MODERN SIGNALS
AND
SYSTEMS

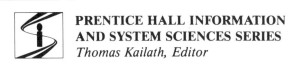

**PRENTICE HALL INFORMATION
AND SYSTEM SCIENCES SERIES**
Thomas Kailath, Editor

MODERN SIGNALS AND SYSTEMS

Huibert Kwakernaak
University of Twente

Raphael Sivan
Technion, Israel Institute of Technology

with software by
Rens C. W. Strijbos
University of Twente

PRENTICE HALL, Englewood Cliffs, NJ 07632

Library of Congress Cataloging-in-Publication Data

KWAKERNAAK, HUIBERT.
    Modern signals and systems/Huibert Kwakernaak, Raphael Sivan: with software by Rens C. W. Strijbos.
        p. cm.
    Includes bibliographical references (p.   )
    ISBN 0-13-809252-4
        1. Signal processing.   2. System analysis.   I. Sivan, Raphael.
II. Strijbos, Rens C. W.   III. Title.
TK5102.5.K895   1991
621.382'2—dc20   89-39977
                                                                CIP

Editorial/production supervision and interior design: Joseph Scordato
Cover design: Annemarie Kwakernaak
Series logo design: A.M. Bruckstein
Manufacturing buyers: Lori Bulwin/Patrice Fraccio

© 1991 by Prentice-Hall, Inc.
A Division of Simon & Schuster
Englewood Cliffs, New Jersey 07632

Printed in the United States of America

10   9   8   7   6   5   4   3   2   1

ISBN 0-13-809252-4

PRENTICE-HALL INTERNATIONAL (UK) LIMITED, *London*
PRENTICE-HALL OF AUSTRALIA PTY. LIMITED, *Sydney*
PRENTICE-HALL CANADA INC., *Toronto*
PRENTICE-HALL HISPANOAMERICANA, S.A., *Mexico*
PRENTICE-HALL OF INDIA PRIVATE LIMITED, *New Delhi*
PRENTICE-HALL OF JAPAN, INC., *Tokyo*
SIMON & SCHUSTER ASIA PTE. LTD., *Singapore*
EDITORA PRENTICE-HALL DO BRASIL, LTDA., *Rio de Janeiro*

To my parents
H. K.

To Ilana, Ori, Ayelet, Keren, and Yael
R. S.

Contents

5 STATE DESCRIPTION OF SYSTEMS   238

Preface

Almost twenty years ago, as young and inexperienced scientists, we dared (where did we take the chutzpah?) to write an advanced graduate textbook, *Linear Optimal Control Systems* (New York: John Wiley, 1972). We found the collaboration enjoyable and rewarding. A few years ago, being not so young anymore, and possibly more mature, we decided to join hands again, this time in what we thought to be a more modest endeavor, to write an undergraduate textbook on modern signals and systems.

Having taught this material over and over again, we had little doubt that it would not be a too difficult or lengthy task to organize what we thought to be the Signals and Systems material. How wrong we were! We ignored the fact that time and again, while standing in front of our classes, we experienced a moment of hesitation, realizing that not all the material we were teaching was totally consistent and completely clear.

It took us nearly four years of work, and almost endless brain racking and soul searching, to integrate the signals and systems material into a unified framework and to produce a book we are satisfied with. We hope that the meticulous care we took in creating the text shows. We also hope that teachers and students alike appreciate the concise and precise style to which we aspired without wanting to compromise the intuitive appeal.

Our *Modern Signals and Systems* is a textbook for a one- or two-semester junior course, which often goes by the name "Signals and Systems", or sometimes by names such as "Linear Systems" or "Dynamical Systems," and is a required course in the curriculum of most electrical engineering departments. The book contains a comprehensive and well-integrated treatment of the basic notions of signal theory and of both the time- and frequency-domain analysis of systems. The discrete- and continuous-time case are treated in parallel, sometimes even typographically in two column format. The book contains many examples, several of which are pursued over many chapters. An extensive collection of homework problems concludes each chapter. In addition, the final section of each chapter provides problems whose solution requires a computer. An unusual feature is that the book comes complete with application software of near-professional quality, provided on a disk, that runs on a personal computer.

It is assumed that the student has a background in basic calculus and algebra, knows how to work with complex numbers, has heard of differential equations and linear algebra, knows the fundamentals of physics and electricity, and possibly has had an introductory course in electrical circuits. To use the software, some familiarity with a personal computer is helpful.

From this book we have taught two-quarter (at the University of Twente) and one-semester (at the Technion) courses to sophomore students. We managed to cover practically all of Chapters 1–8 and a sprinkle of the applications in Chapters 9–11. In departments that devote two semesters to the Signals and Systems course, the entire book may easily be covered. There may even be time to pursue the applications dealt with in Chapters 9–11 in a little more depth, at the discretion of the teacher.

A chapter-by-chapter description of the material follows.

Chapter 1 offers a brief overview of the ideas of signals and systems. By way of motivation, a sketch is given of the application areas that are elaborated in the final three chapters of the book.

Chapter 2 presents an introduction to signals. We describe basic notions such as the time axis of signals, discrete- and continuous-time signals, periodic and harmonic signals, various operations on signals, and signal spaces. The final section is devoted to *generalized* signals. A more elaborate treatment of this material, based on *distribution* theory, is presented in Supplement C.

Chapter 3 deals with a number of fundamental ideas related to systems. Two types of systems are introduced at this point: *input-output* systems and *input-output mapping* systems. The definitions of these systems are *set theoretic,* based on the approach of Jan C. Willems to system theory. These ideas allow an axiomatic approach to system theory even at the level of an undergraduate text. The chapter continues by distinguishing various types of systems. By way of linearity and time-invariance, the discussion arrives at *convolution* systems. The convolution operation is thoroughly treated, and the stability of convolution systems is touched upon. A study of the response of convolution systems to harmonic inputs results in the *frequency response* function. The chapter concludes with a discussion of the response of convolution systems to periodic inputs—leading to *cyclical* convolution—and a brief introduction to the interconnection of systems.

In Chapter 4 we consider systems described by constant coefficient linear difference and differential equations. After a review of elementary material concerning the solution of constant coefficient linear difference and differential equations, we deal with the impulse response, stability, and frequency response of these systems. A central notion is that of the *initially-at-rest system*. In the discussion of stability we introduce *bounded-input-bounded-output* (BIBO) and *converging-input-converging-output* (CICO) stability.

Chapter 5 deals with the state description of systems. Also, here the initial approach is set theoretic. The definition of the state rests upon the *state matching property* formulated by Willems. After introducing the basic ideas, the realization of linear difference and differential systems as state systems is discussed. A further section is devoted to the existence of solutions of state equations and the numerical integration of state differential equations. The chapter continues with the explicit solution of linear state equations and modal analysis. The treatment of the stability of state systems is integrated with that of difference and differential systems.

Chapters 6–8 are devoted to the *frequency domain* description of signals and systems. Chapter 6 begins with a concise but fundamental presentation of signal *expansion*. The theory is made concrete by the introduction of *harmonic* bases for signal spaces and the associated finite and infinite *Fourier series expansions*. Various aspects of these expansions, such as the identities of Plancherel and Parseval, convergence properties, the trigonometric form, symmetry properties, and generalized infinite Fourier series, are carefully presented. The chapter concludes with a treatment of the response of convolution systems to periodic inputs.

Chapter 7 is devoted to Fourier *transforms*. First, transform theory is explained in an abstract setting. The discussion soon focuses on *expansion transforms* (i.e., the transformation from a signal to its expansion coeffcients). In this way, the Fourier series expansions of finite-time and periodic signals of Chapter 6 immediately lead to two of the four Fourier transforms that are considered, namely, the discrete-to-discrete Fourier transform (abbreviated DDFT, more conventionally known as the discrete Fourier transform) and the continuous-to-discrete Fourier transform (CDFT). The properties of these transforms are reviewed. Following this it is shown that the expansion of aperiodic rather than periodic or finite-time signals leads to *Fourier integral expansions*. These result in the discrete-to-continuous Fourier transform (DCFT, also known as the discrete-time Fourier transform) and the continuous-to-continuous Fourier transform (CCFT, commonly known as the Fourier integral transform). Also, the properties of these last two transforms are discussed in detail, emphasizing parallels. The chapter ends with showing how Fourier transform theory is used in the frequency domain analysis of convolution systems.

The third chapter dealing with frequency domain analysis, Chapter 8, is devoted to the *z*-transform and the Laplace transform. First, it is explained that Fourier transform theory cannot handle exponentially increasing signals. To overcome this difficulty, the DCFT and CCFT are modified to the (two-sided) *z*-transform and Laplace transform. To deal with initial value problems, the one-sided *z*-transform and Laplace transform are introduced. The existence, properties and inversion of the *z*-transform and Laplace transform are treated with considerable completeness. Three separate sections of Chapter 8 are devoted to the application of these transforms to the analysis of convolution systems, difference and differential systems, and state systems.

The final chapters of the book present applications of the theory to three important areas for which a course on Signals and Systems forms a prerequisite: signal processing and digital filtering, communication, and feedback and automatic control. Indeed, teachers of courses on these subjects may find that part of their material is covered in this book.

Chapter 9 is the first of the applications chapters. It concerns signal processing and digital filtering. The chapter begins with a discussion of the effect of sampling and interpolation on the frequency content of signals. This leads to a transparent derivation of the sampling theorem. Next on- and off-line signal processing are introduced. After a discussion of windows and windowing, two sections are devoted to an elementary treatment of various methods to design finite and infinite impulse response digital filters. The chapter ends with a derivation of the fast Fourier transform and some considerations about the numerical computation of transforms and convolutions.

Chapter 10 outlines some basic applications to communication theory. The chapter begins with the application of Fourier transforms to the description of narrow-band signals. This is a beautiful theory, which is indispensable in the analysis of modulation and demodulation. Following this, various well-known modulation schemes, including amplitude and frequency modulation, are presented. The chapter ends with a brief discussion of multiplexing.

Chapter 11, finally, contains a concise outline of feedback theory and automatic control. The potential benefits of feedback are demonstrated at a fairly abstract level, illustrated by simple concrete examples. The final section of this chapter reviews various important results on the stability of feedback systems.

The body of the text is complemented with five supplements.

As was mentioned earlier, the book comes with computer software, provided on a disk. The software consists of a powerful interpreter named SIGSYS, which offers a wide and flexible range of operations to generate and handle signals, including Fourier tranformation, convolution, and the integration of differential equations. The software has been stroₗgly inspired by MATLAB, which is a widely used computational tool. The difference is that where the main data type of MATLAB is a matrix, that of SIGSYS is a real- or complex-valued *signal*. In addition, complex and real scalars and polynomials are supported. SIGSYS offers an opportunity to do calculations of a widely varying nature interactively. The graphic support provides instantaneous visualization. An extensive Tutorial, describing and illustrating all the operations and commands of SIGSYS, is included at the end of the book. The READ.ME file on the disk that contains SIGSYS should be consulted before running the program. It contains instructions how to set up and start SIGSYS and also lists corrections and modifications to the Tutorial. The disk furthermore contains a number of demos.

We have used SIGSYS for several years to run a laboratory course in parallel with the Signals and Systems course as follows. During the first weeks of the term, students were instructed to go through the first 16 sections of the Tutorial and to do all the exercises provided by the Tutorial. In addition, the students were assigned two or three Computer Exercises from Chapters 2 and 3, whose main purpose is to become familiar with the computer. Depending on the available time, during the rest of the course students were assigned a number of other problems from the Computer Exercises for the rest of the chapters.

A solutions manual for the Problems and Computer Exercises is available to teachers who adopt the book and may be obtained from the publisher. The manual comes with a disk that provides solutions to all the Computer Exercises in macro form. Also, hard-copy solutions are supplied.

The story of the software is this: Right at the initial stages of planning the book it occurred to us that supporting the text with software would be timely and instructive. The success of MATLAB was a powerful stimulus, and it stood example for SIGSYS in many ways. We were lucky to get Rens Strijbos, a long-time associate, to develop the program for us. The time and effort spent on this work are fully comparable with our own exertions. The software was written in the C language and developed in a UNIX environment by using a variety of programming tools.

Writing this book also got us in other ways involved in high technology. The early versions of the text were prepared by using the text formatting language Troff, and the figures were produced with a drawing program. Being able to keep in daily touch by electronic mail was a forceful incentive. Fax and international courier mail from time to time supplemented communication by ordinary mail. The work on the book took us on numerous trips back and forth between Israel and The Netherlands and also to such places as Amherst, Mass., Berlin, Bern, Englewood Cliffs, N.J., and, last but not least, the Sinai desert.

We are grateful to our teachers, in particular Lotfi Zadeh and Charlie Desoer, who showed us what perspicuity means, and to our students, who keep insisting on getting things clear. Our departments deserve credit for accepting our absorption in the project of writing the book, our periods of absence, footing the bills for computer time, and supporting part of the traveling. In particular, we acknowledge the support of the Fund for the Promotion of Research of the Technion. We thank our secretaries, Marja Langkamp and Annette Berg, for their invaluable help. In conclusion, we wish to express our sincere appreciation to Tim Bozik of Prentice Hall for his continued belief in the project even when we started missing deadlines.

Enschede and Haifa

Huibert Kwakernaak
Raphael Sivan

1

Overview of Signals and Systems

1.1 INTRODUCTION

This textbook is an introductory study of signals and systems. As we develop the subject, we encounter various kinds of signals: discrete-time and continuous-time signals, real- and complex-valued signals, signals of finite and infinite duration, and many more. We also meet a variety of systems: input-output systems, input-output mapping systems, input-output-state systems, linear systems, time-invariant systems, each in discrete- and continuous-time versions. All these notions will be precisely defined as the need arises. In this chapter we first present somewhat loose explanations of what is meant by the terms signal and system, illustrate the definitions with examples, and describe some areas where the theory of signals and systems is applied.

Signals

A *signal,* roughly, is a phenomenon, arising in some environment, that may be described quantitatively. Examples of signals to be discussed in this book are electrical signals, such as electrical voltages or currents in an electrical circuit. Other signals are auditive signals, visual signals, and sequences of bits that emanate from computers.

Systems

A *system* is, more or less, any part of an environment that causes certain signals that exist in that environment to be related. An electrical circuit is a typical example of a system, because the voltages and currents that exist within the circuit are related.

The signals associated with an electrical circuit are not only voltages and currents but are also magnetic fluxes and electrical charges. The signals that go with a mechanical system, such as a car that moves along a highway, are positions and velocities. The signals associated with the national economy of a country, to mention a system of a completely different nature, are quantities such as national income and expenditure, labor force, and capital equipment.

The signals associated with a system are not arbitrary but are interrelated as a result of the internal mechanisms of the system. The electrical voltages and currents in an electrical circuit, for instance, are interrelated because of the electromagnetic laws. The interrelation of the signals imposed by the laws that govern the system is called the *rule* of the system.

Rather than pursuing in this introductory chapter the abstractions of the notions of signals and systems, we devote the chapter to illustrations. In the next section we describe an assortment of systems and associated signals. Next, in Section 1.3, we present some engineering problems whose solution requires a theory of signals and systems. These problems are drawn from the areas of *signal processing, communication engineering,* and *automatic control.*

1.2 EXAMPLES OF SIGNALS AND SYSTEMS

In this section, various examples of signals and systems are presented.

1.2.1.   Example: Resistive electrical circuit element.   Consider a resistive electrical circuit element with two ports as in Fig. 1.1. The signals associated with

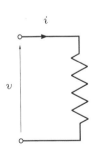

Figure 1.1   A resistive electrical circuit element.

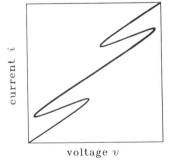

Figure 1.2   Voltage-current characteristic.

the system are the *voltage v* across the circuit element and the *current i* through it. Both the voltage v and the current i are characterized by a real number. The relation between v and i, called the *voltage-current characteristic,* is given by a graph $f(v, i) = 0$, like the one in Fig. 1.2. This graph represents the *rule* of the system: The signals associated with the system are constrained to those voltage-current pairs (v, i) that lie on the graph. Note that for the graph of Fig. 1.2, neither the voltage v determines the current i uniquely nor does the current i uniquely prescribe the voltage v.

∎

Input-Output Systems

Often, though not always, one may designate some of the signals associated with the system as *input* signals, through which the environment influences the system, and some as *output* signals, by which, in turn, the system affects the environment. Figure 1.3 illustrates the situation. Systems of this type are called *input-output* systems. Note that we do *not* require that the input determine the output uniquely. What follows is an example of an input-output system.

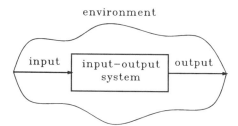

Figure 1.3   Input-output system.

1.2.2.   Example: Resistive element connected to a voltage source.   The resistive circuit element of Example 1.2.1 may be connected to a voltage source that produces a voltage v as in Fig. 1.4. The system we obtain still has the voltage v and current i as associated signals, but it now is natural to designate the voltage v as the *input* signal and the current i as the *output* signal. For some input voltages v, though, there is no unique output current i.

∎

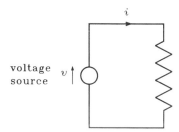

Figure 1.4   Resistive element connected to a voltage source.

Input-Output Mapping Systems

An input-output system whose input *uniquely* determines the output is called an *input-output mapping system*. The following example illustrates this notion.

1.2.3.   Example: Parity bit generator.   A parity bit generator as in Fig. 1.5 is a digital device that accepts words consisting of a fixed number of bits and complements this word with either a 0 or a 1 in such a way that the total number of ones is *even*. Thus, if the input is, say, an eight-bit word of the form $u = u_0 u_1 u_2 u_3 u_4 u_5 u_6 u_7$, with u_k either 0 or 1 for $k = 0, 1, \cdots, 7$, the corresponding output is a nine-bit word $y = y_0 y_1 y_2 y_3 y_4 y_5 y_6 y_7 y_8$, where

$$y_k = \begin{cases} u_k & \text{for } k = 0, 1, \cdots, 7, \\ (u_0 + u_1 + \cdots + u_7) \bmod 2 & \text{for } k = 8. \end{cases}$$

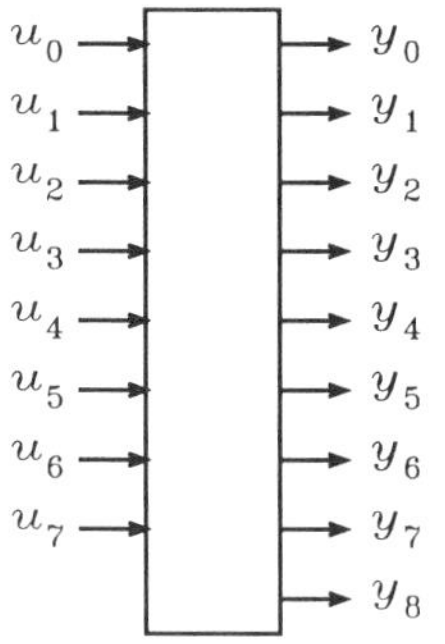

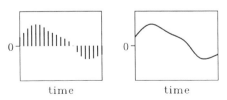

Figure 1.5   Parity bit generator.

If the input is, for instance, $u = 11001100$, the output is $y = 110011000$. This is an input-output mapping system, which maps every eight-bit byte into a nine-bit byte.   ∎

Time Signals

In many instances we need deal with systems whose signals are *functions of time*. Such signals are called *time signals*. The collection of time instants on which the signal is defined is called its *time axis*. Time signals whose time axis is a *finite* or *countable* set of real numbers are called *discrete-time* signals. Time signals whose time axis is a finite or infinite *interval* of real numbers are called *continuous-time* signals. Figure 1.6 illustrates discrete-time and continuous-time signals.

Figure 1.6   Time signals. Left: discrete-time signal. Right: continuous-time signal.

1.2.4.   Example: Dow-Jones averages.   The *Dow-Jones average* is a weighted average of the prices of a selected portfolio of stocks traded at the New York stock exchange, which has been daily computed and recorded since 1897. The sequence of Dow-Jones averages $z = (z(0), z(1), z(2), z(3), \cdots)$, with $z(n)$ denoting the Dow-Jones average on day n, is an example of a discrete-time signal. The time axis of this signal is the countable set of nonnegative integers $\mathbb{Z}_+ = \{0, 1, 2, \cdots\}$.   ■

1.2.5.   Example: Continuous-time signals.   The outdoor temperature at any given place taken as a continuous function of time is a typical example of a continuous-time signal. The physical world abounds with continuous-time signals.   ■

Discrete-Time Systems

Systems associated with discrete-time signals are called *discrete-time systems*. An example of a discrete-time system that is used frequently in the rest of the book follows.

1.2.6.   Example: Exponential smoother.   In signal processing sometimes a procedure called *exponential smoothing* is used to remove unwanted fluctuations from observed time series, such as the Dow-Jones averages in Example 1.2.4. Let $u = (u(0), u(1), u(2), \cdots)$ be a discrete-time signal defined on the time axis $\mathbb{Z}_+ = \{0, 1, 2, \cdots\}$. Then, as we observe the successive values of the time signal u we may form from these values another time signal y, which is a smoothed version of the signal u, according to

$$y(n + 1) = ay(n) + (1 - a)u(n + 1), \qquad (1)$$

for $n = 0, 1, 2, \cdots$. Here, a is a constant such that $0 < a < 1$. This describes a discrete-time system. Equation (1) forms the *rule* of the system. To apply the rule, starting at time 0, we need specify an *initial value* $y(0)$. At each time $n + 1$, the output $y(n + 1)$ is formed as a weighted average of the new input $u(n + 1)$ at time $n + 1$ and the output $y(n)$ at the preceding time instant n. The closer the constant a is to 1, the more the preceding output value is weighted and the "smoother" the output is.

By repeated substitution it is easily found that

$$y(n) = a^n y(0) + (1 - a) \sum_{k=0}^{n-1} a^k u(n - k),$$

for $n = 1, 2, 3, \cdots$. This shows that $y(n)$ is a weighted sum of the present input, all past inputs back to $u(1)$, and the initial value $y(0)$. If $|a| < 1$, the effect of the initial value $y(0)$ asymptotically vanishes as n increases, and the weighting coefficients a^k become exponentially smaller the farther back the input values are in the past. This is why this scheme is called *exponential smoothing*. Figure 1.7 illustrates it.

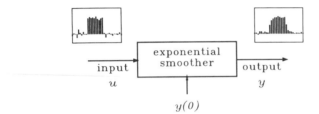

Figure 1.7   Exponential smoothing.

Because the output y is not uniquely determined by the input but also depends on the initial value $y(0)$, the exponential smoother is an input-output system but not an input-output mapping system.   ∎

Continuous-Time Systems

Continuous-time systems are systems that are associated with continuous-time signals. An example of a continuous-time system that will often return is the following RC network.

1.2.7.   Example: RC network.   The electrical circuit of Fig. 1.8 consists of a series connection of a voltage source, a resistor with resistance R, and a capacitor with capacitance C. The voltage u produced by the voltage source varies with time and constitutes the input to the system. The output y is the voltage across the capacitor. Both the input and output are continuous-time signals so that this is an example of a continuous-time system.

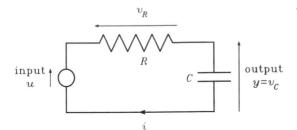

Figure 1.8   An RC network.

By Kirchhoff's voltage law, at each time t

$$u(t) = v_R(t) + v_C(t),$$

where v_R is the voltage across the resistor and v_C that across the capacitor. Denoting the current through the circuit as i, we have $v_R(t) = R\,i(t)$. Since, moreover, $i(t) = C\,dv_C(t)/dt$, it follows that

$$u(t) = RC\frac{dv_C(t)}{dt} + v_C(t).$$

Substitution of $v_C = y$ and division by RC leads to the differential equation

$$\frac{dy(t)}{dt} + \frac{1}{RC}y(t) = \frac{1}{RC}u(t).$$

Any input-output pair (u, y) need satisfy the differential equation, which forms the *rule* of the system.

For a given initial voltage $y(t_o) = v_C(t_o)$ and a given input voltage $u(t)$ for $t \geq t_o$, we may solve for $y(t)$, $t \geq t_o$, as

$$y(t) = e^{-(t-t_o)/RC}\, y(t_o) + \frac{1}{RC}\int_{t_o}^{t} e^{-(t-\tau)/RC}\, u(\tau)\, d\tau, \qquad t \geq t_o.$$

At each time t, the output $y(t)$ is a weighted sum of a term that is determined by the initial condition $y(t_o)$ and a term that is an exponentially weighted integral of past inputs back to the initial time t_o. The effect of the system is much analogous to that of the exponential smoother of Example 1.2.6. Figure 1.9 shows the system's response to the step signal

$$u(t) = \begin{cases} 0 \text{ for } 0 \leq t < a, \\ 1 \text{ for } t \geq a, \end{cases}$$

with $y(0) = 0$. The network turns the step into a smoothed version of a step given by

$$y(t) = \begin{cases} 0 & \text{for } 0 \leq t < a, \\ 1 - e^{-(t-a)/RC} & \text{for } t \geq a. \end{cases} \qquad \blacksquare$$

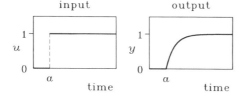

Figure 1.9 Response of the RC network. Left: step input. Right: output.

1.3 APPLICATIONS OF SIGNAL AND SYSTEM THEORY

Signal and system theory is applied in many fields of engineering and science. In this section we describe examples of practical engineering design problems whose solution requires the theory of signals and systems as developed in this book. The design problems are taken from the areas of *signal processing, communication engineering,* and *automatic control.*

Application to Signal Processing

A typical problem in the area of signal processing is the removal of unwanted noise from a signal. Examples are signals received over a telephone line or picked up from the read head of a tape recorder. The model that is frequently used is that the received signal u is given by

$$u = s + n,$$

where s is the transmitted signal and n is *noise* added to the signal. The problem is to design a *filter* as in Fig. 1.10, whose function is to remove or at least attenuate the noise. The input to the filter is the received signal u, and its output y is required to be reasonably similar to the transmitted signal s.

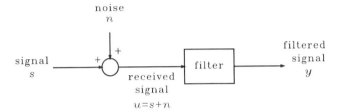

Figure 1.10   The filter design problem.

The exponential smoother of Example 1.2.6 may be viewed as such a filter, because it removes short-term irregularities from the input. The smoother is described by the equation

$$y(t + T) = ay(t) + (1 - a)u(t + T), \tag{1}$$

where we now let t take values on the time axis $\{t_o, \, t_o + T, \, t_o + 2T, \, \cdots \}$. The closer the constant a is to 1, the stronger is the smoothing effect. The stronger the smoothing effect, the more the undesired short-term fluctuations are suppressed. On the other hand, also part of the long-term changes we are actually interested in are smoothed over and eliminated. The choice of the constant a therefore is a crucial aspect of the design of the filter. To make a rational choice, a theory is needed that enables us to study the effect of the filter on both the signal and the noise.

By using modern digital circuitry it is very simple to implement the exponential smoother as a *digital filter,* which accepts a discrete-time signal as input and produces a filtered discrete-time signal as output. Because of their reliability and insensitivity to noise, digital filters are being used on an increasingly large scale in communication, audio and video equipment. Design methods for such filters are discussed in Chapter 9.

Application to Communication Engineering

Communication engineering deals with transferring information, such as audio and video signals and data streams, from one location to another. Since such signals do not propagate over long distances, it is necessary, if the locations are far apart, to "mount" the signal on some *carrier* that may easily be transmitted over great distances.

One much used possibility is to employ high-frequency waves as carrier. At the point of origin the audio, video, or data signal, commonly called the *message* signal, is mounted on the carrier by a process called *modulation*. The modulated carrier propagates through the transmission medium (which may be space, cable, or another medium), and at the destination the message signal is retrieved from the carrier through a process called *demodulation*. Figure 1.11 illustrates such a transmission link.

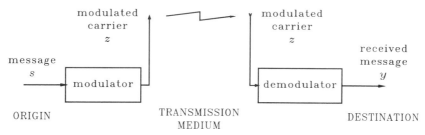

modulated carrier *z*   modulated carrier *z*

message *s* → modulator   demodulator → received message *y*

ORIGIN   TRANSMISSION MEDIUM   DESTINATION

Figure 1.11   Long-distance radio transmission.

Let the continuous-time signal *m* be the message signal that is to be transmitted, and suppose that the continuous-time signal *c* is a high-frequency harmonic carrier with frequency f_c given by

$$c(t) = \cos(2\pi f_c t), \qquad -\infty < t < \infty.$$

Then *amplitude modulation* of the carrier *c* with the signal *m* results in the *amplitude modulated* signal *z* given by

$$z(t) = [m_o + m(t)] \cos(2\pi f_c t), \qquad -\infty < t < \infty.$$

The number m_o is a positive constant such that $m_o + m(t) \geq 0$ for all *t*. Figure 1.12 shows a modulated signal.

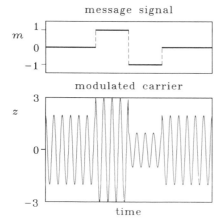

Figure 1.12   Amplitude modulation. Top: message signal. Bottom: modulated carrier.

An important subject in communication theory is the study of modulation schemes such as amplitude and frequency modulation. Chapter 10 presents an introduction to the theory of modulation.

Application to Automatic Control

Control engineering concerns itself with the design of automatic controllers, whose purpose is to govern the dynamic behavior of a given system. By way of example we consider an *automatic cruise controller* in a car. This controller automatically adjusts the engine throttle in such a way that the speed of the car maintains a *reference speed* set by the driver.

The cruise controller is arranged as in the block diagram of Fig. 1.13. The speed v of the car is continuously measured and converted to an electrical voltage. This voltage is compared to a voltage that represents the reference cruise speed v_r that has been set. The error signal $v_r - v$ is transmitted, again in the form of an electrical signal, to an electromechanical device that controls the throttle opening. The operation of this device is such that if the error $v_r - v$ is positive (i.e., the desired speed is *higher* than the actual speed), the throttle opening keeps increasing so that the speed of the car also increases until the speed reaches the desired value v_r. On the other hand, if the reference speed is lower than the actual speed, the throttle opening keeps decreasing. The configuration of Fig. 1.13 is called a *feedback system,* because the output is returned to the input.

The selection of a suitable control mechanism is a central issue in control system design. The theory of feedback and some aspects of control system design are discussed in Chapter 11.

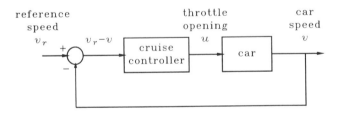

Figure 1.13 Automatic cruise control system.

2

An Introduction to Signals

2.1 INTRODUCTION

In the context of this book, a *signal* is a phenomenon that represents information. Since any signal always is one of a collection of several of many possible signals, signals may mathematically be represented as elements of a set, called the *signal set*. In this chapter we introduce a variety of signal sets. Moreover, operations on and among signals are defined and discussed.

In Section 2.2 signals are defined. We are primarily interested in *time signals*. Section 2.3 deals with elementary operations *on* and *among* signals, such as *signal range* and *signal axis transformation, sampling, interpolation*, and *pointwise addition, multiplication*, and *division*.

In Section 2.4, *signal spaces* are introduced, along with the notions of *norm* and *inner product*. Various important signal spaces, such as spaces of *bounded amplitude, bounded action*, and *bounded energy* signals, are defined.

Section 2.5 is devoted to *generalized* signals. The most prominent example of a generalized signal is the δ-function, which may be viewed as a signal of zero duration, infinite height, and unit area.

In Supplement A at the end of the book various elementary facts about *complex numbers, sets,* and *maps* are reviewed. This material is assumed to be familiar throughout the book.

2.2 SIGNALS

In this section we introduce various kinds of signals, such as discrete- and continuous-time signals, finite- and infinite-time signals, and periodic and harmonic signals.

The signals we are interested in are *functions* of a variable that often is *time*. The *domain* of a signal is a subset $\mathbb{T}$ of the real line and is called the *signal axis*. The signal may take values in any set A, called the *signal range*. Figure 2.1 illustrates the idea. The formal definition of a signal is as follows.

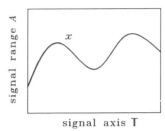

signal axis $\mathbb{T}$

Figure 2.1   A signal is a function $\mathbb{T} \to A$.

2.2.1.   Definition: Signals.   Let A be a set, and suppose that $\mathbb{T}$ is a subset of the reals $\mathbb{R}$. Then, any function $x : \mathbb{T} \to A$ is called a *signal* with *signal axis* $\mathbb{T}$ and *signal range* A.   ∎

The set of all signals with signal axis $\mathbb{T}$ and signal range A thus is the set of all functions from $\mathbb{T}$ to A. This signal set is denoted by $\{x : \mathbb{T} \to A\}$, or by the *power set* notation $A^{\mathbb{T}}$ (see Supplement A).

If the signal axis has the interpretation of time, the signal is called a *time signal*, and the signal axis is called *time axis*. We also meet other signals, in particular *frequency signals*.

2.2.2.   Examples: Signals.

(a) *Dow-Jones averages*. The Dow-Jones averages of Example 1.2.4 form a time signal with time axis $\mathbb{T} = \mathbb{Z}_+ = \{0, 1, 2, \cdots\}$ and signal range $A = \mathbb{R}_+$, the set of nonnegative real numbers.

(b) *Bit stream*. A semi-infinite bit stream such as $1010011 \cdots$ is a signal with signal axis $\mathbb{T} = \mathbb{Z}_+ = \{0, 1, 2, \cdots\}$ and signal range $A = \{0, 1\}$.

(c) *Electrical signal*. The voltage across the capacitor of the RC network of Example 1.2.7 is a time signal with axis $\mathbb{T} = [t_o, \infty)$ and signal range $A = \mathbb{R}$.   ∎

Discrete- and Continuous-Time Signals

Time signals may either be *discrete-* or *continuous-time* signals:

2.2.3.   Definition: Discrete and continuous time axes; discrete- and continuous-time signals.

(a) The time axis $\mathbb{T} \subset \mathbb{R}$ is *discrete* if it consists of a *finite* or *countable* set of time instants. A time signal whose time axis is discrete is called a *discrete-time signal*.

(b) The time axis $\mathbb{T} \subset \mathbb{R}$ is *continuous* if it consists of an interval of $\mathbb{R}$, possibly extending to $-\infty$ or $+\infty$ or to both. A time signal whose time axis is continuous is called a *continuous-time* signal.   ■

In this text, the signal range A usually is the set of real numbers $\mathbb{R}$ or the set of complex numbers $\mathbb{C}$. If $A = \mathbb{R}$, the signal is said to be *real-valued*, while, if $A = \mathbb{C}$, it is called *complex-valued*.

2.2.4.   Examples: Discrete- and continuous-time signals.

(a) *A real-valued discrete-time signal.* The time axis $\mathbb{T} = \{0, 1, 2, 3, 4\}$ is discrete. The signal $x \in \mathbb{R}^{\mathbb{T}}$ shown in Fig. 2.2, defined by

$$x(n) = n + 1, \qquad n \in \{0, 1, 2, 3, 4\},$$

is an example of a real-valued discrete-time signal.

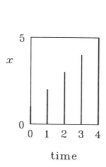

Figure 2.2   A real-valued discrete-time signal.

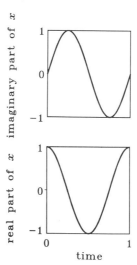

Figure 2.3   A complex-valued continuous-time signal.

(b) *A complex-valued continuous-time signal.* The set $\mathbb{T} = [0, 1]$ is a continuous time axis, and the signal $x \in \mathbb{C}^{\mathbb{T}}$ defined by

$$x(t) = e^{j2\pi t} = \cos(2\pi t) + j \sin(2\pi t), \qquad t \in \mathbb{T},$$

as shown in Fig. 2.3, is a complex-valued continuous-time signal.   ■

Time Sequences and Sampled Signals

Suppose that x is a discrete-time signal defined on the time axis

$$\mathbb{T} = \{t_0, t_1, t_2, \cdots\},$$

with $t_0 < t_1 < t_2 < \cdots$. We may always list the consecutive values

$$x(t_0), x(t_1), x(t_2), \cdots \tag{1}$$

of the signal as a *sequence*. Thus, any discrete-time signal may either be seen as a *map* from a time axis $\mathbb{T}$ to its signal range A or, alternatively, as an *ordered sequence* of values in A. This ordered sequence may be redefined on a time axis consisting of *integers* so that the sequence (1) is rewritten as

$$x(0), x(1), x(2), \cdots.$$

In the sequel, we refer to discrete-time signals defined on time axes that consist of consecutive integers, such as the set $\mathbb{N}$ of all natural numbers, the set $\mathbb{Z}_+$ of all nonnegative integers, and the set $\mathbb{Z}$ of all integers, as *time sequences*.

Often, though not always, a discrete-time signal is obtained by observing a phenomenon that takes place on a continuous time axis at a sequence of discrete time instants $t_0, t_1, t_2, \cdots$, called the *sampling times*. The discrete-time signal obtained in this way is referred to as a *sampled signal*. If the sampling times are *uniformly spaced* (i.e., $t_i = iT$ with i ranging over a set of consecutive integers and $T > 0$), we say that the signal is *uniformly sampled* with *sampling interval T*. The number of samples per unit of time $1/T$ is called the *sampling rate* of the sampled signal.

2.2.5.   Example: Sampled signal and time sequence.   Consider the continuous time axis $[0, \infty)$ and the continuous-time signal x defined on this axis by

$$x(t) = \frac{1}{1 + \dfrac{t}{T}}, \qquad t \in [0, \infty),$$

with $T > 0$, as illustrated in Fig. 2.4(a).

Observing the signal x on the discrete time axis $\{0, T, 2T, \cdots\}$ results in the *uniformly sampled signal x^** defined by

$$x^*(t) = \frac{1}{1 + \dfrac{t}{T}}, \qquad t \in \{0, T, 2T, \cdots\},$$

as shown in Fig. 2.4(b).

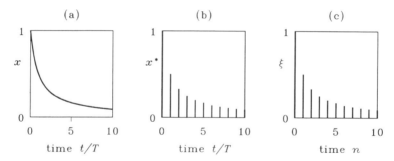

Figure 2.4 Sampled signal and time sequence. Left: a continuous-time signal. Middle: the sampled signal. Right: the resulting time sequence.

The consecutive values taken by the sampled signal $x*$ form the *time sequence* ξ defined by

$$\xi(n) = x*(nT) = x(nT) = \frac{1}{1+n}, \qquad n = 0, 1, 2, \cdots ,$$

as in Fig. 2.4(c). The time sequence ξ is defined on the time axis $\mathbb{Z}_+$. Note that the sampled signal $x*$ and the time sequence ξ consecutively assume the *same* values but have *different* time axes.   ∎

In the sequel, important results that relate to discrete-time signals are first formulated for *time sequences*. Nearly always, the corresponding results for *uniformly sampled signals* are reviewed in a form that emphasizes the parallels between the discrete- and the continuous-time case. The sampled results may be specialized to sequences by simply setting the sampling interval T equal to 1.

Finite-Time, Semi-Infinite-Time, and Infinite-Time Signals

A time signal may be *finite-time, semi-infinite-time,* or *infinite-time,* depending on whether its time axis is finite, semi-infinite, or infinite.

2.2.6.   Definition: Finite, semi-infinite and infinite time axes.
   (a) *Finite time axis.* If a time axis, whether discrete or continuous, is contained in a finite interval, it is called a *finite* time axis.

   (b) *Semi-infinite time axis.* If a time axis is bounded from the left, it is called a *right semi-infinite* time axis, while, if it is bounded from the right, it is said to be *left semi-infinite*.

   (c) *Infinite time axis.* A time axis that is neither bounded from the left nor from the right is called an *infinite* time axis.   ∎

The most common discrete time axes in this text follow. For time sequences we have the *infinite* time axis $\mathbb{Z}$ consisting of all integers, the *right semi-infinite* time axis $\mathbb{Z}_+ = \{0, 1, 2, \cdots\}$ consisting of all nonnegative integers, and the *finite* time axis

$$\underline{N} := \{0, 1, 2, \cdots, N - 1\},$$

with N any natural number. Note that the notation $\underline{N}$ may be used numerically so that $\underline{4}$ denotes the set $\{0, 1, 2, 3\}$.

For uniformly sampled signals with sampling interval T we correspondingly have the infinite, (right) semi-infinite, and finite time axes

$$\mathbb{Z}(T) = \{\cdots, -T, 0, T, 2T, \cdots\},$$
$$\mathbb{Z}_+(T) = \{0, T, 2T, \cdots\},$$
$$\underline{N}(T) := \{0, T, 2T, \cdots, (N - 1)T\},$$

respectively, where the latter is defined for any natural number N.

The usual continuous time axes are the infinite time axis $\mathbb{R}$, the (right) semi-infinite time axis $\mathbb{R}_+$, and finite time axes of the form $[a, b]$, with $-\infty < a < b < \infty$.

In the sequel we occasionally use the following notations for sets of time signals defined on different time axes:

ℓ	all complex-valued time sequences with the infinite time axis $\mathbb{Z}$,
ℓ_+	all complex-valued time sequences with the semi-infinite time axis $\mathbb{Z}_+$,
ℓ_N	all complex-valued time sequences with the finite time axis $\underline{N}$,
$\ell(T)$	all complex-valued uniformly sampled signals with the infinite time axis $\mathbb{Z}(T)$,
$\ell_+(T)$	all complex-valued uniformly sampled signals with the semi-infinite time axis $\mathbb{Z}_+(T)$,
$\ell_N(T)$	all complex-valued uniformly sampled signals with the finite time axis $\underline{N}(T)$,
$\mathcal{L}$	all complex-valued continuous-time signals with the infinite time axis $\mathbb{R}$,
$\mathcal{L}_+$	all complex-valued continuous-time signals with the semi-infinite time axis $\mathbb{R}_+$,
$\mathcal{L}[a, b]$	all complex-valued continuous-time signals with the finite time axis $[a, b]$.

Some Well-Known Signals

In what follows we give examples of some familiar time signals.

2.2.7.   **Example: Some well-known signals.**

(a) *The unit pulse*. The unit pulse of Fig. 2.5 is the time sequence Δ defined by

$$\Delta(n) = \begin{cases} 1 & \text{for } n = 0, \\ 0 & \text{otherwise,} \end{cases} \qquad n \in \mathbb{T},$$

where $\mathbb{T}$ may be any of the discrete time axes $\underline{N}$, $\mathbb{Z}_+$, or $\mathbb{Z}$. Correspondingly, the unit pulse belongs to the signal set $\ell_{\underline{N}}$, ℓ_+, or $\overline{\ell}$.

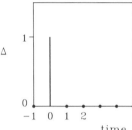

Δ

time   **Figure 2.5**   The unit pulse Δ.

(b) *Rectangular and triangular pulses*. Well-known continuous-time signals are the rectangular and triangular pulse, defined as

$$\text{rect}(t) = \begin{cases} 1 & \text{for } -\tfrac{1}{2} \leq t < \tfrac{1}{2}, \\ 0 & \text{otherwise,} \end{cases} \qquad t \in \mathbb{R},$$

$$\text{trian}(t) = \begin{cases} 1 - |t| & \text{for } |t| < 1, \\ 0 & \text{otherwise,} \end{cases} \qquad t \in \mathbb{R}.$$

The two signals are sketched in Fig. 2.6. They both belong to the signal set $\mathcal{L}$.

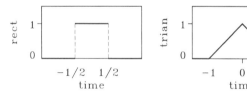

Figure 2.6   Left: rectangular pulse. Right: triangular pulse.

(c) *Unit step and ramp signal*. Two further well-known signals, which exist both in discrete- and continuous-time form, are the unit step $\mathbb{1}$ and the ramp signal, defined as

$$\mathbb{1}(t) = \begin{cases} 0 & \text{for } t < 0, \\ 1 & \text{for } t \geq 0, \end{cases} \qquad t \in \mathbb{T},$$

$$\text{ramp}(t) = \begin{cases} 0 & \text{for } t < 0, \\ t & \text{for } t \geq 0, \end{cases} \qquad t \in \mathbb{T},$$

where $\mathbb{T}$ is any of the infinite time axes $\mathbb{Z}$, $\mathbb{Z}(T)$, or $\mathbb{R}$. The signals are displayed in Fig. 2.7. Depending on the time axis, the signals belong to ℓ, $\ell(T)$, or $\mathcal{L}$. ∎

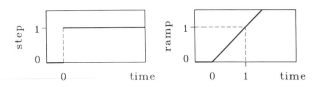

Figure 2.7   Left: unit step. Right: ramp signal.

Periodic Signals

A *periodic signal* is a signal that repeats itself indefinitely. The length of time after which the signal starts repeating itself is called its *period*. The formal definition goes as follows.

2.2.8.   Definition: Periodic time signal.
   (a) *Periodic signal.* A signal $x \in A^\mathbb{T}$ with signal range A and infinite axis $\mathbb{T} = \mathbb{Z}, \mathbb{Z}(T)$ or $\mathbb{R}$ is *periodic* if there exists a $P \in \mathbb{T}$ with $P > 0$ such that

$$x(t + P) = x(t) \qquad \text{for all} \quad t \in \mathbb{T}.$$

   (b) *Period of a periodic signal.* The real number P is called the *period* of a periodic signal x if it is the smallest positive number such that

$$x(t + P) = x(t) \qquad \text{for all} \quad t \in \mathbb{T}. \qquad\qquad ■$$

Examples of a discrete- and a continuous-time periodic signal are given in Fig. 2.8. The reason that in part (b) of the definition we insist on the *smallest* P is that if $x(t) = x(t + P)$ for all $t \in \mathbb{T}$, it follows that $x(t) = x(t + qP)$ for all $t \in \mathbb{T}$ with q any natural number. The definition ensures that the period P of the signal is uniquely determined. Note that if x is a sampled discrete-time periodic signal, its period P necessarily is an integral multiple of the sampling interval T (i.e., $P = pT$, with p a natural number).
   The inverse $1/P$ of the period of a periodic signal is the number of times per unit of time that the signal repeats itself and is called the *repetition rate* of the signal.

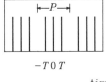
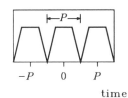

Figure 2.8   Period signals. Left: discrete-time. Right: continuous-time.

Harmonic Signals

The most conspicuous examples of periodic signals are *harmonic* signals, although not *all* harmonic signals are periodic.
   Harmonic signals exist both in the discrete- and the continuous-time case. For any real number f, the (complex) harmonic signal η_f with *frequency f* is defined by

$$\eta_f(t) = e^{j2\pi ft}, \qquad t \in \mathbb{T},$$

where $\mathbb{T}$ is the time axis. The number

$$\omega := 2\pi f$$

is called the *angular frequency* of the harmonic. Also, signals of the form

$$x(t) = a\, e^{j2\pi ft}, \qquad t \in \mathbb{T},$$

with a any complex constant, are said to be harmonic. The constant a is called the *complex amplitude* of the harmonic signal. We may write the complex number a in polar form as

$$a = \alpha\, e^{j\phi},$$

with the magnitude $\alpha = |a| > 0$ and the angle ϕ both real numbers. Then,

$$\begin{aligned}
x(t) &= a\, e^{j2\pi ft} = \alpha\, e^{j\phi}\, e^{j2\pi ft} = \alpha\, e^{j(2\pi ft + \phi)} \\
&= \alpha \cos(2\pi ft + \phi) + j\alpha \sin(2\pi ft + \phi), \qquad t \in \mathbb{T}.
\end{aligned}$$

The real-valued signal c given by

$$\begin{aligned}
c(t) &= \mathrm{Re}(x(t)) \\
&= \alpha \cos(2\pi ft + \phi), \qquad t \in \mathbb{T},
\end{aligned}$$

is said to be a *real harmonic* signal, with frequency f, *amplitude* α, and *phase* ϕ. The complex number $a = \alpha e^{j\phi}$ is called the *phasor* of the real harmonic signal c.

We next discuss the periodicity properties of harmonic signals. First, consider the continuous-time case, with $\mathbb{T} = \mathbb{R}$. Then, for $f \neq 0$, the harmonic signal η_f is periodic with period $P = 1/|f|$. The reason is that $P = 1/|f|$ is the smallest positive value of P such that $e^{j2\pi fP} = 1$. We summarize as follows.

2.2.9.   Summary: Periodicity of continuous-time harmonics.   The continuous-time harmonic η_f, defined on the time axis $\mathbb{R}$, with frequency $f \neq 0$, is periodic with period $P = 1/|f|$. Its repetition rate, hence, is $|f|$.   ■

For $f = 0$, the harmonic signal η_f reduces to $\eta_0(t) = 1$ for all $t \in \mathbb{R}$, which is trivially periodic.

In the discrete-time case, things are a little more complicated. First of all, there are fewer discrete-time harmonic signals than it would seem. The reason is that, as Fig. 2.9 shows, sampling two continuous-time harmonics with *different* frequencies may result in the *same* discrete-time signal. This phenomenon is called *aliasing*.

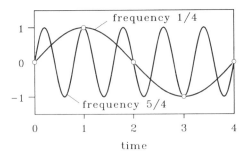

Figure 2.9 Aliasing: on the time axis $\mathbb{Z}$ the real harmonics $\sin(2\pi f_1 t)$ and $\sin(2\pi f_2 t)$ with $f_1 = 1/4$ and $f_2 = 5/4$ coincide.

2.2.10.   Summary: Aliasing.   On the time axis $\mathbb{Z}(T)$, the harmonic signals η_f and $\eta_{f+k/T}$ are identical for any $k \in \mathbb{Z}$.   ∎

2.2.11.   Proof.   For $t \in \mathbb{Z}(T)$ it   follows   that $\eta_{f+k/T}(t) = e^{j2\pi(f+k/T)t} = e^{j2\pi ft} \cdot e^{j2\pi kt/T} = e^{j2\pi ft} = \eta_f(t)$, since kt/T is an integer for all $t \in \mathbb{Z}(T)$.   ∎

As a result, discrete-time harmonics whose frequencies differ by an integral multiple of the sampling rate $1/T$ cannot be distinguished.

One consequence of aliasing is that when studying discrete-time harmonic signals on the time axis $\mathbb{Z}(T)$ we may as well restrict the range of frequencies that are considered to the interval $[0, 1/T)$. Sometimes it is useful to take the interval $[-1/2T, 1/2T)$.

A further fact to bear in mind is that by no means is every discrete-time harmonic signal periodic. The reason is that the sampling interval and the period of a continuous-time harmonic signal that is sampled are not necessarily commensurable.

2.2.12.   Summary: Periodicity of discrete-time harmonic signals.   The discrete-time harmonic signal η_f defined on the time axis $\mathbb{Z}(T)$ is periodic if and only if the frequency f is a *rational* multiple of $1/T$. If $|f| = p/qT$, with p and q coprime natural numbers, the harmonic has period qT.   ∎

2.2.13.   Proof.   As observed before, the period P of any periodic signal on the time axis $\mathbb{Z}(T)$ is of the form $P = qT$, with q a natural number. The harmonic signal η_f, hence, is periodic on the time axis $\mathbb{Z}(T)$ if and only if $\eta_f(t) = \eta_f(t + qT)$ for all $t \in \mathbb{Z}(T)$ for some natural number q, that is, when $e^{j2\pi ft} = e^{j2\pi f(t+qT)} = e^{j2\pi ft} \cdot e^{j2\pi fqT}$ for all $t \in \mathbb{Z}(T)$. This holds if and only if $|fqT|$ is a natural number, say p, which is equivalent to the condition that $|f| = p/qT$. This in turn amounts to the statement that $|f|$ is a rational multiple of $1/T$. If $|f|$ is a rational multiple of $1/T$, the smallest q such that $|f| = p/qT$ is obtained by taking p and q coprime.   ∎

It is easy to see that, although on the time axis $\mathbb{Z}(T)$ the harmonic η_f in general does not repeat itself exactly, it repeats itself *approximately* at a rate $|f|$ if $-1/2T \leq$

$f < 1/2T$. One way of establishing this is by noting that $|f|$ is the average number of maxima and minima of the real and imaginary parts of η_f per unit of time (see Problem 2.6.9). Figure 2.10 illustrates this. If f does not lie between $-1/2T$ and $1/2T$, let $f' = f - k/T$ with $k \in \mathbb{Z}$ such that f' lies in the interval $[-1/2T, 1/2T)$. By 2.2.10, on the time axis $\mathbb{Z}(T)$ the harmonic η_f is indistinguishable from the harmonic $\eta_{f'}$. As a result, the average repetition rate of η_f is $|f'|$.

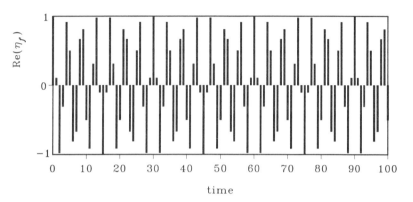

Figure 2.10 Real part of the harmonic η_f on the time axis $\mathbb{Z}$ with $f = 7/30$. The signal repeats itself exactly every 30 sampling intervals and approximately every $30/7 \approx 4.3$ sampling intervals.

2.3 ELEMENTARY OPERATIONS ON AND AMONG TIME SIGNALS

In this section we discuss various operations that *modify* signals. A distinction is made between *unary* and *binary* operations. Unary operations involve a *single* time signal, while binary operations require *two* signals. Important unary operations are *range* and *domain transformation*, which modify the signal range and the signal axis, respectively, and *sampling* and *interpolation*, which convert continuous-time to discrete-time signals and vice-versa. The binary operations we consider are (pointwise) *addition, subtraction, multiplication,* and *division*.

Signal Range Transformation

A signal range transformation changes the *range* of a time signal.

2.3.1.   Definition: Signal range transformation.   Let $\mathbb{T}$ be a signal axis, A_{old} a signal range and $\rho: A_{\text{old}} \to A_{\text{new}}$ a map from A_{old} to another signal range A_{new}. Then, *range transformation* of the signal $x_{\text{old}} \in A_{\text{old}}^{\mathbb{T}}$ under ρ results in the signal $x_{\text{new}} \in A_{\text{new}}^{\mathbb{T}}$ defined by

$$x_{\text{new}}(t) = \rho(x_{\text{old}}(t)), \qquad t \in \mathbb{T}. \qquad \blacksquare$$

Note that range transformation is actually map composition from the *left* because $x_{\text{new}} = \rho \circ x_{\text{old}}$. (See Supplement A for the definition of map composition.) Range transformation is a *pointwise* operation in the sense that $x_{\text{new}}(t)$ for any $t \in \mathbb{T}$ is fully determined by $x_{\text{old}}(t)$.

2.3.2.   Example: Full-wave rectification.   The amplitude transformation ρ: $\mathbb{R} \to \mathbb{R}_+$ defined by $\rho(x) = |x|$ is called *full-wave rectification*. Figure 2.11 illustrates what it does to a real harmonic signal.

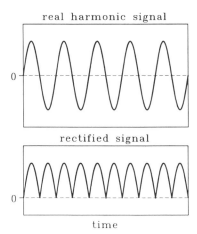

Figure 2.11   Full-wave rectification. Top: before rectification. Bottom: after.

■

Quantization

An important range transformation is *quantization*. This transformation is needed when signals are processed by a digital computer or other digital equipment, because computers can only handle signals that have a *finite* range.

2.3.3.   Definition: Quantization.   Any range transformation such that A_{new} is a finite set is called *quantization*.   ■

The following example illustrates *uniform* quantization.

2.3.4.   Example: Uniform quantization.   Suppose that $A_{\text{old}} = \mathbb{R}$, and let us range transform to the finite signal range $A_{\text{new}} = \underline{N}(H) = \{0, \ H, \ 2H, \ \cdots, \ (N-1)H\}$ by ρ: $\mathbb{R} \to \underline{N}(H)$, where

$$\rho(x) = \begin{cases} 0 & \text{for } x < 0, \\ H \, \text{int}(x/H) & \text{for } 0 \leq x < (N-1)H, \\ (N-1)H & \text{for } x \geq (N-1)H. \end{cases}$$

Here, int is the *entier* function defined such that $\text{int}(x)$ is the greatest integer less than or equal to the real number x. N is a natural number and H a positive number called the *quantization interval*. The graph of ρ is shown in Fig. 2.12. Quantization according to this function implies that signal values below 0 are set equal to 0, signal values between 0 and $(N-1)H$ are rounded down to the nearest value in the set $\underline{N}(H) = \{0, H, 2H, \cdots, (N-1)H\}$, while values above $(N-1)H$ are set equal to $(N-1)H$. We refer to this transformation as *uniform quantization*.

Figure 2.13 illustrates the effect of uniform quantization with $N = 8$ levels and quantization range $\{0, 1/(N-1), 2/(N-1), \cdots, 1\}$ on the discrete-time signal x defined by $x(t) = \frac{1}{2}[1 - \cos(2\pi t)]$, $t \in \mathbb{Z}(T)$, where $T = 1/50$.

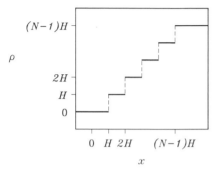

Figure 2.12 Range transformation for uniform quantization.

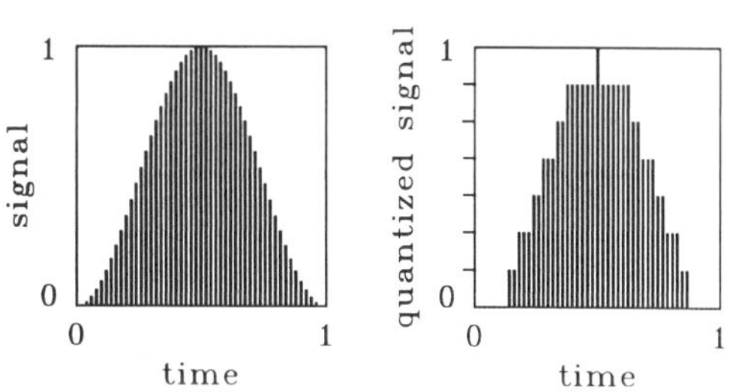

Figure 2.13 Uniform quantization. Left: before quantization. Right: after. ■

Signal Axis Transformation

We continue with defining *signal axis* or *signal domain* transformations.

2.3.5.   Definition: Signal axis transformation.   Let A be a signal range, $\mathbb{T}_{\text{old}}$ a signal axis, and $\tau: \mathbb{T}_{\text{old}} \to \mathbb{T}_{\text{new}}$ a bijective map from the signal axis $\mathbb{T}_{\text{old}}$ to another signal axis $\mathbb{T}_{\text{new}}$, with inverse $\tau^{-1}: \mathbb{T}_{\text{new}} \to \mathbb{T}_{\text{old}}$. Then, *signal axis transformation* of the signal $x_{\text{old}} \in A^{\mathbb{T}_{\text{old}}}$ under τ results in the signal $x_{\text{new}} \in A^{\mathbb{T}_{\text{new}}}$ defined on the signal axis $\mathbb{T}_{\text{new}}$, given by

$$x_{\text{new}}(t) = x_{\text{old}}(\tau^{-1}(t)) \qquad \text{for all} \quad t \in \mathbb{T}_{\text{new}}.$$ ■

For the definition of a bijective map, see Supplement A. Signal axis transformation is map composition from the *right* by τ^{-1}, because $x_{\text{new}} = x_{\text{old}} \circ \tau^{-1}$. The diagram of Fig. 2.14 illustrates this.

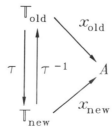

Figure 2.14 Signal axis transformation.

2.3.6.   Example: Time expansion, time compression, and time reversal.
Time *expansion,* time *compression,* and time *reversal* result by application of a time
axis transformation of the form

$$\tau(t) = t/\alpha, \qquad t \in \mathbb{T}_{\text{old}},$$
$$\tau^{-1}(t) = \alpha t, \qquad t \in \mathbb{T}_{\text{new}},$$

so that

$$x_{\text{new}}(t) = x_{\text{old}}(\alpha t), \qquad t \in \mathbb{T}_{\text{new}}.$$

Here, α is a nonzero real constant. If $\alpha > 1$, time is *compressed;* if $0 < \alpha < 1$,
time is *expanded;* while, if $\alpha = -1$, time is *reversed.* Collectively, these transformations are referred to as *time scaling.*
    Let x_{old} be the tooth-shaped pulse defined by

$$x_{\text{old}}(t) = \begin{cases} t/b & \text{for } 0 \le t < b, \\ 0 & \text{otherwise,} \end{cases} \qquad t \in \mathbb{R},$$

where b is a positive real constant. This signal is defined on the time axis $\mathbb{T}_{\text{old}} = \mathbb{R}$.
Time compression, expansion, or reversal result in a new time axis $\mathbb{T}_{\text{new}} = \mathbb{R}$ that
coincides with the old time axis. In Figs. 2.15(c)–(e), plots are given of the time
compressed signal, the time expanded signal, and the time reversed signal.

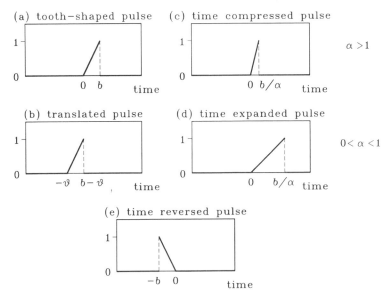

Figure 2.15   Time transformations of the tooth-shaped pulse.   ∎

2.3.7.   Example: Time translation.   Time translation consists of the application of a time axis transformation of the form

$$\tau(t) = t - \theta, \qquad t \in \mathbb{T}_{\text{old}},$$
$$\tau^{-1}(t) = t + \theta, \qquad t \in \mathbb{T}_{\text{new}},$$

where θ is a real constant. Time translation of x_{old} results in the translated signal

$$x_{\text{new}}(t) = x_{\text{old}}(t + \theta), \qquad t \in \mathbb{T}_{\text{new}}.$$

Figure 2.15(b) shows the effect of time translation on the tooth-shaped pulse of Example 2.3.6. The time axis again remains unchanged.   ∎

Sampling and Interpolation

Many physical signals, such as electrical voltages produced by a sound or image recording instrument or a measuring device, are essentially continuous-time signals. Computers and related devices operate on a discrete time axis. Continuous-time signals that are to be processed by such devices therefore first need be converted to discrete-time signals. One way of doing this is *sampling*.

2.3.8.   Definition: Sampling.   Let A be any signal range, $\mathbb{T}_{\text{con}}$ a continuous time axis, and $\mathbb{T}_{\text{dis}} \subset \mathbb{T}_{\text{con}}$ a discrete time axis. Then, *sampling* the continuous time signal $x \in A^{\mathbb{T}_{\text{con}}}$ on the discrete time axis $\mathbb{T}_{\text{dis}}$ results in the sampled signal $x^* \in A^{\mathbb{T}_{\text{dis}}}$ defined by

$$x^*(t) = x(t) \qquad \text{for all} \quad t \in \mathbb{T}_{\text{dis}}.$$ ∎

A device that performs the sampling operation is called a *sampler* and is schematically represented as in Fig. 2.16.

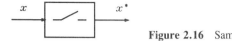

Figure 2.16   Sampler.

2.3.9.   Example: Sampled real harmonic.   Let the continuous-time signal x, given by $x(t) = \frac{1}{2}[1 - \cos(2\pi t)]$, $t \in \mathbb{R}$, be sampled on the uniformly spaced discrete time axis $\mathbb{Z}(T)$. This results in the sampled signal x^* given by

$$x^*(t) = \frac{1}{2}[1 - \cos(2\pi t)], \qquad t \in \mathbb{Z}(T).$$

Figure 2.17 depicts the two signals on the interval $[0, 1]$ for $T = 1/50$.

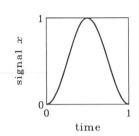
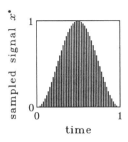

Figure 2.17 Sampling. Left: a continuous-time signal. Right: its sampled version. ■

The converse problem of sampling presents itself when a discrete-time device, such as a computer, produces signals that need drive a physical instrument requiring a continuous-time signal as input. Suppose that a discrete-time signal $x*$ is defined on the discrete time axis $\mathbb{T}_{dis}$ and that we wish to construct from $x*$ a continuous-time signal x defined on the continuous time axis $\mathbb{T}_{con} \supset \mathbb{T}_{dis}$. There obviously are many ways to do this. We introduce a particular class of conversions from discrete-time to continuous-time signals, for which we reserve the term *interpolation*. This type of conversion has the property that the continuous-time signal x agrees with the discrete-time signal $x*$ *at* the sampling times.

2.3.10.   Definition: Interpolation.   Suppose that A is a signal range, $\mathbb{T}_{dis}$ a discrete time axis, and $\mathbb{T}_{con} \supset \mathbb{T}_{dis}$ a continuous time axis. Let $x* \in A^{\mathbb{T}_{dis}}$ be a given discrete-time signal. Then, any continuous-time signal $x \in A^{\mathbb{T}_{con}}$ is called an *interpolation* of $x*$ on $\mathbb{T}_{con}$ if

$$x(t) = x*(t) \qquad \text{for all} \quad t \in \mathbb{T}_{dis}. \qquad\qquad ■$$

Another way of saying that x is an interpolation of $x*$ is the statement that sampling the continuous-time signal x generated by interpolating the discrete-time signal $x*$ on $\mathbb{T}_{dis}$ reproduces the discrete-time signal $x*$.

Clearly, there is no unique interpolation for a given discrete-time signal $x*$. Suppose that $x*$ is defined on the uniformly sampled discrete time axis $\mathbb{Z}(T)$. Then, one possible interpolation method is *step interpolation* as illustrated in Fig. 2.18(a).

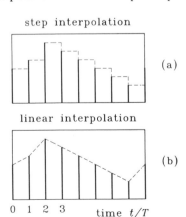

step interpolation

(a)

linear interpolation

(b)

0 1 2 3   time t/T

Figure 2.18 Interpolation. Top: step interpolation. Bottom: linear interpolation.

Given x^*, the step interpolated signal x is given by

$$x(t) = x^*(kT), \qquad kT \leq t < (k+1)T, \qquad k \in \mathbb{Z}.$$

Another interpolation method is *linear interpolation,* as in Fig. 2.18(b). Given x^*, the interpolated signal x is

$$x(t) = \left(1 - \frac{t - kT}{T}\right) x^*(kT) + \frac{t - kT}{T} x^*((k+1)T),$$
$$kT \leq t < (k+1)T, \qquad k \in \mathbb{Z}.$$

Step and linear interpolation are examples of interpolation by *interpolating functions.* An interpolating function is any function $i: \mathbb{R} \to \mathbb{C}$ such that

$$i(t) = \begin{cases} 1 & \text{for } t = 0, \\ 0 & \text{for } t = n, \end{cases} \qquad \text{where} \quad n \neq 0, \qquad n \in \mathbb{Z}, \qquad t \in \mathbb{R}.$$

If $x^* \in \ell(T)$ is a discrete-time signal defined on the time axis $\mathbb{Z}(T)$, and i an interpolating function, the continuous-time signal x given by

$$x(t) = \sum_{n \in \mathbb{Z}} x^*(nT) i\left(\frac{t - nT}{T}\right), \qquad t \in \mathbb{R},$$

is an interpolation of x^*. The reason is that by setting $t = kT$, with k an integer, it follows that

$$x(kT) = \sum_{n \in \mathbb{Z}} x^*(nT) i\left(\frac{kT - nT}{T}\right) = x^*(kT) \qquad \text{for} \quad k \in \mathbb{Z}.$$

Step interpolation is achieved with the *step interpolating function*

$$i_{\text{step}}(t) = \begin{cases} 1 & \text{for } 0 \leq t < 1, \\ 0 & \text{otherwise,} \end{cases} \qquad t \in \mathbb{R},$$

while linear interpolation is obtained with the *linear interpolating function*

$$i_{\text{lin}}(t) = \begin{cases} 1 - |t| & \text{for } |t| < 1, \\ 0 & \text{otherwise,} \end{cases} \qquad t \in \mathbb{R}.$$

Another interpolation function is the *sinc interpolating function*

$$i_{\text{sinc}}(t) = \text{sinc}(\pi t).$$

Here, sinc: $\mathbb{R} \to \mathbb{R}$ is the function defined by

$$\operatorname{sinc}(t) = \begin{cases} \dfrac{\sin(t)}{t} & \text{for } t \neq 0, \\[2mm] 1 & \text{for } t = 0, \end{cases} \qquad t \in \mathbb{R}.$$

Graphs of these interpolating functions are given in Fig. 2.19.

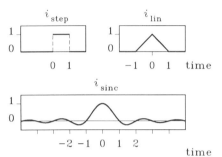

Figure 2.19 Interpolating functions. Top left: step interpolating function. Top right: linear interpolating function. Bottom: sinc interpolating function.

2.3.11.   Example: Interpolation.   Let x^* be the sampled signal

$$x^*(t) = \begin{cases} \cos\left(\dfrac{\pi t}{8T}\right) & \text{for } -4T \leq t < 4T, \\[2mm] 0 & \text{otherwise,} \end{cases} \qquad t \in \mathbb{Z}(T).$$

A plot of x^* is given in Fig. 2.20(a). Interpolation with the step interpolating function results in the staircase-like continuous-time signal of Fig. 2.20(b). Interpolation with the linear interpolating function leads to the signal of Fig. 2.20(c), which is obtained by connecting the sampled values of the original discrete-time signal by straight lines. Interpolation with the sinc interpolating function, finally, yields the signal of Fig. 2.20(d), which is smooth within the interval $[-4T, 4T]$ but shows a "ripple" outside it.   ∎

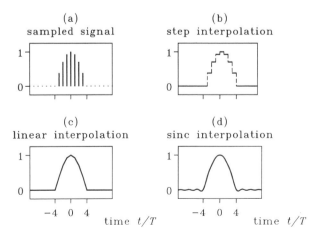

Figure 2.20 Interpolation. (a) A discrete-time signal. (b) Step interpolation. (c) Linear interpolation. (d) Sinc interpolation.

2.3.12.   Remark: Are sampling and interpolation inverses of each other?
Interpolation of a discrete-time signal followed by sampling on the original discrete time axis results in exact reconstruction of the original discrete-time signal. On the other hand, sampling a continuous-time signal followed by interpolation generally does not reproduce the original continuous-time signal. Evidently, sampling is an inverse operation to interpolation, but interpolation is not inverse to sampling. Figure 2.21 illustrates this.   ∎

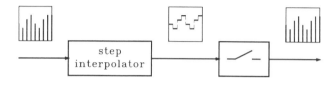

Figure 2.21 Sampling and interpolation. Top: sampling is the inverse of interpolation. Bottom: interpolation is not the inverse of sampling.

2.3.13.   Remark: Analog-to-digital and digital-to-analog conversion.

(a) *A/D conversion.* Computers and other digital equipment do not only operate on a discrete time axis but are also limited to finite signal ranges. Thus, apart from sampling, conversion of real-valued continuous-time signals to input for digital equipment also involves quantization (see 2.3.3 and 2.3.4.) The combined process of sampling and quantization is called *analog-to-digital (A/D) conversion.* Figure 2.22 illustrates how a real-valued continuous-time "analog" signal is converted to a quantized discrete-time "digital" signal.

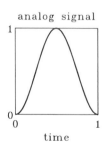

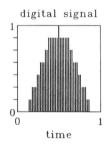

Figure 2.22 Analog-to-digital conversion. Left: analog signal. Right: digital signal.

(b) *D/A conversion.* The inverse process of converting a quantized discrete-time signal to a real-valued continuous-time signal is called *digital-to-analog (D/A) conversion.*

Devices that perform D/A conversion by step interpolation are known as *zero-order hold circuits.* They can function in "real time," meaning that given the sampled signal up to, and including, time t, the continuous-time signal up to and including that same time may be generated.

Devices that perform D/A conversion by linear interpolation are known as *first-order hold circuits*. They cannot precisely function in real time because two successive sampled signal values have to be received before it is known how the continuous-time signal goes. First-order hold circuits therefore introduce a *delay*, equal to the sampling interval T.

Sinc interpolation, finally, cannot be implemented in real time at all because *all* sampled signal values need be received before any point of the continuous-time signal (except at the sampling times) may be computed. Sinc interpolation, however, has great theoretical importance, as will be seen in Chapter 9 when we discuss the sampling theorem.   ∎

Pointwise Binary Operations

We conclude this section by defining several pointwise *binary* operations among complex-valued time signals, namely *addition, subtraction, multiplication,* and *division*. All definitions are obvious extensions of the corresponding complex number operations.

2.3.14.   Definition: Pointwise binary operations on complex-valued time signals.   If x and y are complex-valued time signals with time axis $\mathbb{T}$, their *sum* $x + y$ and *difference* $x - y$ are again complex-valued signals with time axis $\mathbb{T}$, given by

$$(x + y)(t) = x(t) + y(t) \qquad \text{for all} \quad t \in \mathbb{T},$$
$$(x - y)(t) = x(t) - y(t) \qquad \text{for all} \quad t \in \mathbb{T}.$$

Their *product* xy is the complex-valued signal given by

$$(xy)(t) = x(t)y(t) \qquad \text{for all} \quad t \in \mathbb{T},$$

while, if $y(t) \neq 0$ for all $t \in \mathbb{T}$, the *quotient* x/y is the signal given by

$$(x/y)(t) = x(t)/y(t) \qquad \text{for all} \quad t \in \mathbb{T}.$$ ∎

All operations defined in 2.3.14 are *pointwise*, that is, the value of the resulting signal at any given time depends on the values of the two signals that are operated upon at that same time only. Note that the signals that are operated on need be defined on the *same* time axis.

2.3.15.   Example: Sum and product of two time signals.   Let x and y be the continuous-time signals given by

$$x(t) = t \qquad \text{for} \quad t \in \mathbb{R},$$
$$y(t) = \begin{cases} 0 & \text{for } t < 0, \\ 1 & \text{for } t \geq 0, \end{cases} \qquad t \in \mathbb{R},$$

as sketched in Fig. 2.23. Then, $x + y$ and xy are the signals

$$(x + y)(t) = \begin{cases} t & \text{for } t < 0, \\ 1 + t & \text{for } t \geq 0, \end{cases} \quad t \in \mathbb{R},$$

$$(xy)(t) = \begin{cases} 0 & \text{for } t < 0, \\ t & \text{for } t \geq 0, \end{cases} \quad t \in \mathbb{R},$$

also shown in Fig. 2.23.

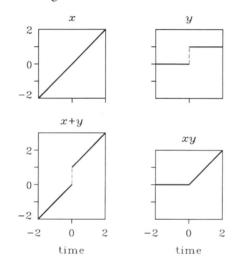

Figure 2.23 Pointwise binary operations. Top: two signals x and y. Bottom left: their sum $x + y$. Bottom right: their product xy. ∎

2.4 SIGNAL SPACES

In this section we introduce various signal sets that are important for the remainder of the book. Each of these sets has the structure of a *linear space*. A linear space, roughly, is a set in which we may *add* any two elements, resulting in another element of the set, and multiply each element by a *scalar*, again resulting in an element in the set. An axiomatic introduction to linear spaces is presented in Supplement B.

The scalars we use as multiplying factors are always *real* or *complex* numbers. In what follows, some important linear spaces are presented.

2.4.1.   Example: Linear spaces.

(a) $\mathbb{R}^N$ *and* $\mathbb{C}^N$. A very well-known linear space is the set $\mathbb{R}^N$ of all N tuples $x = (x_1, x_2, \cdots, x_N)$ with $x_i \in \mathbb{R}$ for $i = 1, 2, \cdots, N$. Addition of two elements $x = (x_1, x_2, \cdots, x_N)$ and $y = (y_1, y_2, \cdots, y_N)$ is defined coordinate-wise as

$$x + y = (x_1 + y_1, x_2 + y_2, \cdots, x_N + y_N),$$

while multiplication of x by the real scalar α is defined as

$$\alpha x = (\alpha x_1, \alpha x_2, \cdots, \alpha x_N).$$

Thus, $\mathbb{R}^N$ is a *linear space* over the *real numbers*. Completely analogously, the set $\mathbb{C}^N$ of complex-valued N tuples is a linear space over the *complex numbers*.

(b) *The space ℓ of all time sequences.* The set ℓ of all complex-valued time sequences $x = (\cdots, x(-1), x(0), x(1), x(2), \cdots)$ is a linear space over the complex numbers if addition and multiplication by a complex number are defined pointwise as

$$(x + y)(n) = x(n) + y(n), \qquad n \in \mathbb{Z},$$

$$(\alpha x)(n) = \alpha x(n), \qquad n \in \mathbb{Z}.$$

This space is an obvious extension of $\mathbb{C}^N$.

(c) *The space $\mathscr{L}$ of all continuous-time signals.* The set of all complex-valued continuous-time signals $\mathscr{L}$ forms a linear space under pointwise addition and pointwise multiplication by a complex scalar:

$$(x + y)(t) = x(t) + y(t), \qquad t \in \mathbb{R},$$

$$(\alpha x)(t) = \alpha x(t), \qquad t \in \mathbb{R}. \qquad \blacksquare$$

Norms

It often is useful to have a measure of the *size* of a signal. The *norm* provides such a measure. The norm $\|x\|$ of an element x of a linear space X is a non-negative real number that is zero if and only if x is the zero element. The norm is *homogeneous with respect to scaling* (i.e., $\|\lambda x\| = |\lambda| \cdot \|x\|$ for every scalar λ and for every x) and satisfies the *triangle inequality* (i.e., $\|x + y\| \le \|x\| + \|y\|$ for every x and y). A brief review of the theory of norms is given in Supplement B.

We first present a well-known family of norms on the spaces $\mathbb{R}^N$ and $\mathbb{C}^N$.

2.4.2.   Example: Norms on $\mathbb{R}^N$ and $\mathbb{C}^N$.   Let p be a real number such that $1 \le p \le \infty$. Then, the *p-norm* of an element $x = (x_1, x_2, \cdots, x_N)$ of $\mathbb{R}^N$ or $\mathbb{C}^N$ is defined as

$$\|x\|_p = \begin{cases} \left(\sum_{i=1}^{N} |x_i|^p \right)^{1/p} & \text{for} \quad 1 \le p < \infty, \\ \max_{1 \le i \le N} |x_i| & \text{for} \quad p = \infty. \end{cases}$$

The most frequently used norms are $\|\cdot\|_1$, $\|\cdot\|_2$, and $\|\cdot\|_\infty$. If, for instance, $x \in \mathbb{C}^2$ is given by $x = (1, j)$, then

$$\|x\|_1 = 1 + 1 = 2,$$

$$\|x\|_2 = \sqrt{1 + 1} = \sqrt{2},$$
$$\|x\|_\infty = \max(1, 1) = 1.$$

In $\mathbb{R}^2$ and $\mathbb{R}^3$, the 2-norm $\|x\|_2$ is actually the *length* of the vector with coordinates x. The 2-norm is often called the *Euclidean* norm. ∎

About the life of the Greek mathematician Euclid (ca 365–ca 300 BC) not much is known for certain. His main work *Elements* has a strong logical structure and was an example for all serious mathematical treatises until the 19th century.

The *p*-norm may be generalized to signal spaces.

2.4.3.   Example: Norms on the signal spaces ℓ and $\mathscr{L}$.

The *p*-norm $\|x\|_p$ of an element x of the space ℓ of time sequences is defined as

The *p*-norm $\|x\|_p$ of an element x of the space $\mathscr{L}$ of continuous-time signals is defined as

$$\|x\|_p = \begin{cases} \left(\sum_{n=-\infty}^{\infty} |x(n)|^p \right)^{1/p}, & 1 \le p < \infty, \\ \sup_{n \in \mathbb{Z}} |x(n)|, & p = \infty. \end{cases}$$

$$\|x\|_p = \begin{cases} \left(\int_{-\infty}^{\infty} |x(t)|^p \, dt \right)^{1/p}, & 1 \le p < \infty, \\ \sup_{t \in \mathbb{R}} |x(t)|, & p = \infty. \end{cases}$$
∎

In 2.4.3, "sup" denotes the *supremum* or *least upper bound*, that is, $\sup_{n \in \mathbb{Z}} |x(n)|$ is the smallest real number α such that $|x(n)| \le \alpha$ for all $n \in \mathbb{Z}$, if any such α exists, and ∞ if no such α exists.

The signal norms that we use in this text are the 1-, 2- and ∞-norms. The ∞-norm, given by

$$\|x\|_\infty = \sup_{n \in \mathbb{Z}} |x(n)|,$$

or

$$\|x\|_\infty = \sup_{t \in \mathbb{R}} |x(t)|,$$

is called the *amplitude* of the signal x. It is the largest magnitude the signal assumes. The square of the 2-norm, given by

$$\|x\|_2^2 = \sum_{n=-\infty}^{\infty} |x(n)|^2,$$

or

$$\|x\|_2^2 = \int_{-\infty}^{\infty} |x(t)|^2 \, dt,$$

is referred to as the *energy* of the signal. The name derives from the fact that in electrical and mechanical systems the physical energy associated with a signal may often be expressed by a quadratic integral. Finally, the 1-norm

$$\|x\|_1 = \sum_{n=-\infty}^{\infty} |x(n)|,$$

or

$$\|x\|_1 = \int_{-\infty}^{\infty} |x(t)| \, dt,$$

is called the *action* of the signal x in this text, because it may be viewed as a measure for the total action associated with the signal. Figure 2.24 illustrates the definitions of the amplitude, energy, and action of a signal.

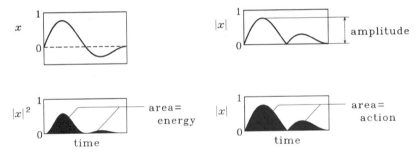

Figure 2.24   Amplitude, energy, and action of a continuous-time signal.

2.4.4.   Remark: Power, rms value, and mean of a signal.   Besides the amplitude, energy, and action, sometimes other quantities are used to represent certain aspects of the signal. If the energy of the signal x is not finite, its (*average*) *power*, if it exists, is defined by

$$\lim_{N \to \infty} \frac{1}{2N} \sum_{n=-N}^{N} |x(n)|^2,$$

or

$$\lim_{T \to \infty} \frac{1}{2T} \int_{-T}^{T} |x(t)|^2 \, dt,$$

in the discrete- and continuous-time case, respectively. The square root of the power of x is the *rms* (root mean square) value of x. The *mean* of x, finally, is defined as

$$\lim_{N \to \infty} \frac{1}{2N} \sum_{n=-N}^{N} x(n),$$

or

$$\lim_{T \to \infty} \frac{1}{2T} \int_{-T}^{T} x(t)\, dt,$$

respectively.

It is easily verified that the continuous-time periodic square wave of Fig. 2.25 with amplitude a has mean $\frac{1}{2}a$, power $\frac{1}{2}a^2$ and rms value $\frac{1}{2}a\sqrt{2}$.

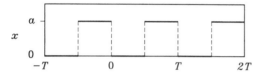

time   **Figure 2.25**   Periodic square wave. ∎

Normed Spaces

The set ℓ of all time sequences is a linear space. Not every element x of ℓ, however, has finite amplitude $\|x\|_\infty$. The subset

$$\ell_\infty := \{x \in \ell \,|\, \|x\|_\infty < \infty\}$$

of ℓ consists of all time sequences with finite amplitude. It is easy to prove that addition of finite-amplitude sequences and multiplication of a finite-amplitude sequence by a scalar result in sequences that have again finite amplitude. Hence, ℓ_∞ is a linear space. Linear spaces whose elements all have finite norms are called *normed spaces*. Thus, ℓ_∞ is a normed space, but ℓ is not. Likewise, the set $\mathscr{L}_\infty$ of all continuous-time signals with finite amplitude is a normed space.

It may be shown that also the subsets of all discrete- and continuous-time signals with *finite energy* or *finite action* are normed spaces. This result, which is elaborated on in Supplement B, may be summarized as follows.

2.4.5.   Summary: Normed signal spaces.

The set

$$\ell_\infty := \{x \in \ell \,|\, \|x\|_\infty < \infty\}$$

The set

$$\mathscr{L}_\infty := \{x \in \mathscr{L} \,|\, \|x\|_\infty < \infty\}$$

of all *finite-amplitude* time sequences, the set

$$\ell_2 := \{x \in \ell \,|\, \|x\|_2 < \infty\}$$

of all *finite-energy* time sequences, and the set

$$\ell_1 := \{x \in \ell \,|\, \|x\|_1 < \infty\}$$

of all *finite-action* time sequences are normed spaces.

of all *finite-amplitude* continuous-time signals, the set

$$\mathcal{L}_2 := \{x \in \mathcal{L} \,|\, \|x\|_2 < \infty\}$$

of all *finite-energy* continuous-time signals, and the set

$$\mathcal{L}_1 := \{x \in \mathcal{L} \,|\, \|x\|_1 < \infty\}$$

of all *finite-action* continuous-time signals are normed spaces.   ∎

Before continuing, we briefly review norms for sampled signals.

2.4.6.   Review: Norms of sampled signals.   The *p*-norm of a sampled signal *x* given on the discrete time axis $\mathbb{Z}(T)$ is defined as

$$\|x\|_p = \begin{cases} \left(T \sum_{t \in \mathbb{Z}(T)} |x(t)|^p\right)^{1/p} & \text{for} \quad 1 \le p < \infty, \\ \sup_{t \in \mathbb{Z}(T)} |x(t)| & \text{for} \quad p = \infty. \end{cases}$$

In particular, the *amplitude, energy,* and *action* of *x* are given by

$$\|x\|_\infty = \sup_{t \in \mathbb{Z}(T)} |x(t)|,$$

$$\|x\|_2^2 = T \sum_{t \in \mathbb{Z}(T)} |x(t)|^2,$$

$$\|x\|_1 = T \sum_{t \in \mathbb{Z}(T)} |x(t)|,$$

respectively. The reason for including the factor *T* in the definition of the energy and action is that the resulting expressions are approximating sums for the integrals

$$\int_{-\infty}^{\infty} |x(t)|^2 \, dt \qquad \text{and} \qquad \int_{-\infty}^{\infty} |x(t)| \, dt$$

that determine the energy and action of the underlying continuous-time signal (see Fig. 2.26). The approximation improves as *T* decreases.

   If the sampling interval *T* equals 1, the sampled signal *x* may be viewed as a time sequence; consistently with this, the *p*-norm of *x* reduces to the *p*-norm of the time sequence as defined in 2.4.3.

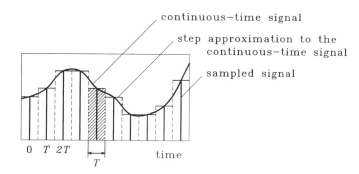

Figure 2.26   The energy and action of a sampled signal are step approximations to the energy and action of the continuous-time signal.

The sets

$$\ell_\infty(T) := \{x \in \ell(T) \,|\, \|x\|_\infty < \infty\},$$

$$\ell_2(T) := \{x \in \ell(T) \,|\, \|x\|_2 < \infty\},$$

$$\ell_1(T) := \{x \in \ell(T) \,|\, \|x\|_1 < \infty\},$$

of all finite-amplitude, finite-energy, and finite-action sampled signals on the time axis $\mathbb{Z}(T)$, respectively, are normed spaces.   ∎

Inner Product

The final topic covered in this section is the notion of *inner product*. We first discuss the idea for vectors in a plane. Let $x = (x_1, x_2)$ and $y = (y_1, y_2)$ be the rectangular coordinates of two vectors in a plane and denote by ϕ_1 and ϕ_2 the angles that the vectors make with the horizontal axis (see Fig. 2.27). Then, the cosine of the angle $\phi_1 - \phi_2$ between the two vectors is

$$\cos(\phi_1 - \phi_2) = \cos(\phi_1)\cos(\phi_2) + \sin(\phi_1)\sin(\phi_2)$$

$$= \frac{x_1}{\sqrt{x_1^2 + x_2^2}} \cdot \frac{y_1}{\sqrt{y_1^2 + y_2^2}} + \frac{x_2}{\sqrt{x_1^2 + x_2^2}} \cdot \frac{y_2}{\sqrt{y_1^2 + y_2^2}}$$

$$= \frac{x_1 y_1 + x_2 y_2}{(x_1^2 + x_2^2)^{1/2}(y_1^2 + y_2^2)^{1/2}}.$$

The denominator of the final expression is the product of the *lengths* of the vectors, while the numerator $x_1 y_1 + x_2 y_2$ is known as the *scalar product* of x and y. Writing $\|x\|_2 = (x_1^2 + x_2^2)^{1/2}$ for the length of x and $\|y\|_2 = (y_1^2 + y_2^2)^{1/2}$ for that of y, we see that the scalar product of x and y is given by

$$x_1 y_1 + x_2 y_2 = \|x\|_2 \cdot \|y\|_2 \cos(\phi_1 - \phi_2).$$

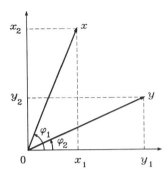

Figure 2.27 Alignment of two vectors in a plane.

For given lengths of the vectors, the scalar product is a measure for the *alignment* of the two vectors. If the scalar product is 0, $\cos(\phi_1 - \phi_2) = 0$ so that the angle between the vectors is 90°, which means that the vectors are perpendicular. The scalar product is maximal when $\cos(\phi_1 - \phi_2) = 1$, which means that the angle between the vectors is 0.

The scalar product of two vectors $x = (x_1, x_2, x_3)$ and $y = (y_1, y_2, y_3)$ in *three*-dimensional space is $x_1y_1 + x_2y_2 + x_3y_3$ and has a geometric interpretation similar to that of the scalar product of two-dimensional vectors.

The *inner product* $\langle x, y \rangle$ of two elements x and y of a linear space is a generalization of the scalar product. An axiomatic definition is given in Supplement B. An *inner product space* is a linear space such that any two elements have a finite inner product.

We first consider inner products on $\mathbb{R}^N$ and $\mathbb{C}^N$.

2.4.7.   Examples: Inner product on $\mathbb{R}^N$ and $\mathbb{C}^N$.

(a) *Inner product on* $\mathbb{R}^N$. The inner product

$$\langle x, y \rangle = \sum_{i=1}^{N} x_i y_i$$

of two elements $x = (x_1, x_2, \cdots, x_N)$ and $y = (y_1, y_2, \cdots, y_N)$ of $\mathbb{R}^N$ is a direct generalization of the scalar product.

(b) *Inner product on* $\mathbb{C}^N$. The elements $x = (x_1, x_2, \cdots, x_N)$ and $y = (y_1, y_2, \cdots, y_N)$ of $\mathbb{C}^N$ in general have complex coordinates. Their inner product is defined as

$$\langle x, y \rangle = \sum_{i=1}^{N} x_i \bar{y}_i,$$

with the overbar denoting the complex conjugate. The introduction of the complex conjugate guarantees the inner product $\langle x, x \rangle$ of any element x of $\mathbb{C}^N$ with itself to be real and positive. In fact, $\langle x, x \rangle = \|x\|_2^2$. ■

The notion of inner product may easily be generalized to signal spaces. The inner product of two signals is defined analogously to the inner product on $\mathbb{C}^N$.

2.4.8.   Example: Inner products on signal spaces.

The inner product of two time sequences x and y belonging to the signal space ℓ is defined as

$$\langle x, y \rangle = \sum_{n \in \mathbb{Z}} x(n) \overline{y(n)}.$$

The inner product of x with itself is

$$\langle x, x \rangle = \sum_{n \in \mathbb{Z}} |x(n)|^2 = \|x\|_2^2.$$

Since two arbitrary elements x and y of ℓ may well have an infinite inner product, such as $x = y = (\cdots, 1, 1, 1, \cdots)$, ℓ is not an inner product space. It follows from the Cauchy-Schwarz inequality below that two *finite-energy* time sequences always have a finite inner product, so that the space ℓ_2 of all finite-energy sequences is an inner product space.

The inner product of two continuous-time signals x and y belonging to the signal space $\mathcal{L}$ is defined as

$$\langle x, y \rangle = \int_{-\infty}^{\infty} x(t) \overline{y(t)} \, dt.$$

The inner product of x with itself is

$$\langle x, x \rangle = \int_{-\infty}^{\infty} |x(t)|^2 \, dt = \|x\|_2^2.$$

Since two arbitrary elements x and y of $\mathcal{L}$ may well have an infinite inner product, $\mathcal{L}$ is not an inner product space. It follows from the Cauchy-Schwarz inequality below that two *finite-energy* continuous-time signals always have a finite inner product, so that the space $\mathcal{L}_2$ of all finite-energy continuous-time signals is an inner product space. ∎

Earlier in this section we saw that if x and y are two elements of $\mathbb{R}^2$ or $\mathbb{R}^3$, then the cosine of the angle between x and y is given by

$$\cos(\phi_1 - \phi_2) = \frac{\langle x, y \rangle}{\|x\|_2 \cdot \|y\|_2}.$$

The right-hand side of this expression may vary from 1 to -1. If it is 1, then the angle between x and y is 0 and x and y are completely aligned. If the right-hand side is 0, then x and y are perpendicular. If the right-hand side is -1, then the vectors are again aligned but in opposed directions.

The *Cauchy-Schwarz inequality* allows a generalization of this argument to any inner product space.

2.4.9.   Summary: The Cauchy-Schwarz inequality.   For any x and y belonging to an inner product space

$$|\langle x, y \rangle|^2 \leq \langle x, x \rangle \langle y, y \rangle.$$

Equality holds if and only if there exists a scalar α such that $x = \alpha y$ or $y = \alpha x$. ∎

The French mathematician Augustin Louis Cauchy (1789–1857) was trained as
a civil engineer and contributed to many fields of mathematics. Hermann
Amandus Schwarz (1843–1921) was a German mathematician.

The proof of the Cauchy-Schwarz inequality may be found in Supplement B. Recalling that in the various inner product spaces we considered the inner product
$\langle x, x \rangle$ of x with itself equals $\|x\|_2^2$, for real spaces we may rewrite the Cauchy-
Schwarz inequality in the form

$$-1 \le \frac{\langle x, y \rangle}{\|x\|_2 \cdot \|y\|_2} \le 1.$$

Interpreting

$$\frac{\langle x, y \rangle}{\|x\|_2 \cdot \|y\|_2}$$

as the "cosine of the angle between x and y" we see that x and y are maximally
aligned (in the same or opposed directions) if their inner product $\langle x, y \rangle$ equals
$\pm \|x\|_2 \cdot \|y\|_2$, while they are least aligned when their inner product is 0.

2.4.10.   Review: Inner product of sampled signals.   The inner product of
two sampled signals x and y on the discrete time axis $\mathbb{Z}(T)$ is defined by

$$\langle x, y \rangle = T \sum_{t \in \mathbb{Z}(T)} x(t)\overline{y(t)}.$$

For $T = 1$ the definition reduces to the inner product of sequences. The set $\ell_2(T)$ of
all finite-energy sampled signals on the time axis $\mathbb{Z}(T)$ is an inner product space. ∎

2.4.11.   Application: Signal recognition.   In this example we show how the
idea of measuring the alignment of two signals by their inner product may be used in
signal recognition. Suppose that we receive a continuous-time signal z along some
transmission channel that is either a "long pulse" or a "short pulse." A "long pulse"
is the signal l defined by

$$l(t) = \begin{cases} \frac{1}{2} & \text{for } -2 \le t < 2, \\ 0 & \text{otherwise,} \end{cases}$$

while a "short pulse" is the signal s given by

$$s(t) = \begin{cases} 1 & \text{for } -\frac{1}{2} \le t < \frac{1}{2}, \\ 0 & \text{otherwise} \end{cases}$$

(see Fig. 2.28). The pulses have been scaled such that their energies and, hence, also their 2-norms, are 1.

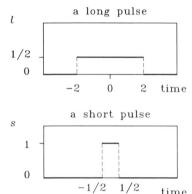

Figure 2.28 Top: a long pulse. Bottom: a short pulse.

One way of detecting whether the received signal z is a long or short pulse is to take the inner product of z with both l and s and to see with which z is best aligned. Suppose that z is a long pulse. Then,

$$\langle z, l \rangle = \langle l, l \rangle = 1,$$
$$\langle z, s \rangle = \langle l, s \rangle = 1/2,$$

so that z is best aligned with the long pulse l, which is not surprising. On the other hand, if z is a short pulse,

$$\langle z, l \rangle = \langle s, l \rangle = 1/2,$$
$$\langle z, s \rangle = \langle s, s \rangle = 1,$$

so that z is best aligned with the short pulse s.

We may thus use the following signal recognition scheme: If z is better aligned with the long pulse l than with the short pulse s, that is, if

$$\langle z, l \rangle > \langle z, s \rangle,$$

then the signal z is recognized as a long pulse, while, if it is better aligned with the short pulse s than with the long pulse l, that is, if

$$\langle z, l \rangle < \langle z, s \rangle,$$

then it is recognized as a short pulse.

This method of detecting which signal was received may also be used if the signal were distorted during transmission. Because taking the inner product is an averaging operation, the method often yields the correct answer. One of the results of advanced communication theory is that under wide assumptions concerning the nature of the distortion taking inner products is the best way of identifying the signal.

∎

2.5 GENERALIZED SIGNALS

All continuous-time signals we encountered so far were real- or complex-valued functions of a real variable. From now on we call such signals *regular* signals, because, in this section, we enlarge our repertoire with what are called *singular* signals. These are *not* functions in the ordinary sense. Together, the regular and singular signals form the *generalized* signals.

The Need for the Delta Function

To explain why singular signals are needed, consider charging a capacitor. In the circuit of Fig. 2.29 the switch initially is in the lower position and the capacitor uncharged. At time 0 the switch is set to the upper position. Because by assumption the resistance in the circuit is very small, a "burst" of current flows from the battery to the capacitor, which is almost instantaneously charged to the voltage V of the battery. Figure 2.30 shows the current i to the capacitor and its voltage. Often it is not important to know the precise shape of the current burst: all that is relevant is that the capacitor has been charged with charge $Q = CV$. Whatever the shape of the current burst is,

$$\int_{-\infty}^{\infty} i(t)\, dt = CV \tag{1}$$

holds.

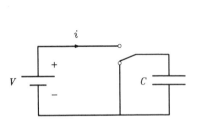

Figure 2.29 Charging a capacitor.

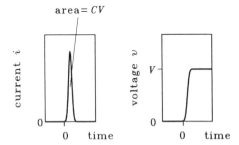

Figure 2.30 Left: the current to the capacitor. Right: the voltage across the capacitor.

If one insists on asking what the values of i are as a function of time under the assumption of *zero* resistance, it is tempting to answer

$$i(t) = \begin{cases} \infty & \text{for } t = 0, \\ 0 & \text{otherwise}, \end{cases} \qquad t \in \mathbb{R}.$$

The difficulty is that

(i) this is *not* a function in the ordinary sense because $\infty \notin \mathbb{C}$, and

(ii) since the current i equals 0 everywhere except at one point of time, its integral must be 0, contradicting (1).

Evidently, we need revise our notion of what a signal is if we wish to include the current burst in our repertoire. To this end, we enlarge the set of regular signals with a *singular* signal δ, called *delta function,* with the properties

$$\delta(t) = 0 \qquad \text{for} \qquad t \neq 0,$$

$$\int_{-\infty}^{\infty} \delta(t)\, dt = 1.$$

This signal is an "infinitely short, infinitely large pulse at time zero, with area one." Figure 2.31 gives a symbolic graphical representation of the δ-function. With this function we may describe the current that charges the capacitor as

$$i(t) = CV\, \delta(t), \qquad t \in \mathbb{R}.$$

The delta function is useful for describing other practical situations as well.

δ

0

time

Figure 2.31 Symbolic representation of a δ-function.

2.5.1.   Example: Mechanical impact.

An example of a mechanical phenomenon for which the δ-function is a useful model is the *impact* of a hammer that hits a nail. The force F exerted on the nail is very large and has very short duration, but the impact

$$P = \int_{-\infty}^{\infty} F(t)\, dt$$

is finite. The penetration of the nail is proportional to the impact. The "time behavior" of the force may be modeled as

$$F(t) = P\delta(t), \qquad t \in \mathbb{R}. \qquad\qquad\blacksquare$$

2.5.2.   **Example: Point masses and point charges.**   If we denote by ρ the mass or charge density along some spatial axis, a convenient model to describe a *point* charge or mass is

$$\rho(x) = A\delta(x), \qquad x \in \mathbb{R},$$

with A denoting the point charge Q or the point mass M, respectively.   ∎

2.5.3.   **Exercise: Approximation to the δ-function.**   Consider the circuit of Fig. 2.32, which is that of Fig. 2.29 with a small resistance R included that represents the resistance of the wires.

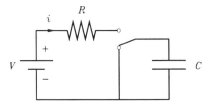

Figure 2.32   Charging a capacitor with a small resistance.

(a) Prove that if the switch is thrown from the lower to the upper position at time 0, the current is given by

$$i_R(t) = \begin{cases} 0 & \text{for } t < 0, \\ \dfrac{V}{R}e^{-t/RC} & \text{for } t \geq 0, \end{cases} \qquad t \in \mathbb{R},$$

as shown in Fig. 2.33, and that

$$\int_{-\infty}^{\infty} i_R(t)\, dt = CV.$$

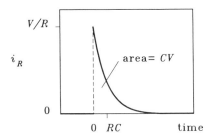

Figure 2.33   The current i_R.

(b) Show that

(i)

$$\lim_{R \to 0} i_R(t) = \begin{cases} 0 & \text{for } t \neq 0, \\ \infty & \text{for } t = 0, \end{cases}$$

(ii)

$$\lim_{R \to 0} \int_{-\infty}^{\infty} i_R(t) \, dt = CV,$$

(iii)

$$\int_{-\infty}^{\infty} \left(\lim_{R \to 0} i_R(t) \right) dt = 0.$$

Evidently, for small R the function i_R approximates the δ-function (within a factor CV), but $\lim_{R \to 0} i_R$ is *not* the δ-function, because of (iii).   ∎

Principles of the Theory of Singular Functions

A delta function is an example of a singular signal. Singular signals are signals that cannot be treated "pointwise," such as ordinary signals. As we have seen, a delta function has zero width and infinite height. Only its area, which is 1, is finite. The correct way to introduce delta functions, therefore, is "under the integral sign." We thus *define* the delta function δ as the entity such that

$$\int_{-\infty}^{\infty} \delta(t)\phi(t) \, dt = \phi(0)$$

for every regular function ϕ that is continuous at 0. We shall see that even though there is *no* such regular function δ, this definition allows us to introduce various operations on the delta function and establish its properties.

The delta function is not the only singular function; in particular, we shall soon meet the *derivatives* of the delta function.

The mathematical theory of singular functions is called *distribution theory*. A brief outline of this theory is given in Supplement C. In what follows we present an abbreviated exposition that introduces all the singular signals we need, and describes how to work with them.

Delta functions originally arose in theoretical physics, but their mathematical justification remained debatable for some time. Their theory was put on a firm mathematical basis in the period 1945–1950 by the French mathematician Laurent Schwartz (b. 1915), who developed distribution theory.

2.5.4.   Definition: Delta function.   The delta function δ is the singular function such that

$$\int_{-\infty}^{\infty} \delta(t)\phi(t) \, dt = \phi(0)$$

for every regular function ϕ that is continuous at 0.   ∎

Thus, for instance,

$$\int_{-\infty}^{\infty} e^{-t^2}\delta(t) \, dt = 1,$$

because the regular function ϕ given by

$$\phi(t) = e^{-t^2}, \qquad t \in \mathbb{R},$$

is continuous at 0 and equals 1 at this point.

In the sequel we encounter various other singular signals, such as shifted delta functions and derivatives of delta functions. Singular signals are defined "under the integral sign." This means that the *equality* of two generalized signals ε_1 and ε_2 need be established by verifying that

$$\int_{-\infty}^{\infty} \varepsilon_1(t)\phi(t) \, dt = \int_{-\infty}^{\infty} \varepsilon_2(t)\phi(t) dt$$

for all regular functions ϕ that are continuous and differentiable wherever and as often as needed.

Linear Combinations of Delta Functions

Equipped with the basic definition of a δ-function and the rule that singular functions are equal if they produce the same answer "under the integral sign," we now set out to define various operations on the delta function, such as linear combination, time scaling and time shifting, and differentiation. These lead to new singular signals. The operations are so defined that they are entirely consistent with the rules for regular functions.

The first of the operations we consider are *addition* of singular signals and *multiplication by a scalar*. For *regular* signals f_1 and f_2 and scalars α_1 and α_2 we have

$$\int_{-\infty}^{\infty} (\alpha_1 f_1 + \alpha_2 f_2)(t)\phi(t) \, dt = \alpha_1 \int_{-\infty}^{\infty} f_1(t)\phi(t) \, dt + \alpha_2 \int_{-\infty}^{\infty} f_2(t)\phi(t) \, dt$$

for every function ϕ such that the integrals exist. Hence, for singular signals ε_1 and ε_2 and scalars α_1 and α_2 we *define*

$$\int_{-\infty}^{\infty} (\alpha_1 \varepsilon_1 + \alpha_2 \varepsilon_2)(t)\phi(t)\, dt = \alpha_1 \int_{-\infty}^{\infty} \varepsilon_1(t)\phi(t)\, dt + \alpha_2 \int_{-\infty}^{\infty} \varepsilon_2(t)\phi(t)\, dt$$

for every ϕ that is continuous and differentiable as often as needed.

In particular, the product $\alpha\delta$ of the scalar α and the δ-function is defined by

$$\int_{-\infty}^{\infty} (\alpha\delta)(t)\phi(t)\, dt = \alpha \int_{-\infty}^{\infty} \delta(t)\phi(t)\, dt = \alpha\phi(0)$$

for every function ϕ that is continuous at 0. ∎

2.5.5.   Exercise: Sum of δ-functions.   Prove that

$$\alpha\delta + \beta\delta = (\alpha + \beta)\delta$$

for any complex scalars α and β. ∎

Time Scaling and Time Translation of Delta Functions

We next study how the δ-function may be scaled in time. Let α be a nonzero real number. If f is a regular function, then it follows by the substitution $\alpha t = \tau$ that

$$\int_{-\infty}^{\infty} f(\alpha t)\phi(t)\, dt = \frac{1}{|\alpha|} \int_{-\infty}^{\infty} f(\tau)\phi\left(\frac{\tau}{\alpha}\right) d\tau$$

for every ϕ such that the integral exists. For singular signals we take this to be the *definition* of time scaling. It follows for the delta function

$$\int_{-\infty}^{\infty} \delta(\alpha t)\phi(t)\, dt = \frac{1}{|\alpha|} \int_{-\infty}^{\infty} \delta(\tau)\phi\left(\frac{\tau}{\alpha}\right) d\tau$$

$$= \frac{1}{|\alpha|}\phi(0),$$

where second step follows by the basic property of the δ-function for any ϕ that is continuous at 0. We recognize that

$$\delta(\alpha t) = \frac{1}{|\alpha|}\delta(t), \qquad t \in \mathbb{R},$$

because application of the left-hand side to any function ϕ that is continuous at 0 gives the same result as applying the right-hand side.

We now consider how the δ-function may be translated in time. Let θ be a real number. Then, if δ were a regular function, it would follow by the substitution

$t - \theta = \tau$ that for every function ϕ that is continuous at θ

$$\int_{-\infty}^{\infty} \delta(t - \theta)\phi(t)\, dt = \int_{-\infty}^{\infty} \delta(\tau)\phi(\tau + \theta)\, d\tau$$
$$= \phi(\theta),$$

where the final step again follows from the basic property of the δ-function. Thus, we define the *translated δ-function* as the singular function such that

$$\int_{-\infty}^{\infty} \delta(t - \theta)\phi(t)\, dt = \phi(\theta)$$

for every regular function ϕ that is continuous at θ. The translated δ-function $\delta(t - \theta)$, $t \in \mathbb{R}$, may graphically be represented as in Fig. 2.34.

translated
delta function

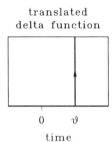

$$0 \qquad \vartheta$$

time   **Figure 2.34**   A translated δ-function.

Multiplication by a Function

We next consider the effect of multiplying the δ-function by a time function $f(t)$, $t \in \mathbb{R}$. If δ were a regular function, we would have for any ϕ that is continuous at 0

$$\int_{-\infty}^{\infty} [f(t)\delta(t)]\phi(t)\, dt = \int_{-\infty}^{\infty} \delta(t)[f(t)\phi(t)]\, dt$$
$$= f(0)\phi(0),$$

where the final step again follows by the basic property of the δ function, *provided f* is continuous at 0. Thus, we *define* the product $f(t)\delta(t)$, $t \in \mathbb{R}$, of a function f that is continuous at 0 and the δ-function as the singular function such that

$$\int_{-\infty}^{\infty} [f(t)\delta(t)]\phi(t)\, dt = f(0)\phi(0)$$

for every function ϕ that is continuous at 0. Inspection shows that

$$f(t)\delta(t) = f(0)\delta(t), \qquad t \in \mathbb{R}.$$

The assumption that f be continuous at 0 is essential. *A fortiori*, f cannot be singular at 0. In particular, the δ-function cannot be multiplied by itself.

Differentiation

We now consider how δ-functions may be differentiated. If f is a regular function, then it follows by partial integration that

$$\int_{-\infty}^{\infty} \frac{df(t)}{dt}\,\phi(t)\,dt = f(t)\phi(t)\Big|_{-\infty}^{\infty} - \int_{-\infty}^{\infty} f(t)\,\frac{d\phi(t)}{dt}\,dt$$

$$= -\int_{-\infty}^{\infty} f(t)\,\frac{d\phi(t)}{dt}\,dt$$

for any ϕ such that the integral exists and such that $f(-\infty)\phi(-\infty) = f(\infty)\phi(\infty) = 0$. If f were the δ-function, the right-hand side would reduce to

$$\int_{-\infty}^{\infty} \delta(t)\,\frac{d\phi(t)}{dt}\,dt = -\phi^{(1)}(0),$$

where $\phi^{(1)}$ stands for the derivative of ϕ. Hence, we *define* the derivative $\delta^{(1)}$ of the delta function by

$$\int_{-\infty}^{\infty} \delta^{(1)}(t)\phi(t)\,dt = -\phi^{(1)}(0)$$

for every function ϕ that is continuously differentiable at 0.

In the same way we may determine the nth derivative $\delta^{(n)}$ of the δ-function. By repeated partial integration it easily follows that $\delta^{(n)}$ need be defined by the requirement that

$$\int_{-\infty}^{\infty} \delta^{(n)}(t)\phi(t)\,dt = (-1)^n\phi^{(n)}(0)$$

for every regular function ϕ that is n times continuously differentiable at 0.

2.5.6.   Exercise. Product of the nth derivative of a δ-function and a function.   Prove that if $f(t)$, $t \in \mathbb{R}$, is a function whose nth derivative is continuous at 0,

$$f(t)\delta^{(n)}(t) = \sum_{k=0}^{n} (-1)^k \binom{n}{k} f^{(k)}(0)\delta^{(n-k)}(t), \qquad t \in \mathbb{R}.$$

In particular,

$$f(t)\delta^{(1)}(t) = f(0)\delta^{(1)}(t) - f^{(1)}(0)\delta(t), \qquad t \in \mathbb{R}.$$

Delta Function as the Derivative of a Step

Another way of looking at the δ-function is to consider it as the derivative of a unit step. In the ordinary sense, the step function $\mathbb{1}$ does not have a derivative, but, if it had, then by partial integration we would have for any differentiable function ϕ

$$\int_{-\infty}^{\infty} \frac{d\mathbb{1}}{dt}(t)\phi(t)\, dt = \mathbb{1}(t)\phi(t)\Big|_{-\infty}^{\infty} - \int_{-\infty}^{\infty} \mathbb{1}(t)\phi^{(1)}(t)\, dt$$

$$= \phi(\infty) - \int_{0}^{\infty} \phi^{(1)}(t)\, dt = \phi(\infty) - \phi(t)\Big|_{0}^{\infty} = \phi(0)$$

$$= \int_{-\infty}^{\infty} \delta(t)\phi(t)\, dt.$$

This proves that in the context of generalized functions the derivative of the unit step is the δ-function. Figure 2.35 illustrates this.

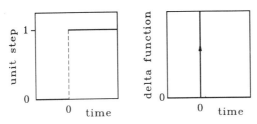

Figure 2.35 The derivative of the unit step is the δ-function. Left: the unit step. Right: its derivative.

The great advantage of the theory of singular signals is that any signal, including signals with discontinuities and singularities, may be differentiated arbitrarily often.

2.5.7.   Example: Acceleration of a mass.   Consider a unit mass that moves along a straight line. Suppose that we wish its position s to depend on time as

$$s(t) = \begin{cases} 0 & \text{for} \quad t < 0, \\ t & \text{for} \quad t \geq 0, \end{cases}$$

that is, the mass is to be at rest until time 0, and to move at a constant speed after time 0. The speed v and acceleration a of the mass follow by successive differentiation as

$$v(t) = \frac{ds(t)}{dt} = \begin{cases} 0 & \text{for} \quad t < 0, \\ 1 & \text{for} \quad t \geq 0, \end{cases}$$

$$= \mathbb{1}(t), \qquad t \in \mathbb{R},$$

$$a(t) = \frac{dv(t)}{dt} = \delta(t), \qquad t \in \mathbb{R}.$$

By Newton's law, this means that the force F applied to the mass need have the impulse-like behavior

$$F(t) = \delta(t), \qquad t \in \mathbb{R}.$$

Figure 2.36 shows the position, velocity, and acceleration of the mass.   ■

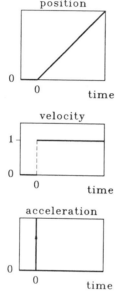

Figure 2.36 A moving mass. Top: its position. Middle: its velocity. Bottom: its acceleration.

Approximations to Delta Functions

The δ-function and its derivatives do not exist as regular functions. They may be *approximated* by sequences of regular functions, however. Define, for instance, the sequence of functions d_n given by

$$d_n(t) = \begin{cases} n & \text{for} \quad -1/2n \leq t < 1/2n, \\ 0 & \text{otherwise}, \end{cases} \qquad t \in \mathbb{R},$$

as shown in Fig. 2.37 (left). This sequence approximates the δ-function in the sense that for every function ϕ that is continuous at 0

$$\lim_{n \to \infty} \int_{-\infty}^{\infty} d_n(t)\phi(t)\, dt = \phi(0).$$

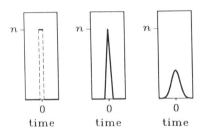

Figure 2.37 Approximations to the δ-function. Left: rectangular approximation. Middle: triangular approximation. Right: bell-shaped approximation.

Note, however, that

$$\lim_{n \to \infty} d_n$$

does *not* exist, which is consistent with the fact that the δ-function is not a regular function.

Other well-known approximations to the δ-function are

$$d_n(t) = n \operatorname{trian}(nt), \qquad t \in \mathbb{R},$$

and

$$d_n(t) = n \operatorname{bell}(nt), \qquad t \in \mathbb{R}.$$

The "bell" function is defined as

$$\operatorname{bell}(t) = \frac{1}{\sqrt{2\pi}} e^{-t^2/2}, \qquad t \in \mathbb{R}.$$

Plots of these approximations are also given in Fig. 2.37.

Approximations to the derivatives $\delta^{(k)}$ of the δ-function may be obtained by differentiating sufficiently smooth approximations to the δ-function, for instance, the bell-shaped approximation. Figure 2.38 shows approximations to the first deriva-

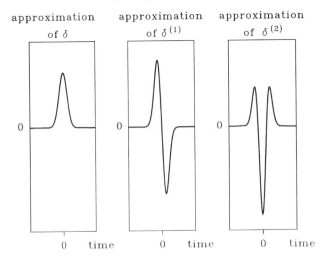

Figure 2.38 Approximations to δ-functions. Left: bell approximation to the δ-function. Middle: approximation to the first derivative $\delta^{(1)}$. Right: approximation to the second derivative $\delta^{(2)}$.

tive $\delta^{(1)}$ and the second derivative $\delta^{(2)}$, obtained by differentiating the bell approximation. Because of the shape of the approximation, $\delta^{(1)}$ is sometimes called a "doublet" function and $\delta^{(2)}$ a "triplet" function. In physics, the conjunction of an infinite positive and negative charge is known as an electrical *dipole*. Its mathematical description is a doublet.

Review

Table 2.1 Summarizes the main properties of the δ-function.

TABLE 2.1   PROPERTIES OF THE δ-FUNCTION

	Property	Conditions
(1)	$\int_{-\infty}^{\infty} \delta(t)\phi(t)\,dt = \phi(0)$	ϕ continuous at 0
(2)	$\delta(\alpha t) = \dfrac{1}{|\alpha|}\delta(t)$	α real and nonzero
(3)	$\int_{-\infty}^{\infty} \delta(t-\theta)\phi(t)\,dt = \phi(\theta)$	θ real, ϕ continuous at θ
(4)	$f(t)\delta(t) = f(0)\delta(t)$	f continuous at 0
(5)	$\dfrac{d}{dt}\mathbb{1}(t) = \delta(t)$	
(6)	$\dfrac{d^n}{dt^n}\delta(t) = \delta^{(n)}(t)$	$n \in \mathbb{N}$
(7)	$\int_{-\infty}^{\infty} \delta^{(n)}(t)\phi(t)\,dt = (-1)^n\phi^{(n)}(0)$	$n \in \mathbb{Z}_+$, ϕ at least n times continuously differentiable at 0
(8)	$f(t)\delta^{(n)}(t) = \sum_{k=0}^{n}(-1)^k\binom{n}{k}f^{(k)}(0)\delta^{(n-k)}(t)$	$n \in \mathbb{Z}_+$, f at least n times continuously differentiable at 0

2.6 PROBLEMS

The first few problems deal with complex numbers, sets, maps, and power sets as reviewed in Supplement A.

2.6.1. Complex numbers. Given the following pairs of complex numbers x and y, compute their sum $x + y$, product xy, quotient x/y as well as the exponential e^x. Represent the results both in Cartesian and in polar form.

(a) $x = 1 + j$, $y = 1 - j$.
(b) $x = e^{j\pi/2}$, $y = e^{-j\pi/2}$.
(c) $x = e^{j\pi/4}$, $y = 1 + j$.

2.6.2. Complex conjugate. If x and y are complex numbers, prove the following:

(a) $\overline{x \pm y} = \overline{x} \pm \overline{y}$.
(b) $\overline{xy} = \overline{x}\,\overline{y}$ and $\overline{(x/y)} = \overline{x}/\overline{y}$.

2.6.3. Product set. Of which sets A and B is the set C = {(1, 1), (1, 2), (1, 3), (2, 1), (2, 2), (2, 3)} the product set?

2.6.4. Power sets.

(a) Give examples of elements of the power sets $\mathbb{B}^{\mathbb{B}}$ and $\mathbb{Z}^{\mathbb{B}}$, where $\mathbb{B}$ is the set $\mathbb{B} = \{0, 1\}$. Show that $\mathbb{B}^{\mathbb{B}}$ has four elements and that $\mathbb{Z}^{\mathbb{B}}$ is countably infinite.

(b) Suppose that X is a finite set containing N elements and Y a finite set with M elements. Prove that the power set X^Y has N^M elements.

(c) Show that the power sets $\ell = \mathbb{C}^{\mathbb{Z}}$ and $\mathscr{L} = \mathbb{C}^{\mathbb{R}}$ are uncountable.

2.6.5. Maps. Verify whether the following maps $\phi\colon X \to Y$ are surjective, injective, or bijective. If the map ϕ is bijective, determine its inverse.

(a) $X = \mathbb{R}$, $Y = \mathbb{R}$, $\phi(x) = \sin(x)$.

(b) $X = \mathbb{R}$, $Y = [-\pi/2, \pi/2]$, $\phi(x) = \operatorname{atan}(x)$.

(c) $X = \mathbb{R}^2$, $Y = \mathbb{R}^2$, $\phi(x_1, x_2) = (2x_1, x_1 + x_2)$.

(d) $X = \mathbb{R}_+$, $Y = \mathbb{R}_+$, $\phi(x) = \sqrt{x}$.

The following problems relate to time signals as introduced in Section 2.2.

2.6.6. Various time signals. Plot the following continuous-time signals x.

(a) $x(t) = \mathbb{1}(t)\mathbb{1}(\theta - t)$, $t \in \mathbb{R}$, for some fixed $\theta \in \mathbb{R}$.

(b) $x(t) = e^{-t}\mathbb{1}(t)$, $t \in \mathbb{R}$.

(c) $x(t) = \sin(\operatorname{ramp}(t))$, $t \in \mathbb{R}$.

(d) $x(t) = \operatorname{ramp}(\sin(t))$, $t \in \mathbb{R}$.

(e) $x(t) = \operatorname{rect}(t/2 - 1)$, $t \in \mathbb{R}$.

Also, plot the following discrete-time signals.

(f) $x(n) = a^n\mathbb{1}(n)$, $n \in \mathbb{Z}$, with a a real number such that $0 < a < 1$.

(g) $x(n) = a^n\mathbb{1}(-n)$, $n \in \mathbb{Z}$, with a a real number such that $a > 1$.

(h) $x(n) = \sum_{k=0}^{\infty} \mathbb{1}(n - kN)$, $n \in \mathbb{Z}$, with N a fixed natural number.

2.6.7. Periodic time signals. Let z_1 and z_2 be the discrete-time signals defined by

$$z_1(n) = (-1)^n, \qquad n \in \mathbb{Z}, \qquad z_2(n) = j^n, \qquad n \in \mathbb{Z}.$$

(a) Plot the time signals.

(b) Show that z_1 and z_2 are periodic, and determine their periods.

2.6.8. Aliasing. Consider the complex harmonic signals η_{f_i}, $i = 1, 2, \cdots, 5$, defined on the discrete time axis $\mathbb{T} = \mathbb{Z}$, where $f_1 = 0$, $f_2 = 1/2$, $f_3 = 1$, $f_4 = 5/4$, and $f_5 = 3/2$. How many distinct harmonic signals do we really have?

2.6.9. Average repetition rate of discrete-time harmonic signals. Let c_f be the discrete-time real harmonic signal with frequency $f \in \mathbb{R}$ defined by

$$c_f(t) = \cos(2\pi f t), \qquad t \in \mathbb{Z}(T).$$

(a) Suppose that $-1/2T \le f < 1/2T$. Define for the natural number N the quantity M_N as the number of *maxima* of the signal c_f on the interval $[-NT, NT)$. Prove that

$$\lim_{N \to \infty} \frac{M_N}{2NT} = |f|,$$

that is, the *average* number of maxima of the harmonic signal per unit of time equals $|f|$.

(b) Suppose that f does not necessarily lie in the interval $[-1/2T, 1/2T)$. Show that the average number of maxima per unit of time now is $|f'|$, where $f' = f - k/T$ with $k \in \mathbb{Z}$ such that f' lies in the interval $[-1/2T, 1/2T)$.

This result shows that the average repetition rate of a discrete-time harmonic signal with frequency f is $|f'|$.

The following problems concern elementary operations on time signals, as presented in Section 2.3.

2.6.10. Signal range transformations. Consider the continuous-time real harmonic signal x given by

$$x(t) = \sin(2\pi ft), \qquad t \in \mathbb{R}.$$

Plot this signal after application of the following signal range transformations ρ: $\mathbb{R} \to \mathbb{R}$.

(a) *Hard limiting*: $\rho(x) = \text{sign}(x)$, $x \in \mathbb{R}$, with the sign function as defined in 3.2.6(b).
(b) *Soft limiting*: $\rho(x) = \text{sat}(x)$, $x \in \mathbb{R}$, with the sat function as defined in 3.2.6(b).
(c) *Half-wave rectification*: $\rho(x) = \text{ramp}(x)$, $x \in \mathbb{R}$.
(d) *Squaring*: $\rho(x) = x^2$, $x \in \mathbb{R}$.

2.6.11. Time transformations. Let x be the continuous-time signal given by

$$x(t) = \sin(2\pi ft)\mathbb{1}(t), \qquad t \in \mathbb{R}.$$

(a) Plot the signal.
(b) Plot the signal after time compression by a factor 2.
(c) Plot the signal after time expansion by a factor 2.
(d) Plot the signal after time reversal.
(e) Plot the signal after translating it in time by 1 time unit.

2.6.12. Sampling and interpolation.

(a) Give an example of a continuous-time signal where sampling followed by step interpolation exactly reconstructs the original continuous-time signal.
(b) Give an example of a continuous-time signal—not that of (a)—where sampling followed by linear interpolation exactly reconstructs the original continuous-time signal.

The next few problems involve various aspects of signal spaces as discussed in Section 2.4 and Supplement B.

2.6.13. Amplitude, energy, and action. Compute the amplitude, energy, and action of the following signals.

(a) $z(t) = \dfrac{1}{1 + t} \mathbb{1}(t), t \in \mathbb{R}$.

(b) $z(t) = \cos(2\pi f t), t \in \mathbb{R}$, for some fixed real frequency f.

(c) $z(n) = a^n \mathbb{1}(-n), n \in \mathbb{Z}$, for some fixed complex number a. Carefully distinguish whether $|a| < 1$, $|a| = 1$, or $|a| > 1$.

(d) $z(n) = a^{|n|}, n \in \mathbb{Z}$, for some fixed complex number a. Carefully distinguish whether $|a| < 1$, $|a| = 1$, or $|a| > 1$.

2.6.14. Effect of time translation, expansion, and compression on amplitude, energy, and action.

(a) Prove that time translation does not change the amplitude, energy, and action of any time signal in $\mathscr{L}$.

(b) Show that time expansion or compression does not affect the amplitude of any signal in $\mathscr{L}$ but multiplies its energy and action by a constant.

2.6.15. Signal spaces. Determine to which of the signal sets ℓ_∞, ℓ_2, and ℓ_1 the following discrete-time signals z belong.

(a) $z(n) = \mathbb{1}(n)\mathbb{1}(N - n), n \in \mathbb{Z}$, for some fixed integer N.

(b) $z(n) = a^{|n|}, n \in \mathbb{Z}$, with a a fixed complex number. Distinguish the cases $|a| < 1$, $|a| = 1$, and $|a| > 1$.

2.6.16. Inner product. Let $x = (1, 1, 1)$ and $y \in \mathbb{R}^3$ represent the rectangular coordinates of two vectors in three-dimensional space. Use the inner product to compute the angle between x and y in the following cases:

(a) $y = (1, 0, 0)$.

(b) $y = (1, 1, 0)$.

(c) $y = (1, 2, 3)$.

2.6.17. Signal recognition. We follow up Example 2.4.11, and suppose that the signal z is a rectangular pulse as in Fig. 2.39, which is nonzero on the interval $[-a, a]$ and has height b.

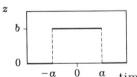

Figure 2.39 A rectangular pulse.

(a) Amplitude scale the pulse such that its energy is 1 (i.e., choose b such that the energy of the signal is 1).

(b) Compute the inner products $\langle z, s \rangle$ and $\langle z, l \rangle$ of the signal with the short pulse s and the long pulse l (carefully distinguish whether a lies between 0 and $1/2$, between $1/2$ and 2, or is greater than 2).

(c) Determine for which values of a the pulse is best aligned with the short pulse and for which values it is best aligned with the long pulse.

(d) Suppose that z is the triangular pulse $z(t) = c \operatorname{trian}(t/2)$, $t \in \mathbb{R}$. Choose c such that the signal z has energy 1 and sketch z.

(e) Is the triangular pulse better aligned with the short or with the long pulse?

The final series of problems for this chapter deals with δ-functions as introduced in Section 2.5 and elaborated upon in Supplement C.

2.6.18. Products with δ-functions. Simplify the following δ-function expressions.

(a) $x(t) = \delta(t)e^{-|t|}$, $t \in \mathbb{R}$.

(b) $x(t) = \delta^{(2)}(t)e^{t^2}$, $t \in \mathbb{R}$.

(c) $x(t) = \delta^{(n)}(t)t^m$, $t \in \mathbb{R}$, with n and m nonnegative integers.

(d) $x(t) = \delta^{(n)}(t) \sin(t)$, $t \in \mathbb{R}$, with n a nonnegative integer.

In (c) and (d) we define $\delta^{(0)} = \delta$.

2.6.19. Symmetry property of derivatives of δ-functions. Prove that

$$\delta^{(n)}(-t) = (-1)^n \delta^{(n)}(t), \qquad t \in \mathbb{R},$$

for $n = 0, 1, 2, \cdots$, where $\delta^{(0)} = \delta$.

2.6.20. Differentiation of generalized signals. Determine the derivative Dx when x is given as follows.

(a) $x(t) = t^2 \mathbb{1}(t)$, $t \in \mathbb{R}$.

(b) $x(t) = \operatorname{rect}(t)$, $t \in \mathbb{R}$.

(c) $x(t) = \sum_{n=-\infty}^{\infty} \mathbb{1}(t - nP)$, $t \in \mathbb{R}$, with P a positive real number.

(d) Determine all derivatives of the signal x given by

$$x(t) = \begin{cases} 0 & \text{for } t < 0, \\ t^N & \text{for } t \geq 0, \end{cases} \qquad t \in \mathbb{R},$$

with N a nonnegative integer.

2.6.21. Differentiation of a jump. Let x be a continuous-time signal that has a jump at time t_o, as shown in Fig. 2.40, but is otherwise continuously differentiable with derivative $\dot{x}$. Show that the derivative Dx of x is given by

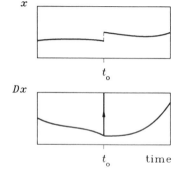

Figure 2.40 Differentiation of a jump. Top: signal with jump at time t_o. Bottom: its derivative.

$$Dx(t) = x'(t) + [x(t_o^+) - x(t_o^-)]\delta(t - t_o), \qquad t \in \mathbb{R},$$

where

$$x'(t) = \dot{x}(t) \quad \text{for} \quad t \in \mathbb{R} \quad \text{with } t \neq t_o,$$
$$x(t_o^+) = \lim_{\varepsilon \downarrow 0} x(t_o + \varepsilon),$$
$$x(t_o^-) = \lim_{\varepsilon \downarrow 0} x(t_o - \varepsilon).$$

As Fig. 2.40 shows, the derivative Dx has a δ-function at the time of the jump. The coefficient of the δ-function equals the size of the jump $x(t_o^+) - x(t_o^-)$.

2.6.22. Composite argument of the δ-function. Suppose that $f: \mathbb{R} \to \mathbb{R}$ is a continuously differentiable monotonically increasing function with derivative f' such that $f(\theta) = 0$ for some real θ. Prove that

$$\delta(f(t)) = \frac{1}{f'(\theta)}\delta(t - \theta), \qquad t \in \mathbb{R}.$$

2.6.23. Impact. An object with mass m moves along a straight line with constant speed v_o. At time 0 the object is subject to an impulse-like force given by

$$F(t) = P\delta(t), \qquad t \in \mathbb{R}.$$

How large should the momentum P be such that the motion of the object is stopped? *Hint:* If v is the speed of the object, by Newton's law $m\,dv(t)/dt = F(t)$.

2.7 COMPUTER EXERCISES

The computer exercises that follow may be carried out with the SIGSYS program. The exercises for Chapter 2 mainly serve to introduce the possibilities of SIGSYS to create and manipulate signals. In addition, the theory of this chapter is illustrated. Many of the exercises may also be done with other software, such as MATLAB.

2.7.1. Complex numbers. (Compare 2.6.1.) Given the following pairs of complex numbers x and y, compute their sum $x + y$, product xy, quotient x/y, as well as the exponential e^x. Represent the results both in Cartesian and in polar form.

(a) $x = 1 + j, y = 1 - j.$ (c) $x = e^{j\pi/4}, y = 1 + j.$
(b) $x = e^{j\pi/2}, y = e^{-j\pi/2}.$

2.7.2. Various time signals. (Compare 2.6.6 and 2.6.7.) Plot the signal x for the following cases. In each case, choose the time axis with care.

(a) $x(t) = \mathbb{1}(t)\mathbb{1}(\theta - t), t \in \mathbb{R}$, with $\theta = 1$.
(b) $x(t) = e^{-1}\mathbb{1}(t), t \in \mathbb{R}.$
(c) $x(t) = \sin(\text{ramp}(t)), t \in \mathbb{R}.$
(d) $x(t) = \text{ramp}(\sin(t)), t \in \mathbb{R}.$

Representation of continous-time signals.

On a digital computer, continuous-time signals necessarily need be represented in sampled form. *Pointwise* operations, such as addition and multiplication, on continuous-time signals that are represented this way yield precise results but only at the sampling instants. Some other operations, however, such as integration, and the computation of the energy, action, and inner product, result in inaccuracies, because integrals are replaced with approximating sums.

When working with continuous-time signals, the values of the sampling interval (inc in SIGSYS) and the number of samples (num in SIGSYS) need be selected with what follows in mind:

1. Sufficient resolution (i.e., the sampling interval small enough and the number of samples large enough) to get good plots.
2. Sufficient resolution for numerical accuracy.
3. The length of the signal axis should be long enough to encompass all events of interest.
4. Choosing the number of samples large slows down computation and requires much memory.

Generating time signals.

In SIGSYS, time signals usually are most conveniently defined by means of signal composition with dot or dotplus as explained in Section 10 of the Tutorial. The time axis of the resulting signals is that of dot or dotplus, depending on which time axis is used. Before creating a time signal it is necessary to decide whether dot is used (for "two-sided" time axes) or dotplus (for "one-sided" time axes), and to select suitable values for the sampling interval inc and the number of sample points num.

Suppose by way of example that one wishes to plot one period of the signal x defined by

$$x(t) = \sin(2\pi t), \qquad t \in \mathbb{R}.$$

To get good resolution for plotting we take the number of samples equal to 100. Because the signal has period 1, we choose the sampling interval as $1/100 = 0.01$. By using a one-sided time axis, the time signal may be generated and displayed by typing

```
num=100
inc=1/num
x=sin(2*pi*dotplus)
plot x
```

(e) $x(t) = \text{rect}(t/2 - 1)$, $t \in \mathbb{R}$.
(f) $x(n) = a^n \mathbb{1}(n)$, $n \in \mathbb{Z}$, with $a = 0.9$.
(g) $x(n) = a^n \mathbb{1}(-n)$, $n \in \mathbb{Z}$, with $a = 1.1$.
(h) $x(n) = \sum_{k=0}^{\infty} \mathbb{1}(n - kN)$, $n \in \mathbb{Z}$, with $N = 4$.
(i) $x(n) = (-1)^n$, $n \in \mathbb{Z}$.
(j) $x(n) = j^n$, $n \in \mathbb{Z}$.

2.7.3. Periodic signals. Generate and plot several periods of the following periodic time signals.

(a) $x(t) = t \bmod 1$, $t \in \mathbb{R}$.
(b) $x(t) = \text{rect}(t \bmod 1)$, $t \in \mathbb{R}$.
(c) The discrete-time *infinite comb*, defined by

$$x(n) = \begin{cases} 1 & \text{for} \quad n \bmod N = 0, \\ 0 & \text{otherwise}, \end{cases} \qquad n \in \mathbb{Z},$$

with N a natural number. Take $N = 4$.

2.7.4. Discrete-time harmonic signals.

(a) *Aliasing.* Figure 2.9 shows that two real harmonic signals with frequencies $f_1 = 1/4$ and $f_2 = 5/4$ coincide on the time axis $\mathbb{T} = \mathbb{Z}$. Reproduce this plot.
(b) *A nonperiodic discrete-time harmonic signal.* According to 2.2.12, discrete-time harmonic signals on the time axis $\mathbb{Z}(T)$ are only periodic if their frequency is a rational multiple of $1/T$. To demonstrate this, plot the real harmonic signal c_f given by

$$c_f(t) = \cos(2\pi f t), \qquad t \in \mathbb{Z}(T),$$

with $f = \pi$ and $T = 1$ on the interval $[0, 50)$. The signal clearly is not periodic, but it seems to be close to periodic with period 7. Explain this. *Hint:* See 2.6.9.

2.7.5. Signal range transformation. (Compare 2.6.10.) In Fig. 2.11 it is shown what

(a) *full-wave rectification*

does to a real harmonic signal. Reproduce this figure. Show also what

(b) *half-wave rectification,*
(c) *hard limiting,*
(d) *soft limiting,* and
(e) *squaring*

do to the signal.

2.7.6. Time transformations. (Compare 2.6.11.) Let x be the signal given by

$$x(t) = \sin(2\pi f t), \qquad t \in \mathbb{R},$$

with f a given frequency.

(a) Choose a convenient time axis and a suitable value of f, and plot the signal.
(b) Plot the signal after time compression by a factor 2.

(c) Plot the signal after time expansion by a factor 2.

(d) Plot the signal after time reversal.

(e) Plot the signal after translating it by a nonzero time.

2.7.7. Some basic operations on signals.

(a) *Sampling and quantization.* Figure 2.17 shows the effect of sampling a continuous-time harmonic, and Fig. 2.13 demonstrates quantization. Reproduce these figures.

(b) *Mean, amplitude, rms value, maximum, minimum, sums, and products of signals.* In Figs. 2.6 and 2.7, plots are given of the rectangular and triangular pulse, the unit step, and the ramp signal. Reproduce these plots. Compute the mean, amplitude, rms value, maximum, and minimum of each of the four signals. Identify the effects of sampling and truncation. Also make plots of the sum and the product of the rectangular and triangular pulse and the sum and the product of the unit step and the ramp signal.

Truncation

Beware of the phenomenon called *truncation*. Some operations on signals involve signal values that lie *outside* the time axis on which the signal has been defined. This is usually resolved by replacing the missing values with zero. Inevitably this sometimes leads to incorrect results. Consider by way of example the problem of back shifting a unit step. The SIGSYS command

    u=step(dot)

results in a unit step defined on a finite time axis. Shifting the result u according to

    y=u(dot+10)

produces a signal y as in Fig. 2.41, which is zero near the end of the time axis. In this particular case the correct result may directly be obtained with the command

    y=step(dot+10)

but this easy way out is not always available.

Another instance of truncation error arises when we compute the mean of the signal *y* by the SIGSYS command mean(y). Even if *y* is correctly generated on the given finite time axis, by truncation this does not result in the correct value 0.5.

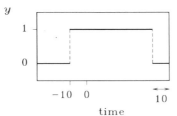

Figure 2.41 A back shifted truncated unit step.

2.7.8. Amplitude, energy and action. (Compare 2.6.13.) Plot the following signals and compute their amplitude, energy and action. Choose in each case the time axis with care. Identify the effects of truncation and sampling.

(a) $z(t) = \dfrac{1}{1 + t} \mathbb{1}(t)$, $t \in \mathbb{R}$.

(b) $z(t) = \cos(2\pi f t)$, $t \in \mathbb{R}$, with $f = 1$.

(c) $z(n) = a^n \mathbb{1}(n)$, $n \in \mathbb{Z}$, with $a = 0.9$.

(d) $z(n) = a^{|n|}$, $n \in \mathbb{Z}$, with $a = 0.9$.

2.7.9. Correlation. In Example 2.4.11 it is shown how the inner product may be used for signal detection. If a signal is received that may be one of several signals, by taking inner products it may be found with which of those signals the received signal is best aligned. This method of signal detection has applications in radar and sonar. A frequent additional complication is that the received signal may have a *delay* so that its exact location in time is not known. To cope with this problem it is useful to introduce the *cross-correlation* $x \square y$ of two signals x and y, which is the signal given by

$$(x \square y)(n) = \sum_{k \in \mathbb{Z}} x(k + n)\overline{y(k)}, \qquad n \in \mathbb{Z},$$

if x and y are discrete-time signals, and by

$$(x \square y)(t) = \int_{-\infty}^{\infty} x(\tau + t)\overline{y(\tau)} \, d\tau, \qquad t \in \mathbb{R},$$

if x and y are continuous-time signals. For sampled signals defined on the time axis $\mathbb{Z}(T)$ we have

$$(x \square y)(t) = T \sum_{k \in \mathbb{Z}} x(kT + t)\overline{y(kT)}, \qquad t \in \mathbb{Z}(T).$$

The value of the cross-correlation $x \square y$ of x and y at time t simply is the inner product of x *back-shifted* by t, and y, that is,

$$(x \square y)(t) = \langle \sigma^t x, y \rangle.$$

Here, the back shift operator σ is defined by $(\sigma^\theta x)(t) = x(t + \theta)$, with t ranging over the appropriate time axis. By looking at the cross-correlation it may be determined for what time shift the signals x and y are best aligned.

By way of example, we again consider the situation of 2.4.11, where the received signal z is either the long pulse l or the short pulse s but is moreover *delayed* by an unknown time θ, that is, we have either

$$z(t) = l(t - \theta), \qquad t \in \mathbb{R},$$

or

$$z(t) = s(t - \theta), \qquad t \in \mathbb{R},$$

(a) Choose a suitable time axis, and generate the long pulse l and the short pulse s.

(b) Compute and plot the two cross-correlations $l \square l$ and $l \square s$. Verify that the maximum of the cross-correlation of l with l is 1 and that this occurs at time 0, and that the maximum of the cross-correlation of l with s is $1/2$.

(c) Generate the signal z given by $z(t) = l(t - \theta)$, with $\theta = 5$, and compute the cross-correlations $z \square l$ and $z \square s$. Verify that the maximum of the cross-correlation of z and l is 1 and the maximum of the cross-correlation of z with s is $1/2$, and that these maxima occur at time θ. Thus, it is reasonable to conclude that the received signal z is the signal l delayed by θ.

(d) Generate the signal x given by

$$x(t) = \text{trian}(t/3 - 1), \qquad t \in \mathbb{R}.$$

Cross-correlate the signal x with l and s to see which signal it resembles most and by how much it is delayed.

(e) To see not only which signal x resembles most but also *how much* it resembles it, it is useful to scale x by a suitable factor such that its energy equals 1 (like that of l and s). After doing this, what is the maximum of the cross-correlation of x with l and s? Use the Cauchy-Schwarz inequality to prove that this maximum is at most equal to 1.

2.7.10. Delta functions. Delta functions and their derivatives are essentially continuous-time signals, but by choosing the sampling interval small enough they may be approximated by discrete-time signals. In what follows, choose the sampling interval small (e.g., equal to 0.01) and the time axis long enough (e.g., extending from -1 to 1) to encompass all events of interest.

(a) Define a signal u as the unit step, and differentiate it numerically to generate a signal d. See what this signal looks like.

(b) Numerically differentiate d twice more to obtain successively signals called d_1 and d_2. Plot them to see what they look like.

(c) The signal d is an approximation to the delta function δ, and d_1 and d_2 are approximations to the derivatives δ' and δ'', respectively. Define the signal ϕ as

$$\phi(t) = \begin{cases} 4(\frac{1}{4} - t^2) & \text{for } -\frac{1}{2} \le t < \frac{1}{2}, \\ 0 & \text{otherwise}, \end{cases} \qquad t \in \mathbb{R},$$

and check numerically in how far

$$\int_{-\infty}^{\infty} \delta^{(n)}(t)\phi(t)\, dt = (-1)^n \phi^{(n)}(0),$$

for $n = 0$, 1 and 2. Comment on the results. *Note:* $\delta^{(0)} = \delta$. *Hint:* Compute the integral as an inner product.

(d) Numerically differentiate the signal ϕ twice on the interval $[-1, 1]$. Explain what you find. *Hint:* Compare 2.6.21.

3

An Introduction to Systems

3.1 INTRODUCTION

As explained in Chapter 1, a *system,* very roughly, is a part of an environment that causes certain signals in that environment to be related. In this chapter we study *input-output systems,* abbreviated as *IO* systems. IO systems are driven by *input* signals and produce *output* signals. IO systems do not necessarily have a single possible output for each input. IO systems that are characterized by a *map* that assigns a *unique* output to each input are called *input-output mapping (IOM) systems. State* systems, which form another class of systems, are studied in Chapter 5.

In Section 3.2 the formal definitions of IO and IOM systems are given. Systems are categorized into *discrete-* and *continuous-time* systems, systems *with memory* and *memoryless* systems, *anticipating* and *non-anticipating* systems, and *time-varying* and *time-invariant* systems. We furthermore present examples of systems from electrical, mechanical, and chemical engineering, signal processing and economics. They are mainly "first-order." Higher-order systems are studied in Chapter 4.

Section 3.3 is devoted to *linear* systems. They derive their importance from the fact that many practical systems may be approximated by linear systems. In Section 3.4 we study *convolution systems*, which are IOM systems that are both linear and time-invariant. Such systems are relatively easy to analyze. In fact, most of this text is devoted to such systems. The convolution operation itself is extensively discussed in Section 3.5. An important notion that is introduced in Section 3.6 is that of the *stability* of convolution systems.

In Section 3.7 the response of convolution systems to harmonic inputs is considered, leading to a discussion of the *frequency response* of convolution systems. Section 3.8 deals with the response of convolution systems to *periodic* inputs, which results in the *cyclical convolution*. In Section 3.9, finally, it is shown how several systems may be *interconnected* to form a larger system.

3.2 INPUT-OUTPUT SYSTEMS AND INPUT-OUTPUT MAPPING SYSTEMS

Input-output systems are systems some of whose signals are designated as *inputs* and some as *outputs*. Usually, but not always, inputs are associated with *causes* and outputs with *effects*. A relationship or *rule* interrelates the input and output signals. Input-output systems are formally defined as follows.

3.2.1.   Definition: Input-output systems.   An *input-output* (IO) system is defined by a signal set $\mathcal{U}$, called the *input set,* a signal set $\mathcal{Y}$, called the *output set,* and a subset $\mathcal{R}$ of the product set $\mathcal{U} \times \mathcal{Y}$, called the *rule* or *relation* of the system. Any pair (u, y) with $u \in \mathcal{U}$, $y \in \mathcal{Y}$, and $(u, y) \in \mathcal{R}$ is said to be an *input-output pair* of the system, with u the *input signal* and y a corresponding *output signal*.   ∎

Figure 3.1 illustrates the definition. It shows that an input-output system is simply defined as a collection $\mathcal{R}$ of input-output pairs (u, y). It also shows that for a given input u in general there is an entire set $\mathcal{Y}_u = \{y \in \mathcal{Y} \mid (u, y) \in \mathcal{R}\}$ of possible outputs that correspond to the input u. If for each input u there exists a *single* corresponding output y, then the system is said to be an *input-output mapping* (IOM) system.

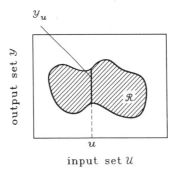

Figure 3.1   Input-output system: $\mathcal{R}$ is the set of possible input-output pairs $(u, y) \in \mathcal{U} \times \mathcal{Y}$.

3.2.2.   Definition: Input-output mapping systems.   Let $\mathcal{U}$ and $\mathcal{Y}$ be the input and output sets of an input-output system with rule $\mathcal{R} \subset \mathcal{U} \times \mathcal{Y}$. Then, if for each input u the fact that $(u, y_1) \in \mathcal{R}$ and $(u, y_2) \in \mathcal{R}$ implies that $y_1 = y_2$ the IO system is called an *input-output mapping* (IOM) system. The map $\phi \colon \mathcal{U} \to \mathcal{Y}$ that assigns a unique output $y \in \mathcal{Y}$ to each input $u \in \mathcal{U}$ is called the *input-output map* (IO map) of the system.   ∎

Figure 3.2 illustrates the input-output relation of an IOM system. Both IO and IOM systems are pictorially often represented by a block as in Fig. 3.3, with sometimes the rule $\mathcal{R}$ written inside the block.

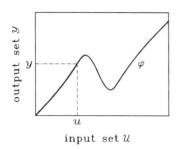

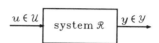

Figure 3.2 Input-output mapping system: For each input u there is a unique output $y = \phi(u)$.

Figure 3.3 IO or IOM system in block diagram form.

Examples of IO and IOM Systems

We illustrate the definitions of IO and IOM systems with some examples.

3.2.3.   Examples: IO and IOM systems.

(a) *Ball tossing system*. Given is a bag with a number of balls inside. Some more balls are tossed into the bag. The following signals may be associated with this system: the number of balls u that have been tossed in and the number of balls y that the bag now contains. Taking u as the input and y as the output, the input set $\mathcal{U}$ and the output set $\mathcal{Y}$ both consist of the set of nonnegative integers $\mathbb{Z}_+$. The rule of the system is

$$y \geq u.$$

In full set notation,

$$\mathcal{R} = \{(u, y) \in \mathbb{Z}_+ \times \mathbb{Z}_+ \,|\, y \geq u\}.$$

Thus, $(5, 6)$ is an IO pair but $(6, 5)$ is not. The ball tossing system is an IO system, but not an IOM system, because to each u there correspond many, in fact infinitely many, possible outputs y.

If, on the other hand, it is known that the bag contained, say, y_o balls before tossing in additional balls, then the system is an *IOM* system, because the input u determines the output y uniquely by the formula

$$y = y_o + u.$$

This equation specifies the input-output map.

(b) *The exponential smoother and the RC network*. The exponential smoother of Example 1.2.6 and the RC network of Example 1.2.7 are input-output systems. In general they are not input-output mapping systems. However, if *fixed* initial conditions are specified once and for all (zero initial conditions, for instance), then the input-output relations of the systems become input-output maps, and consequently the systems are IOM systems.   ∎

Discrete- and Continuous-Time Systems

Most systems we study in this text have *time signals* as input and output. Such systems are called *dynamical* systems. The ball tossing system of Example 3.2.3(a) is not dynamical according to this definition, but the exponential smoother and the RC network are.

Usually, but not always, the input and output of a dynamical system have the *same* time axis $\mathbb{T}$. If both the input and the output of a dynamical system are discrete-time signals, then we call it a *discrete-time* system. If both the input and the output are continuous-time signals, then we speak of a *continuous-time* system. If one of the time axes is continuous and the other discrete, then the system is called a *hybrid* system.

3.2.4.   Examples: Discrete-time, continuous-time, and hybrid systems.
The exponential smoother of Example 1.2.6 is a discrete-time system. The RC network of Example 1.2.7 is a continuous-time system. The sampler of Fig. 2.16 is a hybrid system, because its input is a continuous-time signal and its output a discrete-time signal.   ∎

Memoryless IOM Systems

From Examples 1.2.6 and 1.2.7 it follows that $y(t)$, the output at time t, of the exponential smoother or the RC network depends on the *entire* past input u, and not just on the value $u(t)$ of the input at the same time t. There are also systems, called *memoryless* systems, where the current value of the output is fully determined by the current value of the input alone, and not by the past or future values of the input.

3.2.5.   Definition: Memoryless system.   An IOM system with time axis $\mathbb{T}$, input signal range U, and output signal range Y is memoryless if there exists a *pointwise* map $\psi: \mathbb{T} \times U \to Y$, such that if (u, y) is an input-output pair then

$$y(t) = \psi(t, u(t)), \quad t \in \mathbb{T}.$$ ∎

3.2.6.   Examples: Memoryless systems.
(a) *Resistive circuit*. Consider the resistive circuit of Fig. 3.4, consisting of a voltage source connected to a resistive element with a voltage-current characteristic as in Fig. 3.5. We suppose that the input $u = v$ is a time-varying voltage. Hence,

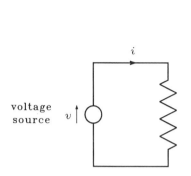

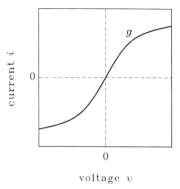

Figure 3.4 A resistive circuit.

Figure 3.5 Voltage-current characteristic of the resistive element.

the time axis is $\mathbb{T} = \mathbb{R}$ and each input u is an element of the continuous-time signal space $\mathcal{L}$. The output $y = i$ is the time-varying current through the network and also an element of $\mathcal{L}$. If the voltage-current characteristic is represented by a function g as in the figure, at each time $t \in \mathbb{R}$ the output current $y(t)$ is uniquely determined by the input voltage $u(t)$ as

$$y(t) = g(u(t)), \quad t \in \mathbb{R}.$$

The resistive circuit therefore is an example of a memoryless system. If the voltage-current characteristic changes with time, for instance by ageing, the system still is memoryless, with an IO map of the form

$$y(t) = g(t, u(t)), \quad t \in \mathbb{R}.$$

(b) *Limiters, rectifiers, squarer.* Some well-known examples of memoryless systems follow:

(1) The *hard limiter* has the IO map

$$y(t) = \text{sign}(u(t)), \qquad t \in \mathbb{T},$$

where sign: $\mathbb{R} \to \mathbb{R}$ is the *signum* or *sign* function defined by

$$\text{sign}(u) = \begin{cases} 1 & \text{for } u > 0, \\ 0 & \text{for } u = 0, \\ -1 & \text{for } u < 0, \end{cases} \quad u \in \mathbb{R}.$$

(2) The *soft limiter* has the IO map

$$y(t) = \text{sat}(u(t)), \qquad t \in \mathbb{R},$$

where the *saturation function* sat: $\mathbb{R} \to \mathbb{R}$ is given by

$$\text{sat}(u) = \begin{cases} 1 \text{ for } u > 1, \\ u \text{ for } |u| \leq 1, \\ -1 \text{ for } u < -1, \end{cases} \quad u \in \mathbb{R}.$$

(3) The *half wave rectifier* is the memoryless system with IO map

$$y(t) = \text{ramp}(u(t)), \quad t \in \mathbb{R},$$

where ramp: $\mathbb{R} \to \mathbb{R}$ is the function defined by

$$\text{ramp}(u) = \begin{cases} 0 \text{ for } u < 0, \\ u \text{ for } u \geq 0, \end{cases} \quad u \in \mathbb{R}.$$

Figure 3.6 shows the graphs of the sign, sat, and ramp functions.

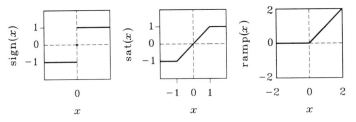

Figure 3.6 Left: the signum function. Middle: the saturation function. Right: the ramp function.

(4) The *full wave rectifier* is the memoryless system with IO map

$$y(t) = |u(t)|, \quad t \in \mathbb{T}.$$

(5) The *squarer,* finally, is the memoryless system described by

$$y(t) = [u(t)]^2, \quad t \in \mathbb{T}. \qquad \blacksquare$$

Non-Anticipating Systems

Real-life systems have the property that their output $y(t)$ at any given time t only depends on the values of the input *before* time t, and not on later values of the input. This property is called *non-anticipativeness*. Theoretically, we may, of course, well hypothesize systems whose current output does not only depend on past values of the input but also on *future* values. Such systems are said to be *anticipative* or *anticipating*.

Formally, non-anticipativeness of IOM systems is defined as follows. In this definition, the system is defined on the time axis $\mathbb{T}$. The input signal has signal range

U, so that the input set consists of all time signals $\mathcal{U} = \{u\colon \mathbb{T} \to U\}$. Likewise, the output signal has signal range Y and hence belongs to the output set $\mathcal{Y} = \{y\colon \mathbb{T} \to Y\}$.

3.2.7.   Definition: Non-anticipating dynamical IOM systems.

Consider an IOM system with time axis $\mathbb{T}$ and rule $\mathcal{R} \subset \mathcal{U} \times \mathcal{Y}$. Let (u_1, y_1) be any IO pair and $t \in \mathbb{T}$ an arbitrary time, and suppose that (u_2, y_2) is any other input-output pair such that $u_1(\tau) = u_2(\tau)$ for all $\tau \leq t$ with $\tau \in \mathbb{T}$. Then the system is *non-anticipating* if $y_1(\tau) = y_2(\tau)$ for all $\tau \leq t$ such that $\tau \in \mathbb{T}$.   ∎

What follows are examples of systems that may or may not be anticipating.

3.2.8.   Examples: Anticipativeness and non-anticipativeness.

(a) *Pure delay and pure predictor*. Consider the discrete-time or continuous-time IOM system whose time axis $\mathbb{T}$ is either $\mathbb{Z}$ or $\mathbb{R}$ and whose output is given by

$$y(t) = u(t - \theta), \qquad t \in \mathbb{T},$$

with $\theta \in \mathbb{T}$ fixed. Figure 3.7 shows that if $\theta \geq 0$ the system *delays* the input by θ, which is why it is called a *pure delay*. The pure delay is non-anticipating. If $\theta < 0$ the system is called a *pure predictor*. Predictors are obviously anticipating.

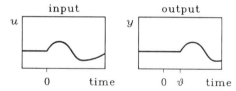

Figure 3.7 IO pair of a pure delay system. Left: input. Right: output.

(b) *Discrete-time sliding window averager*. The rule

$$y(n) = \frac{1}{N + M + 1} \sum_{m=-M}^{N} u(n + m), \qquad n \in \mathbb{Z},$$

with N and M nonnegative integers, describes a discrete-time IOM system, called *sliding window averager*. The system takes an arithmetic average of the input, with a "window" extending from M time intervals before until N intervals after the time n at which the average is taken. The system is non-anticipating if and only if $N = 0$.

(c) *Continuous-time sliding window averager*. The continuous-time equivalent of the discrete-time sliding window averager is the system with rule

$$y(t) = \frac{1}{T_1 + T_2} \int_{t-T_1}^{t+T_2} u(\tau)\, d\tau, \qquad t \in \mathbb{R},$$

with T_1 and T_2 nonnegative real numbers such that $T_1 + T_2 > 0$ (Fig. 3.8.) The system is non-anticipating if and only if $T_2 = 0$.

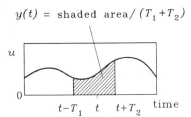

$y(t)$ = shaded area/$(T_1 + T_2)$

Figure 3.8 Continuous-time sliding window averaging. ∎

Time-Invariance

We next discuss the notion of *time-invariance*. A system is time-invariant, roughly, if the system properties do not change with time. Most of the systems studied in this text are time-invariant. To define time-invariance, we introduce the *back shift operator* σ^θ, with θ a real number, which shifts a time signal *backward* if $\theta > 0$ and forward if $\theta < 0$.

3.2.9.   Definition: Back shift operator.   Let x be a discrete- or continuous-time signal defined on an infinite or semi-infinite time axis $\mathbb{T}$. Let θ be a real number such that $t + \theta \in \mathbb{T}$ for all $t \in \mathbb{T}$. Then, the *back shift operator* σ^θ maps the signal x to the back shifted signal $\sigma^\theta x$ given by

$$(\sigma^\theta x)(t) = x(t + \theta), \qquad t \in \mathbb{T}.$$ ∎

Figure 3.9 illustrates the back shift operator. Note that the admissible time shifts θ depend on the time axis. If $\mathbb{T}$ is an infinite time axis such as $\mathbb{Z}$ or $\mathbb{R}$, θ may take any value in $\mathbb{T}$. If $\mathbb{T}$ is right semi-infinite such as $\mathbb{Z}_+$, $\mathbb{R}_+$, or $[t_0, \infty)$, then only *non-negative* values of θ in $\mathbb{T}$ are allowed.

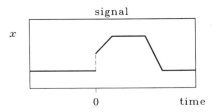

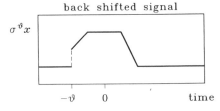

Figure 3.9 Back shift operator. Top: a time signal with time axis $\mathbb{R}$. Bottom: the same time signal after application of the back shift operator with positive parameter θ.

As was noted before, a system is time-invariant if its properties do not change with time: Any input-output pair (u, y) may be arbitrarily shifted in time and *remains* an IO pair. Time-varying systems do not have this property. The formal definition of time-invariance is as follows.

3.2.10.   Definition: Time-invariance.   Consider an IO system with the infinite or right semi-infinite discrete or continuous time axis $\mathbb{T}$. Then the system is *time-invariant* if for every input-output pair (u, y) also the time-shifted pair $(\sigma^\theta u, \sigma^\theta y)$ is an input-output pair for any allowable time shift θ.   ∎

For IOM systems with IO map ϕ, time-invariance reduces to *shift-invariance* of the IO map ϕ, that is,

$$\phi(\sigma^\theta u) = \sigma^\theta \phi(u)$$

for every input u and every allowable time shift θ. Shift-invariance of the IO map is equivalent to the statement that ϕ and σ^θ commute for all θ, as illustrated in Fig. 3.10.

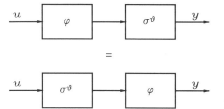

Figure 3.10   Shift invariance.

A system that is not time-invariant is said to be *time-varying*. We illustrate time-invariance with some examples.

3.2.11.   Examples: Time-invariant and time-varying systems.

(a) *The RC network*. The RC network of Example 1.2.7 is a continuous-time system with input u and output y, described by the differential equation

$$\frac{dy(t)}{dt} + \frac{1}{RC}y(t) = \frac{1}{RC}u(t), \qquad t \in \mathbb{R}, \tag{1}$$

where we study the network on the infinite time axis $\mathbb{R}$. Suppose that (u, y) is an IO pair, satisfying the differential equation. The system is time-invariant if we can prove that the back shifted pair (u', y'), with

$$u' = \sigma^\theta u, \qquad y' = \sigma^\theta y,$$

also is an IO pair, i.e., also satisfies the differential equation, for every real θ. Indeed, using the chain rule for differentiation it follows by differentiation with respect to t and using (1) that

$$\frac{d}{dt}\sigma^\theta y(t) = \frac{d}{dt}y(t + \theta) = \frac{dy}{dt}(t + \theta) = -\frac{1}{RC}y(t + \theta) + \frac{1}{RC}u(t + \theta)$$

$$= -\frac{1}{RC}(\sigma^\theta y)(t) + \frac{1}{RC}(\sigma^\theta u)(t), \quad t \in \mathbb{R},$$

which shows that (u', y') is an IO pair.

Suppose now that the resistance of the resistor is not constant, but *varies* with time, perhaps by ageing or as a result of changing environmental temperature. The differential equation (1) then takes the form

$$\frac{dy(t)}{dt} + \frac{1}{R(t)C}y(t) = \frac{1}{R(t)C}u(t), \quad t \in \mathbb{R},$$

where $R(t)$ is the value of the resistance at time t. We assume that $R(t) > 0$ for all t. Let us check whether this IO system is time-invariant. The back shifted pair (u', y') is a solution of the differential equation if

$$\frac{d}{dt}y(t + \theta) + \frac{1}{R(t)C}y(t + \theta) = \frac{1}{R(t)C}u(t + \theta), \quad t \in \mathbb{R}.$$

Replacing t with $\tau - \theta$ we thus require that

$$\frac{dy(\tau)}{d\tau} + \frac{1}{R(\tau - \theta)C}y(\tau) = \frac{1}{R(\tau - \theta)C}u(\tau), \quad \tau \in \mathbb{R}.$$

Even though (u, y) is an IO pair this equation is not necessarily satisfied, unless $R(\tau - \theta) = R(\tau)$ for all $\tau \in \mathbb{R}$ and for any $\theta \in \mathbb{R}$, which implies that the resistance is constant. Thus, if the resistance is not constant, a backshifted IO pair in general is *not* an IO pair, and, as a result, the system is *time-varying*.

(b) *Time-invariance of a memoryless IOM system.* A memoryless IOM system with pointwise IO map given by

$$y(t) = \psi(t, u(t)), \quad t \in \mathbb{T},$$

is time-invariant if and only if $\psi(t_1, u) = \psi(t_2, u)$ for all t_1 and t_2 on the time axis $\mathbb{T}$, and all u belonging to the input signal range of the system. Thus, a memoryless system is time-invariant if and only if the function ψ does not depend on its first argument, which it time. The soft and hard limiter, the half- and full-wave rectifier, and the squarer of Example 3.2.6(b) are all time-invariant memoryless systems. ∎

Additional Examples of Input-Output Systems

In conclusion of this section we present several further examples of input-output systems originating from different areas.

3.2.12.   Example: Heated vessel.   Figure 3.11 shows a vessel containing a continuously stirred fluid, which is heated by an electrical coil. The input to the system is the electrical power u dissipated by the coil, while its output is the temperature T of the fluid. We assume that the vessel is stirred well, so that the fluid in the vessel has the same temperature everywhere. The heat loss to the environment is proportional to the difference $T - T_e$ between the temperature of the fluid and the environment temperature T_e, which is assumed to be constant. Setting up the heat balance of the system we obtain the equation

$$C\frac{dT(t)}{dt} = u(t) - \frac{T(t) - T_e(t)}{R}, \qquad t \in \mathbb{R},$$

where C is the heat capacity of the fluid and R the heat resistance to the environment. The term on the left-hand side is the heat flow needed to increase the temperature of the fluid, the first term on the right-hand side is the external supply of heat, and the second term the loss of heat to the environment. Redefining the output as the temperature *difference* $y = T - T_e$, and substituting $T = y + T_e$ into the differential equation, we easily find that

$$\frac{dy(t)}{dt} = -\frac{1}{RC}y(t) + \frac{1}{C}u(t), \qquad t \in \mathbb{R}.$$

This differential equation is similar to that for the RC circuit of Example 1.2.7. It describes the heated vessel as a continuous-time IO system.

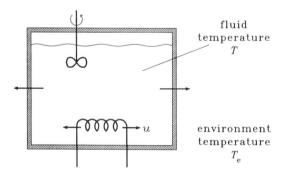

fluid
temperature
T

environment
temperature
T_e

Figure 3.11 A heated vessel.   ∎

3.2.13.   Example: The motion of a car.   In the final part of Section 1.3 we discussed a cruise control system for a car. The motion of the car (without the cruise control system) may be described by the following simple model, which accounts for the major physical effects. By Newton's law,

$$M\frac{dv(t)}{dt} = F_{\text{total}}(t), \qquad t \geq t_o,$$

where M is the mass of the car, dv/dt its acceleration, and F_{total} the total force exerted on the car in forward direction. The total force may be expressed as

$$F_{\text{total}}(t) = cu(t) - Bv^2(t).$$

The first term $cu(t)$ represents the pulling force of the engine, and is proportional to the throttle opening $u(t)$, with proportionality constant c. The second term $Bv^2(t)$ is caused by air resistance. This friction force is proportional to the square of the car speed v, with B representing the friction coefficient. Substitution of F_{total} into Newton's law results in

$$M\frac{dv(t)}{dt} = cu(t) - Bv^2(t), \qquad t \geq t_o.$$

All input-output pairs (u, v) need satisfy this nonlinear first-order differential equation. The system is a continuous-time IO system.   ∎

3.2.14.   Example: Savings account.   Suppose that $y(n)$ represents the balance of a savings account at the beginning of day n, and $u(n)$ the amount deposited during day n. If interest is computed and added daily at a rate of $\alpha \cdot 100\%$, the balance at the beginning of day $n + 1$ is given by

$$y(n + 1) = (1 + \alpha)y(n) + u(n), \qquad n = 0, 1, 2, \cdots.$$

This describes the savings account as a discrete-time system on the time axis $\mathbb{Z}_+$. It is an IO system. If the interest rate α does not change with time, the system is time-invariant; otherwise, it is time-varying.   ∎

3.2.15.   Example: National economy.   A very crude model for the national economy of a country follows. Let $y(n)$ represent the total capital stock of the country accumulated at the beginning of year n. During the year a fraction r of the capital stock is lost owing to ageing and obsolescence. The production during the year is proportional, with productivity coefficient p, to the capital stock. The aggregate consumption during the year is $u(n)$ and is considered the input to the system. The relation

$$y(n + 1) = y(n) - ry(n) + py(n) - u(n), \qquad n = 0, 1, 2, \cdots,$$

expresses that the capital stock at the beginning of the next year equals that of the present year, decreased by the amounts $ry(n)$ lost by depreciation and $u(n)$ used for consumption, and increased by the production $py(n)$. The equation may be rewritten as

$$y(n + 1) = (1 + p - r)y(n) - u(n), \qquad n = 0, 1, 2, \cdots .$$

This difference equation describes the national economy as a discrete-time IO system with time axis $\mathbb{Z}_+$. If the constants p and r do not change with time, the system is time-invariant.   ■

The four systems presented in 3.2.12 to 3.2.15 are all input-output systems but not input output mapping systems. In each case the output y does not depend exclusively on the input but also on the initial value of the output.

3.3 LINEAR SYSTEMS

Linear systems form an important class of systems, because, on the one hand, they lend themselves well to mathematical analysis while, on the other, many practical engineering systems may be accurately modeled as linear systems.

3.3.1.   Definition: Linear IO system.   Let $\mathcal{U}$ and $\mathcal{Y}$ be the input and output sets, respectively, of an input-output system with rule $\mathcal{R} \subset \mathcal{U} \times \mathcal{Y}$. The IO system is *linear* if $\mathcal{U}$ and $\mathcal{Y}$ are *linear spaces* and $\mathcal{R}$ is a *subspace* of $\mathcal{U} \times \mathcal{Y}$.   ■

By definition, R is a subspace if it has the property that if (u_1, y_1) and (u_2, y_2) are any two IO pairs, the linear combination

$$(\alpha u_1 + \beta u_2, \ \alpha y_1 + \beta y_2)$$

is also an IO pair for arbitrary scalars α and β. The scalars usually take values in $\mathbb{R}$ or $\mathbb{C}$, depending on the field over which the spaces $\mathcal{U}$ and $\mathcal{Y}$ are linear.

An IO system that is *not* linear is said to be *nonlinear*.

3.3.2.   Example: Linearity of the exponential smoother.   The exponential smoother, which in Example 1.2.6 is described as an IO system on the time axis $\{0, 1, 2, \cdots \}$, is linear. The reason is that first of all the input and output sets, consisting of all complex-valued sequences on the semi-infinite time axis $\mathbb{Z}_+$, are linear spaces over the complex numbers. To show that the rule $\mathcal{R}$ is linear let (u_1, y_1) and (u_2, y_2) be two input-output pairs, that is, they satisfy

$$y_1(n + 1) = ay_1(n) + (1 - a)u_1(n + 1), \qquad n = 0, 1, 2, \cdots ,$$

$$y_2(n + 1) = ay_2(n) + (1 - a)u_2(n + 1), \qquad n = 0, 1, 2, \cdots .$$

Multiplying the first of these equations by the complex scalar α, the second by β, and adding the two resulting expressions yields

$$[\alpha y_1(n + 1) + \beta y_2(n + 1)] = a[\alpha y_1(n) + \beta y_2(n)]$$
$$+ (1 - a)[\alpha u_1(n + 1) + \beta u_2(n + 1)],$$

$n = 0, 1, 2, \cdots$. This proves that $(\alpha u_1 + \beta u_2, \alpha y_1 + \beta y_2)$ is an IO pair. Hence, $\mathcal{R}$ is a subspace and the system is linear. ∎

3.3.3.   Exercise: Linearity of the RC network.   Prove that the RC circuit of Example 1.2.7, with unspecified initial conditions, is a linear IO system on the time axis $[t_o, \infty)$. Show that it is also linear if the resistor is time-varying, as at the end of Example 3.2.11(a). ∎

Linearity of IOM Systems

Linearity of an input-output mapping system is equivalent to linearity of its IO map.

3.3.4.   Summary: Linearity of IOM systems.   The IOM system with input set $\mathcal{U}$, output set $\mathcal{Y}$ and IO map $\phi: \mathcal{U} \to \mathcal{Y}$ is linear if and only if $\mathcal{U}$ and $\mathcal{Y}$ are linear spaces, and ϕ is a *linear map,* that is,

$$\phi(\alpha u_1 + \beta u_2) = \alpha\phi(u_1) + \beta\phi(u_2) \tag{1}$$

for every u_1 and u_2 in $\mathcal{U}$ and all scalars α and β. ∎

The property (1) is known as the *superposition* property of linear maps. A necessary and sufficient condition for the map ϕ to be linear is that it be both *additive*, that is,

$$\phi(u_1 + u_2) = \phi(u_1) + \phi(u_2)$$

for every u_1 and u_2 in $\mathcal{U}$, and *homogeneous,* that is,

$$\phi(\alpha u) = \alpha\phi(u)$$

for every scalar α and each $u \in \mathcal{U}$. The next exercise shows that both additivity and homogeneity are needed for linearity.

3.3.5.   Exercise: Homogeneity, additivity, and linearity.
    (a) Show that additivity by itself does not imply linearity. *Hint:* Take as a counterexample $U = Y = \mathbb{C}$ and ϕ defined by $\phi(u) = \text{Re}(u)$, and consider linearity over the complex numbers.

    (b) Show that also homogeneity by itself does not imply linearity. *Hint:* Take as a counterexample $\mathcal{U} = \mathbb{R} \times \mathbb{R}$, $\mathcal{Y} = \mathbb{R}$ and ϕ defined by $\phi(u_1, u_2) = (u_1^3 + u_2^3)^{1/3}$, and consider linearity over the real numbers. ∎

We present some examples of linear and nonlinear IOM systems.

3.3.6.   Example: Linearity of the sliding window averager.   The IO map of
the discrete-time sliding window averager of Example 3.2.8(b) is described by the
relation

$$y(n) = \frac{1}{N + M + 1} \sum_{k=-M}^{N} u(n + k), \qquad n \in \mathbb{Z}.$$

By the linearity of the summing operation this IO map is linear, and, hence, the discrete-time sliding window averager is a linear IOM system.

Also the continuous-time sliding window averager of Example 3.2.8(c) is a
linear IOM system.   ∎

3.3.7.   Example: Two nonlinear IOM systems.
(a) *Ball tossing system.* The ball tossing system of Example 3.2.3(a) is nonlinear because the input and output sets $\mathbb{Z}_+$ are not linear spaces.

(b) *Memoryless nonlinear systems.* The memoryless resistive circuit of Example 3.2.6(a) is nonlinear *unless* the voltage-current characteristic is a straight line
crossing through the origin, that is, the function g is of the form

$$g(v) = v/R,$$

with R a constant.
The hard and soft limiter, the half- and full-wave rectifier, and the squarer of
Example 3.2.6(b) are all nonlinear.   ∎

Linearization

Many practical IO systems may be approximated, at least "locally," by a linear IO
system. The approximation procedure is called *linearization*. Linearization is an extension of the idea that a curve may be approximated in the neighborhood of a point
on the curve by the tangent at the point, as illustrated in Fig. 3.12.

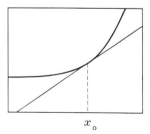

x_o

Figure 3.12 Approximation of a curve
by its tangent at a point x_o.

It is important to note the three main features of linearization:

(i) Linearization always takes place about some fixed "operating point" (the point x_o in Fig. 3.12).
(ii) The approximation is only valid if the system behaves "smoothly" near the operating point.
(iii) The approximation only holds for "small" deviations from the operating point.

We demonstrate linearization of systems with two examples.

3.3.8.   Example: Linearization of the resistive circuit.   The resistive circuit of Example 3.2.6(a) generally is a nonlinear IOM system, described by the IO map

$$i = g(v).$$

Let (v_o, i_o) be a point on the voltage-current characteristic, (i.e., $i_o = g(v_o)$). Any other input voltage v and the corresponding output current i may be written as

$$v = v_o + \tilde{v}, \qquad i = i_o + \tilde{i},$$

with $\tilde{v}$ and $\tilde{i}$ the deviations of the voltage and current, respectively, from the operating point values. It follows that

$$i_o + \tilde{i} = g(v_o + \tilde{v}),$$

as shown in Fig. 3.13. If $\tilde{v}$ is small and g is differentiable at the point v_o with derivative $g'(v_o)$, by using the first term of the Taylor expansion of the right-hand side we may approximate as

$$i_o + \tilde{i} \approx g(v_o) + g'(v_o)\tilde{v}.$$

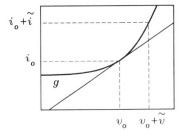

Figure 3.13 Linearization of the voltage-current characteristic of the resistive element at the point (v_o, i_o).

Since $i_o = g(v_o)$, the first terms on both sides of the expression cancel and we have

$$\tilde{i} \approx g'(v_o)\tilde{v}.$$

Let us denote the deviation of the voltage $u = \tilde{v}$ as the input of the system, and the deviation of the current $y = \tilde{i}$ as its output. Then, replacing the approximate equality with equality, we obtain the IO map

$$y = \frac{u}{R_o},$$

where $R_o = 1/g'(v_o)$. This *linear* IO map describes the *linearized* or *variational* system about the operating point (v_o, i_o). R_o is the *equivalent resistance* of the nonlinear resistive element at the operating point. In particular, if, for instance, $g(v) = \alpha v + \beta v^3$, then $g'(v_o) = \alpha + 3\beta v_o^2$ and

$$R_o = \frac{1}{\alpha + 3\beta v_o^2}. \qquad \blacksquare$$

3.3.9.   Example: Linearization of the differential equation for the motion of a car.   In Example 3.2.13 we saw that the motion of a car may be described by the differential equation

$$M\frac{dv(t)}{dt} = cu(t) - Bv^2(t), \qquad t \geq t_o,$$

with the input u the throttle position and the output v the car speed. It is simple to verify that if (u_1, v_1) is any solution pair to the differential equation and (u_2, v_2) another, an arbitrary linear combination $(\alpha u_1 + \beta u_2, \alpha v_1 + \beta v_2)$, with α and β real, in general does *not* satisfy the differential equation. The culprit is the quadratic term in the differential equation, and the result is that the IO system is nonlinear.

To linearize, we first choose an "operating point," which in principle may be any IO pair (u_o, y_o) satisfying the differential equation. A convenient choice is to take for u_o the constant signal $u_o(t) = U_o$, and for v_o the corresponding constant solution $v_o(t) = V_o$ of the differential equation. Substitution into the equation yields $0 = cU_o - BV_o^2$, so that $V_o = \sqrt{cU_o/B}$. The constant input U_o, of course, represents a constant throttle position, and V_o is the constant cruising speed corresponding to this throttle position.

The next step is to write

$$u(t) = U_o + \tilde{u}(t), \qquad v(t) = V_o + \tilde{v}(t), \qquad t \geq t_o,$$

with $\tilde{u}$ and $\tilde{v}$ the deviations of u and v from U_o and V_o, respectively. Substitution into the differential equation yields

$$M\frac{d}{dt}[V_o + \tilde{v}(t)] = c[U_o + \tilde{u}(t)] - B[V_o + \tilde{v}(t)]^2, \qquad t \geq t_o.$$

By using the equality $cU_o = BV_o^2$, this reduces to

$$M\frac{d}{dt}\tilde{v}(t) = c\tilde{u}(t) - 2BV_o\tilde{v}(t) - B\tilde{v}^2(t), \qquad t \geq t_o.$$

If $\tilde{v}$ is small, we may neglect the quadratic term on the right-hand side and obtain the *linear* differential equation

$$M\frac{d}{dt}\tilde{v}(t) = c\tilde{u}(t) - 2BV_o\tilde{v}(t), \qquad t \geq t_o. \tag{2}$$

We may now redefine the input and output to be $\tilde{u}$ and $\tilde{v}$, respectively, and call the IO system described by (2) the *linearized* or *variational* system about the solution (u_o, v_o). It is easy to verify that the variational system is linear because the differential equation (2) is linear.

By using the variational system we may obtain approximations to the original nonlinear system that are quite good as long as $\tilde{v}$ is small.   ∎

The IO Map of Linear IOM Systems

We continue by showing that the IO map of linear dynamical IOM systems may be expressed explicitly in the form of a sum (in the discrete-time case) or an integral (in the continuous-time case.)

Consider first a discrete-time IOM system whose IO map may be represented as

$$y(n) = \sum_{m=-\infty}^{\infty} k(n, m)\, u(m), \qquad n \in \mathbb{Z}, \tag{3}$$

with k a given function of two variables and u and y real- or complex-valued signals. The function k is called the *kernel* of the system. An example of a system whose IO map takes this form is the discrete-time sliding window averager of Example 3.2.8(b), where

$$k(n, m) = \begin{cases} \dfrac{1}{N + M + 1} & \text{for } n - M \leq m \leq n + N, \\ 0 & \text{otherwise,} \end{cases} \qquad n, m \in \mathbb{Z}.$$

It is easy to show that the IO map (3) represents a *linear* system because it has the superposition property.

The kernel k has the following interpretation. Suppose that the input u to the system described by (3) is

$$u(n) = \Delta(n - n_o), \qquad n \in \mathbb{Z},$$

that is, the input u is the unit pulse shifted to the time $n_o \in \mathbb{Z}$. Then, from (3) the corresponding output is

$$y(n) = \sum_{m=-\infty}^{\infty} k(n, m)u(m) = \sum_{m=-\infty}^{\infty} k(n, m)\Delta(m - n_o)$$

$$= k(n, n_o), \quad n \in \mathbb{Z}.$$

Thus, the value $k(n, m)$ of k at the point (n, m) is the *response of the system at time n if the input is a unit pulse applied at time m*. Figure 3.14 illustrates this for an arbitrary system.

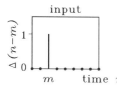
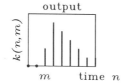

input   output

Figure 3.14 Kernel of a discrete-time linear system. Left: unit pulse at time m. Right: the response is $k(n, m)$, $n \in \mathbb{Z}$.

Similar to the discrete-time case, continuous-time IOM systems whose IO map may be represented in the form

$$y(t) = \int_{-\infty}^{\infty} k(t, \tau)u(\tau)\, d\tau, \quad t \in \mathbb{R}, \tag{4}$$

are linear. The function k is again called the *kernel* of the system. The continuous-time sliding window averager is such a system, with kernel

$$k(t, \tau) = \begin{cases} \dfrac{1}{T_1 + T_2} & \text{for } t - T_1 \le \tau < t + T_2, \\ 0 & \text{otherwise,} \end{cases} \quad t, \tau \in \mathbb{R}.$$

Suppose that the input u to the system described by (4) is a δ-function shifted to the time τ_o, that is,

$$u(\tau) = \delta(\tau - \tau_o), \quad \tau \in \mathbb{R}.$$

Then the output is

$$y(t) = \int_{-\infty}^{\infty} k(t, \tau)u(\tau)\, d\tau = \int_{-\infty}^{\infty} k(t, \tau)\delta(\tau - \tau_o)\, d\tau$$

$$= k(t, \tau_o), \quad t \in \mathbb{R}.$$

Thus, $k(t, \tau)$ is the response at time t when the input is a delta function shifted to time τ.

So far we have seen that an IO map given in the form (3) or (4) defines a linear IOM system. What is perhaps surprising is that the IO map of *every* linear IOM system on the infinite time axis $\mathbb{Z}$ or $\mathbb{R}$ may be written in the form (3) or (4). We first state the complete result.

3.3.10.   Summary: The IO map of linear IOM systems.

A discrete-time IOM system with time axis $\mathbb{Z}$ and real- or complex-valued input and output is linear if and only if its IO map is of the form

$$y(n) = \sum_{m=-\infty}^{\infty} k(n, m)u(m),$$

$n \in \mathbb{Z}$. The function k, called the *kernel* of the system, is any function $\mathbb{Z} \times \mathbb{Z} \to \mathbb{C}$.

A continuous-time IOM system with time axis $\mathbb{R}$ and real- or complex-valued input and output is linear if and only if its IO map is of the form

$$y(t) = \int_{-\infty}^{\infty} k(t, \tau)u(\tau)\, d\tau,$$

$t \in \mathbb{R}$. The function k, called the *kernel* of the system, is any function $\mathbb{R} \times \mathbb{R} \to \mathbb{C}$, possibly singular.   ∎

To see that any linear IOM system may be represented as in 3.3.10, first consider an arbitrary discrete-time linear IOM system and denote the response of the system at time n to a unit pulse at time m as $k(n, m)$. An arbitrary input u may be represented as a linear combination of shifted unit pulses in the form

$$u(n) = \sum_{m=-\infty}^{\infty} u(m)\Delta(n - m), \qquad n \in \mathbb{Z}. \tag{5}$$

By *homogeneity* the response of the system to the shifted and scaled pulse $u(m)\Delta(n - m)$, $n \in \mathbb{Z}$, is $k(n, m)u(m)$, $n \in \mathbb{Z}$, and by *additivity* the response to the composite input (5) is

$$y(n) = \sum_{m=-\infty}^{\infty} k(n, m)u(m), \qquad n \in \mathbb{Z}.$$

This proves that the IO map of *every* linear discrete-time IOM system may be represented in the form (3).

A similar argument holds for continuous-time linear IOM systems. Denote the response of the system to the shifted δ-function $\delta(t - \tau)$, $t \in \mathbb{R}$, as $k(t, \tau)$, $t \in \mathbb{R}$. Any input u to the system may be decomposed as an infinite linear combination of shifted δ-functions of the form

$$u(t) = \int_{-\infty}^{\infty} u(\tau)\delta(t - \tau)\, d\tau, \qquad t \in \mathbb{R}.$$

By homogeneity and addivity it is plausible that the response to this input is

$$y(t) = \int_{-\infty}^{\infty} k(t, \tau)u(\tau) \, d\tau, \qquad t \in \mathbb{R}.$$

The discrete- and continuous-time sliding window averager were already mentioned as examples of systems whose IO map may be represented in sum or integral form. A further example is the following.

3.3.11.   Example: The kernel of a delay.   The pure delay of Example 3.2.8(a) is characterized by

$$y(t) = u(t - \theta), \qquad t \in \mathbb{T},$$

with θ the delay.

(a) *Discrete-time case.* When the time axis is $\mathbb{T} = \mathbb{Z}$, the system is discrete-time. The response of the discrete-time delay to the shifted unit pulse $u(n) = \Delta(n - m)$, $n \in \mathbb{Z}$, is

$$y(n) = \Delta(n - \theta - m), \qquad n \in \mathbb{Z}.$$

Hence, the kernel of the system is

$$k(n, m) = \Delta(n - \theta - m) = \begin{cases} 1 \text{ for } n = m + \theta, \\ 0 \text{ otherwise,} \end{cases} \quad n, m \in \mathbb{Z}.$$

(b) *Continuous-time case.* If $\mathbb{T} = \mathbb{R}$ the system is continuous-time. The response to the shifted delta function $u(t) = \delta(t - \tau)$, $t \in \mathbb{R}$, then is

$$y(t) = \delta(t - \theta - \tau), \qquad t \in \mathbb{R}.$$

As a result, the kernel of the system is

$$k(t, \tau) = \delta(t - \theta - \tau), \qquad t, \tau \in \mathbb{R}.$$

Thus, the continuous-time delay system has a *singular* kernel.   ∎

Non-Anticipating and Real Linear IOM Systems

In the discrete-time case the interpretation of the kernel k of a linear IOM system is that $k(n, m)$ is the response of the system at time n to a unit pulse at time m. In the continuous-time case $k(t, \tau)$ is the response at time t to a delta function at time τ. A non-anticipating system clearly cannot respond to a pulse or delta function before its arrival. Thus, a necessary condition for non-anticipativeness is both in the discrete- and the continuous-time case that $k(t, \tau) = 0$ for $t < \tau$. It is easy to see that this is also a sufficient condition.

3.3.12.   Summary: Non-anticipativeness of linear IOM systems.   A discrete- or continuous-time linear IOM system with kernel k is non-anticipating if and only if

$$k(t, \tau) = 0 \qquad \text{for} \qquad t < \tau$$

for all t and τ belonging to the time axis on which the system is defined.   ∎

3.3.13.   Example: Non-anticipativeness of the sliding window averager.
The continuous-time sliding window averager of Example 3.2.8(c) is described by the IO relation

$$y(t) = \frac{1}{T_1 + T_2} \int_{t-T_1}^{t+T_2} u(\tau) \, d\tau, \qquad t \in \mathbb{R}.$$

The kernel of this system is

$$k(t, \tau) = \begin{cases} \dfrac{1}{T_1 + T_2} & \text{for } t - T_1 \leq \tau < t + T_2, \\ 0 & \text{otherwise,} \end{cases}$$

$$= \begin{cases} \dfrac{1}{T_1 + T_2} & \text{for } \tau - T_2 < t \leq \tau + T_1, \\ 0 & \text{otherwise,} \end{cases} \qquad t, \tau \in \mathbb{R}.$$

A plot of k is given in Fig. 3.15. Inspection shows that $k(t, \tau) = 0$ for $t < \tau$ if and only if $T_2 = 0$, so that by 3.3.12 the averager is non-anticipating if and only if $T_2 = 0$. This agrees with what we concluded in 3.2.8(c).

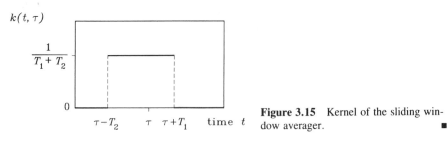

Figure 3.15   Kernel of the sliding window averager.   ∎

3.3.14.   Remark: Real systems.   A *real* IOM system is an IOM system with complex-valued input and output signals, with the property that if the input signal is real, the corresponding output signal is also real. It is easy to see that linear IOM systems are real if and only if their kernel k is real-valued. The pure delay and the sliding window averager are both real systems.   ∎

3.3.15.   Review: The IO map of sampled discrete-time systems.   A *sampled* discrete-time system is a discrete-time system whose input and output signals are defined on the time axis $\mathbb{Z}(T)$. Sampled IOM systems are linear if and only if their IO map may be represented in the form

$$y(t) = T \sum_{\tau \in \mathbb{Z}(T)} k(t, \tau)u(\tau), \qquad t \in \mathbb{Z}(T),$$

where the function k is again called the *kernel* of the sampled system. The value $k(t, \tau)$ of the kernel k at the point (t, τ) is the response of the system at time t if the input is the shifted pulse

$$u(t) = \frac{1}{T} \Delta\left(\frac{t - \tau}{T}\right), \qquad t \in \mathbb{Z}(T).$$

The necessary and sufficient conditions for a linear IOM system to be non-anticipating or real as stated in 3.3.12 and 3.3.14 also apply to sampled systems.   ∎

3.4 CONVOLUTION SYSTEMS

In Section 3.3 it was found that the IO map of any discrete-time linear IOM system may be represented in the form

$$y(n) = \sum_{m=-\infty}^{\infty} k(n, m)u(m), \qquad n \in \mathbb{Z}, \tag{1}$$

with k the kernel of the system. The value $k(n, m)$ of the kernel k at (n, m) is the response of the system at time n when the input is a unit pulse shifted to time m.

Suppose that besides linear the system is also *time-invariant*. Then by time-invariance for every $d \in \mathbb{Z}$ the response at time $n + d$ to a pulse at time $m + d$ is the same as the response at time n to a pulse at time m, that is,

$$k(n + d, m + d) = k(n, m) \qquad \text{for all} \quad n, m, d \in \mathbf{Z}.$$

Evidently, adding the same number d to each of the arguments of the kernel k does not change its value, which means that k actually is a function of the *difference* of its arguments. Thus, there exists a function h of a *single* variable such that

$$k(n, m) = h(n - m) \qquad \text{for all} \quad n, m \in \mathbb{Z}. \tag{2}$$

This shows that the response at time n of a discrete-time time-invariant system to a pulse at time m is $h(n - m)$ and, hence, only depends on the time *elapsed* since the arrival of the pulse. Substitution of (2) into (1) shows that the response of a discrete-time linear time-invariant IOM system may be expressed as

$$y(n) = \sum_{m=-\infty}^{\infty} h(n - m)u(m), \qquad n \in \mathbb{Z}. \tag{3}$$

This expression defines the output y as the result of an operation on the discrete-time signals h and u that is called (discrete-time) *convolution* and written as

$$y = h * u. \tag{4}$$

We next consider continuous-time systems. It is easy to see that also the kernel k of a continuous-time linear IOM system that is time-invariant is a function

$$k(t, \tau) = h(h - \tau) \qquad \text{for all} \quad t, \tau \in \mathbb{R},$$

of the difference of its arguments only. The response of the system may thus be expressed as

$$y(t) = \int_{-\infty}^{\infty} h(t - \tau)u(\tau)\, d\tau, \qquad t \in \mathbb{R}. \tag{5}$$

This operation on the signals h and u is similar to that for the discrete-time case and is called (continuous-time) *convolution*. Again we write

$$y = h * u. \tag{6}$$

For an overview of different types of convolution we refer to Table 3.2 in Section 3.8. The convolution operation is discussed in detail in Section 3.5.

IO maps of the form (4) and (6) are called *convolution maps,* and the corresponding IOM systems are *convolution systems.*

We have shown that every linear time-invariant system is a convolution system. Conversely, every convolution system is linear and time-invariant. We summarize these facts as follows.

3.4.1.   Summary: Convolution systems.

A discrete-time linear IOM system with time axis $\mathbb{Z}$ and real- or complex-valued input and output is time-invariant if and only if it is a convolution system.

A continuous-time linear IOM system with time axis $\mathbb{R}$ and real- or complex-valued input and output is time-invariant if and only if it is a convolution system.   ∎

Impulse Response

The function h occurring in the convolution maps (3) and (5) is called the *impulse response* of the discrete- or continuous-time system. The reason for this name is that the value $h(t)$ of the function h at time t is the response of the system at time t to a unit pulse (in the discrete-time case) or a delta function (in the continuous-time case) at time 0. It follows that for linear time-invariant systems it is sufficient to know the response h of the system to a pulse or delta function at time 0 to determine the response to *any* input.

   Note that we use the name *impulse response* for the function h that describes the response of both discrete-time and continuous-time convolution systems.

   Figure 3.16 shows diagrammatically how the output of a convolution system follows by convolution. It also demonstrates how, in particular, the impulse response h is the response to the unit pulse Δ or the delta function δ.

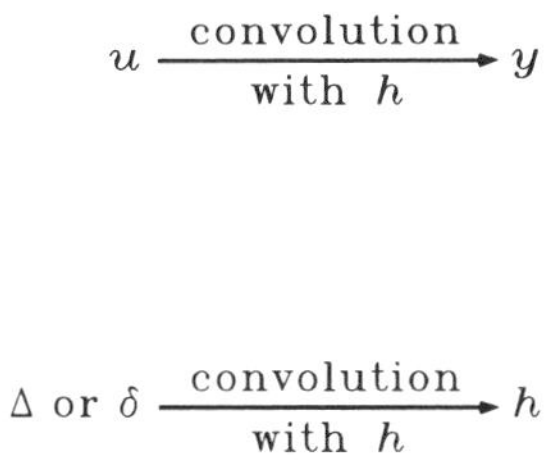

$$u \xrightarrow[\text{with } h]{\text{convolution}} y$$

$$\Delta \text{ or } \delta \xrightarrow[\text{with } h]{\text{convolution}} h$$

Figure 3.16   The output of a convolution system follows by convolving the input with the impulse response h. Top: general input. Bottom: unit input Δ or δ.

Examples

Convolution systems form a major topic in this text. Before starting a detailed discussion of the convolution operation, we give several examples of convolution systems and their impulse responses.

3.4.2.   Example: Convolution systems.
   (a) *The exponential smoother*. The exponential smoother of Example 1.2.6 is described by the difference equation

$$y(n + 1) = ay(n) + (1 - a)u(n + 1), \qquad n \in \mathbb{Z},$$

where we take the time axis as $\mathbb{Z}$. By repeated substitution it easily follows that if $y(n_o)$ is known for some initial time n_o, the output of the system is given by

$$y(n) = a^{n-n_o}y(n_o) + (1 - a) \sum_{m=n_o+1}^{n} a^{n-m} u(m), \qquad n \geq n_o, \qquad n \in \mathbb{Z}.$$

Suppose that the initial condition is $y(n_o) = 0$, and let the initial time n_o approach $-\infty$. Then the response of the system takes the form

$$y(n) = (1 - a) \sum_{m=-\infty}^{n} a^{n-m} u(m), \qquad n \in \mathbb{Z},$$

assuming that the input u is such that the infinite sum converges. Defining the function h by

$$h(n) = \begin{cases} 0 & \text{for } n < 0, \\ (1 - a)a^n & \text{for } n \geq 0, \end{cases}$$

$$= (1 - a)a^n 1(n), \qquad n \in \mathbb{Z},$$

we see that on the infinite time axis $\mathbb{Z}$ the system may be represented as the *convolution system* with IO map

$$y(n) = \sum_{m=-\infty}^{\infty} h(n - m)u(m), \qquad n \in \mathbb{Z}.$$

A plot of the impulse response h of the system is given in Fig. 3.17.

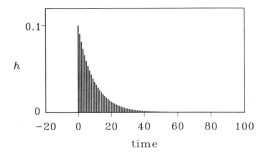

Figure 3.17 The impulse response of the exponential smoother for $a = 0.9$.

(b) *The RC network.* In Example 1.2.7 we found that the solution of the differential equation

$$\frac{dy}{dt}(t) + \frac{1}{RC} y(t) = \frac{1}{RC} u(t), \qquad t \in \mathbb{R},$$

which describes the RC network, for given initial condition $y(t_o)$ at time t_o may be written as

$$y(t) = e^{-(t-t_o)/RC} y(t_o) + \frac{1}{RC} \int_{t_o}^{t} e^{-(t-\tau)/RC} u(\tau)\, d\tau, \qquad t \geq t_o.$$

Keeping the initial condition $y(t_o)$ fixed and letting the initial time t_o go to $-\infty$ we see that the output is given by

$$y(t) = \frac{1}{RC} \int_{-\infty}^{t} e^{-(t-\tau)/RC} u(\tau)\, d\tau, \qquad t \in \mathbb{R},$$

where we assume that the input u is such that the integral converges. Defining the function

$$h(t) = \begin{cases} 0 & \text{for } t < 0, \\ \dfrac{1}{RC} e^{-t/RC} & \text{for } t \geq 0, \end{cases}$$

$$= \frac{1}{RC} e^{-t/RC} \, \mathbb{1}(t), \qquad t \in \mathbb{R},$$

we may represent the system as the continuous-time *convolution system* with IO map

$$y(t) = \int_{-\infty}^{\infty} h(t - \tau) u(\tau) \, d\tau, \qquad t \in \mathbb{R}.$$

The impulse response h of the system is plotted in Fig. 3.18.

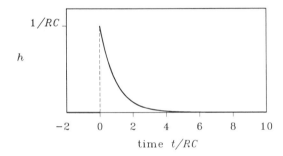

Figure 3.18   The impulse response of the RC network.

(c) *Continuous-time pure delay.* The IO map of the pure delay of Example 3.2.8(a), which is given by

$$y(t) = u(t - \theta), \qquad t \in \mathbb{R},$$

may be rewritten in the form

$$y(t) = \int_{-\infty}^{\infty} \delta(t - \theta - \tau) u(\tau) \, d\tau, \qquad t \in \mathbb{R}.$$

This shows that the delay is a convolution system with the *singular* impulse response

$$h(t) = \delta(t - \theta), \qquad t \in \mathbb{R}. \qquad\qquad \blacksquare$$

Non-Anticipativeness of Convolution Systems

Anticipativeness or non-anticipativeness of convolution systems is easily established from the impulse response.

3.4.3.   Summary: Non-anticipativeness of convolution systems.   A discrete- or continuous-time convolution system with impulse response h is non-anticipating if and only if

$$h(t) = 0 \quad \text{for} \quad t < 0,$$

with t ranging over the time axis on which the system is defined.   ■

This result is an immediate consequence of 3.3.12 and the fact that $k(t, \tau) = h(t - \tau)$.

3.4.4.   Examples: Non-anticipating and anticipating convolution systems. Inspection of the impulse responses of the systems of Example 3.4.2 shows that the exponential smoother and the RC network both are non-anticipating. The pure delay is non-anticipating if and only if the delay θ is nonnegative.   ■

Step Response

The *step response* of a convolution system is its response s when the input u is the unit step $\mathbb{1}$. The step response is closely related to the impulse response. Substitution of $u = \mathbb{1}$ into the IO map yields for the discrete-time case

$$s(n) = (h * \mathbb{1})(n) = \sum_{m=-\infty}^{\infty} h(n - m)\mathbb{1}(m) = \sum_{m=0}^{\infty} h(n - m), \quad n \in \mathbb{Z}.$$

By the change of variable $k = n - m$ this assumes the form

$$s(n) = \sum_{k=-\infty}^{n} h(k), \quad n \in \mathbb{Z}.$$

This shows that the step response is the "running sum" of the impulse response. Conversely, by differencing the step response we find

$$s(n) - s(n - 1) = h(n), \quad n \in \mathbb{Z}.$$

This shows how to retrieve the impulse response from the step response.
    For the continuous-time case it is easy to obtain the corresponding results.

3.4.5.   Summary: Step response of convolution systems.

The step response s of a discrete-time convolution system is the response of the system to the unit step $u = \mathbb{1}$. It is related to the impulse response h by

The step response s of a continuous-time convolution system is the response of the system to the unit step $u = \mathbb{1}$. It is related to the impulse response h by

$$s(n) = \sum_{m=-\infty}^{n} h(m),$$

$$h(n) = s(n) - s(n-1),$$

$$n \in \mathbb{Z}.$$

$$s(t) = \int_{-\infty}^{t} h(\tau)\, d\tau,$$

$$h(t) = \frac{ds(t)}{dt},$$

$$t \in \mathbb{R}.$$

■

The step response is sometimes easier to determine than the impulse response. The impulse response may then be obtained from the step response by differencing (in the discrete-time case) or differentiating (in the continuous-time case).

3.4.6.   Example: Step responses.

(a) *Exponential smoother*. Given the impulse response of the exponential smoother as found in Example 3.4.2(a), we may determine the step response of the smoother as

$$s(n) = \sum_{m=-\infty}^{n} h(m) = (1-a) \sum_{m=0}^{n} a^m$$

$$= \begin{cases} 0 & \text{for } n < 0, \\ 1 - a^{n+1} & \text{for } n \geq 0, \end{cases} \quad n \in \mathbb{Z}.$$

A plot of the step response s is given in Fig. 3.19.

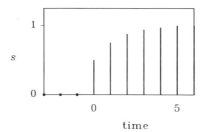

Figure 3.19 Step response of the exponential smoother for $a = 1/2$.

(b) *Pure delay*. The IO map of the continuous-time pure delay of Example 3.2.8(a) is characterized by the relation

$$y(t) = u(t - \theta), \quad t \in \mathbb{R}.$$

It follows that the step response of the system is

$$s(t) = \mathbb{1}(t - \theta), \quad t \in \mathbb{R},$$

as sketched in Fig. 3.20. By 3.4.5, the impulse response h of the system follows by differentiation of s and, hence, is given by

$$h(t) = \frac{d}{dt}\mathbb{1}(t - \theta) = \delta(t - \theta), \quad t \in \mathbb{R}.$$

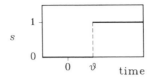

Figure 3.20 Pure delay system. Top: step response. Bottom: impulse response.

This confirms what we found in Example 3.4.2(c).   ∎

In conclusion, we briefly review the results of this section for sampled systems.

3.4.7.   Review: Sampled convolution systems.   The IO map of a linear time-invariant sampled system (see 3.3.15), defined on the time axis $\mathbb{Z}(T)$, may be written as

$$y(t) = T \sum_{\tau \in \mathbb{Z}(T)} h(t - \tau)y(\tau), \qquad t \in \mathbb{Z}(T), \tag{7}$$

with h the *impulse response* of the sampled system. The impulse response is related to the kernel k of the system as $h(t - \tau) = k(t, \tau)$, for t and τ belonging to $\mathbb{Z}(T)$. The expression (7) defines the output y as the sampled convolution

$$y = h * u$$

of the impulse response h and the input u. The impulse response h is the response to the unit pulse

$$u(t) = \frac{1}{T}\Delta\left(\frac{t}{T}\right), \qquad t \in \mathbb{Z}(T).$$

A sampled convolution system is non-anticipating if and only if its impulse response satisfies $h(t) = 0$ for $t < 0$ and $t \in \mathbb{Z}(T)$. The step response s of a sampled convolution system is its response to a unit step $u = \mathbb{1}$. The relation of the step response s to the impulse response h is given by

$$s(t) = T \sum_{\tau \leq t, \tau \in \mathbb{Z}(T)} h(\tau), \qquad t \in \mathbb{Z}(T),$$

$$h(t) = \frac{s(t) - s(t - T)}{T}, \qquad t \in \mathbb{Z}(T).$$

Suppose that the sampled convolution system given by (7) is described in *sequence* form by defining the time sequences u^* and y^* as

$$u^*(n) = u(nT), \qquad y^*(n) = y(nT), \qquad n \in \mathbb{Z}(T).$$

Then (7) may be rewritten as

$$y^*(n) = \sum_{m=-\infty}^{\infty} h^*(n-m)u^*(m), \qquad n \in \mathbb{Z},$$

where h^* is given by

$$h^*(n) = T \cdot h(nT), \qquad n \in \mathbb{Z}.$$

This defines a discrete-time convolution system in sequence form. We refer to the impulse response h^* of this system as the *pulse response* of the system (7). For discrete-time convolution systems that are originally described in sequence form the impulse response h and pulse response h^* are identical.   ∎

3.5 CONVOLUTION

In this section we present a detailed discussion of the *convolution* operation. Convolution is a binary operation among two discrete- or continuous-time signals, which results in another signal. If x and y are real- or complex-valued time *sequences* defined on the infinite time axis $\mathbb{Z}$, the convolution $z = x * y$ of x and y is the signal given by

$$z(n) = (x * y)(n) = \sum_{m=-\infty}^{\infty} x(n-m)y(m), \qquad n \in \mathbb{Z}, \tag{1}$$

provided the infinite sum exists for all $n \in \mathbb{Z}$. If x and y are two real- or complex-valued *continuous-time* signals defined on the infinite time-axis $\mathbb{R}$, their convolution $z = x * y$ is defined as

$$z(t) = (x * y)(t) = \int_{-\infty}^{\infty} x(t - \tau)y(\tau)d\tau, \qquad t \in \mathbb{R}, \tag{2}$$

provided the integral exists for all $t \in \mathbb{R}$. With this notation, the IO map of a discrete- or continuous-time convolution system may be represented as

$$y = h * u,$$

which explains the name convolution system.

Convolution is *not* a pointwise operation, so that the value of $x * y$ at any given time depends on the *entire* time behavior of the signals x and y. For a good understanding of the convolution it is worthwhile to look at the details of the operation. We consider continuous-time convolution but the same arguments apply to discrete-time convolution. From $z = x * y$ and the definition

$$z(t) = (x * y)(t) = \int_{-\infty}^{\infty} x(t - \tau) y(\tau) d\tau, \qquad t \in \mathbb{R},$$

we see that computing the convolution $z(t)$ at some fixed time $t \in \mathbb{R}$ involves the following steps:

(a) Time reverse the signal x and shift the result forward by t.
(b) Multiply y pointwise by the shifted time reversed signal x and integrate the result to obtain $z(t)$.

These steps need be repeated for every value of t. The procedure shows that the convolution $z = x * y$ may be seen as a *local averaging operation* on y with weights obtained by time reversing and shifting x.

We illustrate the procedure with two examples.

3.5.1.   Examples: Convolution.

(a) *Discrete-time convolution.* In Example 3.4.2(a) we found that if we choose the initial condition zero and let the initial time n_o approach $-\infty$, the output of the exponential smoother is given by

$$y = h * u,$$

where h is the one-sided exponential signal plotted in Fig. 3.21(a). Suppose that u is the discrete-time unit step, that is, $u = \mathbb{1}$, as depicted in Fig. 3.21(b) "Locally averaging" u with the time reversed and shifted impulse response h, as shown in Fig. 3.21(c), in this case amounts to taking *past* values of the input u only and weighting them exponentially while averaging. Analytical computation results in

$$y(n) = (h * \mathbb{1})(n) = \begin{cases} 0 & \text{for } n < 0, \\ (1-a) \sum_{m=0}^{n} a^{n-m} = 1 - a^{n+1} & \text{for } n \geq 0, \end{cases} \quad n \in \mathbb{Z},$$

as shown in Fig. 3.21(d). Because of the shape of h, convolution has the effect of smoothing: u undergoes a step change at time 0, while y gradually changes from 0 to the final value 1.

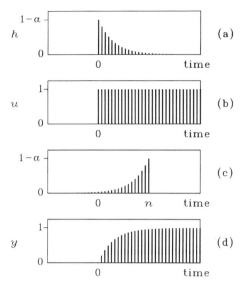

Figure 3.21 Discrete-time convolution of two signals h and u. (a) The signal h. (b) The signal u. (c) The time reversed signal h, shifted by n. (d) the convolution $h * u$.

(b) *Continuous-time convolution.* In Example 3.4.2(b) we found that if the initial time t_o approaches $-\infty$, the output of the RC network is given by the convolution

$$y = h * u,$$

of h and u, with h as shown in Fig. 3.22(a). Since h is a one-sided exponential, convolution of h and u results in exponential weighting of past values of the input. Suppose that u is a rectangular pulse of the form

$$u(t) = \begin{cases} 1 \text{ for } 0 \le t < a, \\ 0 \text{ otherwise,} \end{cases} \qquad t \in \mathbb{R}.$$

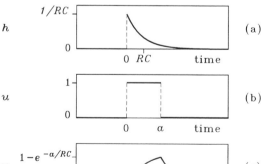

Figure 3.22 Continuous-time convolution of h and u. Top: the signal h. Middle: the signal u. Bottom: the convolution $h * u$.

It follows that

$$y(t) = \frac{1}{RC} \int_{-\infty}^{t} e^{-(t-\tau)/RC} \, u(\tau) \, d\tau$$

$$= \begin{cases} 0 & \text{for } t < 0, \\ 1 - e^{-t/RC} & \text{for } 0 \leq t < a, \\ (1 - e^{-a/RC})e^{-(t-a)/RC} & \text{for } t \geq a. \end{cases}$$

Plots of h, u, and y are given in Fig. 3.22. If one thinks of $h * u$ as the action of h on u, h makes u smoother. The larger RC is, the stronger the effect. ∎

Both in Examples 3.5.1(a) and (b) the effect of convolution is to smooth the input u. Depending on the shape of the impulse response h, also other effects may be achieved. In 3.5.7, for instance, we see that differentiation, which is the opposite of a smoothing operation, may be represented as a convolution.

Properties and Existence of the Convolution

The convolution is a binary operation among time signals. In contrast to all previous binary operations we considered it is not a pointwise operation. Even so, convolution has certain aspects in common with (pointwise) multiplication. In what follows we list the most important properties of the convolution operation.

3.5.2.   Summary: Properties of the convolution   Let $*$ denote discrete-time or continuous-time convolution, and suppose that x, y, and z are discrete- or continuous-time signals defined on the infinite time axis $\mathbb{Z}$ or $\mathbb{R}$, respectively. Then the following holds.

(a) *Commutativity:* If $x * y$ exists, then

$$x * y = y * x.$$

(b) *Associativity:* If $(x * y) * z$ exists, then

$$(x * y) * z = x * (y * z).$$

(c) *Distributivity:* If $x * y$ and $x * z$ exist, then

$$x * (y + z) = x * y + x * z.$$

(d) *Commutativity of scalar multiplication and convolution:* If $x * y$ exists, then

$$\alpha (x * y) = (\alpha x) * y = x * (\alpha y)$$

for any $\alpha \in \mathbb{C}$.

(e) *Shift property*. Let σ denote the back shift operator. Then, if $x * y$ exists,

$$\sigma^t(x * y) = (\sigma^t x) * y = x * (\sigma^t y)$$

for any $t \in \mathbb{Z}$ in the discrete-time case and any $t \in \mathbb{R}$ in the continuous-time case.

(f) *Differentiation property*. Let D denote the differentiation operator, that is, if the continuous-time signal z is differentiable, $Dz(t) = dz(t)/dt$, $t \in \mathbb{R}$. Suppose that the continuous-time convolution $x * y$ exists and is differentiable. Then,

$$D(x * y) = (Dx) * y$$

if x is differentiable and

$$D(x * y) = x * (Dy),$$

if y is differentiable.   ∎

The proof of these properties is not difficult.

The caveat in the definition of the convolution (1) or (2) that the infinite sum or integral may not exist is not superfluous. For instance, convolution of two signals x and y that both are constant leads to diverging infinite sums or integrals so that $x * y$ does not exist. We consider some helpful sufficient conditions for the existence of convolutions.

Before stating the existence results we need introduce some terminology. The *support* of a signal x defined on the discrete or continuous time axis $\mathbb{T}$ is the set

$$\{t \in \mathbb{T} \mid x(t) \neq 0\}.$$

Thus, the support is the set of time instants on which the system is nonzero. (Actually, the support is the *closure* of this set but we do not elaborate on this.) A signal has *bounded support* if its support is contained in some finite interval. A signal is *right one-sided* if its support is contained in a right semi-infinite interval and *left one-sided* if its support is contained in a left semi-infinite interval. Figure 3.23 gives plots of a signal with bounded support and a right one-sided signal.

A continuous-time signal x is called *locally integrable* if

$$\int_a^b |x(t)| \, dt$$

exists and is finite for every finite a and b.

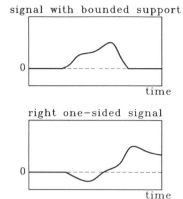

Figure 3.23 Top: a signal with bounded support. Bottom: a right one-sided signal.

3.5.3. Summary: Sufficient conditions for the existence of the convolution.

In what follows, x and y are either both time sequences with time axis $\mathbb{Z}$ or both locally integrable continuous-time signals with time axis $\mathbb{R}$.

(a) If x *or* y has bounded support, then $x * y$ exists. If *both* x and y have bounded support, then also $x * y$ has bounded support.

(b) If x *and* y are one-sided (both right or both left), then $x * y$ exists and is also one-sided (in the same direction as x and y).

(c) If $\|x\|_2$ and $\|y\|_2$ are both finite, then $x * y$ exists, and $\|x * y\|_\infty$ is finite (but $\|x * y\|_2$ is not necessarily finite!).

(d) If $\|x\|_1$ and $\|y\|_\infty$ exist, then $x * y$ exists and $\|x * y\|_\infty$ is finite.   ∎

3.5.4. Exercise: Proof of 3.5.3. Prove 3.5.3. *Hints:* The proofs of (a), (b), and (d) are straightforward. To prove (c), use the Cauchy-Schwartz inequality.   ∎

The statements in parts (a) and (b) concerning the support of the convolution may be made much more specific. The following result makes it possible to delimit the support of a convolution, which is quite useful when doing actual calculations.

3.5.5. Summary: Support of the convolution. Let x and y either be both time-sequences or both continuous-time signals. Suppose that the support of x is contained in some interval $[a, b]$ and that of y in $[c, d]$, with a and c possibly $-\infty$ and b and d possibly ∞. Then the support of $x * y$ is contained in $[a + c, b + d]$.
    ∎

The proof is left to the reader. Note that if x and y both have bounded support, it immediately follows from 3.5.5 that also $x * y$ has bounded support, as stated in 3.5.3(a). Moreover, if x and y are both one-sided in a certain direction, so is $x * y$, as claimed in 3.5.3(b).

3.5.6.   Example: Support of convolutions.   In Examples 3.5.1(a) and (b) one
of the signals that are convolved always has its support in $[0, \infty)$, while the other either has that same interval as its support or a finite interval of the form $[0, a]$. It follows from 3.5.5 that in both cases the convolution of the two signals has its support
in $[0, \infty)$, which agrees with the results that were found.   ∎

Convolution With the Unit Functions Δ and δ

Convolving a discrete-time signal x with the unit pulse Δ leaves x unchanged, because

$$\sum_{m=-\infty}^{\infty} x(n - m)\Delta(m) = x(n), \qquad n \in \mathbb{Z},$$

so that $x * \Delta = x$.
   Similarly, in the continuous-time case convolving a signal x with the δ-func-
tion leaves x unchanged, because

$$\int_{-\infty}^{\infty} x(t - \tau)\delta(\tau) \, d\tau = x(t), \qquad t \in \mathbb{R},$$

so that $x * \delta = x$. In particular,

$$\delta * \delta = \delta.$$

These facts are summarized by the statement that the unit pulse Δ is the *unit* of the
discrete-time convolution and the delta function δ that of the continuous-time convolution.

Convolution With Derivatives of the Delta Function

We continue by briefly discussing convolutions with *derivatives* of the delta function.
The full development of convolutions of generalized functions is given in Supplement C. Suppose that f is an n times continuously differentiable regular function,
with $n \in \mathbb{Z}_+$. Then we have by Property (7) of Table 2.1

$$(f * \delta^{(n)})(t) = \int_{-\infty}^{\infty} f(t - \tau)\delta^{(n)}(\tau) \, d\tau = (-1)^n \frac{d^n}{d\tau^n} f(t - \tau)\Big|_{\tau=0}$$

$$= f^{(n)}(t), \qquad t \in \mathbb{R},$$

with $f^{(n)}$ denoting the nth derivative of f. It follows that

$$f * \delta^{(n)} = f^{(n)}. \tag{3}$$

If f is not an n times differentiable regular function, but *any* generalized function, we still take the convolution of f and $\delta^{(n)}$ to be given by (3). In particular, by letting $f = \delta^{(m)}$, with $m \in \mathbb{Z}_+$, we obtain

$$\delta^{(m)} * \delta^{(n)} = \delta^{(n+m)}, \qquad n, m \in \mathbb{Z}_+.$$

Table 3.1 shows the results of this discussion. The generalized convolution possesses *all* the properties of the continuous-time convolution listed in 3.5.2, including commutativity and the shift and differentiation properties.

TABLE 3.1 CONVOLUTIONS WITH δ-FUNCTIONS

Property	Conditions
(1) $f * \delta^{(n)} = f^{(n)}$	f any regular or generalized function, $n \in \mathbb{Z}_+$
(2) $\delta^{(n)} * \delta^{(m)} = \delta^{(n+m)}$	$n, m \in \mathbb{Z}_+$

3.5.7.   Example: The differentiator.   The *differentiator* is the continuous-time IOM system with IO relation $y = Du$, with D the differentiation operator. In full,

$$y(t) = \frac{du(t)}{dt}, \qquad t \in \mathbb{R}.$$

It is easily verified that the system is linear and time-invariant and, hence, is a convolution system. To find its impulse response, choose the input as $u = \delta$. The response y is the impulse response h, so that

$$h = D\delta = \delta^{(1)}.$$

By property (1) of Table 3.1 the response s of the system to a unit step may be found as

$$s = h * 1 = \delta^{(1)} * 1 = D1$$
$$= \delta.$$

The same result is of course obtained by direct use of the IO relationship

$$s(t) = \frac{d}{dt} 1(t) = \delta(t), \qquad t \in \mathbb{R}. \qquad\qquad \blacksquare$$

We conclude this section with a brief review of the sampled convolution.

3.5.8.   Review: Sampled convolution.   The convolution $z = x * y$ of two sampled signals x and y on the time axis $\mathbb{Z}(T)$ is defined as

$$z(t) = T \sum_{\tau \in \mathbb{Z}(T)} x(t - \tau) y(\tau), \qquad t \in \mathbb{Z}(T).$$

The sampled convolution has all the properties 3.5.2(a)–(e) of the discrete-time convolution, exists under the same conditions 3.5.3 as the discrete-time convolution, and its support may be delimited as in 3.5.5. The unit of the sampled convolution is the unit pulse $\Delta(t/T)/T$, $t \in \mathbb{Z}(T)$. ∎

3.6 STABILITY OF CONVOLUTION SYSTEMS

Stability is an important subject in system theory. It is also a complex subject and we deal only with some of the many forms and definitions of stability in this book. Roughly, a system is *stable* if *"small"* or *bounded* inputs and initial conditions result in bounded outputs. If a small or bounded input or initial condition produces a response that grows indefinitely the system is *unstable*.

We develop the subject of stability step by step over the next few chapters. At this point we introduce the notion of *bounded-input bounded-output* (BIBO) *stability* of convolution systems.

3.6.1.   Definition: BIBO stability of convolution systems.   A discrete- or continuous-time convolution system is *bounded-input bounded-output* (BIBO) *stable* if the response y to every input u with finite amplitude has finite amplitude, that is, if $\| u \|_\infty < \infty$ implies $\| y \|_\infty < \infty$. ∎

Thus, a convolution system that is BIBO stable is well-behaved in the sense that if the input is bounded, so is the output. Because the response of a convolution system is fully determined by its impulse response h, it is not surprising that the BIBO stability of a convolution system may be verified from its impulse response.

3.6.2.   Summary: BIBO stability of convolution systems.
    (a) A discrete- or continuous-time convolution system with impulse response h is BIBO stable if and only if the impulse response has finite action, that is, $\| h \|_1 < \infty$.

    (b) If the convolution system is BIBO stable, then

$$\| y \|_\infty \leq \| h \|_1 \cdot \| u \|_\infty$$

for every input u with finite amplitude, where y is the corresponding output. Equality may always be achieved, for instance by the input given by

$$u(t) = \begin{cases} \overline{h(-t)}/|h(-t)| & \text{if } h(-t) \neq 0, \\ 0 & \text{otherwise,} \end{cases} \quad t \in \mathbb{T},$$

with $\mathbb{T}$ the time axis of the system.   ∎

3.6.3.   Proof.   We only give the proof for the discrete-time case. That for the continuous-time case is analogous. The output y of the system is given by

$$y(n) = \sum_{m=-\infty}^{\infty} h(n-m)u(m), \quad n \in \mathbb{Z}.$$

By the triangle inequality for complex numbers

$$|y(n)| = \left| \sum_{m=-\infty}^{\infty} h(n-m)u(m) \right| \leq \sum_{m=-\infty}^{\infty} |h(n-m)u(m)|$$

$$\leq \left(\sum_{m=-\infty}^{\infty} |h(n-m)| \right) \|u\|_\infty, \quad n \in \mathbb{Z},$$

where we use the fact that $|u(m)| \leq \sup_m |u(m)| = \|u\|_\infty$ for all $m \in \mathbb{Z}$. Substituting $n - m = k$ it follows that

$$|y(n)| \leq \left(\sum_{k=-\infty}^{\infty} |h(k)| \right) \|u\|_\infty = \|h\|_1 \cdot \|u\|_\infty, \quad n \in \mathbb{Z}.$$

As a result,

$$\|y\|_\infty \leq \|h\|_1 \cdot \|u\|_\infty.$$

Thus, if the action $\|h\|_1$ of the impulse response is finite, the amplitude $\|y\|_\infty$ of the output corresponding to any input u with finite amplitude $\|u\|_\infty$ is bounded from above by $\|h\|_1 \cdot \|u\|_\infty$ and, hence, is finite. This shows that a *sufficient* condition for BIBO stability is that the impulse response h has finite action, and also proves the inequality in (b). Note that it also proves 3.5.3(d).

To prove that $\|h\|_1 < \infty$ is also a *necessary* condition, choose the input as

$$u(m) = \begin{cases} \overline{h(-m)}/|h(-m)| & \text{if } h(-m) \neq 0, \\ 0 & \text{otherwise,} \end{cases} \quad m \in \mathbb{Z}.$$

If h is real, u reduces to $u(m) = \text{sign}[h(-m)]$, $m \in \mathbb{Z}$, where sign is the sign function introduced in 3.2.6(b). The amplitude of this input is 1 and, hence, the input is certainly bounded. For the corresponding output y we have at time 0

$$y(0) = \sum_{m=-\infty}^{\infty} h(-m)u(m) = \sum_{m=-\infty}^{\infty} |h(-m)| = \|h\|_1.$$

This proves that in (b) equality may be achieved. It also shows that if $\|h\|_1 = \infty$, the output is unbounded and, hence, $\|h\|_1 < \infty$ is not only sufficient but also necessary for BIBO stability.   ∎

We illustrate BIBO stability with three examples.

3.6.4.   Examples: BIBO stability.

(a) *Exponential smoother*. For the purpose of this example we change the difference equation of the exponential smoother of 1.2.6 to

$$y(n + 1) = ay(n) + bu(n + 1), \qquad n \in \mathbb{Z},$$

which has the solution

$$y(n) = a^{n-n_o}y(n_o) + b \sum_{m=0}^{n-n_o-1} a^m u(n - m), \qquad n \geq n_o, n \in \mathbb{Z}.$$

For $y(n_o) = 0$ and $n_o \to -\infty$ we obtain

$$y = h * u,$$

where the impulse response h is given by

$$h(n) = \begin{cases} 0 & \text{for } n < 0, \\ ba^n & \text{for } n \geq 0, \end{cases} \qquad n \in \mathbb{Z}.$$

Assuming that $b \neq 0$, the action of the impulse response is given by

$$\|h\|_1 = \sum_{k=0}^{\infty} |b| \cdot |a|^k = \begin{cases} \dfrac{|b|}{1 - |a|} & \text{for } |a| < 1, \\ \infty & \text{for } |a| \geq 1. \end{cases}$$

It follows from 3.6.2 that the system is BIBO stable if and only if $|a| < 1$.

Indeed, from Example 3.4.6(a) it easily follows that if $a \neq 1$ the step response s of the system is

$$s(n) = \frac{b}{1 - a}(1 - a^{n+1})\mathbb{1}(n), \qquad n \in \mathbb{Z}.$$

If $|a| > 1$ the step response increases indefinitely with time, confirming that the system is not BIBO stable for these values of the parameter a.

Exercise. Verify that also for $a = 1$ the step response is unbounded. Determine a bounded input that results in an unbounded output if $a = -1$.

(b) *Integrator*. An *integrator*, when defined on the infinite time axis $\mathbb{R}$, is a continuous-time system whose IO map is given by

$$y(t) = \int_{-\infty}^{t} u(\tau)\, d\tau, \qquad t \in \mathbb{R}.$$

Taking the input u as δ it follows that the impulse response of the system is

$$h(t) = \mathbb{1}(t), \qquad t \in \mathbb{R}.$$

Obviously the action of this system is $\|h\|_1 = \infty$, and, hence, the system is not BIBO stable. Indeed, the step response of the system is

$$s(t) = \int_{-\infty}^{t} \mathbb{1}(\tau)\, d\tau = \mathrm{ramp}(t), \qquad t \in \mathbb{R},$$

which is unbounded.

(c) *Differentiator*. In Example 3.5.7 we saw that the impulse response of the differentiator is

$$h(t) = \delta^{(1)}(t), \qquad t \in \mathbb{R}.$$

The response of the differentiator to the unit step is

$$s(t) = \frac{d}{dt}\mathbb{1}(t) = \delta(t), \qquad t \in \mathbb{R},$$

which is unbounded at time 0. Hence, the differentiator is not BIBO stable.

To see that the impulse response $h = \delta^{(1)}$ of the differentiator has infinite action, one may compute the action of an *approximation* of the derivative of the δ-function. For instance, if we use the triangular approximation $d_n(t) = n\,\mathrm{trian}(nt)$, $t \in \mathbb{R}$, of the δ-function, as proposed in Section 2.5, the action of the derivative $d_n^{(1)}$ of d_n is $2n$, which approaches ∞ as $n \to \infty$. ∎

We conclude with a review of the BIBO stability of sampled convolution systems.

3.6.5.   Review: BIBO stability of sampled convolution systems.   A sampled convolution system (see 3.4.7) is BIBO stable if and only if the action

$$\|h\|_1 = T \sum_{\tau \in \mathbb{Z}(T)} |h(\tau)|$$

of its impulse response h is finite. ∎

3.7 HARMONIC INPUTS

In this section we discuss the response of convolution systems to *harmonic* inputs. The reason for our interest is that many signals may be *expanded* as a finite or infinite linear combination of harmonic signals, as we show later. If the response to a single harmonic input is known, the linearity of convolution systems may be exploited to find their response to a linear combination of harmonics.

Harmonic Inputs and Frequency Response

As introduced in Chapter 2, the *harmonic* η_f with frequency $f \in \mathbb{R}$ is the complex-valued signal defined by

$$\eta_f(t) = e^{j2\pi ft}, \qquad t \in \mathbb{T},$$

with $\mathbb{T}$ the time axis on which the signal is defined. We consider the continuous-time case, where $\mathbb{T} = \mathbb{R}$. Then the response of a convolution system with impulse response h to the harmonic $u = \eta_f$ is given by

$$y(t) = \int_{-\infty}^{\infty} h(t - \tau)u(\tau)\, d\tau = \int_{-\infty}^{\infty} h(t - \tau)e^{j2\pi f\tau}\, d\tau, \qquad t \in \mathbb{R},$$

provided the integral exists. By substitution of $t - \tau = \theta$ it follows that

$$\begin{aligned}
y(t) &= \int_{-\infty}^{\infty} h(\theta)e^{j2\pi f(t-\theta)}\, d\theta = \left(\int_{-\infty}^{\infty} h(\theta)e^{-j2\pi f\theta}\, d\theta \right) e^{j2\pi ft} \\
&= \hat{h}(f)e^{j2\pi ft}, \qquad t \in \mathbb{R}.
\end{aligned} \tag{1}$$

Here $\hat{h}$ is the function given by

$$\hat{h}(f) = \int_{-\infty}^{\infty} h(\theta)e^{-j2\pi f\theta}\, d\theta, \qquad f \in \mathbb{R}, \tag{2}$$

provided the integral exists. We see from (1) that the response of the convolution system to the harmonic η_f, if it exists, is *again* harmonic and of the form

$$y = \hat{h}(f)\eta_f,$$

where $\hat{h}$ is defined by (2). The function $\hat{h}$ is called the *frequency response function* of the convolution system. Note that $\hat{h}$ usually is a *complex-valued* function.

In the discrete-time case it follows similarly that the response of a convolution system with impulse response h to the harmonic signal $u = \eta_f$ on the time axis $\mathbb{T} = \mathbb{Z}$ also is of the form

$$y = \hat{h}(f)\eta_f,$$

where the frequency response function $\hat{h}$ now is given by

$$\hat{h}(f) = \sum_{n=-\infty}^{\infty} h(n)e^{-j2\pi fn}, \qquad f \in \mathbb{R}.$$

Figure 3.24 illustrates the results. We summarize as follows.

$$\eta_f \xrightarrow[\text{with } h]{\text{convolution}} \hat{h}(f)\,\eta_f$$

Figure 3.24   Response of a convolution system to a harmonic input.

3.7.1.   Summary: Response of convolution systems to harmonic inputs.

The response of a discrete-time convolution system with time axis $\mathbb{Z}$ and impulse response h to the harmonic input

$$u(n) = e^{j2\pi fn}, \qquad n \in \mathbb{Z},$$

with real frequency f, is

$$y(n) = \hat{h}(f)e^{j2\pi fn}, \qquad n \in \mathbb{Z}.$$

The factor $\hat{h}(f)$ is the value at f of the *frequency response function* $\hat{h}$ of the system, given by

$$\hat{h}(f) = \sum_{n=-\infty}^{\infty} h(n)e^{-j2\pi fn}, \qquad f \in \mathbb{R},$$

provided it exists.

The response of a continuous-time convolution system with time axis $\mathbb{R}$ and impulse response h to the harmonic input

$$u(t) = e^{j2\pi ft}, \qquad t \in \mathbb{R},$$

with real frequency f, is

$$y(t) = \hat{h}(f)e^{j2\pi ft}, \qquad t \in \mathbb{R}.$$

The factor $\hat{h}(f)$ is the value at f of the *frequency response function* $\hat{h}$ of the system, given by

$$\hat{h}(f) = \int_{-\infty}^{\infty} h(t)e^{-j2\pi ft}\, dt, \qquad f \in \mathbb{R},$$

provided it exists.   ∎

What follows gives a sufficient condition for the existence of the frequency response function.

3.7.2.   Summary: Existence of frequency response functions.   The frequency response function $\hat{h}$ of a discrete- or continuous-time convolution system with impulse response h exists and is bounded if the impulse response has finite action, that is, $\|h\|_1 < \infty$.   ∎

3.7.3.   Proof.   We give the proof for the discrete-time case only; that for the continuous-time case is similar. Suppose that $\|h\|_1$ is finite. Then

$$|\hat{h}(f)| = \left| \sum_{n=-\infty}^{\infty} h(n)e^{-j2\pi fn} \right| \leq \sum_{n=-\infty}^{\infty} |h(n)| \cdot |e^{-j2\pi fn}| = \sum_{n=-\infty}^{\infty} |h(n)| = \|h\|_1 < \infty,$$

which shows that $\hat{h}(f)$ exists for every $f \in \mathbb{R}$, and moreover is bounded.   ∎

Note that the condition for the existence of the frequency response function is the same as the necessary and sufficient condition of 3.6.2 for the BIBO stability of the convolution system. Hence, if the system is BIBO stable, its frequency response function exists.

We saw that in the discrete-time case the frequency response function is defined by

$$\hat{h}(f) = \sum_{n=-\infty}^{\infty} h(n)e^{-j2\pi fn}, \qquad f \in \mathbb{R}.$$

Because the complex exponential $e^{-j2\pi fn}$, with n an integer, is periodic in f with period 1, so is the frequency response function $\hat{h}$. Hence, the response to a harmonic input with frequency f is the same as the response to a harmonic input with frequency $f + m$, with m any integer. This is, of course, an immediate consequence of the *aliasing* effect of 2.2.10, which implies that on the time axis $\mathbb{Z}$ the harmonics η_f and η_{f+m} are indistinguishable for any integer m.

3.7.4.   Summary: Periodicity of discrete-time frequency response functions.
The frequency response function of a discrete-time convolution system defined on the time axis $\mathbb{Z}$ is periodic with period 1.   ∎

Because of the periodicity of the frequency response function of a discrete-time convolution system, it is only necessary to specify or display it on a single period, such as the interval $[-\frac{1}{2}, \frac{1}{2})$. Frequency response functions of continuous-time systems generally are *not* periodic.

We consider a few examples of frequency response functions.

3.7.5.   Examples: Frequency response functions.
(a) *Exponential smoother*. From Example 3.4.2(a) the impulse response of the exponential smoother is

$$h(n) = (1 - a)a^n \mathbb{1}(n), \qquad n \in \mathbb{Z}.$$

If $|a| < 1$, the impulse response has finite action, so that by 3.7.2 the system has a frequency response function. It follows that

$$\hat{h}(f) = \sum_{n=-\infty}^{\infty} h(n)e^{-j2\pi fn} = (1 - a) \sum_{n=0}^{\infty} a^n e^{-j2\pi fn} = (1 - a) \sum_{n=0}^{\infty} (ae^{-j2\pi f})^n$$

$$= \frac{1 - a}{1 - ae^{-j2\pi f}}, \qquad f \in \mathbb{R}.$$

The assumption $|a| < 1$ guarantees the convergence of the infinite sum. Plots of the magnitude and phase of $\hat{h}$ are given in Fig. 3.25 for $a = 1/2$. The periodicity with period 1 is evident.

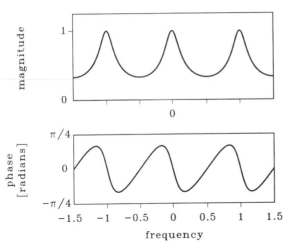

Figure 3.25 Frequency response function of the exponential smoother. Top: magnitude. Bottom: phase.

(b) *RC network*. Next consider the RC network. In Example 3.4.2(b) we saw how the RC network may be considered as a continuous-time convolution system with impulse response

$$h(t) = \frac{1}{RC} e^{-t/RC} \mathbb{1}(t), \qquad t \in \mathbb{R}.$$

The impulse response function has finite action for all $RC > 0$. The frequency response function of the system is

$$\hat{h}(f) = \int_{-\infty}^{\infty} h(t) e^{-j2\pi f t}\, dt = \frac{1}{RC} \int_{0}^{\infty} e^{-t/RC} e^{-j2\pi f t}\, dt$$

$$= \frac{1}{RC} \int_{0}^{\infty} e^{-(j2\pi f + 1/RC)t}\, dt = -\frac{\dfrac{1}{RC}}{\dfrac{1}{RC} + j2\pi f} e^{-\left(\frac{1}{RC} + j2\pi f\right)t} \Bigg|_{0}^{\infty}$$

$$= \frac{1}{1 + RC j 2\pi f}, \qquad f \in \mathbb{R}.$$

Plots of the magnitude and phase of $\hat{h}$ are shown in Fig. 3.26.

(c) *Integrator*. As we have seen in Example 3.6.4(b), the impulse response of the integrator is

$$h(t) = \mathbb{1}(t), \qquad t \in \mathbb{R}.$$

The unit step has infinite action, so that the condition of 3.7.2 for the existence of the frequency response function $\hat{h}$ is not satisfied. Indeed, the integral

$$\int_{0}^{\infty} e^{-j2\pi f t}\, dt$$

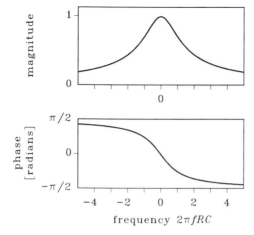

Figure 3.26 Frequency response function of the RC network. Top: magnitude. Bottom: phase.

diverges for every real f, so that the integrator does *not* have a frequency response function in the regular sense. Later (in 7.5.5(a)) we see that the frequency response does exist, but only in the sense of generalized functions.

(d) *Differentiator*. In Example 3.5.7 the impulse response of the differentiator was found to be

$$h(t) = \delta^{(1)}(t), \qquad t \in \mathbb{R}.$$

Although the impulse response has infinite action, it still has a frequency response function, given by

$$\hat{h}(f) = \int_{-\infty}^{\infty} \delta^{(1)}(t)e^{-j2\pi ft}\, dt = j2\pi f \cdot e^{-j2\pi ft}\big|_{t=0}$$
$$= j2\pi f, \qquad f \in \mathbb{R}.$$

This result may be obtained more easily by substituting $u(t) = e^{j2\pi ft}$, $t \in \mathbb{R}$, into the IO relationship $y(t) = du(t)/dt$ of the system. This results in

$$y(t) = j2\pi f \cdot e^{j2\pi ft}, \qquad t \in \mathbb{R},$$

confirming that the frequency response function of the system is $\hat{h}(f) = j2\pi f$, $f \in \mathbb{R}$.

The magnitude and argument of $\hat{h}$ are given by

$$|\hat{h}(f)| = 2\pi |f|, \qquad f \in \mathbb{R},$$
$$\arg(\hat{h}(f)) = \begin{cases} -\pi/2 & \text{for } f < 0, \\ \pi/2 & \text{for } f \geq 0, \end{cases} \qquad f \in \mathbb{R}.$$

as plotted in Fig. 3.27.

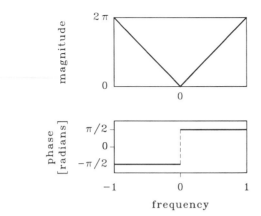

Figure 3.27 Frequency response function of the differentiator. Top: magnitude. Bottom: phase. ∎

Response to Real Harmonic Signals

In the remainder of this section we restrict the discussion to *real* convolution systems. By 3.3.14, real convolution systems are convolution systems whose impulse response h is real-valued. Physical and engineering convolution systems usually are real.

Even if the impulse response is real, the frequency response function generally is a complex-valued function. It has several symmetry properties, enumerated in the following.

3.7.6.   Summary: Symmetry properties of the frequency response function of real convolution systems.   The frequency response function $\hat{h}$ of a *real* discrete- or continuous-time convolution system is *conjugate symmetric,* that is,

$$\hat{h}(-f) = \overline{\hat{h}(f)} \qquad \text{for all} \quad f \in \mathbb{R},$$

with the overbar denoting the complex conjugate. This implies that

(a) the magnitude $|\hat{h}|$ of $\hat{h}$ is *even,* that is,

$$|\hat{h}(-f)| = |\hat{h}(f)| \qquad \text{for all} \quad f \in \mathbb{R},$$

(b) the phase $\arg(\hat{h})$ of $\hat{h}$ is *odd,* that is,

$$\arg(\hat{h}(-f)) = -\arg(\hat{h}(f)) \qquad \text{for all} \quad f \in \mathbb{R}. \qquad ∎$$

These properties are easy to prove. The plots of Figures 3.25, 3.26, and 3.27 show clearly that the frequency response functions of the exponential smoother, RC network, and differentiator, which are real systems, have even magnitude and odd phase.

Because of the symmetry properties, it is enough to specify the frequency response function of a real system for *nonnegative* frequencies only. For this reason the negative frequency part is often omitted in plots. Because the frequency response function of *discrete-time* systems according to 3.7.4 is periodic with period 1, it is sufficient to plot the frequency response of discrete-time real convolution systems on the frequency interval $[0, \frac{1}{2}]$ only.

For real discrete- and continuous-time convolution systems we may give an illuminating interpretation of the *magnitude* $|\hat{h}(f)|$ and *phase* $\arg(\hat{h}(f))$ of the frequency response function at some frequency f in terms of the response of the system to a *real* harmonic signal with frequency f. Let x be the real harmonic signal

$$x(t) = \alpha_x \cos(2\pi f t + \phi_x), \qquad t \in \mathbb{T},$$

with amplitude α_x and phase ϕ_x. We recall from Section 2.2 that the *phasor* of this signal is the complex number

$$a_x = \alpha_x e^{j\phi_x},$$

and that the relation between the real harmonic signal x and its phasor a_x is

$$x(t) = \mathrm{Re}[a_x e^{j2\pi f t}], \qquad t \in \mathbb{R}.$$

3.7.7.   Summary: Response of real convolution systems to real harmonics.
The response of a real discrete- or continuous-time convolution system with frequency response function $\hat{h}$ to the real harmonic input

$$u(t) = \alpha_u \cos(2\pi f t + \phi_u) = \mathrm{Re}[a_u e^{j2\pi f t}], \qquad t \in \mathbb{T},$$

with $\mathbb{T} = \mathbb{Z}$ or $\mathbb{T} = \mathbb{R}$, respectively, $\alpha_u \geq 0$ and ϕ_u real numbers and $a_u = \alpha_u e^{j\phi_u}$ the phasor of u, is the real harmonic signal

$$y(t) = \alpha_y \cos(2\pi f t + \phi_y) = \mathrm{Re}[a_y e^{j2\pi f t}], \qquad t \in \mathbb{T}.$$

The phasor a_y of y is given by

$$a_y = \hat{h}(f)a_u.$$

It follows that the amplitude α_y and phase ϕ_y of the output are

$$\alpha_y = |\hat{h}(f)| \cdot \alpha_u,$$
$$\phi_y = \phi_u + \arg(\hat{h}(f)). \qquad\qquad\qquad\qquad\qquad \blacksquare$$

3.7.8.   Proof.
The proof is not difficult. The idea is to write the real harmonic input u as a sum of complex harmonics and then use linearity to obtain the response of the system. The phasor of the input is $a_u = \alpha_u e^{j\phi_u}$, and, hence, we may write the input as

$$u = \mathrm{Re}(a_u \eta_f) = \frac{1}{2}(a_u \eta_f + \overline{a_u \eta_f}) = \frac{1}{2}(a_u \eta_f + \overline{a}_u \eta_{-f}).$$

This shows that the input is the sum of two complex harmonic signals, one with frequency f and the other with frequency $-f$. By linearity, the response to this input is

$$y = \tfrac{1}{2}[\hat{h}(f)a_u\eta_f + \hat{h}(-f)\overline{a}_u\eta_{-f}].$$

Because by assumption the system is real, by conjugate symmetry $\hat{h}(-f) = \overline{\hat{h}(f)}$, and, hence,

$$y = \tfrac{1}{2}[\hat{h}(f)a_u\eta_f + \overline{\hat{h}(f)a_u\eta_f}] = \mathrm{Re}[\hat{h}(f)a_u\eta_f].$$

This shows that the output is again real harmonic, with phasor $a_y = \hat{h}(f)a_u$. It follows that the amplitude and phase of the output are given by

$$\alpha_y = |a_y| = |\hat{h}(f)|\cdot|a_u| = |\hat{h}(f)|\cdot\alpha_u,$$
$$\phi_y = \arg(a_y) = \arg(\hat{h}(f)) + \arg(a_u) = \arg(\hat{h}(f)) + \phi_u.$$

This completes the proof.   ∎

This result shows the interpretation of the magnitude and phase plots of Figs. 3.25–3.27. The magnitude plots indicate how much the amplitudes of real harmonic signals are amplified or attenuated, depending on their frequency. The phase plots give the corresponding phase shifts.

3.7.9.   Example: Response of the RC network to a real harmonic.   From Example 3.7.5(b) the frequency response function of the RC network is

$$\hat{h}(f) = \frac{1}{1 + RCj2\pi f}, \qquad f \in \mathbb{R}.$$

The magnitude and phase of $\hat{h}$ are given by

$$|\hat{h}(f)| = \frac{1}{\sqrt{1 + R^2C^24\pi^2f^2}}, \qquad f \in \mathbb{R},$$
$$\arg(\hat{h}(f)) = -\mathrm{atan}(RC2\pi f), \qquad f \in \mathbb{R}.$$

If the input is a real harmonic with frequency $f_o = 1/RC2\pi$, we have $\hat{h}(f_o) = 1/(1 + j)$, and the magnitude and phase are

$$|\hat{h}(f_o)| = \tfrac{1}{2}\sqrt{2}, \qquad \arg(\hat{h}(f_o)) = -\frac{\pi}{4}.$$

Thus, the response to the input

$$u(t) = \cos(2\pi f_o t), \qquad t \in \mathbb{R},$$

is

$$y(t) = \tfrac{1}{2}\sqrt{2}\,\cos\!\left(2\pi f_o t - \frac{\pi}{4}\right), \qquad t \in \mathbb{R}.$$

Input and output are shown in Fig. 3.28. The input is attenuated by a factor $\tfrac{1}{2}\sqrt{2} = 0.707$ and delayed by one eighth of a period (because $\pi/4$ is one eighth of 2π).

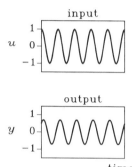

Figure 3.28   Response of the RC network to a real harmonic input. Top: input. Bottom: output. ■

Engineering Significance of Frequency Response

Convolution systems are often used as *filters*. Filters modify harmonic signals by certain desired frequency response functions. A filter that attenuates high-frequency harmonics relative to low-frequency harmonics is called a *low-pass* filter. Low-pass filters have the general effect of *smoothing* the input. Conversely, if low-frequency harmonics are attenuated compared with high-frequency harmonics, the filter is *high-pass*. High-pass filters remove slowly varying components of the input. A *band-pass* filter passes harmonic signals with frequencies in a certain band and attenuates harmonics with all other frequencies. A *band-stop* filter, finally, rejects harmonics with frequencies in a certain band, and passes all other harmonics. Figure 3.29 illustrates these notions.

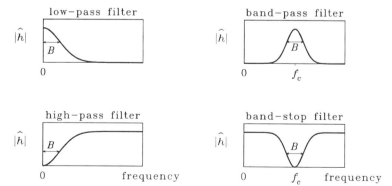

Figure 3.29   Low-pass, high-pass, band-pass, and band-stop filters. B is the bandwidth, f_c the center frequency.

The *bandwidth B* of a low-pass filter specifies the band $[0, B]$ transmitted by the filter. If the filter does not cut off sharply, the bandwidth is not well defined. In such a case it is usually taken as the frequency where the magnitude of the frequency response function has fallen off to a certain fraction, often taken as $\frac{1}{2}\sqrt{2}$—approximately 70%—of the peak value. The bandwidth of a high-pass filter is the width of the band that is suppressed by the filter. To characterize band-pass and band-stop filters it is customary to distinguish the *center frequency* f_c, which more or less specifies the middle of the band, and the bandwidth B. These quantities are shown in Fig. 3.29.

The frequency response functions of Figures 3.25 and 3.26 both represent low-pass filters. For a correct interpretation of the low-, high- or band-pass character of discrete-time filters, their frequency response function should be considered on the frequency interval $[0, \frac{1}{2}]$.

Since the range of the magnitude of frequency response functions usually encompasses several orders of magnitude, the magnitude is often plotted *logarithmically*. In engineering applications it is common to plot the logarithm of the *square* $|\hat{h}|^2$ of the magnitude, because this is a measure for the transfer of the *power* of the input signal. If $\hat{h}$ is physically dimensionless, $\log_{10}(|\hat{h}|^2)$ is expressed in *bel*. Usually this number is converted to *decibel* (dB), which is one tenth of a bel. Thus, the magnitude of $\hat{h}$ is expressed as $10 \log_{10}(|\hat{h}|^2) = 20 \log_{10}(|\hat{h}|)$ dB. The following little table compares how factors of different magnitudes are expressed in dB:

Absolute value	dB
100	40
10	20
1	0
0.1	−20
0.01	−40

Decibels are often also used when $\hat{h}$ is not physically dimensionless. The dB scale then refers to the ratio of $|\hat{h}|^2$ and the relevant physical unit. An advantage of expressing the magnitude of $\hat{h}$ logarithmically is that *multiplication* of the magnitudes of frequency response functions simplifies to *addition*. As seen in Section 3.9, frequency response functions need be multiplied whenever systems are connected in series.

> The bel and decibel are named after the American inventor Alexander Graham Bell (1847–1922.)

Because commonly frequency also ranges over several orders of magnitude, in the continuous-time case frequency is also often plotted logarithmically. The engineering unit corresponding to a logarithmic frequency scale is the *octave*, which measures a factor *two* between two frequencies. Thus, 3 Hz and 6 Hz are one octave apart. The scientific unit is *decade*, which measures a factor *ten*. The frequencies 1.1 Hz and 11 Hz are one decade apart.

We defined the frequency response function $\hat{h}$ as a function of frequency f. Sometimes, however, it is convenient to consider and plot $\hat{h}$ as a function $\hat{h}(f) = \hat{h}(\omega/2\pi)$ of the *angular frequency* $\omega = 2\pi f$.

3.7.10.   Example: RC network as low- and high-pass filter.

(a) *Low-pass.* So far we considered the RC network with the voltage v_C across the capacitor as output, as shown in Fig. 3.30. In Example 3.7.5(b) we found that its frequency response function (here denoted as $\hat{h}_C$) is

$$\hat{h}_C(f) = \frac{1}{1 + RCj2\pi f}, \qquad f \in \mathbb{R}.$$

In Fig. 3.26 the magnitude and phase of $\hat{h}$ are plotted as a function of the frequency f with linear scales. The plot shows that the RC network is a low-pass filter.

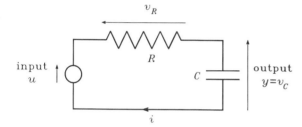

Figure 3.30   RC network.

In Fig. 3.31 the same plot is given with logarithmic scales for frequency and magnitude. It confirms the low-pass nature. The plot is repeated in Fig. 3.32 with the modifications that the frequency response is plotted as a function

$$\hat{h}(\omega/2\pi) = \frac{1}{1 + RCj\omega}, \qquad \omega \in \mathbb{R},$$

of the angular frequency ω on a logarithmic frequency scale, while the magnitude is plotted in dB and the phase in degrees.

The magnitude plot has a low-frequency asymptote 1, corresponding to 0 dB. The high-frequency asymptote of the magnitude is $1/RC\omega$, which corresponds to $-20\log_{10}(\omega) - 20\log_{10}(RC)$ dB, resulting in a straight line with slope -20 dB/ decade in the magnitude plot of Fig. 3.32. The low- and high-frequency asymptote intersect at the angular frequency $\omega = 1/RC$, which may be taken as the bandwidth of the low-pass filter.

The low-frequency asymptote of the phase is $0°$ and the high-frequency asymptote $-90°$.

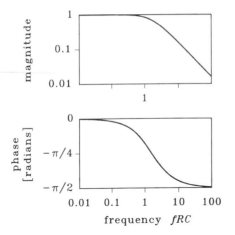

Figure 3.31 Frequency response of the RC network. Top: magnitude. Bottom: phase.

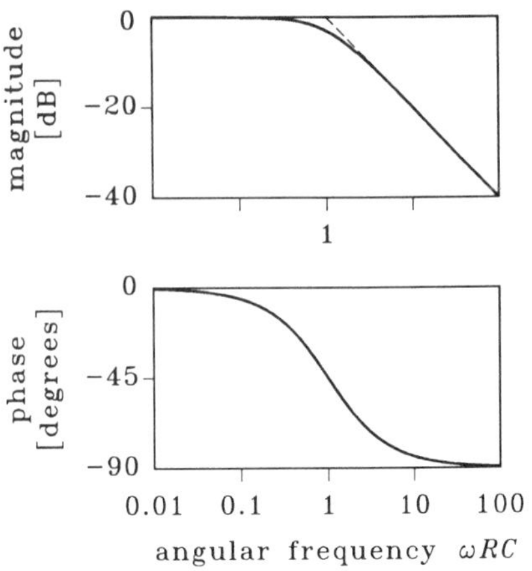

Figure 3.32 Frequency response of the RC network. Top: magnitude. Bottom: phase.

(b) *High-pass.* We also consider the situation that the voltage v_R across the resistor is the output of the network. Since $u = v_R + v_C$, it follows that

$$v_R = u - v_C.$$

If the input is the harmonic η_f we have $u = \eta_f$ and $v_C = \hat{h}_C(f)\eta_f$, so that

$$v_R = [1 - \hat{h}_C(f)]\eta_f.$$

This shows that if the voltage across the resistor is the output of the network, the frequency response function $\hat{h}_R$ is

$$\hat{h}_R(f) = 1 - \hat{h}_C(f) = 1 - \frac{1}{1 + RCj2\pi f} = \frac{RCj2\pi f}{1 + RCj2\pi f}, \qquad f \in \mathbb{R}.$$

The magnitude and phase of this frequency response function are given by

$$|\hat{h}_R(f)| = \frac{RC2\pi|f|}{\sqrt{1 + R^2C^24\pi^2f^2}},$$

$$\arg(\hat{h}_R(f)) = \frac{\pi}{2} - \operatorname{atan}(RC2\pi f), \qquad f \in \mathbb{R}.$$

The plot of the frequency response function of $\hat{h}_R$ given in Fig. 3.33 shows that the system is a *high-pass* filter. Figure 3.34 shows the same plot as a function

$$\hat{h}_R(\omega/2\pi) = \frac{RCj\omega}{1 + RCj\omega}, \qquad \omega \in \mathbb{R},$$

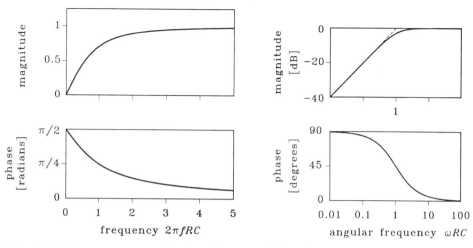

Figure 3.33 Frequency response function of the RC network as a high-pass filter. Top: magnitude. Bottom: phase.

Figure 3.34 Frequency response of the RC network as a high-pass filter. Top: magnitude. Bottom: phase.

of angular frequency, with logarithmic frequency and amplitude scales. The low frequency asymptote of the magnitude is $RC\omega$, corresponding to $20\log_{10}(\omega) + 20\log_{10}(RC)$ dB, which is a straight line with slope 20 dB/decade. The high-frequency asymptote is 1, corresponding to 0 dB. The low- and high-frequency asymptotes intersect at the angular frequency $\omega = 1/RC$, which may be taken as the bandwidth. The low frequency asymptote of the phase is 90° and the high-frequency asymptote 0°.

The quality of the low- and high-pass filters represented by the RC network is low, in the sense that the filter does not cut off sharply. To obtain better filters, higher-order networks are needed. The construction of band-pass and band-stop filters also requires higher-order systems.   ∎

As usual, we conclude with a review of the results of this section for sampled systems.

3.7.11.   Review: Frequency response of sampled convolution systems.   The response of a sampled convolution system with time axis $\mathbb{Z}(T)$ to the harmonic input $u(t) = e^{j2\pi ft}$, $t \in \mathbb{Z}(T)$, is given by

$$y(t) = \hat{h}(f)e^{j2\pi ft}, \qquad t \in \mathbb{Z}(T).$$

The frequency response $\hat{h}$ follows from the impulse response h of the system as

$$\hat{h}(f) = T \sum_{t\in\mathbb{Z}(T)} h(t)e^{-j2\pi ft}, \qquad f \in \mathbb{R}.$$

The frequency response function $\hat{h}$ exists if the impulse response h has finite action $\|h\|_1$. The frequency response function is *periodic* with period $1/T$. If the system is *real* (i.e., the impulse response h is real-valued), then the frequency response function possesses the symmetry properties of 3.7.6. The response of the system to real harmonic inputs may be obtained as in 3.7.7. ∎

3.7.12. Examples: Sampled version of the exponential smoother. In Example 3.7.5(a) we considered the exponential smoother on the time axis $\mathbb{Z}$. On the time axis $\mathbb{Z}(T)$ the exponential smoother is described by the difference equation

$$y(t + T) = ay(t) + (1 - a)u(t + T), \qquad t \in \mathbb{Z}(T).$$

For given initial condition $y(t_o)$, with $t_o \in \mathbb{Z}(T)$, the solution of the difference equation is

$$y(t) = a^{\frac{t-t_o}{T}}y(t_o) + (1 - a) \sum_{\substack{t_o < \tau \leq t, \\ \tau \in \mathbb{Z}(T)}} a^{\frac{t-\tau}{T}}u(\tau), \qquad t \geq t_o, \qquad t \in \mathbb{Z}(T).$$

Taking $y(t_o) = 0$ and letting $t_o \to -\infty$ we obtain

$$y(t) = (1 - a) \sum_{\tau \leq t, \tau \in \mathbb{Z}(T)} a^{\frac{t-\tau}{T}}u(\tau), \qquad t \in \mathbb{Z}(T).$$

This shows that the smoother is a sampled convolution system with impulse response

$$h(t) = \frac{1 - a}{T}a^{\frac{t}{T}}1(t), \qquad t \in \mathbb{Z}(T).$$

The impulse response has finite action $\|h\|_1$ if and only if $|a| \leq 1$. If $|a| \leq 1$, the frequency response function is given by

$$\hat{h}(f) = (1 - a) \sum_{t \geq 0, t \in \mathbb{Z}(T)} a^{\frac{t}{T}}e^{-j2\pi ft}$$

$$= (1 - a) \sum_{k=0}^{\infty} a^k e^{-j2\pi fkT} = (1 - a) \sum_{k=0}^{\infty} (ae^{-j2\pi fT})^k$$

$$= \frac{1 - a}{1 - ae^{-j2\pi fT}}, \qquad f \in \mathbb{R}.$$

The frequency response function is periodic in the frequency f with period $1/T$. Because the impulse response h is real, $\hat{h}$ is conjugate symmetric, so that it is sufficient to consider the frequency response on the frequency interval $[0, 1/2T]$. A plot of the magnitude and phase of $\hat{h}$ is given in Fig. 3.35 for $a = 1/2$.

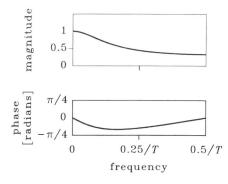

Figure 3.35 Frequency response of the exponential smoother as a sampled convolution system. Top: magnitude. Bottom: phase.

The exponential smoother is a low-pass filter of poor quality. Figure 3.35 shows that for $a = 1/2$ the 70% cut-off frequency (i.e., the frequency at which the magnitude of $\hat{h}$ is 70% of the peak value), is about $0.125/T$. The cut-off frequency thus is proportional to the sampling rate $1/T$. ∎

3.8 PERIODIC INPUTS

In the preceding section we saw that the response of convolution systems to a harmonic input is also harmonic. We now go a step farther, and show that the response of convolution systems to *any* periodic input is periodic. Furthermore, to compute the response to a periodic input, the convolution may be reduced to a special form, called *cyclical* convolution. The cyclical convolution is defined on a *finite* rather than an infinite time axis.

We start with the continuous-time case. Let the continuous-time periodic signal u with period P be the input to a convolution system with impulse response h. Then if the output y exists, we may write

$$y(t) = \int_{-\infty}^{\infty} h(t - \tau)u(\tau)\, d\tau$$

$$= \cdots + \int_{-P}^{0} h(t - \tau)u(\tau)\, d\tau + \int_{0}^{P} h(t - \tau)u(\tau)\, d\tau$$

$$+ \int_{P}^{2P} h(t - \tau)u(\tau)\, d\tau + \cdots, \qquad t \in \mathbb{R}.$$

By suitable changes of variable this may be brought into the form

$$y(t) = \cdots + \int_{0}^{P} h(t - \tau + P)u(\tau - P)\, d\tau + \int_{0}^{P} h(t - \tau)u(\tau)\, d\tau$$

$$+ \int_{0}^{P} h(t - \tau - P)u(\tau + P)\, d\tau + \cdots$$

$$= \sum_{k=-\infty}^{\infty} \int_{0}^{P} h(t - \tau - kP)u(\tau + kP)\, d\tau, \qquad t \in \mathbb{R}.$$

By using the fact that u is periodic with period P, this may be rewritten as

$$y(t) = \int_0^P \left(\sum_{k=-\infty}^{\infty} h(t - \tau - kP) \right) u(\tau) \, d\tau$$

$$= \int_0^P h_{\text{per}}(t - \tau)u(\tau) \, d\tau, \qquad t \in \mathbb{R}, \tag{1}$$

where

$$h_{\text{per}}(t) = \sum_{k=-\infty}^{\infty} h(t - kP), \qquad t \in \mathbb{R}.$$

The signal h_{per} is called the *periodic extension* of h. It is easy to see that h_{per} is periodic with period P. Also y is periodic with period P, because by the periodicity of h_{per} we have from (1)

$$y(t + P) = \int_0^P h_{\text{per}}(t + P - \tau)u(\tau) \, d\tau$$

$$= \int_0^P h_{\text{per}}(t - \tau)u(\tau) \, d\tau = y(t), \qquad t \in \mathbb{R}.$$

In (1) the integration is carried out over a finite interval rather than the entire time axis $\mathbb{R}$. Moreover, because the output y is periodic with period P, it is sufficient to evaluate it over one period, say on the interval $[0, P)$. By the periodicity of h_{per} we may write

$$y(t) = \int_0^P h_{\text{per}}((t - \tau) \bmod P)u(\tau) \, d\tau, \qquad t \in [0, P). \tag{2}$$

Mod is the modulo operator. If a and b are real numbers with b positive, then $a \bmod b = a - kb$, with the integer k such that $0 \le a - kb < b$.

In (2), the three signals y, h_{per} and u that are involved need only be given on the interval $[0, P)$. To emphasize this, let U be a signal defined on the *finite* time axis $[0, P)$ such that

$$U(t) = u(t) \quad \text{for } 0 \le t < P.$$

U is called the *one-period restriction* of the periodic signal u. Similarly, let H be the one-period restriction of h_{per} and Y that of y. Then it follows from (2) that

$$Y(t) = \int_0^P H((t - \tau) \bmod P)U(\tau) \, d\tau, \qquad t \in [0, P).$$

This operation on the finite-time signals H and U is called the (continuous-time) *cyclical convolution* of H and U. We write

$$Y = H \odot U.$$

Table 3.2 reviews the different types of convolutions encountered so far. It also includes the sampled cyclical convolution introduced in 3.8.13.

TABLE 3.2 SUMMARY OF CONVOLUTIONS

Type of convolution	Time axis	Convolution	
		shorthand	*longhand*
discrete-time cyclical convolution	$\underline{N}$	$x \odot y$	$\displaystyle\sum_{m=0}^{N-1} x((n-m) \bmod N)y(m)$
discrete-time (regular) convolution	$\mathbb{Z}$	$x * y$	$\displaystyle\sum_{m=-\infty}^{\infty} x(n-m)y(m)$
sampled cyclical convolution	$\underline{N}(T)$	$x \odot y$	$\displaystyle T \sum_{\tau \in \underline{N}(T)} x((t-\tau) \bmod NT)y(\tau)$
sampled (regular) convolution	$\mathbb{Z}(T)$	$x * y$	$\displaystyle T \sum_{\tau \in \mathbb{Z}(T)} x(t-\tau)y(\tau)$
continuous-time cyclical convolution	$[0^-, P)$	$x \odot y$	$\displaystyle\int_{0^-}^{P} x((t-\tau) \bmod P)y(\tau)\, d\tau$
continuous-time (regular) convolution	$\mathbb{R}$	$x * y$	$\displaystyle\int_{-\infty}^{\infty} x(t-\tau)y(\tau)\, d\tau$

This introduction shows how the response of continuous-time convolution systems to periodic inputs may be obtained by cyclical convolution. A very similar derivation, with sums replacing integrals, applies to discrete-time systems. Again, one period of the response of the system to a periodic input may be obtained by cyclical convolution of one period of the periodic extension of the impulse response and one period of the input.

The results obtained here are summarized in 3.8.7(a) and 3.8.9. Before arriving there we introduce the periodic extension, one-period restriction, and cyclical convolution more formally.

Periodic Extension and One-Period Restriction

By periodic extension we manufacture a periodic signal from an infinite-time signal. One-period restriction turns one period of a periodic signal into a finite-time signal.

3.8.1.   Definitions: Periodic extension and one-period restriction.

(a) *Periodic extension.* Let x be a complex-valued signal on the discrete time axis $\mathbb{T} = \mathbb{Z}$ or the continuous time axis $\mathbb{T} = \mathbb{R}$. Then for given $P \in \mathbb{T}$, with $P > 0$, the *periodic extension with period P* of x is the signal x_{per} defined by

$$x_{\text{per}}(t) = \sum_{k=-\infty}^{\infty} x(t - kP), \qquad t \in \mathbb{T},$$

provided the sum exists for all $t \in \mathbb{T}$.

(b) *One-period restriction.* Let x be a periodic signal with period $P \in \mathbb{T}$ defined on the infinite discrete time axis $\mathbb{T} = \mathbb{Z}$ or the infinite continuous time axis $\mathbb{T} = \mathbb{R}$. Then the *one-period restriction* of x is the signal X defined on the finite time axis $\mathbb{T}_P = [0, P) \cap \mathbb{T}$ given by

$$X(t) = x(t) \qquad \text{for} \quad t \in \mathbb{T}_P. \qquad\qquad \blacksquare$$

Periodic extension of a continuous-time signal is illustrated in Fig. 3.36.

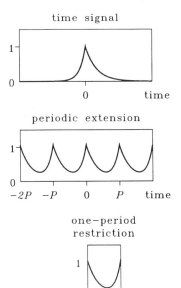

Figure 3.36   Top: an infinite-time signal. Middle: its periodic extension with period P. Bottom: the one-period restriction of the periodic extension.

3.8.2.   Exercise: Properties of the periodic extension.

(a) *Existence.* Show that if the infinite-time signal x has finite action its periodic extension exists for any period $P > 0$.

(b) *Periodic extension of a finite-support signal.* Suppose that the support of the infinite-time signal x with time axis $\mathbb{T}$ is contained in the finite interval $[0, P)$. Prove that the periodic extension of x with period P always exists and coincides with x on $[0, P) \cap \mathbb{T}$. $\blacksquare$

Cyclical Convolution

We continue with the definitions of the discrete- and continuous-time cyclical convolutions.

3.8.3.   Definition: Cyclical convolution.

Let x and y be two complex-valued discrete-time signals defined on the finite time axis $\underline{N} = \{0, \ 1, \ \cdots, \ N - 1\}$. Then, the (discrete-time) *cyclical convolution* $x \odot y$ of x and y is defined on the same time axis and given by

$$(x \odot y)(n) = \sum_{m \in \underline{N}} x((n - m) \bmod N) y(m),$$

for $n \in \underline{N}$.

Let x and y be two complex-valued continuous-time signals defined on the finite time axis $[0, P)$. Then, the (continuous-time) *cyclical convolution* $x \odot y$ of x and y is defined on the same time axis and given by

$$(x \odot y)(t) = \int_0^P x((t - \tau) \bmod P) y(\tau) \, d\tau,$$

for $t \in [0, P)$.   ∎

The cyclical convolution is quite similar to ordinary convolution. The principal difference is that if the argument of one of the two signals that are convolved cyclically moves out of the signal axis, it is cyclically shifted back into the signal axis.

3.8.4.   Example: Cyclical convolution.   Consider the signals x and y defined on the time axis $\underline{4} = \{0, 1, 2, 3\}$, given by $x = (0, 1, 2, 3)$ and $y = (4, 5, 6, 7)$. Then, $z = x \odot y$ is defined on the same time axis, with

$$z(0) = 0 \cdot 4 + 3 \cdot 5 + 2 \cdot 6 + 1 \cdot 7 = 34,$$

$$z(1) = 1 \cdot 4 + 0 \cdot 5 + 3 \cdot 6 + 2 \cdot 7 = 36,$$

$$z(2) = 2 \cdot 4 + 1 \cdot 5 + 0 \cdot 6 + 3 \cdot 7 = 34,$$

$$z(3) = 3 \cdot 4 + 2 \cdot 5 + 1 \cdot 6 + 0 \cdot 7 = 28.$$ ∎

3.8.5.   Exercise: Matrix representation of the discrete-time cyclical convolution.   If u is a complex-valued discrete-time signal with time axis $\underline{N}$, denote by $\vec{u}$ the N-dimensional column vector

$$\vec{u} = \text{col}[u(0), u(1), \cdots, u(N - 1)]$$

and by C_u the $N \times N$ matrix

$$C_u = \begin{bmatrix} u(0) & u(N - 1) & \cdots & u(1) \\ u(1) & u(0) & \cdots & u(2) \\ \cdots & \cdots & \cdots & \cdots \\ u(N - 1) & u(N - 2) & \cdots & u(0) \end{bmatrix}.$$

Note that each column of C_u is a cyclical shift of the preceding column. A matrix with this structure is called a *circulant* matrix. Prove for any two signals x and y defined on the time axis $\underline{N}$ and $z = x \odot y$ that

$$\vec{z} = C_x \vec{y} = C_y \vec{x}.$$

Suppose for instance that the signals x and y are represented by $\vec{x} = \mathrm{col}(1, 1, 0, 0)$ and $\vec{y} = \mathrm{col}(8/15, 4/15, 2/15, 1/15)$. Then if $z = x \odot y$ we have

$$\vec{z} = \begin{bmatrix} 1 & 0 & 0 & 1 \\ 1 & 1 & 0 & 0 \\ 0 & 1 & 1 & 0 \\ 0 & 0 & 1 & 1 \end{bmatrix} \begin{bmatrix} 8/15 \\ 4/15 \\ 2/15 \\ 1/15 \end{bmatrix} = \begin{bmatrix} 3/5 \\ 4/5 \\ 2/5 \\ 1/5 \end{bmatrix}.$$

■

The properties of the cyclical convolution parallel those of the ordinary convolution.

3.8.6.   Summary: Properties of the cyclical convolution.   Let $\odot$ denote the cyclical convolution of signals defined on the finite discrete time axis $\underline{N}$ or the finite continuous time axis $[0, P)$.

(a) All the properties listed in 3.5.2(a)–(d) for the regular convolution, namely *commutativity, associativity, distributivity,* and *commutativity of scalar multiplication and convolution* also apply to the cyclical convolution.

(b) *Shift property.* Define the *cyclical back shift operator* σ in the discrete-time case by

$$(\sigma^\theta x)(n) = x((n + \theta) \bmod N), \qquad n \in \underline{N},$$

and in the continuous-time case by

$$(\sigma^\theta x)(t) = x((t + \theta) \bmod P), \qquad t \in [0, P).$$

Then, if $x \odot y$ exists,

$$\sigma^\theta(x \odot y) = (\sigma^\theta x) \odot y = x \odot (\sigma^\theta y)$$

for any $\theta \in \underline{N}$ in the discrete-time case and any $\theta \in [0, P)$ in the continuous-time case.

■

The proof is not difficult.

Cyclical and Regular Convolution

In what follows we formulate two important connections between the regular and the cyclical convolution. The first result is that if one of the operands of a regular convolution is periodic, the result is also periodic (if it exists). One period of the re-sulting signal may be obtained by periodic extension and cyclical convolution. The other result is that the regular convolution of *finite support* signals may be obtained by cyclical convolution. This is an important result for signal processing, because, as seen in Chapter 9, there exist very efficient numerical algorithms for cyclical con-volution.

3.8.7.   Summary: Cyclical and regular convolution.

(a) *Regular convolution with a periodic signal.* Suppose that x and y are dis-crete- or continuous-time infinite-time signals such that y is periodic with period P and has finite amplitude. Then, if the action $\|x\|_1$ of x is finite, the convolution

$$z = x * y$$

exists and is periodic with period P. Define the finite-time signals, X, Y, and Z as follows:

> X is the one-period restriction of the periodic extension x_{per} of x with period P,
> Y is the one-period restriction of y, and
> Z is the one-period restriction of z.

Then,

$$Z = X \odot Y.$$

(b) *Convolution of finite support signals by cyclical convolution.* Suppose that x and y are discrete- or continuous-time signals defined on the infinite time axis $\mathbb{T}$ such that x, y and their convolution $z = x * y$ all have their support inside the inter-val $[0, P)$ for some $P \in \mathbb{T}$. Define $\mathbb{T}_P$ as the finite time axis $\mathbb{T}_P = [0, P) \cap \mathbb{T}$. Let Z be the restriction of z to $\mathbb{T}_P$, i.e., Z is defined on $\mathbb{T}_P$, and

$$Z(t) = z(t) \qquad \text{for} \quad t \in \mathbb{T}_P.$$

Similarly, X is the restriction of x to $\mathbb{T}_P$ and Y that of y. Then,

$$Z = X \odot Y. \qquad\qquad\qquad\qquad\qquad\qquad\qquad \blacksquare$$

3.8.8.   Proof.   The proof of (a) for the continuous-time case is given in the in-troduction to this section. The existence of z follows from 3.5.3(d). The proof for the discrete-time case is similar.

To prove (b), let y_{per} be the periodic extension of y with period P. By (a), $z = x * y_{\text{per}}$ is periodic with period P. Following (a), define X as the one-period re-striction of the periodic extension x_{per} of x. Because by assumption x has finite sup-port within the interval $[0, P)$, by 3.8.2(b) X and x coincide on $\mathbb{T}_P$. Hence, X is as given in (b). Again, following (a), define Y as the one-period restriction of the periodic extension y_{per} of y. Because again by assumption y has finite support within $[0, P)$, also Y is as defined in (b). It follows from (a) that $Z = X \odot Y$, where Z is the one-period restriction of z. By the shift property of the convolution,

$$z = x * y_{per} = x * \sum_{k=-\infty}^{\infty} \sigma^{-kP} y = \sum_{k=-\infty}^{\infty} \sigma^{-kP}(x * y) = (x * y)_{per},$$

where $(x * y)_{per}$ is the periodic extension of $x * y$. Because by assumption $x * y$ has its support within $[0, P)$, by 3.8.2(b) Z and $x * y$ coincide on $\mathbb{T}_P$, which proves (b).
∎

Response of Convolution Systems of Periodic Inputs

We are now fully prepared to express the response of an infinite-time linear time-invariant system to a periodic input in terms of *finite-time* signals.

3.8.9.   Summary: Response of a convolution system to a periodic input. Suppose that the impulse response h of a discrete- or continuous-time convolution system has finite action.

(a) If the input u is periodic with period P and has finite amplitude the output y exists and is again periodic with period P.

(b) Let Y denote the one-period restriction of the output y, H the one-period restriction of the periodic extension h_{per} of h with period P, and U the one-period restriction of u. Then,

$Y = H \odot U.$ ∎

This result is an immediate consequence of 3.8.7(a). It allows obtaining the response of a convolution system to a periodic input by cyclical convolution.

Figure 3.37 shows how the response to periodic inputs on the one hand follows by regular convolution, and on the other by cyclical convolution. Figure 3.38 is the specialization of the diagram of Fig. 3.37 when the input is the "unit periodic input" Δ_{per} or δ_{per}, successively defined by

$$\Delta_{per}(n) = \sum_{k=-\infty}^{\infty} \Delta(n + kN), \qquad n \in \mathbb{Z},$$

and

$$\delta_{per}(t) = \sum_{k=-\infty}^{\infty} \delta(t + kP), \qquad t \in \mathbb{R}.$$

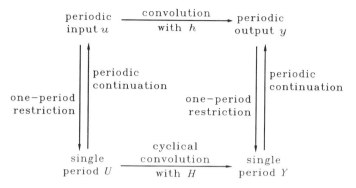

Figure 3.37 The response of a convolution system to a periodic input may be found by regular or by cyclical convolution.

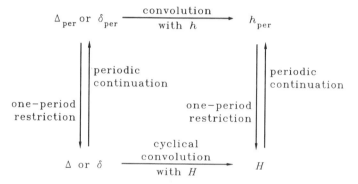

Figure 3.38 Response of a convolution system to the unit periodic input Δ_{per} or δ_{per}.

The result summarized in 3.8.9 shows that the response of a convolution system to a periodic input with a given, fixed period may be obtained by the cyclical convolution

$$Y = H \odot U.$$

This cyclical convolution (with H given) may be thought of as the IO map of a finite-time system, called a *cyclical convolution system*.

3.8.10.   Examples: Response to periodic inputs.

(a) *Exponential smoother*. We consider the exponential smoother with a periodic input u with period 4, one period of which is given by

$$u(0) = U(0) = u(1) = U(1) = 1,$$
$$u(2) = U(2) = u(3) = U(3) = 0.$$

U is the one-period restriction of u. From Example 3.4.2(a) we know that the impulse response h of the exponential smoother is given by

$$h(n) = (1 - a)a^n 1(n), \quad n \in \mathbb{Z}.$$

To apply 3.8.9 we first compute the periodic extension h_{per} of h with period $P = N$. The impulse response h has finite action if $|a| < 1$. For $0 \leq n < N$ we have

$$h_{per}(n) = \sum_{k=-\infty}^{\infty} h(n - kN)$$

$$= (1 - a) \sum_{k=-\infty}^{\infty} a^{n-kN} 1(n - kN) = (1 - a)a^n \sum_{k=-\infty}^{0} a^{-kN}$$

$$= \frac{1 - a}{1 - a^N} a^n, \quad 0 \leq n < N, \quad n \in \mathbb{Z}.$$

Plots of the impulse response h and its periodic extension h_{per} with period $N = 4$ are given in Fig. 3.39 for $a = 1/2$. The numerical values are

$$h_{per}(0) = H(0) = \frac{8}{15},$$

$$h_{per}(1) = H(1) = \frac{4}{15},$$

$$h_{per}(2) = H(2) = \frac{2}{15},$$

$$h_{per}(3) = H(3) = \frac{1}{15},$$

with H the one-period restriction of h_{per}. The cyclical convolution $Y = H \odot U$ is the cyclical convolution of the signals $(1, 1, 0, 0)$ and $(8/15, 4/15, 2/15, 1/15)$ that is computed in Example 3.8.5. It follows from 3.8.9 that one period of the periodic output is given by

$$y(0) = Y(0) = \frac{3}{5},$$

$$y(1) = Y(1) = \frac{4}{5},$$

$$y(2) = Y(2) = \frac{2}{5},$$

$$y(3) = Y(3) = \frac{1}{5}.$$

Plots of the input together with the output are given in Fig. 3.40.

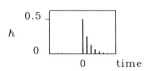

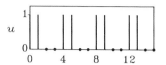

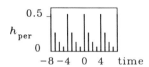

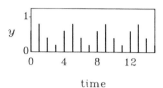

Figure 3.39   Top: the impulse response h of the exponential smoother. Bottom: its periodic extension h_{per} with period 4.

Figure 3.40   Top: periodic input u to the exponential smoother. Bottom: the corresponding output y.

(b) *Echo system.* Consider a continuous-time system whose impulse response is given by

$$h(t) = \sum_{n=0}^{\infty} (\tfrac{1}{2})^n \delta(t - nP), \qquad t \in \mathbb{R},$$

with P a positive number. The response of this system to a delta function is a series of regularly spaced delta functions ("echos") with exponentially decreasing coefficients. We study the response of the system to a periodic input whose period P is the same as the time interval between echos. To apply 3.8.9 we first compute the periodic extension h_{per} of the impulse response h. It follows that

$$h_{\text{per}}(t) = \sum_{k=-\infty}^{\infty} \sum_{n=0}^{\infty} (\tfrac{1}{2})^n \delta(t + kP - nP), \qquad t \in \mathbb{R}.$$

Interchanging the order of summation and substituting $n - k = m$ we obtain

$$h_{\text{per}}(t) = \sum_{n=0}^{\infty} (\tfrac{1}{2})^n \sum_{k=-\infty}^{\infty} \delta(t + kP - nP) = \sum_{n=0}^{\infty} (\tfrac{1}{2})^n \sum_{m=-\infty}^{\infty} \delta(t - mP)$$

$$= 2 \sum_{m=-\infty}^{\infty} \delta(t - mP), \qquad t \in \mathbb{R}.$$

It follows that the one-period restriction H of h_{per} is

$$H(t) = 2\delta(t), \qquad t \in [0, P).$$

As a result, by 3.8.9 the one-period restriction Y of the response y of the echo system to the periodic input u is

$$Y(t) = (H \odot U)(t) = \int_0^P 2\delta((t - \tau) \bmod P)U(\tau) \, d\tau = 2U(t), \qquad t \in [0, P),$$

with U the one-period restriction of the input u. Hence, if the input is periodic with period P, the output of the echo system is

$$y(t) = 2u(t), \qquad t \in \mathbb{R},$$

independently of the precise behavior of the input. *Exercise:* Obtain the same solution by regular convolution. Show that this interesting result is the effect of superposition of the echos of successive periods.   ∎

Response of Convolution Systems to Harmonic Periodic Inputs

We conclude this section with some comments on the response of convolution systems to *harmonic* periodic inputs. As seen in Chapter 2, the harmonic signal η_f defined on the discrete time axis $\mathbb{Z}$ repeats itself periodically with period N if the frequency f satisfies $f \in \mathbb{Z}(F)$, with $F = 1/N$. By aliasing, we actually only need consider frequencies $f \in \underline{N}(F)$. If defined on the continuous time axis $\mathbb{R}$, the harmonic η_f repeats itself periodically with period P if $f \in \mathbb{Z}(F)$, where $F = 1/P$.

As seen earlier in this section, the response of a discrete- or continuous-time convolution system with impulse response h to a periodic input may be obtained by considering the cyclical convolution system with IO map

$$Y = H \odot U.$$

Y is one period of the output and U one period of the input, while H is the one-period restriction (for the given period) of the periodic extension h_{per} (again with the given period) of the impulse response h.

In the discrete-time case, the response of the cyclical convolution system to the harmonic input $U = \eta_f$, with $f \in \underline{N}(F)$, $F = 1/N$, is

$$Y(n) = \sum_{m=0}^{N-1} H((n - m) \bmod N)u(m)$$

$$= \sum_{m=0}^{N-1} H((n - m) \bmod N)e^{j2\pi fm}, \qquad n = 0, 1, \cdots, N - 1.$$

By the substitution $n - m = k$ it easily follows that

$$Y(n) = \hat{H}(f)e^{j2\pi fn}, \qquad n = 0, 1, \cdots, N - 1, \tag{3}$$

where $\hat{H}$ is defined by

$$\hat{H}(f) = \sum_{n=0}^{N-1} H(n)e^{-j2\pi fn}, \qquad f \in \underline{N}(F).$$

The function $\hat{H}$ is called the *frequency response function* of the cyclical convolution system.

In the continuous-time case a similar result holds. We summarize as follows.

3.8.11.   Summary: Response of cyclical convolution systems to harmonic inputs.

The response of the discrete-time cyclical convolution system

$$Y = H \odot U,$$

defined on the time axis $\underline{N}$, to the harmonic input U given by

$$U(n) = \eta_f(n) = e^{j2\pi nf}, \qquad n \in \underline{N},$$

with $f \in \underline{N}(F)$, $F = 1/N$, is

$$Y = \hat{H}(f)\eta_f.$$

The *frequency response function* $\hat{H}$ of the cyclical convolution system is defined by

$$\hat{H}(f) = \sum_{n=0}^{N-1} H(n)e^{-j2\pi nf}, \qquad f \in \underline{N}(F).$$

The response of the continuous-time cyclical convolution system

$$Y = H \odot U,$$

defined on the time axis $[0, P)$, to the harmonic input U given by

$$U(t) = \eta_f(t) = e^{j2\pi ft}, \qquad t \in [0, P),$$

with $f \in \mathbb{Z}(F)$, $F = 1/P$, is

$$Y = \hat{H}(f)\eta_f.$$

The *frequency response function* $\hat{H}$ of the cyclical convolution system is defined by

$$\hat{H}(f) = \int_{0^-}^{P} H(t)e^{-j2\pi ft}dt, \quad f \in \mathbb{Z}(F). \quad \blacksquare$$

We next investigate what this result means for the response of a regular discrete-time convolution system with impulse response h. By periodic continuation of Y as given by (3) it follows that the response of the regular convolution system to the harmonic input η_f, is

$$y = \hat{H}(f)\eta_f, \qquad f \in \underline{N}(F), \tag{4}$$

with $F = 1/N$. On the other hand, we know from Section 3.7 that the response of the system to the harmonic input $u = \eta_f$ is

$$y = \hat{h}(f)\eta_f,$$

where $\hat{h}$ is the frequency response function of the regular convolution system with IO map $y = h * u$, given by

$$\hat{h}(f) = \sum_{n=-\infty}^{\infty} h(n)e^{-j2\pi fn}, \qquad f \in \mathbb{R}. \tag{5}$$

Comparison of (4) and (5) shows that

$$\hat{H}(f) = \hat{h}(f), \qquad f \in \underline{N}(F).$$

Indeed, this may be proved directly. Similarly, in the continuous-time case

$$\hat{H}(f) = \hat{h}(f), \qquad f \in \mathbb{Z}(F),$$

$F = 1/P$, where the frequency response function now is given by

$$\hat{h}(f) = \int_{-\infty}^{\infty} h(t)e^{-j2\pi ft}\, dt, \qquad f \in \mathbb{R}.$$

3.8.12.   Exercise: Connection between $\hat{h}$ and $\hat{H}$.

Let H be the one-period restriction of the periodic extension h_{per} with period N of a finite action signal h defined on the discrete time axis $\mathbb{Z}$. Define

$$\hat{H}(f) = \sum_{n=0}^{N-1} H(n)e^{-j2\pi fn}, \qquad f \in \underline{N}(F),$$

with $F = 1/N$, and

$$\hat{h}(f) = \sum_{n=-\infty}^{\infty} h(n)e^{-j2\pi fn}, \qquad f \in \mathbb{R}.$$

Prove that

$$\hat{H}(f) = \hat{h}(f) \qquad \text{for } f \in \underline{N}(F).$$

Let H be the one-period restriction of the periodic extension h_{per} with period P of a finite action signal h defined on the continuous time axis $\mathbb{R}$. Define

$$\hat{H}(f) = \int_0^P H(t)e^{-j2\pi ft}\, dt, \qquad f \in \mathbb{Z}(F),$$

with $F = 1/P$, and

$$\hat{h}(f) = \int_{-\infty}^{\infty} h(t)e^{-j2\pi ft}\, dt, \qquad f \in \mathbb{R}.$$

Prove that

$$\hat{H}(f) = \hat{h}(f) \qquad \text{for } f \in \mathbb{Z}(F). \qquad \blacksquare$$

Figure 3.41 shows the specialization of Fig. 3.37 to periodic harmonic inputs. It clearly demonstrates the equivalence of $\hat{h}$ and $\hat{H}$ on the appropriate set of frequencies.

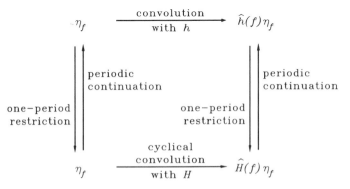

Figure 3.41   The response of convolution systems to a periodic harmonic input with frequency f. In the discrete-time case $f \in \underline{N}(F)$, $F = 1/N$, while in the continuous-time $f \in \mathbb{Z}(F)$, $F = 1/P$.

We conclude with a review of the main result of this section for sampled systems.

3.8.13.   Review: Response of sampled convolution systems to periodic inputs.
The periodic extension z_{per} of a sampled signal z defined on the time axis $\mathbb{Z}(T)$ with period $P \in \mathbb{Z}(T)$ is

$$z_{per}(t) = \sum_{k=-\infty}^{\infty} z(t - kP), \qquad t \in \mathbb{Z}(T).$$

The cyclical convolution $x \odot y$ of two sampled signals x and y given on the time axis $\underline{N}(T)$ is defined as

$$(x \odot y)(t) = T \sum_{\tau \in \underline{N}(T)} x((t - \tau) \bmod NT)\, y(\tau), \qquad t \in \underline{N}(T).$$

Then the one-period restriction Y of the response y of an infinite-time sampled convolution system with impulse response h to a bounded periodic input u with period $P = NT$ is given by

$$Y = H \odot U,$$

where H is the one-period restriction of the periodic extension h_{per} of h and U the one-period restriction of u.

The response of the infinite-time convolution system to the harmonic periodic input η_f, with $f \in \mathbb{Z}(F)$, $F = 1/NT$, is

$$y = \hat{H}(f)\eta_f,$$

where the frequency response function $\hat{H}$ of the sampled cyclical convolution system $Y = H \odot U$ follows from

$$\hat{H}(f) = T \sum_{t \in \underline{N}(T)} H(t)e^{-j2\pi ft}, \qquad f \in \underline{N}(F).$$

$\hat{H}$ is related to the frequency response function

$$\hat{h}(f) = T \sum_{t \in \mathbb{Z}(T)} h(t)e^{-j2\pi ft}, \qquad f \in \mathbb{R},$$

of the infinite-time sampled convolution system $y = h * u$ as

$$\hat{H}(f) = \hat{h}(f), \qquad f \in \underline{N}(F). \qquad\qquad\blacksquare$$

3.9 INTERCONNECTIONS OF SYSTEMS

Two or more IO systems may be *interconnected* by arranging that the outputs of some of the systems serve as inputs to other systems. One way of representing such interconnections is by means of a *block diagram*.

We have already met instances of block diagrams in Figs. 1.10, 1.13, and 2.21. The basic element of a block diagram is a box or block as in Fig. 3.42. A block generally has a number of incoming connections and outgoing connections. If the block has K incoming connections, marked, say, $u_1, u_2, \cdots, u_K$, this signifies that the input of the system may be decomposed into K components, $u_1 \in \mathcal{U}_1$, $u_2 \in \mathcal{U}_2, \cdots, u_K \in \mathcal{U}_K$. As a result, the input set of the system is a product set of the form $\mathcal{U} = \mathcal{U}_1 \times \mathcal{U}_2 \times \cdots \times \mathcal{U}_K$. If the block has M outgoing connections, marked, say, $y_1, y_2, \cdots, y_M$, the system output has M components and the output set is also a product set, $\mathcal{Y} = \mathcal{Y}_1 \times \mathcal{Y}_2 \times \cdots \times \mathcal{Y}_M$.

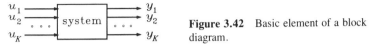

Figure 3.42   Basic element of a block diagram.

A block diagram usually comprises several blocks. In the block diagram it is graphically indicated which component outputs of each subsystem are connected to which component inputs of other subsystems. A connection is allowed only if the associated component output set is contained in the associated component input set. The input to the interconnected (overall) system consists of all unattached incoming connections taken together, while the output of the interconnected (overall) system consists of all unattached outgoing connections. The following example clarifies this.

3.9.1.   Example: A simple interconnection.   In the block diagram of Fig. 3.43, system 2 has the input u with components u_1 and u_2 and a single output y_1. System 1 has the input v_1 and output z with components z_1 and z_2. Component output z_1 of system 1 is connected to component input u_2 of system 2. The overall input to the interconnected system is (u_1, v_1) and the overall output is (y_1, z_2). If the individual systems are IOM systems with IO maps represented in the form

$$y_1 = \phi(u_1, u_2)$$

and

$$z_1 = \psi_1(v_1),$$
$$z_2 = \psi_2(v_1),$$

the IO map of the interconnected system is given by

$$y_1 = \phi(u_1, \psi_1(v_1)),$$
$$z_2 = \psi_2(v_1).$$

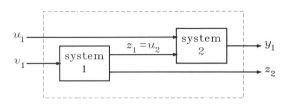

Figure 3.43 Example of a block diagram. ■

Some common subsystems have special graphical representations. Two well-known ones are depicted in Fig. 3.44. The system on the left is a *branch*. It has input u and two component outputs y_1 and y_2, and its IO map is specified by

$$y_1 = u,$$

$$y_2 = u.$$

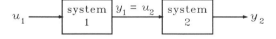

Figure 3.44 Two common subsystems. Left: branch. Right: adder.

The system on the right in Fig. 3.44 is an *adder*. It has two component inputs u_1 and u_2 that belong to the same linear space. The adder has output y given by

$$y = u_1 + u_2.$$

The plus signs may be replaced with minus signs with obvious corresponding changes in the definition of the system.

Series and Parallel Connections

Two well-known connections of IOM systems are the *series* or *cascade connection* and the *parallel connection* as shown in Fig. 3.45. The parallel connection has a branch as well as an adder.

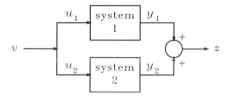

Figure 3.45 Two well-known interconnections of two systems. Top: series connection. Bottom: parallel connection.

In what follows, some facts about the series connection of two IOM systems are collected.

3.9.2. Summary: Series connection.

(a) *IO map.* The series connection of two IOM systems consisting of a system with IO map $\phi_1: \mathcal{U}_1 \to \mathcal{Y}_1$ followed by a system with IO map $\phi_2: \mathcal{U}_2 \to \mathcal{Y}_2$, is well-defined if $\mathcal{Y}_1 \subset \mathcal{U}_2$. It has the IO map $\phi: \mathcal{U}_1 \to \mathcal{Y}_2$ given by

$$\phi = \phi_2 \circ \phi_1,$$

where $\circ$ denotes map composition.

(b) *Impulse response.* If the systems are discrete- or continuous-time convolution systems with impulse responses h_1 and h_2, respectively, the series connection again is a convolution system with impulse response

$$h = h_2 * h_1.$$

(c) *Frequency response function.* If the systems are discrete- or continuous-time convolution systems with frequency response functions $\hat{h}_1$ and $\hat{h}_2$, respectively, the series connection has the frequency response function

$$\hat{h} = \hat{h}_2 \hat{h}_1. \qquad \blacksquare$$

Figure 3.46 illustrates the results. Their proof is simple:

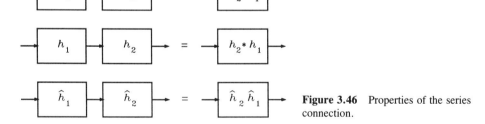

Figure 3.46 Properties of the series connection.

3.9.3. Proof.

(a) By substitution it follows that $y_2 = \phi_2(u_2) = \phi_2(\phi_1(u_1)) = (\phi_2 \circ \phi_1)(u_1)$, so that $\phi = \phi_2 \circ \phi_1$.

(b) Substituting $u_2 = y_1 = h_1 * u_1$ into $y_2 = h_2 * u_2$ it follows by the associativity of the convolution that $y_2 = h_2 * (h_1 * u_1) = (h_2 * h_1) * u_1$. Hence, the series connection is a convolution system with impulse response $h = h_2 * h_1$.

(c) If $u_1 = \eta_f$ (both in the discrete- and the continuous-time case) then $u_2 = y_1 = \hat{h}_1(f)\eta_f$, so that $y_2 = \hat{h}_2(f)\hat{h}_1(f)\eta_f$. It follows that the series connection has the frequency response function $\hat{h} = \hat{h}_2\hat{h}_1$. $\qquad \blacksquare$

For the parallel connection we have what follows.

3.9.4.   Summary: Parallel connection.

(a) *IO map.* The parallel connection of two IOM systems consisting of a system with IO map $\phi_1\colon \mathcal{U} \to \mathcal{Y}$ in parallel with a system with IO map $\phi_2\colon \mathcal{U} \to \mathcal{Y}$, is well-defined if $\mathcal{Y}$ is a linear space, and has the IO map $\phi\colon \mathcal{U} \to \mathcal{Y}$ given by

$$\phi = \phi_1 + \phi_2.$$

Here, $\phi_1 + \phi_2$ is the map defined by $(\phi_1 + \phi_2)(u) = \phi_1(u) + \phi_2(u)$ for all $u \in \mathcal{U}$.

(b) *Impulse response.* If the systems are discrete- or continuous-time convolution systems with impulse responses h_1 and h_2, the parallel connection again is a convolution system with impulse response

$$h = h_1 + h_2.$$

(c) *Frequency response function.* If the systems are discrete- or continuous-time convolution systems with frequency response functions $\hat{h}_1$ and $\hat{h}_2$, the parallel connection has the frequency response function

$$\hat{h} = \hat{h}_1 + \hat{h}_2. \qquad\blacksquare$$

Figure 3.47 illustrates the results, whose proof is left as an exercise.

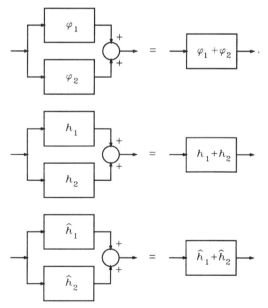

Figure 3.47   Properties of the parallel connection.

3.9.5.   Example: Series connection of hifi audio components.   As shown in Fig. 3.48, home audio equipment often consists of a number of components connected in series: tuner, cassette deck, record or CD player, followed by an amplifier,

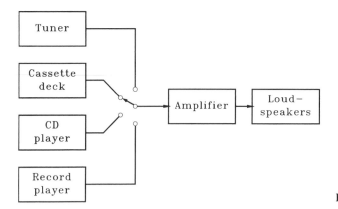

Figure 3.48 Home audio system.

which in turn is connected to the loudspeaker boxes. Each of the three components in the series connection may be characterized by a frequency response function. The frequency response function of the overall system is the product of the three functions. Ideally the overall frequency response function is flat over a wide frequency range, as in Fig. 3.49. The weakest link in the system is the component whose frequency response function starts dropping off soonest on the low- and high-frequency sides. Usually this is the loudspeaker.

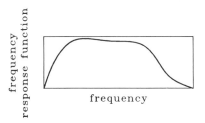

Figure 3.49 Frequency response function of the audio system. ∎

3.9.6.   Example: Audio equalizer.   Audio amplifiers are often provided with an *equalizer*, by means of which the frequency response characteristics may be adjusted. Its purpose is to compensate for weaknesses in the frequency responses of other components, such as the loudspeakers, or to correct for acoustical peculiarities of the room. An N-band equalizer may be realized as the parallel connection of N band-pass filters as in Fig. 3.50, each with a different center frequency but with partly overlapping bands, and provided with an adjustable gain. A *gain* is a system with IO map $y = ku$, with k a constant. If the ith band-pass filter has frequency response function $\hat{h}_i$ and the corresponding gain is k_i, the overall frequency response function is

$$\hat{h} = k_1\hat{h}_1 + k_2\hat{h}_2 + \cdots + k_N\hat{h}_N,$$

as illustrated in Fig. 3.51.

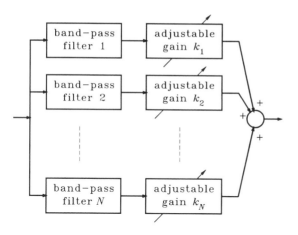

Figure 3.50 *N*-band equalizer.

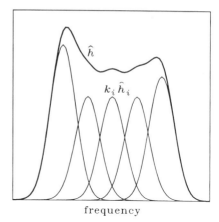

Figure 3.51 Frequency response function of a five-band equalizer. ■

3.10 PROBLEMS

The first problems for this chapter deal with several elementary notions concerning systems as introduced in Section 3.2.

3.10.1. Adder. A calculating device accepts two real numbers as input and produces their sum as output. Describe this as an input-output mapping system by specifying the input set $\mathcal{U}$, the output set $\mathcal{Y}$ and the input-output map $\phi: \mathcal{U} \to \mathcal{Y}$.

3.10.2. Sorting. A computer routine sorts alphanumeric input data $u_1, u_2, \cdots, u_N$ alphabetically. Each input item u_i is a word of variable but finite length consisting of a sequence of characters from the ASCII character set $\mathbb{A}$, and sorting is lexicographically according to the order of the ASCII characters. Describe this routine as an IOM system by specifying the input and output set $\mathcal{U}$ and $\mathcal{Y}$ and the IO map $\phi: \mathcal{U} \to \mathcal{Y}$.

3.10.3. Memoryless system. Prove that a memoryless IOM system is non-anticipating.

3.10.4. Time scaler. A time scaler is a continuous-time IOM system with input and output sets $\mathcal{U} = \mathcal{Y} = \mathcal{L}$, whose IO map is defined by

$$y(t) = u(\alpha t), \qquad t \in \mathbb{R},$$

with α a real constant.

(a) Prove that this system is time-invariant if and only if $\alpha = 1$.
(b) Similarly, prove that the system is non-anticipating if and only if $\alpha = 1$.

Linear systems, which form the subject of the next series of problems, are introduced in Section 3.3.

3.10.5. Backward and forward differencer. A *backward differencer* is a discrete-time IOM system with time axis $\mathbb{Z}$ and input and output sets $\mathcal{U} = \mathcal{Y} = \ell$, whose IO map is defined by

$$y(n) = u(n) - u(n - 1), \qquad n \in \mathbb{Z}.$$

(a) Prove that this system is linear.
(b) Determine its kernel.
(c) Compute and plot the response of the system to the unit step

$$u(n) = \mathbb{1}(n), \qquad n \in \mathbb{Z}.$$

(d) Compute and plot the response of the system to the ramp signal

$$u(n) = n\,\mathbb{1}(n), \qquad n \in \mathbb{Z}.$$

(e) Is the system non-anticipating?
(f) Repeat (b), (c), and (e) for the *forward differencer,* which is the system defined by

$$y(n) = u(n + 1) - u(n), \qquad n \in \mathbb{Z}.$$

3.10.6. Time scaler. A *time scaler,* as introduced in Problem 3.10.4, is a continuous-time IOM system with input and output sets $\mathcal{U} = \mathcal{Y} = \mathcal{L}$, whose IO map is defined by

$$y(t) = u(\alpha t), \qquad t \in \mathbb{R},$$

with α a real constant.

(a) Prove that this system is linear.
(b) Determine the kernel of the system.
(c) Compute and plot the response of the system to a step input

$$u(t) = \mathbb{1}(t), \qquad t \in \mathbb{R}.$$

Distinguish various values of α.

3.10.7. Amplitude modulator. An *amplitude modulator* is a continuous-time system with input and output sets $\mathcal{U} = \mathcal{Y} = \mathcal{L}$ and IO map characterized by

$$y(t) = u(t) \cdot \sin(2\pi f_o t), \qquad f \in \mathbb{R},$$

with f_o a fixed real constant.

(a) Prove that the system is linear and time-varying.
(b) What is the kernel of the system?
(c) Is the system non-anticipating?

3.10.8. Linearization of a "robot arm." Figure 3.52 shows a much simplified "robot arm," consisting of a point mass m at the end of a weightless inflexible rod of length l that rotates about a fixed pivot O. The input to the system is an external torque u, while its output is the angle ϕ the rod makes with the vertical. Gravity exerts a vertical force mg, with g the acceleration of gravity, on the mass. Since the component of this force perpendicular to the rod is $mg \sin(\phi)$, gravitation results in a torque $mgl \sin(\phi)$ on the rod. Thus, from Newton's law we obtain the equation

$$J\frac{d^2\phi(t)}{dt^2} = T_{\text{total}}(t) = u(t) - mgl \sin[\phi(t)], \qquad t \geq t_o.$$

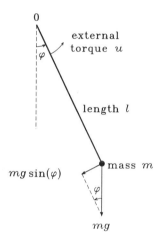

0

external
torque u

length l

mass m

$mg \sin(\varphi)$

mg

Figure 3.52 Robot arm.

Here, T_{total} is the total torque exerted on the arm, and $J = ml^2$ the moment of inertia of the arm. Dividing by J and substituting $J = ml^2$ we obtain after rearrangement

$$\frac{d^2\phi(t)}{dt^2} + \frac{g}{l} \sin[\phi(t)] = \frac{1}{ml^2}u(t), \qquad t \geq t_o.$$

Consider an equilibrium solution corresponding to a constant input torque $u(t) = U_o, t \geq t_o$.

(a) Show that there exists a constant equilibrium solution $\phi(t) = \Phi_o, t \geq t_o$ if and only if $|U_o| \leq mgl$. Also show that for each U_o in this range there in fact exist *two* equilibrium solutions. Clarify these geometrically.

(b) Show that for a given equilibrium solution (U_o, Φ_o) the variational system is described by the linear differential equation

$$\frac{d^2\bar{\phi}(t)}{dt^2} + \frac{g}{l} \cos(\Phi_o)\bar{\phi}(t) = \frac{1}{ml^2}\bar{u}(t), \qquad t \geq t_o.$$

Hint: See 3.3.9.

The purpose of the following two problems is to identify a number of basic system properties.

3.10.9. A discrete-time system. Consider the discrete-time system with input signal range $U = \mathbb{R}$ and output signal range $Y = \mathbb{B} = \{0, 1\}$, whose input u and output y are related by

$$y(n) = \begin{cases} 0 \text{ if } \sum_{k=-\infty}^{n} u(k) < 100, \\ 1 \text{ otherwise,} \end{cases} \quad n \in \mathbb{Z}.$$

Is this system

(a) an IOM system,
(b) linear,
(c) time-invariant,
(d) non-anticipating?

3.10.10. A continuous-time system. Consider a continuous-time system whose input and output sets are subsets of $\mathcal{L}$, and whose input u and output y are related by

$$\dot{y}(t) + y(t) = u(t^2), \quad t \in \mathbb{R}.$$

Is this system

(a) an IOM system,
(b) linear,
(c) time-invariant?

We continue with some problems on convolution systems, defined in Section 3.4.

3.10.11. Backward and forward differencer. The backward differencer introduced in Problem 3.10.5 is a discrete-time system described by

$$y(n) = u(n) - u(n - 1), \quad n \in \mathbb{Z}.$$

(a) Prove that this IOM system is linear and time-invariant, and hence is a convolution system.
(b) Determine and plot the impulse response of the system.
(c) Also determine the impulse response of the forward differencer, described by

$$y(n) = u(n + 1) - u(n), \quad n \in \mathbb{Z}.$$

3.10.12. Finite and infinite summer. A *finite summer* is a discrete-time system with input and output sets $\mathcal{U} = \mathcal{Y} = \ell$, and IO map defined by

$$y(n) = u(n) + u(n - 1) + \cdots + u(n - M + 1), \quad n \in \mathbb{Z}.$$

with M a fixed natural number.

(a) Prove that the system is linear and time-invariant, and hence is a convolution system.

(b) Determine and plot the impulse response of the system.

(c) Determine and plot its step response.

(d) Repeat (b) and (c) for the *infinite summer*, which follows by taking $M = \infty$.

3.10.13. Continuous-time sliding window averager. A continuous-time *sliding window averager* is a system whose input u and output y are continuous-time complex-valued signals that are related by

$$y(t) = \frac{1}{T_1 + T_2} \int_{t-T_1}^{t+T_2} u(\tau)\, d\tau, \qquad t \in \mathbb{R},$$

with T_1 and T_2 nonnegative real numbers such that $T_1 + T_2 \neq 0$.

(a) Prove that this relation can be expressed as a continuous-time convolution $y = h * u$ and determine the impulse response h.

(b) Suppose that the input u is a unit step, that is, $u(t) = 1(t)$, $t \in \mathbb{R}$. Compute the output y.

(c) Use 3.4.3 to establish under what conditions on T_1 and T_2 the system is nonanticipating.

3.10.14. Step response of a continuous-time convolution system. Prove that the step response s and the impulse response h of a continuous-time convolution system are related as

$$s(t) = \int_{-\infty}^{t} h(\tau)\, d\tau, \qquad h(t) = \frac{ds(t)}{dt}, \qquad t \in \mathbb{R}.$$

3.10.15. Discrete-time convolution system with given step response. A discrete-time convolution system has the step response

$$s(n) = n\,1(n), \qquad n \in \mathbb{Z}.$$

(a) What is the impulse response of the system?

(b) Determine the response y of the system to the input

$$u(n) = \begin{cases} 0 & \text{for } n < 0, \\ 1 & \text{for } n = 0, \\ -1 & \text{for } n = 1, \\ 0 & \text{for } n \geq 2, \end{cases} \qquad n \in \mathbb{Z}.$$

Plot both u and y.

3.10.16. Continuous-time convolution system with given step response. The step response of a continuous-time convolution system is given by

$$s(t) = \begin{cases} 1 & \text{for } a \leq t < b, \\ 0 & \text{otherwise,} \end{cases} \qquad t \in \mathbb{R},$$

with a and b real constants such that $a < b$.

(a) Determine the IO map of the system.
(b) For which values of a and b is the system non-anticipating?

3.10.17. Continuous-time system with a given IO pair. A continuous-time convolution system has the IO pair (u^o, y^o) of Fig. 3.53.

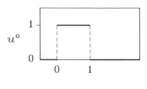

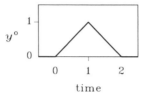

Figure 3.53 An input-output pair of a convolution system.

(a) Find the step response of the system. *Hint:* Use linearity and time-invariance to determine the response of the system to the input u given by $u(t) = u^o(t) + u^o(t - 1) + u^o(t - 2) + \cdots, t \in \mathbb{R}$.
(b) Determine and plot the impulse response of the system.

The next series of problems concern convolution, as discussed in Section 3.5.

3.10.18. Convolutions. Compute the convolutions $x * y$ of the following signal pairs.

(a) $x(n) = \mathbb{1}(n)$, $n \in \mathbb{Z}$, and $y = x$.
(b) $x = \Delta$ is the unit pulse and y is any discrete-time signal defined on the time axis $\mathbb{Z}$.
(c) $x(t) = \text{rect}(t)$, $t \in \mathbb{R}$, and $y = x$.
(d) $x(t) = e^{\alpha t}\mathbb{1}(t)$, $t \in \mathbb{R}$, and $y(t) = e^{\beta t}\mathbb{1}(t)$, $t \in \mathbb{R}$, with α and β complex numbers. Consider the case $\alpha = \beta$ separately.

3.10.19. Convolutions with generalized functions. Determine the convolution $x * y$ when x and y are the following generalized functions.

(a) $x(t) = \delta(t)$ and $y(t) = t$, $t \in \mathbb{R}$.
(b) $x(t) = \delta^{(1)}(t)$ and $y(t) = t$, $t \in \mathbb{R}$.
(c) $x(t) = \mathbb{1}(t - 1)$ and $y(t) = \mathbb{1}(t) + \delta^{(1)}(t - 1)$, $t \in \mathbb{R}$.
(d) $x(t) = w_P(t)$ and $y(t) = \text{trian}(t/P)$, $t \in \mathbb{R}$, where w_P is the infinite comb given by

$$w_p(t) = \sum_{n=\infty}^{\infty} \delta(t - nP), \quad t \in \mathbb{R}.$$

P is a positive real number.

3.10.20. Shift property of the convolution. Prove the shift property of both the discrete-time and the continuous-time convolution as stated in 3.5.2(e).

3.10.21. Support of the convolution. Suppose that x and y are discrete- or continuous-time signals (both defined on the same time axis) such that the support of x is contained in some interval $[a, b]$ and that of y in some other interval $[c, d]$. Prove the statement of 3.5.5 that the support of $x * y$ is contained in $[a + c, b + d]$.

3.10.22. Convolution of truncated signals. To convolve signals with infinite support numerically it is necessary to *truncate* them to signals with finite support (i.e., to set the signals equal to zero outside an interval of finite length). The result of truncation is that the convolution is computed incorrectly, but not always over the entire support of the convolution. In what follows x and y are either both continuous-time signals on the time axis $\mathbb{T} = \mathbb{R}$ or both sampled or discrete-time signals defined on the time axis $\mathbb{T} = \mathbb{Z}(T)$.

(a) *Truncation on the right of right one-sided signals.* Suppose that the support of x lies inside the interval $[a, b]$ and that of y inside the interval $[c, d]$, with a and c finite and b and d possibly infinite. Let x be truncated on the right to the interval $[a, B]$, that is, the truncation x_{tr} of x is defined as

$$x_{\mathrm{tr}}(t) = \begin{cases} x(t) & \text{for } a \leq t \leq B, \\ 0 & \text{otherwise,} \end{cases} \qquad t \in \mathbb{T},$$

with $B \leq b$. Similarly, let y be truncated to the interval $[c, D]$, with $D \leq d$. Prove that the convolution $x_{\mathrm{tr}} * y_{\mathrm{tr}}$ of the truncated signals x_{tr} and y_{tr} agrees with $x * y$ on the interval $(-\infty, e]$, where

$$e = \min(a + D, c + B).$$

Hint: Write the truncation of x as $x_{\mathrm{tr}} = x + x_B$ and that of y as $y_{\mathrm{tr}} = y + y_D$, and prove that $x_{\mathrm{tr}} * y_{\mathrm{tr}} = x * y + \tilde{z}$, where $\tilde{z}$ has support inside (e, ∞).

(b) *Truncation on the left of left one-sided signals.* Find a result similar to (a) for the effect of truncation on the left of left one-sided signals.

(c) *Truncation on both sides.* If both x and y have two-sided infinite support, truncation of the two signals leads to convolutions that are in error over the entire time axis. If the signals are very small outside the truncation intervals it may be hoped that the truncation errors are also small. Prove that if x has two-sided infinite support but is truncated to the interval $[A, B]$, and y has finite support inside $[c, d]$, the convolution of the truncation x_{tr} of x with the (untruncated) signal y coincides with $x * y$ on $[A + d, B + c]$.

3.10.23. Convolution as matrix multiplication. If $u \in \ell$ is a discrete-time signal such that $u(n) = 0$ for $n < 0$, and N is any nonnegative integer, define the column vector $\vec{u}$ and the matrix M_u as follows:

$$\vec{u} = \begin{bmatrix} u(0) \\ u(1) \\ u(2) \\ \cdots \\ u(N-1) \end{bmatrix}, \quad M_u = \begin{bmatrix} u(0) & 0 & 0 & \cdots & 0 \\ u(1) & u(0) & 0 & \cdots & 0 \\ u(2) & u(1) & u(0) & \cdots & 0 \\ \cdots & \cdots & \cdots & \cdots & \cdots \\ u(N-1) & u(N-2) & u(N-3) & \cdots & u(0) \end{bmatrix}.$$

The matrix M_u has the property that the elements along each diagonal are all equal. Such a matrix is called a *Toeplitz* matrix. Because moreover all elements above the main diagonal are zero, the matrix is also *lower triangular*.

Let x and y be discrete-time signals in ℓ such that $x(n) = y(n) = 0$ for $n < 0$. By 3.5.3(b) their convolution $z = x * y$ exists.

(a) Show that $z(n) = 0$ for $n < 0$.
(b) Prove that

$$\vec{z} = M_x \vec{y} = M_y \vec{x}.$$

(c) Suppose that x and y are given by

$$x(n) = \begin{cases} 0 & \text{for } n < 0, \\ 1 & \text{for } n \geq 0, \end{cases} \qquad n \in \mathbb{Z}.$$

$$y(n) = \begin{cases} 0 & \text{for } n < 0 \text{ and } n \geq 3, \\ 1 & \text{for } 0 \leq n < 3, \end{cases} \qquad n \in \mathbb{Z}.$$

Use (b) to compute $z = x * y$ on the time interval $0 \leq n \leq 5$, $n \in \mathbb{Z}$.

Otto Toeplitz (1881–1940) was a German mathematician who emigrated to Palestine in 1938. He contributed to the theory of quadratic forms.

The following problems concern BIBO stability as introduced in Section 3.6.

3.10.24. "Impulsive" systems. Consider the continuous-time system with input set $\mathcal{U} \subset \mathcal{L}$ and output set $\mathcal{Y} \subset \mathcal{L}$, whose IO map is of the form

$$y(t) = \int_{-\infty}^{\infty} h^o(t - \tau)u(\tau)\, d\tau + \sum_{i \in \mathbb{Z}} a_i u(t - t_i), \qquad t \in \mathbb{R}.$$

Here, $h^o \in \mathcal{L}$ is a given function, while the a_i, $i \in \mathbb{Z}$, are real or complex coefficients and the t_i, $i \in \mathbb{Z}$, distinct real time instants.

(a) Prove that the system is linear and time-invariant.
(b) Determine the impulse response of the system.
(c) Prove that the system is non-anticipating if and only if $h^o(t) = 0$ for $t < 0$ and $a_i = 0$ for $t_i < 0$.
(d) Prove that the system is BIBO stable if and only if both h^o and a have finite action, where a is the sequence $\{\cdots, a_{-1}, a_0, a_1, a_2, \cdots\}$.

3.10.25. Echo system. A continuous-time system defined on the infinite time axis $\mathbb{T} = \mathbb{R}$ is described by the IO relation

$$y(t) = u(t) + \alpha y(t - \theta), \qquad t \in \mathbb{R},$$

with θ a fixed positive real constant, and α a real constant. The first term is the direct effect of the input on the output, while the second represents an *echo*.

(a) Show that this system is linear and time-invariant, and find its impulse response.
(b) For which values of α is the system BIBO stable?

3.10.26. BIBO stability. Verify whether the following systems are BIBO stable. If the system is not BIBO stable, show an example of a bounded input that results in an unbounded output.

(a) The backward and forward differencers of Problem 3.10.5.
(b) The finite summer of Problem 3.10.12.
(c) The infinite summer of Problem 3.10.12.
(d) The continuous-time sliding window averager of Problem 3.10.13.
(e) The system of Problem 3.10.15.
(f) The system of Problem 3.10.16.

3.10.27. Output maximization. Consider a continuous-time non-anticipating convolution system with impulse response h such that

$$h(t) = 0 \qquad \text{for} \quad t \geq T,$$

with T a fixed positive time.

(a) Determine an input u with $|u(t)| \leq 1$ for all $t \in \mathbb{R}$ such that $|y(T)|$ is maximal.
(b) Show that the maximal value of $|y(T)|$ equals the action $\|h\|_1$ of h.

Hint: See 3.6.3.

The following problems deal with the frequency response of convolution systems, as discussed in Section 3.7.

3.10.28. Frequency response functions. Determine whether the following systems have a frequency response function. If the system has a frequency response function, calculate it and sketch its magnitude and phase as a function of frequency.

(a) The backward differencer of Problem 3.10.5.
(b) The forward differencer of Problem 3.10.5.
(c) The finite summer of Problem 3.10.12.
(d) The infinite summer of Problem 3.10.12.
(e) The continuous-time delay, which is the continuous-time system described by

$$y(t) = u(t - \theta), \qquad t \in \mathbb{R},$$

with θ a positive real constant.

3.10.29. Response to real harmonic inputs.

(a) Given the frequency response function of the backward differencer as obtained in Problem 3.10.28(a), determine the response of the system to the real harmonic input

$$u(n) = u_{\max} \cos(2\pi f_o n), \qquad n \in \mathbb{Z}.$$

Are there any frequencies f_o for which the output of the system is identical to zero? Answer the same questions for the forward differencer.

(b) Determine the response of the continuous-time system with frequency response function

$$\hat{h}(f) = \frac{1}{(1 + j2\pi f)^2}, \qquad f \in \mathbb{R},$$

to the real harmonic input

$$u(t) = \sin(t), \qquad t \in \mathbb{R}.$$

3.10.30. Symmetry properties of the frequency response function. Let $\hat{h}$ be the frequency response function of a *real* discrete- or continuous-time linear time-invariant system, i.e., a system whose impulse response h is a real-valued function. Prove the symmetry properties of 3.7.6:

(i) *Conjugate symmetry:* $\hat{h}(-f) = \overline{\hat{h}(f)}$ for all $f \in \mathbb{R}$.
(ii) *Evenness of the magnitude:* $|\hat{h}|$ is an even function.
(iii) *Oddness of the phase:* $\arg(\hat{h})$ is an odd function.

The next series of problems relate to cyclical convolution, treated in Section 3.8.

3.10.31. Periodic extension. Determine the periodic extension with period P of the following continuous-time signals:

(a) The one-sided exponential signal

$$x(t) = e^{-t}1(t), \qquad t \in \mathbb{R}.$$

(b) The rectangular pulse

$$x(t) = \begin{cases} 1 & \text{for } 0 \le t < a, \\ 0 & \text{otherwise}, \end{cases} \qquad t \in \mathbb{R},$$

with (b.1) $0 < a \le P$, and (b.2) $P < a \le 2P$.

3.10.32. Cyclical convolution. Determine the cyclical convolution of the following pairs of signals x and y:

(a) The discrete-time signals defined on the time axis $\underline{8}$ given by $x = (1, 1, 1, 1, 0, 0, 0, 0)$ and $y = (1, 1, 0, 0, 0, 0, 0, 0)$.
(b) The continuous-time signals defined on the time axis $[0, 1)$ by

$$x(t) = y(t) = \begin{cases} 1 & \text{for } 0 \le t < \frac{1}{2}, \\ 0 & \text{for } \frac{1}{2} \le t < 1. \end{cases}$$

3.10.33. Shift property of the cyclical convolution. Prove the shift property 3.8.6(b) of the continuous-time cyclical convolution.

3.10.34. Response of the RC network to the periodic rectangular pulse. If $RC = 1$, the impulse response of the RC network is given by

$$h(t) = e^{-t}\mathbb{1}(t), \qquad t \in \mathbb{R}.$$

(a) Compute the periodic extension h_{per} of h with period 1.
(b) Use cyclical convolution to determine the response of the RC network to the periodic rectangular pulse u, one period of which is given by

$$u(t) = \begin{cases} 1 & \text{for } 0 \le t < \frac{1}{2}, \\ 0 & \text{for } \frac{1}{2} \le t < 1. \end{cases}$$

The final problems concern interconnections of systems, dealt with in Section 3.9.

3.10.35. An interconnection. Three linear time-invariant systems with impulse responses h_1, h_2 and h_3 are interconnected as in Fig. 3.54. The systems are either all three discrete-time or all three continuous-time.

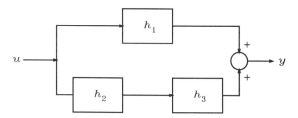

Figure 3.54   An interconnection of three linear time-invariant systems.

(a) Determine the impulse response of the interconnection.
(b) Suppose that the systems have frequency response functions $\hat{h}_1$, $\hat{h}_2$, and $\hat{h}_3$, respectively. Determine the frequency response function of the interconnection.

3.10.36. Series connection of two differencers. Consider the series connection of a backward and a forward differencer (see Problem 3.10.5), as in Fig. 3.55. Determine

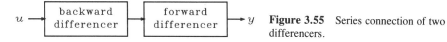

Figure 3.55   Series connection of two differencers.

(a) the impulse response
(b) the frequency response function

of the series connection.

3.10.37. Series connection. Consider two IOM systems S_1 and S_2 connected in series as in Fig. 3.56. We denote the series connection as $S_2 \circ S_1$. Prove the following four statements. In each case give an example to show that the *converse* statement is not true.

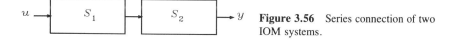

Figure 3.56   Series connection of two IOM systems.

(a) S_1 is linear and S_2 linear $\Rightarrow$ $S_2 \circ S_1$ is linear.
(b) S_1 is time-invariant and S_2 is time-invariant $\Rightarrow$ $S_2 \circ S_1$ is time-invariant.
(c) S_1 is non-anticipating and S_2 is non-anticipating $\Rightarrow$ $S_2 \circ S_1$ is non-anticipating.
(d) S_1 is memoryless and S_2 is memoryless $\Rightarrow$ $S_2 \circ S_1$ is memoryless.

3.11 COMPUTER EXERCISES

The computer exercises for Chapter 3 serve to introduce various elementary ways to handle linear time-invariant systems numerically. The first exercises deal with convolution.

3.11.1. Convolution. (Compare 3.10.18.) Compute and plot the convolutions $x * y$ of the following signals. Identify the effect of truncation, which may make some of the results partially incorrect. For continuous-time convolutions be sure to choose the sampling interval small enough.

(a) $x(n) = \mathbb{1}(n)$, $n \in \mathbb{Z}$, and $y = x$.
(b) $x = \Delta$ is the unit pulse and y is defined by $y(n) = a^{|n|}$, $n \in \mathbb{Z}$, with $a = 0.9$.
(c) $x(t) = \text{rect}(t)$, $t \in \mathbb{R}$, and $y = x$.
(d) $x(t) = e^{\alpha t}\mathbb{1}(t)$ and $y(t) = e^{\beta t}\mathbb{1}(t)$, $t \in \mathbb{R}$, with $\alpha = \beta = -1$.

3.11.2. Convolutions with δ-functions. Generate the sampled approximations d, d_1, and d_2 of the δ-function and its first two derivatives as in 2.7.10.

(a) Check the equalities $\delta * \delta' = \delta'$ and $\delta' * \delta' = \delta''$ numerically. Explain any discrepancies.
(b) Define the function ϕ as in 2.7.10(c), and compute the convolutions $\delta * \phi$, $\delta' * \phi$, and $\delta'' * \phi$ numerically. Verify the results and explain any discrepancies.

Numerical convolution.

Numerical convolution involves several kinds of errors.

Truncation errors. For numerical computation time signals with infinite support necessarily need be *truncated* to signals with finite support. Convolving truncated signals leads to truncation errors. In Problem 3.10.22 it is shown that if the signals are one-sided or one of the signals has finite support the convolution of the signals is correct inside a well-defined interval, and hence is free of truncation errors on this interval. Outside this interval the truncation errors may be expected to be small if the signals are small outside their truncation intervals.

Sampling errors. To compute the convolution of continuous-time signals they need be sampled, so that the convolution integral may be replaced by an approximating convolution sum. To keep the resulting error small the sampling interval should be chosen sufficiently small.

Rounding errors. On top of the truncation and sampling errors there are rounding errors. They may be kept small by using sufficient numerical precision.

3.11.3. Interpolation. As discussed in Section 2.3, the interpolation of a sampled signal $x*$, defined on the time axis $\mathbb{Z}(T)$, using the interpolation function i is the continuous-time signal given by

$$x(t) = \sum_{n \in \mathbb{Z}} x*(nT) i\left(\frac{t - nT}{T}\right), \qquad t \in \mathbb{R}.$$

Define the continuous-time signals $x^\dagger$ and i_T by

$$x^\dagger(t) = \sum_{n \in \mathbb{Z}} x*(nT)\delta(t - nT), \qquad t \in \mathbb{R},$$

$$i_T(t) = i\left(\frac{t}{T}\right), \qquad t \in \mathbb{R}.$$

(a) Prove that the interpolated signal x may be expressed as

$$x = x^\dagger * i_T.$$

(b) In Fig. 2.20 it is shown how a discrete-time signal is interpolated with three different interpolating functions. Reproduce this figure taking $T = 1/8$. *Hint:* Define a discrete time axis with increment $\text{inc} \ll T$ to approximate the continuous time axis, and approximate the signal $x^\dagger$ as

$$x^\dagger(t) = \begin{cases} \dfrac{x*(t)}{\text{inc}} & \text{for } t \in \mathbb{Z}(T), \\ 0 & \text{otherwise,} \end{cases} \qquad t \in \mathbb{Z}(\text{inc}).$$

It is convenient to choose T as an integral multiple of inc.

3.11.4. Response of the sliding window averager. In 3.2.8(b) the discrete-time sliding window averager is described as a system whose output y corresponding to the input u is given by

$$y(n) = \frac{1}{N + M + 1} \sum_{m=-M}^{N} u(n + m), \qquad n \in \mathbb{Z},$$

with N and M nonnegative integers.

(a) Define a function "slide" that implements the IO map, that is, if u is a discrete-time signal, then $y = \text{slide}(u, M, N)$ is the output of the sliding window averager with parameters M and N. *Hint:* Write

$$\sum_{m=-M}^{N} u(n + m) = \sum_{k=n-M}^{n+N} u(k) = \sum_{k=-\infty}^{n+N} u(k) - \sum_{k=-\infty}^{n-M-1} u(k)$$

$$= s(n + N) - s(n - M - 1), \qquad n \in \mathbb{Z},$$

where s is the "summed" signal defined by

$$s(n) = \sum_{k=-\infty}^{n} u(k), \qquad n \in \mathbb{Z}.$$

Use a suitable standard function to compute the summed signal.

(b) Determine and plot the impulse response of the averager by computing its response to the unit pulse $u = \Delta$. Take $N = 10$ and $M = 5$. *Hint:* Note that the result is in error near the end of the time axis owing to truncation. The error may be corrected by restricting the impulse response as computed to a slightly smaller interval.

(c) Determine and plot the response of the system to a step and to a ramp

   (c.1) directly using the IO map
   (c.2) by convolution with the impulse response.

   Hint: Again there are differences near the end of the time axis. Explain and correct.

3.11.5. Response of the RC network. In 3.4.2(b) it is found that if the output y of the RC network is the voltage v_C across the capacitor, the initially-at-rest network may be described as the continuous-time convolution system

$$y = h * u,$$

where the impulse response h is given by

$$h(t) = \frac{1}{RC} e^{-t/RC} \mathbb{1}(t), \qquad t \in \mathbb{R}.$$

Suppose that rather than the voltage $y = v_C$ across the capacitor we take the voltage $z = v_R$ across the resistor as output. Then since by Kirchhoff's voltage law $u = v_C + v_R$ or $u = y + z$, the IO map of the system now is

$$z = u - y = u - h * u.$$

We may rewrite this as the convolution system

$$z = g * u,$$

whose impulse response g is given by

$$g = \delta - h.$$

Take $RC = 1$, and let u be the rectangular input as in 3.5.1(b) with $a = 2$.

(a) Determine and plot the response of the voltage y across the capacitor to this input by convolution of the input with the impulse response h.

(b) Determine and plot the response of the voltage z across the resistor to this input by convolution of the input with the impulse response g. Verify that $y + z = u$. Explain any discrepancies.

(c) Compute and plot the step response of each of the two systems considered.

Discrete-time approximation of the δ-function.

Suppose that a discrete time axis with small increment inc is used to approximate a continuous time axis. Then the δ-function may be approximated by the discrete-time signal d given by

$$d(t) = \begin{cases} \dfrac{1}{\text{inc}} & \text{for } t = 0, \\ 0 & \text{otherwise,} \end{cases} \qquad t \in \mathbb{Z}(\text{inc}).$$

3.11.6. Frequency response of the exponential smoother. According to 3.7.5(a) the frequency response function of the exponential smoother is

$$\hat{h}(f) = \frac{1 - a}{1 - ae^{-j2\pi f}}, \qquad f \in \mathbb{R}.$$

(a) Let $a = 0.5$. Reproduce the plots of the magnitude and phase of $\hat{h}$ of Fig. 3.23. Include in the same figure plots for $a = 0.2$ and $a = 0.9$, and comment on the differences.

(b) Again let $a = 0.5$, and suppose that the input to the exponential smoother is the real harmonic signal

$$u(t) = \cos(2\pi f t), \qquad t \in \mathbb{Z},$$

with frequency $f = 0.25$. Use 3.7.7 to determine the response to this input. Plot both the harmonic input and the response.

3.11.7. Frequency response of the RC network. According to 3.7.5(b) the frequency response function of the RC network is given by

$$\hat{h}(f) = \frac{1}{1 + RCj2\pi f}, \qquad f \in \mathbb{R}.$$

Assume that $RC = 1$.

(a) Reproduce the magnitude and phase plots of $\hat{h}$ as shown in Fig. 3.24.

(b) Use 3.7.7 to compute the response of the RC network to the real harmonic input

$$u(t) = \cos(2\pi f_o t), \qquad t \in \mathbb{R},$$

for $2\pi f_o = 1/RC$. Plot the input u and the corresponding output y.

(c) Use 3.7.7 to compute the response of the RC network to the composite real harmonic input u given by

$$u(t) = \cos(2\pi f_o t) + \tfrac{1}{4} \cos(2\pi f_1 t), \qquad t \in \mathbb{R},$$

with f_o as in (b), and $2\pi f_1 = 3/RC$. Plot both the input u and the output y. *Hint:* Use the linearity of the system.

3.11.8. Series connection of two differencers. According to 3.10.5 a (backward) differencer is a discrete-time IOM system whose IO map is specified by

$$y(n) = u(n) - u(n - 1), \qquad n \in \mathbb{Z}.$$

(a) It is simple to program the IO map. Determine the impulse response of the system by determining its response to the unit pulse Δ.
(b) Use 3.9.2 to determine the impulse response of the series connection of the differencer with itself in the following two ways:
   (b.1) by map composition, that is, by determining the response of the second system to the response of the first system to the unit pulse;
   (b.2) by convolution of the impulse response of the differencer with itself.
(c) Confirm numerically that the step response of the series connection equals the impulse response of a single differencer. Explain.
(d) Show (theoretically) that the frequency response of a single backward differencer is given by

$$\hat{h}(f) = 1 - e^{-j2\pi f}, \qquad f \in \mathbb{R}.$$

Plot the magnitude and phase of $\hat{h}$ on the frequency interval $[0, \tfrac{1}{2}]$. Why is it sufficient to plot $\hat{h}$ on this interval?
(e) Use 3.9.2 to compute the frequency response function of the series connection of two backward differencers. Plot its magnitude and phase.

4

Difference and Differential Systems

4.1 INTRODUCTION

Most input-output systems dealt with in this text are *difference systems* or *differential systems*. These are input-output systems whose IO relationship consists of a *difference* or *differential* equation. The importance of difference and differential systems stems from the fact that application of the natural laws that govern the behavior of a system over a small time interval often leads to difference equations in the discrete-time case and differential equations in the continuous-time case. The more complex the phenomena are, the higher is the order of the difference or differential equation.

Examples of discrete-time systems described by a difference equation are the exponential smoother of Example 1.2.6, the sliding window averager of Example 3.2.8(b), the savings account of Example 3.2.14, and the national economy of Example 3.2.15. We have also encountered instances of continuous-time systems described by differential equations, such as the RC circuit of Example 1.2.7, the heated vessel of Example 3.2.12, and the moving car of Example 3.2.13. All these systems are described by *first-order* difference and differential equations.

In Section 4.2 we define *difference systems* as discrete-time IO systems whose IO relationship consists of a difference equation involving the input and output. Similarly, *differential systems* are continuous-time systems whose IO relationship takes the form of a differential equation relating input and output. The section includes several examples of difference and differential systems described by difference or differential equations of higher order.

Section 4.3 deals with a number of basic issues. First it is shown how given an input, corresponding outputs follow from the difference and differential equation, given suitable *initial conditions*. Next we consider the *non-anticipativity, time-invariance,* and *linearity* of difference and differential systems. Difference and differential systems that are both linear and time-invariant are described by linear difference and differential equations with constant coefficients. The assumption that the initial conditions are zero leads to a *unique* output for any given input. The IO *mapping* system thus defined is called the *initially-at-rest* difference or differential system. Linear time-invariant initially-at-rest difference and differential systems are convolution systems.

The rest of the chapter is concerned with linear time-invariant difference and differential systems. Section 4.4 presents an explicit analysis of the response of linear time-invariant difference and differential systems. First *homogeneous* difference and differential equations are studied. These may easily be solved. Given the general solution of the homogeneous equation together with a *particular* solution we may find the general solution of the nonhomogeneous equation, which gives us all possible outputs corresponding to a given input. In Section 4.5 it is studied how to find the *impulse response* of the initially-at-rest system defined by the difference or differential equation and how to use this to determine the response of the system when not initially at rest.

In Section 4.6 we discuss the *stability* of difference and differential systems. Besides BIBO stability, which was introduced for convolution systems in Section 3.6, we define the notion of *CICO* (converging-input converging-output) stability. The BIBO and CICO stability of constant coefficient difference and differential systems may easily be established from the locations of the *characteristic roots* and the *poles* of the system.

Section 4.7, finally, is devoted to an analysis of the *frequency response* function of difference and differential systems. It may directly be obtained from the coefficients of the difference and differential equation. Its existence depends on the locations of the poles of the system. We, furthermore, consider the *transient* and *steady-state* response to harmonic inputs and show how the frequency response functions of electrical networks may conveniently be obtained by using *impedances*.

4.2 DIFFERENCE AND DIFFERENTIAL SYSTEMS: DEFINITION AND EXAMPLES

Difference systems are discrete-time systems whose IO relation consists of a difference equation. Similarly, differential systems are continuous-time systems whose rule consists of a differential equation. We consider time axes $\mathbb{T}$ that are either *right semi-infinite,* such as $\{n_o,\ n_o + 1,\ n_o + 2,\ \cdots\}$ and $[t_o, \infty)$, or *infinite,* namely, $\mathbb{Z}$ and $\mathbb{R}$.

4.2.1.   Definition: Difference and differential systems.

A *difference system* is a discrete-time IO system whose time axis $\mathbb{T}$ is right semi-infinite or infinite, such that any IO pair (u, y) satisfies a difference equation of the form

$$F[\,y(n), y(n + 1), \cdots, y(n + N),$$
$$u(n), u(n + 1), \cdots, u(n + M), n\,]$$
$$= 0 \quad (1)$$

for all $n \in \mathbb{T}$, with N and M nonnegative integers and F a given map $F\colon \mathbb{C}^{N+M+2} \times \mathbb{T} \to \mathbb{C}$.

A *differential system* is a continuous-time IO system whose time axis $\mathbb{T}$ is right semi-infinite or infinite, such that any IO pair (u, y) satisfies a differential equation of the form

$$F\left[\,y(t), \frac{dy(t)}{dt}, \cdots, \frac{d^N y(t)}{dt^N},\right.$$
$$\left. u(t), \frac{du(t)}{dt}, \cdots, \frac{du^M(t)}{dt^M}, t\,\right]$$
$$= 0 \quad (1')$$

for all $t \in \mathbb{T}$, with N and M nonnegative integers and F a given map $F\colon \mathbb{C}^{N+M+2} \times \mathbb{T} \to \mathbb{C}$.   ∎

Usually, given an input u to a difference or differential system, there are *many* corresponding outputs y satisfying the difference or differential equation. We shall learn that by imposing extra conditions, in particular *initial conditions,* a unique output y may be established.

The difference equation (1) describing difference systems has been put into a standard form such that the values of the input u and output y at time n and a (finite) number of *later* times $n + 1, n + 2, n + 3, \cdots$, are interrelated.

We assume in the following that for any $n \in \mathbb{T}$ the difference equation (1) may uniquely be solved for $y(n + N)$ in terms of the other arguments. The number N is called the *order* of the difference equation. Also, the difference system is said to have order N. Similarly, we assume that the differential equation $(1')$ may be uniquely solved for $y^{(N)}(t)$ for any $t \in \mathbb{T}$. Again, the differential equation and corresponding system are said to be of *order N*.

4.2.2.   Review: Sampled difference systems.   Sampled difference systems are sampled systems defined on the semi-infinite time axis $\mathbb{T} = \{t_o,\ t_o + T,\ t_o + 2T, \cdots\}$, with $t_o \in \mathbb{Z}(T)$, or the infinite time axis $\mathbb{T} = \mathbb{Z}(T)$. They are described by a difference equation of the form

$$F[\,y(t), y(t + T), \cdots, y(t + NT),$$
$$u(t), u(t + T), \cdots, u(t + MT), t\,] = 0, \qquad t \in \mathbb{T}.$$

We assume that for any $t \in \mathbb{T}$ the difference equation may uniquely be solved for $y(t + NT)$, and call N the *order* of the difference equation and the corresponding IO system.   ∎

Examples of Difference and Differential Systems

We have already met several examples of difference and differential systems.

4.2.3.   Examples: First-order difference and differential systems.

(a) *Exponential smoother*. The exponential smoother of Example 1.2.6 is described by the difference equation

$$y(n + 1) - ay(n) - (1 - a)u(n + 1) = 0, \qquad n \in \mathbb{T},$$

with time axis $\mathbb{T} = \{n_o, n_o + 1, n_o + 2, \cdots\}$, with n_o an integer. If $n_o = -\infty$, the time axis is $\mathbb{Z}$. The sampled version of the exponential smoother (see 3.7.12 and 4.2.2) is described by the difference equation

$$y(t + T) - ay(t) - (1 - a)u(t + T) = 0, \qquad t \in \mathbb{T},$$

with time axis $\mathbb{T} = \{t_o, t_o + T, t_o + 2T, \cdots\}$, with $t_o \in \mathbb{Z}(T)$. If $t_o = -\infty$, the time axis is $\mathbb{Z}(T)$.

(b) *Savings account*. The savings account of Example 3.2.14 is a difference system described by the difference equation

$$y(n + 1) - (1 + \alpha)y(n) - u(n) = 0, \qquad n = 0, 1, 2, \cdots.$$

(c) *RC network*. The RC circuit, as introduced in Example 1.2.7, is a differential system characterized by the differential equation

$$\frac{dy(t)}{dt} + \frac{1}{RC}y(t) - \frac{1}{RC}u(t) = 0, \qquad t \in [t_0, \infty).$$

(d) *Moving car*. The moving car of Example 3.2.13 is also a differential system, whose differential equation is

$$M\frac{dv(t)}{dt} + Bv^2(t) - cu(t) = 0, \qquad t \in [t_o, \infty). \qquad ∎$$

The orders of the systems of 4.2.3 are all one. We next look at some slightly more complex systems.

4.2.4.   Examples: Higher-order systems.

(a) *Discrete-time sliding window averager*. The discrete-time sliding window averager of Example 3.2.8(b) is described by the difference equation

$$y(n) - \frac{1}{N+M+1} \sum_{m=-M}^{N} u(n+m) = 0, \qquad n \in \mathbb{Z},$$

with N and M nonnegative integers. If $M > 0$ this equation is not in the standard form (1), but it may be transformed to it by the simple substitution $n = n' + M$. It follows that

$$y(n'+M) - \frac{1}{N+M+1} \sum_{m=-M}^{N} u(n'+M+m) = 0, \qquad n' \in \mathbb{Z}.$$

By the change of variable $m' = m + M$ this equation takes the form

$$y(n'+M) - \frac{1}{N+M+1} \sum_{m'=0}^{N+M} u(n'+m') = 0, \qquad n' \in \mathbb{Z}.$$

This in turn may be rewritten in full as

$$y(n+M) - \frac{1}{N+M+1}[u(n) + u(n+1) + \cdots + u(n+N+M)] = 0,$$

for $n \in \mathbb{Z}$, where we drop the prime on n. The system has order M.

(b) *Second-order smoother*. The difference equation

$$y(n) = ay(n-1) + (1-a)u(n), \qquad n \in \mathbb{Z},$$

obtained by replacing n with $n - 1$ in the equation for the exponential smoother, shows that the current output $y(n)$ of the smoother is a weighted sum of the immediate past output $y(n-1)$ and the current input $u(n)$. Inspection of the frequency response function of the smoother, as obtained in 3.7.5(a), shows that it is a low-pass filter. The filter does not cut off high frequencies very sharply, however. We could contemplate including extra terms in the difference equation involving further past outputs and also past inputs, in the hope that this improves the filtering effect. Taking one more past output value and one past input value, for instance, we obtain

$$y(n) = a_1 y(n-1) + a_0 y(n-2) + b_2 u(n) + b_1 u(n-1), \qquad n \in \mathbb{Z},$$

with a_1, a_0, b_2, and b_1 constants to be determined. The constants need be selected carefully for the given filtering task, which is a problem we can only deal with later. To obtain the difference equation in the standard form (1) we replace n with $n + 2$ and thus find

$$y(n + 2) - a_1 y(n + 1) - a_0 y(n) - b_2 u(n + 2) - b_1 u(n + 1) = 0, \qquad n \in \mathbb{Z}.$$

Comparison with the standard form (1) shows that $N = 2$, so that the order of the system is two. Also, M equals 2.

(c) *RCL circuit*. The input to the RCL circuit of Fig. 4.1 is the voltage u produced by the voltage source. We consider two different choices for the output, namely,

  (i) the output is the current i through the circuit, and

  (ii) the output is the voltage v_L across the inductor.

The differential equations that describe the system in these two cases may be derived as follows. By Kirchhoff's voltage law,

$$u(t) = v_R(t) + v_C(t) + v_L(t), \qquad t \in \mathbb{R}.$$

> The voltage and current laws are the first two of the four laws found by the German physicist Gustav Robert Kirchhoff (1824–1887.) The other two laws concern the spectral emission of glowing bodies.

For the resistor, capacitor, and inductor we have

$$v_R(t) = Ri(t), \qquad v_C(t) = \frac{q(t)}{C}, \qquad v_L(t) = L\frac{di(t)}{dt}, \qquad t \in \mathbb{R},$$

respectively, where q represents the charge of the capacitor. It follows that

$$u(t) = Ri(t) + \frac{q(t)}{C} + L\frac{di(t)}{dt}, \qquad t \in \mathbb{R}.$$

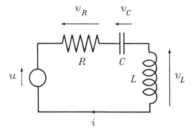

Figure 4.1.   An RCL network.

Because $dq(t)/dt = i(t)$, we may differentiate to eliminate q. This results in

$$\frac{du(t)}{dt} = R\frac{di(t)}{dt} + \frac{i(t)}{C} + L\frac{d^2i(t)}{dt^2}, \qquad t \in \mathbb{R}. \tag{2}$$

(i) *Current i as output*. Suppose that the output is $y = i$. It immediately follows from (2) that

$$\frac{1}{LC}y(t) + \frac{R}{L}\frac{dy(t)}{dt} + \frac{d^2y(t)}{dt^2} - \frac{1}{L}\frac{du(t)}{dt} = 0, \qquad t \in \mathbb{R}.$$

This shows that the RCL network with the current as output is a second-order differential system.

(ii) *Voltage v_L as output*. If the output is $y = v_L = L\,di(t)/dt$, we need differentiate (2) once more. It follows that

$$\frac{d^2u(t)}{dt^2} = R\frac{d}{dt}\left(\frac{di(t)}{dt}\right) + \frac{1}{C}\left(\frac{di(t)}{dt}\right) + L\frac{d^2}{dt^2}\left(\frac{di(t)}{dt}\right), \qquad t \in \mathbb{R}.$$

By the substitution $L\,di/dt = y$ we find that the system is described by the differential equation

$$\frac{1}{LC}y(t) + \frac{R}{L}\frac{dy(t)}{dt} + \frac{d^2y(t)}{dt^2} - \frac{d^2u(t)}{dt^2} = 0, \qquad t \in \mathbb{R}.$$

This represents another second-order differential system.

(d) *Moving car*. The moving car of Example 3.2.13 is described by the first-order differential equation

$$M\frac{dv(t)}{dt} = cu(t) - Bv^2(t), \qquad t \in [t_o, \infty), \tag{3}$$

with v representing the car speed and the input u the throttle position. Suppose that we take as output y the *distance* traveled by the car. Because

$$v(t) = \frac{dy(t)}{dt},$$

we may find the differential equation that describes the system with this output by the simple substitution $v = dy/dt$. This results in the differential equation

$$M \frac{d^2 y(t)}{dt^2} + B\left(\frac{dy(t)}{dt}\right)^2 - cu(t) = 0, \qquad t \in \mathbb{R},$$

which represents a second-order differential system.   ∎

4.3 BASICS OF DIFFERENCE AND DIFFERENTIAL SYSTEMS

In this section we deal with a number of basic issues concerning difference and differential systems. First it is explained how by specifying *initial conditions* the output uniquely follows from the input, given the difference or differential equation. After studying non-anticipativeness, time-invariance, and linearity of difference and differential systems, we arrive at the systems that occupy our attention for most of the rest of this chapter, namely systems described by *constant coefficient linear difference* and *differential equations*. Finally, we consider *initially-at-rest* systems, which are the IOM systems that result by setting the initial conditions of the difference and differential equations equal to zero. This is a natural way of reducing IO difference and differential systems to IOM systems.

Solutions to Difference and Differential Equations

We devote some attention to the question how difference and differential equations as introduced in 4.2.1 define IO pairs.

The situation for difference systems is simple. Suppose that for each time n the equation

$$F[y(n), y(n + 1), \cdots, y(n + N), u(n), u(n + 1), \cdots, u(n + M), n] = 0$$

may uniquely be solved for $y(n + N)$ in the form

$$\begin{aligned} y(n + N) = G[\,&y(n), y(n + 1), \cdots, y(n + N - 1), \\ &u(n), u(n + 1), \cdots, u(n + M), n], \qquad n \in \mathbb{T}. \end{aligned} \tag{1}$$

Take the time axis as $\mathbb{T} = \{n_o, n_o + 1, n_o + 2, \cdots\}$, and assume that the input u on this time axis is known. Then for $n = n_o$ the relation (1) yields $y(n_o + N)$, provided

$$y(n_o), y(n_o + 1), \cdots, y(n_o + N - 1)$$

are given. These numbers are called the *initial conditions*. Once $y(n_o + N)$ has been obtained, application of (1) for $n = n_o + 1$ yields $y(n_o + N + 1)$. Continuing in this way, the entire output $y(n)$ for $n \geq n_o + N$ may be constructed by what is called *successive substitution*.

Given the initial conditions and the input u, the output y is fully determined. Hence, by specifying the initial conditions the IO system becomes an input-output *mapping* system.

In the continuous-time case matters are more complicated. For the sake of the argument it is helpful to assume that the input u is M times continuously differentiable. Suppose that for each time t the relation

$$F\left[y(t), \frac{dy(t)}{dt}, \cdots, \frac{d^N y(t)}{dt^N}, u(t), \frac{du(t)}{dt}, \cdots, \frac{d^M u(t)}{dt^M}, t\right] = 0$$

may uniquely be solved for $d^N y(t)/dt^N$ in the form

$$\frac{d^N y(t)}{dt^N} = G\left[y(t), \frac{dy(t)}{dt}, \cdots, \frac{d^{N-1} y(t)}{dt^{N-1}}, u(t), \frac{du(t)}{dt}, \cdots, \frac{d^M u(t)}{dt^M}, t\right]. \quad (2)$$

This relation shows that given the input u, at any time t the Nth derivative $d^N y(t)/dt^N = y^{(N)}(t)$ of the output y is determined by $y(t)$, $y^{(1)}(t)$, $\cdots$, $y^{(N-1)}(t)$. Thus, given the input u and the N numbers $y(t_o)$, $y^{(1)}(t_o)$, $\cdots$, $y^{(N-1)}(t_o)$ at some initial time t_o we may compute, at least approximately, the quantities y, $y^{(1)}$, $\cdots$, $y^{(N-1)}$ at a short time h later, because

$$y(t_o + h) = y(t_o) + \int_{t_o}^{t_o + h} y^{(1)}(\tau) \, d\tau \approx y(t_o) + y^{(1)}(t_o)h,$$

$$y^{(1)}(t_o + h) = y^{(1)}(t_o) + \int_{t_o}^{t_o + h} y^{(2)}(\tau) \, d\tau \approx y^{(1)}(t_o) + y^{(2)}(t_o)h,$$

$$\cdots$$

$$y^{(N-1)}(t_o + h) = y^{(N-1)}(t_o) + \int_{t_o}^{t_o + h} y^{(N)}(\tau) \, d\tau \approx y^{(N-1)}(t_o) + y^{(N)}(t_o)h,$$

where $y^{(N)}(t_o)$ in the latter expression is obtained from (2). By repeating this procedure at the time $t_o + h$ we may find, at least approximately, y and its derivatives $y^{(1)}$, $\cdots$, $y^{(N-1)}$ at time $t_o + 2h$, and continuing, at the times $t_o + 3h$, $t_o + 4h$, $\cdots$. It is plausible that if the time interval h is small enough we may obtain an arbitrarily accurate approximation to the solution of the differential equation for $t \geq t_o$.

The role of the *initial conditions* at time t_o is taken by the N numbers

$$y(t_o), y^{(1)}(t_o), \cdots, y^{(N-1)}(t_o).$$

It may be proved that under certain hypotheses on the function G, which in the cases we are interested in are usually satisfied, the solution of the differential equation

(2) for $t \geq t_o$ indeed is uniquely defined by the input together with the initial conditions. Like difference systems, differential systems are IO systems that become input-output *mapping* systems once the initial conditions have been specified.

4.3.1. Summary: Difference and differential IO and IOM systems.

A discrete-time system with time axis $\mathbb{T} = \{n_o, n_o + 1, \cdots\}$ and rule

$$F[\, y(n), y(n+1), \cdots, y(n+N),$$
$$u(n), u(n+1), \cdots, \ u(n+M), n]$$
$$= 0, \quad (3)$$

$n \in \mathbb{T}$, represents an IO system. Suppose that (3) may be written in the form

$$y(n+N) =$$
$$G[\, y(n), y(n+1), \cdots, y(n+N-1),$$
$$u(n), u(n+1), \cdots, u(n+M), n],$$

$n \in \mathbb{T}$. Then for given *initial conditions*

$$y(n_o), y(n_o + 1), \cdots, y(n_o + N - 1)$$

the system becomes an IOM system, which has a unique output y for any given input u.

A continuous-time system with time axis $\mathbb{T} = [t_o, \infty)$ and rule

$$F\left[y(t), \frac{dy(t)}{dt}, \cdots, \frac{d^N y(t)}{dt^N}, \right.$$
$$\left. u(t), \frac{du(t)}{dt}, \cdots, \frac{d^M u(t)}{dt^M}, t \right]$$
$$= 0 \quad (3')$$

$t \in \mathbb{T}$, represents an IO system. Suppose that (3') may be written in the form

$$\frac{d^N y(t)}{dt^N} = G\left[y(t), \frac{dy(t)}{dt}, \cdots, \frac{d^{N-1} y(t)}{dt^{N-1}}, \right.$$
$$\left. u(t), \frac{du(t)}{dt}, \cdots, \frac{d^M u(t)}{dt^M}, t \right],$$

$t \in \mathbb{T}$. Then for given *initial conditions*

$$y(t_o), y^{(1)}(t_o), \cdots, y^{(N-1)}(t_o)$$

the system becomes an IOM system, which has a unique output y for any given input u.   ∎

If in the continuous-time case the input u is not M times continuously differentiable, its derivatives contain δ-functions and derivatives of δ-functions. As a result, the corresponding output y may also contain singular functions, and the differential equation is required to hold in the *distribution sense*. For linear differential systems, discussed in the next section, this is the usual situation.

Non-Anticipativity of Difference and Differential IOM Systems

The non-anticipativity of a difference IOM system with given initial conditions may easily be checked from its difference equation. Differential IOM systems are always non-anticipating.

4.3.2.   Summary: Anticipativity of difference and differential IOM systems.

The difference system described by the difference equation

$$y(n + N) =$$
$$G[\, y(n), y(n + 1), \cdots, y(n + N - 1),$$
$$u(n), u(n + 1), \cdots, u(n + M), n],$$

for $n \in \mathbb{T} = \{n_o, n_o + 1, \cdots\}$, with given initial conditions $y(n_o)$, $y(n_o + 1)$, $\cdots$, $y(n_o + N - 1)$, is non-anticipating if $M \leq N$. If $M > N$ and the function G depends nontrivially on $u(n + N + 1)$, $u(n + N + 2), \cdots, u(n + M)$, the system is anticipating.

The differential system described by the differential equation

$$\frac{dy^N(t)}{dt^N} = G\left[y(t), \frac{dy(t)}{dt}, \cdots, \frac{d^{N-1}y(t)}{dt^{N-1}},\right.$$
$$\left. u(t), \frac{du(t)}{dt}, \cdots, \frac{du^M(t)}{dt^M}, t \right],$$

for $t \in \mathbb{T} = [t_o, \infty)$, with given initial conditions $y(t_0)$, $y^{(1)}(t_o)$, $\cdots$, $y^{(N-1)}(t_o)$, is non-anticipating.   ∎

The proof is simple for the discrete-time case. The continuous-time result is plausible because the differential equation shows that the instantaneous changes in y and its derivatives at any time t depend on u and its derivatives at that same time and not on those at later times.

4.3.3.   Examples: Anticipativeness.

(a) *Exponential smoother and RC network*. Both the exponential smoother of Example 4.2.3(a) and the RC network of 4.2.3(c) are non-anticipating.

(b) *Sliding window averager*. As we saw in Example 4.2.4(a), the discrete-time sliding window averager is described by the difference equation

$$y(n + M) = \frac{1}{N + M + 1} \sum_{k=0}^{N+M} u(n + k), \qquad t \in \mathbb{T}.$$

It follows by 4.3.2 that the system is anticipating if and only if $N > 0$. This confirms what we found in 3.2.8(b).   ∎

Time-Invariance of Difference and Differential Systems

We next discuss the time-invariance of difference and differential systems. Time-invariance may usually immediately be verified from the difference or differential equation by checking whether this equation only depends on time via u or y.

4.3.4.   Summary: Time-invariance of difference and differential systems.

Consider the difference IO system described by the difference equation

$$F[y(n), y(n + 1), \cdots, y(n + N),$$
$$u(n), u(n + 1), \cdots, u(n + M), n]$$
$$= 0,$$

$n \in \mathbb{T}$, with $\mathbb{T} = \{n_o, n_o + 1, \cdots\}$ or $\mathbb{T} = \mathbb{Z}$. Then if the function F does not depend on its last argument, that is,

$$F(y_o, y_1, \cdots, y_N, u_0, u_1, \cdots, u_M, n)$$
$$= F(y_0, y_1, \cdots, y_N, u_0, u_1, \cdots, u_M, n')$$

for all n, $n' \in \mathbb{T}$ and for all $y_0, y_1, \cdots$, $y_N, u_0, u_1, \cdots, u_M$ in $\mathbb{C}$, the system is time-invariant on the time axis $\mathbb{T}$.

Consider the differential IO system described by the differential equation

$$F\left[y(t), \frac{dy(t)}{dt}, \cdots, \frac{d^N y(t)}{dt^N},\right.$$
$$\left. u(t), \frac{du(t)}{dt}, \cdots, \frac{du^M(t)}{dt^M}, t\right] = 0,$$

$t \in \mathbb{T}$, with $\mathbb{T} = [t_o, \infty)$ or $\mathbb{T} = \mathbb{R}$. Then if the function F does not depend on its last argument, that is,

$$F(y_0, y_1, \cdots, y_N, u_0, u_1, \cdots, u_M, t)$$
$$= F(y_0, y_1, \cdots, y_N, u_0, u_1, \cdots, u_M, t')$$

for all t, $t' \in \mathbb{T}$ and for all $y_0, y_1, \cdots$, $y_N, u_0, u_1, \cdots, u_M$ in $\mathbb{C}$, the system is time-invariant on the time axis $\mathbb{T}$.   ∎

Note that the condition that is given is *sufficient* for time-invariance, but not necessary. The trivial discrete- or continuous-time system described by $\alpha(t)y(t) - \alpha(t)u(t) = 0$, $t \in \mathbb{T}$, with α a nonzero time-dependent coefficient, for instance, does not satisfy the condition of 4.3.4. Nevertheless the system is time-invariant, because division by $\alpha(t)$ results in the equivalent equation $y(t) - u(t) = 0$, $t \in \mathbb{T}$, which by 4.3.4 represents a time-invariant system.

4.3.5.   Proof.   We only give the proof for the discrete-time case. That for the continuous-time case is similar. Suppose that (u, y) is an IO pair. Then

$$F[y(n), y(n + 1), \cdots, y(n + N), u(n), u(n + 1), \cdots, u(n + M), n] = 0,$$

$n \in \mathbb{T}$. Let θ be such that $n + \theta \in \mathbb{T}$ for all $n \in \mathbb{T}$. It follows that

$$F[y(n + \theta), y(n + \theta + 1), \cdots, y(n + \theta + N),$$
$$u(n + \theta), u(n + \theta + 1), \cdots, u(n + \theta + M), n + \theta] = 0, \qquad n \in \mathbb{T}.$$

Because F does not depend on its last argument, we may also write

$$F[y(n + \theta), y(n + \theta + 1), \cdots, y(n + \theta + N),$$
$$u(n + \theta), u(n + \theta + 1), \cdots, u(n + \theta + M), n] = 0, \qquad n \in \mathbb{T},$$

which shows that $(\sigma^\theta y,\ \sigma^\theta u)$ is an IO pair. Hence, the system is time-invariant.   ■

We consider some examples.

4.3.6.   Examples: Time-invariant difference and differential systems.

(a) *Time-invariant systems.* The difference and differential systems we encountered so far, such as the exponential smoother of Example 4.2.3(a), the savings account of 4.2.3(b), the discrete-time sliding window averager of 4.2.4(a), the second-order smoother of 4.2.4(b), the RC network of 4.2.3(c), the moving car of 4.2.3(d) and 4.2.4(d), and the RCL network of 4.2.4(c) are all time-invariant.

(b) *RC network with time-dependent resistor.* For an example of a time-varying system suppose that the resistance of the resistor of the RC network of 1.2.7 is not constant, but a function $R(t)$, $t \in \mathbb{R}$, of time. Referring to Fig. 4.2 we have as in 1.2.7

$$\frac{dy(t)}{dt} = -\frac{1}{R(t)C}\,y(t) + \frac{1}{R(t)C}u(t), \qquad t \in \mathbb{R},$$

provided $R(t) > 0$ for all t. Because the coefficients of this differential equation vary with time, 4.3.4 is not satisfied. Indeed it may be verified that the system is time-invariant if and only if the resistance R is *constant*.

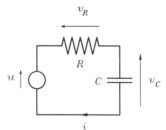

Figure 4.2.   RC network.

(c) *RC network with time-dependent capacitor.* Next suppose that the resistance R is constant, but the capacitance $C(t)$, $t \in \mathbb{R}$, of the capacitor of the RC network varies with time. The equations

$$u(t) = v_R(t) + v_C(t), \tag{4}$$

and

$$v_R(t) = Ri(t) \tag{5}$$

remain valid, but the relation $i(t) = Cdv_C(t)/dt$ need be replaced with $i(t) = dq(t)/dt$, where the charge q of the capacitor is given by $q(t) = C(t)v_C(t)$. It follows that $i(t) = \dot{C}(t)v_C(t) + C(t)dv_C(t)/dt$, where $\dot{C}$ is the derivative of C with respect to time. As a result, we obtain from (4) and (5) that

$$u(t) = R\left[\dot{C}(t)v_C(t) + C(t)\frac{dv_C(t)}{dt}\right] + v_C(t).$$

The substitution $v_C = y$ and rearrangement yield the differential equation

$$\frac{dy(t)}{dt} + \left(\frac{1}{RC(t)} + \frac{\dot{C}(t)}{C(t)}\right)y(t) = \frac{1}{RC(t)}u(t), \qquad t \in \mathbb{R},$$

provided $C(t) > 0$ for all t. This differential equation has time-dependent coefficients and represents a time-varying system. Note that the differential equation does *not* simply follow by replacing the constant C in the differential equation

$$\frac{dy(t)}{dt} + \frac{1}{RC}y(t) = \frac{1}{RC}u(t), \qquad t \in \mathbb{R}.$$

with the time-dependent parameter $C(t)$.   ∎

Linearity of Difference and Differential Systems

Also the linearity of a difference or differential system may often immediately be established by inspection of the difference or differential equation. If this equation is linear, so is the system.

4.3.7.   Linearity of difference and differential systems.

The difference system described by the linear difference equation

$$q_0(n)y(n) + q_1(n)y(n + 1) + \cdots$$
$$+ q_N(n)y(n + N)$$
$$= p_0(n)u(n) + p_1(n)u(n + 1) + \cdots$$
$$+ p_M(n)u(n + M),$$

$n \in \mathbb{T}$, with the q_k and p_k real or complex possibly time-dependent coefficients, is linear.

The differential system described by the linear differential equation

$$q_0(t)y(t) + q_1(t)\frac{dy(t)}{dt} + \cdots$$
$$+ q_N(t)\frac{d^Ny(t)}{dt^N}$$
$$= p_0(t)u(t) + p_1(t)\frac{du(t)}{dt} + \cdots$$
$$+ p_M(t)\frac{d^Mu(t)}{dt^M},$$

$t \in \mathbb{T}$, with the q_k and p_k real or complex possibly time-dependent coefficients, is linear.   ∎

Again, 4.3.7 gives a *sufficient* condition for linearity. An example of a trivial difference or differential equation that does not satisfy the condition but nevertheless represents a linear system is $[1 + |y(t)|^2]y(t) = [1 + |y(t)|^2]u(t)$, $t \in \mathbb{T}$.

4.3.8.   Proof of 4.3.7.   This time we present the proof for the continuous-time case. That for the discrete-time case is analogous. To prove linearity we need show that if (u, y) and (u', y') are two IO pairs, any linear combination of these pairs is also an IO pair. Because (u, y) and (u', y') are IO pairs we have

$$
\begin{aligned}
q_0(t)y(t) &+ q_1(t)\frac{dy(t)}{dt} + \cdots + q_N(t)\frac{d^N y(t)}{dt^N} \\
&= p_0(t)u(t) + p_1(t)\frac{du(t)}{dt} + \cdots + p_M(t)\frac{d^M u(t)}{dt^M},
\end{aligned}
\tag{6}
$$

for $t \in \mathbb{T}$, and

$$
\begin{aligned}
q_0(t)y'(t) &+ q_1(t)\frac{dy'(t)}{dt} + \cdots + q_N(t)\frac{d^N y'(t)}{dt^N} \\
&= p_0(t)u'(t) + p_1(t)\frac{du'(t)}{dt} + \cdots + p_M(t)\frac{d^M u'(t)}{dt^M},
\end{aligned}
\tag{7}
$$

for $t \in \mathbb{T}$. Multiplying (6) by any complex scalar α, (7) by α' and adding the results it follows that

$$
\begin{aligned}
q_0(t)[\alpha y(t) + \alpha'y'(t)] &+ q_1(t)\frac{d[\alpha y(t) + \alpha'y'(t)]}{dt} + \cdots \\
&+ q_N(t)\frac{d^N[\alpha y(t) + \alpha'y'(t)]}{dt^N} \\
= p_0(t)[\alpha u(t) + \alpha'u'(t)] &+ p_1(t)\frac{d[\alpha u(t) + \alpha'u'(t)]}{dt} + \cdots \\
&+ p_M(t)\frac{d^M[\alpha u(t) + \alpha'u'(t)]}{dt^M},
\end{aligned}
$$

for $t \in \mathbb{T}$, which proves that $(\alpha u + \alpha'u', \alpha y + \alpha'y')$ is an IO pair. Hence, the system is linear.   ∎

We look at some examples.

4.3.9.   Examples: Linearity.

(a) *Linear difference and differential systems.* Most difference and differential systems we have seen so far are linear, notably the exponential smoother of Example

4.2.3(a), the savings account of 4.2.3(b), the discrete-time sliding window averager of 4.2.4(a), the second-order smoother of 4.2.4(b), the RC network of 4.2.3(c), and the RCL network of 4.2.4(c). Also, the time-varying versions of the RC network of 4.3.6(b) and 4.3.6(c) are linear systems.

(b) *Moving car*. The moving car of Example 4.2.3(d) is described by the differential equation

$$M\frac{dv(t)}{dt} = cu(t) - Bv^2(t), \qquad t \geq t_o.$$

The differential equation is not linear. Indeed, the system is not linear. For instance, if u_o is a constant input, $v_o = \sqrt{cu_o/B}$ is a corresponding output, but $(\alpha u_o, \alpha v_o)$ is not an IO pair unless $\alpha = 0$, $\alpha = 1$ or $u_o = 0$.

Also when we take the position of the car as output, as in Example 4.2.4(d), the system is nonlinear.   ∎

Linear Time-Invariant Difference and Differential Systems

An important and useful class of difference and differential systems consists of those difference and differential systems that are both linear and time-invariant. By combining 4.3.4 and 4.3.7 we obtain the following result.

4.3.10.   Summary: Linear time-invariant difference and differential systems.

The difference system described by the linear constant coefficient difference equation

$$q_0 y(n) + q_1 y(n + 1) + \cdots$$
$$+ q_N y(n + N)$$
$$= p_0 u(n) + p_1 u(n + 1) + \cdots$$
$$+ p_M u(n + M), \qquad (8)$$

$n \in \mathbb{T}$, with the q_k and p_k real or complex coefficients, is linear and time-invariant.

The differential system described by the linear constant coefficient differential equation

$$q_0 y(t) + q_1 \frac{dy(t)}{dt} + \cdots + q_N \frac{d^N y(t)}{dt^N}$$
$$= p_0 u(t) + p_1 \frac{du(t)}{dt} + \cdots + p_M \frac{d^M u(t)}{dt^M},$$
$$(8')$$

$t \in \mathbb{T}$, with the q_k and p_k real or complex coefficients, is linear and time-invariant. ∎

We next introduce a useful notation. Let Q and P be the polynomials

$$Q(\lambda) = q_0 + q_1\lambda + \cdots + q_N\lambda^N,$$
$$P(\lambda) = p_0 + p_1\lambda + \cdots + p_M\lambda^M.$$

We may then write the difference equation (8) and the differential equation (8') in the more compact forms

$$Q(\sigma)y = P(\sigma)u,$$

or

$$Q(D)y = P(D)u,$$

respectively. Here σ is the *back shift operator* defined by

$$\sigma y(n) = y(n + 1), \qquad n \in \mathbb{T},$$

and D the *differential operator* defined by

$$Dy(t) = \frac{dy(t)}{dt}, \qquad t \in \mathbb{T}.$$

4.3.11.   Examples: Linear time-invariant difference and differential systems.   Many of the difference and differential systems we have seen are both linear and time-invariant. To this category belong the exponential smoother of 4.2.3(a), the savings account of 4.2.3(b), the discrete-time sliding window averager of 4.2.4(a), the second-order smoother of 4.2.4(b), the RC circuit of 4.2.3(c), and the RCL network of 4.2.4(c). The moving car of 4.2.3(d) and 4.2.4(d) is time-invariant but nonlinear. The RC network with time-varying resistor of 4.3.4(b) or time-varying capacitor of 4.3.4(c) is linear but time-varying.   ■

The Initially-At-Rest System

Suppose that the RCL network of 4.2.4(c) at some given time contains electrical or magnetic energy. If the input voltage is kept at zero, the energy dissipates, and eventually reduces to zero. Such an energy-free circuit is said to be *at rest*. A system that is at rest has the property that its response to a zero input is also zero. Generally, if the system is at rest it has a unique response to any given (nonzero) input. The input-output mapping system thus defined is called the *initially-at-rest* system.

We now present a precise definition of initially-at-rest difference and differential systems. It is simplest to consider only systems defined on the infinite time axis $\mathbb{Z}$ or $\mathbb{R}$. First we need establish conditions such that the zero input and zero output actually form an IO pair. Consider the difference system described by the difference equation

$$y(n + N) = G[\, y(n), y(n + 1), \cdots, y(n + N - 1),$$
$$u(n), u(n + 1), \cdots, u(n + M), n], \qquad n \in \mathbb{Z}, \tag{9}$$

or the differential system described by the differential equation

$$\frac{dy^N(t)}{dt^N} = G[\, y(t), \frac{dy(t)}{dt}, \cdots, \frac{dy^{(N-1)}(t)}{dt^{N-1}},$$

$$u(t), \frac{du(t)}{dt}, \cdots, \frac{du^M(t)}{dt^M}, t], \qquad t \in \mathbb{R}. \tag{10}$$

Then if the function G in (9) or (10) has the property

$$G(0, 0, \cdots, 0, t) = 0 \qquad \text{for all} \quad t \in \mathbb{Z} \text{ or } \mathbb{R},$$

the difference system (9) or the differential system (10) has $(0, 0)$ as an IO pair.

We next look for conditions such that the zero input $u = 0$ has the zero output $y = 0$ as its *unique* response. First, consider the difference system. Let n_o be some instant of time on the discrete time axis $\mathbb{Z}$, and suppose that

$$y(n_o - N) = y(n_o - N + 1) = \cdots = y(n_o - 1) = 0,$$

and

$$u(n_o - N) = u(n_o - N + 1) = \cdots = u(n_o - 1) = 0.$$

It easily follows by successive substitution that the unique response to the input $u(n) = 0$ for $n = n_o, n_o + 1, \cdots$, is $y(n) = 0$ for $n = n_o, n_o + 1, \cdots$. Moreover, to any *nonzero* input u, restricted to $n_o, n_o + 1, \cdots$, there corresponds a unique response $y_{n_o}(n)$, $n = n_o, n_o + 1, \cdots$.

Now, let $n_o \to -\infty$. The limit of y_{n_o}, if it exists, is denoted as y. We take the limit *pointwise*, that is,

$$y(n) = \lim_{n_o \to -\infty} y_{n_o}(n)$$

for any fixed n. The IOM system consisting of all IO pairs (u, y) thus defined is called the *initially-at-rest* system. Note that $(0, 0)$ is an IO pair. Also note that any right one-sided input maps into a right one-sided output.

Before going on to the continuous-time case we present an example.

4.3.12.   Example: Initially-at-rest exponential smoother.   The exponential smoother is described by the first-order difference equation

$$y(n + 1) - ay(n) = (1 - a)u(n + 1), \qquad n \in \mathbb{Z}.$$

Because $N = 1$, the conditions that ensure the system to be at rest at time n_o are

$$y(n_o - 1) = 0, \qquad u(n_o - 1) = 0.$$

The latter condition happens to be superfluous for this particular example because the difference equation has no term with $u(n)$ on the right hand side. By 3.4.2(a), for $y(n_o - 1) = 0$ the solution to the difference equation is

$$y(n) = (1 - a) \sum_{m=n_o}^{n} a^{n-m}u(m), \qquad n = n_o, n_o + 1, n_o + 2, \cdots.$$

Letting n_o approach $-\infty$ we obtain

$$y(n) = (1 - a) \sum_{m=-\infty}^{n} a^{n-m}u(m), \qquad n \in \mathbb{Z},$$

which is defined for all u for which the infinite sum exists. This is the IO map of the initially-at-rest exponential smoother. In fact, this is a *convolution map* of the form

$$y = h * u,$$

with

$$h(n) = (1 - a)a^n 1(n), \qquad n \in \mathbb{Z}.$$

The initially-at-rest exponential smoother thus is a *convolution system* with impulse response h.   ∎

We next discuss initially-at-rest differential systems. Suppose that for the differential system defined by (10) we have

$$y(t_o^-) = y^{(1)}(t_o^-) = \cdots = y^{(N-1)}(t_o^-) = 0, \tag{11}$$

and

$$u(t_o^-) = u^{(1)}(t_o^-) = \cdots = u^{(M-1)}(t_o^-) = 0, \tag{12}$$

for some $t_o \in \mathbb{R}$. Then the unique response to the input $u(t) = 0$ for $t \geq t_o$ is $y(t) = 0$, $t \geq t_o$. Note that we need condition (12) on the derivatives of u at time t_o^- (i.e., just before time t_o) to prevent any delta functions or derivatives of delta functions at time t_o to enter into the right-hand side of (10). For any *nonzero* input u, restricted to $[t_o, \infty)$, the initial conditions (11) and (12) define a unique response $y_{t_o}(t)$, $t \geq t_o$. Letting $t_o \to -\infty$ we denote the limit of y_{t_o}, if it exists, as y. We say that y is the response to the input u of the *initially-at-rest* differential system derived from (10). As in the discrete-time case, the initially-at-rest system has $(0, 0)$ as an IO pair. Right one-sided inputs results in right one-sided outputs.

Both in the discrete- and continuous-time case inputs for which the response does not exist as the initial time approaches $-\infty$ are excluded from the input set.

4.3.13.   Example: Initially-at-rest RC network.   The RC network is a first-order differential system described by the differential equation

$$\frac{dy(t)}{dt} + \frac{1}{RC}y(t) = \frac{1}{RC}u(t), \qquad t \in \mathbb{R}.$$

The system is at rest at time t_o if $y(t_o^-) = 0$. In 3.4.2(b) we saw that the output of the RC network may be expressed as

$$y(t) = e^{-(t-t_o)/RC}y(t_o) + \frac{1}{RC}\int_{t_o}^{t} e^{-(t-\tau)/RC}u(\tau)\,d\tau, \qquad t \geq t_o.$$

Replacing t_o with t_o^- and setting $y(t_o^-) = 0$ we obtain

$$y(t) = \frac{1}{RC}\int_{t_o^-}^{t} e^{-(t-\tau)/RC}u(\tau)\,d\tau, \qquad t \geq t_o^-.$$

By letting t_o approach $-\infty$ we see that the IO map of the initially-at-rest RC network is given by

$$y(t) = \frac{1}{RC}\int_{-\infty}^{t} e^{-(t-\tau)/RC}u(\tau)\,d\tau, \qquad t \in \mathbb{R},$$

for those inputs u for which the integral exists. Inspection shows that the initially-at-rest RC network is a continuous-time convolution system

$$y = h * u$$

with impulse response

$$h(t) = \frac{1}{RC}e^{-\frac{t}{RC}}\,\mathbb{1}(t), \qquad t \in \mathbb{R}. \qquad\qquad \blacksquare$$

The examples show that both the initially-at-rest exponential smoother and the initially-at-rest RC network are convolution systems. In fact, *all* initially-at-rest difference and differential systems defined by constant coefficient difference and differential systems are convolution systems. The reason is that initially-at-rest difference and differential systems are (1) linear, (2) time-invariant, and (3) IOM systems. Both (1) and (2) may easily be proved. By 3.4.1, linear time-invariant IOM systems on the time axis $\mathbb{Z}$ or $\mathbb{R}$ are convolution systems.

4.3.14.   Summary: Initially-at-rest linear time-invariant difference and differential systems are convolution systems.   The initially-at-rest system defined by the constant coefficient linear difference equation

$$Q(\sigma)y = P(\sigma)u$$

on the time axis $\mathbb{Z}$, or the differential equation

$$Q(D)y = P(D)u$$

on the time axis $\mathbb{R}$, is a linear time-invariant IOM system, and, hence, a convolution system. ∎

We conclude by reviewing the main results of this section for sampled systems.

4.3.15.   Sampled difference systems.   Sampled difference systems are defined by difference equations of the form

$$F[y(t), y(t + T), \cdots, y(t + NT), u(t), u(t + T), \cdots, u(t + MT), t] = 0$$

for $t \in \mathbb{T}$, where $\mathbb{T} = \{t_o, t_o + T, t_o + 2T, \cdots\}$ with $t_o \in \mathbb{Z}(T)$, or $\mathbb{T} = \mathbb{Z}(T)$. We assume that for each $t \in \mathbb{T}$ this equation may uniquely be solved for $y(t + NT)$ in the form

$$y(t + NT) = G[y(t), y(t + T), \cdots, y(t + (N - 1)T),$$
$$u(t), u(t + T), \cdots, u(t + MT), t], \qquad t \in \mathbb{T}.$$

N is the *order* of the system. The solution from time t_o on is fully determined by the initial conditions

$$y(t_o), y(t_o + T), \cdots, y(t_o + (N - 1)T)$$

together with the input u from time t_o on. The resulting IOM system is *anticipating* if and only if both $M > N$ and G depends nontrivially on $u(t + (N + 1)T)$, $u(t + (N + 2)T), \cdots, u(t + MT)$.

The sampled difference system is *time-invariant* if the function F does not depend on its last argument, and *linear* if the function F is linear in its first $N + M + 2$ arguments.

If the difference equation is the constant coefficient linear difference equation

$$q_0 y(t) + q_1 y(t + T) + \cdots + q_N y(t + NT)$$
$$= p_0 u(t) + p_1 u(t + T) + \cdots + p_M u(t + MT), \qquad t \in \mathbb{T},$$

with the q_k and p_k real or complex coefficients, the system is both linear and time-invariant. Defining the polynomials

$$Q(\lambda) = q_0 + q_1 \lambda + \cdots + q_N \lambda^N, \qquad P(\lambda) = p_0 + p_1 \lambda + \cdots + p_M \lambda^M,$$

the difference equation may be written in the compact form

$$Q(\sigma^T)y = P(\sigma^T)u,$$

with σ the back shift operator.

    Initially-at-rest sampled difference systems are defined analogously to initially-at-rest difference systems. Initially-at-rest sampled difference systems defined by constant coefficient linear difference equations are linear time-invariant and, hence, are sampled convolution systems.   ■

4.4 RESPONSE OF LINEAR TIME-INVARIANT DIFFERENCE AND DIFFERENTIAL SYSTEMS

In this section we study linear time-invariant difference and differential systems described by constant coefficient linear difference and differential equations. The difference equations are of the form

$$Q(\sigma)y = P(\sigma)u,$$

while the differential equations look like

$$Q(D)y = P(D)u.$$

Q and P are the polynomials

$$Q(\lambda) = q_0 + q_1\lambda + \cdots + q_N\lambda^N, \tag{1a}$$

$$P(\lambda) = p_0 + p_1\lambda + \cdots + p_M\lambda^M. \tag{1b}$$

Without loss of generality we assume in the rest of this chapter that the *leading coefficients* q_N and p_M are both nonzero.

    The time axis $\mathbb{T}$ is taken to be *right semi-infinite* or *infinite*. In the discrete-time case the appropriate time axes are

$$\text{Semi-infinite:} \quad \mathbb{T}_+ = \{n_o, \, n_o + 1, \, n_o + 2, \, \cdots \},$$
$$\text{Infinite:} \quad \mathbb{T}_\infty = \mathbb{Z},$$

with $n_o \in \mathbb{Z}$. In the continuous-time case we use the time axes

$$\text{Semi-infinite:} \quad \mathbb{T}_+ = [t_o, \, \infty),$$
$$\text{Infinite:} \quad \mathbb{T}_\infty = \mathbb{R},$$

with $t_o \in \mathbb{R}$. Because in the continuous-time case we wish to allow the input to have δ-functions at time t_o, the semi-infinite time axis $\mathbb{T}_+$ is understood to include the time t_o^-. This means that the time axis really is the interval $\mathbb{T}_+ = (t_o - \epsilon, \infty)$ with $\epsilon \downarrow 0$.

Difference and differential systems are IO systems but not IOM systems. In the sequel we study how to find the set of *all* possible outputs y corresponding to a given input u. We also consider difference and differential systems with *given* initial conditions, so that the IO system becomes an IOM system, and see how we may determine the unique output y corresponding to any given input u. In Section 4.5 we investigate in particular initially-at-rest constant coefficient linear difference and differential systems.

The plan of the present section is as follows:

   (i) First we study the *homogeneous* solution of linear constant coefficient difference and differential equations. This gives us *all* input-output pairs of the form $(0, y)$, that is, the set of all *zero-input responses* of the system.

  (ii) Next we introduce the idea of a *particular solution* of the difference or differential equation for a given input u. A particular solution y_{part} is *any* solution of the difference or differential equation such that (u, y_{part}) forms an IO pair.

 (iii) Given a particular solution y_{part} corresponding to some input u, and given the set of homogeneous solutions we may determine the *general* solution of the difference or differential equation. The general solution defines the set of all input-output pairs (u, y) for a given input u.

  (iv) Given the set of all IO pairs for a given input u we may use initial conditions to select a special solution.

In Section 4.5, where we consider initially-at-rest linear time-invariant difference and differential systems, it will be seen how particular solutions of the difference or differential equation may be constructed if the impulse response of the intially-at-rest system is known.

Solution of the Homogeneous Equation

The *homogeneous* equation corresponding to the difference equation $Q(\sigma)y = P(\sigma)u$ or the differential equation $Q(D)y = P(D)u$ is obtained by setting the right-hand side equal to zero, resulting in

$$Q(\sigma)y = 0$$

or

$$Q(D)y = 0,$$

respectively. The solution of the homogeneous equation is entirely determined by the roots of the polynomial Q. These roots are called the *characteristic roots* of the difference or differential system.

The following facts are well-known from the elementary theory of difference and differential equations.

4.4.1.   Summary: Solution of the homogeneous equation.

(a) On the time axis $\mathbb{T}_+ = \{n_o, n_o + 1, n_o + 2, \cdots\}$, with $n_o \in \mathbb{Z}$, the homogeneous difference equation $Q(\sigma)y = 0$ has N linearly independent solutions $y_1, y_2, \cdots, y_N$, called *basis solutions*.

On the time axis $\mathbb{T}_\infty = \mathbb{Z}$ the homogeneous equation has N_o basis solutions $y_1, y_2, \cdots, y_{N_o}$, where N_o is the number of *nonzero* roots of Q.

(b) Any solution of the homogeneous difference equation is a linear combination of the basis solutions.

(c) One way of selecting basis solutions is to determine corresponding to each *nonzero* root λ of multiplicity m of the polynomial Q the m basis solutions given by

$$n^i \lambda^n, \qquad n \in \mathbb{T}_+ \text{ or } \mathbb{T}_\infty,$$

for $i = 0, 1, \cdots, m - 1$. On the time axis $\mathbb{T}_+$ the equation has the m_o *additional* basis solutions

$$\Delta(n - n_o - i), \qquad n \in \mathbb{T}_+,$$

for $i = 0, 1, \cdots, m_o - 1$, where $m_o = N - N_o$ is the number of *zero* roots of Q.

(a') Both on the time axis $\mathbb{T}_+ = [t_o, \infty)$, with $t_o \in \mathbb{R}$, and $\mathbb{T}_\infty = \mathbb{R}$ the homogeneous differential equation $Q(D)y = 0$ has N linearly independent solutions $y_1, y_2, \cdots, y_N$, called *basis solutions*.

(b') Any solution of the homogeneous differential equation is a linear combination of the basis solutions.

(c') One way of selecting basis solutions is to determine corresponding to each root λ of multiplicity m of the polynomial Q the m basis solutions given by

$$t^i e^{\lambda t}, \qquad t \in \mathbb{T}_+ \text{ or } \mathbb{T}_\infty,$$

for $i = 0, 1, \cdots, m - 1$.

■

The basis solutions $y_1, y_2, \cdots, y_N$ or $y_1, y_2, \cdots, y_{N_o}$ are said to be linearly independent if no nontrivial linear combination of the basis solutions is identical to zero. In (c), Δ is the pulse of 2.2.7(a), defined by

$$\Delta(n) = \begin{cases} 1 & \text{for } n = 0, \\ 0 & \text{otherwise,} \end{cases} \qquad n \in \mathbb{Z}.$$

The main result of 4.4.1 is that any solution of the homogeneous difference or differential equation is a suitable linear combination of basis solutions. The basis solutions are not unique, but according to (c) and (c') a convenient set of basis solutions may be found from the characteristic roots (i.e., the roots of the polynomial Q). For each root we find as many basis solutions as its multiplicity.

In the discrete-time case, zero characteristic roots play a special role. If the time axis is $\mathbb{T}_+ = \{n_o, n_o + 1, n_o + 2, \cdots\}$ and Q has m_o zero roots, we obtain m_o basis solutions consisting of shifted unit pulses, as shown in Fig. 4.3. If the time axis is $\mathbb{T}_\infty = \mathbb{Z}$, these basis solutions are missing (because they move "out of view" towards $-\infty$).

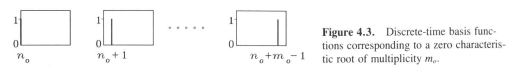

Figure 4.3. Discrete-time basis functions corresponding to a zero characteristic root of multiplicity m_o.

The result of 4.4.1. allows us to construct the set $\mathcal{Y}_0$ of all *zero-input* responses of the difference or differential system. It is given by

$$\mathcal{Y}_0 = \left\{ y \mid y = \sum_i \alpha_i y_i, \; \alpha_i \in \mathbb{C} \text{ for each } i \right\},$$

with the y_i the basis solutions of the homogeneous difference or differential equation.

We consider some examples.

4.4.2.   Examples: Solutions of homogeneous equations.

(a) *Exponential smoother*. The exponential smoother is described by the difference equation

$$y(n + 1) - ay(n) = (1 - a)u(n + 1), \qquad n \in \mathbb{T},$$

where $\mathbb{T}$ is either $\mathbb{T}_+ = \{n_o, n_o + 1, \cdots\}$ or $\mathbb{T}_\infty = \mathbb{Z}$. The polynomial Q is given by $Q(\lambda) = \lambda - a$, which has the single root $\lambda_1 = a$.

If $a \neq 0$ all solutions of the homogeneous equation are of the form

$$y_{\text{hom}}(n) = \alpha a^n, \qquad n \in \mathbb{T},$$

with α an arbitrary constant, possibly complex.

If $a = 0$ and the time axis is $\mathbb{T}_\infty$, the homogeneous equation is $y(n + 1) = 0$, $n \in \mathbb{T}_\infty$, which has the unique trivial solution

$$y_{\text{hom}}(n) = 0, \qquad t \in \mathbb{T}_\infty.$$

If $a = 0$ and the time axis is $\mathbb{T}_+$, all solutions of the homogeneous equation $y(n + 1) = 0$, $n \in \mathbb{T}_+$, are of the form

$$y_{\text{hom}}(n) = \alpha \, \Delta(n - n_o), \qquad n \in \mathbb{T}_+,$$

with α an arbitrary constant, possibly complex.

(b) *RCL network*. The RCL network of Example 4.2.4.(c) is both in case (i) and case (ii) described by a differential equation of the form $Q(D)y = P(D)u$, where Q is given by

$$Q(\lambda) = \frac{1}{LC} + \frac{R}{L}\lambda + \lambda^2,$$

while P depends on the choice of the output.

If $R^2 \neq 4L/C$, the system has two different characteristic roots, given by

$$\lambda_{1,2} = \frac{-R \pm \sqrt{R^2 - 4L/C}}{2L}.$$

All solutions of the homogeneous equation hence are of the form

$$y_{\text{hom}}(t) = \alpha_1 e^{\lambda_1 t} + \alpha_2 e^{\lambda_2 t}, \qquad t \in \mathbb{T},$$

with α_1 and α_2 arbitrary constants, both on the time axis $\mathbb{T} = \mathbb{T}_+ = [t_o,\infty)$ and on $\mathbb{T} = \mathbb{T}_\infty = \mathbb{R}$.

If $R^2 > 4L/C$, the roots λ_1 and λ_2 are both real and y_{hom} consists of a linear combination of two real-valued decaying exponentials.

If $R^2 < 4L/C$, the roots form a complex conjugate pair, and the exponentials are complex-valued. In Example 4.4.4 we see that in this case the exponentials may be replaced by two real-valued decaying harmonics.

If $R^2 = 4L/C$, the polynomials Q has a root $-R/2L = -1/\sqrt{LC}$ of multiplicity 2. As a result, the homogeneous solution is of the form

$$y_{\text{hom}}(t) = \alpha_1 e^{-t/\sqrt{LC}} + \alpha_2 t e^{-t/\sqrt{LC}}, \qquad t \in \mathbb{T}.$$

When this solution applies the system is said to be *critically damped*.   ∎

When a characteristic root λ of a difference or differential equation is real, the corresponding basis solution is real. Even if the polynomial Q has real coefficients, it may easily have complex roots, which make the corresponding basis solutions also complex. Often we are interested in real solutions only. If the polynomial Q has real coefficients (which is the usual situation), its complex roots always occur in complex conjugate pairs. The corresponding basis solutions may be combined to construct pairs of *real* basis solutions as follows.

4.4.3.   Summary: Real basis solutions.

If the homogeneous difference equation $Q(\sigma)y = 0$ has a pair of complex conjugate roots λ, $\bar\lambda$ of multiplicity m, the equation has the $2m$ corresponding *real* basis solutions

$$n^i \rho^n \cos(\psi n),$$
$$n^i \rho^n \sin(\psi n), \qquad n \in \mathbb{T},$$

for $i = 0, 1, \cdots, m - 1$, where

$$\rho := |\lambda|, \qquad \psi := \arg(\lambda).$$

If the homogeneous differential equation $Q(D)y = 0$ has a pair of complex conjugate roots λ, $\bar\lambda$ of multiplicity m, the equation has the $2m$ corresponding *real* basis solutions

$$t^i e^{\sigma t} \cos(\omega t),$$
$$t^i e^{\sigma t} \sin(\omega t), \qquad t \in \mathbb{T},$$

for $i = 0, 1, \cdots, m - 1$, where

$$\sigma := \operatorname{Re}(\lambda), \qquad \omega := \operatorname{Im}(\lambda). \qquad \blacksquare$$

The proof is left as an exercise.

4.4.4.   Example: Real basis solutions for the RCL network.

In Example 4.4.2(b) we saw that if $R^2 \neq 4L/C$, then the polynomial Q of the RCL network has the two roots

$$\lambda_{1,2} = \frac{-R \pm \sqrt{R^2 - 4L/C}}{2L}.$$

We define the *resonance frequency* ω_r and the *quality factor* q of the network as

$$\omega_r = \frac{1}{\sqrt{LC}}, \qquad q = \frac{\omega_r L}{R}.$$

Then the characteristic roots may be expressed as

$$\lambda_{1,2} = \frac{\omega_r}{2q}(-1 \pm \sqrt{1 - 4q^2}).$$

For $q \leq 1/2$ the characteristic roots are both real, but if $q > 1/2$ they form a complex conjugate pair λ, $\bar\lambda$ with real and imaginary parts

$$\sigma = \operatorname{Re}(\lambda) = -\frac{\omega_r}{2q}, \qquad \omega = \operatorname{Im}(\lambda) = \omega_r\sqrt{1 - 1/4q^2}.$$

From 4.4.3 all solutions of the homogeneous equation may be expressed as

$$y_{\text{hom}}(t) = \alpha_1 e^{\sigma t} \cos(\omega t) + \alpha_2 e^{\sigma t} \sin(\omega t), \qquad t \in \mathbb{T}, \qquad (2)$$

with α_1 and α_2 arbitrary constants. All real solutions of the homogeneous equation follow by taking α_1 and α_2 real. If α_1 and α_2 are real, (2) may alternatively be written as

$$
\begin{aligned}
y_{\text{hom}}(t) &= \alpha e^{\sigma t} \cos(\omega t + \phi), \\
&= \alpha e^{-\omega_r t/2q} \cos(\omega_r t \sqrt{1 - 1/4q^2} + \phi), \qquad t \in \mathbb{T}, \tag{3}
\end{aligned}
$$

with α and ϕ arbitrary real constants. This solution represents a damped harmonic with frequency $\omega = \omega_r \sqrt{1 - 1/4q^2}$ and time constant $2q/\omega_r$. If the quality factor q is large, the frequency ω is close to the resonance frequency ω_r and the damping is small.

Figure 4.4 shows the real basis solutions of the RCL network we obtained in 4.4.2(b) and those of the present example.

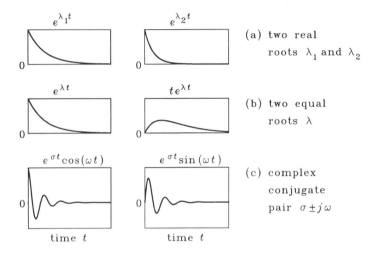

Figure 4.4.   Real basis solutions of the RCL network. Top: two real roots. Middle: two equal roots. Bottom: complex conjugate pair of roots.

Exercise. Prove that if α_1 and α_2 are real, then (2) may be rewritten in the form (3) with α and ϕ real.   ∎

Particular Solutions

A *particular solution* y_{part} of the difference equation

$$
Q(\sigma)y = P(\sigma)u
$$

or the differential equation

$$
Q(D)y = P(D)u
$$

corresponding to a given input u simply is *any* output y_{part} such that $Q(\sigma)y_{part} = P(\sigma)u$ or $Q(D)y_{part} = P(D)u$, respectively.

For some input signals u particular solutions are easy to guess. If the input is an exponential signal, for instance, there usually exists an exponential particular solution with the same parameter.

4.4.5.   Example: Particular solutions for exponential inputs to the exponential smoother.   The exponential smoother is described by the difference equation

$$y(n + 1) - ay(n) = (1 - a)u(n + 1), \qquad n \in \mathbb{T},$$

with $\mathbb{T}$ a right semi-infinite or infinite time axis. Suppose that the input is the exponential signal

$$u(n) = u_o z^n, \qquad n \in \mathbb{T},$$

with u_o and z given real or complex constants. We try to find a particular solution of a similar form $y(n) = y_o z^n$, $n \in \mathbb{T}$, with the constant y_o to be determined. Substitution of u and y into the difference equation yields

$$y_o z^{n+1} - ay_o z^n = (1 - a)u_o z^{n+1}, \qquad n \in \mathbb{T}.$$

Cancellation of the common factor z^n and solution for y_o results in

$$y_o = \frac{1 - a}{z - a} u_o z,$$

provided $z \neq a$. This shows that if $z \neq a$ for the given exponential input the difference equation has an exponential particular solution

$$y_{part}(n) = \frac{1 - a}{z - a} u_o z^{n+1}, \qquad n \in \mathbb{T}.$$

Exercise. Show that if $z = a$, the difference equation has the particular solution

$$y_{part}(n) = (1 - a)n u_o a^n, \qquad n \in \mathbb{T}. \qquad\qquad \blacksquare$$

General Solution of Difference and Differential Equations

We finally study the *general solution* of constant coefficient linear difference and differential equations. The general solution is a form in which every solution may be written.

4.4.6.   Summary: General solution of the nonhomogeneous equation. Every
solution of the constant coefficient linear difference equation $Q(\sigma)y = P(\sigma)u$ or the
differential equation $Q(D)y = P(D)u$ for a given input u may be written as

$$y = y_{\text{part}} + y_{\text{hom}},$$

where y_{part} is a particular solution of the nonhomogeneous difference or the differential equation corresponding to the input u, and y_{hom} is a suitable linear combination

$$y_{\text{hom}} = \sum_i \alpha_i y_i, \qquad \alpha_i \in \mathbb{C} \text{ for each } i,$$

of the basis solutions y_i of the homogeneous equation.   ■

In other words, if we happen to know some solution y_{part} of the nonhomogeneous
difference or differential equation (on the time axis of interest), any other solution
may be obtained by adding a suitable solution of the homogeneous equation. It follows that the set of solutions $\mathcal{Y}_u$ of the difference or differential equations corresponding to the input u is given by

$$\mathcal{Y}_u = \left\{ y \mid y = y_{\text{part}} + \sum_i \alpha_i y_i, \qquad \alpha_i \in \mathbb{C} \text{ for each } i \right\}.$$

The arbitrary constants α_i occurring in the general solution may be used to satisfy N initial or boundary conditions, such as

$$y(0) = y_0, \qquad y(1) = y_1, \qquad \cdots, \qquad y(N - 1) = y_{N-1},$$

in the discrete-time case, or

$$y(0) = y_0, \qquad y^{(1)}(0) = y_1, \qquad \cdots, \qquad y^{(N-1)}(0) = y_{N-1},$$

in the continuous-time case, where $y_0, y_1, \cdots, y_{N-1}$, are N given real or complex
numbers.

4.4.7.   Example: General solution and initial conditions.
(a) *Exponential smoother*. The exponential smoother is described by the difference equation

$$y(n + 1) = ay(n) + (1 - a)u(n + 1), \qquad n = 0, 1, 2, \cdots,$$

where we take the time axis as $\mathbb{T} = \mathbb{Z}_+$. Suppose that the input is constant and
given by

$$u(n) = 1, \qquad n = 0, 1, 2, \cdots .$$

We conjecture that there exists a constant particular solution $y(n) = y_o$, $n = 0, 1, 2, \cdots$. Substitution into the difference equation results in

$$y_o = ay_o + (1 - a), \qquad n = 0, 1, 2, \cdots ,$$

which is satisfied for $y_o = 1$.

In Example 4.4.2(a) we found that for $a \neq 0$ the homogeneous equation has the single basis solution $y_1(n) = a^n$, $n = 0, 1, 2, \cdots$. It follows from 4.4.6 that corresponding to the given input the difference equation has the general solution

$$y(n) = 1 + \alpha a^n, \qquad n = 0, 1, 2, \cdots , \tag{4}$$

with α an arbitrary constant.

The arbitrary constant α may be determined if an initial condition is given. Suppose that $y(0) = 0$. Substitution of $n = 0$ into (4) then yields

$$0 = 1 + \alpha,$$

so that $\alpha = -1$. It follows from (4) that the response of the system to the constant input $u(n) = 1$, $n = 0, 1, 2, \cdots$, with the initial condition $y(0) = 0$ is

$$y(n) = 1 - a^n, \qquad n = 0, 1, 2, \cdots .$$

(b) *RCL network*. As a second, more complicated example we consider the RCL network of Example 4.2.4(c). If we take the current i through the network as the output y, the network is described by the differential equation

$$y^{(2)}(t) + \frac{R}{L} y^{(1)}(t) + \frac{1}{LC} y(t) = \frac{1}{L} u^{(1)}(t), \qquad t \in \mathbb{T}.$$

Suppose that the input is constant and given by

$$u(t) = 1 \qquad \text{for} \quad t \geq 0^-.$$

It is easy to verify that correspondingly the differential equation has the particular solution

$$y_{\text{part}}(t) = 0, \qquad t \geq 0^-.$$

In Example 4.4.2(b) we found that if $R^2 \neq 4L/C$ the homogeneous equation has the basis solutions $y_i(t) = \exp(\lambda_i t)$, $t \in \mathbb{T}$, $i = 1, 2$, where λ_1 and λ_2 are the characteristic roots. Thus, corresponding to the given constant input the general solution of the differential equation is

$$y(t) = \alpha_1 e^{\lambda_1 t} + \alpha_2 e^{\lambda_2 t}, \qquad t \geq 0^-. \tag{5}$$

Suppose now that the initial conditions of the RCL network (see Fig. 4.1) are specified by the requirement that at time zero both the charge q of the capacitor and the flux ϕ contained in the inductor are zero. We determine the resulting initial conditions on y. Since the flux through the inductor equals $\phi = Li = Ly$ it follows immediately from $\phi(0) = 0$ that $y(0) = 0$, which provides us with the first initial condition. Substitution of $t = 0$ into (5) yields

$$0 = \alpha_1 + \alpha_2. \tag{6}$$

The second initial condition follows from the equality $u = v_R + v_C + v_L$ (see Example 4.4.2(b)), or

$$v_L(0) = u(0) - v_R(0) - v_C(0).$$

Since $u(0) = 1$, $v_R(0) = Ri(0) = Ry(0) = 0$, and because the charge of the capacitor at time zero $q(0) = Cv_C(0)$ is zero we have also $v_C(0) = 0$, it follows that

$$v_L(0) = 1.$$

Since $v_L = Ldi/dt = Ldy/dt$ it follows that the second initial condition we are looking for is

$$y^{(1)}(0) = \frac{1}{L} v_L(0) = \frac{1}{L}.$$

Differentiation of (5) with respect to time and setting $t = 0$ accordingly leads to

$$\frac{1}{L} = \alpha_1 \lambda_1 + \alpha_2 \lambda_2. \tag{7}$$

The two equations (6) and (7) may be solved for the constants α_1 and α_2, resulting in

$$\alpha_1 = -\alpha_2 = \frac{\dfrac{1}{L}}{\lambda_1 - \lambda_2}.$$

As a result, the solution of the differential equation for the given input and initial conditions is

$$y(t) = \frac{\frac{1}{L}}{\lambda_1 - \lambda_2}(e^{\lambda_1 t} - e^{\lambda_2 t}), \qquad t \geq 0.$$

To make this result more concrete, we adopt the numerical values $R = 11 \ \Omega$, $L = 0.01$ H, and $C = 0.001$ F. As a result, the polynomial Q is given by $Q = \lambda^2 + 1100\lambda + 100000 = (\lambda + 100)(\lambda + 1000)$, so that the characteristic roots are $\lambda_1 = -100$ and $\lambda_2 = -1000$. After substitution of the numerical values we find that the solution of the differential equation is

$$y(t) = \frac{1}{9}(e^{-100t} - e^{-1000t}), \qquad t \geq 0.$$

A plot is given in Fig. 4.5.

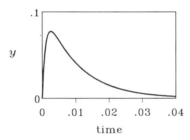

Figure 4.5. Response of the current through the RCL network when the input is a constant voltage equal to 1 and the initial charge of the capacitor and the initial flux through the inductor are zero.

∎

We review the results of this section for sampled difference systems.

4.4.8.   Review: Homogeneous solutions of sampled linear time-invariant difference systems.   Sampled linear time-invariant difference systems are described by difference equations of the form

$$Q(\sigma^T)y = P(\sigma^T)u,$$

where Q and P are polynomials as in (1). We study such systems on the semi-infinite time axis $\mathbb{T}_+ = \{t_o, t_o + T, t_o + 2T, \cdots\}$, with $t_o \in \mathbb{Z}(T)$, or on the infinite time axis $\mathbb{T}_\infty = \mathbb{Z}(T)$.

On the time axis $\mathbb{T}_+$ the homogeneous equation

$$Q(\sigma^T)y = 0$$

has N basis solutions $y_1, y_2, \cdots, y_N$, with N the degree of Q. On the time axis $\mathbb{T}_\infty$, the homogeneous equation has N_o basis solutions, with N_o the number of nonzero roots of Q. Every solution of the homogeneous equation may be expressed as a linear combination of the basis solutions.

Corresponding to each *nonzero* characteristic root λ of multiplicity m of the polynomial Q the homogeneous equation has m basis solutions given by

$$(t/T)^i \lambda^{t/T}, \qquad t \in \mathbb{T}_+ \text{ or } \mathbb{T}_\infty,$$

for $i = 0, 1, \cdots, m - 1$. On the time axis $\mathbb{T}_+$ the homogeneous equation has the m_o *additional* basis solutions

$$\frac{1}{T}\Delta(t - t_o - iT), \qquad t \in \mathbb{T}_+,$$

for $i = 0, 1, \cdots, m_o - 1$, where $m_o = N - N_o$ is the number of *zero* roots of Q.

Any solution of the nonhomogeneous difference equation may be written as

$$y = y_{\text{part}} + y_{\text{hom}},$$

where y_{part} is a particular solution, and y_{hom} a suitable solution of the homogeneous equation.   ■

4.5 INITIALLY-AT-REST LINEAR TIME-INVARIANT DIFFERENTIAL AND DIFFERENCE SYSTEMS

In Section 4.3 we defined initially-at-rest difference and differential systems as difference and differential systems that are "at rest" at time $-\infty$. The system is at rest if a zero input from the initial time on results in a unique output that is identical to zero.

Initially-at-rest linear time-invariant difference and differential systems are input-output mapping systems. According to 4.3.14 they are *convolution* systems, characterized by an IO map of the form

$$y = h * u,$$

with h the *impulse response* of the system.

In this section we first consider how to determine the impulse response h of initially-at-rest linear difference and differential systems. Next it is seen how the impulse response may be used to generate particular solutions of difference and differential systems defined on infinite and semi-infinite time axes. Given a particular solution, the general solution of the difference or differential equation, or the response of the system to given initial conditions, may be found as in Section 4.4.

Impulse Response

We first discuss the impulse response of initially-at-rest difference systems described by a difference equation of the form

$$Q(\sigma)y = P(\sigma)u,$$

with Q and P the polynomials

$$Q(\lambda) = q_0 + q_1\lambda + \cdots + q_N\lambda^N,$$
$$P(\lambda) = p_0 + p_1\lambda + \cdots + p_M\lambda^M.$$

As before, the leading coefficients q_N and p_M are assumed to be nonzero.

The impulse response h of the initially-at-rest system is the response of the system if the input u is the unit pulse Δ. Because this input is zero for positive times, for those times the impulse response h is a solution of the *homogeneous* equation, and hence may be written as a linear combination of basis solutions on the time axis $\{1, 2, \cdots\}$. Because the input is zero for negative times as well and the system is initially at rest, the impulse response is zero for negative times up to some finite time, which is negative if the system is anticipating, and nonnegative if it is not anticipating. It may be shown that the impulse response of the difference system may actually be written in the form

$$h(n) = \sum_{i=1}^{N} \alpha_i y_i(n-1)1(n-1) + \sum_{i=0}^{M-N} \beta_i\Delta(n+i), \qquad n \in \mathbb{Z}.$$

The α_i and β_i are suitable constants, and the y_i are basis solutions of the homogeneous difference equation on the semi-infinite time axis $\mathbb{Z}_+$, arbitrarily extended for negative times. If the lower limit of the second sum exceeds the upper limit (i.e., if $M - N < 0$), the sum is canceled.

In the continuous-time case, the system is described by the differential equation

$$Q(D)y = P(D)u,$$

and the impulse response h of the initially-at-rest system is its response to the input $u = \delta$. Again, because this input is zero for positive times, for those times the impulse response is a solution of the homogeneous equation, which means that for positive times it is a linear combination of the basis solutions. Also, because the input is zero for negative times and the system is initially at rest, the impulse response is zero for negative times. *At* time zero the impulse response has some delta functions, depending on whether $M - N \geq 0$. The general form of the impulse response for the continuous-time case is

$$h(t) = \sum_{i=1}^{N} \alpha_i y_i(t)1(t) + \sum_{i=0}^{M-N} \beta_i\delta^{(i)}(t), \qquad t \in \mathbb{R},$$

where again the α_i and β_i are suitable constants, the y_i are basis solutions of the homogeneous differential equation, and the second sum is canceled if $M - N < 0$.

4.5.1.   Summary: Impulse response of linear constant coefficient initially-at-rest difference and differential systems.

The impulse response of the initially-at-rest discrete-time system described by the constant coefficient linear difference equation $Q(\sigma)y = P(\sigma)u$ is of the form

$$h(n) = \sum_{i=1}^{N} \alpha_i y_i(n-1)\mathbb{1}(n-1)$$
$$+ \sum_{i=0}^{M-N} \beta_i \Delta(n+i),$$

$n \in \mathbb{Z}$, where the α_i and β_i are suitable constants, N is the degree of Q, M that of P, and the y_i are N basis solutions of the homogeneous equation $Q(\sigma)y = 0$ on the time axis $\mathbb{Z}_+$ and arbitrarily extended for negative times.

The impulse response of the initially-at-rest continuous-time system described by the constant coefficient linear differential equation $Q(D)y = P(D)u$ is of the form

$$h(t) = \sum_{i=1}^{N} \alpha_i y_i(t)\mathbb{1}(t)$$
$$+ \sum_{i=0}^{M-N} \beta_i \delta^{(i)}(t),$$

$t \in \mathbb{R}$, where the α_i and β_i are suitable constants, N is the degree of Q, M that of P, and the y_i are N basis solutions of the homogeneous equation $Q(D)y = 0$ on $\mathbb{R}$. ∎

A comment about anticipativeness and non-anticipativeness is in order. From 3.4.13 we know that convolution systems are non-anticipating if and only if their impulse response h satisfies $h(t) = 0$ for $t < 0$. Inspection of the impulse responses as given in 4.5.1 confirms what we know from 4.3.2, namely, that difference systems are non-anticipating if and only if $M \leq N$, and that differential systems are always non-anticipating.

We do not present completely general ways for computing the impulse response, but content ourselves with some examples. Often it is easier to determine the impulse response using z- or Laplace transforms, which are the subject of Chapter 8.

4.5.2.   Examples: Impulse response of difference and differential systems.

(a) *Second-order anticipating difference system.* This example illustrates how to find the impulse response of a difference system by constructing its response from rest to the unit pulse Δ. Consider the difference equation

$$\frac{1}{6}y(n) - \frac{5}{6}y(n+1) + y(n+2) = u(n+3), \qquad n \in \mathbb{Z}, \tag{1}$$

which represents a second-order system with $Q(\lambda) = 1/6 - 5\lambda/6 + \lambda^2 = (\lambda - 1/2)(\lambda - 1/3)$ and $P(\lambda) = \lambda^3$. The impulse response h may quite straightforwardly be obtained from the difference equation

$$\frac{1}{6}h(n) - \frac{5}{6}h(n + 1) + h(n + 2) = \Delta(n + 3), \qquad n \in \mathbb{Z},$$

which we rewrite as

$$h(n + 2) = -\frac{1}{6}h(n) + \frac{5}{6}h(n + 1) + \Delta(n + 3), \qquad n \in \mathbb{Z}. \tag{2}$$

Because the impulse response is the response from rest, h is certainly zero for large negative times. Inspection of (2) shows that as n increases from $-\infty$ the first time the last term on the right-hand side is nonzero occurs when $n = -3$. This is also the first time the left-hand side is nonzero, so that $h(n + 2) = 0$ for $n < -3$ (i.e., $h(n) = 0$ for $n < -1$). Substitution of $n = -3$ yields

$$h(-1) = -\frac{1}{6}h(-3) + \frac{5}{6}h(-2) + \Delta(0) = 1,$$

since $h(-3) = h(-2) = 0$ and $\Delta(0) = 1$. Further inspection shows that for $n \geq -2$ the equation (1) reduces to the homogeneous equation

$$\frac{1}{6}h(n) - \frac{5}{6}h(n + 1) + h(n + 2) = 0, \qquad n \geq -2, \qquad n \in \mathbb{Z}.$$

Because Q has the roots $1/2$ and $1/3$, this equation has the general solution

$$h(n) = \alpha_1 \left(\frac{1}{2}\right)^n + \alpha_2 \left(\frac{1}{3}\right)^n, \qquad n \geq -2, \qquad n \in \mathbb{Z}.$$

From the initial conditions $h(-2) = 0$ and $h(-1) = 1$, we obtain after substitution of $n = -2$ and $n = -1$

$$h(-2) = 4\alpha_1 + 9\alpha_2 = 0,$$
$$h(-1) = 2\alpha_1 + 3\alpha_2 = 1.$$

It follows that $\alpha_1 = 3/2$ and $\alpha_2 = -2/3$, and hence the impulse response is given by

$$h(n) = \left[\frac{3}{2}\left(\frac{1}{2}\right)^n - \frac{2}{3}\left(\frac{1}{3}\right)^n\right]\mathbb{1}(n + 1), \qquad n \in \mathbb{Z}.$$

A plot is given in Fig. 4.6. It is seen that the impulse response anticipates the unit pulse Δ by one time interval, which agrees with the fact that $M - N = 1$.

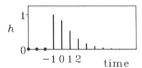

Figure 4.6. Impulse response of the anticipating difference system.

The impulse response as obtained may be rewritten as

$$h(n) = \left[\frac{3}{2} \cdot \frac{1}{2}\left(\frac{1}{2}\right)^{n-1} - \frac{2}{3} \cdot \frac{1}{3}\left(\frac{1}{3}\right)^{n-1} \right] \mathbb{1}(n-1) + \Delta(n+1) + \frac{5}{6}\Delta(n)$$

$$= \left[\frac{3}{4}\left(\frac{1}{2}\right)^{n-1} - \frac{2}{9}\left(\frac{1}{3}\right)^{n-1} \right] \mathbb{1}(n-1) + \Delta(n+1) + \frac{5}{6}\Delta(n), \qquad n \in \mathbb{Z},$$

which is of the form given in 4.5.1. An alternative way of finding h is to assume that it is of the form of 4.5.1, and substituting it into (2) to determine the coefficients α_1, α_2, β_1 and β_2 it contains.

(b) *RCL network*. In the continuous-time case the impulse response may likewise be constructed by determining the initial conditions for the homogeneous equation for $t > 0$. The RCL network of Example 4.2.4(c), when taking the voltage across the inductor as its output, is described by the differential equation

$$\frac{1}{LC}y(t) + \frac{R}{L}\frac{dy(t)}{dt} + \frac{d^2y(t)}{dt^2} = \frac{d^2u(t)}{dt^2}, \qquad t \in \mathbb{R}.$$

The impulse response h is the response to the δ-function from rest, and hence satisfies

$$Q(D)h = P(D)\delta,$$

while $h(t) = 0$ for $t < 0$. In this particular case we have

$$\frac{1}{LC}h(t) + \frac{R}{L}\frac{dh(t)}{dt} + \frac{d^2h(t)}{dt^2} = \delta^{(2)}(t), \qquad t \in \mathbb{R}.$$

Because $M - N = 0$, according to 4.5.1 the impulse response has the form

$$h = h_o\mathbb{1} + \beta\delta,$$

where h_o is a solution of the homogeneous equation on $\mathbb{R}$ and β a constant to be determined. Differentiating twice with application of the product rule we obtain

$$h^{(1)} = h_o^{(1)}\mathbb{1} + h_o(0)\delta + \beta\delta^{(1)},$$

$$h^{(2)} = h_o^{(2)}\mathbb{1} + h_o^{(1)}(0)\delta + h_o(0)\delta^{(1)} + \beta\delta^{(2)}.$$

Substitution into the differential equation results in

$$\frac{1}{LC}[h_o \mathbb{1} + \beta\delta] + \frac{R}{L}[h_o^{(1)}\mathbb{1} + h_o(0)\delta + \beta\delta^{(1)}]$$

$$+ [h_o^{(2)}\mathbb{1} + h_o^{(1)}(0)\delta + h_o(0)\delta^{(1)} + \beta\delta^{(2)}] = \delta^{(2)}.$$

Equality is obtained if the δ-functions on the left and the right have equal coefficients. This results in the equations

$$\delta^{(2)}: \quad \beta = 1,$$

$$\delta^{(1)}: \quad \frac{R}{L}\beta + h_o(0) = 0,$$

$$\delta: \quad \frac{1}{LC}\beta + \frac{R}{L}h_o(0) + h_o^{(1)}(0) = 0,$$

$$\mathbb{1}: \quad \frac{1}{LC}h_o + \frac{R}{L}h_o^{(1)} + h_o^{(2)} = 0.$$

The latter equation is satisfied because by assumption h_o is a solution of the homogeneous equation. The first three equations may be solved for the coefficient β and the initial conditions $h_o(0)$ and $h_o^{(1)}(0)$. We find

$$\beta = 1,$$

$$h_o(0) = -\frac{R}{L},$$

$$h_o^{(1)}(0) = -\frac{1}{LC}\left(1 - \frac{R^2C}{L}\right).$$

The homogeneous equation may now be solved for h_o with the initial conditions $h_o(0)$ and $h_o^{(1)}(0)$ thus found. Suppose in particular that the network is critically damped, that is, $R^2 = 4L/C$. Then $h_o(0) = -2/\sqrt{LC}$ and $h_o^{(1)}(0) = 3/LC$. According to Example 4.4.2(b) the general solution of the homogeneous equation is

$$h_o(t) = \alpha_1 e^{-t/\sqrt{LC}} + \alpha_2 t e^{-t/\sqrt{LC}}, \qquad t \geq 0,$$

with α_1 and α_2 to be determined. It follows that

$$h_o(0) = \alpha_1 = -\frac{2}{\sqrt{LC}},$$

$$h_o^{(1)}(0) = -\frac{\alpha_1}{\sqrt{LC}} + \alpha_2 = \frac{3}{LC}.$$

Solution for the unknown constants gives $\alpha_1 = -2/\sqrt{LC}$ and $\alpha_2 = 1/LC$, which yields the desired homogeneous solution. The impulse response follows as

$$h(t) = \left[-\frac{2}{\sqrt{LC}} e^{-t/\sqrt{LC}} + \frac{t}{LC} e^{-t/\sqrt{LC}} \right] \mathbb{1}(t) + \delta(t), \qquad t \in \mathbb{R}.$$

A plot of the impulse response is given in Fig. 4.7.

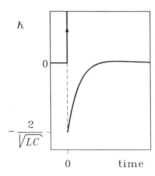

Figure 4.7. Impulse response of the RCL network for critical damping.

■

Particular Solutions of Linear Time-Invariant Difference and Differential Systems

In Section 4.4 we introduced the notion of a *particular solution* of the difference equation $Q(\sigma)y = P(\sigma)u$ or the differential equation $Q(D)y = P(D)u$ corresponding to an input u. Given the impulse response h of the initially-at-rest system defined by the difference or differential equation, the following result makes it possible to construct particular solutions for a large class of inputs. Once a particular solution is available it is a simple matter to determine the general solution and the response to given initial conditions.

4.5.3.   Summary: Particular solutions of linear constant coefficient difference and differential equations.

Let h be the impulse response of the initially-at-rest system described by the difference equation $Q(\sigma)y = P(\sigma)u$.

    (a) On the infinite time axis $\mathbb{Z}$ the difference equation has the particular solution

$$y_{\text{part}} = h * u,$$

provided the convolution exists.

Let h be the impulse response of the initially-at-rest system described by the differential equation $Q(D)y = P(D)u$.

    (a') On the infinite time axis $\mathbb{R}$ the differential equation has the particular solution

$$y_{\text{part}} = h * u,$$

provided the convolution exists.

(b) On the semi-infinite time axis $\mathbb{T}_+ = \{n_o, \ n_o + 1, \cdots\}$ the difference equation has a particular solution which on $\mathbb{T}_+$ coincides with the signal

$$y_{\text{part}} = h * u_+,$$

provided the convolution exists, where u_+ is defined on the time axis $\mathbb{Z}$ by

$$u_+(n) = \begin{cases} u(n) & \text{for } n \geq n_o, \\ 0 & \text{otherwise,} \end{cases} \qquad n \in \mathbb{Z}.$$

(b') On the semi-infinite time axis $\mathbb{T}_+ = [t_o, \infty)$ the differential equation has a particular solution which on $\mathbb{T}_+$ coincides with the signal

$$y_{\text{part}} = h * u_+,$$

provided the convolution exists, where u_+ is defined on the time axis $\mathbb{R}$ by

$$u_+(t) = \begin{cases} u(t) & \text{for } t \geq t_o, \\ 0 & \text{otherwise,} \end{cases} \qquad t \in \mathbb{R}. \ ∎$$

4.5.4.   Remark: Particular solutions.   If the convolutions in 4.5.3 do not exist, both in the discrete- and the continuous-time case one may try to find particular solutions of the form

$$y_{\text{part}} = \left(h + \sum_i \alpha_i y_i \right) * u$$

on the infinite time axis $\mathbb{T}$, or

$$y_{\text{part}} = \left(h + \sum_i \alpha_i y_i \right) * u_+$$

on the semi-infinite time axis $\mathbb{T}_+$. The y_i are basis solutions on the infinite time axis $\mathbb{T}_\infty$ and the α_i constants that are chosen such that the convolutions exist.   ∎

4.5.5.   Proof of 4.5.3 and 4.5.4.   The proof is given for the discrete-time case and a semi-infinite time axis. The proof for the continuous-time case follows parallel lines and the proof for infinite time axes is obtained by replacing n_o or t_o with $-\infty$.

Using the linearity and shift properties of the convolution it follows that

$$Q(\sigma)y_{\text{part}} = Q(\sigma)\left(\left(h + \sum_i \alpha_i y_i \right) * u_+ \right) = \left(Q(\sigma)h + \sum_i \alpha_i Q(\sigma)y_i \right) * u_+.$$

Because y_i is a basis solution of the homogeneous equation, $Q(\sigma)y_i = 0$. Furthermore, since h is the impulse response, $Q(\sigma)h = P(\sigma)\Delta$. As a result,

$$Q(\sigma)y_{\text{part}} = (P(\sigma)\Delta) * u_+ = P(\sigma)(\Delta * u_+) = P(\sigma)u_+.$$

By restricting this equality to the time axis $\mathbb{T}_+$ we may replace u_+ with u, which proves that y_{part} satisfies the difference equation on $\mathbb{T}_+$.   ∎

It follows from 4.5.3 that both in the discrete- and the continuous-time case the response of a difference or differential system on the infinite time axis $\mathbb{T}$ may be expressed as

$$y = h * u + y_{\text{hom}},$$

where y_{hom} is a suitable solution of the homogeneous equation. The first term $h * u$ is the *initially-at-rest* response of the system to the input u. The second term y_{hom} is the response if the input were zero and, hence, is the *zero-input* response. Thus we may write

$$y = y_{\text{initially-at-rest}} + y_{\text{zero-input}}.$$

The following example illustrates how 4.5.3 may be used to construct particular and general solutions.

4.5.6.   Example: Particular solutions for the exponential smoother.   As found in 4.4.2(a), for $a \neq 0$ the exponential smoother has the single basis solution

$$y_1(n) = a^n, \qquad n \in \mathbb{T}_+ \text{ or } \mathbb{T}_\infty,$$

while from Example 4.3.12 its impulse response is

$$h(n) = (1 - a)a^n \mathbb{1}(n), \qquad n \in \mathbb{Z}.$$

We consider the problem of finding a particular solution corresponding to the constant input

$$u(n) = 1, \qquad n \in \mathbb{T},$$

for different choices of $\mathbb{T}$ and a.

(a) $\mathbb{T} = \mathbb{T}_+ = \{n_o, n_o + 1, n_o + 2, \cdots\}$. We first consider the smoother on the semi-infinite time axis $\mathbb{T}_+ = \{n_0, n_0 + 1, n_0 + 2, \cdots\}$. Defining u_+ as

$$u_+(n) = \begin{cases} 1 & \text{for } n \geq n_o, \\ 0 & \text{otherwise}, \end{cases} \qquad n \in \mathbb{Z},$$

it follows from 4.5.3 that a particular solution corresponding to the input u coincides on $\mathbb{T}_+$ with

$$y_{\text{part}}(n) = (h * u_+)(n) = \sum_{m=-\infty}^{\infty} h(n - m)u_+(m)$$

$$= (1 - a) \sum_{m=n_o}^{\infty} a^{n-m} \mathbb{1}(n - m), \qquad n \in \mathbb{Z}.$$

For $n \geq n_o$ we have

$$y_{\text{part}}(n) = (1 - a) \sum_{m=n_o}^{n} a^{n-m} = 1 - a^{n-n_o+1}, \qquad n \geq n_o, n \in \mathbb{Z}. \tag{3}$$

This is the desired particular solution. It follows that the *general* solution of the difference equation on the time axis $\mathbb{T}_+$ is given by

$$y(n) = 1 - a^{n-n_o+1} + \alpha a^{n-n_o}, \qquad n = n_o, n_o + 1, n_o + 2, \cdots, \tag{4}$$

with α an arbitrary constant.

(b) $\mathbb{T} = \mathbb{Z}$ *and* $|a| < 1$. On the infinite time axis $\mathbb{T}_\infty = \mathbb{Z}$ a particular solution follows by taking $n_o = -\infty$ in (3). This is well-defined if $|a| < 1$ and results in

$$y_{\text{part}}(n) = 1, \qquad n \in \mathbb{Z}. \tag{5}$$

(c) $\mathbb{T} = \mathbb{Z}$ *and* $|a| \geq 1$. When the time axis is $\mathbb{Z}$, but $|a| \geq 1$, (3) has no limit for $n_o \to -\infty$. Obviously, though, the constant solution (5) is a correct particular solution of the difference equation $y(n + 1) - ay(n) = (1 - a)$, $n \in \mathbb{Z}$, for *any* a. To find this solution constructively when $|a| \geq 1$, we use 4.5.4. Consider $h' = h + \alpha_1 y_1$, with α_1 to be chosen. We have

$$h'(n) = (1 - a)a^n \mathbb{1}(n) + \alpha_1 a^n, \qquad n \in \mathbb{Z}.$$

By choosing $\alpha_1 = a - 1$ this reduces to

$$h'(n) = \begin{cases} (a - 1)a^n & \text{for } n < 0, \\ 0 & \text{for } n \geq 0 \end{cases}$$

$$= (a - 1)a^n \mathbb{1}(-n - 1), \qquad n \in \mathbb{Z}.$$

For $|a| \geq 1$ the function h' has finite action, so that its convolution with the constant input u exists. It is easily found that

$$(h' * u)(n) = 1, \qquad n \in \mathbb{Z},$$

which confirms that also for $|a| \geq 1$ the constant 1 is a particular solution of the difference equation.   ∎

We very briefly review the results of this section for sampled systems.

4.5.7.   Review: Impulse response and particular solutions of sampled difference systems. The impulse response of initially-at-rest sampled difference systems defined on the time axis $\mathbb{Z}(T)$ and described by the difference equation $Q(\sigma^T)y = P(\sigma^T)u$ is of the form

$$h(t) = \sum_{i=1}^{N} \alpha_i y_i(t - T) \mathbb{1}(t - T) + \sum_{i=0}^{M-N} \beta_i \Delta\left(\frac{t + iT}{T}\right), \qquad t \in \mathbb{Z}(T),$$

where the y_i are basis solutions on the time axis $\mathbb{Z}(T)$, and the α_i and β_i suitable constants. Particular solutions of the difference equation may be generated by convolution with the impulse response as in 4.5.3(a) and (b) and 4.5.4.   ∎

4.6 STABILITY OF DIFFERENCE AND DIFFERENTIAL SYSTEMS

In 3.6.1 we defined *BIBO* (bounded-input bounded-output) stability of *convolution* systems. A convolution system is BIBO stable if every bounded input results in a bounded output. In the present section we extend the notion of BIBO stability first to IOM systems, and then to IO systems, which do not have a unique output for each input. We also introduce a stronger form of stability, namely, *CICO* (converging-input converging-output) stability. An IO or IOM system is CICO stable if first of all it is BIBO stable, and furthermore its responses to two inputs that converge to each other also converge to each other.

For linear time-invariant difference and differential systems simple conditions for the various types of stability may be obtained in terms of their characteristic roots and *poles*. First, the latter notion is defined.

4.6.1.   Definition: Poles and zeros.   Consider the constant coefficient difference system

$$Q(\sigma)y = P(\sigma)u$$

or the constant coefficient differential system

$$Q(D)y = P(D)u.$$

The roots of the polynomial Q are the *characteristic roots* of the system. Let P' and Q' be the polynomials that are obtained by canceling all common polynomial factors of P and Q. Then the roots of Q' are called the *poles* of the system, and those of Q' the *zeros* of the system.   ∎

We note that the rational function H defined by

$$H = \frac{P}{Q}$$

is *infinite* at the poles of the system, while at the zeros of the system it is *zero*. In Chapter 8 the function H is re-encountered as the *transfer function* of the system.

4.6.2.   Example: Poles and zeros.   Consider the difference or differential system characterized by the polynomials

$$Q(\lambda) = \lambda^2 + \lambda - 2 = (\lambda - 1)(\lambda + 2),$$
$$P(\lambda) = \lambda^2 + 2\lambda - 3 = (\lambda - 1)(\lambda + 3).$$

From Q we see that the system has the characteristic roots 1 and -2. After cancellation of the common factor $\lambda - 1$ we obtain the polynomials

$$Q'(\lambda) = \lambda + 2, \qquad P'(\lambda) = \lambda + 3.$$

As a result, the system has a single pole -2 and a single zero -3.   ∎

BIBO Stability of Initially-At-Rest Linear Time-Invariant Difference and Differential Systems

Initially-at-rest difference and differential systems are convolution systems defined on the infinite time axis $\mathbb{Z}$ or $\mathbb{R}$, respectively. BIBO stability of convolution systems was defined in 3.6.1. In 3.6.2 we saw that a convolution system is BIBO stable if and only if its impulse response h has finite action $\|h\|_1$. For initially-at-rest difference and differential systems we may formulate necessary and sufficient conditions for BIBO stability in terms of their poles.

4.6.3.   Summary: BIBO stability of initially-at-rest linear time-invariant difference and differential systems.

The initially-at-rest difference system characterized by the constant coefficient difference equation

$$Q(\sigma)y = P(\sigma)u,$$

is BIBO stable if and only if all its poles have magnitude strictly less than 1.

The initially-at-rest differential system characterized by the constant coefficient differential equation

$$Q(D)y = P(D)u,$$

is BIBO stable if and only if
    (i) the degree of P is less than or equal to that of Q, and
    (ii) all the poles of the system have strictly negative real part.   ∎

The proof follows by inspection of the impulse response h of the initially-at-rest system, which by 4.5.1 depends on the basis solutions of the homogeneous equation. If some of the poles have magnitude greater than or equal to 1 (in the discrete-time case) or real part greater than or equal to zero (in the continuous-time case) the corresponding basis solutions do not converge to zero. As a result the impulse response has infinite action $\|h\|_1$, and by 3.4.15 the system is BIBO unstable. Characteristic roots that are *not* poles, that is, characteristic roots that cancel against roots of P, do *not* appear in the impulse response, and therefore do not affect the BIBO stability of the system. The details are given in E.1 in Supplement E. In this supplement a number of proofs have been collected. If in the continuous-time case the degree of P exceeds that of Q, the impulse response contains derivative δ-functions, which make the action infinite and hence render the convolution system BIBO unstable.

4.6.4.   Example: BIBO stability of an initially-at-rest differential system.
From Example 4.6.2 it follows that the differential system

$$y^{(2)}(t) + y^{(1)}(t) - 2y(t) = u^{(2)}(t) + 2u^{(1)}(t) - 3u(t), \qquad t \in \mathbb{R},$$

has the two characteristic roots 1 and -2 but a single pole -2. By 4.6.3 the corresponding initially-at-rest system is BIBO stable even though one characteristic root is positive. Indeed, it may be found that the impulse response of the system is

$$h(t) = \delta(t) + e^{-2t}\mathbb{1}(t), \qquad t \in \mathbb{R}, \tag{1}$$

which does not contain a term e^t corresponding to the characteristic root 1, and has finite action. *Exercise:* Prove that the system has the impulse response (1).   ∎

Boundedness and Convergence of Zero Input Responses

Before plunging into a detailed exposition of the stability of input-output systems we consider the boundedness and convergence of their *zero-input response,* that is, the solution of the homogeneous difference or differential equation $Q(\sigma)y = 0$ or $Q(D)y = 0$, respectively. The results apply both to semi-infinite and infinite time axes.

4.6.5.   Summary: Boundedness and convergence of homogeneous solutions.

(a) Necessary and sufficient conditions for any solution of the homogeneous difference equation

$$Q(\sigma)y = 0$$

to remain bounded from any finite time on are that
  (i) all characteristic roots have magnitude less than or equal to 1 and
  (ii) any characteristic root with magnitude equal to 1 has multiplicity one.
(b) Necessary and sufficient conditions for any solution of the homogeneous difference equation

(a′) Necessary and sufficient conditions for any solution of the homogeneous differential equation

$$Q(D)y = 0$$

to remain bounded from any finite time on are that
  (i) all characteristic roots have nonpositive real part and
  (ii) any characteristic root with zero real part has multiplicity one.
(b′) Necessary and sufficient conditions for any solution of the homogeneous differential equation

$$Q(\sigma)y = 0 \qquad\qquad\qquad Q(D)y = 0$$

to converge to zero as time increases to infinity are that all characteristic roots have magnitude strictly less than 1.

to converge to zero as time increases to infinity are that all characteristic roots have strictly negative real part.   ∎

The proof follows by the fact that any solution of the homogeneous equation is a linear combination of basis solutions. By 4.4.1 we may select basis solutions that in the discrete-time case are of the form

$$n^k \lambda^n = n^k \rho^n [\cos(\psi n) + j \sin(\psi n)], \qquad n \in \mathbb{Z},$$

where λ is a characteristic root, $\rho = |\lambda|$ and $\psi = \arg(\lambda)$. In the continuous-time case we choose the basis solutions

$$t^k e^{\lambda t} = t^k e^{\sigma t} [\cos(\omega t) + j \sin(\omega t)], \qquad t \in \mathbb{R},$$

where again λ is a characteristic root, $\sigma = \operatorname{Re}(\lambda)$ and $\omega = \operatorname{Im}(\lambda)$. The basis solutions, and hence any linear combination of the basis solutions, are bounded or converge to zero as time increases under the conditions stated in 4.6.5.

4.6.6.   Example: Convergence of the zero-input response of a differential system.   The system of Example 4.6.4, which is described by the differential equation

$$y^{(2)}(t) + y^{(1)}(t) - 2y(t) = u^{(2)}(t) + 2u^{(1)}(t) - 3u(t), \qquad t \in \mathbb{R},$$

has the characteristic roots 1 and -2. Because the first of these has positive real part, by 4.6.5 the zero-input response generally does not converge to zero. Indeed, the zero-input response is any linear combination of the exponentials e^t and e^{-2t}, $t \in \mathbb{R}$, the first of which goes to infinity as time increases.   ∎

We are now in a position to discuss the stability of *IO* systems, beginning with their BIBO stability and proceeding to CICO stability.

BIBO Stability of IO Systems

BIBO stability of convolution systems is the property that bounded inputs result in bounded outputs. For general input-output systems the definition need be slightly modified, as an example will illustrate. We first give the definition.

4.6.7.   Definition: BIBO stability of IO systems.   An input-output system with scalar input and output defined on the infinite or right semi-infinite time axis $\mathbb{T}$ is *BIBO stable* if for any input-output pair (u, y)

$$\|u\|_\infty < \infty \qquad \text{implies} \qquad \|y_\theta\|_\infty < \infty$$

for every $\theta \in \mathbb{T}$, where

$$y_\theta(t) = \begin{cases} y(t) & \text{for } t \geq \theta, \\ 0 & \text{otherwise}, \end{cases} \qquad t \in \mathbb{T}. \qquad \blacksquare$$

Figure 4.8 illustrates the signal y_θ. The definition implies that an IO system is BIBO stable if any bounded input always produces an output that is bounded *from any finite time on*.

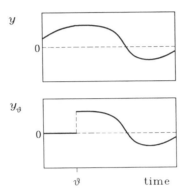

Figure 4.8. The signals y and y_θ.

First we show why the definition of BIBO stability must be modified for IO systems.

4.6.8.   Example: BIBO stability of the exponential smoother as an IO system.   The exponential smoother is described by the difference equation

$$y(n + 1) = ay(n) + (1 - a)u(n + 1), \qquad n \in \mathbb{Z}.$$

The impulse response h of the smoother is given by

$$h(n) = (1 - a)a^n \mathbb{1}(n), \qquad n \in \mathbb{Z}.$$

Suppose that $|a| < 1$; then h has finite action and the convolution $h * u$ exists for any input u with finite amplitude. By 4.5.3(a) any output y corresponding to a bounded input u may be written as

$$y(n) = (h * u)(n) + \alpha a^n, \qquad n \in \mathbb{Z},$$

with α an arbitrary constant. Because h has finite action and u is bounded, the first term on the right-hand side is bounded. The second term is always bounded from any finite time on, even though it increases to infinity in the negative time direction and hence is not bounded. By Definition 4.6.7 this system is still called BIBO stable.

$\blacksquare$

Because IOM systems are IO systems, Definition 4.6.7 also applies to IOM systems and, in particular, also to convolution systems. It is easily seen that BIBO stability of convolution systems as defined in 3.6.1 implies BIBO stability in the sense of 4.6.7. By a slight modification of the proof of 3.6.2, it may be shown that convolution systems are BIBO stable in the sense of 4.6.7 if and only if their impulse response h has finite action $\|h\|_1$. Hence, BIBO stability of convolution systems in the sense of 3.6.1 is equivalent to BIBO stability in the sense of 4.6.7.

4.6.9.   Summary: BIBO stability of convolution systems.   A discrete- or continuous-time convolution system with impulse response h is BIBO stable (in the sense of 4.6.7) if and only if h has finite action, that is, $\|h\|_1 < \infty$.   ∎

We next consider the BIBO stability of IO systems described by constant coefficient difference and difference equations. Obviously, a necessary condition for the BIBO stability of such systems is that their impulse response has bounded action; otherwise there exist bounded inputs that result in unbounded initially-at-rest responses. From 4.6.3 we know that a necessary and sufficient condition for the impulse response to have bounded action in the discrete-time case is that all the poles of the system have magnitude strictly less than one. In the continuous-time case the degree of Q should not be less than that of P, and the poles should have strictly negative real part.

Suppose that these conditions are satisfied, so that the impulse response h of the system has finite action. Then by 4.5.3 the response to a bounded input u may be written as

$$y = h * u + \sum_i \alpha_i y_i,$$

where the y_i are basis solutions and the α_i arbitrary constants. Because by assumption h has finite action and u is bounded, the first term is bounded. Necessary and sufficient conditions for the second term to be bounded are given in 4.6.5. Since by assumption the poles of the system have magnitude strictly less than one (in the discrete-time case) or strictly negative real part (in the continuous-time case), for the boundedness of the second term we need only look at those characteristic roots that are *not* poles (i.e., that cancel against roots of P). Combining the results we conclude what follows.

4.6.10.   Summary: BIBO stability of constant coefficient linear difference and differential systems.

Necessary and sufficient conditions for the IO system described by the constant coefficient linear difference equation

$$Q(\sigma)y = P(\sigma)u$$

to be BIBO stable are the following:

Necessary and sufficient conditions for the IO system described by the constant coefficient linear differential equation

$$Q(D)y = P(D)u$$

to be BIBO stable are the following:

(i) all the poles of the system have magnitude strictly less than 1,

(ii) the system has no canceled characteristic roots with magnitude strictly greater than 1, and

(iii) any canceled characteristic root with magnitude equal to 1 has multiplicity one.

(i) the degree of P is less than or equal to that of Q,

(ii) all the poles of the system have strictly negative real part,

(iii) the system has no canceled characteristic roots with strictly positive real part, and

(iv) any canceled characteristic root with zero real part has multiplicity one. ∎

This result applies to systems defined on both semi-infinite and infinite time axes.

4.6.11.   Example: BIBO unstable differential system.   The initially-at-rest differential system described by the differential equation

$$y^{(2)}(t) + y^{(1)}(t) - 2y(t) = u^{(2)}(t) + 2u^{(1)}(t) - 3u(t), \qquad t \in \mathbb{R},$$

was in Example 4.6.4 found to be BIBO stable. Because the impulse response h of the system has finite action, by 4.5.3 for bounded inputs the general solution of the differential equation is given by

$$y(t) = (h * u)(t) + \alpha_1 e^t + \alpha_2 e^{-2t}, \qquad t \in \mathbb{R},$$

with h the impulse response (1), and α_1 and α_2 arbitrary constants. The first term on the right-hand side is bounded if the input u is bounded, but the second term generally is unbounded in the positive time direction. Hence, the system is not BIBO stable. Indeed, the system has the canceled characteristic root 1, which violates the conditions of 4.6.10. ∎

CICO Stability

A form of stability that is stronger than BIBO stability is *converging-input converging-output* (CICO) stability. The idea of CICO stability may be explained as follows. Suppose that an IO or IOM system is subjected to a bounded input u that approaches zero as time increases, that is, $u(t) \to 0$ as $t \to \infty$. Then, if the system is BIBO stable, all we know is that the corresponding output is bounded. It is easy to think of systems, however, that have the additional property that if the input approaches zero, any corresponding output y also approaches zero, (i.e., if the input eventually comes to rest, so does the output). Such systems are called CICO stable.

The formal definition of CICO stability is slightly more involved.

4.6.12.   Definition: CICO stability.   Let (u_1, y_1) and (u_2, y_2) be two input-output pairs of an IO system, which is defined on the infinite or right semi-infinite time axis $\mathbb{T}$, such that $\| u_1 - u_2 \|_\infty < \infty$. Then the system is *converging-input converging-output* (CICO) *stable* if

(i) $\| (y_1 - y_2)_\theta \|_\infty < \infty$ for every $\theta \in \mathbb{T}$,

(ii) $|u_1(t) - u_2(t)| \to 0$ as $t \to \infty$ implies $|y_1(t) - y_2(t)| \to 0$ as $t \to \infty$. ∎

The notation $(y_1 - y_2)_\theta$ is as in 4.6.7. Fig. 4.9 illustrates CICO stability. Note that (i) implies that any CICO system is also BIBO stable (by setting $u_2 = 0$.) The converse, namely, that BIBO stable IO systems are also CICO stable, is not true in general. It does hold for BIBO stable *convolution* systems, however.

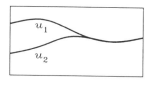

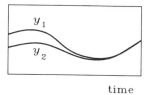

time

Figure 4.9. CICO stability. Top: two inputs approach each other. Bottom: the two corresponding outputs also approach each other.

4.6.13.   Summary: CICO stability of convolution systems.   A discrete- or continuous-time convolution system is CICO stable if and only if it is BIBO stable. ∎

The proof is presented in E.2 in Supplement E.

We consider the CICO stability of constant coefficient difference and differential systems. Let (u_1, y_1) and (u_2, y_2) be two input-output pairs such that $u_1 - u_2$ is bounded. If the impulse response h of the system has finite action, by linearity,

$$y_1 - y_2 = h * (u_1 - u_2) + \sum_i \alpha_i y_i,$$

where the y_i are basis solutions, and the α_i constants. By 4.6.13, the first term on the right-hand side converges to zero as time increases if $u_1 - u_2$ converges to zero. By 4.6.5, the second term converges to zero as time increases if and only if all characteristic roots of the system have magnitude strictly less than 1 (in the discrete-time case) or strictly negative real part (in the continuous-time case.) These conditions are also sufficient for the impulse response to have finite action (plus, in the continuous-time case, the requirement that the impulse response contain no derivatives of delta functions) and, hence, imply CICO stability.

The conclusions apply to systems defined both on semi-infinite and infinite time axes, and may be summarized as follows.

4.6.14.   Summary: CICO stability of constant coefficient difference and differential systems.

Necessary and sufficient conditions for the IO system described by the constant coefficient difference equation

$$Q(\sigma)y = P(\sigma)u$$

to be CICO stable are that all the characteristic roots of the system have magnitude strictly less than 1.

Necessary and sufficient conditions for the IO system described by the constant coefficient differential equation

$$Q(D)y = P(D)u$$

to be CICO stable are that
    (i) the degree of Q is greater than or equal to that of P and
    (ii) all the characteristic roots of the system have strictly negative real part.   ■

Note that the conditions for CICO stability of constant coefficient difference and differential systems are the same as the conditions of 4.6.5 for the convergence of the zero-input response.

CICO stability is the strongest form of stability of constant coefficient linear difference and differential systems. It implies, but is not implied by, BIBO stability.

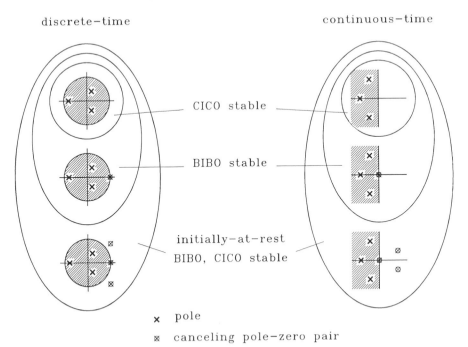

discrete-time   continuous-time

CICO stable

BIBO stable

initially-at-rest
BIBO, CICO stable

×   pole

⊠   canceling pole-zero pair

Figure 4.10.   Interrelation of various forms of stability of constant coefficient linear difference and differential systems, and examples of pole-zero patterns.

BIBO stability, in turn, implies but is not implied by BIBO and CICO stability of the *initially-at-rest* system. Figure 4.10 illustrates that the set of all initially-at-rest BIBO and CICO stable constant coefficient linear difference and differential systems contains the set of all BIBO stable systems, which in turn contains the set of all CICO stable systems. The figure also shows examples of pole-zero patterns for each form of stability.

4.6.15.   Example: CICO stability.

(a) *Exponential smoother*. From the difference equation

$$y(n + 1) - ay(n) = (1-a)u(n + 1), \qquad n \in \mathbb{Z},$$

that describes the exponential smoother we see that the system has a single characteristic root a. Hence, the smoother is CICO stable if and only if $|a| < 1$.

(b) *Differential system*. In Example 4.6.4 we saw that the differential system

$$y^{(2)}(t) + y^{(1)}(t) - 2y(t) = u^{(2)}(t) + 2u^{(1)}(t) - 3u(t), \qquad t \in \mathbb{R},$$

has the two characteristic roots 1 and -2. Hence, by 4.6.14 the system is not CICO stable. We note that according to 4.6.4 the *initially-at-rest* differential system is BIBO and hence also CICO stable. The *IO* system, however, is neither BIBO nor CICO stable. The reason is that because of the characteristic root 1 the response of the IO system generally has an unstable component of the form e^t, $t \in \mathbb{R}$.   ∎

We conclude, as usual, with a review of the results of this section applied to sampled difference systems.

4.6.16.   Review: Stability of sampled constant coefficient difference systems.
Sampled constant coefficient difference systems are discrete-time systems on the time axis $\mathbb{Z}(T)$ described by the difference equation

$$Q(\sigma^T)y = P(\sigma^T)u, \tag{2}$$

with Q and P polynomials. The *characteristic roots* of the system are the roots of Q. The *poles* of the system are those characteristic roots that do not cancel against roots of P.

The initially-at-rest system defined by the difference equation (2) is BIBO stable if and only if all the poles of the system have magnitude strictly less than 1.

BIBO and CICO stability of sampled discrete-time systems are defined as in 4.6.7 and 4.6.12, respectively. The conditions for BIBO and CICO stability of sampled constant coefficient difference IO systems are identical to those given in 4.6.10 and 4.6.14 for difference systems defined on the time axis $\mathbb{Z}$.   ∎

4.7 FREQUENCY RESPONSE OF DIFFERENCE AND DIFFERENTIAL SYSTEMS

This section is devoted to an investigation of the response of linear constant coefficient difference and differential systems to *harmonic* inputs. As before, the initially-at-rest system plays a central role, and is analyzed first.

The initially-at-rest system is a convolution system. As shown in Section 3.7, the response of a convolution system to harmonic inputs is determined by its *frequency response function*. It is seen in the present section that the frequency response function of difference and differential systems may directly be obtained from the polynomials Q and P that define the difference or differential equation.

Once the frequency response function is known it is not difficult to find the general response of difference and differential IO systems to harmonic inputs.

Frequency Response Function of Difference and Differential Systems

In Section 3.7 we found that the response of a convolution system to a harmonic input, if it exists, is again harmonic. If the system has the impulse response h with finite action $\|h\|_1$, then its frequency response function $\hat{h}$ is given by

$$\hat{h}(f) = \sum_{n=-\infty}^{\infty} h(n)e^{-j2\pi fn}, \qquad f \in \mathbb{R},$$

in the discrete-time case, and

$$\hat{h}(f) = \int_{-\infty}^{\infty} h(t)e^{-j2\pi ft}\, dt, \qquad f \in \mathbb{R},$$

in the continuous-time case. Given the frequency response function, we have the respective harmonic input-output pairs

discrete-time: $\quad u(n) = e^{j2\pi fn}, \quad y(n) = \hat{h}(f)e^{j2\pi fn}, \qquad n \in \mathbb{Z},$

continuous-time: $\quad u(t) = e^{j2\pi ft}, \quad y(t) = \hat{h}(f)e^{j2\pi ft}, \qquad t \in \mathbb{R}.$

The frequency response function of initially-at-rest difference and differential systems may immediately be determined from the difference or differential equation, without first computing the impulse response.

4.7.1.   Summary: Frequency response function of initially-at-rest difference and differential systems.

The frequency response function $\hat{h}$ of the initially-at-rest difference system with difference equation

$$Q(\sigma)y = P(\sigma)u$$

exists if and only if all the poles of the system have magnitude strictly less than one and is given by

$$\hat{h}(f) = \frac{P(e^{j2\pi f})}{Q(e^{j2\pi f})}, \qquad f \in \mathbb{R}.$$

The frequency response function $\hat{h}$ of the initially-at-rest differential system with differential equation

$$Q(D)y = P(D)u \ ,$$

exists if and only if all the poles of the system have strictly negative real part, and is given by

$$\hat{h}(f) = \frac{P(j2\pi f)}{Q(j2\pi f)}, \qquad f \in \mathbb{R}. \qquad ■$$

4.7.2   Proof.   The proof is given for the continuous-time case and consists of two steps. First we establish the existence of the frequency response function, and then we derive the function itself.

(i) *Existence.* From 4.5.1 the impulse response h of the initially-at-rest system is the sum of a δ-function and some derivative δ-functions and a linear combination of basis solutions multiplied by the unit step. The δ-functions do not affect the existence of the frequency response function. The basis solutions that are involved are of the form $t^k e^{\lambda t}$, $t \in \mathbb{R}$, where λ is a pole of the system. Characteristic roots λ that are not poles do not appear in h. If the poles all have strictly negative real parts, each of the basis solutions decays exponentially. As a result, the nonsingular part of the impulse response has finite action so that by 3.7.2 the frequency response function of the system exists. On the other hand, if one or several of the poles have zero or positive real parts the impulse response does not converge to zero, the integral that defines the frequency response function diverges, and the frequency response function does not exist, at least not in the regular sense.

(ii) *Derivation of the frequency response function.* From Section 3.7 we know that the response of a continuous-time convolution system to the harmonic input $u(t) = e^{j2\pi ft}$, $t \in \mathbb{R}$, if it exists, is again harmonic and of the form $y(t) = y_o e^{j2\pi ft}$, with $y_o = \hat{h}(f)$. The existence of y is guaranteed if all the poles of the system have strictly negative real parts, as established in (i). All we have to do to find y_o is to substitute this solution into the differential equation. Since $n$fold differentiation of harmonic signals amounts to multiplication by $(j2\pi f)^n$, it follows after substitution of u and y into the differential equation $Q(D)y = P(D)u$ that

$$Q(j2\pi f)y_o e^{j2\pi ft} = P(j2\pi f)e^{j2\pi ft}, \qquad t \in \mathbb{R}.$$

Solution for y_o yields $y_o = \hat{h}(f) = P(j2\pi f)/Q(j2\pi f)$.

The proof for the discrete-time case follows the same lines, with the difference that the factor $j2\pi f$ in (ii) is replaced with $e^{j2\pi f}$.   ■

4.7.3.   Example: Frequency response function of difference and differential systems.

(a) *Exponential smoother*. The exponential smoother is described by the difference equation

$$y(n+1) - ay(n) = (1-a)u(n+1), \qquad n \in \mathbb{Z},$$

so that the polynomials Q and P are given by

$$Q(\lambda) = \lambda - a, \qquad P(\lambda) = (1-a)\lambda.$$

If $a \neq 1$, the smoother has the single pole a. By 4.7.1, the frequency response function of the smoother exists if $|a| < 1$, and then is given by

$$\hat{h}(f) = \frac{P(e^{j2\pi f})}{Q(e^{j2\pi f})} = \frac{(1-a)e^{j2\pi f}}{e^{j2\pi f} - a}$$

$$= \frac{1-a}{1 - a\,e^{-j2\pi f}}, \qquad f \in \mathbb{R}.$$

This is the same as the result we found in Example 3.7.5(a) with considerably more effort after first determining the impulse response h of the system. *Exercise:* Show that if $a = 1$ the frequency response function is $\hat{h} = 0$.

(b) *RCL network*. From Example 4.2.4(c) the differential equation that describes the RCL network with the voltage across the inductor as output is

$$\frac{d^2y(t)}{dt^2} + \frac{R}{L}\frac{dy(t)}{dt} + \frac{1}{LC}y(t) = \frac{d^2u(t)}{dt^2}, \qquad t \in \mathbb{R}.$$

The polynomials Q and P for this differential equation are given by

$$Q(\lambda) = \lambda^2 + \frac{R}{L}\lambda + \frac{1}{LC},$$

$$P(\lambda) = \lambda^2.$$

The poles of the system are the characteristic roots

$$\lambda_{1,2} = \frac{-R \pm \sqrt{R^2 - 4L/C}}{2L}.$$

If $R^2 \geq 4L/C$, both poles are real and negative (assuming that R, C and L are all positive.) If $R^2 < 4L/C$, the poles form a conjugate pair with real part $-R/2L$. In either case, the poles have strictly negative real parts, so that the frequency response function $\hat{h}$ exists. It is given by

$$\hat{h}(f) = \frac{P(j2\pi f)}{Q(j2\pi f)} = \frac{-4\pi^2 f^2}{\left(\dfrac{1}{LC} - 4\pi^2 f^2\right) + \dfrac{R}{L} j2\pi f}, \qquad f \in \mathbb{R}.$$

In terms of the resonance frequency $\omega_r = 1/\sqrt{LC}$ and the quality factor $q = \omega_r L/R$ of the network this may be rewritten as

$$\hat{h}(\omega/2\pi) = \frac{-\omega^2/\omega_r^2}{(1 - \omega^2/\omega_r^2) + j\omega/\omega_r q}, \qquad \omega \in \mathbb{R},$$

where for convenience we use the angular frequency $\omega = 2\pi f$ rather than the frequency f. The magnitude and phase of $\hat{h}$ determine the response of the initially-at-rest RCL network to real harmonics, and are given by

$$|\hat{h}(\omega/2\pi)|^2 = \frac{\omega^4/\omega_1^4}{(1 - \omega^2/\omega_r^2)^2 + \omega^2/\omega_r^2 q^2},$$

$$\arg\left(\hat{h}(\omega/2\pi)\right) = \begin{cases} -\mathrm{atan}\left(\dfrac{\omega/\omega_r q}{1 - \omega^2/\omega_r^2}\right) + \pi & \text{for } 0 < \omega/\omega_r \leq 1, \\[3mm] -\mathrm{atan}\left(\dfrac{\omega/\omega_r q}{1 - \omega^2/\omega_r^2}\right) & \text{for } \omega/\omega_r > 1, \end{cases} \qquad \omega \in \mathbb{R}_+.$$

Plots of the magnitude and phase are given in Fig. 4.11 both for $q > 1/2$ and $q < 1/2$ on the normalized angular frequency scale ω/ω_r, $\omega \geq 0$. For $q > 1/2$ the characteristic roots of the network form a complex conjugate pair and as a result the impulse response has a damped oscillatory behavior. This is reflected by the presence of a peak in the magnitude of the frequency response function near the resonance frequency ω_r.

4.7.4.   Remark: Poles on the unit circle or imaginary axis.   If a constant coefficient linear difference system $Q(\sigma)y = P(\sigma)u$ has all its poles strictly inside the unit circle, the poles all have magnitude strictly less than one, so that by 4.7.1 the frequency response function of the system exists. If besides poles inside the unit circle the system also has poles *on* the unit circle, the frequency response function $\hat{h}$ of the system only exists in the *generalized* sense. It may be found by application of the generalized DCFT (see Section 7.4) to the impulse response of the system.

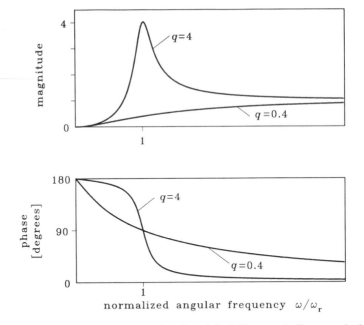

Figure 4.11. Frequency response function of the RCL network. Top: magnitude. Bottom: phase.

Likewise, a constant coefficient linear differential system has a frequency response function in the generalized sense if besides poles to the left of the imaginary axis it also has poles *on* the imaginary axis. The frequency response function may then be found by application of the generalized CCFT (see Section 7.5) to the impulse response. An *integrator*, for instance, is the differential system

$$y^{(1)}(t) = u(t), \qquad t \in \mathbb{R}.$$

Because $P(\lambda) = 1$ and $Q(\lambda) = \lambda$, the system has a single pole at 0 and, hence, by 4.7.1 the system does not have a frequency response function. Indeed, the corresponding initially-at-rest system has the IO map

$$y(t) = \int_{-\infty}^{t} u(\tau) \, d\tau, \qquad t \in \mathbb{R},$$

so that the impulse response of the system is

$$h(t) = \int_{-\infty}^{t} \delta(\tau) \, d\tau = \mathbb{1}(t), \qquad t \in \mathbb{R}.$$

The impulse response has infinite action, and hence the frequency response function does not exist in the ordinary sense.

As found in 7.5.5, the system has the *generalized* frequency response function

$$\hat{h}(f) = \frac{1}{j2\pi f} + \tfrac{1}{2}\delta(f), \qquad f \in \mathbb{R}.$$

The first term on the right-hand side is the frequency response function that one would expect from 4.7.1, but the frequency response function has an additional singularity at frequency zero (corresponding to the pole at 0).   ∎

Steady-State and Transient Response to Harmonic Inputs

We continue with a discussion of the response of difference and differential *IO* systems to harmonic inputs. It is easy to establish the following result.

4.7.5.   Summary: Response of difference and differential systems to harmonic inputs.   Any output of a CICO stable constant coefficient difference or differential system with frequency response function $\hat{h}$ corresponding to

(a) the complex harmonic input

$$u(t) = u_o e^{j2\pi ft}, \qquad t \in \mathbb{T},$$

or

(b) the real harmonic input

$$u(t) = u_o \cos(2\pi ft + \phi), \qquad t \in \mathbb{T},$$

with the time axis $\mathbb{T}$ infinite or right semi-infinite, is of the form

$$y = y_{\text{steady-state}} + y_{\text{transient}}.$$

The steady-state response $y_{\text{steady-state}}$ is given by

(a)

$$y_{\text{steady-state}}(t) = \hat{h}(f)u_o e^{j2\pi ft}, \qquad t \in \mathbb{T},$$

or

(b)

$$y_{\text{steady-state}}(t) = |\hat{h}(f)|u_o \cos(2\pi ft + \phi + \psi(f)), \qquad t \in \mathbb{T},$$

respectively, with $\psi(f) = \arg(\hat{h}(f))$. The transient response $y_{\text{transient}}$ is a solution of the homogeneous equation and satisfies

$$y_{\text{transient}}(t) \to 0 \qquad \text{as} \quad t \to \infty. \qquad\qquad ■$$

The proof for case (a) follows by observing that $\hat{h}(f)u_o e^{j2\pi ft}$, $t \in \mathbb{T}$, is a particular solution of the difference or differential equation for the given harmonic input. The general solution thus is given by

$$y(t) = \hat{h}(f)u_o e^{j2\pi ft} + y_{\text{hom}}(t), \qquad t \in \mathbb{T},$$

where y_{hom} is a solution of the homogeneous equation whose coefficients are determined by the initial conditions. Since by assumption the system is CICO stable, any solution of the homogeneous equation tends to zero. The proof for the real-valued harmonic input (b) is similar.

4.7.6.   Example: Steady-state and transient response of the RC network. The RC network is described by the differential equation

$$\frac{dy(t)}{dt} + \frac{1}{RC}y(t) = \frac{1}{RC}u(t), \qquad t \in \mathbb{R}.$$

Because the system has the single pole $-1/RC$ the frequency response function of the system is well-defined and given by

$$\hat{h}(f) = \frac{1/RC}{j2\pi f + 1/RC} = \frac{1}{j2\pi fRC + 1}, \qquad f \in \mathbb{R}.$$

Suppose that the input is the real harmonic signal

$$u(t) = u_o \cos(2\pi f_o t), \qquad t \geq 0,$$

with f_o a fixed positive frequency. The general solution of the differential equation corresponding to this input is

$$y(t) = |\hat{h}(f_o)|u_o \cos(2\pi f_o t + \psi_o) + \alpha e^{-t/RC}, \qquad t \geq 0, \qquad (1)$$

where

$$|\hat{h}(f_o)| = \frac{1}{\sqrt{4\pi^2 f_o^2 R^2 C^2 + 1}},$$

$$\psi_o = \arg(\hat{h}(f_o)) = -\text{atan}(2\pi f_o RC).$$

The arbitrary constant α in (1) may be found from the initial condition $y(0) = y_o$. It follows that

$$y(t) = |\hat{h}(f_o)|u_o \cos(2\pi f_o t + \psi_o) + [y_o - |h(f_o)|u_o \cos(\psi_o)]e^{-t/RC}, \qquad t \geq 0.$$

The first term is the steady-state response to the real harmonic input, the second the transient response. The responses are shown in Fig. 4.12 for $y_o = 0$.

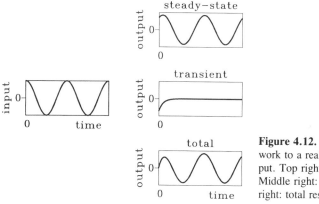

Figure 4.12. Response of the RC network to a real harmonic input. Left: input. Top right: steady-state response. Middle right: transient response. Bottom right: total response. ∎

We review the results of this section for sampled linear time-invariant difference systems.

4.7.7.   Review: Frequency response function of sampled constant coefficient difference systems.

The frequency response function of sampled systems defined on the time axis $\mathbb{Z}(T)$ described by a constant coefficient linear difference equation of the form

$$Q(\sigma^T)y = P(\sigma^T)u$$

exists if and only if the magnitudes of all the poles of the system are strictly less than 1. Its form may easily be established by determining a particular solution of the form $y(t) = y_o e^{j2\pi ft}$, $t \in \mathbb{Z}(T)$, given the harmonic input $u(t) = e^{j2\pi ft}$, $t \in \mathbb{Z}(T)$. It follows that the frequency response function is given by

$$\hat{h}(f) = \frac{P(e^{j2\pi fT})}{Q(e^{j2\pi fT})}, \qquad f \in \mathbb{R}. \qquad ∎$$

Frequency Response Functions of Electrical Networks

We conclude this section by describing a simple and elegant way of finding frequency response functions of *electrical networks*. Consider an interconnection of basic electrical network elements such as resistors, capacitors and inductors (see Fig. 4.13.) We assume that the input u to the system originates from a single voltage or current source, and that the output y is a single voltage or current somewhere in the network (see for an example 4.7.8.)

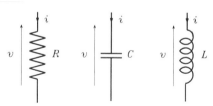

Figure 4.13. Three elementary electrical network elements. Left: resistor. Middle: capacitor. Right: inductor.

From the theory of this section it follows that we can determine the frequency-response function $\hat{h}$ of the network if corresponding to the harmonic input $u(t) = u_o e^{j2\pi ft}$, $t \in \mathbb{R}$, we can find for the output a particular solution $y(t) = y_o e^{j2\pi ft}$, $t \in \mathbb{R}$, with y_o a suitable scalar constant; then $y_o = \hat{h}(f)u_o$.

We exploit the fact that corresponding to the input $u(t) = u_o e^{j2\pi ft}$, $t \in \mathbb{R}$, *every* current and voltage within the network has a particular solution of the form $c\, e^{j2\pi ft}$, $t \in \mathbb{R}$, with c a suitable constant, and first consider each network element individually.

Resistor. The current i through and the voltage v across a resistor with resistance R are related by Ohm's law as $v(t) = Ri(t)$. Suppose that both the current and the voltage are harmonic signals of the form $i(t) = \hat{i}\, e^{j2\pi ft}$, $v(t) = \hat{v}\, e^{j2\pi ft}$, $t \in \mathbb{R}$, with $\hat{i}$ and $\hat{v}$ constants to be determined. By substitution it follows that $\hat{i}$ and $\hat{v}$ are related as $\hat{v} = R\hat{i}$, or

$$Z_R(f) = \frac{\hat{v}}{\hat{i}} = R.$$

The ratio $Z_R(f) := \hat{v}/\hat{i}$ is called the *impedance* of the network element. As the notation indicates, the impedance generally is frequency dependent. The impedance Z_R of the resistor is precisely its resistance R and, hence, is constant.

George Simon Ohm (1787–1854) was a German physicist.

Capacitor. The voltage v across a capacitor and the current i flowing into it are related as $i(t) = C\, dv(t)/dt$, with C its capacitance. Substituting $i(t) = \hat{i}\, e^{j2\pi ft}$ and $v(t) = \hat{v}\, e^{j2\pi ft}$, $t \in \mathbb{R}$, it follows that $\hat{i} = C\, j2\pi f\hat{v}$, so that the impedance of the capacitor is

$$Z_C(f) = \frac{\hat{v}}{\hat{i}} = \frac{1}{j2\pi fC}, \qquad f \neq 0.$$

In terms of the *angular frequency* $\omega = 2\pi f$ the impedance of the capacitor is $Z_C(\omega/2\pi) = 1/j\omega C$.

Inductor.   The voltage v across an inductor and the current i through it, finally, are related by $v(t) = L\, di(t)/dt$, with L its inductance. It easily follows with $v(t) = \hat{v}\, e^{j2\pi ft}$ and $i(t) = \hat{i}\, e^{j2\pi ft}$, $t \in \mathbb{R}$, that $\hat{v} = L\, j2\pi f\hat{i}$, and, hence, the impedance of the inductor is

$$Z_L(f) = j2\pi fL.$$

In terms of the angular frequency it follows that $Z_L(\omega/2\pi) = j\omega L$.

For elementary interconnections of network elements it is very simple to find their *replacement impedances,* that is, the ratio $Z(f) = \hat{v}/\hat{i}$ when the voltage across and the current through the interconnection are $\hat{v}\, e^{j2\pi ft}$ and $\hat{i}\, e^{j2\pi ft}$, $t \in \mathbb{R}$, respectively. The results are completely analogous to those obtained when analyzing DC networks that only contain resistors.

Consider for instance the series connection of Fig. 4.14 (top), where the individual network elements have impedances Z_1 and Z_2, respectively. It follows that $\hat{v} = \hat{v}_1 + \hat{v}_2 = Z_1\hat{i} + Z_2\hat{i} = (Z_1 + Z_2)\hat{i}$, so that the replacement impedance of the series connection is $Z = Z_1 + Z_2$. For convenience we omit the argument f of the impedances.

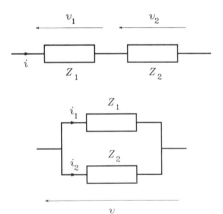

Figure 4.14.   Connection of two electrical network elements. Top: series. Bottom: parallel.

For the parallel connection of Fig. 4.14 (bottom), on the other hand, we have $\hat{i} = \hat{i}_1 + \hat{i}_2 = \hat{v}/Z_1 + \hat{v}/Z_2 = (1/Z_1 + 1/Z_2)\hat{v}$. Consequently, the replacement impedance Z of the parallel connection follows from $1/Z = 1/Z_1 + 1/Z_2$.

In this way, given a harmonic input current or voltage of the form $u(t) = \hat{u}\, e^{j2\pi ft}$, $t \in \mathbb{R}$, one may usually easily determine the corresponding output voltage or current $y(t) = \hat{y}\, e^{j2\pi ft}$, $t \in \mathbb{R}$, and thus establish the frequency response function of the network.

4.7.8.   Example: RCL network. By way of example we consider the RCL series network, whose diagram is repeated in Fig. 4.15. The input is the voltage produced by the voltage source, and the output is the voltage across the inductance. The first step to find the frequency response function is to construct the replacement network of Fig. 4.16. The impedance Z_1 is the series connection of the resistor (with impedance R) and the capacitor (with impedance $1/j\omega C$), so that $Z_1 = R + 1/j\omega C$. The impedance Z_2 is that of the inductor, so that $Z_2 = j\omega L$. We use the angular frequency ω throughout this example.

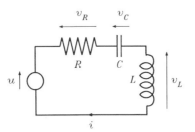

Figure 4.15. RCL network.

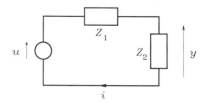

Figure 4.16. Replacement network.

Given the input voltage $u(t) = \hat{u}\, e^{j\omega t}$, $t \in \mathbb{R}$, the current through the network is $i(t) = \hat{i}\, e^{j\omega t}$, $t \in \mathbb{R}$, where

$$\hat{i} = \frac{\hat{u}}{Z_1 + Z_2}.$$

It follows that the voltage across the impedance Z_2 is $y(t) = \hat{y}\, e^{j\omega t}$, $t \in \mathbb{R}$, where

$$\hat{y} = Z_2 \hat{i} = \frac{Z_2}{Z_1 + Z_2}\hat{u}.$$

Thus, the system frequency response function is

$$\hat{h}(f) = \hat{h}(\omega/2\pi) = \frac{\hat{y}}{\hat{u}} = \frac{Z_2}{Z_1 + Z_2} = \frac{j\omega L}{R + \dfrac{1}{j\omega C} + j\omega L}$$

$$= \frac{(j\omega)^2}{(j\omega)^2 + \dfrac{R}{L}j\omega + \dfrac{1}{LC}}.$$

This agrees with what we found in Example 4.7.3(b).   ∎

4.8 PROBLEMS

In the first problem we follow up Example 3.2.13 as an instance of a nonlinear differential system.

4.8.1. Response of the car. As found in Example 3.2.13, the speed of a car may be described by the differential equation

$$M\frac{dv(t)}{dt} = cu(t) - Bv^2(t), \qquad t \in [0, \infty).$$

Here v is the car speed, u (ranging between 0 and 1) the throttle position, and M, c and B physical constants. If the throttle position has the constant value 1, the speed has a corresponding stationary value v_{max} that satisfies $0 = c - Bv_{max}^2$ and, hence, is given by $v_{max} = \sqrt{c/B}$.

(a) Define

$$w = \frac{v}{v_{max}}$$

as the speed expressed as fraction of the top speed, and show that w satisfies the differential equation

$$\frac{dw(t)}{dt} = \alpha[u(t) - w^2(t)], \qquad t \geq 0,$$

where $\alpha = \sqrt{Bc}/M$.

(b) Take the initial time equal to 0, assume that at this time the car has speed w_o (as fraction of the top speed) and that the throttle position remains constant at the value u_o. Use separation of variables to show that the solution of the differential equation is

$$w(t) = w_o \frac{1 + \dfrac{w_\infty}{w_o}\tanh{(\alpha w_\infty t)}}{1 + \dfrac{w_o}{w_\infty}\tanh{(\alpha w_\infty t)}}, \qquad t \geq 0,$$

where $w_\infty = \sqrt{u_o}$ is the stationary speed corresponding to the constant throttle setting u_o, and tanh is the hyperbolic tangent given by

$$\tanh{(x)} = \frac{e^x - e^{-x}}{e^x + e^{-x}}, \qquad -\infty < x < \infty.$$

(c) Plot the response of the speed of the car w if $\alpha = 1/10 \ [s^{-1}]$, $w_o = 0.5$ and $w_\infty = 0.6$.

(d) What is the behavior of the speed w if $w_o = 0$ and $u_o = 1$?

The next problems deal with linear difference and differential systems as introduced in Section 4.3.

4.8.2. Various difference systems. Show that the following discrete-time systems may be described by a constant coefficient linear difference equation of the form

$$Q(\sigma)y = P(\sigma)u,$$

and determine the polynomials Q and P.

(a) The *delay* system described by

$$y(n) = u(n - M), \qquad n \in \mathbb{Z},$$

with M a nonnegative integer.

(b) The *tapped delay-line*, which is a system described by an expression of the form

$$y(n) = a_M u(n) + a_{M-1} u(n - 1) + \cdots + a_0 u(n - M), \qquad n \in \mathbb{Z},$$

with $a_0, a_1, \cdots, a_M$ real coefficients, and M a nonnegative integer. The reason for the name is that the system may be implemented with a delay line as shown in Fig. 4.17. The system is also known as a *moving averager* (MA system), or *transversal filter*.

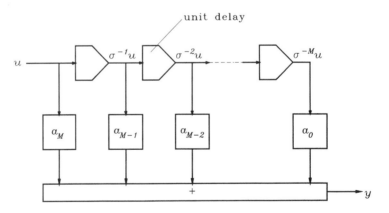

Figure 4.17.   A tapped delay line.

(c) An *auto-regressive* (AR) system, which is a system described by an expression of the form

$$y(n) = b_{N-1} y(n - 1) + b_{N-2} y(n - 2) + \cdots + b_0 y(n - N) + a_0 u(n),$$

$n \in \mathbb{Z}$, with $a_0, b_0, b_1, \cdots, b_{N-1}$ real coefficients, and N a nonnegative integer. The reason for the name is that the current value of the output *regresses* on the previous values.

(d) An *ARMA* (auto-regressive moving-average) system, which is a system whose output is given by an expression of the form

$$y(n) = a_M u(n) + a_{M-1} u(n-1) + \cdots + a_0 u(n-M)$$
$$+ b_{N-1} y(n-1) + b_{N-2} y(n-2) + \cdots + b_0 y(n-N),$$

$n \in \mathbb{Z}$, where $a_0, a_1, \cdots, a_M, b_0, b_1, \cdots, b_{N-1}$ are real coefficients and M and N nonnegative integers. An ARMA system is a combination of an auto-regressive and a moving-average system.

4.8.3. Some differential systems. For the following systems, determine the differential equation that relates the continuous-time input u and output y. If the system is linear and time-invariant bring the equation in the form $Q(D)y = P(D)u$ and exhibit the polynomials P and Q.

(a) *RL network*. The input to the network of Fig. 4.18 is the voltage u of the voltage source, while the output of the system is the voltage y across the resistor.

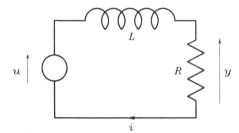

Figure 4.18.   An RL circuit.

(b) *Double spring-mass system*. In the double spring-mass system of Fig. 4.19, the springs have spring constants k_1 and k_2, while the two blocks have masses m_1 and m_2, respectively. The blocks move in one dimension only. The input to the system is the displacement u of the left-hand side of the leftmost spring, which is measured relative to a fixed point. The displacements of the blocks are z_1 and z_2, respectively, both taken relative to fixed points that correspond to the positions of the blocks when the system is at rest. There is no friction. The output y of the system is the position z_2 of the second block. *Hints:* Argue that the system satisfies the equations

$$m_1 \ddot{z}_1 = -k_1(z_1 - u) + k_2(z_2 - z_1),$$
$$m_2 \ddot{z}_2 = \qquad\qquad - k_2(z_2 - z_1).$$

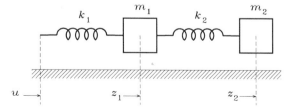

Figure 4.19.   A double spring-mass system.

Differentiate the second equation twice and eliminate z_1 and $\ddot{z}_1$ from the three equations that now have been obtained. This leads to a differential equation of order 4.

Linearity, time-invariance, and initially-at-rest-systems are introduced in Section 4.3.

4.8.4. Linearity and time-invariance. Determine whether the following differential systems are (i) linear, (ii) time-invariant.

(a) $t\dfrac{dy(t)}{dt} + y(t) = u(t), \qquad t \in \mathbb{R}.$

(b) $\dfrac{dy(t)}{dt} + y(t) = [u(t)]^2, \qquad t \in \mathbb{R}.$

(c) $\dfrac{dy(t)}{dt} + y(t) = t[u(t)]^2, \qquad t \in \mathbb{R}.$

(d) $\left[\dfrac{dy(t)}{dt}\right]^2 = [u(t)]^2, \qquad t \in \mathbb{R}.$

4.8.5. Initially-at-rest systems. Determine the initially-at-rest response of each of the following systems by first assuming the system to be at rest at some initial time and next letting the initial time approach $-\infty$. What is the impulse response h of the initially-at-rest system?

(a) The savings account of Example 4.2.3(b).
(b) The delay line of Problem 4.8.2(a).
(c) The tapped delay line of Problem 4.8.2(b).
(d) The continuous-time differential system

$$\frac{dy(t)}{dt} = u(t), \qquad t \in \mathbb{R}.$$

In the next series of problems we deal with the response of linear time-invariant difference and differential systems, as discussed in Section 4.4.

4.8.6. Basis solutions. Determine the basis solutions, both in complex and in real form, of the linear constant coefficient difference and differential equations describing the following systems.

(a) The auto-regressive system

$$y(n + 2) - \frac{1}{4}y(n) = u(n + 2), \qquad n \in \mathbb{T},$$

both when $\mathbb{T} = \mathbb{Z}$ and $\mathbb{T} = \mathbb{Z}_+$.

(b) The auto-regressive system

$$y(n + 2) - y(n + 1) + \frac{5}{16} y(n) = u(n + 2), \qquad n \in \mathbb{T},$$

both when $\mathbb{T} = \mathbb{Z}$ and $\mathbb{T} = \mathbb{Z}_+$.

(c) The tapped delay line (or moving averager)

$$y(n + M) = a_0 u(n) + a_1 u(n + 1) + \cdots + a_M u(n + M), \qquad n \in \mathbb{T},$$

both when $\mathbb{T} = \mathbb{Z}$ and $\mathbb{T} = \mathbb{Z}_+$.

(d) The ARMA system

$$y(n + 2) - \frac{1}{4} y(n) = u(n + 2) + u(n + 1), \qquad n \in \mathbb{T},$$

both when $\mathbb{T} = \mathbb{Z}$ and $\mathbb{T} = \mathbb{Z}_+$.

(e) The LR network of Problem 4.8.3(a), which is described by the differential equation

$$\frac{L}{R} \frac{dy(t)}{dt} + y(t) = u(t), \qquad t \in \mathbb{R}.$$

(f) The double spring-mass system of Problem 4.8.3(b), which is described by the differential equation

$$y^{(4)}(t) + \left(\omega_1^2 + \omega_2^2 + \frac{m_2}{m_1} \omega_2^2 \right) y^{(2)}(t) + \omega_1^2 \omega_2^2 y(t) = \omega_1^2 \omega_2^2 u(t), \qquad t \in \mathbb{R},$$

where $\omega_1^2 = k_1/m_1$ and $\omega_2^2 = k_2/m_2$. Assume that

$$\omega_1^2 = 3/2, \quad \omega_2^2 = 4/3, \quad m_2/m_1 = 1/8.$$

(g) The "n fold integrator," which is the system described by the differential equation

$$y^{(n)}(t) = u(t), \qquad t \in \mathbb{R}.$$

(h) The system described by the differential equation

$$y''(t) - y(t) = u'(t) - u(t), \qquad t \in \mathbb{R}.$$

4.8.7. Initial value problems. Solve the following initial value problems.

(a) *Fibonacci numbers.* The Fibonacci numbers are the solution of the difference equation

$$y(n + 2) = y(n + 1) + y(n), \qquad n = 0, 1, 2, \cdots,$$

with $y(0) = 0$ and $y(1) = 1$.

Fibonacci was the nickname of Leonardo of Pisa (ca. 1180–ca. 1250). Born of the trading class, Fibonacci learned Arabic mathematics on his travels and wrote about it in his book *Liber Abaci* (1202).

(b) *ARMA system:*

$$y(n + 2) - \frac{1}{4}y(n) = u(n + 2) + u(n + 1), \qquad n = 0, 1, 2, \cdots,$$

with $u(n) = 1$ for $n \geq 0$, and $y(0) = y(1) = 1$.

(c) *LR network:*

$$\frac{L}{R}\frac{dy(t)}{dt} + y(t) = u(t), \qquad t \geq 0,$$

with $u(t) = t$ for $t \geq 0$ and $y(0) = 0$.

(d) *Double spring-mass system:*

$$y^{(4)}(t) + \left(\omega_1^2 + \omega_2^2 + \frac{m_2}{m_1}\omega_2^2\right)y^{(2)}(t) + \omega_1^2\omega_2^2\, y(t) = \omega_1^2\omega_2^2\, u(t), \qquad t \geq 0,$$

where $\omega_1^2 = 3/2$, $\omega_2^2 = 4/3$, $m_2/m_1 = 1/8$, with $u(t) = 0$ for $t \geq 0$, and $y(0) = 1$, $y^{(1)}(0) = y^{(2)}(0) = y^{(3)}(0) = 0$.

(e) *Multiple integrator:*

$$y^{(n)}(t) = u(t), \qquad t \geq 0,$$

with $u(t) = t^k$, $t \geq 0$, and $y^{(i)}(0) = 0$ for $i = 0, 1, \cdots, n - 1$. Both n and k are nonnegative integers.

Initially-at-rest linear time-invariant difference and differential systems are convolution systems. In Section 4.5 it is shown how their impulse response may be found.

4.8.8. Impulse and step responses. Determine the impulse and step responses of each of the following systems. Sketch them. Establish for each system whether it is non-anticipating.

(a) The auto-regressive system

$$y(n + 2) - \frac{1}{4}y(n) = u(n + 2), \qquad n \in \mathbb{Z}.$$

(b) The auto-regressive system

$$y(n + 2) - y(n + 1) + \frac{5}{16} y(n) = u(n + 2), \qquad n \in \mathbb{Z}.$$

(c) The tapped delay line (or moving averager)

$$y(n + M) = a_0 u(n) + a_1 u(n + 1) + \cdots + a_M u(n + M), \qquad n \in \mathbb{Z}.$$

(d) The ARMA system

$$y(n + 2) - \frac{1}{4} y(n) = u(n + 2) + u(n + 1), \qquad n \in \mathbb{Z}.$$

(e) The LR network of Problem 4.8.3(a), which is described by the differential equation

$$\frac{L}{R} \frac{dy(t)}{dt} + y(t) = u(t), \qquad t \in \mathbb{R}.$$

(f) The double spring-mass system of Problem 4.8.3(b), which is described by the differential equation

$$y^{(4)}(t) + \left(\omega_1^2 + \omega_2^2 + \frac{m_2}{m_1} \omega_2^2 \right) y^{(2)}(t) + \omega_1^2 \omega_2^2 \, y(t) = \omega_1^2 \omega_2^2 \, u(t), \qquad t \in \mathbb{R},$$

where $\omega_1^2 = k_1/m_1$ and $\omega_2^2 = k_2/m_2$. Assume that $\omega_1^2 = 3/2$, $\omega_2^2 = 4/3$, $m_2/m_1 = 1/8$.

(g) The "n fold integrator," which is the system described by the differential equation

$$y^{(n)}(t) = u(t), \qquad t \in \mathbb{R}.$$

(h) The system described by the differential equation

$$y''(t) - y(t) = u'(t) - u(t), \qquad t \in \mathbb{R}.$$

The next problem deals with BIBO and CICO stability of difference and differential systems as discussed in Section 4.6.

4.8.9. BIBO and CICO stability. For the following difference and differential systems, investigate whether (i) the initially-at-rest system is BIBO stable, (ii) the IO system described by the difference or differential equation is BIBO stable, and (iii) the IO system is CICO stable.

(a) The delay of Problem 4.8.2(a).
(b) The "Fibonacci system" of 4.8.7(a).
(c) The auto-regressive system of Problem 4.8.8(a).

(d) The auto-regressive system of Problem 4.8.8(b).
(e) The tapped delay line of Problem 4.8.8(c).
(f) The ARMA system of Problem 4.8.8(d).
(g) The LR network of Problem 4.8.8(e), with L and R both positive.
(h) The RCL network of Example 4.2.4(c), with R, C and L all positive.
(i) The double spring-mass system of Problem 4.8.8(f).
(j) The $n$fold integrator of Problem 4.8.8(g).
(k) The second-order system of Problem 4.8.8(h).

The frequency response of difference and differential systems is treated in Section 4.7.

4.8.10. Frequency response of difference and differential systems. Determine the characteristic roots, the poles and the zeros, if any, of the following systems. Establish whether each system has a frequency response function, and determine it if it exists.

(a) The delay of Problem 4.8.2(a).
(b) The auto-regressive system of Problem 4.8.8(a).
(c) The auto-regressive system of Problem 4.8.8(b).
(d) The tapped delay line of Problem 4.8.8(c), with $M = 2$, $a_0 = 4$, $a_1 = -4$, $a_2 = 1$.
(e) The ARMA system of Problem 4.8.8(d).
(f) The LR network of Problem 4.8.8(e).
(g) The double spring-mass system of Problem 4.8.8(f).
(h) The $n$fold integrator of Problem 4.8.8(g).

4.8.11. Steady-state and transient response to harmonic inputs.

(a) *Exponential smoother.* Consider the exponential smoother $y(n + 1) = ay(n) + (1 - a)u(n + 1)$, $n \in \mathbb{Z}_+$, with $a = \sqrt{3}/3$. Determine the steady-state and transient response of the smoother to the real harmonic input

$$u(n) = \cos(2\pi fn), \qquad n \in \mathbb{Z}_+,$$

with the initial condition $y(0) = 0$, for $f = 1/4$.

(b) *Second-order differential system.* Determine the steady-state and transient response of the system $y'' + 2y' + y = u' - u$ to the real harmonic input

$$u(t) = \cos(2\pi ft), \qquad t \geq 0,$$

with the initial conditions $y(0) = y'(0) = 0$, where $2\pi f = 1$.

4.8.12. Frequency response functions of electrical networks. Determine the frequency response functions of the following electrical networks.

(a) The second-order RC network of Fig. 4.20, whose input is the current u produced by the current source, and whose output y is the current through the capacitor C_2.
(b) The second-order LC network of Fig. 4.21, whose input is the voltage u of the voltage source, and whose output y is the voltage across the capacitor C_2.

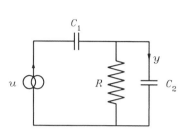

Figure 4.20. An RC network.

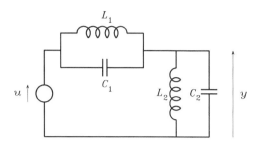

Figure 4.21. An LC network.

4.9 COMPUTER EXERCISES

The computer exercises for Chapter 4 deal with the numerical solution of difference and differential equations, computing characteristic roots, poles and zeros, and Bode plots.

4.9.1. Solution of difference equations.

(a) *Fibonacci equation.* The Fibonacci equation is the second-order difference equation

$$y(n + 2) = y(n) + y(n + 1), \qquad n = 0, 1, 2, \cdots .$$

The *Fibonacci numbers* are its solution for the initial conditions $y(0) = 0$, $y(1) = 1$. Solve the Fibonacci equation numerically for $0 \leq n \leq 20$ with the given initial conditions, and plot the solution y. What is $y(20)$? *Hint:* The equation may be converted into a *pair* of first-order difference equations, which are easier to program, by defining $x_1(n) = y(n - 1)$ and $x_2(n) = y(n)$ for $n = 1$, $2, \cdots$. Determine difference equations for x_1 and x_2 and program these.

(b) *Exponential smoother.* The exponential smoother is described by the first-order difference equation

$$y(n + 1) = ay(n) + (1 - a)u(n + 1), \qquad n = 0, 1, 2, \cdots .$$

(b.1) Write a macro or simple program to solve the difference equation numerically on the time axis $\{0, 1, 2, \cdots, 100\}$, for a given input signal u and given initial condition $y(0) = 0$, with the constant a externally defined.

(b.2) Let $a = 0.5$, and compute and plot the response of the exponential smoother to a pure noise input. Amplitude scale the response such that its root mean square (rms) value is 1, and use the resulting signal as the input for (b.3). *Hint:* In SIGSYS, generate the noise input as the standard signal `noiseplus`.

(b.3) Use the signal computed in (b.2) as input to the exponential smoother for $a = 0.2$, $a = 0.5$ and $a = 0.9$. Plot the corresponding outputs and observe the different smoothing effects.

Numerical solution of differential equations. The numerical solution of a set of differential equations is often referred to as the numerical *integration* of the equations. Numerical integration involves a *step size* along the time axis. In integration routines where the step size is not automatically selected or adjusted a practical approach to determine the step size is the following. First choose the step size tentatively as, say, 1/100 of the time interval over which the differential equations are to be integrated. If the solution becomes unbounded (which often happens if the step size is too large, and is manifested by numerical overflow), then reduce the step size to, say, half the size. After a solution has been obtained, recompute with a smaller (again, say, halved) step size. If reducing the step size produces no significant change, the step size is small enough. An alternative to reducing the step size is switching to a more powerful integration routine, say from Runge-Kutta 2 to Runge-Kutta 4.

4.9.2. Numerical solution of differential equations. Differential equations of the form

$$F[y(t), \frac{dy(t)}{dt}, \cdots, \frac{d^N y(t)}{dt^N}, t] = 0, \qquad t \geq 0,$$

most conveniently may be solved numerically by converting them to a *set* of *N first-order* differential equations. First, assume that for each t the given equation may uniquely be solved for $d^N y(t)/dt^N$ in the form

$$\frac{d^N y(t)}{dt^N} = G\left[y(t), \frac{dy(t)}{dt}, \cdots, \frac{d^{N-1} y(t)}{dt^{N-1}}, t\right], \qquad t \geq 0.$$

Next, define the N auxiliary signals

$$x_1(t) = y(t),$$
$$x_2(t) = y^{(1)}(t),$$
$$\cdots \quad \cdots$$
$$x_N(t) = y^{(N-1)}(t),$$

all for $t \geq 0$. It is easily seen that $x_1, x_2, \cdots, x_N$ satisfy the set of first-order simultaneous differential equations

$$\dot{x}_1(t) = x_2(t),$$

$$\dot{x}_2(t) = x_3(t),$$

$$\cdots \quad \cdots$$

$$\dot{x}_{N-1}(t) = x_N(t),$$

$$\dot{x}_N(t) = G[x_1(t), x_2(t), \cdots, x_N(t), t],$$

all for $t \geq 0$. The initial conditions for this set of differential equations follow directly from the given initial conditions $y(0)$, $y^{(1)}(0)$, $\cdots$, $y^{(N-1)}(0)$ for the original differential equation. For the numerical solution of sets of first-order differential equations many methods exist (see Section 5.4), and a plethora of standard software is available.

(a) *RC network*. The RC network (with the voltage across the capacitor as output), is described by the differential equation

$$\frac{dy(t)}{dt} + \frac{1}{RC} y(t) = u(t), \qquad t \geq 0.$$

Take $RC = 1$, and solve the differential equation numerically on the interval $[0, 20]$ for the input

$$u(t) = \cos(2\pi f_o t), \qquad t \geq 0,$$

with $2\pi f_o = 1$ and the initial condition $y(0) = 0$.

(b) *Moving car* (compare 4.8.1.) The moving car is described by the normalized differential equation

$$\frac{dw(t)}{dt} = \alpha[u(t) - w^2(t)], \qquad t \geq 0,$$

with w the speed of the car as fraction of the top speed, and u the throttle position. Integrate the differential equation numerically on the interval $[0, 100]$ for the input

$$u(t) = 0.36, \qquad t \geq 0,$$

with the initial condition $w(0) = 0.5$, while $\alpha = 0.1 \ [\text{s}^{-1}]$. Compare the solution numerically with the analytical result of 4.8.1.

(c) *RCL network*. The RCL circuit of 4.2.4(c) with the current as output is described by the differential equation

$$y^{(2)}(t) + \frac{R}{L} y^{(1)}(t) + \frac{1}{LC} y(t) = \frac{1}{L} u^{(1)}(t), \qquad t \geq 0.$$

Suppose that the input u is identical to zero, and convert the differential equation to a set of two first-order equations. Choose $R = 1/4$, $C = 1$ and $L = 1$ and integrate the differential equations over the interval $[0, 20]$ for the initial conditions $y(0) = 1$, $y^{(1)}(0) = 0$.

4.9.3. Basis solutions of constant coefficient linear difference and differential equations. (Compare 4.8.6.) Compute the characteristic roots of the linear constant coefficient difference and differential equations that describe the following systems. Determine and plot the basis solutions of the homogeneous equations.

(a) The auto-regressive system

$$y(n + 2) - y(n + 1) + \frac{5}{16} y(n) = u(n + 2), \qquad n \in \mathbb{Z}_+.$$

(b) The double spring-mass system, described by the differential equation

$$y^{(4)}(t) + \left(\omega_1^2 + \omega_2^2 + \frac{m_2}{m_1}\omega_2^2\right)y^{(2)}(t) + \omega_1^2\omega_2^2\, y(t) = \omega_1^2\omega_2^2\, u(t), \qquad t \in \mathbb{R}_+,$$

with $\omega_1^2 = 3/2$, $\omega_2^2 = 4/3$ and $m_2/m_1 = 1/8$.

(c) The discrete-time "echo" system, described by the difference equation

$$y(n) = u(n) + \alpha y(n - N), \qquad n = N, N + 1, N + 2, \cdots,$$

with N a nonnegative integer and α a real constant. Let $N = 4$ and $\alpha = 7/8$.

4.9.4. Stability of constant coefficient linear difference and differential systems. Compute the characteristic values, poles and zeros of the following linear time-invariant difference and differential systems. Determine the BIBO and CICO stability of each system.

(a) The difference system $Q(\sigma)y = P(\sigma)u$, where P and Q are the polynomials

$$Q(\lambda) = 1 + 2\lambda^4, \qquad P(\lambda) = 1 + \lambda.$$

(b) The differential system $Q(D)y = P(D)u$, where the polynomials P and Q are given as in (a).

(c) The difference system $Q(\sigma)y = P(\sigma)u$, where P and Q are the polynomials

$$Q(\lambda) = 1 - \lambda^4, \qquad P(\lambda) = 1 - \lambda + \lambda^2 - \lambda^3.$$

(d) The differential system $Q(D)y = P(D)u$, where the polynomials P and Q are given as in (c).

4.9.5. Asymptotic Bode plots. *Bode plots* graphically represent frequency response functions of continuous-time systems as follows:

(i) the logarithm of the magnitude is plotted versus the logarithm of frequency, and

(ii) the phase is plotted linearly versus the logarithm of frequency.

This often is a convenient way of displaying frequency response functions.

Suppose that the frequency response function $\hat{h}$ is the rational function given by

$$\hat{h}(\omega/2\pi) = \frac{P(j\omega)}{Q(j\omega)}, \qquad \omega \in \mathbb{R},$$

with P and Q polynomials with real coefficients. Because $\hat{h}$ is conjugate symmetric it is sufficient to plot it for positive angular frequencies only.

According to the fundamental theorem of algebra, the polynomials P and Q may be factored as

$$P(\lambda) = p_0 \prod_k (\lambda - \zeta_k), \qquad Q(\lambda) = q_0 \prod_k (\lambda - \pi_k),$$

where the ζ_k are the roots of P, π_k those of Q, and p_0 and q_0 constants. It follows that

$$\log|\hat{h}(\omega/2\pi)| = \log(|p_0/q_0|) + \sum_k \log|j\omega - \zeta_i| - \sum_k \log|j\omega - \pi_k|. \qquad (1)$$

Let us look at one term individually, for instance $\log|j\omega - \alpha|$, with α a complex number. We have

$$\log|j\omega - \alpha| \approx \begin{cases} \log|\alpha| & \text{for } \omega \ll |\alpha|, \\ \log(\omega) & \text{for } \omega \ll |\alpha|, \end{cases} \qquad \omega \in \mathbb{R}_+.$$

This shows that on a logarithmic frequency scale for *low* frequencies the term $\log|j\omega - \alpha|$ has a constant as asymptote, while for *high* frequencies it has a straight line $\log(\omega)$ with slope 1 as asymptote. Figure 4.22 shows the position of the asymptotes. The asymptotes *intersect* at the angular frequency $\omega = |\alpha|$, which is called the *break frequency* corresponding to the factor $j\omega - \alpha$. Near the break frequency the plot of $|j\omega - \alpha|$ deviates most significantly from the asymptotes.

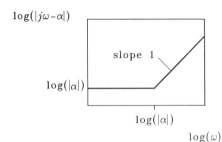

Figure 4.22. Asymptotes of the magnitude of $j\omega - \alpha$ in a double-logarithmic plot.

The asymptotes of the various terms in (1) may be used to obtain a quick sketch of the general behavior of the Bode magnitude plot. Consider for instance the frequency response function

$$\hat{h}(\omega/2\pi) = \frac{P(j\omega)}{Q(j\omega)},$$

where

$$P(\lambda) = 4(\lambda + 1),$$
$$Q(\lambda) = \lambda^2 + \lambda + 4.$$

Bode plots.

Bode plots graphically represent frequency response functions of continuous-time systems as follows

  (i) the logarithm of the magnitude is plotted versus the logarithm of frequency, and
  (ii) the phase is plotted linearly versus the logarithm of frequency.

This often is a convenient way of displaying frequency response functions.

A Bode plot of a frequency response function $\hat{h}(\omega/2\pi)$, $\omega \in \mathbb{R}_+$, may be prepared as follows. We assume that $\hat{h}$ is specified in analytic form. First determine the frequency interval over which the plot is made. A good way of doing this is by first sketching an asymptotic magnitude plot as explained in 4.9.5. The frequency interval may then be chosen, say, starting from one or two frequency decades down from the lower break frequency of interest and extending to one or two decades up from the highest break frequency. (A *decade* is a factor of 10.)

Suppose that it is decided to plot over the angular frequency range extending from 10^a to 10^b. To plot the angular frequency logarithmically, first define an underlying linear signal axis with domain $[a, b]$ and a sufficient number of points to make nice plots (say 100). Next, compute an exponentially sampled angular frequency axis by evaluating 10^r with r ranging over the underlying linear axis. Finally, evaluate the frequency response function $\hat{h}$ on the angular frequency axis, and from this obtain the $\log_{10}$ of its magnitude and its phase.

In SIGSYS, the underlying linear axis could be created as LOGOM = a:(b-a)/100:b. Next, the exponentially spaced angular frequency axis may be obtained as OM = 10^LOGOM. Suppose that the frequency response function to be evaluated is given by $\hat{h}(\omega/2\pi) = P(j\omega)/Q(j\omega)$, with P and Q given polynomials. Then we may evaluate $\hat{h}$ on the angular frequency axis as hhat = P(j*OM)/Q(j*OM). The logarithm of the magnitude and phase follow as ahhat = log10(abs(hhat)) and phhat = arg(hhat), respectively.

Hendrik Wade Bode (1905–1982) was an American research engineer who did important work on network design during his long association with the Bell Telephone Laboratories.

The numerator polynomial P has a single root -1, whose magnitude is 1. The denominator Q has a complex pair of roots, each of which has magnitude $\sqrt{4} = 2$. At low frequencies $\hat{h}$ has the low frequency asymptote $\hat{h}(0) = P(0)/Q(0) = 1$. As frequency increases, the first break frequency that is encountered is 1. Because this break frequency is contributed by the numerator, at this point an asymptote with slope 1 sets in, as indicated in Fig. 4.23. The next break frequency 2 represents a double root contributed by the denominator. As a consequence, an asymptote with slope -2 sets in, which results in a slope change of the asymptotic magnitude plot from 1 to -1. Figure 4.23 shows the resulting asymptotic plot.

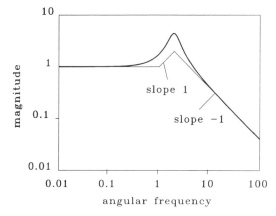

Figure 4.23. Asymptotic and actual Bode magnitude plots.

The figure also gives the full Bode magnitude plot. The agreement with the asymptotic plot is very good at high and low frequencies, but less so near the break frequencies.

(a) Reproduce Fig. 4.23 by plotting the Bode magnitude plot and the asymptotic plot of the given frequency response function in one frame.
(b) Compute the break frequencies, sketch the asymptotic Bode magnitude plot, and display the Bode magnitude and phase plots of the frequency response function

$$\hat{h}_c(\omega/2\pi) = \frac{\dfrac{1}{LC}}{(j\omega)^2 + j\omega\dfrac{R}{L} + \dfrac{1}{LC}}, \qquad \omega \in \mathbb{R}_+,$$

with $R = 1/4$, $C = 1$ and $L = 1$. This is the frequency response function of the RCL network of 4.2.4(c) when the output is the voltage across the capacitor.

(c) Repeat (b) for the frequency response function

$$\hat{h}_R(\omega/2\pi) = \frac{\dfrac{R}{L}j\omega}{(j\omega)^2 + j\omega\dfrac{R}{L} + \dfrac{1}{LC}}, \qquad \omega \in \mathbb{R}_+,$$

with the numerical values of (b).

(d) Repeat (b) for the frequency response function

$$\hat{h}_L(\omega/2\pi) = \frac{(j\omega)^2}{(j\omega)^2 + j\omega\dfrac{R}{L} + \dfrac{1}{LC}}, \qquad \omega \in \mathbb{R}_+,$$

with the numerical values of (b).

4.9.6. Butterworth filters. A normalized *Butterworth filter* of order N is a continuous-time differential system described by a differential equation of the form

$$Q_N(D)y = u,$$

where the polynomial Q_N is obtained as follows. The polynomial

$$\phi_N(\lambda) = \lambda^{2N} + (-1)^N,$$

with N a positive integer, has $2N$ roots λ_i, $i = 1, 2, \cdots, 2N$, that may be arranged such that λ_i, $i = 1, 2, \cdots, N$, all have strictly negative real parts, while the remaining N roots all have strictly positive real parts. Then

$$Q_N(\lambda) = \prod_{i=1}^{N} (\lambda - \lambda_i).$$

The frequency response function of the Nth order Butterworth filter is

$$\hat{h}_N(\omega/2\pi) = \frac{1}{Q_N(j\omega)}, \qquad \omega \in \mathbb{R}_+.$$

Butterworth filters are low-pass filters. The angular cut-off frequency of a normalized Butterworth filter is 1. The cut-off frequency may be shifted to ω_o by modifying the frequency response function to

$$\frac{1}{Q_N(j\omega/\omega_o)}, \qquad \omega \in \mathbb{R}_+,$$

which is the frequency response function of the differential system

$$Q(D/\omega_o)y = u.$$

(a) Find the Butterworth polynomials Q_N for $N = 1, 2, 3, 4$, and 5.
(b) Make Bode plots of the magnitude and phase of the normalized Butterworth frequency response functions $\hat{h}_N$ as a function of angular frequency for $N = 1, 2, 3, 4$, and 5. Observe how the low-pass nature of the frequency response function improves as N increases.

Butterworth filters date back to the early days of electronics: S. Butterworth, "On the theory of filter amplifiers," *Wireless Engineer*, London, vol. 7, 1930, pp. 536–541.

5

State Description of Systems

5.1 INTRODUCTION

In Chapters 3 and 4 we considered systems described by their input-output relationship. Such a description reduces the system to a "black box," whose internal mechanisms are ignored. In the present chapter we study the *state description* of systems, which encompasses not only their external but also their internal behavior. Such a description may often be obtained by direct application of the physical or other fundamental laws that govern the system. The state description is of eminent importance for understanding, analyzing and simulating systems.

In Section 5.2 state systems are first informally and then formally introduced. By a progression of assumptions this section leads to linear time-invariant discrete- and continuous-time systems described by *state difference* and *state differential* equations, respectively, together with an *output equation*. These systems form the focus for the remainder of the chapter. In Section 5.3 we show how discrete-time difference systems and continuous-time differential systems may be represented as state difference systems and state differential systems, respectively.

In Section 5.4 the existence of solutions of state difference and differential equations is discussed and methods are presented for their numerical solution. For linear systems the solution of state difference and differential equations is given in Section 5.5 using the state transition matrix.

Section 5.6 presents the details of the *modal* analysis of linear time-invariant state systems, based on the diagonal form of the system matrix. This section leads to a considerable improvement of the understanding of the internal behavior of linear time-invariant state difference and differential systems. Section 5.7 is devoted to the definition and analysis of the *stability* of state systems. In Section 5.8, finally, the frequency response of linear time-invariant state systems is analyzed.

> The idea of state, though not under this name, has been around in physics for some time. What is called "phase" in statistical mechanics, for instance, is precisely the notion state. In system theory the state description broke through in the late 1950s.

5.2 STATE DESCRIPTION OF SYSTEMS

In this section we first introduce the idea of the *state* of a system, next present a few examples, and then proceed to a formal definition of *input-output-state* systems.

The Notion of State

Input-output systems are described by the *input-output rule,* which amounts to a specification of all possible input-output pairs (u, y). Such a description is adequate for many purposes, but suffers from the following defect: Suppose that we are interested in the response $y(t)$, $t \geq t_o$, from some fixed time t_o on, given the input $u(t)$, $t \geq t_o$, from that same time on. It then is necessary to know the *entire past input* $u(t)$, $t < t_o$, as well. In many instances, this information is much more than is required to determine the output from time t_o on. What is really needed is to know the "state" in which the system is. The state of the system summarizes the past of the system insofar as relevant for its future behavior.

What the "state" of the system is may often be ascertained by physical or other fundamental considerations. For electrical circuits, for instance, we know from physics that if the charges of the capacitors and the magnetic fluxes contained by the inductors in the network are known at some fixed time we may predict the future behavior of the system without any information about past values of the network variables. The charges of the capacitors together with the fluxes contained by the inductors constitute the *state* of the network. The electromagnetic laws that apply to the network determine the *evolution* of the state. We illustrate this by a simple example.

5.2.1.   Example: RCL network.   Consider the RCL network of Example 4.2.4(c), whose diagram is repeated in Fig. 5.1. Rather than obtaining a single differential equation of order two that relates input and output, as in Example 4.2.4(c), we look for a set of *first-order* differential equations that determine the rate of

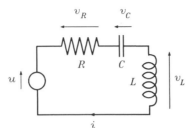

Figure 5.1.   An RCL network.

change of the charge q of the capacitor and the flux ϕ contained by the inductor. These two quantities determine the energy that is stored in the network and hence together constitute its state. By Kirchhoff's current law we have

$$\frac{dq(t)}{dt} = i(t),$$

with i the current through the circuit. Since $i(t) = \phi(t)/L$ this may be rewritten as

$$\frac{dq(t)}{dt} = \frac{1}{L}\phi(t). \tag{1}$$

On the other hand we have for the inductor

$$\frac{d\phi(t)}{dt} = v_L(t),$$

with v_L the voltage across the inductor. By Kirchhoff's voltage law it follows that

$$u(t) = v_R(t) + v_C(t) + v_L(t),$$

with v_R the voltage across the resistor and v_C that across the capacitor. Because $v_R = Ri = R\phi/L$ and $v_C = q/C$ this may be rewritten as

$$v_L(t) = u(t) - \frac{R}{L}\phi(t) - \frac{1}{C}q(t),$$

so that

$$\frac{d\phi(t)}{dt} = u(t) - \frac{R}{L}\phi(t) - \frac{1}{C}q(t). \tag{2}$$

This equation together with (1) shows that the rates of change of the charge q and the flux ϕ at time t are determined by $q(t)$, $\phi(t)$ and $u(t)$, which are the values of the charge, flux and input *at time t*. Thus, if the initial charge $q(t_o)$ and the initial flux $\phi(t_o)$ are known together with the input $u(t)$, $t \geq t_o$, from time t_o on, the further evolution of the charge and flux is fully determined. This is more fully discussed in Section 5.4.

It is easy to see that if the charge, flux and input are given at any time t, also the output at that time is known. If the output y is the current i through the network, we have

$$y(t) = i(t) = \frac{1}{L}\phi(t).$$

If the output is the voltage v_L across the inductor it follows that

$$y(t) = v_L(t) = u(t) - \frac{R}{L}\phi(t) - \frac{1}{C}q(t).$$

Thus, if the initial charge and flux are known together with the subsequent behavior of the input, we do not only know the further behavior of the charge and the flux, but also that of the output.

Note that in this case the state is not a single, scalar quantity, but a vector-valued quantity with two components. Indeed, we may rewrite the differential equations (1) and (2) in matrix notation as

$$\frac{d}{dt}\begin{bmatrix} q(t) \\ \phi(t) \end{bmatrix} = \begin{bmatrix} 0 & \dfrac{1}{L} \\ -\dfrac{1}{C} & -\dfrac{R}{L} \end{bmatrix} \begin{bmatrix} q(t) \\ \phi(t) \end{bmatrix} + \begin{bmatrix} 0 \\ 1 \end{bmatrix} u(t). \tag{3}$$

Here

$$\frac{d}{dt}\begin{bmatrix} q(t) \\ \phi(t) \end{bmatrix} := \begin{bmatrix} \dfrac{dq(t)}{dt} \\ \dfrac{d\phi(t)}{dt} \end{bmatrix}.$$

The equation (3) is known as the *state differential equation* of the system. If the output of the network is the current,

$$y(t) = \begin{bmatrix} 0 & \dfrac{1}{L} \end{bmatrix}\begin{bmatrix} q(t) \\ \phi(t) \end{bmatrix}, \tag{4}$$

while if the output is the voltage across the inductor,

$$y(t) = \begin{bmatrix} -\dfrac{1}{C} & -\dfrac{R}{L} \end{bmatrix} \begin{bmatrix} q(t) \\ \phi(t) \end{bmatrix} + u(t). \tag{5}$$

The equation (4) or (5), respectively, is known as the *output equation* of the system.
Defining the *state vector*

$$x(t) := \begin{bmatrix} q(t) \\ \phi(t) \end{bmatrix}$$

the state differential equation and output equation of the system may be rewritten in the form

$$\dot{x}(t) = Ax(t) + Bu(t), \tag{6a}$$

$$y(t) = Dx(t) + Eu(t), \tag{6b}$$

where the overdot denotes differentiation, and the matrices A and B are given by

$$A = \begin{bmatrix} 0 & \dfrac{1}{L} \\ -\dfrac{1}{C} & -\dfrac{R}{L} \end{bmatrix}, \qquad B = \begin{bmatrix} 0 \\ 1 \end{bmatrix}.$$

For the output equation (4) the matrices D and E in (6b) are

$$D = \begin{bmatrix} 0 & \dfrac{1}{L} \end{bmatrix}, \qquad E = 0,$$

while for the output equation (5)

$$D = \begin{bmatrix} -\dfrac{1}{C} & -\dfrac{R}{L} \end{bmatrix}, \qquad E = 1.$$

Later in this chapter systems described by state differential and output equations of the form (6) are extensively analyzed.

We conclude by pointing out that the choice of the state often is not unique. For the RCL network, for instance, rather than the charge of the capacitor we may choose its voltage v_C, and instead of the flux contained by the inductor we may use the current i through it.

Exercise: Rederive the state differential and output equations of the RCL network when the state consists of v_C and i.   ■

The state of *mechanical* systems is described by the *positions* and *velocities* of all the masses: If these are given, then the further motion of the system may be determined without requiring information on the past behavior of the system. Again we illustrate this with a specific system.

5.2.2   Example: Moon rocket.

A rocket descends vertically to the surface of the moon, as sketched in Fig. 5.2. The elevation of the center of gravity of the rocket above the surface at time t (measured in upward direction) is $h(t)$, while the speed of ascent is $v(t)$. The thrust force of the rocket engine is in upward direction and given by $cu(t)$, where $u(t)$ is the amount of mass expelled per unit time and c the (constant) expulsion speed.

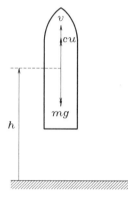

Figure 5.2.   A rocket descends on the moon.

Given the time behavior of the thrust force from a fixed time t on, the descent of the rocket is determined by its initial position and velocity. These are the components of the state of the system. It is easy to find their rates of changes. For the elevation we have

$$\dot{h}(t) = v(t).$$

By Newton's law it follows that $m\dot{v}(t) = cu(t) - mg$, with m the mass of the rocket and g the acceleration of gravity on the moon, so that

$$\dot{v}(t) = \frac{c}{m}u(t) - g.$$

The English mathematician and natural philosopher Sir Isaac Newton (1642–1727) formulated the laws of gravity and motion and the elements of differential calculus.

Together these two equations form the state differential equation. Denoting $x_1(t) = h(t)$ and $x_2(t) = v(t)$ we have

$$\begin{cases} \dot{x}_1(t) = x_2(t), \\ \dot{x}_2(t) = \dfrac{c}{m}\,u(t) - g. \end{cases}$$

These equations hold under the assumption that the mass of the rocket is *constant*. Actually, it is not, because of the expulsion of mass. A more accurate model therefore also includes the mass $m(t)$ as a component of the state. Because of the variable mass it is appropriate to replace the velocity $v(t)$ by the momentum $p(t) = m(t)v(t)$ as state variable. The rates of change of the state variables may now be expressed as

$$\dot{h}(t) = v(t) = \frac{p(t)}{m(t)},$$

$$\dot{p}(t) = cu(t) - m(t)g,$$

$$\dot{m}(t) = -u(t).$$

Defining $x_1(t) = h(t)$, $x_2(t) = p(t)$ and $x_3(t) = m(t)$, the state differential equation thus takes the form

$$\begin{cases} \dot{x}_1(t) = \dfrac{x_2(t)}{x_3(t)}, \\ \dot{x}_2(t) = cu(t) - gx_3(t), \\ \dot{x}_3(t) = -u(t). \end{cases}$$

If we consider the elevation of the rocket as its output, the output equation of the system is $y(t) = h(t)$, or

$$y(t) = x_1(t),$$

both when the mass is taken constant and time-varying.   ∎

As a final example to illustrate the notion of state we consider a *digital network* consisting of delays, adders, and gains. If at any fixed time the contents of all the *delay elements* are known, it is not necessary to know anything about the past behavior. The contents of the delays thus constitute the *state* of the system. We consider a specific example.

5.2.3.   Example: A tapped delay line.   A tapped delay line, also called a *transversal filter*, is the digital network schematically represented in Fig. 5.3. The discrete-time input u is fed into a series connection of M unit delays. After multipli-

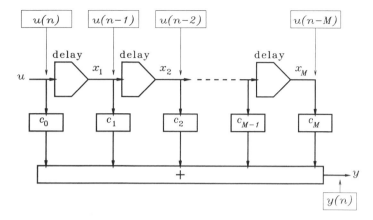

Figure 5.3.   A tapped delay line. The boxed quantities are the signal values at time n.

cation by scalar coefficients the outputs of the delays are added to form the output y of the circuit.

To find the state of the system at time n we need know which quantities determine the current and future output values $y(n)$, $y(n + 1)$, $\cdots$, if the current and future values of the input $u(n)$, $u(n + 1)$, $\cdots$, are given. From the block diagram the required quantities may be seen to be the outputs $u(n - 1)$, $u(n - 2)$, $\cdots$, $u(n - M)$ of the delays. Thus, the M components of the state $x(n)$ at time n are

$$x_1(n) = u(n - 1),$$
$$x_2(n) = u(n - 2),$$
$$\cdots \quad \cdots \quad \cdots$$
$$x_M(n) = u(n - M).$$

The evolution of the state is determined by a set of first-order difference equations that is easily found as follows:

$$x_1(n + 1) = u(n),$$
$$x_2(n + 1) = u(n - 1) = x_1(n),$$
$$x_3(n + 1) = u(n - 2) = x_2(n),$$
$$\cdots \quad \cdots \quad \cdots$$
$$x_M(n + 1) = u(n - M + 1) = x_{M-1}(n).$$

Defining the state vector

$$x(n) := \begin{bmatrix} x_1(n) \\ x_2(n) \\ \cdots \\ x_M(n) \end{bmatrix},$$

we may rewrite these equations in the form of the *state difference equation*

$$x(n + 1) = Ax(n) + Bu(n), \tag{7}$$

where the matrices A and B are given by

$$A = \begin{bmatrix} 0 & 0 & 0 & \cdots & 0 & 0 & 0 \\ 1 & 0 & 0 & \cdots & 0 & 0 & 0 \\ 0 & 1 & 0 & \cdots & 0 & 0 & 0 \\ \cdots & \cdots & \cdots & \cdots & \cdots & \cdots & \cdots \\ 0 & 0 & 0 & \cdots & 0 & 1 & 0 \end{bmatrix}, \ B = \begin{bmatrix} 1 \\ 0 \\ 0 \\ \cdots \\ 0 \end{bmatrix}.$$

Given the state $x(n)$ and the input $u(n)$ at time n the output $y(n)$ at that same time follows as

$$\begin{aligned} y(n) &= c_0 u(n) + c_1 u(n - 1) + \cdots + c_M u(n - M) \\ &= c_1 x_1(n) + c_2 x_2(n) + \cdots + c_M x_M(n) + c_0 u(n). \end{aligned}$$

This *output equation* may be rewritten as

$$y(n) = Cx(n) + Du(n), \tag{8}$$

where

$$C = [c_1 \quad c_2 \quad \cdots \quad c_M], \qquad D = c_0.$$

In the sequel we pay considerable attention to state systems described by a state difference equation of the form (7) and output equation (8). ∎

The procedure of the examples may be summarized as follows:

1. First determine those variables that summarize the past, referred to as the *state variables*.
2. Next find first-order difference or differential equations that describe the evolution of the state.
3. Finally express the output at any given time in the state and input at that same time.

State Systems

After having introduced the notion of state intuitively and by examples we turn to a formal definition of *input-output-state* (IOS) systems. The idea is to introduce, in addition to the input and output, a third signal—the state—that has the property that if the state at a fixed time t is given, no values of the input prior to t are needed to determine the output and state after time t.

5.2.4.   Definition: Input-output-state system.   Let $\mathbb{T}$ be a time axis, and $\mathcal{U}$, $\mathcal{Y}$ and $\mathcal{X}$ sets of time signals defined on the time axis $\mathbb{T}$. Elements of $\mathcal{U}$, $\mathcal{Y}$ and $\mathcal{X}$ are called *input signals, output signals* and *state signals,* respectively. Then an *input-output-state system* is defined by a subset $\mathcal{R}$ of $\mathcal{U} \times \mathcal{Y} \times \mathcal{X}$, called the *rule* or *relation* of the system, that has the following property.

Suppose that (u', y', x') and (u'', y'', x'') both belong to $\mathcal{R}$ such that

$$x'(t_o) = x''(t_o) \tag{9}$$

for some $t_o \in \mathbb{T}$. Then the following hold:

(a) *State matching property:* Define the *concatenated* signals u, y and x by

$$u(\tau) = \begin{cases} u'(\tau) & \text{for } \tau < t_o, \\ u''(\tau) & \text{for } \tau \geq t_o, \end{cases} \qquad y(\tau) = \begin{cases} y'(\tau) & \text{for } \tau < t_o, \\ y''(\tau) & \text{for } \tau \geq t_o, \end{cases}$$

$$x(\tau) = \begin{cases} x'(\tau) & \text{for } \tau < t_o, \\ x''(\tau) & \text{for } \tau \geq t_o, \end{cases}$$

for $\tau \in \mathbb{T}$. Then if (9) holds (u, y, x) forms an input-output-state triple, that is, $(u, y, x) \in \mathcal{R}$.

(b) *Causality:* If $u'(\tau) = u''(\tau)$ for $t_o \leq \tau < t$ and $\tau \in \mathbb{T}$, then if (9) holds

$$x'(\tau) = x''(\tau) \quad \text{for } t_o \leq \tau \leq t,$$

$$y'(\tau) = y''(\tau) \quad \text{for } t_o \leq \tau < t,$$

for $\tau \in \mathbb{T}$.   ∎

In the sequel we often simply write *state system* for input-output-state system.

The matching property implies that if two different "pasts" of input, output and state lead to the same state at some given time, both "futures" constitute a valid continuation. Causality means that if the initial state is the same and the inputs coincide during a certain initial time interval, the outputs and states also coincide on this interval.

The signal range X of the state signal is called the *state space* of the system.

5.2.5   Example: State Systems.

(a) *RCL network.* The RCL network of Example 5.2.1 is a state system whose state has the components $x_1 = q$ and $x_2 = \phi$. The state space of the system is $X = \mathbb{R}^2$.

(b) *Moon rocket*. The moon rocket of Example 5.2.2 is a state system. Accounting for the time-dependent behavior of the mass of the rocket the state of the system has the three components $x_1 = h$, $x_2 = p$ and $x_3 = m$. The state space of the system is $X = \mathbb{R}^3$.

(c) *Tapped delay line*. Also the tapped delay line of Example 5.2.3 is a state system. Its state x has M components, so that the state space of the system is $X = \mathbb{R}^M$.

(d) *Binary tapped delay line*. Suppose that in the tapped delay line of 5.2.3 all variables, including the coefficients, are *binary*, i.e., take their values in $\mathbb{B} = \{0, 1\}$, and that the additions and multiplications are defined modulo 2. Then the state x still has the M components as in Fig. 5.3, but the state space now is $\mathbb{B}^M$. ∎

The examples we saw so far all had a state vector with a finite number of components (i.e., their state space was of the form $X = \mathbb{R}^N$). The state space of the following example consists of *functions*.

5.2.6.   Example: Continuous-time delay.   Consider the continuous-time system whose IO relation is given by

$$y(t) = u(t - \theta), \qquad t \in \mathbb{R},$$

with θ a positive number. Given the input $u(\tau)$ for $\tau \geq t$, we also need know the preceding input segment $u(\tau)$, $t - \theta \leq \tau < t$, to determine the output y from time t on. Hence, the state at time t is the function segment $x(t)$ defined by

$$(x(t))(\theta) = u(t - \theta + \tau), \qquad 0 \leq \tau < \theta,$$

as illustrated in Fig. 5.4. The state space X of this system consists of all functions defined on the interval $[0, \theta)$. ∎

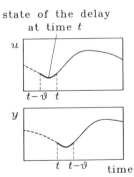

state of the delay
at time t

Figure 5.4.   State of the continuous-time delay.

State Difference and Differential Systems

In the sequel we restrict ourselves to two special types of state systems, one discrete-time, the other continuous-time. They are the most common although by no means the only important types of state systems, and are characterized by the fact that their state space X is *finite-dimensional,* with $X = \mathbb{R}^N$ or $X = \mathbb{C}^N$. The RCL network, the moon rocket and the digital network of Examples 5.2.1, 5.2.2, and 5.2.3 all belong to this class of state systems.

First consider a state system as in 5.2.4 defined on the discrete time axis $\mathbb{T} = \mathbb{Z}$. Then if for some $n \in \mathbb{Z}$ the state $x(n)$ and input $u(n)$, both at time n, are given, by property 5.2.4(b) the state $x(n + 1)$ at time $n + 1$ and the output $y(n)$ at time n are fully determined. Hence, there exist functions f and g such that

$$x(n + 1) = f(n, x(n), u(n)),$$
$$y(n) = g(n, x(n), u(n)), \qquad n \in \mathbb{Z}.$$

The first of these equations is the *state difference equation* of the system and the second the *output equation*. If the state difference equation is known, the evolution of the state from any initial state is fully determined by successive application of the difference equation. Once the behavior of the state has been found, that of the output follows by the output equation.

The state difference equation describes the evolution of the state over the shortest possible time span on the discrete time axis $\mathbb{Z}$. The state difference and output equations often may directly be obtained from the physical or other fundamental laws that govern the system. Systems of this type where at any given time the state belongs to the finite-dimensional state space $X = \mathbb{R}^N$ or $X = \mathbb{C}^N$, and the input and output signal ranges are also finite-dimensional spaces, are called *state difference systems*. The state difference and output equations for such systems may more explicitly be rendered as the *set* of equations

$$\begin{cases} x_1(n + 1) &= f_1(n, x_1(n), x_2(n), \cdots, x_N(n), u_1(n), u_2(n), \cdots, u_K(n)), \\ x_2(n + 1) &= f_2(n, x_1(n), x_2(n), \cdots, x_N(n), u_1(n), u_2(n), \cdots, u_K(n)), \\ \cdots & \quad \cdots \cdots \\ x_N(n + 1) &= f_N(n, x_1(n), x_2(n), \cdots, x_N(n), u_1(n), u_2(n), \cdots, u_K(n)), \end{cases}$$

$$\begin{cases} y_1(n) &= g_1(n, x_1(n), x_2(n), \cdots, x_N(n), u_1(n), u_2(n), \cdots, u_K(n)), \\ y_2(n) &= g_2(n, x_1(n), x_2(n), \cdots, x_N(n), u_1(n), u_2(n), \cdots, u_K(n)), \\ \cdots \cdots & \quad \cdots \\ y_M(n) &= g_M(n, x_1(n), x_2(n), \cdots, x_N(n), u_1(n), u_2(n), \cdots, u_K(n)), \end{cases}$$

for $n \in \mathbb{Z}$. Here the x_i are the components of x, the u_i those of u and the y_i those of y. Also, the f_i are the components of the vector-valued function f, and the g_i those of g. The tapped delay line of Example 5.2.3 is an example of a state difference system.

The continuous-time equivalent of a state difference system is a *state differential system*. This is a state system defined on the continuous time axis $\mathbb{T} = \mathbb{R}$ with state space $X = \mathbb{R}^N$ or $X = \mathbb{C}^N$ and finite-dimensional input and output signal ranges, and is described by the *state differential* and *output* equations

$$\dot{x}(t) = f(t, x(t), u(t)),$$

$$y(t) = g(t, x(t), u(t)), \qquad t \in \mathbb{R}.$$

As before, the overdot denotes differentiation. The state differential equation again describes the evolution of the state on the shortest possible time span. Under suitable conditions on the (vector-valued) function f, which are briefly touched upon in Section 5.4, the state differential equation uniquely determines the behavior of the state starting from any initial state if the input is known. The state differential equation and continuous-time output equation may be expressed in component form as

$$
\begin{cases}
\dot{x}_1(t) &=& f_1(t, x_1(t), x_2(t), \cdots, x_N(t), u_1(t), u_2(t), \cdots, u_K(t)), \\
\dot{x}_2(t) &=& f_2(t, x_1(t), x_2(t), \cdots, x_N(t), u_1(t), u_2(t), \cdots, u_K(t)), \\
\cdots & \cdots & \cdots \\
\dot{x}_N(t) &=& f_N(t, x_1(t), x_2(t), \cdots, x_N(t), u_1(t), u_2(t), \cdots, u_K(t)),
\end{cases}
$$

$$
\begin{cases}
y_1(t) &=& g_1(t, x_1(t), x_2(t), \cdots, x_N(t), u_1(t), u_2(t), \cdots, u_K(t)), \\
y_2(t) &=& g_2(t, x_1(t), x_2(t), \cdots, x_N(t), u_1(t), u_2(t), \cdots, u_K(t)), \\
\cdots & \cdots & \cdots \\
y_M(t) &=& g_M(t, x_1(t), x_2(t), \cdots, x_N(t), u_1(t), u_2(t), \cdots, u_K(t)).
\end{cases}
$$

Again, the x_i are the components of x, the u_i those of u, and the y_i those of y, while the f_i are the components of the vector-valued function f and the g_i those of g. The RCL network of Example 5.2.1 and the moon rocket of Example 5.2.2 are state differential systems. The continuous-time delay of Example 5.2.6 is *not* a state differential system because its state space is not finite-dimensional.

State Transition Map

Consider a general input-output-state system as defined in 5.2.4. Part (b) of the definition implies that if the state $x(t_o)$ of the system at some initial time t_o and the input u are given, the state $x(t)$ at time t with $t > t_o$ is uniquely determined. Hence, there exists a map, denoted s, such that

$$x(t) = s(t, t_o, x(t_o), u).$$

Since the map s describes how the input u takes the system from the state $x(t_o)$ at time t_o to the state $x(t)$ at time t, it is called the *state transition map* of the system.

The state transition map has the following properties, which may easily be verified:

(a) *Consistency:* For any $t \in \mathbb{T}$, any $x \in X$ and any $u \in \mathcal{U}$

$$s(t, t, x, u) = x.$$

Consistency simply means that if no time elapses, the system remains in the same state.

(b) *Semigroup property:* Let t_0, t_1 and t_2 be three time instants belonging to the time axis $\mathbb{T}$ such that $t_0 \leq t_1 \leq t_2$. Then for all $x_o \in X$ and all $u \in \mathcal{U}$

$$s(t_2, t_0, x_o, u) = s(t_2, t_1, s(t_1, t_0, x_o, u), u).$$

This property implies that taking the system from some initial state at time t_0 to some intermediate state at time t_1 and from this to a final state at time t_2 results in the same final state as taking the system directly from the initial state at time t_0 to the final state at time t_2.

(c) *Non-anticipativity:* Let t_o and t be two time instants such that $t_o < t$, and u' and u'' two inputs that coincide on the interval $[t_o, t)$, i.e., $u'(\tau) = u''(\tau)$ for $t_o \leq \tau < t$ and $\tau \in \mathbb{T}$. Then for any $x_o \in X$

$$s(t, t_o, x_o, u') = s(t, t_o, x_o, u'').$$

Non-anticipativity means that the transition of the state from time t_o to time t only depends on the input during the intervening interval, and *not* on the input at any other time.

We illustrate the state transition map for the RC network.

5.2.7.   Example: State transition map of the RC network.   The circuit diagram of the RC network is repeated in Fig. 5.5. The state of the system consists of the charge q of the capacitor. By Kirchhoff's current law $\dot{q}(t) = i(t)$, with i the current through the network. Since by Kirchhoff's voltage law $u(t) = v_R(t) + v_C(t) = Ri(t) + q(t)/C$ it follows that $i(t) = u(t)/R - q(t)/RC$, so that

$$\dot{q}(t) = -\frac{1}{RC}q(t) + \frac{1}{R}u(t), \qquad t \in \mathbb{R}.$$

Replacing q with x results in the state differential equation

$$\dot{x}(t) = -\frac{1}{RC}x(t) + \frac{1}{R}u(t), \qquad t \in \mathbb{R}. \tag{10}$$

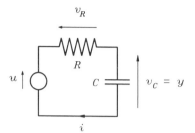

Figure 5.5. RC network.

The corresponding output equation is

$$y(t) = \frac{q(t)}{C} = \frac{1}{C}x(t).$$

For a given initial condition $x(t_o)$ and given input u the solution of the scalar first-order state differential equation (10) may be verified to be given by

$$x(t) = e^{-\frac{t-t_o}{RC}}x(t_o) + \frac{1}{R}\int_{t_o}^{t} e^{-\frac{t-\tau}{RC}}u(\tau)\,d\tau, \qquad t \geq t_o.$$

This expression explicitly represents the state transition map. The consistency, semi-group and non-anticipativity properties are easily seen to hold.   ∎

Linearity of State Systems

We conclude this section with brief discussions of the *linearity* and *time-invariance* of input-output-state systems. Linearity, as in the case of input-output systems, is equivalent to linearity of the relation $\mathcal{R}$ in 5.2.4 that defines the system.

5.2.8.   Definition: Linearity of input-output-state systems.   The input-output-state system with relation $\mathcal{R} \subset \mathcal{U} \times \mathcal{Y} \times \mathcal{X}$ is *linear* if $\mathcal{U}$, $\mathcal{Y}$ and $\mathcal{X}$ are linear spaces and the relation $\mathcal{R}$ is a linear subspace.   ∎

Linearity of a state system means that if $(u', y', x') \in \mathcal{R}$ and $(u'', y'', x'') \in \mathcal{R}$ are two input-output-state triples, any linear combination of these is also an input-output-state triple.

It may easily be proved that state difference systems are linear if their state difference and output equations are both linear, and that state differential systems are linear if their state differential and output equations are both linear:

5.2.9.   Summary: Linearity of state difference systems and state differential systems.

The state difference system with state difference and output equations

$$x(n + 1) = A(n)x(n) + B(n)u(n),$$

$$y(n) = C(n)x(n) + D(n)u(n),$$

$n \in \mathbb{Z}$, with $A(n)$, $B(n)$, $C(n)$, $D(n)$, $n \in \mathbb{Z}$, matrices of appropriate dimensions whose elements are real- or complex-valued functions of time, is linear.

The state differential system with state differential and output equations

$$\dot{x}(t) = A(t)x(t) + B(t)u(t),$$

$$y(t) = C(t)x(t) + D(t)u(t),$$

$t \in \mathbb{R}$, with $A(t)$, $B(t)$, $C(t)$, $D(t)$, $t \in \mathbb{R}$, matrices of appropriate dimensions whose elements are real- or complex-valued bounded and continuous functions of time, is linear.   ∎

As seen in Section 5.4, the boundedness and continuity of the matrix functions A, B, C and D guarantee the existence of solutions to the state differential equation.

5.2.10.   Examples: Linear state difference and differential systems.   It is easily recognized that the state difference and output equations of the tapped delay line of Example 5.2.3 are of the type $x(n + 1) = Ax(n) + Bu(n)$, $y(n) = Cx(n) + Du(n)$, with A, B, C and D constant matrices. Hence, the tapped delay line is linear.

The RCL network of Example 5.2.1 and the RC network of Example 5.2.7 have a state differential equation and output equation of the form $\dot{x}(t) = Ax(t) + Bu(t)$, $y(t) = Cx(t) + Du(t)$, with A, B, C and D constant matrices. Hence, also these systems are linear.

The state differential and output equations of the moon rocket of Example 5.2.2 (accounting for the variable mass) are *not* linear. Indeed, it may easily be shown that the system is nonlinear.   ∎

Time-Invariance of State Systems

It remains to discuss time-invariance of state systems. As in the case of input-output systems, time-invariance amounts to shift-invariance of the relation that defines the system.

5.2.11.   Definition: Time-invariance of input-output-state systems.   The input-output-state system with relation $\mathcal{R} \subset \mathcal{U} \times \mathcal{Y} \times \mathcal{X}$ defined on an infinite or right semi-infinite time axis is time-invariant if $\mathcal{R}$ is shift-invariant, that is, if $(u, y, x) \in \mathcal{R}$ then $(\sigma^\theta u, \sigma^\theta y, \sigma^\theta x) \in \mathcal{R}$ for any allowable time shift θ.   ∎

Time-invariance of the state system implies that if any input-output-state triple is shifted over any amount of time, it still is an input-output-state triple. Time-invariance means that the dynamic properties of the system do not change with time.

State difference and differential systems are time-invariant if the right-hand sides of the state difference or differential equation, respectively, and that of the output equation depend on time only through the state x and the input u and not directly:

5.2.12.   Summary: Time-invariance of state difference and differential systems.

The discrete-time system described by the state difference and output equations

$$x(n+1) = f(x(n), u(n)),$$
$$y(n) = g(x(n), u(n)), \quad n \in \mathbb{Z},$$

is time-invariant.

The continuous-time system described by the state differential and output equations

$$\dot{x}(t) = f(x(t), u(t)),$$
$$y(t) = g(x(t), u(t)), \quad t \in \mathbb{R},$$

is time-invariant.   ∎

Combining 5.2.11 and 5.2.12 we obtain the following result.

5.2.13.   Summary: Linear time-invariant state difference and differential systems.

The state difference system with state difference and output equations

$$x(n+1) = Ax(n) + Bu(n),$$
$$y(n) = Cx(n) + Du(n), \quad n \in \mathbb{Z},$$

with A, B, C and D constant matrices, is both linear and time-invariant.

The state differential system with state differential and output equations

$$\dot{x}(t) = Ax(t) + Bu(t),$$
$$y(t) = Cx(t) + Du(t), \quad t \in \mathbb{R},$$

with A, B, C and D constant matrices, is both linear and time-invariant.   ∎

5.2.14.   Example: Time-invariant state difference and differential systems. The tapped delay line of Example 5.2.3 has a state difference and output equation of the form $x(n+1) = Ax(n) + Bu(n)$, $y(n) = Cx(n) + Du(n)$, $n \in \mathbb{Z}$, with A, B, C and D constant matrices, Hence, this system is both time-invariant and linear.

The RCL network of Example 5.2.1 and the RC network of Example 5.2.7 have state differential and output equations of the form $\dot{x}(t) = Ax(t) + Bu(t)$, $y(t) = Cx(t) + Du(t)$, $t \in \mathbb{R}$, with A, B, C and D constant matrices, and hence are also both linear and time-invariant.

The moon rocket of Example 5.2.2 is time-invariant but not linear. If the constant c would change with time the system would not only be nonlinear but also time-varying.   ∎

5.2.15.   Review: Sampled state difference systems.   Sampled state difference systems are input-output-state systems defined on the time axis $\mathbb{Z}(T)$. They are described by a state difference and output equation of the form

$$x(t + T) = f(t, x(t), u(t)),$$
$$y(t) = g(t, x(t), u(t)), \qquad t \in \mathbb{Z}(T).$$

Sampled state difference systems are linear if their state difference and output equations are of the form

$$x(t + T) = A(t)x(t) + B(t)u(t),$$
$$y(t) = C(t)x(t) + D(t)u(t), \qquad t \in \mathbb{Z}(T),$$

with A, B, C, and D possibly time-dependent matrices of suitable dimensions. Sampled state difference systems described by

$$x(t + T) = f(x(t), u(t)),$$
$$y(t) = g(x(t), u(t)), \qquad t \in \mathbb{Z}(T),$$

are time-invariant. Linear time-invariant sampled state difference systems, finally, have a state difference and output equation of the form

$$x(t + T) = Ax(t) + Bu(t),$$
$$y(t) = Cx(t) + Du(t), \qquad t \in \mathbb{Z}(T),$$

with A, B, C, and D *constant* matrices.   ∎

5.3 REALIZATION OF DIFFERENCE AND DIFFERENTIAL SYSTEMS AS STATE SYSTEMS

In this section we study how linear difference and differential systems, which are described by a difference or differential equation that relates input and output, may be represented as state systems. Because as we shall shortly see state difference and differential systems may easily be implemented in hardware or software, these results are important when it comes to actually building difference and differential systems. Moreover, constructing a state representation helps to study and understand its properties.

Implementation of State Difference and Differential Systems

We first discuss the implementation of *state difference systems*. For this we assume the availability of three basic building blocks, namely, *unit time delays, function generators,* and *time clocks.* These building blocks may exist as *hardware* (i.e., in the form of physical devices), or as *software* (i.e., computer routines that emulate the physical function).

A *unit time delay* may be represented as in Fig. 5.6(a). It accepts a scalar or vector-valued discrete-time signal u as input and delays it by one time unit, so that its output is

$$y(n) = u(n - 1), \qquad n \geq n_o, \qquad n \in \mathbb{Z}.$$

At start-up time n_o the unit delay requires an initial value $y(n_o) = y_o$, which is also indicated in Fig. 5.6(a).

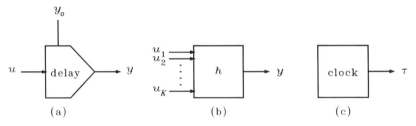

Figure 5.6. Building blocks for discrete-time state systems. Left: unit delay. Middle: function generator. Right: time clock.

A *function generator* may be depicted as in Fig. 5.6(b). It accepts a number of scalar or vector-valued discrete-time signals $u_1, u_2, \cdots, u_K$ as inputs, and produces the output y given by

$$y(n) = h(u_1(n), u_2(n), \cdots, u_K(n)), \qquad n \in \mathbb{Z},$$

with h a given function.

A *time clock,* finally, as shown in Fig. 5.6(c), is a system without input whose output is a discrete-time signal τ such that

$$\tau(n) = n, \qquad n \in \mathbb{Z}.$$

Given these building blocks, a state difference system with state difference and output equations

$$x(n + 1) = f(n, x(n), u(n)),$$
$$y(n) = g(n, x(n), u(n)), \qquad n = n_o, n_o + 1, \cdots,$$

may be realized as in the block diagram of Fig. 5.7. At any instant of time n one may measure at each point in the block diagram the value of the corresponding signal at that time. From the block diagram we see that at time n

$$(\sigma x)(n) = f(\tau(n), x(n), u(n)),$$

or

$$x(n + 1) = f(n, x(n), u(n)).$$

Moreover,

$$y(n) = g(\tau(n), x(n), u(n)) = g(n, x(n), u(n)).$$

In practice, since the input to the unit delay is a vector-valued signal x with components $x_1, x_2, \cdots, x_N$, the unit delay consists of a *parallel bank of delays,* each of which accepts a scalar signal as input. Similarly, the function generators f and possibly also g are parallel banks of function generators, each accepting several signals simultaneously as input and producing one of the components of x or y, respectively, as output.

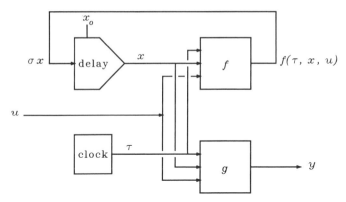

Figure 5.7.   Implementation of a state difference system.

Sampled state difference equations as discussed in 5.2.15 are defined on the time axis $\mathbb{T} = \mathbb{Z}(T)$, and may be realized as in Fig. 5.7 with delays that retard by one sampling interval T.

For the realization of state differential systems the role of the unit delay is taken by an *integrator*. An integrator accepts a scalar or vector-valued signal u as input and produces the output

$$y(t) = y(t_o) + \int_{t_o}^{t} u(\theta)\, d\theta, \qquad t \geq t_o.$$

The initial value $y(t_o)$ need be set at the initial time t_o. Differentiating, we find that input and output are related as

$$\dot{y}(t) = u(t), \qquad t \geq t_o.$$

Integrators are represented in a block diagram as in Fig. 5.8.

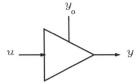

Figure 5.8. An integrator.

The state differential system with state differential and output equations

$$\dot{x}(t) = f(t, x(t), u(t)),$$
$$y(t) = g(t, x(t), u(t)), \qquad t \geq t_o, \qquad t \in \mathbb{R},$$

may be implemented using an integrator, function generators and a time clock as in Fig. 5.9. The time clock now runs in continuous time. Also here, an integrator that operates on vector-valued signals in practice is constructed as a parallel bank of scalar integrators.

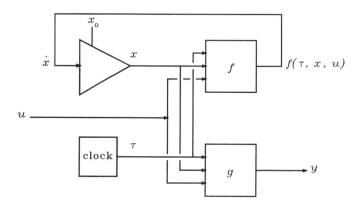

Figure 5.9. Implementation of a state differential system.

5.3.1.   Example: Simulation of the moon rocket.   Suppose that we wish to build an implementation of the equations that describe the moon rocket of Example 5.2.2. Such an implementation is called a *simulation,* and allows studying the dynamic behavior of the rocket without constructing the rocket itself. The state differential equation is given by

$$\dot{x}_1(t) = \frac{x_2(t)}{x_3(t)},$$

$$\dot{x}_2(t) = cu(t) - gx_3(t),$$

$$\dot{x}_3(t) = -u(t),$$

while the output equation is

$$y(t) = x_1(t).$$

Figure 5.10 shows a block diagram of the implementation. It uses three scalar integrators and three function generators. Two of these function generators are "gains," which simply multiply their scalar input by a fixed constant. The third is a "divider," whose output is the quotient x_2/x_3 of its scalar inputs x_2 and x_3. The block diagram also includes an adder, which is another special type of function generator. Because the implementation has three integrators, three initial values are needed at start-up time.   ■

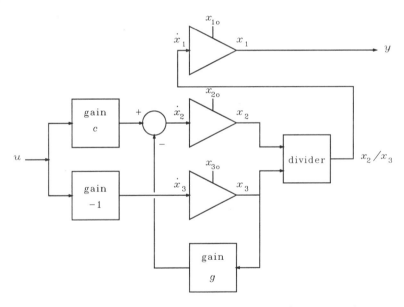

Figure 5.10.   Implementation of the state equations of the moon rocket.

State Realization of Linear Difference and Differential Systems: Examples

To implement a difference or differential system we need a state realization of the system. A state realization of a linear constant coefficient difference or differential system may be found by the *matching property* of the state as defined in 5.2.4. We illustrate this by means of two examples.

5.3.2.   Examples: State realization of difference and differential systems.

(a) *Exponential smoother*. The exponential smoother is described by the difference equation

$$y(n + 1) = ay(n) + (1 - a)u(n + 1), \qquad n \in \mathbb{Z}. \tag{1}$$

To determine the state of the smoother we use the *state matching* property of 5.2.4. This property implies the following. Suppose that (u', y') and (u'', y'') are two input-output pairs of the smoother. Then for any $n_o \in \mathbb{Z}$ the concatenated pair (u, y) defined by

$$u(n) = \begin{cases} u'(n) & \text{for } n < n_o, \\ u''(n) & \text{for } n \geq n_o, \end{cases} \qquad n \in \mathbb{Z}, \tag{2a}$$

$$y(n) = \begin{cases} y'(n) & \text{for } n < n_o \\ y''(n) & \text{for } n \geq n_o, \end{cases} \qquad n \in \mathbb{Z}, \tag{2b}$$

is an IO pair provided $x'(n_o) = x''(n_o)$, where x' is the state signal corresponding to the pair (u', y') and x'' that corresponding to (u'', y''). Hence, we may identify the state by recognizing what constraint the requirement that (u, y) be an IO pair imposes on (u', y') and (u'', y'').

Substitution of (u, y) as given by (2) into the difference equation (1) requires distinguishing the cases $n < n_o - 1$, $n = n_o - 1$ and $n > n_o$. We successively obtain

$$
\begin{aligned}
n < n_o - 1: &\quad y'(n + 1) = ay'(n) + (1 - a)u'(n + 1) \\
n = n_o - 1: &\quad y''(n_o) = ay'(n_o - 1) + (1 - a)u''(n_o), \\
n > n_o - 1: &\quad y''(n + 1) = ay''(n) + (1 - a)u''(n + 1).
\end{aligned} \tag{3}
$$

The equality for $n < n_o - 1$ is satisfied because by assumption (u', y') is an IO pair. Similarly, the equality for $n > n_o - 1$ is satisfied because by assumption also (u'', y'') is an IO pair. From (3) it follows that the two IO pairs need be related by

$$ay'(n_o - 1) = y''(n_o) - (1 - a)u''(n_o).$$

Since (u'', y'') satisfies the difference equation this may be rewritten in the form

$$ay'(n_o - 1) = ay''(n_o - 1).$$

Now let

$$x'(n_o) = ay'(n_o - 1), \qquad x''(n_o) = ay''(n_o - 1).$$

Then the condition $x'(n_o) = x''(n_o)$ guarantees (u, y) to be an IO pair. It follows that the state of the exponential smoother (1) is given by

$$x(n) = ay(n-1), \qquad n \in \mathbb{Z}.$$

The resulting state difference and output equations for the exponential smoother may easily be established. We have from (1)

$$\begin{aligned} x(n+1) = ay(n) &= a^2 y(n-1) + a(1-a)u(n) \\ &= ax(n) + a(1-a)u(n), \qquad n \in \mathbb{Z}, \end{aligned}$$

which constitutes the state difference equation. On the other hand, it follows from the input-output difference equation that

$$\begin{aligned} y(n) &= ay(n-1) + (1-a)u(n) \\ &= x(n) + (1-a)u(n), \qquad n \in \mathbb{Z}, \end{aligned}$$

which is the output equation of the state representation of the smoother. An implementation of the state representation is given in Fig. 5.11.

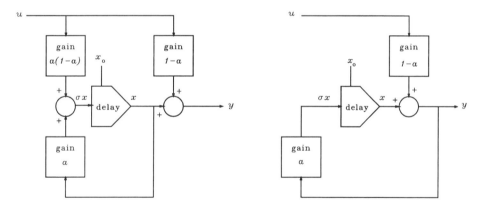

Figure 5.11.   Implementation of the exponential smoother. Left: a direct implementation of the state difference and output equations. Right: a more efficient implementation using fewer coefficients.

(b) *RC network*. In Example 1.2.7 we have seen that if the output y of the RC network of Fig. 5.12 is the voltage across the capacitor, the network is described by the differential equation

$$\frac{dy(t)}{dt} + \frac{1}{RC} y(t) = \frac{1}{RC} u(t), \qquad t \in \mathbb{R}. \tag{4}$$

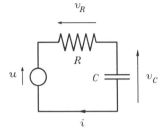

Figure 5.12. RC network.

Suppose now that the output y is not the voltage v_C across the capacitor but the voltage v_R across the resistor. Since $v_C = u - v_R$, the differential equation for this output follows by replacing y in (4) by $u - y$, resulting in the differential equation

$$\frac{dy(t)}{dt} + \frac{1}{RC}y(t) = \frac{du(t)}{dt}, \qquad t \in \mathbb{R}. \tag{5}$$

Again we use the state matching property to identify the state of this system. Suppose that (u', y') and (u'', y'') are two IO pairs satisfying the differential equation (5). Denoting the corresponding state signals as x' and x'', the concatenated pair (u, y) defined by

$$u(t) = \begin{cases} u'(t) & \text{for } t < t_o, \\ u''(t) & \text{for } t \geq t_o, \end{cases} \qquad t \in \mathbb{R},$$

$$y(t) = \begin{cases} y'(t) & \text{for } t < t_o, \\ y''(t) & \text{for } t \geq t_o, \end{cases} \qquad t \in \mathbb{R},$$

is an IO pair if $x'(t_o) = x''(t_o)$. Rewriting y as

$$y(t) = [1 - \mathbb{1}(t - t_o)]y'(t) + \mathbb{1}(t - t_o)y''(t), \qquad t \in \mathbb{R},$$

it follows after differentiation that

$$\frac{dy(t)}{dt} = [1 - \mathbb{1}(t - t_o)]\frac{dy'(t)}{dt} + \mathbb{1}(t - t_o)\frac{dy''(t)}{dt}$$
$$+ [y''(t_o) - y'(t_o)]\delta(t - t_o), \qquad t \in \mathbb{R}.$$

Similarly we have

$$\frac{du(t)}{dt} = [1 - \mathbb{1}(t - t_o)]\frac{du'(t)}{dt} + \mathbb{1}(t - t_o)\frac{du''(t)}{dt}$$
$$+ [u''(t_o) - u'(t_o)]\delta(t - t_o), \qquad t \in \mathbb{R}.$$

Substitution of y and u into the differential equation (5) and using the fact that both (u', y') and (u'', y'') satisfy the differential equation we find that (u, y) is an IO pair if

$$[y''(t_o) - y'(t_o)]\delta(t - t_o) = [u''(t_o) - u'(t_o)]\delta(t - t_o),$$

or

$$y''(t_o) - y'(t_o) = u''(t_o) - u'(t_o).$$

Rearrangement shows that (u, y) is an IO pair, that is, $x'(t_o) = x''(t_o)$, if

$$y'(t_o) - u'(t_o) = y''(t_o) - u''(t_o).$$

Hence, the state of the system may be chosen as

$$x(t) = y(t) - u(t), \qquad t \in \mathbb{R},$$

because then $x'(t_o) = x''(t_o)$ guarantees the concatenated pair (u, y) to be an IO pair. The state differential equation follows from (5) as

$$\dot{x}(t) = \frac{dy(t)}{dt} - \frac{du(t)}{dt} = -\frac{1}{RC}y(t) = -\frac{1}{RC}[x(t) + u(t)]$$

$$= -\frac{1}{RC}x(t) - \frac{1}{RC}u(t), \qquad t \in \mathbb{R},$$

while the output equation is

$$y(t) = x(t) + u(t), \qquad t \in \mathbb{R}.$$

An implementation of the system is given in Fig. 5.13.

We observe from the circuit diagram of Fig. 5.12 that $x = y - u = v_R - u = -v_C$. Hence, within a factor -1 the state equals the voltage across the capacitor, which in turn within a factor $1/C$ equals the charge q of the capacitor. This confirms the physical argument of 5.2.7 that the charge q may be chosen as the state of the network. ∎

State Realization of Linear Difference and Differential Systems

We now consider quite generally the state realization of difference systems described by the linear constant coefficient difference equation

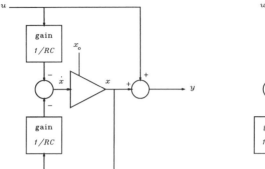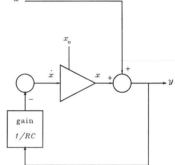

Figure 5.13. Implementation of the RC network with the output across the resistor as output. Left: a direct implementation of the state differential and output equations. Right: a more efficient implementation using fewer coefficients.

$$Q(\sigma)y = P(\sigma)u,$$

and differential systems described by

$$Q(D)y = P(D)u.$$

Q and P are the polynomials

$$Q(\lambda) = q_0 + q_1\lambda + \cdots + q_N\lambda^N,$$
$$P(\lambda) = p_0 + p_1\lambda + \cdots + p_M\lambda^M.$$

Without loss of generality we assume that $q_N = 1$. We furthermore suppose that $M \leq N$. By the latter assumption, without loss of generality we may take $M = N$, where in contrast to Chapter 4 the leading coefficient p_N may be zero.

The main result may be formulated as follows.

5.3.3.   Summary: State realization of difference and differential systems.

The difference system described by the difference equation

$$Q(\sigma)y = P(\sigma)u,$$

where

$$Q(\lambda) = q_0 + q_1\lambda + \cdots + q_N\lambda^N,$$
$$P(\lambda) = p_0 + p_1\lambda + \cdots + p_N\lambda^N,$$

The differential system described by the differential equation

$$Q(D)y = P(D)u,$$

where

$$Q(\lambda) = q_0 + q_1\lambda + \cdots + q_N\lambda^N,$$
$$P(\lambda) = p_0 + p_1\lambda + \cdots + p_N\lambda^N,$$

with $q_N = 1$, has a state realization

$$x(n + 1) = Ax(n) + Bu(n),$$
$$y(n) = Cx(n) + Du(n), \quad n \in \mathbb{Z},$$

such that

$$A = \begin{bmatrix} -q_{N-1} & 1 & 0 & \cdots & 0 \\ -q_{N-2} & 0 & 1 & \cdots & 0 \\ \cdots & \cdots & \cdots & \cdots & \cdots \\ -q_1 & 0 & 0 & \cdots & 1 \\ -q_0 & 0 & 0 & \cdots & 0 \end{bmatrix},$$

$$C = [1 \quad 0 \quad 0 \quad \cdots \quad 0],$$

with $q_N = 1$, has a state realization

$$\dot{x}(t) = Ax(t) + Bu(t),$$
$$y(t) = Cx(t) + Du(t), \quad t \in \mathbb{R},$$

such that

$$B = \begin{bmatrix} p_{N-1} - q_{N-1}p_N \\ p_{N-2} - q_{N-2}p_N \\ \cdots \\ p_1 - q_1 p_N \\ p_0 - q_0 p_N \end{bmatrix},$$

$$D = p_N. \qquad \blacksquare$$

In Figures 5.14 and 5.15 implementations are given of the state realizations using gains, adders, unit delays and integrators. The number of unit delays or integrators needed for the implementation equals the order N of the system.

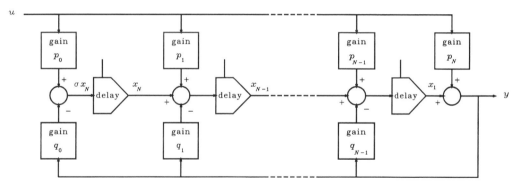

Figure 5.14.   State realization of a difference system.

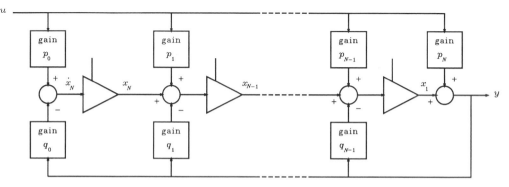

Figure 5.15.   State realization of a differential system.

5.3.4. Proof of 5.3.3.   We only give an outline of the proof. The details are straightforward but lengthy.

(a) *Discrete-time case*. To identify the state of the system described by the difference equation $Q(\sigma)y = P(\sigma)u$ we employ the state matching property of 5.2.4. N conditions are needed to make sure that the concatenation of two IO pairs at time n is again an IO pair. Inspection of the conditions shows that the corresponding N components of the state are

$$x_1(n) = p_0 u(n - N) + p_1 u(n - N + 1) + \cdots + p_{N-1} u(n - 1)$$
$$\qquad - q_0 y(n - N) - q_1 y(n - N + 1) - \cdots - q_{N-1} y(n - 1),$$
$$x_2(n) = p_0 u(n - N + 1) + p_1 u(n - N + 2) + \cdots + p_{N-2} u(n - 1)$$
$$\qquad - q_0 y(n - N + 1) - q_1 y(n - N + 2) - \cdots - q_{N-2} y(n - 1),$$
$$\cdots \quad \cdots \quad \cdots \quad \cdots$$
$$x_N(n) = p_0 u(n - 1) - q_0 y(n - 1).$$

From these definitions the state difference and output equations follow easily with the help of the input-output difference equation.

(b) *Continuous-time case*. Also in the continuous-time case the matching property is used. Concatenation of two IO pairs normally causes a step change, which after substitution into the input-output differential equation results in δ-functions and derivatives of δ-functions. Matching the coefficients of the δ-functions leads to N conditions, from which it follows that the components of the state at time t may be chosen as

$$x_1(t) = y(t) - p_N u(t),$$
$$x_2(t) = q_{N-1} y(t) + Dy(t) - p_{N-1} u(t) - p_N Du(t),$$
$$\cdots \quad \cdots \quad \cdots$$
$$x_N(t) = q_1 y(t) + q_2 Dy(t) + \cdots + D^{N-1} y(t)$$
$$\qquad - p_1 u(t) - p_2 Du(t) - \cdots - p_N D^{N-1} u(t).$$

The state differential and output equations follow easily from this.   ∎

Given a difference system $Q(\sigma)y = P(\sigma)u$ or a differential system $Q(D)y = P(D)u$, the realizations of 5.3.3 are by no means the only state representations. In fact, as seen in Section 5.6, any nonsingular transformation of the state may serve as a state for the system.

We close this section with some further examples of state realizations.

5.3.5. Examples: State realizations of difference and differential systems.

(a) *Second-order smoother*. Consider the second-order smoother described by the difference equation

$$y(n + 2) - a_1 y(n + 1) - a_0 y(n) = b_2 u(n + 2) + b_1 u(n + 1), \qquad n \in \mathbb{Z},$$

introduced in Example 4.2.4(b). It follows that

$$Q(\lambda) = -a_0 - a_1\lambda + \lambda^2, \qquad P(\lambda) = b_1\lambda + b_2\lambda^2.$$

Since the order of the system is $N = 2$, the dimension of the state is also 2. Identifying $q_1 = -a_1$, $q_0 = -a_0$, $p_2 = b_2$, $p_1 = b_1$ and $p_0 = 0$, by 5.3.3 the system may be represented in state form as

$$x(n + 1) = \begin{bmatrix} a_1 & 1 \\ a_0 & 0 \end{bmatrix} x(n) + \begin{bmatrix} b_1 + a_1 b_2 \\ a_0 b_2 \end{bmatrix} u(n),$$

$$y(n) = [1 \quad 0]x(n) + b_2 u(n), \qquad n \in \mathbb{Z}.$$

A block diagram of the implementation of the smoother is given in Fig. 5.16.

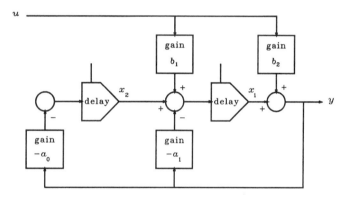

Figure 5.16.   State space realization of the second-order smoother.

(b) *Approximate differentiator*. A differentiator is a continuous-time system with IO map $y = Du$. Its frequency response function is

$$\hat{h}(f) = j2\pi f, \qquad f \in \mathbb{R}.$$

As frequency increases, the magnitude of the frequency response function goes to infinity. This causes large amplification of any noise added to the input, which is always present and usually contains high frequencies. This difficulty may be avoided by using an *approximate* differentiator, whose frequency response function behaves like that of the differentiator at low frequencies, but stops increasing or even starts decreasing at high frequencies. An example of an approximate differentiator is a system with frequency response function

$$\hat{h}(f) = \frac{j2\pi f}{(1 + j2\pi f/\omega_0)^2}, \qquad f \in \mathbb{R},$$

with ω_0 a positive number. The system behaves like a differentiator more or less up to the frequency $\omega_0/2\pi$. The function $\hat{h}$ is the frequency response function

$$\hat{h}(f) = \frac{P(j2\pi f)}{Q(j2\pi f)} = \frac{\omega_o^2 j2\pi f}{\omega_o^2 + 2\omega_o j2\pi f + (j2\pi f)^2}, \qquad f \in \mathbb{R},$$

of a differential system $Q(D)y = P(D)u$, whose denominator and numerator polynomials are

$$Q(\lambda) = \omega_o^2 + 2\omega_o \lambda + \lambda^2, \qquad P(\lambda) = \omega_o^2 \lambda.$$

Identifying $q_0 = \omega_o^2$, $q_1 = 2\omega_o$, $p_0 = p_2 = 0$ and $p_1 = \omega_o^2$, by 5.3.3 the system may be realized in state form as

$$\dot{x}(t) = \begin{bmatrix} -2\omega_o & 1 \\ -\omega_0^2 & 0 \end{bmatrix} x(t) + \begin{bmatrix} \omega_o^2 \\ 0 \end{bmatrix} u(t),$$

$$y(t) = [1 \quad 0]x(t), \qquad t \in \mathbb{R}.$$

A block diagram of the realization is given in Fig. 5.17.   ■

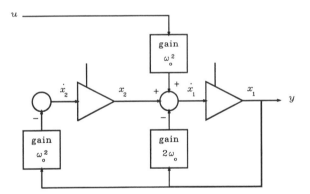

Figure 5.17.   Realization of the approximate differentiator.

5.4 SOLUTION OF STATE EQUATIONS

In Section 5.2 we introduced state systems, and in Section 5.3 we showed how such systems may be implemented. In the present section we discuss the existence of solutions to state difference and differential equations, and their numerical solution.

The starting point for the description of discrete- and continuous-time state systems, with finite-dimensional state space $X = \mathbb{R}^N$ or $\mathbb{C}^N$ and finite-dimensional input and output signal ranges $U = \mathbb{R}^K$ or $\mathbb{C}^K$ and $Y = \mathbb{R}^M$ or $\mathbb{C}^M$, consists of their state difference or differential equation, respectively, and output equation:

Discrete-time: For $n \in \mathbb{Z}$

$$x(n + 1) = f(n, x(n), u(n)),$$
$$y(n) = g(n, x(n), u(n)).$$

Continuous-time: For $t \in \mathbb{R}$

$$\dot{x}(t) = f(t, x(t), u(t)),$$
$$y(t) = g(t, x(t), u(t)).$$

Existence of Solutions

In the discrete-time case the solution of the state difference equation is clear. Given an initial time n_o, an initial state $x(n_o)$ and an input $u(n)$ for $n \geq n_o$, $n \in \mathbb{Z}$, the state difference equation

$$x(n + 1) = f(n, x(n), u(n)), \qquad n \geq n_o, n \in \mathbb{Z},$$

may be solved for $x(n_o + 1)$, $x(n_o + 2)$, $\cdots$, by *successive substitution*. It is easy to program a digital computer to do this numerically, and sometimes it may be done analytically. Once the state trajectory $x(n)$, $n \geq n_o$, $n \in \mathbb{Z}$, has been found, it is straightforward to determine the corresponding output y using the output equation

$$y(n) = g(n, x(n), u(n)), \qquad n \geq n_o, \qquad n \in \mathbb{Z}.$$

For linear time-invariant state difference systems much can be inferred about the solution of the state difference equation before resorting to the computer. This is discussed in Section 5.5.

In the continuous-time case the situation is not quite so simple. First of all, the state differential equation does not always have a unique solution for a given initial state $x(t_o)$. Second, even if it does, it is not as easy as in the discrete-time case to compute the state trajectory. In the remainder of this section we first discuss the existence of solutions of the state differential equation, and then review some basic numerical methods for solving it.

The following theorem gives a well-known sufficient condition for the existence of unique solutions of a set of differential equations of the form $\dot{x}(t) = f(t, x(t))$, $t \in \mathbb{R}$. State differential equations are of this type when the input is a fixed given time function.

5.4.1.   Summary: Existence of solutions to state differential equations.
Suppose that the map $f: \mathbb{R} \times \mathbb{R}^N \to \mathbb{R}^N$ satisfies the following conditions:

   (a) f is continuous on the set

$$Q = \{(t, x) \in \mathbb{R} \times \mathbb{R}^N \mid \|x - x_o\| \leq \alpha, |t - t_o| \leq \beta\},$$

with α and β some positive real constants, and

$$\|f(t, x)\| < \gamma \text{ for all } (t, x) \in Q$$

for some positive real constant γ;

   (b) f satisfies the *Lipschitz condition*

$$\|f(t, x_1) - f(t, x_2)\| \le \kappa \|x_1 - x_2\| \qquad \text{for all } (t, x_1), (t, x_2) \in Q$$

for some positive real constant κ.

   Then if $\eta = \min(\alpha, \beta/\gamma)$ there exists a unique continuous function x defined on $[t_o - \eta, t_o + \eta]$ such that

   (i) $x(t_o) = x_o$ and
   (ii) x satisfies the differential equation $\dot{x}(t) = f(t, x(t))$ for $t \in [t_o - \eta, t_o + \eta]$.
   ∎

> Rudolf Lipschitz (1832–1903) was a German mathematician who is well-known for his work on differential equations and Fourier series.

The norm $\|\cdot\|$ denotes any natural norm on $\mathbb{R}^N$.

5.4.2.   Example: Nonunique solution.   The standard example to show that if the conditions of 5.4.1 are not satisfied the initial value problem may have no unique solution is the scalar initial value problem

$$\dot{x}(t) = \sqrt{|x(t)|}, \qquad x(0) = 0.$$

The function $f : \mathbb{R} \times \mathbb{R} \to \mathbb{R}$ defined by $f(t, x) = \sqrt{|x|}$ is continuous everywhere but does not satisfy a Lipschitz condition on any set that includes the origin. Both

$$x(t) = 0, \qquad t \in \mathbb{R},$$

and

$$x(t) = \begin{cases} -\frac{1}{4}t^2 & \text{for } t < 0, \\ \frac{1}{4}t^2 & \text{for } t \ge 0, \end{cases} \qquad t \in \mathbb{R},$$

solve the initial value problem, which therefore has no unique solution.   ∎

Numerical Integration of State Differential Equations

Even if the solution of a state differential equation is known to exist, it may be not easy or even impossible to find a closed-form solution. In this case numerical solution is called for. There exist many efficient methods for the numerical solution of differential equations. We describe some of these for the solution of initial value problems of the form $\dot{x}(t) = f(x(t))$, $x(0) = x_o$. Initial value problems for time-varying differential equations of the form $\dot{x}(t) = f(t, x(t))$ may easily be reduced to this form.

5.4.3.   Exercise: Reduction of a time-varying to a time-invariant equation.
Show that by defining an additional state variable $x_{N+1}(t) = t$ the initial value problem

$$\dot{x}(t) = f(t, x(t)), \qquad x(t_o) = x_o,$$

is equivalent to the initial value problem

$$\dot{z}(\tau) = g(z(\tau)), \qquad z(0) = z_o,$$

where $\tau = t - t_o$, and

$$z(\tau) = \begin{bmatrix} x(\tau) \\ x_{N+1}(\tau) \end{bmatrix}, \qquad g(z) = \begin{bmatrix} f(x_{N+1}, x) \\ 1 \end{bmatrix}, \qquad z_o = \begin{bmatrix} x_o \\ 0 \end{bmatrix}. \qquad \blacksquare$$

The simplest but usually not the most practical technique for solving the initial value problem $\dot{x} = f(x)$, $x(0) = x_o$, is *Euler's method*.

The Swiss mathematician Leonhard Euler (1707–1783) lived in Berlin and St. Petersburg. He was the central figure in the mathematical activities of the 18th century and the most prolific mathematician of all time.

5.4.4.   Summary: Euler's method for approximating the solution of differential equations.
Euler's method for approximating the solution of the initial value problem $\dot{x} = f(x)$, $x(0) = x_o$ consists of solving the approximating difference equation

$$x(t + T) = x(t) + T \cdot f(x(t)), \qquad t = 0, T, 2T, \cdots,$$

and interpolating linearly between the points $x(0)$, $x(T)$, $x(2T)$, $\cdots$ thus found according to

$$x(t) = x(iT) + \frac{t - iT}{T} x((i + 1)T), \qquad iT \le t \le (i + 1)T,$$

$i = 0, 1, 2, \cdots$. If f satisfies the conditions of 5.4.1, as $T \to 0$ the function x thus obtained uniformly approaches the solution of the initial value problem on $[0, \eta]$. $\blacksquare$

The number $T > 0$ is known as the *step size* of the integration method. Uniform convergence of the solution on $[0, \eta]$ means that the *largest* deviation between the approximate solution and the exact solution on this interval converges to zero as T approaches 0.

The idea behind the method is clear: Given any intermediate solution point $x(iT)$, compute the rate of change $\dot{x}(iT) = f(x(iT))$, and estimate the solution at an interval of length T later as $x((i + 1)T) = x(iT) + T \cdot \dot{x}(iT)$.

In practice for many problems the step size T in Euler's method has to be chosen very small to obtain sufficiently accurate solutions. There are other methods that do better and require less computational effort for most problems. A scheme that is well known for its simplicity and reliability is that of *Runge-Kutta*. The method exists in infinitely many variants but we only describe the three most common ones. The Runge-Kutta order 1 scheme is in fact Euler's method.

5.4.5. Summary: Runge-Kutta order 1, 2, and 4 schemes for the integration of differential equations.
Consider the initial value problem $\dot{x} = f(x)$ with $x(0)$ given. An approximate solution at the time points $0, T, 2T, \cdots$ is obtained as follows.

(a) The *first-order Runge-Kutta method* for obtaining an approximate solution x at time $(i + 1)T$, given an approximate solution x_o at time iT, is to compute successively

$$r = f(x_o),$$

$$x = x_o + Tr.$$

(b) The *standard second-order Runge-Kutta method* for obtaining an approximate solution x at time $(i + 1)T$, given an approximate solution x_o at time iT, is to compute successively

$$r_1 = f(x_o),$$

$$x_1 = x_o + \tfrac{1}{2}Tr_1, \qquad r_2 = f(x_1),$$

$$x = x_o + Tr_2.$$

(c) The *standard fourth-order Runge-Kutta method* for obtaining an approximate solution x at time $(i + 1)T$, given an approximate solution x_o at time iT, is to compute successively

$$r_1 = f(x_o),$$

$$x_1 = x_o + \tfrac{1}{2}Tr_1, \qquad r_2 = f(x_1),$$

$$x_2 = x_o + \tfrac{1}{2}Tr_2, \qquad r_3 = f(x_2),$$

$$x_3 = x_o + Tr_3, \qquad r_4 = f(x_3),$$

$$x = x_o + \tfrac{1}{6}T(r_1 + 2r_2 + 2r_3 + r_4).$$

> Carl Runge (1856–1927) and Wilhelm Kutta (1867–1944) both were German applied mathematicians.

The idea of the second-order method is first to compute the rate of change $\dot{x} = r_1$ at the given initial point, to use this to estimate the solution at mid-interval, and to obtain from this in turn a better estimate r_2 of the rate of change over the entire interval. The fourth-order scheme is a refinement of this procedure. There is of course a rationale for it, which may be found in the literature.

We present two examples.

5.4.6. Examples: Numerical solution of differential equations.

(a) *First-order equation.* By way of illustration we consider the numerical solution of the simple initial value problem

$$\dot{x}(t) = x(t), \qquad x(0) = 1.$$

It is clear that the solution is $x(t) = e^t$, $t \in \mathbb{R}$. In Table 5.1 the exact solution on the interval [0, 5] is compared with the numerical solution obtained by the Runge-Kutta schemes of orders 1, 2 and 4 with step size $T = 0.5$. This step size actually is quite large, because the relative change of the solution over one step is

$$\left| \frac{e^{t+T} - e^t}{e^t} \right| = e^T - 1 = 0.4687.$$

For this large step size Euler's method gives poor results, while the second-order Runge-Kutta scheme is better but also inaccurate. The fourth-order scheme yields a solution that is accurate to almost three decimal places on the interval [0, 5]. By taking a smaller step size much more accurate solutions may be obtained.

(b) *Moon rocket.* If we account for the expulsion of mass, the moon rocket of Example 5.2.2 is described by the state differential equation

$$\dot{x}_1(t) = x_2(t)/x_3(t),$$
$$\dot{x}_2(t) = cu(t) - gx_3(t),$$
$$\dot{x}_3(t) = -u(t).$$

We wish to solve these equations for $t \geq 0$ with the following data:

$$
\begin{aligned}
c &= 2000 \text{ m/s}, & x_1(0) &= 1000 \text{ m}, \\
g &= 1.7 \text{ m/s}^2, & x_2(0) &= 0 \text{ kg} \cdot \text{m/s}, \\
u(t) &= 0.5 \text{ kg/s for } t \geq 0, & x_3(0) &= 1000 \text{ kg}.
\end{aligned}
$$

TABLE 5.1 NUMERICAL SOLUTION OF THE DIFFERENTIAL EQUATION $\dot{x}(t) = x(t)$, $x(0) = 1$
ACCORDING TO THE RUNGE-KUTTA ORDER 1, 2, AND 4 SCHEMES WITH STEP SIZE $T = 0.5$

| | | | x | |
t	Order 1 (Euler)	Order 2	Order 4	Exact solution
0	1	1	1	1
0.5	1.5	1.625	1.648	1.649
1	2.25	2.641	2.717	2.718
1.5	3.375	4.291	4.479	4.482
2	5.063	6.973	7.384	7.389
2.5	7.594	11.33	12.17	12.18
3	11.39	18.41	20.06	20.09
3.5	17.09	29.92	33.08	33.12
4	25.63	48.62	54.42	54.60
4.5	38.44	79.01	89.88	90.02
5	57.67	128.4	148.2	148.4

Figure 5.18 shows the result of integrating the differential equation according to the standard second-order Runge-Kutta scheme with step size $T = 1$ s. Halving the step size results in a relative change of the various outcomes of less than 10^{-6} so that the accuracy seems sufficient.

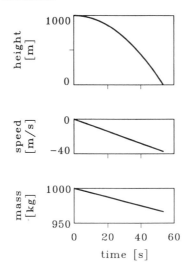

Figure 5.18. Flight trajectory of the moon rocket. Top: elevation. Middle: speed. Bottom: mass.

The initial state corresponds to the rocket being momentarily stationary at 1000 m above the moon surface with zero velocity. The thrust needed to keep the rocket stationary follows from $0 = cu - gx_3$ and would be provided by a constant mass expulsion rate of $u = gx_3/c = 1.7 \times 1000/2000 = 0.85$ kg/s. Because the actual mass expulsion rate u is less than this, the rocket starts descending. After almost 54 s the rocket crashes into the surface of the moon with a speed of nearly 37.5 m/s. Its initial mass of 1000 kg meanwhile has decreased to about 973 kg.   ∎

5.5 SOLUTION OF LINEAR STATE EQUATIONS

Quite explicit results may be obtained for the solution of the state equations of the following *linear* finite-dimensional state difference and differential systems:

Discrete-time   *Continuous-time*

$$x(n + 1) = A(n)x(n) + B(n)u(n),$$ $$\dot{x}(t) = A(t)x(t) + B(t)u(t),$$

$$y(n) = C(n)x(n) + D(n)u(n),$$ $$y(t) = C(t)x(t) + D(t)u(t)$$

for $n \in \mathbb{T}$.   for $t \in \mathbb{T}$.

In both cases, the time axis $\mathbb{T}$ may be right semi-infinite, such as $\mathbb{Z}_+$ or $\mathbb{R}_+$, or infinite.

For mathematical convenience we allow the input, state and output of linear state systems to be *complex-valued,* so that the state space is $X = \mathbb{C}^N$, while the input and output signal ranges are $U = \mathbb{C}^K$ and $Y = \mathbb{C}^M$, respectively. A is an $N \times N$ matrix with time-dependent complex or real elements, B an $N \times K$ time-dependent matrix, C an $M \times N$ time-dependent matrix, and D, finally, an $M \times K$ time-dependent matrix.

It is assumed in the following that in the continuous-time case the entries of the matrix functions A, B, C, and D are bounded and continuous on any finite time interval. These conditions ensure the existence of the solutions of the state differential and output equations.

Homogeneous State Difference and Differential Equations

We begin by studying the solution of the *homogeneous* state difference equation

$$x(n + 1) = A(n)x(n), \qquad n \in \mathbb{T},$$

and the homogeneous state differential equation

$$\dot{x}(t) = A(t)x(t), \qquad t \in \mathbb{T}.$$

First consider the discrete-time case. If $n_o \in \mathbb{T}$ is some initial time, the solution of the homogeneous equation for $n \geq n_o$ is fully determined by the initial condition $x(n_o)$. Indeed by successive substitution it easily follows that

$$x(n) = A(n - 1)A(n - 2) \cdots A(n_o)x(n_o), \qquad n = n_o + 1, n_o + 2, \cdots.$$

We write this more compactly as

$$x(n) = \Phi(n, n_o)x(n_o), \qquad n = n_o + 1, n_o + 2, \cdots. \tag{1}$$

where Φ is the $N \times N$ matrix given by $\Phi(n, n_o) = A(n - 1)A(n - 2) \cdots A(n_o)$, for $n = n_o + 1, n_o + 2, \cdots$. The matrix Φ is called the *state transition matrix* of the system, because it describes the transition of the state from time n_o to time n (for the homogeneous system.)

The relation (1) shows that the transition from the initial state $x(n_o)$ to the state $x(n)$ at the fixed time n is a *linear* map. Indeed, this follows directly from the linearity of the state difference equation.

Also in the continuous-time case the map from the initial state $x(t_o)$, with t_o some initial time, to the state $x(t)$, with t some other time, is linear. We thus may write

$$x(t) = \Phi(t, t_o)x(t_o), \qquad t, t_o \in \mathbb{T},$$

where again the matrix function Φ is called the *state transition matrix* of the continuous-time system. By 5.4.1 the solution of continuous-time initial value problems, if it exists, also exists for time instants *before* the time instant t_o. Hence, in contrast to the discrete-time case, the continuous-time transition matrix Φ generally is also defined for $t < t_o$.

The continuous-time transition matrix is less easy to find than for the discrete-time case. Before turning to this problem we summarize the results found so far.

5.5.1.   Summary: Solution of the homogeneous state difference and differential equation.

The solution of the homogeneous state difference equation

$$x(n + 1) = A(n)x(n),$$

with $n \geq n_o$ and n and n_o in $\mathbb{T}$, may be expressed as

$$x(n) = \Phi(n, n_o)x(n_o), \qquad n \geq n_o.$$

The $N \times N$ *transition matrix* $\Phi(n, n_o)$ is defined for all n and n_o in $\mathbb{T}$ such that $n \geq n_o$.

The solution of the homogeneous state differential equation

$$\dot{x}(t) = A(t)x(t),$$

$t \in \mathbb{T}$, may be expressed as

$$x(t) = \Phi(t, t_o)x(t_o)$$

for all and t and t_o in $\mathbb{T}$, where the $N \times N$ *transition matrix* $\Phi(t, t_o)$ is defined for all t and t_o in $\mathbb{T}$.   ■

State Transition Matrix

We continue with a discussion of the state transition matrix Φ. As seen previously, in the discrete-time case we have an explicit expression for Φ:

5.5.2.   Summary: Discrete-time transition matrix.   The transition matrix of the homogeneous state difference equation $x(n + 1) = A(n)x(n)$, $n \in \mathbb{T}$, is given by

$$\Phi(n, n_o) = \begin{cases} I & \text{for } n = n_o, \\ A(n-1)A(n-2)\cdots A(n_o) & \text{for } n > n_o, \end{cases} \qquad n, n_o \in \mathbb{T}. \quad \blacksquare$$

In the continuous-time case such an explicit expression generally is not available, except if $N = 1$ (i.e., the state is one-dimensional so that the state differential equation $\dot{x} = Ax$ reduces to a *scalar* equation).

5.5.3.   Exercise: Continuous-time transition matrix for the scalar case. Prove that if the state differential equation $\dot{x}(t) = A(t)x(t)$, $t \in \mathbb{T}$, is a *scalar* equation, the transition matrix may be expressed as

$$\Phi(t, t_o) = \exp\left(\int_{t_o}^{t} A(\tau)\,d\tau\right), \qquad t, t_o \in \mathbb{T}.$$

Hint: Use separation of variables to solve the differential equation $dx(t)/dt = A(t)x(t)$, $t \in \mathbb{T}$. $\qquad \blacksquare$

Both in the discrete-time and the continuous-time case the transition matrix has a number of important properties, which may be summarized as follows.

5.5.4.   Summary: Properties of the transition matrix.

(a) The transition matrix Φ of a state difference system satisfies the matrix difference equation

$$\Phi(n+1, n_o) = A(n)\Phi(n, n_o),$$

for $n \geq n_o$ with n and n_o in $\mathbb{T}$, with the initial condition

$$\Phi(n_o, n_o) = I.$$

(b) Φ has the *consistency property*

$$\Phi(n, n) = I \qquad \text{for all} \quad n \in \mathbb{T}.$$

(c) Φ possesses the *semigroup property*

$$\Phi(n_2, n_1)\Phi(n_1, n_o) = \Phi(n_2, n_o)$$

for all n_o, n_1 and n_2 in $\mathbb{T}$ such that $n_o \leq n_1 \leq n_2$.

(a') The transition matrix of a state differential system satisfies the matrix differential equation

$$\frac{\partial}{\partial t}\Phi(t, t_o) = A(t)\Phi(t, t_o)$$

for all t and t_o in $\mathbb{T}$, with the initial condition

$$\Phi(t_o, t_o) = I.$$

(b') Φ has the *consistency property*

$$\Phi(t, t) = I \qquad \text{for all} \quad t \in \mathbb{T}.$$

(c') Φ possesses the *semigroup property*

$$\Phi(t_2, t_1)\Phi(t_1, t_o) = \Phi(t_2, t_o)$$

for all t_o, t_1 and t_2 in $\mathbb{T}$.

(d') In particular, $\Phi(t, t_o)$ is nonsingular for every t and t_o in $\mathbb{R}$, and

$$\Phi^{-1}(t, t_o) = \Phi(t_o, t). \qquad \blacksquare$$

5.5.5.   Proof.   We indicate the proof for the continuous-time case only. That for the discrete-time case is similar.

The proof of (a′) follows by substituting $x(t) = \Phi(t, t_o)x(t_o)$ into the state differential equation. It follows that $[\partial\Phi(t, t_o)/\partial t]x(t_o) = A(t)\Phi(t, t_o)x(t_o)$. Since this holds identical in $x(t_o)$, the differential equation for Φ immediately follows. The initial condition $\Phi(t_o, t_o) = I$ is obtained by substituting $t = t_o$ into $x(t) = \Phi(t, t_o)x(t_o)$. This at the same time yields the consistency property (b′).

The semigroup property (c′) is proved by writing $x(t_2) = \Phi(t_2, t_1)x(t_1) = \Phi(t_2, t_1)\Phi(t_1, t_o)x(t_o)$, while, on the other hand, $x(t_2) = \Phi(t_2, t_o)x(t_o)$. The property follows because $\Phi(t_2, t_o)x(t_o) = \Phi(t_2, t_1)\Phi(t_1, t_o)x(t_o)$ holds identical in $x(t_o)$.

The inversion formula of (d′), finally, follows from (c′) by setting $t_2 = t_o$. ∎

From 5.5.4(a) and (a′) we may give the following interpretation of the transition matrix Φ: The ith column of Φ is the solution of the homogeneous state difference or differential equation for the initial condition $col(0, \cdots, 0, 1, 0, \cdots, 0)$, with the 1 in the ith position. In the continuous-time case this provides a way to find the transition matrix numerically by solution of the homogeneous state differential equation with appropriate initial conditions.

In the continuous-time case the semigroup property holds for any t_o, t_1 and t_2, while in the discrete-time case we need the ordering $n_o \leq n_1 \leq n_2$. Because of this, and because moreover the discrete-time transition matrix $\Phi(n, n_o)$ is not defined for $n < n_o$, the inversion formula (d′) has no discrete-time equivalent.

By way of example we consider two first-order systems.

5.5.6.   Examples: Transition matrices.

(a) *Exponential smoother*. In 5.3.2(a) we found that the exponential smoother may be represented by the state difference and output equations

$$x(n + 1) = ax(n) + a(1 - a)u(n),$$
$$y(n) = x(n) + (1 - a)u(n), \qquad n \in \mathbb{Z}.$$

The homogeneous state difference equation is

$$x(n + 1) = ax(n), \qquad n \in \mathbb{Z}.$$

It follows that the 1×1 state transition matrix is given by

$$\Phi(n, n_o) = A(n - 1)A(n - 2) \cdots A(n_o) = a \cdot a \cdots a$$
$$= a^{n-n_o}, \qquad n \geq n_o, \qquad n, n_o \in \mathbb{Z}.$$

The consistency and semigroup properties are easily seen to hold.

(b) *RC network*. In 5.2.7 we found that the RC network may be described by the state differential equation

$$\dot{x}(t) = -\frac{1}{RC}x(t) + \frac{1}{R}u(t), \qquad t \in \mathbb{R},$$

and output equation

$$y_1(t) = \frac{1}{C}x(t), \qquad t \in \mathbb{R},$$

where the state x is the charge of the capacitor. The homogeneous state differential equation is

$$\dot{x}(t) = -\frac{1}{RC}x(t), \qquad t \in \mathbb{R}.$$

Given the initial condition $x(t_o)$, the solution of the homogeneous equation is

$$x(t) = e^{-\frac{t-t_o}{RC}}x(t_o), \qquad t \in \mathbb{R}.$$

It follows that the state transition matrix is

$$\Phi(t, t_o) = e^{-\frac{t-t_o}{RC}}, \qquad t, t_o \in \mathbb{R}.$$

Also here it is simple to see that the consistency and semigroup properties hold.   ∎

Transition Matrix of Time-Invariant Systems

We now consider the important case that the state difference or differential system besides linear is also time-invariant. This results in quite explicit formulas for the state transition matrix.

5.5.7.   Summary: Transition matrix of linear time-invariant state difference and differential systems.

The state transition matrix of the homogeneous linear time-invariant finite-dimensional state difference equation

$$x(n + 1) = Ax(n), \quad n \geq n_o, \quad n \in \mathbb{T},$$

with A a constant matrix, is given by

$$\Phi(n, n_o) = A^{n-n_o},$$

The state transition matrix of the homogeneous linear time-invariant finite-dimensional state differential equation

$$\dot{x}(t) = Ax(t), \qquad t \in \mathbb{T},$$

with A a constant matrix, is given by

$$\Phi(t, t_o) = e^{A(t-t_o)},$$

for $n \geq n_o$, with n and n_o in $\mathbb{T}$. By defini-   for all t and t_o in $\mathbb{T}$.   ∎
tion, $A^0 = I$.

For every square matrix M the *exponential matrix* e^M is defined by the converging infinite sum

$$e^M = I + M + \frac{1}{2!}M^2 + \frac{1}{3!}M^3 + \cdots .$$

The result 5.5.7 shows that in the discrete-time case the transition matrix of a time-invariant system simply follows by taking suitable powers of the matrix A. In the continuous-time case the transition matrix follows as the exponential of the matrix $A(t - t_o)$. The exponential of a matrix is defined as an infinite matrix power series whose coefficients are those of the exponential power series. Before showing some examples we discuss the proof of 5.5.7.

5.5.8.   Proof of 5.5.7.

Discrete-time case.   In the discrete-time case the transition matrix follows immediately from 5.5.2.

Continuous-time case.   The proof for the continuous-time relies on *Picard's algorithm* for the solution of differential equations, which may be stated as follows: Given the differential equation

$$\dot{x}(t) = f(t, x(t)), \qquad t \in \mathbb{R},$$

with the initial condition $x(t_o) = x_o$, construct the sequence of functions $x_{(k)}(t)$, $t_o - \eta \leq t \leq t_o + \eta$, for $k = 0, 1, 2, \cdots$, as follows:

$$x_{(0)}(t) = x_o,$$

$$x_{(k+1)}(t) = x_o + \int_{t_o}^{t} f(\tau, x_{(k)}(\tau)) \, d\tau, \qquad t_o - \eta \leq t \leq t_o + \eta,$$

for $k = 0, 1, 2, \cdots$. Then if f satisfies the conditions of 5.4.1 the sequence $x_{(k)}$, $k = 0, 1, 2, \cdots$, converges uniformly to the solution of the initial value problem $\dot{x}(t) = f(t, x(t)), t \in [t_o - \eta, t_o + \eta], x(t_o) = x_o$.

The differential equation $\dot{x}(t) = Ax(t), t \in \mathbb{R}$, satisfies the existence condition of 5.4.1 on any time interval. We find successively

$$x_{(0)}(t) = x_o,$$

$$x_{(1)}(t) = x_o + \int_{t_o}^{t} Ax_o \, d\tau = x_o + A(t - t_o)x_o,$$

$$x_{(2)}(t) = x_o + \int_{t_o}^{t} A[x_o + A(\tau - t_o)x_o] \, d\tau$$

$$= x_o + A(t - t_o)x_o + \frac{1}{2!}A^2(t - t_o)^2 x_o.$$

By induction it is easily proved that in general

$$x_{(k)}(t) = \sum_{j=0}^{k} \frac{1}{j!}A^j(t - t_o)^j x_o, \qquad k = 0, 1, 2, \cdots, \qquad t \in \mathbb{R}.$$

Because the sequence $x_{(k)}$ converges uniformly to the solution x of the initial value problem as $k \to \infty$ for any x_o, the infinite matrix sum

$$\sum_{j=0}^{\infty} \frac{1}{j!}A^j(t - t_o)^j$$

converges for every $t - t_o$ and every A, and hence equals the state transition matrix $\Phi(t, t_o)$. The coefficients of the infinite matrix sum are precisely the coefficients of the power series for the exponential function, so that we denote this sum as $e^{A(t-t_o)}$. It follows that

$$\Phi(t, t_o) = e^{A(t-t_o)}. \qquad\qquad\qquad \blacksquare$$

We consider a simple example.

5.5.9.   Examples: Transition matrix of linear time-invariant state differential systems.

(a) *RC network*. In 5.2.7 we obtained the state differential equation

$$\dot{x}(t) = -\frac{1}{RC}x(t) + \frac{1}{R}u(t), \qquad t \in \mathbb{R},$$

for the RC network. Since $A = -1/RC$, it follows that

$$e^{At} = 1 - \frac{t}{RC} + \frac{1}{2!}\left(\frac{t}{RC}\right)^2 - \frac{1}{3!}\left(\frac{t}{RC}\right)^3 + \cdots$$

$$= e^{-\frac{t}{RC}}, \qquad t \in \mathbb{R}.$$

As a result, the transition matrix (which in this case is 1×1) is

$$\Phi(t, t_o) = e^{A(t-t_o)} = e^{-\frac{t-t_o}{RC}}, \qquad t, t_o \in \mathbb{R}.$$

This agrees with what we found in 5.5.6(b).

(b) *Moon rocket.* In 5.2.2 we saw that if the mass of the moon rocket is assumed to be *constant*, the system is described by the state differential equation

$$\dot{x}_1(t) = x_2(t),$$

$$\dot{x}_2(t) = \frac{c}{m} u(t) - g,$$

while the output equation is

$$y(t) = x_1(t).$$

Because of the presence of the term $-g$ in the second component of the state differential equation the moon rocket is not precisely a linear time-invariant differential system. This may be remedied by not taking the mass flow u as input, but the difference

$$w = u - \frac{mg}{c}.$$

The quantity mg/c indicates the size of the mass flow whose thrust compensates gravity. The state differential equation now becomes $\dot{x}_1(t) = x_2(t)$, $\dot{x}_2(t) = cw(t)/m$, so that in matrix form we have

$$\dot{x}(t) = \begin{bmatrix} 0 & 1 \\ 0 & 0 \end{bmatrix} x(t) + \begin{bmatrix} 0 \\ c/m \end{bmatrix} w(t),$$

$$y(t) = [1 \quad 0]x(t).$$

Since

$$A = \begin{bmatrix} 0 & 1 \\ 0 & 0 \end{bmatrix}, \qquad A^2 = \begin{bmatrix} 0 & 0 \\ 0 & 0 \end{bmatrix},$$

it follows that $A^j = 0$ for $j = 2, 3, \cdots$, so that

$$e^{At} = I + At + \frac{1}{2!}A^2t^2 + \frac{1}{3!}A^3t^3 + \cdots$$

$$= \begin{bmatrix} 1 & 0 \\ 0 & 1 \end{bmatrix} + \begin{bmatrix} 0 & 1 \\ 0 & 0 \end{bmatrix} t = \begin{bmatrix} 1 & t \\ 0 & 1 \end{bmatrix}, \qquad t \in \mathbb{R}.$$

As a result, the transition matrix of the system is

$$\Phi(t, t_o) = \begin{bmatrix} 1 & t - t_o \\ 0 & 1 \end{bmatrix}, \qquad t, t_o \in \mathbb{R}. \qquad \blacksquare$$

In the last example, fortuitously the infinite matrix sum for the exponential e^{At} reduces to a finite sum, consisting of only two terms. Normally this does not happen. In Section 5.6 it is seen how by *diagonalization* of the matrix A convenient expressions may be obtained for the transition matrix, both in the discrete- and the continuous-time case.

Solution of the Inhomogeneous Equation

Once the transition matrix Φ of a linear state difference or differential system has been found, the solution of the *inhomogeneous* equation may also be obtained.

5.5.10.   Summary: Solution of the inhomogeneous state difference and differential equation.

The solution of the state difference equation

$$x(n + 1) = A(n)x(n) + B(n)u(n),$$

$n \in \mathbb{T}$, is

$$x(n) = \Phi(n, n_o)x(n_o)$$
$$+ \sum_{k=n_o}^{n-1} \Phi(n, k + 1)B(k)u(k),$$

for $n \geq n_o$ with n and n_o in $\mathbb{Z}$. If the lower limit of the sum exceeds the upper limit the sum is canceled.

The solution of the state differential equation

$$\dot{x}(t) = A(t)x(t) + B(t)u(t),$$

$t \in \mathbb{T}$, is

$$x(t) = \Phi(t, t_o)x(t_o)$$
$$+ \int_{t_o}^{t} \Phi(t, \tau)B(\tau)u(\tau)d\tau,$$

for any t and t_o in $\mathbb{T}$. $\blacksquare$

5.5.11.   Proof.

(a) *Discrete-time case.* The proof of the discrete-time solution follows by induction on n.

(b) *Continuous-time case.* The continuous-time solution is usually proved by the "variation of constant" argument. Given that any solution of the *homogeneous* equation has the form $x(t) = \Phi(t, t_o)a$, with a a suitable constant vector, we attempt to find a solution of the *inhomogeneous* equation of the form $x(t) = \Phi(t, t_o)a(t)$, with $a(t)$, $t \in \mathbb{R}$, a time-varying vector to be determined. Note that $a(t_o) = x(t_o)$. Differentiation of $x(t) = \Phi(t, t_o)a(t)$ with respect to time yields by application of the product rule

$$\dot{x}(t) = \left[\frac{\partial}{\partial t}\Phi(t, t_o)\right]a(t) + \Phi(t, t_o)\dot{a}(t) = A(t)\Phi(t, t_o)a(t) + \Phi(t, t_o)\dot{a}(t).$$

Substitution into the nonhomogeneous state differential equation results in $A(t)\Phi(t, t_o)a(t) + \Phi(t, t_o)\dot{a}(t) = A(t)\Phi(t, t_o)a(t) + B(t)u(t)$, which shows that

$$\Phi(t, t_o)\dot{a}(t) = B(t)u(t), \qquad t \in \mathbb{R}.$$

Using the inversion formula for the transition matrix (see 5.5.4(d')) we have $\dot{a}(t) = \Phi(t_o, t)B(t)u(t)$, which after integration from t_o results in

$$a(t) = a(t_o) + \int_{t_o}^{t} \Phi(t_o, \tau)B(\tau)u(\tau)\, d\tau$$

$$= x(t_o) + \int_{t_o}^{t} \Phi(t_o, \tau)B(\tau)u(\tau)\, d\tau, \qquad t \in \mathbb{R}.$$

Substitution into $x(t) = \Phi(t, t_o)a(t)$ finally yields with the use of the semi-group property (see 5.5.4(c')) that

$$x(t) = \Phi(t, t_o)\left[x(t_o) + \int_{t_o}^{t} \Phi(t_o, \tau)B(\tau)u(\tau)\, d\tau\right]$$

$$= \Phi(t, t_o)x(t_o) + \int_{t_o}^{t} \Phi(t, t_o)\Phi(t_o, \tau)B(\tau)u(\tau)\, d\tau$$

$$= \Phi(t, t_o)x(t_o) + \int_{t_o}^{t} \Phi(t, \tau)B(\tau)u(\tau)\, d\tau, \qquad t \in \mathbb{R},$$

which is what we set out to prove.   ∎

We elaborate a little on the output y that results from the solution of the inhomogeneous state difference or differential equation. With the state expressed as in 5.5.10, it immediately follows with the output equation that in the discrete-time case the output of the system is

$$y(n) = C(n)\Phi(n, n_o)x(n_o) + \sum_{k=n_o}^{n-1} C(n)\Phi(n, k+1)B(k)u(k) + D(n)u(n),$$

for $n \geq n_o$ with n and n_o in $\mathbb{Z}$. Similarly, in the continuous-time case

$$y(t) = C(t)\Phi(t, t_o)x(t_o) + \int_{t_o}^{t} C(t)\Phi(t, \tau)B(\tau)u(\tau)\, d\tau + D(t)u(t),$$

for all t and t_o in $\mathbb{R}$. We may write both expressions in the form

$$y = y_{\text{zero-input}} + y_{\text{zero-state}},$$

where in the discrete-time case

$$y_{\text{zero-input}}(n) = C(n)\Phi(n, n_o)x(n_o),$$

$$y_{\text{zero-state}}(n) = \sum_{k=n_o}^{n-1} C(n)\Phi(n, k+1)B(k)u(k) + D(n)u(n),$$

and in the continuous-time case

$$y_{\text{zero-input}}(t) = C(t)\Phi(t, t_o)x(t_o),$$

$$y_{\text{zero-state}}(t) = \int_{t_o}^{t} C(t)\Phi(t, \tau)B(\tau)u(\tau)\, d\tau + D(t)u(t).$$

The term $y_{\text{zero-input}}$ is called the *zero-input response*, because it is the response to the initial state if the input is zero. The term $y_{\text{zero-state}}$ constitutes the *zero-state response*, because it forms the response if the initial state is zero.

In the time-invariant case these expressions reduce to

$$y_{\text{zero-input}}(n) = CA^{n-n_o}x(n_o),$$

$$y_{\text{zero-state}}(n) = \sum_{k=n_o}^{n-1} CA^{n-k-1}Bu(k) + Du(n),$$

in the discrete-time case, while in the continuous-time case

$$y_{\text{zero-input}}(t) = Ce^{A(t-t_o)}x(t_o),$$

$$y_{\text{zero-state}}(t) = \int_{t_o}^{t} Ce^{A(t-\tau)}Bu(\tau)\, d\tau + Du(t).$$

The zero-state response may be rewritten as

$$y_{\text{zero-state}}(n) = \sum_{k=n_o}^{\infty} h(n-k)u(k),$$

in the discrete-time case, and in the continuous-time case as

$$y_{\text{zero-state}}(t) = \int_{t_o}^{\infty} h(t-\tau)u(\tau)\, d\tau,$$

where the *impulse response matrix h* is given by

$$h(n) = CA^{n-1}B\mathbb{1}(n-1) + D\Delta(n), \qquad n \in \mathbb{Z},$$

in the discrete-time case, and in the continuous-time case as

$$h(t) = Ce^{At}B\,\mathbb{1}(t) + D\delta(t), \qquad t \in \mathbb{R}.$$

If the input u and the output y are both scalar (i.e., $K = M = 1$), the impulse response h is a scalar function. It may be interpreted as the zero-state response to the unit pulse Δ in the discrete-time case and the delta function δ in the continuous-time case. In the multi-input multi-output case (i.e., K and M are greater than 1) the (i, j) element h_{ij} is the zero-state response of the ith component of the output when the jth component of the input is Δ or δ and the remaining components of the input are zero.

The following example illustrates how the solution to the inhomogeneous equation may be found if the state transition matrix is known.

5.5.12.   Example: Moon rocket.   In Example 5.5.9(b) we found that if the mass of the moon rocket is assumed to be constant, and its input is the difference $w = u - mg/c$ of the mass expulsion rate u and the rate mg/c that keeps the rocket stationary, the rocket may be described by the state differential equation $\dot{x} = Ax + Bw$, where

$$A = \begin{bmatrix} 0 & 1 \\ 0 & 0 \end{bmatrix}, \qquad B = \begin{bmatrix} 0 \\ c/m \end{bmatrix}.$$

Also in 5.5.9(b) we found that

$$e^{At} = \begin{bmatrix} 1 & t \\ 0 & 1 \end{bmatrix}.$$

It follows that the solution of the state differential equation with given initial state at time 0 is

$$x(t) = e^{At}x(0) + \int_0^t e^{A(t-\tau)}Bw(\tau)\, d\tau$$

$$= \begin{bmatrix} 1 & t \\ 0 & 1 \end{bmatrix} x(0) + \int_0^t \begin{bmatrix} 1 & t - \tau \\ 0 & 1 \end{bmatrix} \begin{bmatrix} 0 \\ c/m \end{bmatrix} w(\tau)\, d\tau$$

$$= \begin{bmatrix} 1 & t \\ 0 & 1 \end{bmatrix} x(0) + \frac{c}{m} \int_0^t \begin{bmatrix} t - \tau \\ 1 \end{bmatrix} w(\tau)\, d\tau, \qquad t \geq 0.$$

Suppose as in Example 5.4.6(b), where we integrated the state differential equation of the rocket numerically, that the input w is constant, say $w(t) = w_o$ for $t \geq 0$. It easily follows that

$$x_1(t) = x_1(0) + x_2(0)t + \frac{c}{2m}w_o t^2,$$

$$t \geq 0.$$

$$x_2(t) = x_2(0) + \frac{c}{m}w_o t,$$

With the numerical values as in 5.4.6(b), namely, $c/m = 2000/1000 = 2$ m/kg·s, $x_1(0) = 1000$ m, $x_2(0) = 0$ m/s, and $w_o = 0.5 - 0.85 = -0.35$ kg/s, it follows that

$$x_1(t) = 1000 - 0.35t^2 \text{ [m]},$$

$$t \geq 0.$$

$$x_2(t) = -0.7t \text{ [m/s]},$$

The corresponding trajectories are plotted in Fig. 5.19. They are very similar to those of Fig. 5.18, where they were computed while accounting for the change of mass of the rocket during the flight. It follows from the present solution that the rocket crashes onto the surface of the moon at the time t_f given by

$$0 = 1000 - 0.35t_f^2,$$

so that $t_f = \sqrt{1000/0.35} = 53.45$ s. The velocity of the rocket at the time of impact is $v(t_f) = -0.7t_f = -37.44$ m/s. Because these calculations do not account for the change of mass of the rocket, the results are only approximate.

Suppose that we take the output of the system as

$$y(t) = \begin{bmatrix} x_1(t) \\ x_2(t) \end{bmatrix},$$

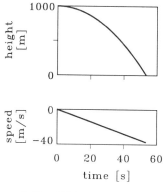

Figure 5.19.   Flight trajectory of the moon rocket assuming constant mass. Top: elevation. Bottom: speed.

that is, the elevation x_1 is the first component of the output, and the speed x_2 the second component. Since $C = I$ and $D = 0$ the impulse response matrix h of the system is

$$h(t) = Ce^{At}B\mathbb{1}(t) + D\delta(t)$$

$$= \frac{c}{m}\begin{bmatrix} t \\ 1 \end{bmatrix}\mathbb{1}(t), \qquad t \in \mathbb{R}.$$

This shows that the zero-state response of the elevation to a delta function at time zero is a ramp, while that of the speed is a step. ∎

5.5.13.   Review: Sampled state difference systems.   Sampled linear state difference systems are described by a state difference and output equation of the form

$$x(t + T) = A(t)x(t) + B(t)u(t),$$

$$y(t) = C(t)x(t) + D(t)u(t), \qquad t \in \mathbb{T},$$

where the time axis $\mathbb{T}$ may be semi-infinite, such as $\mathbb{Z}_+(T)$, or $\mathbb{Z}(T)$. Given the initial state $x(t_o)$, the solution of the homogeneous state difference equation

$$x(t + 1) = A(t)x(t), \qquad t \in \mathbb{T},$$

may be written as

$$x(t) = \Phi(t, t_o)x(t_o), \qquad t \in \mathbb{T},$$

where the *transition matrix* Φ is given by

$$\Phi(t, t_o) = \begin{cases} I & \text{for } t = t_o, \\ A(t - T)A(t - 2T) \cdots A(t_o) & \text{for } t > t_o, \end{cases} \qquad t \in \mathbb{T}.$$

The transition matrix Φ has the following properties. First of all it satisfies the matrix difference equation

$$\Phi(t + T, t_o) = A(t)\Phi(t, t_o), \qquad t \geq t_o, \qquad t, t_o \in \mathbb{T},$$

$$\Phi(t_o, t_o) = I.$$

Besides the *consistency* property $\Phi(t, t) = I$ for all $t \in \mathbb{T}$ the transition matrix possesses the *semigroup* property

$$\Phi(t_2, t_1)\Phi(t_1, t_0) = \Phi(t_2, t_0)$$

for all t_2, t_1 and t_0 in $\mathbb{T}$ such that $t_2 \geq t_1 \geq t_0$. In the time-invariant case, i.e., when the matrix function A is a constant matrix,

$$\Phi(t, t_o) = A^{\frac{t - t_o}{T}}, \qquad t \geq t_o, \qquad t, t_o \in \mathbb{T}.$$

The solution of the inhomogeneous state difference equation may be expressed as

$$x(t) = \Phi(t, t_o)x(t_o) + \sum_{\substack{t_o \leq \tau < t, \\ \tau \in \mathbb{Z}(T)}} \Phi(t, \tau + T)B(\tau)u(\tau),$$

for $t \geq t_o$ and $t \in \mathbb{T}$. The zero-input and zero-state response of the system hence are

$$y_{\text{zero-input}}(t) = C(t)\Phi(t, t_o)x(t_o), \qquad t \geq t_o, \qquad t, t_o \in \mathbb{Z}(T),$$

$$y_{\text{zero-state}}(t) = T \sum_{\substack{\tau \geq t_o, \\ \tau \in \mathbb{Z}(T)}} k(t, \tau)u(\tau), \qquad t \geq t_o, \qquad t, t_o \in \mathbb{Z}(T),$$

where the kernel k is given by

$$k(t, \tau) = \begin{cases} \dfrac{1}{T}C(t)\Phi(t, \tau + T)B(\tau) & \text{for } t \geq \tau + T, \\ \dfrac{1}{T}D(t) & \text{for } t = \tau, \qquad t, \tau \in \mathbb{Z}(T). \\ 0 & \text{for } t < \tau, \end{cases}$$

In the time-invariant case the zero-state response is

$$y_{\text{zero-state}}(t) = T \sum_{\substack{\tau \geq t_o, \\ \tau \in \mathbb{Z}(T)}} h(t - \tau)u(\tau), \qquad t \geq t_o, \qquad t, t_o \in \mathbb{Z}(T),$$

where the impulse response h is given by

$$h(t) = \frac{1}{T}[CA^{\frac{t - T}{T}}B\mathbb{1}(t - T) + D\Delta(t)], \qquad t \in \mathbb{Z}(T). \qquad \blacksquare$$

5.6 MODAL ANALYSIS OF LINEAR TIME-INVARIANT STATE SYSTEMS

The state of a system describes the internal situation of the system. The choice of the state is not unique, however, in the sense that there may be many different descriptions of the same system, each with different state variables, that *externally* show the same behavior. Any nonsingular transformation of the state, for instance, is also a state.

In this section we consider state transformations for linear time-invariant difference and differential systems that bring the system in *modal* form. *Modes* are special motions of the system that are decoupled and facilitate the analysis of the dynamic behavior.

All discrete-time systems in this section are considered on the infinite time axis $\mathbb{Z}$ and all continuous-time systems on $\mathbb{R}$.

Time-Invariant State Transformations

First *linear time-invariant state transformations* are discussed. By writing the state as

$$x = Vx',$$

with V a nonsingular constant matrix and x' the transformed state, the state differential and output equations assume a different form, as summarized in the following.

5.6.1.   Summary: Linear time-invariant state transformations.

Let V be a constant nonsingular $N \times N$ matrix. Then the transformation

$$x = Vx'$$

transforms the equations

$$x(n + 1) = Ax(n) + Bu(n),$$
$$y(n) = Cx(n) + Du(n), \qquad n \in \mathbb{Z}.$$

into

$$x'(n + 1) = A'x'(n) + B'u(n),$$
$$y(n) = C'x'(n) + D'u(n), \qquad n \in \mathbb{Z},$$

where

$$A' = V^{-1}AV, \qquad B' = V^{-1}B,$$
$$C' = CV, \qquad D' = D.$$

Let V be a constant nonsingular $N \times N$ matrix. Then the transformation

$$x = Vx'$$

transforms the equations

$$\dot{x}(t) = Ax(t) + Bu(t),$$
$$y(t) = Cx(t) + Du(t), \qquad t \in \mathbb{R},$$

into

$$\dot{x}'(t) = A'x'(t) + B'u(t),$$
$$y(t) = C'x'(t) + D'u(t), \qquad t \in \mathbb{R},$$

where

$$A' = V^{-1}AV, \qquad B' = V^{-1}B,$$
$$C' = CV, \qquad D' = D. \qquad \blacksquare$$

The proof is not difficult.

5.6.2.   Proof.   We consider the proof only for the discrete-time case. That for the continuous-time case is very similar. Substitution of $x(n) = Vx'(n)$ into the state difference equation $x(n + 1) = Ax(n) + Bu(n)$ yields $Vx'(n + 1) = AVx'(n) + Bu(n)$, which, after premultiplication by V^{-1}, results in $x'(n + 1) = V^{-1}AVx'(n) + V^{-1}Bu(n) = A'x'(n) + B'u(n)$. For the output equation we have $y(n) = Cx(n) + Du(n) = CVx'(n) + Du(n) = C'x'(n) + D'u(n)$, which completes the proof.   ∎

A question of interest is how the state transition matrix and impulse response matrix behave under state transformation.

5.6.3.   Summary: Effect of state transformation on the state transition and impulse response matrices.

(a) The transition matrix Φ' of a discrete-time system that has been subjected to a state transformation $x = Vx'$ is related to the transition matrix Φ of the system before transformation as

$$\Phi'(n, n_o) = V^{-1}\Phi(n, n_o)V,$$

for all n and n_o in $\mathbb{Z}$ such that $n \geq n_o$.

(b) The impulse response matrix of the system remains unchanged under the state transformation.

(a') The transition matrix Φ' of a continuous-time system that has been subjected to a state transformation $x = Vx'$ is related to the transition matrix Φ of the system before transformation as

$$\Phi'(t, t_o) = V^{-1}\Phi(t, t_o)V,$$

for all t and t_o in $\mathbb{R}$.

(b') The impulse response matrix of the system remains unchanged under the transformation.   ∎

5.6.4.   Proof.
(a) Again for the discrete-time case, if the input is zero we have $x'(n) = V^{-1}x(n) = V^{-1}\Phi(n, n_o)x(n_o) = V^{-1}\Phi(n, n_o)Vx'(n_o)$, which shows that the transformed system has the transition matrix $V^{-1}\Phi(n, n_o)V$.

(b) Because the zero state transforms into the zero state, the zero state response of the transformed system is the same as the zero state response of the system before transformation, which means that the impulse response matrix of the system is not changed by the state transformation.   ∎

It follows from 5.6.3 that the state transition matrix is subject to the same transformation as the A-matrix. The impulse response matrix is *not* changed by the state transformation. The impulse response is an *external* property of the system, which is not affected by its internal representation.

Modal Transformations

The particular state transformation that is of interest in this section is the *modal transformation*. This transformation is closely related to the eigenvectors and eigenvalues of the matrix A occurring in the state difference or differential equation.

We briefly review some notions from linear algebra. The complex number λ is an *eigenvalue* of the $N \times N$ matrix A if

$$\det (\lambda I - A) = 0,$$

with I the $N \times N$ unit matrix. The eigenvalues of A hence are the N roots of the polynomial $\det (\lambda I - A)$ of degree N, which is called the *characteristic polynomial* of the matrix A. The nonzero vector $v \in \mathbb{C}^N$ is an *eigenvector* of A corresponding to the eigenvalue λ if

$$Av = \lambda v.$$

If the N eigenvalues of A are all *different*, A always has N linearly independent eigenvectors. If some of the eigenvalues of A are equal to each other, A may or may not have N linearly independent eigenvectors. If it does not, A is called *defective*.

5.6.5.   Exercise: Similarity transformation.   The transformation of A to $A' = V^{-1}AV$ is called a *similarity transformation* of the matrix A. Prove that a similarity transformation does not affect the eigenvalues, i.e., A and A' have the same eigenvalues.   ∎

A modal transformation is a state transformation such that the columns of the transform matrix V are the eigenvectors of the matrix A. This is only possible if A is not defective. The case where A is defective is discussed in Supplement D. If A is not defective, modal transformation brings it into *diagonal* form. We summarize as follows.

5.6.6.   Summary: Modal transformation.   Suppose that A is an $N \times N$ matrix with eigenvalues $\lambda_1, \lambda_2, \cdots, \lambda_N$, and N corresponding linearly independent eigenvectors $v_1, v_2, \cdots, v_N$. Let V be the $N \times N$ matrix whose columns are the eigenvectors $v_1, v_2, \cdots, v_N$, that is,

$$V = [v_1 \quad v_2 \quad \cdots \quad v_N],$$

and let Λ be the $N \times N$ diagonal matrix whose diagonal elements are the eigenvalues $\lambda_1, \lambda_2, \cdots, \lambda_N$, that is,

$$\Lambda = \begin{bmatrix} \lambda_1 & 0 & 0 & 0 & 0 \\ 0 & \lambda_2 & 0 & 0 & 0 \\ \cdots & \cdots & \cdots & \cdots & \cdots \\ 0 & 0 & 0 & 0 & \lambda_N \end{bmatrix} = \text{diag}(\lambda_1, \lambda_2, \cdots, \lambda_N).$$

Then the state transformation

$$x = Vx'$$

results in the transformations

$$V^{-1}AV = \Lambda, \tag{1a}$$

$$V^{-1}A^iV = \Lambda^i \qquad \text{for all} \quad i \in \mathbb{Z}_+, \tag{1b}$$

$$V^{-1}e^{At}V = e^{\Lambda t} \qquad \text{for all} \quad t \in \mathbb{R}, \tag{1c}$$

where

$$\Lambda^i = \text{diag}(\lambda_1^i, \lambda_2^i, \cdots, \lambda_N^i), \qquad i \in \mathbb{Z}_+,$$

$$e^{\Lambda t} = \text{diag}(e^{\lambda_1 t}, e^{\lambda_2 t}, \cdots, e^{\lambda_N t}), \qquad t \in \mathbb{R}.$$

Conversely,

$$A = V\Lambda V^{-1},$$

$$A^i = V\Lambda^i V^{-1} \qquad \text{for all} \qquad i \in \mathbb{Z}_+,$$

$$e^{At} = Ve^{\Lambda t}V^{-1} \qquad \text{for all} \qquad t \in \mathbb{R}. \qquad \blacksquare$$

Note that if $\lambda = 0$ we define

$$\lambda^i = \begin{cases} 1 & \text{for } i = 0, \\ 0 & \text{otherwise,} \end{cases} \qquad i \in \mathbb{Z}_+.$$

A matrix V such that $V^{-1}AV$ is diagonal is said to *diagonalize* the matrix A. Diagonalization of A is only possible if A is not defective, that is, if A has N linearly independent eigenvectors.

It follows from 5.6.6 that once A has been diagonalized, it is a simple matter to determine powers of A and its exponential, which is what we need if we wish to find the transition matrix of the system.

5.6.7. **Proof of 5.6.6.** By the definition of V it follows that

$$AV = A[v_1 \quad v_2 \quad \cdots \quad v_N] = [Av_1 \quad Av_2 \quad \cdots \quad Av_N]$$

$$= [\lambda_1 v_1 \quad \lambda_2 v_2 \quad \cdots \quad \lambda_N v_n]$$

$$= [v_1 \quad v_2 \quad \cdots \quad v_N] \begin{bmatrix} \lambda_1 & 0 & 0 & 0 & 0 \\ 0 & \lambda_2 & 0 & 0 & 0 \\ \cdots & \cdots & \cdots & \cdots & \cdots \\ 0 & 0 & 0 & 0 & \lambda_N \end{bmatrix} = V\Lambda.$$

Because by assumption the eigenvectors $v_1, v_2, \cdots, v_N$ are linearly independent, the matrix V is nonsingular. Premultiplication of $AV = V\Lambda$ by V^{-1} yields $V^{-1}AV = \Lambda$, which proves (1a).

The equality (1b) follows by writing

$$\Lambda^i = (V^{-1}AV)^i = V^{-1}AV \cdot V^{-1}AV \cdots V^{-1}AV = V^{-1}A \cdots AV$$

$$= V^{-1}A^i V.$$

To prove (1c), we write $e^{\Lambda t}$ in the form of an exponential series and use (1b). It follows that

$$e^{\Lambda t} = I + \Lambda t + \frac{1}{2!}\Lambda^2 t^2 + \frac{1}{3!}\Lambda^3 t^3 + \cdots$$

$$= I + V^{-1}AVt + \frac{1}{2!}V^{-1}A^2Vt^2 + \frac{1}{3!}V^{-1}A^3Vt^3 + \cdots$$

$$= V^{-1}\left[I + At + \frac{1}{2!}A^2 t^2 + \frac{1}{3!}A^3 t^3 + \cdots\right]V$$

$$= V^{-1}e^{At}V, \qquad t \in \mathbb{R}.$$

It is a simple matter to derive the remaining equalities of 5.6.6.   ∎

Once a modal transformation of a discrete- or continuous-time system has been performed, the state transition matrix of the original system from the latter part of 5.6.6 may be obtained as

$$\Phi(n, n_o) = A^{n-n_o} = V\Lambda^{n-n_o}V^{-1}, \qquad n \geq n_o, \qquad n, n_o \in \mathbb{Z},$$

in the discrete-time case, and

$$\Phi(t, t_o) = e^{A(t-t_o)} = Ve^{\Lambda(t-t_o)}V^{-1}, \qquad t, t_o \in \mathbb{R},$$

in the continuous-time case. The following examples illustrate the modal transformation.

5.6.8.   Examples: Modal transformation.

(a) *Second-order smoother*. In 5.3.5(a) we found that the second-order smoother of 4.2.4(b) may be described by the state difference system $x(n + 1) = Ax(n) + Bu(n)$, $y(n) = Cx(n) + Du(n)$, $n \in \mathbb{Z}$, with

$$A = \begin{bmatrix} a_1 & 1 \\ a_o & 0 \end{bmatrix}, \qquad B = \begin{bmatrix} b_1 + a_1 b_2 \\ a_o b_2 \end{bmatrix},$$

$$C = [1 \quad 0], \qquad D = b_2.$$

Adopting the numerical values $a_0 = 0$, $a_1 = \frac{1}{2}$, $b_1 = 1$ and $b_2 = 0$ this reduces to

$$A = \begin{bmatrix} \frac{1}{2} & 1 \\ 0 & 0 \end{bmatrix}, \qquad B = \begin{bmatrix} 1 \\ 0 \end{bmatrix},$$

$$C = [1 \quad 0], \qquad D = 0.$$

The characteristic polynomial of the matrix A is

$$\det (\lambda I - A) = \det \left(\begin{bmatrix} \lambda - \frac{1}{2} & -1 \\ 0 & \lambda \end{bmatrix} \right) = \lambda (\lambda - \frac{1}{2}).$$

As a result, the eigenvalues of A are $\lambda_1 = 0$ and $\lambda_2 = \frac{1}{2}$. It is easily found that the corresponding eigenvectors are

$$v_1 = \begin{bmatrix} 1 \\ -\frac{1}{2} \end{bmatrix}, \qquad v_2 = \begin{bmatrix} 1 \\ 0 \end{bmatrix}.$$

It follows that the modal transformation matrix V, its inverse V^{-1} and the diagonal matrix Λ are

$$V = \begin{bmatrix} 1 & 1 \\ -\frac{1}{2} & 0 \end{bmatrix}, \qquad V^{-1} = \begin{bmatrix} 0 & -2 \\ 1 & 2 \end{bmatrix}, \qquad \Lambda = \begin{bmatrix} 0 & 0 \\ 0 & \frac{1}{2} \end{bmatrix}.$$

Thus, after modal transformation the system is represented as $x'(n + 1) = A'x'(n) + B'u(n)$, $y(n) = C'x'(n) + Du(n)$, $n \in \mathbb{Z}$, where

$$A' = V^{-1}AV = \Lambda = \begin{bmatrix} 0 & 0 \\ 0 & \frac{1}{2} \end{bmatrix}, \qquad B' = V^{-1}B = \begin{bmatrix} 0 \\ 1 \end{bmatrix},$$

$$C' = CV = [1 \quad 1].$$

The state transition matrix of the transformed system is

$$\Phi'(n, n_o) = \Lambda^{n-n_o} = \begin{bmatrix} \Delta(n - n_o) & 0 \\ 0 & (\tfrac{1}{2})^{n-n_o} \end{bmatrix}, \qquad n \geq n_o.$$

It follows for the impulse response of the transformed system, which is also the impulse response of the untransformed system,

$$h(n) = C'\Lambda^{n-1}B'1(n - 1) + D\Delta(n) = \begin{bmatrix} 1 & 1 \end{bmatrix} \begin{bmatrix} \Delta(n - 1) & 0 \\ 0 & (\tfrac{1}{2})^{n-1} \end{bmatrix} \begin{bmatrix} 0 \\ 1 \end{bmatrix}$$

$$= (\tfrac{1}{2})^{n-1}1(n - 1), \qquad n \in \mathbb{Z}.$$

We may also compute the transition matrix Φ of the untransformed system as

$$\Phi(n, n_o) = A^{n-n_o} = V\Lambda^{n-n_o}V^{-1}$$

$$= \begin{bmatrix} 1 & 1 \\ -\tfrac{1}{2} & 0 \end{bmatrix} \begin{bmatrix} \Delta(n - n_o) & 0 \\ 0 & (\tfrac{1}{2})^{n-n_o} \end{bmatrix} \begin{bmatrix} 0 & -2 \\ 1 & 2 \end{bmatrix}$$

$$= \begin{bmatrix} (\tfrac{1}{2})^{n-n_o} & -2\Delta(n - n_o) + 2(\tfrac{1}{2})^{n-n_o} \\ 0 & \Delta(n - n_o) \end{bmatrix}, \qquad n \geq n_o,$$

with n and n_o both belonging to $\mathbb{Z}$. This may be simplified to $\Phi(n_o, n_o) = I$ and

$$\Phi(n, n_o) = \begin{bmatrix} (\tfrac{1}{2})^{n-n_o} & 2(\tfrac{1}{2})^{n-n_o} \\ 0 & 0 \end{bmatrix}, \qquad n > n_o.$$

(b) *RCL network.* In 5.2.1 we considered a series connection of a resistor, a capacitor and an inductor, which may be described by the state differential equation $\dot{x}(t) = Ax(t) + Bu(t)$, where

$$A = \begin{bmatrix} 0 & 1/L \\ -1/C & -R/L \end{bmatrix}, \qquad B = \begin{bmatrix} 0 \\ 1 \end{bmatrix}.$$

The characteristic polynomial of the matrix A is

$$\det(\lambda I - A) = \det\left(\begin{bmatrix} \lambda & -1/L \\ 1/C & \lambda + R/L \end{bmatrix}\right) = \lambda^2 + \frac{R}{L}\lambda + \frac{1}{LC}.$$

Assume that $R = 11\ \Omega$, $L = 0.01$ H and $C = 0.001$ F, so that the characteristic polynomial is $\lambda^2 + 1100\lambda + 100000 = (\lambda + 100)(\lambda + 1000)$. As a result, A has two eigenvalues given by $\lambda_1 = -100$ and $\lambda_2 = -1000$. It is easily found that two corresponding eigenvectors of the matrix

$$A = \begin{bmatrix} 0 & 100 \\ -1000 & -1100 \end{bmatrix}$$

are

$$v_1 = \begin{bmatrix} 1 \\ -1 \end{bmatrix}, \qquad v_2 = \begin{bmatrix} 1 \\ -10 \end{bmatrix}.$$

It follows that the modal transformation matrix V, its inverse V^{-1} and the diagonal matrix Λ are

$$V = \begin{bmatrix} 1 & 1 \\ -1 & -10 \end{bmatrix}, \qquad V^{-1} = \frac{1}{9}\begin{bmatrix} 10 & 1 \\ -1 & -1 \end{bmatrix}, \qquad \Lambda = \begin{bmatrix} -100 & 0 \\ 0 & -1000 \end{bmatrix}.$$

Hence, the transition matrix Φ of the system *before* modal transformation is given by

$$\Phi(t, t_o) = e^{A(t-t_o)} = Ve^{\Lambda(t-t_o)}V^{-1}$$

$$= \begin{bmatrix} 1 & 1 \\ -1 & -10 \end{bmatrix}\begin{bmatrix} e^{-100(t-t_o)} & 0 \\ 0 & e^{-1000(t-t_o)} \end{bmatrix}\frac{1}{9}\begin{bmatrix} 10 & 1 \\ -1 & -1 \end{bmatrix}$$

$$= \begin{bmatrix} \dfrac{10}{9}e^{-100(t-t_o)} - \dfrac{1}{9}e^{-1000(t-t_o)} & \dfrac{1}{9}e^{-100(t-t_o)} - \dfrac{1}{9}e^{-1000(t-t_o)} \\ -\dfrac{10}{9}e^{-100(t-t_o)} + \dfrac{10}{9}e^{-1000(t-t_o)} & -\dfrac{1}{9}e^{-100(t-t_o)} + \dfrac{10}{9}e^{-1000(t-t_o)} \end{bmatrix},$$

for t and $t_o \in \mathbb{R}$.   ∎

Modes

By using the modal transform, the zero-input response of the state of the discrete-time system $x(n + 1) = Ax(n) + Bu(n)$, $n \in \mathbb{Z}$, may be written as

$$x(n) = A^n x(0) = V\Lambda^n V^{-1}x(0), \qquad n = 0, 1, 2, \cdots .$$

We may expand this expression as follows. The columns of V are the eigenvectors $v_1, v_2, \cdots, v_N$ of A. Denote the *rows* of the inverse matrix V^{-1} as $w_1, w_2, \cdots, w_N$. Then we have

$$x(n) = [v_1 \ v_2 \ \cdots \ v_N] \begin{bmatrix} \lambda_1^n & 0 & 0 & 0 & 0 \\ 0 & \lambda_2^n & 0 & 0 & 0 \\ \cdots & \cdots & \cdots & \cdots & \cdots \\ 0 & 0 & 0 & 0 & \lambda_N^n \end{bmatrix} \begin{bmatrix} w_1 \\ w_2 \\ \cdots \\ w_N \end{bmatrix} x(0)$$

$$= \sum_{i=1}^{N} v_i \lambda_i^n w_i x(0), \qquad n = 0, 1, \cdots .$$

Because w_i is a row vector and $x(0)$ a column vector (of the same dimension), the quantity $w_i x(0)$, which we denote as α_i, is a *scalar*. We thus may write

$$x(n) = \sum_{i=1}^{N} \alpha_i \lambda_i^n v_i, \qquad n = 0, 1, \cdots .$$

This shows that the zero-input response is a *linear combination* of the solutions

$$\lambda_i^n v_i, \qquad n = 0, 1, \cdots ,$$

for $i = 1, 2, \cdots , N$. Each of these solutions is called a *mode*. Because

$$x(0) = \sum_{i=1}^{N} \alpha_i v_i,$$

the numbers α_i simply are the coefficients of the expansion of the initial state $x(0)$ in the vectors $v_1, v_2, \cdots , v_N$. A mode is said to be *excited* if the corresponding expansion coefficient α_i is nonzero.

In the continuous-time case we similarly have for the zero-input response

$$x(t) = e^{At}x(0) = Ve^{\Lambda t}V^{-1}x(0) = \sum_{i=1}^{N} v_i e^{\lambda_i t} w_i x(0)$$

$$= \sum_{i=1}^{N} \alpha_i e^{\lambda_i t} v_i, \qquad t \geq 0.$$

Here the modes are solutions of the form $e^{\lambda_i t} v_i$, $t \geq 0$.

We summarize as follows.

5.6.9. Summary: Modes of a linear time-invariant system.

Suppose that the $N \times N$ matrix A has the eigenvalues

$$\lambda_1, \lambda_2, \cdots , \lambda_N,$$

Suppose that the $N \times N$ matrix A has the eigenvalues

$$\lambda_1, \lambda_2, \cdots , \lambda_N,$$

and N corresponding linearly independent eigenvectors

$$v_1, v_2, \cdots, v_N.$$

Then the solution of the homogeneous state difference equation

$$x(n + 1) = Ax(n), \qquad n \in \mathbb{Z}_+,$$

is a linear combination of the N modes

$$\lambda_i^n v_i, \qquad n \in \mathbb{Z}_+,$$

$i = 1, 2, \cdots, N$. The coefficient α_i of the ith mode is the ith component of the expansion of $x(0)$ in the vectors $v_1, v_2, \cdots, v_N$.

and N corresponding linearly independent eigenvectors

$$v_1, v_2, \cdots, v_N.$$

Then the solution of the homogeneous state differential equation

$$\dot{x}(t) = Ax(t), \qquad t \in \mathbb{R}_+,$$

is a linear combination of the N modes

$$e^{\lambda_i t} v_i, \qquad t \in \mathbb{R}_+,$$

$i = 1, 2, \cdots, N$. The coefficient α_i of the ith mode is the ith component of the expansion of $x(0)$ in the vectors $v_1, v_2, \cdots, v_N$. ∎

The modal expansion of the zero-input response gives a good idea of the general behavior of the zero-input response. If the eigenvalue λ_i is *real,* for instance, also the corresponding eigenvector and hence the corresponding mode may be chosen real. The time behavior of the mode is in this case exponential. In the discrete-time case the mode *decreases exponentially* if $|\lambda_i| < 1$, *increases exponentially* if $|\lambda_i| > 1$, and remains *bounded* without decreasing to zero if $|\lambda_i| = 1$. In the continuous-time case, the mode decreases exponentially if $\lambda_i < 0$, increases exponentially if $\lambda_i > 0$, and remains bounded without decreasing to zero if $\lambda_i = 0$.

When A has *complex* eigenvalues, the corresponding modes are complex-valued functions of time. If for physical reasons the state of the system is essentially real-valued, it is useful to have a real representation for the complex modes.

5.6.10.   Summary: Real representation of complex modes.
(a) Suppose that the real matrix A has a complex eigenvalue λ with corresponding eigenvector v. Then the complex conjugate $\bar{\lambda}$ of λ is also an eigenvalue of A and the complex conjugate $\bar{v}$ of v is a corresponding eigenvector.

(b) Let

$$\lambda = \rho e^{j\psi},$$

with $\rho = |\lambda|$ and ψ real, and

$$v = r + js,$$

(b') Let

$$\lambda = \sigma + j\omega,$$

with σ and ω real, and

$$v = r + js,$$

with r and s real vectors. Then if the initial state $x(0)$ lies in the space spanned by r and s, the solution of the homogeneous state difference equation

$$x(n + 1) = Ax(n), \qquad n \in \mathbb{Z}_+,$$

may be expressed as

$$x(n) = \alpha \rho^n [r \cos(\psi n + \phi) \\ - s \sin(\psi n + \phi)]$$

$n \in \mathbb{Z}_+$, with α and ϕ real constants that are determined by $x(0)$.

with r and s real vectors. Then if the initial state $x(0)$ lies in the space spanned by r and s, the solution of the homogeneous state differential equation

$$\dot{x}(t) = Ax(t), \qquad t \in \mathbb{R}_+,$$

may be expressed as

$$x(t) = \alpha e^{\sigma t} [r \cos(\omega t + \phi) \\ - s \sin(\omega t + \phi)],$$

$t \in \mathbb{R}_+$, with α and ϕ real constants that are determined by $x(0)$.   ∎

5.6.11.   Proof.

(a) This fact is well-known from linear algebra.

(b) Since $x(0)$ lies in the space spanned by r and s, there exist real constants c_1 and c_2 such that $x(0) = c_1 r + c_2 s = \frac{1}{2}(\mu v + \overline{\mu v})$, where $\mu = c_1 - jc_2$. Because v and $\overline{v}$ are eigenvectors of A corresponding to the eigenvalues λ and $\overline{\lambda}$, respectively, the associated solution of the homogeneous difference equation is

$$x(n) = \frac{1}{2}(\mu \lambda^n v + \overline{\mu}\,\overline{\lambda}^n \overline{v}), \qquad n \in \mathbb{Z}_+.$$

Substitution of $\lambda = \rho e^{j\psi}$, $v = r + js$ and $\mu = \alpha e^{j\phi}$, with α and ϕ suitable real numbers, completes the proof.

(b′) The proof of the continuous-time result is similar.   ∎

We see from 5.6.10 that the modes corresponding to a complex conjugate pair of eigenvalues λ, $\overline{\lambda}$ may be combined to a real solution that contains two arbitrary constants.

In the discrete-time case, this solution is *harmonically damped* if the magnitude $\rho = |\lambda|$ of the eigenvalue is less than 1, *purely harmonic* if $\rho = 1$, and *harmonically increasing* if $\rho > 1$. The number $\psi = \arg(\lambda)$ is the *angular frequency* of the harmonic solution.

In the continuous-time case, the solution is harmonically damped if the real part $\sigma = \mathrm{Re}(\lambda)$ of the eigenvalue is less than 0, purely harmonic if $\sigma = 0$, and harmonically increasing if $\sigma > 0$. The imaginary part $\omega = \mathrm{Im}(\lambda)$ is the angular frequency of the harmonic.

We consider two examples.

5.6.12.   Examples: Modes.

(a) *Second-order smoother*. In Example 5.6.8(a) we found that the second-order smoother has the eigenvalues $\lambda_1 = 0$ and $\lambda_2 = \frac{1}{2}$, with corresponding eigenvectors $v_1 = \text{col } (1, -\frac{1}{2})$ and $v_2 = \text{col } (1, 0)$. As a result, the system has the two real modes

$$\lambda_1^n v_1 = \begin{bmatrix} 1 \\ -\frac{1}{2} \end{bmatrix} \Delta(n), \qquad \lambda_2^n v_2 = \begin{bmatrix} 1 \\ 0 \end{bmatrix} (\tfrac{1}{2})^n, \qquad n \in \mathbb{Z}_+.$$

Any solution of the homogeneous equation on $\mathbb{Z}_+$ is a linear combination of these modes. The first of the modes is what is called a "dead-beat" mode, because any initial component along this mode reduces exactly to zero after a finite number of time instants (in this case after one time instant). The second mode decays exponentially. It decreases quickly, because after, say, five time instants this mode reduces to $(\frac{1}{2})^5 = 1/32$ of its initial value.

(b) *Fourth-order RCL network*. As a continuous-time example we consider the electrical network of Fig. 5.20. It contains a current source, which provides the input u to the system, two capacitors, two inductors, and two resistors. For the resistors, capacitors, and inductors we may write the following ten element equations:

Resistors:

$$v_{R_1} = R_1 i_{R_1}, \qquad v_{R_2} = R_2 i_{R_2}.$$

Capacitors:

$$q_{C_1} = C_1 v_{C_1}, \qquad \dot{q}_{C_1} = i_{C_1}, \qquad q_{C_2} = C_2 v_{C_2}, \qquad \dot{q}_{C_2} = i_{C_2}.$$

Inductors:

$$\phi_{L_1} = L_1 i_{L_1}, \qquad \dot{\phi}_{L_1} = v_{L_1}, \qquad \phi_{L_2} = L_2 i_{L_2}, \qquad \dot{\phi}_{L_2} = v_{L_2}.$$

In these equations v denotes voltage, i current, q charge, and ϕ flux, while the subscripts refer to the various circuit elements. The directions in which the currents are taken positive are indicated in the diagram. Also the current i in the connecting

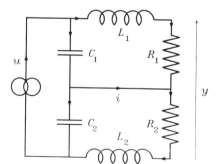

Figure 5.20. A fourth-order RCL network.

branch is shown. Application of Kirchhoff's voltage law to each of the three meshes yields successively

$$v_{C_1} = v_{L_1} + v_{R_1}, \qquad v_{C_2} = v_{R_2} + v_{L_2}, \qquad y = v_{R_1} + v_{R_2}.$$

Application of Kirchhoff's current law to each of the five nodes yields

$$u = i_{C_1} + i_{L_1}, \qquad i_{L_1} = i_{R_1}, \qquad i_{C_1} = i + i_{C_2}, \qquad i_{R_1} + i = i_{R_2}, \qquad i_{R_2} = i_{L_2}.$$

Because the charges and the fluxes determine the amount of energy that is stored in the network, the four state variables are ϕ_{L_1}, q_{C_1}, ϕ_{L_2} and q_{C_2}. The six element voltages (excluding the output voltage y) and the seven currents may be eliminated from the 18 equations. Defining the state as

$$x = \begin{bmatrix} \phi_{L_1} \\ q_{C_1} \\ \phi_{L_2} \\ q_{C_2} \end{bmatrix},$$

this leads to the state differential equation $\dot{x}(t) = Ax(t) + Bu(t)$ and output equation $y(t) = Cx(t)$, where

$$A = \begin{bmatrix} -R_1/L_1 & 1/C_1 & 0 & 0 \\ -1/L_1 & 0 & 0 & 0 \\ 0 & 0 & -R_2/L_2 & 1/C_2 \\ 0 & 0 & -1/L_2 & 0 \end{bmatrix}, \qquad B = \begin{bmatrix} 0 \\ 1 \\ 0 \\ 1 \end{bmatrix},$$

$$B = [R_1/L_1 \quad 0 \quad R_2/L_2 \quad 0].$$

To perform a modal analysis we first determine the eigenvectors and eigenvalues of the matrix A. The characteristic polynomial of A is easily found to be

$$\det(\lambda I - A) = \left(\lambda^2 + \frac{R_1}{L_1}\lambda + \frac{1}{L_1 C_1}\right)\left(\lambda^2 + \frac{R_2}{L_2}\lambda + \frac{1}{L_2 C_2}\right).$$

Suppose that $R_1 = 1$, $C_1 = 1$, $L_1 = 1/2$, $R_2 = 1$, $C_2 = 5/6$, and $L_2 = 1/5$. Then the characteristic polynomial is $(\lambda^2 + 2\lambda + 2)(\lambda^2 + 5\lambda + 6)$, whose roots are

$$\lambda_{1,2} = -1 \pm j, \qquad \lambda_3 = -2, \qquad \lambda_4 = -3.$$

It is not difficult to find that

$$A = \begin{bmatrix} -2 & 1 & 0 & 0 \\ -2 & 0 & 0 & 0 \\ 0 & 0 & -5 & -5 \\ 0 & 0 & -5 & 0 \end{bmatrix}$$

has the corresponding eigenvectors

$$v_{1,2} = \begin{bmatrix} \frac{1}{2} \pm (-\frac{1}{2}j) \\ 1 \\ 0 \\ 0 \end{bmatrix}, \qquad v_3 = \begin{bmatrix} 0 \\ 0 \\ 2 \\ 5 \end{bmatrix}, \qquad v_4 = \begin{bmatrix} 0 \\ 0 \\ 3 \\ 5 \end{bmatrix}.$$

For the eigenvalue pair $\lambda_{1,2} = -1 \pm j$ and corresponding eigenvector pair $v_{1,2}$ we have

$$\sigma_1 = \mathrm{Re}\,(\lambda_1) = -1, \qquad \omega_1 = \mathrm{Im}\,(\lambda_1) = 1,$$

$$r_1 = \mathrm{Re}\,(v_1) = \begin{bmatrix} \frac{1}{2} \\ 1 \\ 0 \\ 0 \end{bmatrix}, \qquad s_1 = \mathrm{Im}\,(v_1) = \begin{bmatrix} -\frac{1}{2} \\ 0 \\ 0 \\ 0 \end{bmatrix}.$$

Thus, according to 5.6.10 corresponding to the eigenvalue pair $\lambda_{1,2} = -1 \pm j$ the homogeneous state differential equation has the solution

$$\alpha e^{-t} \left(\begin{bmatrix} \frac{1}{2} \\ 1 \\ 0 \\ 0 \end{bmatrix} \cos\,(t + \phi) - \begin{bmatrix} -\frac{1}{2} \\ 0 \\ 0 \\ 0 \end{bmatrix} \sin\,(t + \phi) \right)$$

for $t \geq 0$, with α and ϕ arbitrary real constants. This solution represents damped harmonics with angular frequency 1. Corresponding to the eigenvalues $\lambda_3 = -2$ and $\lambda_4 = -3$ the equation has the solutions

$$\alpha_3 e^{-2t} \begin{bmatrix} 0 \\ 0 \\ 2 \\ 5 \end{bmatrix}, \qquad \alpha_4 e^{-3t} \begin{bmatrix} 0 \\ 0 \\ 3 \\ 5 \end{bmatrix},$$

for $t \geq 0$, with α_3 and α_4 arbitrary real constants. Both solutions decay exponentially, the second faster than the first, and both faster than the harmonically damped solutions. The first solution pair corresponds to excitation of the top RCL mesh in the circuit, and the second to excitation of the bottom mesh.   ∎

5.6.13.   Review: Modal analysis of sampled linear time-invariant state difference systems.
We briefly discuss the modal analysis of sampled linear time-invariant state difference systems with the homogeneous state difference equation

$$x(t + T) = Ax(t), \qquad t \in \mathbb{Z}(T).$$

If A has the eigenvalues $\lambda_1, \lambda_2, \cdots, \lambda_N$, with corresponding linearly independent eigenvectors $v_1, v_2, \cdots, v_N$, we may define the modal transformation matrix V and the diagonal matrix Λ as in 5.6.6. It follows for the transition matrix of the system

$$\Phi(t, t_o) = A^{\frac{t-t_o}{T}} = V\Lambda^{\frac{t-t_o}{T}} V^{-1}, \qquad t \geq t_o, \qquad t, t_o \in \mathbb{Z}(T).$$

The modes of the system are the solutions

$$\lambda_i^{t/T} v_i, \qquad t \in \mathbb{Z}_+(T), \qquad i = 1, 2, \cdots, N.$$

If λ is a complex eigenvalue with corresponding complex eigenvector v, we may write

$$\lambda = \rho e^{j\psi T}, \qquad v = r + js,$$

with $\rho = |\lambda|$ and ψ, r and s real. Then corresponding to the eigenvalue pair λ, $\bar{\lambda}$ the homogeneous state difference equation has the real solution

$$x(t) = \alpha\rho^{t/T}[r \cos (\psi t + \phi) - s \sin (\psi t + \phi)], \qquad t \in \mathbb{Z}_+(T). \qquad \blacksquare$$

5.7 STABILITY OF STATE SYSTEMS

In Section 4.6 we discussed the *stability* of input-output systems. Two types of stability were distinguished: *BIBO* and *CICO* stability. These notions also apply to state systems:

(a) *BIBO stability:* Bounded-Input Bounded-Output stability of a state system implies that if the input u of the system is bounded, any corresponding output y is also bounded from any finite time on.

(b) *CICO stability:* Converging-Input Converging-Output stability of a state system signifies that if two inputs u_1 and u_2 converge to each other as time increases, so do any two corresponding outputs y_1 and y_2.

In the case of state systems, a situation of particular interest arises when in these definitions the output is replaced with the state. This leads to the notion of *BIBS* (bounded-input bounded-state) stability, and that of CICS (converging-input converging-state) stability.

BIBS stability implies that in particular the zero-input response remains bounded. This is closely related to the notion of *Lyapunov stability*, which is extensively discussed in many more specialized texts. CICS stability implies that in particular the zero-input response converges to zero. This is connected to the idea of *asymptotic stability*, found in these same texts.

The discussion in this section is restricted to linear time-invariant state difference and differential systems.

Boundedness and Convergence of the Zero-Input State Response

Before entering a more complete discussion of the stability of state systems we consider the asymptotic behavior of the zero-input state response. The conditions for boundedness and convergence exhibit considerable similarity to those for difference and differential IO systems as given in 4.6.5.

5.7.1.   Summary: Boundedness and convergence of the zero-input state response.

(a) Necessary and sufficient conditions for any solution of the homogeneous state difference equation

$$x(n + 1) = Ax(n), \qquad n \in \mathbb{Z},$$

to remain bounded from any finite time on are that
   (i) all eigenvalues of A have magnitude less than or equal to 1, and
   (ii) to any eigenvalue with magnitude 1 and multiplicity m there correspond m linearly independent eigenvectors.
   (b) Necessary and sufficient conditions for any solution of the homogeneous state difference equation to converge to zero as time increases are that all the eigenvalues of A have magnitude strictly less than 1.

(a′) Necessary and sufficient conditions for any solution of the homogeneous state differential equation

$$\dot{x}(t) = Ax(t), \qquad t \in \mathbb{R},$$

to remain bounded from any finite time on are that
   (i′) all eigenvalues of A have real part less than or equal to 0, and
   (ii′) to any eigenvalue with real part 0 and multiplicity m there correspond m linearly independent eigenvectors.
   (b′) Necessary and sufficient conditions for any solution of the homogeneous state differential equation to converge to zero as time increases are that all the eigenvalues of A have strictly negative real part.   ∎

If A is nondefective, the proof follows by inspection of the modal expansion of the zero-input response, as discussed in Section 5.6. The proof when A is defective relies on the Jordan normal form discussed in Supplement E.
   We illustrate the result with a simple example.

5.7.2.   Example: Boundedness and convergence.   Consider the state differential equation $\dot{x} = Ax$, with

$$A = \begin{bmatrix} -2 & 0 & 0 & 0 \\ 0 & -3 & 0 & 0 \\ 0 & 0 & 0 & 0 \\ 0 & 0 & 0 & 0 \end{bmatrix}.$$

A has the eigenvalues -2, -3, 0, and 0. Corresponding to the double eigenvalue 0 the matrix A has the two linearly independent eigenvectors col(0, 0, 1, 0) and col(0, 0, 0, 1). A thus satisfies the conditions of 5.7.1(a') but not those of (b'). Hence, all solutions are bounded but they do not all converge to zero. Indeed, the solution of the state differential equation is easily seen to be given by

$$x_1(t) = e^{-2t}x_1(0),$$

$$x_2(t) = e^{-3t}x_2(0),$$

$$x_3(t) = x_3(0), \qquad t \geq 0.$$

$$x_4(t) = x_4(0),$$

Inspection shows that each of the components of x and hence x itself is bounded. The last two components do not converge to 0, however.

Suppose now that A is modified to

$$A = \begin{bmatrix} -2 & 0 & 0 & 0 \\ 0 & -3 & 0 & 0 \\ 0 & 0 & 0 & 1 \\ 0 & 0 & 0 & 0 \end{bmatrix}$$

The eigenvalues still are -2, -3, 0, and 0, but the matrix only has a single eigenvector col(0, 0, 1, 0) corresponding to the double eigenvalue 0. As a result, the conditions of neither 5.7.1(a') nor (b') are satisfied, and, hence, not all solutions are bounded and converge to zero. Indeed, it may easily be verified that the solution is given by

$$x_1(t) = e^{-2t}x_1(0),$$

$$x_2(t) = e^{-3t}x_2(0),$$

$$x_3(t) = x_3(0) + x_4(0)t, \qquad t \geq 0.$$

$$x_4(t) = x_4(0),$$

The third component of x, and, hence, x itself, is unbounded when $x_4(0) \neq 0$. ∎

BIBO, CICO, BIBS, and CICS Stability of State Systems

We limit ourselves to a rather simple-minded discussion of the stability of state systems. The full story covering all contingencies is quite intricate. The solution of the state difference equation

$$x(n + 1) = Ax(n) + Bu(n), \qquad n \in \mathbb{Z}_+, \tag{1}$$

may be written as

$$x(n) = A^n x(0) + \sum_{k=0}^{n} h(n - k)u(k), \qquad n \in \mathbb{Z}_+,$$

where

$$h(n) = A^{n-1}B\mathbb{1}(n-1), \qquad n \in \mathbb{Z}.$$

The matrix function h is the impulse response matrix of the system when the state is taken as the output. Suppose that the eigenvalues of the matrix A all have magnitude strictly less than 1. Then,

(i) by 5.7.1 the zero-input response $A^n x(0)$, $n \in \mathbb{Z}_+$, is bounded and converges exponentially to zero, and

(ii) each of the entries of the impulse response matrix h has finite action, so that by 4.6.9 the zero-state response to any bounded input is bounded, and by 4.6.13 the response to any input that converges to zero also converges to zero.

It follows that the response of the system (1) to any bounded input is bounded, so that the system is BIBS stable. Moreover, by linearity (ii) implies that the responses to any two inputs that converge to each other also converge to each other. Hence, the system is also CICS stable.

This shows that the condition that all eigenvalues have magnitude strictly less than 1 is *sufficient* for BIBS and CICS stability. By 5.7.1(b) the condition is also *necessary* for CICS stability, but not for BIBS stability (take, for instance, the system $x(n+1) = x(n)$, $n \in \mathbb{Z}$).

Suppose now that the state difference equation (1) is complemented with an output equation

$$y(n) = Cx(n) + Du(n), \qquad n \in \mathbb{Z}.$$

Clearly, if the system is BIBS stable it is also BIBO stable, and if it is CICS stable it is also CICO stable. Hence, the condition that all eigenvalues of A have magnitude strictly less than 1 is not only sufficient for BIBS and CICS stability, but also for both BIBO and CICO stability. It is necessary for neither (think of the system $x(n+1) = 2x(n)$, $n \in \mathbb{Z}$ with the trivial output equation $y(n) = 0$, $n \in \mathbb{Z}$).

Similar considerations apply to state differential systems. They lead to the conclusion that the condition that all eigenvalues of A have strictly negative real part is sufficient for BIBS, BIBO and CICO stability and necessary and sufficient for CICS stability.

We summarize our findings as follows.

5.7.3.   Summary: BIBO, CICO, BIBS, and CICS stability of linear time-invariant state difference and differential systems.

A sufficient condition for the state difference system

$$x(n+1) = Ax(n) + Bu(n),$$
$$y(n) = Cx(n) + Du(n), \qquad n \in \mathbb{Z},$$

A sufficient condition for the state differential system

$$\dot{x}(t) = Ax(t) + Bu(t),$$
$$y(t) = Cx(t) + Du(t), \qquad t \in \mathbb{R},$$

to be BIBO, CICO, and BIBS stable is that all the eigenvalues of the matrix A have magnitude strictly less than one. The condition is necessary and sufficient for CICS stability.

to be BIBO, CICO, and BIBS stable is that all the eigenvalues of the matrix A have strictly negative real part. The condition is necessary and sufficient for CICS stability.   ∎

In many applications, CICS stability is the preferred form of stability. The condition given in 5.7.3 is both necessary and sufficient for CICS stability.

The following examples illustrate the results.

5.7.4.   Examples: Stability of state systems.

(a) *RCL network*. In Example 5.6.8(b) we found that for the assumed numerical values the eigenvalues of the RCL network are -100 and -1000. Hence, by 5.7.3 the network is BIBO, CICO, BIBS, and CICS stable.

(b) *A digital filter*. We show an example of a system for which the sufficient condition of 5.7.3 is not satisfied, but which still is both BIBO and CICO stable though not BIBS and CICS stable. This demonstrates that the sufficient condition is not always necessary.

Consider a second-order digital filter described by the state difference and output equations

$$A = \begin{bmatrix} \frac{1}{2} & 0 \\ 0 & \rho \end{bmatrix}, \qquad B = \begin{bmatrix} 1 \\ 0 \end{bmatrix}, \qquad C = [1 \quad 0], \qquad D = 0,$$

with ρ a real number to be chosen. The eigenvalues of the matrix A are $\frac{1}{2}$ and ρ. If the magnitude $|\rho|$ of the second eigenvalue is not strictly less than 1, the sufficient condition of 5.7.3 is not satisfied.

However, the zero-input response of the system is given by

$$y_{\text{zero-input}}(n) = CA^n x(0) = [1 \quad 0] \begin{bmatrix} (\frac{1}{2})^n & 0 \\ 0 & \rho^n \end{bmatrix} x(0)$$

$$= (\tfrac{1}{2})^n x_1(0), \qquad n \in \mathbb{Z}_+,$$

with x_1 the first component of the state x. Clearly, the zero-input response is bounded and converges to zero for any initial state. Furthermore, the impulse response h of the system is

$$h(n) = CA^{n-1}B\mathbb{1}(n-1) + D\Delta(n) = [1 \quad 0] \begin{bmatrix} (\frac{1}{2})^{n-1} & 0 \\ 0 & \rho^{n-1} \end{bmatrix} \begin{bmatrix} 1 \\ 0 \end{bmatrix} \mathbb{1}(n-1)$$

$$= (\tfrac{1}{2})^{n-1}\mathbb{1}(n-1), \qquad n \in \mathbb{Z}.$$

Because the impulse response has finite action, the zero-state response of the system to any bounded input is bounded and that to any input that converges to zero also converges to zero.

As a result, the system is both BIBO and CICO stable, even if the second eigenvalue does not have magnitude strictly less than 1. The reason is that by the particular structure of the matrices A, B, and C the eigenvalue ρ, which potentially may result in a nonconverging zero-input response and make the action of the impulse response infinite, in fact appears neither in the zero-input response nor in the impulse response.

The realization of the system of Fig. 5.21 clarifies the situation. The system consists of two separate subsystems, the second of which is not affected by the input and does not affect the output. The input-output stability of the overall system only depends on the stability of the first subsystem.

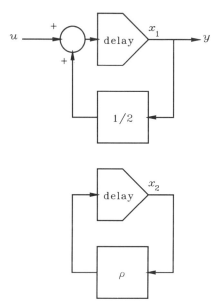

Figure 5.21.   Realization of the digital filter.

On the other hand, it is clear that if $|\rho|$ is greater than 1, the system is both BIBS and CICS unstable. If the initial value of the second component x_2 of the state is nonzero, this component becomes larger with every time step and very soon overflow occurs.

The phenomenon that a particular eigenvalue of the system may not affect the zero-input response or the impulse response, or both, explains why certain systems do not satisfy the sufficient condition of 5.7.3 and still are BIBO stable or even CICO stable.

(c) *Moon rocket*. In Example 5.4.6(b) we saw that a constant input to the moon rocket causes both its elevation and its speed to increase without bound. Hence, the rocket is not BIBS stable.

To see that the rocket neither is CICS stable, consider two initial states such that the initial elevations are different but the initial speeds and masses are the same. Then if the input is zero, in the resulting free drops of the rocket the difference between the elevations remains constant and never diminishes.   ■

5.7.5.   Review: Sampled state difference systems.   The various stability definitions apply without modification to sampled state difference systems. A sufficient condition for the linear time-invariant sampled state difference system with state difference and output equation

$$x(t + T) = Ax(t) + Bu(t),$$
$$y(t) = Cx(t) + Du(t), \qquad t \in \mathbb{Z}(T),$$

to be BIBO, CICO, and BIBS stable is that the eigenvalues of A all have magnitude strictly less than one. This condition is sufficient *and* necessary for the system to be CICS stable.   ■

5.8 FREQUENCY RESPONSE OF STATE SYSTEMS

In Section 3.7 we found that if a continuous- or discrete-time linear time-invariant input-output mapping system has a harmonic signal as input, the corresponding output signal is again harmonic, with the same frequency. All the system does is to multiply the harmonic input by a scaling function, called the *frequency response function* of the system.

In the present section we study the response of linear time-invariant state difference and differential systems to harmonic inputs. To define their frequency response function we consider the zero-state response. According to Section 5.5, the zero-state response with initial time $n_o = -\infty$ of the state difference system

$$x(n + 1) = Ax(n) + Bu(n), \tag{1a}$$
$$y(n) = Cx(n) + Du(n), \qquad n \in \mathbb{Z}, \tag{1b}$$

is

$$y(n) = \sum_{k=-\infty}^{\infty} h(n - k)u(k), \qquad n \in \mathbb{Z}, \tag{2}$$

where h is the impulse response matrix

$$h(n) = CA^{n-1}B\mathbb{1}(n - 1) + D\Delta(n), \qquad n \in \mathbb{Z}.$$

The zero-state response with initial time $t_o = -\infty$ of the state differential system

$$\dot{x}(t) = Ax(t) + Bu(t), \tag{3a}$$
$$y(t) = Cx(t) + Du(t), \qquad t \in \mathbb{R}, \tag{3b}$$

is

$$y(t) = \int_{-\infty}^{\infty} h(t - \tau)u(\tau)\,d\tau, \qquad t \in \mathbb{R},$$

where the impulse response matrix h is

$$h(t) = Ce^{At}B\,\mathbb{1}(t) + D\delta(t), \qquad t \in \mathbb{R}.$$

Because the input u and the output y of state difference or differential systems may be vector-valued signals, the notion of frequency response function need be generalized to that of *frequency response matrix*.

Frequency Response Matrix

To extend the idea of frequency response function to that of frequency response matrix, we suppose that the input to the state difference system (1) is the *vector-valued* harmonic signal

$$u(n) = u_o e^{j2\pi fn}, \qquad n \in \mathbb{Z}.$$

Here, $f \in \mathbb{R}$ is a real frequency and $u_o \in \mathbb{C}^K$ a constant vector whose dimension is that of the input. Then, by (2) the zero-state response to this harmonic input, if it exists, is

$$y(n) = \sum_{k=-\infty}^{\infty} h(n - k)u_o e^{j2\pi fk}, \qquad n \in \mathbb{Z}.$$

Replacement of the summation variable k by $n - m$ results in

$$y(n) = \sum_{m=-\infty}^{\infty} h(m)u_o e^{j2\pi f(n-m)} = \left(\sum_{m=-\infty}^{\infty} h(m)e^{-j2\pi fm} \right) u_o e^{j2\pi fn}$$

$$= \hat{h}(j2\pi f)u_o e^{j2\pi fn}, \qquad n \in \mathbb{Z},$$

where the *frequency response matrix* $\hat{h}$ is a matrix function of the same dimensions as the impulse response matrix h, and is given by

$$\hat{h}(f) = \sum_{m=-\infty}^{\infty} h(m)e^{-j2\pi fm}, \qquad f \in \mathbb{R}.$$

From Section 5.7 we know that if all the eigenvalues of the system matrix A have magnitude strictly less than one, all the elements of the impulse response matrix h have finite action. This is a sufficient condition for the existence of the frequency response matrix $\hat{h}$ of the system.

The frequency response matrix of the linear time-invariant state differential system (3) follows by considering vector-valued harmonic inputs of the form

$$u(t) = u_o e^{j2\pi ft}, \qquad t \in \mathbb{R}.$$

The zero-state response to this input with initial time $t_o = -\infty$, if it exists, is again harmonic of the form

$$y(t) = \hat{h}(f)u_o e^{j2\pi ft}, \qquad t \in \mathbb{R},$$

where the frequency response matrix $\hat{h}$ has the same dimensions as the impulse response matrix h, and is given by

$$\hat{h}(f) = \int_{-\infty}^{\infty} h(t)e^{-j2\pi ft}\, dt, \qquad f \in \mathbb{R}.$$

A sufficient condition for the existence of the frequency response matrix is that all the eigenvalues of the system matrix A have strictly negative real parts.
    We summarize the results found so far.

5.8.1.   Summary. Frequency response of linear time-invariant state difference and differential systems.

Suppose that the eigenvalues of the system matrix A of the linear time-invariant state difference system

$$x(n + 1) = Ax(n) + Bu(n),$$
$$y(n) = Cx(n) + Du(n), \qquad n \in \mathbb{Z},$$

all have magnitude strictly less than one.
    Then, the system has a well-defined zero-state response to the harmonic input

$$u(n) = u_o e^{j2\pi fn}, \qquad n \in \mathbb{Z},$$

with $u_o \in \mathbb{C}^K$ a constant vector and f a constant real frequency, of the form

$$y(n) = \hat{h}(f)u_o e^{j2\pi fn}, \qquad n \in \mathbb{Z}.$$

The *frequency response matrix* $\hat{h}$ is given by

Suppose that the eigenvalues of the system matrix A of the linear time-invariant state differential system

$$\dot{x}(t) = Ax(t) + Bu(t),$$
$$y(t) = Cx(t) + Du(t), \qquad t \in \mathbb{R},$$

all have strictly negative real part.
    Then, the system has a well-defined zero-state response to the harmonic input

$$u(t) = u_o e^{j2\pi ft}, \qquad t \in \mathbb{R},$$

with $u_o \in \mathbb{C}^K$ a constant vector and f a constant real frequency, of the form

$$y(t) = \hat{h}(f)u_o e^{j2\pi ft}, \qquad t \in \mathbb{R}.$$

The *frequency response matrix* $\hat{h}$ is given by

$$\hat{h}(f) = \sum_{n=-\infty}^{\infty} h(n)e^{-j2\pi fn}, \qquad f \in \mathbb{R}, \qquad\qquad \hat{h}(f) = \int_{-\infty}^{\infty} h(t)e^{-j2\pi ft}\, dt, \qquad f \in \mathbb{R},$$

with h the impulse response matrix of the system.

where h is the impulse response matrix of the system.   ∎

To determine the frequency response matrix explicitly we observe that if the input is harmonic the state is likely also to be harmonic of the form

$$x(n) = x_o e^{j2\pi fn}, \qquad n \in \mathbb{Z},$$

with x_o a constant vector to be determined. Substitution of this conjectured particular solution into the state difference equation results in

$$x_o e^{j2\pi f(n+1)} = Ax_o e^{j2\pi fn} + Bu_o e^{j2\pi fn}, \qquad n \in \mathbb{Z},$$

which after cancellation of the common factor $e^{j2\pi fn}$ simplifies to

$$x_o e^{j2\pi f} = Ax_o + Bu_o.$$

This expression is independent of time. After rewriting it as

$$(e^{j2\pi f}I - A)x_o = Bu_o$$

with I the $N \times N$ unit matrix, we may solve for the unknown constant vector x_o as

$$x_o = (e^{j2\pi f}I - A)^{-1}Bu_o.$$

It may be shown that under the condition that all eigenvalues of A have magnitude strictly less than one the inverse matrix in this expression always exists. Substitution of the particular solution $x(n) = x_o e^{j2\pi fn}$, $n \in \mathbb{Z}$, into the output equation results in

$$y(n) = (Cx_o + Du_o)e^{j2\pi fn}$$
$$= [C(e^{j2\pi f}I - A)^{-1}B + D]u_o e^{j2\pi fn}, \qquad n \in \mathbb{Z}.$$

This shows that the frequency response matrix $\hat{h}$ of the state difference system is given by

$$\hat{h}(f) = C(e^{j2\pi f}I - A)^{-1}B + D, \qquad f \in \mathbb{R}.$$

The frequency response matrix may thus directly be found from the coefficient matrices A, B, C, and D.

Analogously, by considering a particular solution of the state differential equation of the form $x(t) = x_o e^{j2\pi ft}$, $t \in \mathbb{R}$, it is easily found that the frequency response matrix of the linear time-invariant state differential system (3) may explicitly be expressed in the coefficient matrices as

$$\hat{h}(f) = C(j2\pi fI - A)^{-1}B + D, \qquad f \in \mathbb{R}.$$

We summarize as follows.

5.8.2.   Summary: Frequency response matrix of linear time-invariant state difference and differential systems.

The frequency response matrix $\hat{h}$ of the linear time-invariant state difference system

$$x(n + 1) = Ax(n) + Bu(n),$$
$$y(n) = Cx(n) + Du(n), \qquad n \in \mathbb{Z},$$

exists if the eigenvalues of A all have magnitude strictly less than one, and is given by

$$\hat{h}(f) = C(e^{j2\pi f}I - A)^{-1}B + D, \quad f \in \mathbb{R}.$$

The frequency response matrix of the linear time-invariant state differential system

$$\dot{x}(t) = Ax(t) + Bu(t),$$
$$y(t) = Cx(t) + Du(t), \qquad t \in \mathbb{R},$$

exists if the eigenvalues of the matrix A all have strictly negative real part, and is given by

$$\hat{h}(f) = C(j2\pi fI - A)^{-1}B + D, \quad f \in \mathbb{R}.$$

∎

For multi-input multi-output (MIMO) systems with K-dimensional input and M-dimensional output the frequency response matrix $\hat{h}$ is an $M \times K$ matrix function. The (i, j)th element $\hat{h}_{ij}$ of the frequency response matrix $\hat{h}$ may be interpreted as follows. Suppose that the jth component of the input is harmonic with frequency f and all other components of the input are zero. Then the response of the ith component of the output equals $\hat{h}_{ij}(f)$ multiplied by the jth component of the input. For single-input single-output (SISO) systems $\hat{h}$ is a 1×1 function and is called, as before, the frequency response *function* of the system.

5.8.3.   Example: RCL network.   In Example 5.2.1 it was found that if the state of the RCL network is chosen to consist of the charge q of the capacitor and the flux ϕ of the inductor, the system has the state differential equation $\dot{x}(t) = Ax(t) + Bu(t)$, $t \in \mathbb{R}$, with

$$A = \begin{bmatrix} 0 & \dfrac{1}{L} \\ -\dfrac{1}{C} & -\dfrac{R}{L} \end{bmatrix}, \qquad B = \begin{bmatrix} 0 \\ 1 \end{bmatrix}.$$

Suppose now that we consider *two* outputs simultaneously. Taking the first component y_1 of the output as the current through the network we have from 5.2.1

$$y_1(t) = \frac{1}{L}x_2(t), \qquad t \in \mathbb{R}.$$

If we choose as the second component y_2 of the output the voltage v_L across the inductor we have as seen in 5.2.1

$$y_2(t) = -\frac{1}{C}x_1(t) - \frac{R}{L}x_2(t) + u(t), \qquad t \in \mathbb{R}.$$

Hence, the output equation is

$$y(t) = \begin{bmatrix} 0 & \dfrac{1}{L} \\ -\dfrac{1}{C} & -\dfrac{R}{L} \end{bmatrix} x(t) + \begin{bmatrix} 0 \\ 1 \end{bmatrix} u(t), \qquad t \in \mathbb{R}.$$

This is an output equation of the form

$$y(t) = Dx(t) + Eu(t),$$

with

$$D = \begin{bmatrix} 0 & \dfrac{1}{L} \\ -\dfrac{1}{C} & -\dfrac{R}{L} \end{bmatrix}, \qquad E = \begin{bmatrix} 0 \\ 1 \end{bmatrix}.$$

We determine the frequency response matrix $\hat{h}$ of the system. Because the output is two-dimensional and the input one-dimensional, the frequency response matrix is a 2×1 matrix function. The characteristic polynomial of the matrix A is $\lambda^2 + (R/L)\lambda + 1/LC$. It may be verified that if R, C, and L are positive constants, the two roots always have strictly negative real part. Hence, by 5.8.2 the frequency response matrix is well defined. Temporarily writing s for $j2\pi f$, we have for the matrix $sI - A$ and its inverse:

$$sI - A = \begin{bmatrix} s & -\dfrac{1}{L} \\ \dfrac{1}{C} & s + \dfrac{R}{L} \end{bmatrix}, \qquad (sI - A)^{-1} = \frac{1}{s^2 + \dfrac{R}{L}s + \dfrac{1}{LC}} \begin{bmatrix} s + \dfrac{R}{L} & \dfrac{1}{L} \\ -\dfrac{1}{C} & s \end{bmatrix}.$$

It follows that

$$D(sI - A)^{-1}B + E = \begin{bmatrix} 0 & \dfrac{1}{L} \\[2mm] -\dfrac{1}{C} & -\dfrac{R}{L} \end{bmatrix} \dfrac{1}{s^2 + \dfrac{R}{L}s + \dfrac{1}{LC}} \begin{bmatrix} s + \dfrac{R}{L} & \dfrac{1}{L} \\[2mm] -\dfrac{1}{C} & s \end{bmatrix} \begin{bmatrix} 0 \\[1mm] 1 \end{bmatrix} + \begin{bmatrix} 0 \\[1mm] 1 \end{bmatrix}$$

$$= \dfrac{1}{s^2 + \dfrac{R}{L}s + \dfrac{1}{LC}} \begin{bmatrix} \dfrac{s}{L} \\[2mm] s^2 \end{bmatrix}.$$

Back substituting $s = j2\pi f$ we thus find for the frequency response matrix of the system

$$\hat{h}(f) = \begin{bmatrix} \dfrac{\dfrac{j2\pi f}{L}}{(j2\pi f)^2 + \dfrac{R}{L}j2\pi f + \dfrac{1}{LC}} \\[8mm] \dfrac{(j2\pi f)^2}{(j2\pi f)^2 + \dfrac{R}{L}j2\pi f + \dfrac{1}{LC}} \end{bmatrix}, \qquad f \in \mathbb{R}.$$

The frequency response matrix has two entries. The top entry is the frequency response function from the input to the current through the network. The bottom entry is that from the input to the voltage across the inductor. ∎

5.8.4. Review: Frequency response of sampled systems.

The zero-state response of the linear time-invariant sampled state difference system

$$x(t + T) = Ax(t) + Bu(t),$$
$$y(t) = Cx(t) + Du(t), \qquad t \in \mathbb{Z}(T),$$

to the harmonic input

$$u(t) = u_o e^{j2\pi ft}, \qquad t \in \mathbb{Z}(T),$$

with f a real frequency, is well-defined if all the eigenvalues of A have magnitude strictly less than one. The response is given by

$$y(t) = \hat{h}(f)u_o e^{j2\pi ft}, \qquad t \in \mathbb{Z}(T),$$

where the frequency response matrix $\hat{h}$ may be expressed in terms of the impulse response matrix h of the system (see 5.5.13) as

$$\hat{h}(f) = T \sum_{t \in \mathbb{Z}(T)} h(t)e^{-j2\pi ft}, \qquad f \in \mathbb{R}.$$

The frequency response matrix may be given directly in terms of the coefficient matrices as

$$\hat{h}(f) = C(e^{j2\pi fT}I - A)^{-1}B + D, \qquad f \in \mathbb{R}. \qquad \blacksquare$$

5.9 PROBLEMS

The first series of problems deals with the state description of systems, as introduced in Section 5.2. Realizations are discussed in Section 5.3.

5.9.1. **State description of a multiple delay.** Consider the discrete-time multiple delay system with IO map

$$y(n) = u(n - M), \qquad n \in \mathbb{Z},$$

with M a positive integer.

(a) Make it plausible that to determine the system behavior from time n on given the input $u(k)$ for $k \geq n$ it is necessary to know the M past values $u(n - M)$, $u(n - M + 1), \cdots, u(n - 1)$ of the input.

(b) For this reason, choose the state $x(n)$ at time n as $x(n) = \text{col } (u(n - M), u(n - M + 1), \cdots, u(n - 1))$, and determine the state difference and output equations of the system.

(c) Is the system linear? Is it time-invariant? If the answer is yes to both questions, then represent the system in the standard form $x(n + 1) = Ax(n) + Bu(n)$, $y(n) = Cx(n) + Du(n), n \in \mathbb{Z}$.

(d) Set up a block diagram for the realization of the system using unit delays, adders, and gains.

5.9.2. **Fibonacci equation as a state system.** Consider a system described by the Fibonacci equation

$$y(n + 2) = y(n) + y(n + 1), \qquad n \in \mathbb{Z}_+.$$

(a) Show that the state at time n may be chosen as $x(n) = \text{col } (y(n - 1), y(n))$.

(b) Derive the state difference and output equations of the system.

(c) Is the system linear? Is it time-invariant? If the answer to both questions is yes, represent the system in the standard form $x(n + 1) = Ax(n) + Bu(n)$, $y(n) = Cx(n) + Du(n), n \in \mathbb{Z}_+$.

(d) Determine the block diagram for the realization of the system by using unit delays, adders, and gains.

5.9.3. **State of a binary shift register.** The binary shift register of Fig. 5.22 consists of a sequence of N binary memory elements that are connected in series as indicated. The input u consists of a sequence of bits $u(n), n \in \mathbb{Z}_+$, that one by one shift into the left-most memory element. Each time a bit is entered, the entire contents of the register move one position to the right. The final bit moves out and is the output at that time.

(a) What is the state of this system?
(b) Determine the state difference and output equations.
(c) Is the system time-invariant?

$u=(1101001...)$ $y=(0001101...)$ **Figure 5.22.**   A binary shift register.

5.9.4. State description of a nonlinear RC network. The electrical network of Fig. 5.23 consists of a voltage source, which produces the input u to the system, two resistors, a capacitor, and an ideal diode. The output is the voltage across the capacitor.

(a) Make it plausible by an argument involving initial conditions that the state of the system is the voltage across the capacitor.
(b) Determine the state differential and output equations of the network.
(c) Show that the system is time-invariant but not linear.
(d) Give a block diagram for the realization of the system using integrators, adders, and function generators.

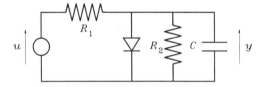

Figure 5.23.   An RC network with a diode.

5.9.5. State representation of the van der Pol equation. The *van der Pol equation* is the nonlinear differential equation

$$y'' + \mu(y^2 - 1)y' + y = 0,$$

with μ a nonnegative constant. For $\mu = 0$ the equation reduces to what is often called the *harmonic oscillator*. If μ is nonzero, then the second term of the equation introduces damping, which is negative for $|y| < 1$ and positive if $|y| > 1$. The negative damping for small y keeps the circuit oscillating, while the positive damping for $|y| > 1$ more or less stabilizes the amplitude of the oscillation.

(a) Show that if $\mu = 0$ the basis solutions of the resulting linear differential equation are harmonics.
(b) Make it plausible by an argument involving initial conditions that the state at time t consists of $y(t)$ and its derivative $y'(t)$.
(c) Derive the corresponding state differential and output equations.
(d) Is the system linear? Is it time-invariant?
(e) Give a block diagram for the realization of the system by using integrators, adders, gains, and multipliers.

Balthasar van der Pol was a Dutch physicist. He discovered his equation in the study of nonlinear effects in oscillator circuits with electronic vacuum tubes and described it in a classic paper in 1926.

5.9.6. State description of the double spring-mass system. The double spring-mass system of 4.8.3(b) (Fig. 4.18) comprises two masses and two springs. The output of the system is now assumed to consist of the positions z_1 and z_2 of the two masses.

(a) The state variables of mechanical systems typically are *positions* and *velocities*. Therefore, choose the state of the system as $x = \text{col}\,(z_1,\, z_1',\, z_2,\, z_2')$, and determine the state differential and output equations.

(b) Is the system linear? Is is time-invariant? If the answer is affirmative to both questions, represent the system in the standard form $\dot{x} = Ax + Bu$, $y = Cx + Du$.

(c) Give a block diagram for the realization of the system using integrators, adders, and gains.

5.9.7. Double integrator. The block diagram of Fig. 5.24 shows a *double integrator*, consisting of a series connection of two integrators. Determine a state representation of this system, including the state differential and output equations. Find the state transition map of the system.

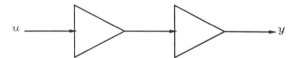

Figure 5.24.   A double integrator.

5.9.8. Series and parallel connection of two state systems. Consider two continuous-time systems, whose state differential and output equations are of the form

$$\dot{x}_1(t) = A_1 x_1(t) + B_1 u_1(t),$$
$$y_1(t) = C_1 x_1(t) + D_1 u_1(t), \qquad t \in \mathbb{R},$$

and

$$\dot{x}_2(t) = A_2 x_2(t) + B_2 u_2(t),$$
$$y_2(t) = C_2 x_2(t) + D_2 u_2(t), \qquad t \in \mathbb{R},$$

respectively.

(a) Suppose that the systems are connected in series as in Fig. 5.25 (a), so that $u_2 = y_1$. For this to make sense we need u_2 and y_1 to have the same dimensions. Prove that the series connection is described by the combined state differential and output equations

$$\begin{bmatrix} \dot{x}_1(t) \\ \dot{x}_2(t) \end{bmatrix} = \begin{bmatrix} A_1 & 0 \\ B_2 C_1 & A_2 \end{bmatrix} \begin{bmatrix} x_1(t) \\ x_2(t) \end{bmatrix} + \begin{bmatrix} B_1 \\ B_2 D_1 \end{bmatrix} u(t),$$

$$y_2(t) = [D_2 C_1 \quad C_2] \begin{bmatrix} x_1(t) \\ x_2(t) \end{bmatrix} + D_2 D_1 u(t), \qquad t \in \mathbb{R}.$$

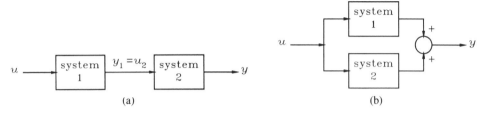

(a)   (b)

Figure 5.25.   (a) series connection of two systems. (b) parallel connection.

(b) Find the combined state differential and output equations for the parallel connection of Fig. 5.25 (b). For this to make sense we need on the one hand u_1 and u_2 to have the same dimensions, and on the other y_1 and y_2 to have equal dimensions.

The state representation of difference and differential systems is treated in Section 5.3.

5.9.9. State representation of difference and differential systems.

(a) *Difference system without input differences.* Consider the difference system

$$y(n) + q_{N-1} y(n-1) + \cdots + q_0 y(n-N) = p_0 u(n), \qquad n \in \mathbb{Z},$$

with $q_0, q_1, \cdots, q_{N-1}$ and p_0 constant coefficients. Show that the state of the system at time t may be chosen as $x(n) = \mathrm{col}\,(y(n-1), y(n-2), \cdots, y(n-N))$, and find the state difference and output equations.

(b) *Differential system without input derivatives.* Consider the differential system

$$y^{(N)}(t) + q_{N-1} y^{(N-1)}(t) + \cdots + q_0 y(t) = p_0 u(t), \qquad t \in \mathbb{R},$$

with $q_0, q_1, \cdots, q_{N-1}$ and p_0 constant coefficients. Show that the state of the system at time t may be chosen as $x(t) = \mathrm{col}\,(y(t), y^{(1)}(t), \cdots, y^{(N-1)}(t))$, and find the state differential equation and output equations.

(c) *Backward differencer.* Determine a state representation of the backward differencer of Problem 3.10.5, described by

$$y(n) = u(n) - u(n-1), \qquad n \in \mathbb{Z}.$$

(d) Differential system. Find a state representation of the differential system of Problem 4.8.6(h), which is given by

$$y''(t) - y(t) = u'(t) - u(t), \qquad t \in \mathbb{R}.$$

The solution of state difference and differential equations is the subject of Section 5.4.

5.9.10. Existence of the solution of a state differential equation. Consider the system with
state differential equation

$$\dot{x}(t) = \sqrt{|x(t)|}, \qquad t \in \mathbb{R},$$

with the initial condition $x(0) = x_0 \neq 0$.

(a) Use 5.4.1 to show that the differential equation has a unique solution on some
interval $[-\eta, \eta]$ with $\eta > 0$.

(b) Use separation of variables to find this solution.

*The solution of linear state difference and differential equations is extensively discussed in
Section 5.5.*

5.9.11. State transition matrix of the harmonic oscillator. The harmonic oscillator (see
Problem 5.9.5) is described by the state differential equation

$$\dot{x}(t) = \begin{bmatrix} 0 & 1 \\ -1 & 0 \end{bmatrix} x(t), \qquad t \in \mathbb{R}.$$

Determine the state transition matrix Φ of the system by summing the infinite sum
for e^{At}.

5.9.12. Solution of the state equations of the discrete-time multiple delay system. The
state equations of the discrete-time delay system of Problem 5.9.1 are of the form
$x(n + 1) = Ax(n) + Bu(n)$, $y(n) = Cx(n) + Du(n)$, $n \in \mathbb{Z}$, where A is an
$M \times M$ matrix, B an $M \times 1$ matrix, C a $1 \times M$ matrix and D a 1×1 matrix as
follows

$$A = \begin{bmatrix} 0 & 1 & 0 & 0 & \cdots & 0 & 0 \\ 0 & 0 & 1 & 0 & \cdots & 0 & 0 \\ \cdots & \cdots & \cdots & \cdots & \cdots & \cdots & \cdots \\ 0 & 0 & 0 & 0 & \cdots & 0 & 1 \\ 0 & 0 & 0 & 0 & \cdots & 0 & 0 \end{bmatrix}, \quad B = \begin{bmatrix} 0 \\ 0 \\ \cdots \\ 0 \\ 1 \end{bmatrix},$$

$$C = [1 \ \ 0 \ \ 0 \ \ 0 \ \ \cdots \ \ 0 \ \ 0], \qquad D = 0.$$

(a) Find the state transition matrix Φ of the system.

(b) Given the state transition matrix, find the impulse response of the system. Is the
result surprising?

*In Section 5.6 it is shown that modal analysis, or diagonalization of the matrix A, may sim-
plify the analysis of time-invariant state difference and differential systems.*

5.9.13. Modal analysis of the Fibonacci system. The state representation of the Fibonacci
system of Problem 5.9.2 is of the form $x(n + 1) = Ax(n)$, $n \in \mathbb{Z}_+$, with

$$A = \begin{bmatrix} 1 & 1 \\ 1 & 0 \end{bmatrix}.$$

(a) Find the eigenvalues and corresponding eigenvectors of the matrix A.
(b) Determine the modes of the system.
(c) For each mode, find initial conditions such that the mode is excited without exciting the other modes.
(d) Compute the transition matrix Φ of the system using the modal transformation.

5.9.14. Modal analysis of the harmonic oscillator. The harmonic oscillator (see Problem 5.9.11) is described by the state differential equation

$$\dot{x}(t) = \begin{bmatrix} 0 & 1 \\ -1 & 0 \end{bmatrix} x(t), \qquad t \in \mathbb{R}.$$

(a) Find the eigenvalues and corresponding eigenvectors of the matrix A.
(b) Determine the modes of the system, in real form.
(c) Compute the transition matrix Φ of the system by using the modal transformation.

5.9.15. Modal analysis of the double spring-mass system. The state differential and output equations of the double spring-mass system of Problem 5.9.6 are of the form $\dot{x} = Ax + Bu$, $y = Cx + Du$, where A is a 4×4 matrix, B a 4×1 matrix, C a 1×4 matrix, and D a 1×1 matrix as follows

$$A = \begin{bmatrix} 0 & 1 & 0 & 0 \\ -\left(\omega_1^2 + \dfrac{m_2}{m_1}\omega_2^2\right) & 0 & \dfrac{m_2}{m_1}\omega_2^2 & 0 \\ 0 & 0 & 0 & 1 \\ \omega_2^2 & 0 & -\omega_2^2 & 0 \end{bmatrix}, \qquad B = \begin{bmatrix} 0 \\ \omega_1^2 \\ 0 \\ 0 \end{bmatrix},$$

$$C = \begin{bmatrix} 1 & 0 & 0 & 0 \\ 0 & 0 & 1 & 0 \end{bmatrix}, \qquad D = 0,$$

where $\omega_1^2 = k_1/m_1$ and $\omega_2^2 = k_2/m_2$. Assume the numerical values $\omega_1^2 = 3/2$, $\omega_2^2 = 4/3$ and $m_2/m_1 = 1/8$.

(a) Find the eigenvalues and corresponding eigenvectors of the matrix A.
(b) Determine the modes of the system, in real form.
(c) For each mode, find initial conditions such that the mode is excited without exciting the other modes. Which physical motion of the system corresponds to each of the modes?
(d) Compute the transition matrix Φ of the system by using the modal transformation.
(e) Given the state transition matrix, determine the impulse response matrix of the system.

5.9.16. RCL network. Use the modal transformation to compute the transition matrix of the fourth-order RCL system of Example 5.6.12(b) for the given numerical values.

BIBO, CICO, BIBS, and CICS stability of state systems are discussed in Section 5.7.

5.9.17. BIBO, CICO, BIBS, and CICS stability. Determine whether the following state systems are BIBO, CICO, BIBS, and CICS stable.

    (a) The harmonic oscillator of Problem 5.9.11.
    (b) The double spring-mass system of Problem 5.9.15.
    (c) The multiple delay system of Problem 5.9.12.
    (d) The nonlinear system with state differential and output equations

$$\dot{x}(t) = \alpha[u(t) - x^2(t)],$$

$$y(t) = x(t), \qquad t \in \mathbb{R},$$

    with α a positive constant. This is the moving car as represented in Problem 4.8.1.
    (e) The state differential system $\dot{x}(t) = Ax(t) + Bu(t)$, $y(t) = Cx(t)$, with

$$A = \begin{bmatrix} -1 & 1 \\ 0 & 0 \end{bmatrix}, \qquad B = \begin{bmatrix} 2 \\ 1 \end{bmatrix}, \qquad C = \begin{bmatrix} 1 & -1 \end{bmatrix}.$$

5.9.18. Stability of the parallel and series connections of two linear time-invariant state differential systems.

    (a) Prove that the parallel connection of two linear time-invariant state differential systems, as discussed in Problem 5.9.8(b), is BIBS stable if and only if each of the two subsystems is BIBS stable. Likewise, prove that the parallel connection is CICS stable if and only each of the component systems is CICS stable.
    (b) What can be said about the BIBS and CICS stability of the *series* connection in terms of the BIBS or CICS stability of the component systems?

The frequency response of linear time-invariant state difference and differential systems, finally, is treated in Section 5.8.

5.9.19. Frequency response.

    (a) Consider the multiple delay system as represented in Problem 5.9.12, and let $M = 3$. Verify that the sufficient conditions for the existence of the frequency response function are satisfied, and determine the frequency response function. Is the result surprising? What is the frequency response function for arbitrary M?
    (b) Show that the system of Problem 5.9.17(e) has a well-defined steady-state response to any harmonic input $u(t) = e^{j2\pi ft}$, $t \geq 0$, of the form $y(t) = \hat{h}(f)e^{j2\pi ft}$, $t \geq 0$. Determine the frequency response function $\hat{h}$.

5.9.20. Electrical circuit. The electrical network of Fig. 5.26 contains one inductor and two capacitors. Its input is the voltage u of the voltage source, and its output y is the voltage across the capacitor C_2.

    (a) Take the charges of the capacitors and the flux contained by the inductor as components of the state, and derive the state differential and output equations of the network. (Alternatively, take the voltages across the capacitors and the current through the inductor as components of the state.)

(b) Determine the frequency response function of the network.

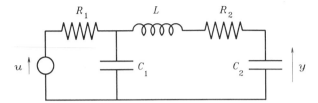

Figure 5.26. An electrical network.

5.10 COMPUTER EXERCISES

The computer exercises for this chapter involve the simulation of various discrete- and continuous-time state systems.

5.10.1. Discrete-time delay and moving averager. Consider the discrete-time delay, described by

$$y(n) = u(n - N), \qquad n \in \mathbb{Z},$$

and the simple moving averager

$$y(n) = u(n) + u(n - 1) + \cdots + u(n - N), \qquad n \in \mathbb{Z}.$$

(a) Find the state difference equations describing these two systems.
(b) Let $N = 3$ and assume that the systems are at rest at time 0. Use the state difference equations to compute and plot the response of these two systems on the time axis $\{0, 1, \cdots, 100\}$ to a pure noise input. *Hint:* In SIGSYS, generate the input by the command u=noiseplus after having defined the appropriate time axis.

5.10.2. Hénon's equation. The solutions of certain nonlinear (state) difference equations (without input) exhibit a highly irregular behavior, which is said to be "chaotic." A well-known example is *Hénon's equation,* which is given by

$$x_1(n + 1) = 1 - ax_1(n)^2 + x_2(n),$$
$$x_2(n + 1) = bx_1(n), \qquad n \in \mathbb{Z}_+,$$

with $a = 1.4$ and $b = 0.3$. (Chaotic behavior is not obtained for all values of a and b.)

(a) Solve the difference equations numerically for $n \in \{0, 1, \cdots, N\}$, with $N = 100$ or 200. Observe that if the initial conditions $x_1(0)$ and $x_2(0)$ are chosen too large (one of the two or both in absolute value greater than 1 or 2), then the solution is unstable, but that for small initial conditions chaotic behavior is obtained. Plot the solutions as a function of time.
(b) Next plot the solution in the (x_1, x_2)-plane rather than as a function of time. Observe that the points form an interesting pattern, which is not at all discernible in the chaotic behavior of the time signals. The pattern is clearer when the (x_1, x_2)

pairs are plotted individually, without connecting lines. *Hint:* In SIGSYS, a real signal x_2 may be plotted against a real signal x_1 by the command polarplot x1+j*x2.

Chaos is a popular subject among physicists and mathematicians, because it is thought to be a model for important physical phenomena such as turbulence. Hénon's equation was published in 1976. (M. Hénon, "A Two-Dimensional Mapping with a Strange Attractor." *Comm. Math. Phys.,* Vol. **50,** 69–77, 1976.)

5.10.3. Lotka-Volterra prey-predator model. The *Lotka-Volterra* equations model an ecological system consisting of two populations that inhabit an isolated environment. The first population, the *preys,* live on a vegetarian diet. The other population, the *predators,* feed entirely on the preys. The growths or declines of the populations are mutually dependent. Let x denote the size of the predator population, and y that of the prey population. Then the Lotka-Volterra model states that x and y change with time according to the equations

$$\dot{x} = -ax + bxy,$$
$$\dot{y} = cy - dxy, \tag{1}$$

with a, b, c, and d constants. The quantity $\dot{x}$ is the rate of increase of the predator population. The term $-ax$ on the right-hand side of the first equation indicates the natural decline of the predator population in the absence of prey. The term bxy is the growth of the predator population, which is proportional both to the size of the predator population itself and to that of the prey population. The term cy on the right-hand side of the second equation is the natural growth of the prey population when left to itself, while the term $-dxy$ represents its decline due to the predators. The equations (1) form the state differential equation for the model.

Lotka and Volterra independently proposed their population model in the 1920s. Vito Volterra (1860–1940) was an Italian mathematican who worked on differential and integral equations. The work of the actuarian and demographer Alfred J. Lotka extends from the first decennium of this century until the 1940s.

(a) Show that the system has two equilibrium states in the (x, y)-plane, one at the origin and one in the first quadrant (assuming that the constants a, b, c, and d are all positive). The following will show that the equilibrium state at the origin is unstable (meaning that all solutions starting close to it tend to move away), while the other equilibrium state is stable (i.e., solutions starting close to it stay close).

(b) Assume $a = b = c = d = 1$, and determine the stable equilibrium state. Solve the Lotka-Volterra equations numerically on the time interval $[0, 10]$ for the following initial conditions:

(b.1) $x(0) = y(0) = 2$,

(b.2) $x(0) = y(0) = 1.1$,

(b.3) $x(0) = y(0) = 0.1$.

Plot the population sizes x and y both against time and against each other. The populations exhibit a periodic behavior. Explain this behavior qualitatively. *Hint:* To plot a real signal y versus a real signal x, in SIGSYS the command polarplot x+j*y may be used.

(c) Ascertain the effect of changing the parameter c from 1 to 2 on the equilibrium state and the trajectories. The change means that the natural growth of the prey population is twice as large as before. Explain the results of this change qualitatively.

Numerical solution of differential equations.

Refer to the box with this title in Section 4.9 for recommendations on the numerical integration of differential equations.

5.10.4. Van der Pol's equation. (Compare 5.9.5.) Van der Pol's equation

$$y'' + \mu(y^2 - 1)y' = 0,$$

is a model for a nonlinear oscillator. The nonlinearity causes the oscillator to have negative damping for small values of y, and an increasingly large positive damping if y exceeds the value 1. The effect of the nonlinearity depends on the magnitude of the constant μ. Defining $x_1 = y$ and $x_2 = y'$ the second-order equation may be rewritten in the form of the state differential equation

$$\dot{x}_1 = x_2,$$
$$\dot{x}_2 = -\mu(x_1^2 - 1)x_2 - x_1.$$

(a) For $\mu = 0$ the system reduces to a harmonic oscillator. Determine the period P of the harmonic.

(b) Integrate the differential equations for $\mu = 0$ on the interval $[0, a]$ with the initial conditions $y(0) = 2.1$ and $y'(0) = 0$. Here a is $2P$ rounded up to a "nice" value. Plot $x_1 = y$ and $x_2 = y'$ both against time and against each other. *Hint:* To plot a real signal x_2 versus a real signal x_1, in SIGSYS the command polarplot x1+j*x2 may be used.

(c) Repeat (b) for $\mu = 0.1$, 1 and 10. Use the same initial conditions, and increase the length of the integration interval as much as needed to allow the behavior of the solution to become stationary. The larger μ becomes, the "stiffer" the differential equation is (see box), so that it may be necessary to reduce the step size of the integration.

(d) For $\mu = 1$, study the effect of changing the initial conditions. Observe that in the (x_1, x_2)-plane whatever the initial conditions are the solution eventually keeps moving on a closed trajectory, which is called the *limit cycle* of the system. In this particular case the initial condition (2, 0) is always on the limit cycle.

5.10.5. Simulation of a simple demodulation circuit. A well-known technique in communication is *amplitude modulation*. Modulation is needed to convert low-frequency signals to high-frequency signals, which may be transmitted as radio waves or along a transmission line. The idea of amplitude modulation is the following. Suppose that the continuous-time message signal m is to be transmitted, and let the *carrier* c be the real harmonic signal given by

$$c(t) = \cos (2\pi f_c t), \qquad t \in \mathbb{R},$$

"Stiff" differential equations.

The numerical solution of the differential equation describing the demodulator of Exercise 5.10.5 is made difficult by the presence of the time constant $R_1 C$. For the demodulator to function adequately, this time constant should be small, necessitating a small step size for the numerical integration. Differential equations containing time constants that widely differ in magnitude are said to be *stiff*. This name derives from mechanical systems, where stiff springs result in small time constants. Stiffness and nonlinearities are characteristic for digital electronic networks. The simulation of such networks, especially large scale networks, is numerically very demanding, and special software exists for this purpose.

---

with the carrier frequency f_c large. Then amplitude modulation of the carrier c with the message signal m results in the modulated signal u given by

$$u(t) = [m_o + m(t)] \cos (2\pi f_c t), \qquad t \in \mathbb{R}.$$

The number m_o is a positive constant such that $m_o + m(t) \geq 0$ for all t.

After the modulated signal u has been transmitted and received, *demodulation* is required to recover the message signal m. A common approximate demodulation scheme that is used in simple AM receivers is implemented by the network of Fig. 5.27. In Fig. 5.28 the output y of the network corresponding to a modulated input signal u is plotted.

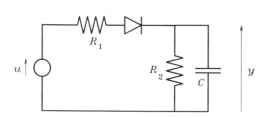

Figure 5.27. A simple demodulator circuit.

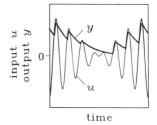

Figure 5.28. Input u and output y of the demodulator circuit.

The demodulated output exhibits a high-frequency ripple owing to imperfect removal of the carrier. The ripple may be reduced by choosing the carrier frequency
large. Choosing the carrier frequency large also allows to make the time constant
$R_2 C$ small which results in a closer approximation of the original waveform.

(a) Describe the operation of the demodulation network qualitatively.
(b) Generate the message signal m as

$$m(t) = \begin{cases} 1 & \text{for } 0.25 \leq t < 0.5 \text{ and } 0.75 \leq t < 1, \\ 0 & \text{otherwise,} \end{cases} \qquad t \in [0, 2).$$

Take the carrier frequency as $f_c = 25$ and generate the modulated signal u with
$m_o = 0.25$. Plot u. *Hint:* Generate the continuous-time signals on a discrete time
axis with sampling interval 0.0025 or less.

(c) Show that the demodulation network is described by the differential equation

$$\dot{y} = -\frac{1}{R_2 C} y + \frac{1}{R_1 C} \text{ramp } (u - y).$$

(d) Compute the response of the network to the input u by integrating the differential equation with the initial condition $y(0) = 0$ for $R_1 C = 0.002$ and
$R_2 C = 0.05$. Plot the input u and the output y. Compute the response if $R_2 C$ is
changed to 0.2 and discuss the difference. *Hints:* Consult the Tutorial to see how
in SIGSYS time t need be integrated along with y to include the external input u
in the differential equation. A Runge-Kutta integration scheme of order two with
step size equal to the sampling interval yields adequate results.

5.10.6. Cruise control system. In 4.8.1 it was found that the moving car of Example 3.2.13
is described by the differential equation

$$\frac{dw(t)}{dt} = \alpha[u(t) - w^2(t)], \qquad t \geq 0, \tag{2}$$

where w is the normalized speed of the car, u the throttle position, and α a physical
constant. Assume that the car is equipped with a cruise control system as in Fig.
1.13, which adjusts the throttle position according to

$$\frac{du(t)}{dt} = k[w_r(t) - w(t)], \qquad t \geq 0, \tag{3}$$

where w_r is the reference speed (again as fraction of the top speed), and k the "gain"
of the cruise controller. The function of the cruise controller is to keep increasing
the throttle position as long as the reference speed exceeds the actual speed, and to
decrease the throttle position in the other case. Together, (2) and (3) form the state
differential equation for the controlled system, with the reference speed w_r as external input.

The purpose of this exercise is to see whether this scheme works, and what the
best value for the gain k is. In the following, take $\alpha = 1/10$.

(a) First choose $k = 0$ (i.e., the controller is inactive). Solve the state differential equation numerically on the interval $[0, 100]$ with the initial conditions $w(0) = 0.5$ and $u(0) = 0.36$. For $k = 0$ the throttle position keeps the constant value $u_o = 0.36$. Observe how the car speed reaches the steady-state constant value $w_\infty = \sqrt{u_o} = \sqrt{0.36} = 0.6$.

(b) Determine the constant throttle position u_o that corresponds to the constant cruising speed $w_o = 0.5$. Solve the differential equations numerically with $w_r(t) = 0.6$ for $t \geq 0$ and the initial conditions $w(0) = w_o = 0.5$ and $u(0) = u_o$ for $k = 0.02, 0.05,$ and 0.1. Compare the responses of the system to that of (a). What is the best value of k?

6

Expansion Theory and Fourier Series

6.1 INTRODUCTION

The next three chapters are devoted to *frequency domain* methods for the analysis of linear time-invariant systems. The methods for obtaining the response of a linear time-invariant system we studied so far rely on *convolutions*. Because the summation or integration implicit in convolution is over time, these methods constitute what is called *time domain analysis*.

Convolution suffers from the drawback that it is a complicated operation, which makes it difficult to assess the system behavior. Also, performing (discrete-time) convolution on a digital computer involves many multiplications and additions, which means that it is computationally expensive.

Both difficulties may to a large extent be avoided when using *frequency domain* analysis. Frequency domain analysis is based on the following two facts: First, surprisingly, practically every time signal may be written as a linear combination of (complex) harmonic signals with different frequencies. Second, the response of a linear time-invariant system to a complex harmonic input signal is the same harmonic multiplied by a gain, which is the value of the *frequency response function* of the system at that frequency.

Frequency domain methods exploit these two facts as follows. To determine the response of a linear time-invariant system we first *decompose* the input as a linear combination of harmonics with different frequencies. Next, the responses to the individual harmonics are obtained, which again are harmonic. Finally, by the superposition property of linear systems, the harmonic responses are linearly combined to obtain the total response.

Frequency domain analysis has considerable advantages. It has intuitive appeal, and moreover leads to efficient computer algorithms, notably the *fast Fourier transform* (FFT), for signal processing.

We start this chapter by developing in Section 6.2 the theory of *signal expansion* in an abstract setting. The idea is to decompose a given signal as a linear combination of *basis signals*. In particular, *orthogonal* expansions are emphasized. In Section 6.3 it is shown that for the analysis of the response of a linear system the *spectral basis* is the most suitable basis. For linear time-invariant systems the spectral basis turns out to consist of harmonics.

In Section 6.4 we develop the expansion of *periodic* signals in harmonics. This leads to the finite and infinite *Fourier series* expansions. Section 6.5 describes how the Fourier series expansion may be used to analyze the response of discrete- and continuous-time linear time-invariant systems to periodic inputs.

In Chapter 7 the frequency domain approach is extended to *aperiodic* signals based on *Fourier integral* theory. The *Laplace* and *z-transforms,* introduced in Chapter 8, are further developments that allow the application of frequency domain methods to constant coefficient linear difference and differential systems.

6.2 SIGNAL EXPANSION

In this section we explore how signals may be represented as linear combinations of a number of fixed signals, together called a *basis*. Special attention is given to expansions in *orthogonal* bases.

Linear Independence

We first define linear *dependence* and *independence*.

6.2.1.   Definition: Linear dependence and independence.   Suppose that X is a linear space over the field $\mathscr{F}$ of real or complex scalars, and that S is a subset of X.

(a) The element $y \in X$ is a *finite linear combination* of the elements $x_1, x_2, \cdots, x_N$ of X if there exist scalars $\alpha_1, \alpha_2, \cdots, \alpha_N \in \mathscr{F}$ such that

$$y = \sum_{i=1}^{N} \alpha_i x_i.$$

(b) The element $y \in X$ is *linearly dependent* on S if y can be expressed as a finite linear combination of elements in S. Otherwise, y is said to be *linearly independent* of S.

(c) The subset S is a *linearly dependent set* if there exist elements x_1, x_2, $\cdots$, $x_N \in S$, with N a natural number, and scalars α_1, α_2, $\cdots$, $\alpha_N \in \mathcal{F}$ such that $\alpha_k x_k \neq 0$ for at least one value of k and

$$\sum_{i=1}^{N} \alpha_i x_i = 0.$$

Otherwise, S is called a *linearly independent set*.   ∎

6.2.2. **Example: Linear dependence and independence in $\mathbb{C}^N$.** Let a_1, a_2, $\cdots$, a_K be vectors in $\mathbb{C}^N$ such that $a_i = \text{col}(\alpha_{1i}, \alpha_{2i}, \cdots, \alpha_{Ni})$, $i = 1, 2, \cdots, K$. Form the $N \times K$ matrix A from the column vectors a_1, a_2, $\cdots$, a_K as

$$A = [a_1, a_2, \cdots, a_K] = \begin{bmatrix} \alpha_{11} & \alpha_{12} & \cdots & \alpha_{1K} \\ \alpha_{21} & \alpha_{22} & \cdots & \alpha_{2K} \\ \cdots & \cdots & \cdots & \cdots \\ a_{N1} & \alpha_{N2} & \cdots & \alpha_{NK} \end{bmatrix}.$$

It is well-known from linear algebra that a_1, a_2, $\cdots$, a_K are linearly independent if and only if the rank of the matrix A equals K. If $\text{rank}(A) < K$, the vectors a_1, a_2, $\cdots$, a_K are linearly dependent. If $K = N$, the vectors a_1, a_2, $\cdots$, a_K are linearly independent if and only if $\det(A) \neq 0$.

Let, for instance, $a_1 = \text{col}(1, 0, 0)$, $a_2 = \text{col}(1, 1, 0)$, $a_3 = \text{col}(1, 1, 1)$ and $b_1 = \text{col}(1, 0, 0)$, $b_2 = \text{col}(1, 1, 0)$, $b_3 = \text{col}(0, 1, 0)$. Forming from these vectors the matrices

$$A = \begin{bmatrix} 1 & 1 & 1 \\ 0 & 1 & 1 \\ 0 & 0 & 1 \end{bmatrix} \qquad B = \begin{bmatrix} 1 & 1 & 0 \\ 0 & 1 & 1 \\ 0 & 0 & 0 \end{bmatrix},$$

we see that $\text{rank}(A) = 3$ and $\text{rank}(B) = 2$, so that a_1, a_2, and a_3 are linearly independent and b_1, b_2, and b_3 linearly dependent. In fact, $b_1 - b_2 + b_3 = 0$.   ∎

Basis

We next introduce the important notion of a *basis* of a linear space.

6.2.3. **Definition: Basis.** Let X be a linear space over the field $\mathcal{F}$ of real or complex scalars.

(a) The set $S = \{v_1, v_2, \cdots, v_N\} \subset X$ is called a *finite basis* for X if for every $y \in X$ there exist *unique* scalars α_1, α_2, $\cdots$, $\alpha_N \in \mathcal{F}$ such that

$$y = \sum_{i=1}^{N} \alpha_i v_i.$$

(b) Suppose that X is a normed space. Then the countably infinite set $S = \{v_i, i \in \mathbb{I}\}$, with $\mathbb{I}$ a countably infinite index set, is a *countably infinite basis* for X if for every $y \in X$ there exist unique scalars α_i, $i \in \mathbb{I}$, such that

$$y = \sum_{i \in \mathbb{I}} \alpha_i v_i.$$ ∎

If the set $\mathbb{I}$ is for instance $\mathbb{Z}$, the equality in 6.2.3(b) is to be read in the sense that

$$\lim_{N \to \infty} \left\| y - \sum_{i=-N}^{N} \alpha_i v_i \right\| = 0,$$

with $\| \cdot \|$ the norm defined on X. If $\mathbb{I}$ is another countably infinite set the definition of equality is accordingly modified.

It may be shown that a basis, whether finite or countably infinite, is a linearly independent set. Also, if a linear space X has a finite basis with N elements, every basis for X has N elements. In fact, if X has a finite basis with N elements, *every* linearly independent subset of X with N elements is a basis for X. Finally, if the normed space X has a countably infinite basis, then *every* basis for X is countably infinite.

These facts allow us to introduce the notion of *dimension*.

6.2.4.   Definition: Dimension.   A linear space X that has a finite basis with N elements is called a *finite-dimensional* linear space with *dimension N*. Otherwise, X is called an *infinite-dimensional* space.   ∎

6.2.5.   Examples: Bases.
(a) *The Euclidean space* $\mathbb{C}^N$. Any N linearly independent vectors in $\mathbb{C}^N$ form a basis for $\mathbb{C}^N$, so that $\mathbb{C}^N$ has dimension N. A particularly convenient basis in $\mathbb{C}^N$ is the *natural basis*, which consists of the N vectors

$$e_1 = \begin{bmatrix} 1 \\ 0 \\ 0 \\ \cdots \\ 0 \end{bmatrix}, \quad e_2 = \begin{bmatrix} 0 \\ 1 \\ 0 \\ \cdots \\ 0 \end{bmatrix}, \quad \cdots, \quad e_N = \begin{bmatrix} 0 \\ 0 \\ \cdots \\ 0 \\ 1 \end{bmatrix}.$$

It is easy to see that this set is linearly independent.

(b) *The signal space* $\ell_{\underline{N}}$. The signal space $\ell_{\underline{N}} = \mathbb{C}^{\underline{N}}$ is the space of all complex-valued signals on the finite time axis $\underline{N} = \{0, 1, 2, \cdots, N - 1\}$. The natural basis for this signal space consists of the signals e_i, $i = 0, 1, \cdots, N - 1$, where

$$e_i(n) = \begin{cases} 1 & \text{for } n = i, \\ 0 & \text{otherwise,} \end{cases} \qquad n \in \underline{N}.$$

The signal space is finite-dimensional with dimension N.

(c) *The signal space ℓ_2.* The signal space ℓ_2 consists of all complex-valued signals with finite energy defined on the infinite discrete time axis $\mathbb{Z}$. It has the countable natural basis $\{e_i, i \in \mathbb{Z}\}$, with e_i the signal

$$e_i(n) = \begin{cases} 1 & \text{for } n = i, \\ 0 & \text{otherwise,} \end{cases} \qquad n \in \mathbb{Z}.$$

The space is infinite-dimensional.   ∎

Given a basis for an *inner product space X*, with inner product $\langle \cdot, \cdot \rangle$, expansions may most conveniently be obtained by introducing another basis, called the *reciprocal basis*.

6.2.6.   Summary: Reciprocal basis.   Let $\{v_i, i \in \mathbb{I}\}$, with $\mathbb{I}$ a finite or countable index set, be a basis for the inner product space X. Then there exists a unique set $\{w_i, i \in \mathbb{I}\}$ of elements in X such that

$$\langle v_i, w_k \rangle = \begin{cases} 1 & \text{for } i = k, \\ 0 & \text{otherwise,} \end{cases} \qquad i, k \in \mathbb{I}. \tag{1}$$

The set $\{w_i, i \in \mathbb{I}\}$ is linearly independent and forms a basis for X, called the *reciprocal basis* to $\{v_i, i \in \mathbb{I}\}$.   ∎

If $\{w_i, i \in \mathbb{I}\}$ is a reciprocal basis to $\{v_i, i \in \mathbb{I}\}$, then $\{v_i, i \in \mathbb{I}\}$ is a reciprocal basis to $\{w_i, i \in \mathbb{I}\}$.

In $\mathbb{C}^N$, reciprocal bases may conveniently be obtained by matrix inversion. In the following the superscript ^{H} denotes the *Hermitian* of a matrix or vector, that is,

$$A^H = \overline{A}^T,$$

with the overbar indicating the complex-conjugate and the superscript ^{T} the transpose.

Charles Hermite (1822–1901) was a French mathematician who taught at the Sorbonne.

6.2.7.   Summary: Reciprocal basis in $\mathbb{C}^N$.   Let $\{v_1, v_2, \cdots, v_N\}$ be a basis for $\mathbb{C}^N$. Then the reciprocal basis $\{w_1, w_2, \cdots, w_N\}$ may be found as follows.

(a) Form the $N \times N$ matrix V whose columns are the basis vectors $v_1, v_2, \cdots, v_N$.

(b) Invert the matrix V, and let the $N \times N$ matrix W be the Hermitian of the matrix V^{-1}, that is,

$$W = (V^{-1})^{\mathrm{H}}.$$

(c) The columns of W form the reciprocal basis $\{w_1, w_2, \cdots, w_N\}$.   ∎

The proof of this result is not difficult.

6.2.8.   Example: Reciprocal basis in $\mathbb{C}^2$.   By way of example we determine the reciprocal basis to the basis for $\mathbb{C}^2$ formed by the vectors $v_1 = \mathrm{col}(1, 1)$ and $v_2 = \mathrm{col}(0, j)$. We have

$$V = \begin{bmatrix} 1 & 0 \\ 1 & j \end{bmatrix}, \qquad V^{-1} = \begin{bmatrix} 1 & 0 \\ j & -j \end{bmatrix},$$

so that

$$W = \begin{bmatrix} 1 & 0 \\ j & -j \end{bmatrix}^{\mathrm{H}} = \begin{bmatrix} 1 & -j \\ 0 & j \end{bmatrix}.$$

It follows that the reciprocal basis consists of $w_1 = \mathrm{col}(1, 0)$ and $w_2 = \mathrm{col}(-j, j)$. ∎

We now have all that is needed to expand a given vector as a linear combination of basis vectors.

6.2.9.   Summary: Expansion in a basis.   Let $S = \{v_i, i \in \mathbb{I}\}$, with $\mathbb{I}$ a finite or countable index set, be a basis for the inner product space X, and $\{w_i, i \in \mathbb{I}\}$ the corresponding reciprocal basis.

(a) The expansion of any $x \in X$ in the basis S is given by

$$x = \sum_{i \in \mathbb{I}} \langle x, w_i \rangle v_i. \tag{2}$$

(b) The inner product of any x and y in X may be expressed as

$$\langle x, y \rangle = \sum_{i \in \mathbb{I}} \langle x, w_i \rangle \langle v_i, y \rangle, \tag{3}$$

while the norm of x equals

$$\|x\| = \left(\sum_{i \in \mathbb{I}} \langle x, w_i \rangle \langle v_i, x \rangle \right)^{1/2}. \tag{4}$$

∎

The proof of 6.2.9 for infinite-dimensional spaces is beyond the scope of this text, but that for the finite-dimensional case is not difficult.

6.2.10.   **Proof of 6.2.9 for finite-dimensional spaces.**   Assume that the space X has dimension N and possesses the basis $\{v_1, v_2, \cdots, v_N\}$. It follows that

$$x = \sum_{i=1}^{N} \alpha_i \, v_i, \tag{5}$$

with the constants α_i to be determined. To find α_k, with $k \in \{1, 2, \cdots, N\}$, we take the inner product of x as given by (5) with the element w_k of the reciprocal basis. This results in

$$\langle x, w_k \rangle = \sum_{i=1}^{N} \alpha_i \, \langle v_i, w_k \rangle = \alpha_k,$$

which proves (a). To obtain (3), take the inner product of the right-hand side of (2) with y. To prove (4), simply set $y = x$ in (3).   ∎

6.2.11.   **Example: Expansion in $\mathbb{C}^2$.**   For an example, refer back to Example 6.2.8, where we found that the basis $v_1 = \mathrm{col}(1, 1)$, $v_2 = \mathrm{col}(0, j)$ for $\mathbb{C}^2$ results in the reciprocal basis $w_1 = \mathrm{col}(1, 0)$, $w_2 = \mathrm{col}(-j, j)$. We expand $x = \mathrm{col}(1, j)$. It follows that

$$\langle x, w_1 \rangle = 1 \cdot 1 + j \cdot 0 = 1,$$
$$\langle x, w_2 \rangle = 1 \cdot j + j \cdot (-j) = 1 + j,$$

so that $x = 1 \cdot v_1 + (1 + j) \cdot v_2$. Indeed,

$$x = \begin{bmatrix} 1 \\ j \end{bmatrix} = \begin{bmatrix} 1 \\ 1 \end{bmatrix} + (1 + j) \begin{bmatrix} 0 \\ j \end{bmatrix}.$$ ∎

Orthogonal and Orthonormal Bases

It is much easier to expand in a given basis when the basis vectors are *orthogonal*, that is, when $\langle v_i, v_k \rangle = 0$ for $i \neq k$. Orthogonality is a generalization of the notion of *perpendicularity* in geometry.

6.2.12.   **Definition: Orthogonality and orthonormality.**   Let X be an inner product space.

(a) Two elements x and y of X are *orthogonal* if $\langle x, y \rangle = 0$.

(b) The element $x \in X$ is *orthogonal* to the set $S \subset X$ if $\langle x, s \rangle = 0$ for every $s \in S$.

(c) The sets $R \subset X$ and $S \subset X$ are orthogonal if $\langle r, s \rangle = 0$ for every $r \in R$ and every $s \in S$.

(d) The set $S \subset X$ forms an *orthogonal set* if $\langle s_1, s_2 \rangle = 0$ for all s_1 and s_2 in S with $s_1 \neq s_2$. If in addition $\langle s, s \rangle = 1$ for every $s \in S$, then S is said to be *orthonormal*.   ∎

6.2.13.   Exercise: Pythagorean equality.   Prove that if x and y are orthogonal, the *Pythagorean equality* holds, that is,

$$\| x + y \|^2 = \| x \|^2 + \| y \|^2,$$

with $\| \cdot \|$ denoting the natural norm for the inner product space.   ∎

About the life of Pythagoras of Samos (ca 570–497 BC) little is known for certain. He left Samos for political reasons and lived in southern Italy, where he was occupied with religious and ethical mysticism. The theorem named after him was known long before in Babylon.

The natural norm

$$\| x \| = \sqrt{\langle x, x \rangle}$$

is used throughout the rest of the chapter.

6.2.14.   Examples: Orthogonality and orthonormality.
    (a) *Three-dimensional Euclidean space* $\mathbb{C}^3$. The vectors $a = \mathrm{col}(1, j, 0)$ and $b = \mathrm{col}(1, -j, 1)$ in $\mathbb{C}^3$ are orthogonal since

$$\langle a, b \rangle = 1 \cdot 1 + j \cdot \overline{(-j)} + 0 \cdot 1 = 0.$$

The vectors become orthonormal if they are *normalized* to

$$\frac{1}{\|a\|} a = \frac{1}{\sqrt{2}} \begin{bmatrix} 1 \\ j \\ 0 \end{bmatrix}, \qquad \frac{1}{\|b\|} b = \frac{1}{\sqrt{3}} \begin{bmatrix} 1 \\ -j \\ 1 \end{bmatrix}.$$

    (b) *N-dimensional Euclidean space* $\mathbb{C}^N$. The natural basis defined in 6.2.5(a) for $\mathbb{C}^N$ is an orthonormal set.

    (c) *Signal space* ℓ_2. The signal space ℓ_2 is an inner product space with inner product as in 2.4.8. The natural basis defined in 6.2.5(c) constitutes an orthonormal set.   ∎

If a basis for an inner product space is an orthogonal set, then it is called an *orthogonal basis*. If the elements of an orthogonal basis are normalized such that their norms are 1, then the basis is an orthonormal set and, hence, is called an *orthonormal basis*.

Because an orthonormal basis is reciprocal to itself, expansion in an orthonormal basis becomes very simple.

6.2.15.   Summary: Expansion in an orthonormal basis. Let $\{v_i,\ i \in \mathbb{I}\}$, with $\mathbb{I}$ a finite or countable index set, be an orthonormal basis for the inner product space X.

(a) Every $x \in X$ may be expanded as

$$x = \sum_{i \in \mathbb{I}} \hat{x}_i v_i,$$

where $\hat{x}_i = \langle x, v_i \rangle$ for $i \in \mathbb{I}$.

(b) The inner product of any x and y in X with expansion coefficients $\hat{x}_i$ and $\hat{y}_i$, $i \in \mathbb{I}$, respectively, may be written as

$$\langle x, y \rangle = \sum_{i \in \mathbb{I}} \hat{x}_i \bar{\hat{y}}_i.$$

In particular,

$$\|x\|^2 = \sum_{i \in \mathbb{I}} |\hat{x}_i|^2. \qquad\qquad\blacksquare$$

6.2.16.   Proof of 6.2.15. Since the reciprocal basis to an orthonormal basis is the basis itself, the formulas of 6.2.15 follow from those of 6.2.9 by simply replacing w_i with v_i throughout. $\qquad\blacksquare$

6.2.17.   Examples: Orthonormal expansion.

(a) *Two-dimensional space* $\mathbb{C}^2$. The vectors $v_1 = \text{col}(1, 1)/\sqrt{2}$ and $v_2 = \text{col}(1, -1)/\sqrt{2}$ form an orthonormal basis for $\mathbb{C}^2$. The expansion of $x = \text{col}(1, j)$ in this basis is

$$x = \hat{x}_1 v_1 + \hat{x}_2 v_2,$$

where

$$\hat{x}_1 = \langle x, v_1 \rangle = \frac{1}{\sqrt{2}}(1 \cdot 1 + j \cdot 1) = \frac{1 + j}{\sqrt{2}},$$

$$\hat{x}_2 = \langle x, v_2 \rangle = \frac{1}{\sqrt{2}}(1 \cdot 1 + j \cdot (-1)) = \frac{1 - j}{\sqrt{2}}.$$

(b) *N-dimensional space* $\mathbb{C}^N$. The natural basis e_i, $i = 1, 2, \cdots, N$, of 6.2.5(a) is an orthonormal basis for $\mathbb{C}^N$. The expansion of $x = \text{col}(x_1, x_2, \cdots, x_N)$ in this basis is

$$x = \sum_{i=1}^{N} x_i e_i. \qquad\qquad \blacksquare$$

Best Approximation and the Projection Theorem

Given an expansion

$$x = \sum_{i \in \mathbb{I}} \hat{x}_i v_i$$

of a vector x in the orthogonal or orthonormal basis $\{v_i, i \in \mathbb{I}\}$, it is of interest to know what may be said about the *partial sum*

$$\sum_{i \in \mathbb{I}'} \hat{x}_i v_i,$$

with $\mathbb{I}'$ a subset of the index set $\mathbb{I}$. Such a partial sum arises when some of the terms in the expansion are omitted. To answer this question, we present a geometric argument explaining the idea of *best approximation* and its relation to expansions in orthogonal bases.

In geometry, there is a close connection between perpendicularity and distance. If P is a plane and A a point outside the plane, the distance from the point A to the plane P is found by passing a line through A that is perpendicular to the plane (see Fig. 6.1.) The point of intersection of this line with the plane (the point C in Fig. 6.1) is the (perpendicular) *projection* of A on the plane. The distance of the point A to the plane P is the length of the line from C to A. C is the point in the plane P that is *closest* to the point A.

The *projection theorem* generalizes these concepts to inner products spaces.

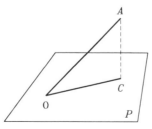

Figure 6.1. The closest point in the plane P to the point A is the projection C of A on P.

6.2.18.   Summary: Projection theorem.   Let X be a complete inner product space and S a closed subspace of X.

(a) For every $x \in X$ there exists a unique $x^* \in S$ such that $x - x^*$ is orthogonal to S. The vector x^* is called the *orthogonal projection* of x on S.

(b) For every $x \in X$ there exists a unique $x^* \in S$ that is closest to x, that is,

$$\| x - x^* \| = \min_{y \in S} \| x - y \|.$$

The vector x^* is called the *best approximation* to x in S.

(c) The orthogonal projection of x on S and the best approximation to x in S are one and the same vector.   ∎

The projection theorem implies that the best approximation to x in S may be determined by finding its orthogonal projection on S. It is illustrated in Fig. 6.2.

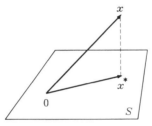

Figure 6.2.   The closest point in the subspace S to x is the projection x^* of x on S.

6.2.19.   Remarks: Closed and complete spaces.

(a) *Closed spaces.* The projection theorem requires the subspace S to be closed, which means that S contains all its "boundary points." A finite-dimensional subspace S is always closed.

(b) *Complete spaces.* A normed linear space X is complete if every "Cauchy sequence" in X converges to an element of X. For us it is enough to know that both ℓ_2 and $\mathscr{L}_2$ are complete inner product spaces.   ∎

An important consequence of the projection theorem is that a *truncated* expansion in an orthogonal basis, that is, the expansion obtained by omitting any number of terms from the expansion, is still the best expansion in the remaining basis vectors. Expansions in bases that are not orthogonal do not have this remarkable property.

6.2.20.   Summary: Best truncated orthogonal expansion.   Let $\{v_i, \ i \in \mathbb{I}\}$, with $\mathbb{I}$ a finite or countable index set, be an orthogonal basis for the inner product space X, and suppose that the expansion of $x \in X$ in this basis is given by

$$x = \sum_{i \in \mathbb{I}} \hat{x}_i v_i.$$

(a) The best approximation x^* to x in the subspace spanned by the basis vectors v_i, $i \in \mathbb{I}'$, with $\mathbb{I}'$ a subset of $\mathbb{I}$, is

$$x^* = \sum_{i \in \mathbb{I}'} \hat{x}_i v_i,$$

where the coefficients $\hat{x}_i$ are the same as those of the full expansion.

(b) If the basis is not only orthogonal but also orthonormal, the *approximation error* is given by

$$\|x^* - x\| = \left(\sum_{i \in \mathbb{I}, i \notin \mathbb{I}'} |\hat{x}_i|^2 \right)^{\frac{1}{2}}. \tag{6}$$

∎

In numerical computations infinite expansions are always truncated. The result of 6.2.20(a) shows that if the basis is orthogonal there is no need to adjust the coefficients to improve accuracy. Moreover, 6.2.20(b) gives an explicit expression for the truncation error in terms of the coefficients of the omitted terms.

6.2.21.   Proof of 6.2.20. The proof of 6.2.20 follows directly from the projection theorem since x^* is the projection of x on the subspace spanned by v_k, $k \in \mathbb{I}'$, and, hence, is the best approximation to x in that subspace. The error $\|x - x^*\|$ is the norm of the vector

$$x - x^* = \sum_{i \in \mathbb{I}, i \notin \mathbb{I}'} \hat{x}_i v_i.$$

If the basis is orthonormal this norm is given by (6). ∎

6.2.22.   Example: Truncated expansion in $\mathbb{C}^N$. The best approximation of $x = \mathrm{col}(x_1, x_2, \cdots, x_N) \in \mathbb{C}^N$ by an element of the subspace S spanned by the first K elements of the natural basis, with $K \leq N$, is the vector

$$x^* = \sum_{i=1}^{K} x_i e_i = \begin{bmatrix} x_1 \\ x_2 \\ \cdots \\ x_K \\ 0 \\ \cdots \\ 0 \end{bmatrix}.$$

∎

Uncountable and Harmonic Bases

We conclude this section with two remarks.

6.2.23.   Remark: Uncountable bases.   In this section we presented the theory of the expansion of signals in a *finite* or *countable* basis. This theory is applied in Section 6.4. In Chapter 7 we encounter "expansions" in a *continuum* of basis signals, together forming an *uncountable* "basis." Such expansions are treated as *transforms*.   ■

6.2.24.   Remark: Harmonic bases.   The most important orthogonal bases for time signals that we deal with in the sequel are *harmonic* bases, consisting of signals of the form η_f, $f \in \mathbb{I}$, with $\mathbb{I}$ the index set. In Section 6.4 we discuss *finite* and *countable infinite* harmonic bases. In Chapter 7, where Fourier transform theory is used to expand signals in a continuum of harmonic signals, we are confronted with *uncountable* harmonic "bases."   ■

6.3 SIGNAL EXPANSION FOR LINEAR SYSTEMS

In this section we apply expansion theory to the analysis of linear IOM systems. The idea is to expand the input to the system and to use the superposition property of the IO map to determine the response of the system.

6.3.1.   Examples: Linear systems with finite and countable bases.
    (a) *Linear system with a finite basis.* The simplest example of a linear system whose input and output sets are finite-dimensional spaces is a discrete-time linear system whose input u and output y are signals on a finite time axis. Suppose that the input and output signals are real- or complex-valued signals on the time axis $\underline{N} = \{0, 1, \cdots, N - 1\}$. According to 6.2.5(b), these signals belong to the N-dimensional signal space $\ell_{\underline{N}}$. Analogous to 3.3.10, the IO map of such a system may be characterized by

$$y(n) = \sum_{m=0}^{N-1} k(n, m)u(m), \qquad n \in \underline{N},$$

where the function k is the *kernel* of the system.
    To emphasize the similarity to results from linear algebra, we represent the input u and output y as the column vectors

$$u = \begin{bmatrix} u(0) \\ u(1) \\ \cdots \\ u(N - 1) \end{bmatrix}, \qquad y = \begin{bmatrix} y(0) \\ y(1) \\ \cdots \\ y(N - 1) \end{bmatrix}.$$

The IO map may then be expressed in matrix form as

$$y = Ku,$$

where K is the $N \times N$ matrix

$$K = \begin{bmatrix} k(0, 0) & k(0, 1) & \cdots & k(0, N-1) \\ k(1, 0) & k(1, 1) & \cdots & k(1, N-1) \\ \cdots & \cdots & \cdots & \cdots \\ k(N-1, 0) & k(N-1, 1) & \cdots & k(N-1, N-1) \end{bmatrix}.$$

(b) *Linear system with a countable basis.* As a second example we consider linear systems with real- or complex-valued inputs defined on the infinite discrete time axis $\mathbb{Z}$. The input and output sets of this system may be taken to be subspaces of the signal space ℓ. As we saw in 3.3.10, the IO map of such systems may be put into the form

$$y(n) = \sum_{m \in \mathbb{Z}} k(n, m) u(m), \qquad n \in \mathbb{Z}.$$

From 6.2.5(c) we know that if the input and output sets both are ℓ_2, they are infinite-dimensional with a countable basis. One choice for the basis is the natural basis.   ∎

Expansion of Input and Output

We study the response of a linear IOM system with IO map $\phi \colon X \to X$, whose input and output sets are the same linear space X with the countable basis $\{v_i, i \in \mathbb{I}\}$. Expanding the input u as

$$u = \sum_{i \in \mathbb{I}} \hat{u}_i v_i,$$

it follows from the superposition property of linear maps that the output may be expressed as

$$y = \phi\left(\sum_{i \in \mathbb{I}} \hat{u}_i v_i\right) = \sum_{i \in \mathbb{I}} \hat{u}_i \phi(v_i).$$

This shows that the output y is a linear combination of the vectors $\phi(v_i)$, $i \in \mathbb{I}$. Note that if $\mathbb{I}$ is not finite we need assume that superposition also holds for infinite sums. By expanding each of the vectors $\phi(v_i)$ in the basis $\{v_k, k \in \mathbb{I}\}$ we obtain what follows.

6.3.2.   Summary: Input and output signal expansion for linear systems.
Suppose that the input and output set X of the linear IOM system with IO map $\phi: X \to X$ is a linear space with a finite or countable basis $\{v_i, i \in \mathbb{I}\}$. Then if the input u to the system is expanded as

$$u = \sum_{i \in \mathbb{I}} \hat{u}_i v_i,$$

the corresponding output may be expanded as

$$y = \sum_{i \in \mathbb{I}} \hat{y}_i v_i.$$

The expansion coefficients $\hat{y}_i$ are given by

$$\hat{y}_i = \sum_{j \in \mathbb{I}} \hat{u}_j \hat{\phi}_{ij}, \qquad i \in \mathbb{I},$$

where the $\hat{\phi}_{ij}$ are the coefficients in the expansions

$$\phi(v_i) = \sum_{j \in \mathbb{I}} \hat{\phi}_{ji} v_j, \qquad i \in \mathbb{I}. \tag{1}$$

■

6.3.3.   Proof.   As we have already seen, after expanding the input in the given basis with coefficients $\hat{u}_i$, the corresponding output may be expressed as

$$y = \sum_{i \in \mathbb{I}} \hat{u}_i \phi(v_i).$$

Expanding each vector $\phi(v_i)$ according to (1), substitution and reversal of the order of the summation results in

$$y = \sum_{i \in \mathbb{I}} \hat{u}_i \sum_{j \in \mathbb{I}} \hat{\phi}_{ji} v_j = \sum_{j \in \mathbb{I}} \left(\sum_{i \in \mathbb{I}} \hat{u}_i \hat{\phi}_{ji} \right) v_j = \sum_{j \in \mathbb{I}} \hat{y}_j v_j.$$

■

The result 6.3.2 is mainly useful for two special choices of the basis $\{v_i, i \in \mathbb{I}\}$. The first basis is the *natural basis,* if it exists, while the second is the one that results from the so-called *spectral expansion.* The latter is explained in the next subsection, while we illustrate the former for finite-time discrete-time systems.

6.3.4.   Example: Expansion in the natural basis for linear discrete-time systems.   The natural basis for the input and output space of a linear discrete-time

IOM system with finite time axis $\underline{N} = \{0, 1, 2, \cdots, N - 1\}$ consists of the signals e_i, $i = 0, 1, \cdots, N - 1$, defined by

$$e_i(n) = \begin{cases} 1 & \text{for } n = i, \\ 0 & \text{otherwise,} \end{cases} \qquad n \in \underline{N}.$$

Since

$$u(n) = \sum_{i=0}^{N-1} u(i)e_i(n), \qquad n \in \underline{N},$$

the coefficients of the expansion of the input u in the natural basis are

$$\hat{u}_i = u(i), \qquad i = 0, 1, \cdots, N - 1.$$

Similarly, the expansion coefficients of the output y in this basis are

$$\hat{y}_i = y(i), \qquad i = 0, 1, \cdots, N - 1.$$

According to 6.3.2, the expansion coefficients are related by

$$y(i) = \sum_{j=0}^{N-1} u(j)\hat{\phi}_{ij}, \qquad i = 0, 1, \cdots, N - 1. \tag{2}$$

On the other hand, we know that if k is the kernel of the system

$$y(i) = \sum_{j=0}^{N-1} k(i, j)u(j), \qquad i = 0, 1, \cdots, N - 1. \tag{3}$$

Comparison of (2) and (3) shows that

$$\hat{\phi}_{ij} = k(i, j), \qquad i, j = 0, 1, \cdots, N - 1.$$

We thus see that the representation of the IO map in terms of the system kernel is equivalent to expanding the input and output in the natural basis. ∎

Spectral Expansion

Linear system analysis by signal expansion becomes particularly simple when the IO map $\phi: X \to X$ has a set of *eigensignals* $\{v_i, i \in \underline{I}\}$ that forms a basis for the space X.

The vector v_i is an eigensignal of the map ϕ if there exists a real or complex number λ_i such that

$$\phi(v_i) = \lambda_i v_i.$$

The number λ_i is called the *eigenvalue* corresponding to the eigensignal v_i. If there exists a basis consisting of eigensignals, then the expansion coefficients of the output of the system simply follow from those of the input by multiplication by the eigenvalues. We summarize as follows.

6.3.5.   Summary: Spectral expansion.   Suppose that the IO map $\phi: X \rightarrow X$ of a linear IOM system has a finite or countable set of eigensignals $\{v_i,\ i \in \mathbb{I}\}$ that forms a basis for the input and output space X, called a *spectral basis*. Let λ_i be the eigenvalue corresponding to the eigensignal v_i. Then, if the input u to the system is expanded as

$$u = \sum_{i \in \mathbb{I}} \hat{u}_i v_i,$$

the corresponding output $y = \phi(u)$ may be expanded as

$$y = \sum_{i \in \mathbb{I}} \hat{y}_i v_i,$$

where the expansion coefficients are given by

$$\hat{y}_i = \lambda_i \hat{u}_i, \qquad i \in \mathbb{I}. \qquad\qquad\qquad \blacksquare$$

Spectral expansion turns out to be very useful in the sequel, because *harmonic* signals form a spectral basis for convolution systems. At this point we content ourselves with showing the application of spectral expansion to finite-time linear discrete-time systems, whose spectral basis, if it exists, is *finite*.

6.3.6.   Example: Spectral expansion of finite-time linear discrete-time systems.   Consider a linear discrete-time system with the finite time axis $\underline{N} = \{0, 1, 2, \cdots, N - 1\}$, whose IO map is given by

$$y(n) = \sum_{m=0}^{N-1} k(n, m) u(m), \qquad n \in \underline{N}.$$

As in 6.3.1(a) we may write the IO map in matrix form as

$$y = Ku,$$

with u and y finite-dimensional column vectors formed from u and y, and K an $N \times N$ matrix formed from the kernel k.

A signal v (represented as a column vector) is an eigensignal of the IO map of the system if there exists a constant λ such that

$$Kv = \lambda v$$

(i.e., v is an eigenvector of the matrix K and λ the corresponding eigenvalue). Thus, we may find the eigensignals of the system by determining the eigenvectors of the matrix K. If the matrix K is not defective (i.e., has N linearly independent eigenvectors), then the resulting eigensignals of the IO map ϕ form a spectral basis for the system.

In particular, consider a very simple discrete-time system with time axis $\underline{2} = \{0, 1\}$, whose IO map is given by

$$y(0) = \frac{3}{2}u(0) + \frac{1}{2}u(1),$$

$$y(1) = \frac{1}{2}u(0) + \frac{3}{2}u(1).$$

In matrix form the system is represented by the IO map $y = Ku$, with K the 2×2 matrix

$$K = \begin{bmatrix} 3/2 & 1/2 \\ 1/2 & 3/2 \end{bmatrix}.$$

The eigenvalues of the matrix K are the roots of its characteristic polynomial

$$\det(\lambda I - K) = \det\left(\begin{bmatrix} \lambda - 3/2 & -1/2 \\ -1/2 & \lambda - 3/2 \end{bmatrix}\right) = \lambda^2 - 3\lambda + 2$$

$$= (\lambda - 2)(\lambda - 1).$$

It follows that K has the two eigenvalues $\lambda_1 = 2$ and $\lambda_2 = 1$. Since the two eigenvalues are different, we expect K to have two linearly independent eigenvectors v_1 and v_2. It is easily found that these may be chosen as

$$v_1 = \frac{1}{\sqrt{2}}\begin{bmatrix} 1 \\ 1 \end{bmatrix}, \qquad v_2 = \frac{1}{\sqrt{2}}\begin{bmatrix} -1 \\ 1 \end{bmatrix}.$$

We note that the eigenvectors are orthogonal and have been normalized so that they are orthonormal. This makes it particularly simple to expand signals in the basis.

Let u be the input $u = \text{col}(0, 1)$. Then its expansion coefficients are $\hat{u}_1 = \langle u, v_1 \rangle = \frac{1}{2}\sqrt{2}$ and $\hat{u}_2 = \langle u, v_2 \rangle = \frac{1}{2}\sqrt{2}$. It follows that the corresponding output is

$$y = \lambda_1 \hat{u}_1 v_1 + \lambda_2 \hat{u}_2 v_2 = 2 \cdot \tfrac{1}{2}\sqrt{2} \cdot \frac{1}{\sqrt{2}}\begin{bmatrix} 1 \\ 1 \end{bmatrix} + 1 \cdot \tfrac{1}{2}\sqrt{2} \cdot \frac{1}{\sqrt{2}}\begin{bmatrix} -1 \\ 1 \end{bmatrix}$$

$$= \begin{bmatrix} 1/2 \\ 3/2 \end{bmatrix}.$$

In this simple example it is actually much easier to obtain this result directly from the IO map, but this is not the case in general.   ∎

Orthogonality of Spectral Bases

In Section 6.2 we have seen that signal expansions are most convenient when the basis is orthogonal or orthonormal. Linear maps that possess an orthogonal spectral basis are called *normal.*

The following result gives a necessary and sufficient condition for the normality of maps represented by square matrices.

6.3.7.   Summary: Normal matrix.   The linear map $\phi: \mathbb{C}^N \to \mathbb{C}^N$ given by

$$y = Ku,$$

with K an $N \times N$ matrix, possesses an orthogonal spectral basis if and only if K is *normal,* that is,

$$K^H K = K K^H,$$

with the superscript H denoting the Hermitian.   ∎

6.3.8.   Example: Normal matrix.   In Example 6.3.6 we considered a simple system whose IO map is represented by the 2×2 matrix

$$K = \begin{bmatrix} 3/2 & 1/2 \\ 1/2 & 3/2 \end{bmatrix},$$

and found that it has an orthonormal spectral basis. Indeed it is easily verified that

$$K^H K = K K^H = \begin{bmatrix} 5/2 & 3/2 \\ 3/2 & 5/2 \end{bmatrix},$$

so that the matrix K is normal.   ∎

6.3.9.   Exercise: A Hermitian matrix is normal.   Prove that if the square matrix K is Hermitian (i.e., $K^H = K$), then it is normal.   ∎

6.4 FOURIER EXPANSION

In this section we show how finite-time and periodic signals may be expanded in a harmonic basis. *Discrete-time* signals of this type require a *finite* harmonic basis, while *continuous-time* signals need a *countably infinite* basis. The resulting expansions are called the *finite* and *infinite Fourier series expansion*, respectively.

> Jean Baptiste Joseph Fourier (1768–1830) was a French mathematician who did research on heat transmission. In a prize-winning solution to a problem put forth by the French Academy of Sciences he stated in 1811 without proof that an arbitrary function may be represented by a trigonometric series.

The reason why we are interested in expanding in harmonic signals is that as we found in Section 3.7 linear time-invariant systems pass harmonic signals through as harmonic signals so that they are *eigensignals* of linear time-invariant IOM systems. The Fourier series expansion thus constitutes a *spectral expansion* for linear time-invariant systems, which greatly facilitates the analysis of the response of linear time-invariant systems to periodic inputs. This analysis is the subject of Section 6.5.

Fourier Series Expansion

Recall that the harmonic η_f with real frequency f is the time signal defined by

$$\eta_f(t) = e^{j2\pi ft}, \qquad t \in \mathbb{T},$$

where $\mathbb{T}$ is the appropriate time axis.

Our first result is that suitably chosen harmonic signals form an orthogonal basis for (finite-energy) signals defined on a finite time axis.

6.4.1.   Summary: Finite and countably infinite harmonic bases.

(a) The space $\ell_{\underline{N}}$ of complex-valued discrete-time signals with finite time axis $\underline{N} = \{0,\ 1,\ 2,\ \cdots,\ N-1\}$ is an N-dimensional inner product space with inner product

$$\langle x, y \rangle = \sum_{n=0}^{N-1} x(n)\overline{y(n)}.$$

(b) The set of N harmonic signals

$$\{F\eta_f, f \in \underline{N}\,(F)\},$$

(a′) The space $\mathcal{L}_2[0,\ P)$ of finite-energy complex-valued continuous-time signals with finite time axis $[0,\ P)$ is an infinite-dimensional inner product space with inner product

$$\langle x, y \rangle = \int_0^P x(t)\overline{y(t)}\ dt.$$

(b′) The countably infinite set of harmonic signals

$$\{F\eta_f, f \in \mathbb{Z}(F)\},$$

with   with

$$\underline{N}(F) = \{0, F, 2F, \cdots, (N-1)F\} \qquad \mathbb{Z}(F) = \{kF, k \in \mathbb{Z}\}$$

and $F = 1/N$, forms an orthogonal basis for ℓ_N, called the *finite harmonic basis*.

and $F = 1/P$, forms an orthogonal basis for $\mathscr{L}_2[0, P)$, called the *countably infinite harmonic basis*. ∎

6.4.2.   Proof.

(a) The space ℓ_N clearly is linear. It has the natural basis $\{e_i, i = 0, 1, \cdots, N-1\}$ with $e_i(n) = 1$ for $n = i$ and $e_i(n) = 0$ for $n \neq i$, and hence is N-dimensional. It is an inner product space with the inner product as given.

(b) Let

$$w_N = e^{j2\pi/N}.$$

Then we have for $i, k = 0, 1, \cdots, N-1$,

$$\langle F\eta_{iF}, F\eta_{kF} \rangle = F^2 \sum_{n=0}^{N-1} e^{j2\pi iFn} e^{-j2\pi kFn} = F^2 \sum_{n=0}^{N-1} e^{j2\pi(i-k)nF}$$

$$= F^2 \sum_{n=0}^{N-1} e^{j2\pi(i-k)n/N} = F^2 \sum_{n=0}^{N-1} w_N^{(i-k)n}$$

$$= \begin{cases} F^2 \dfrac{1 - w_N^{(i-k)N}}{1 - w_N^{i-k}} = 0 & \text{for } i \neq k, \\[2mm] F & \text{for } i = k. \end{cases}$$

This proves that the set $\{F\eta_f, f \in \underline{N}(F)\}$ is orthogonal. Suppose that the set is linearly dependent. Then there exist constants $\alpha_0, \alpha_1, \cdots, \alpha_{N-1}$ that are not all zero such that

$$\sum_{i=0}^{N-1} \alpha_i F\eta_{iF} = 0.$$

Taking the inner product of both the left- and right-hand sides with $F\eta_{kF}$, $k \in \{0, 1, \cdots, N-1\}$, yields by the properties of the inner product and the fact that the set is orthogonal $\alpha_k F = 0$, so that $\alpha_k = 0$ for $k \in \{0, 1, \cdots, N-1\}$. This contradicts the assumption that the set is linearly dependent. Hence, the set is linearly independent and thus constitutes a basis.

(a′) and (b′). The proof that the countable set $\{F\eta_f, f \in \mathbb{Z}(F)$ is a basis for $\mathscr{L}_2[0, P)$ is outside the scope of this book. The proof that they form a linearly independent orthogonal set is straightforward and parallels that for the discrete-time case. ∎

It follows from 6.4.1 that the space of discrete-time complex-valued time signals on the finite time axis $\{0, 1, \cdots, N - 1\}$ has a basis consisting of a finite set of harmonics with frequencies $0, F, 2F, \cdots, (N - 1)F$, where the *fundamental frequency F* equals $1/N$.

In the continuous-time case, the space of finite-energy signals on the finite time axis $[0, P)$ has a *countably infinite* basis consisting of harmonics whose frequencies are integral multiples of the fundamental frequency $F = 1/P$.

Because both the finite and the countably infinite harmonic basis are orthogonal, expansions in these basis may easily be obtained. They result in the *Fourier series expansion*.

6.4.3.   Summary: Finite and infinite Fourier series expansions.

The expansion of any signal $x \in \ell_N$ in the finite harmonic basis of 6.4.1(b) is given by the *finite Fourier series expansion*

$$x(n) = F \sum_{f \in \underline{N}(F)} \hat{x}(f)e^{j2\pi fn}, \qquad n \in \underline{N},$$

$F = 1/N$, where $\hat{x} \in \ell_N(F)$, called the *Fourier coefficients* of x, is given by

$$\hat{x}(f) = \sum_{n=0}^{N-1} x(n)e^{-j2\pi fn}, \qquad f \in \underline{N}(F).$$

The expansion of any signal $x \in \mathcal{L}_2[0, P)$ in the infinite harmonic basis of 6.4.1(b′) is given by the *infinite Fourier series expansion*

$$x(t) = F \sum_{f \in \mathbb{Z}(F)} \hat{x}(f)e^{j2\pi ft}, \qquad t \in [0, P),$$

$F = 1/P$, where $\hat{x} \in \ell_2(F)$, called the *Fourier coefficients of x*, is given by

$$\hat{x}(f) = \int_0^P x(t)e^{-j2\pi ft}\, dt, \qquad f \in \mathbb{Z}(F).$$

∎

We recall from Section 2.2 that $\ell_N(F)$ is the space of complex-valued signals defined on the signal axis $\underline{N}(F)$. The space $\ell_2(F)$ is the space of finite-energy signals defined on the signal axis $\mathbb{Z}(F)$, introduced in 2.4.6.

Note that the Fourier coefficients $\hat{x}$ are not time signals but *frequency signals,* defined on the finite frequency axis $\underline{N}(F)$ in the discrete-time case and the countably infinite frequency axis $\mathbb{Z}(F)$ in the continuous-time case.

6.4.4.   Proof of 6.4.3.   First consider the discrete-time case. The reciprocal basis to the orthogonal basis $\{F\eta_f, f \in \underline{N}(F)\}$ is easily seen to be the set $\{\eta_f, f \in \underline{N}(F)\}$. Hence, by 6.2.9 the coefficient of $F\eta_f$ in the expansion with respect to the basis $\{F\eta_f, f \in \underline{N}(F)\}$ is

$$\hat{x}(f) = \langle x, \eta_f \rangle = \sum_{n=0}^{N-1} x(n)e^{-j2\pi fn}, \qquad f \in \underline{N}(F).$$

This proves the finite Fourier series expansion. The proof of the infinite Fourier series expansion is similar. The fact that in the continuous-time case the Fourier coefficients $\hat{x}$ belong to $\ell_2(F)$ (and, hence, have finite energy) follows from Parseval's identity as stated in 6.4.7.   ∎

6.4.5.   Remark: Alternative forms of the Fourier series.   In many texts the basis signals for the finite and infinite Fourier series expansion are taken as η_f rather than $F\eta_f$. From this one obtains the *alternative forms* of the Fourier series expansions:

(a) *Discrete-time case*. With the alternative choice of the basis the finite Fourier series expansion takes the form

$$x(n) = \sum_{f \in \underline{N}(F)} \hat{x}(f) e^{j2\pi fn}, \qquad n \in \underline{N},$$

where

$$\hat{x}(f) = \frac{1}{N} \sum_{n=0}^{N-1} x(n) e^{-j2\pi fn}, \qquad f \in \underline{N}(F).$$

If we write

$$x(n) =: x_n, \qquad n = 0, 1, \cdots, N - 1,$$
$$\hat{x}(kF) =: \hat{x}_k, \qquad k = 0, 1, \cdots, N - 1,$$

the alternative finite Fourier series expansion takes the commonly encountered appearance

$$x_n = \sum_{k=0}^{N-1} \hat{x}_k e^{j2\pi kn/N}, \qquad n = 0, 1, \cdots, N - 1,$$

where

$$\hat{x}_k = \frac{1}{N} \sum_{n=0}^{N-1} x_n e^{-j2\pi kn/N}, \qquad k = 0, 1, \cdots, N - 1.$$

(b) *Continuous-time case*. With the alternative choice of the basis the infinite Fourier series expansion takes the form

$$x(t) = \sum_{f \in \mathbb{Z}(F)} \hat{x}(f) e^{j2\pi ft}, \qquad t \in [0, P),$$

where

$$\hat{x}(f) = \frac{1}{P} \int_0^P x(t) e^{-j2\pi ft} \, dt, \qquad f \in \mathbb{Z}(F).$$

Denoting $\hat{x}(kF) =: \hat{x}_k$, $k \in \mathbb{Z}$, this may be rewritten as

$$x(t) = \sum_{k=-\infty}^{\infty} \hat{x}_k e^{j2\pi tk/P}, \qquad t \in [0, P),$$

where

$$\hat{x}_k = \frac{1}{P} \int_0^P x(t) e^{-j2\pi tk/P} \, dt, \qquad k \in \mathbb{Z}. \qquad\blacksquare$$

Examples

We illustrate the Fourier series expansion by some examples.

6.4.6.   Example: Fourier series expansions.
    (a) *Finite Fourier series expansion.* For practical computation it is usually convenient to rearrange the formulas for the Fourier expansion and its coefficients as

$$x(n) = F \sum_{k=0}^{N-1} \hat{x}(kF) \, w_N^{kn}, \qquad n = 0, 1, \cdots, N-1, \tag{1}$$

$$\hat{x}(kF) = \sum_{n=0}^{N-1} x(n) \, w_N^{-kn}, \qquad k = 0, 1, \cdots, N-1, \tag{2}$$

where $w_N := e^{j2\pi/N}$. Let the column vector

$$x = \mathrm{col}(x(0), x(1), \cdots, x(N-1))$$

represent the signal x in vector form and

$$\hat{x} = \mathrm{col}(\hat{x}(0), \hat{x}(F), \cdots, \hat{x}((N-1)F))$$

the Fourier coefficients. Then (1) may be rewritten in matrix form as

$$x = FW_N \hat{x},$$

where W_N is the $N \times N$ matrix

$$W_N = \begin{bmatrix} 1 & 1 & 1 & \cdots & 1 \\ 1 & w_N & w_N^2 & \cdots & w_N^{N-1} \\ 1 & w_N^2 & w_N^4 & \cdots & w_N^{2(N-1)} \\ \cdots & \cdots & \cdots & \cdots & \cdots \\ 1 & w_N^{N-1} & w_N^{2(N-1)} & \cdots & w_N^{(N-1)(N-1)} \end{bmatrix}.$$

Conversely, (2) takes the form

$$\hat{x} = W_N^H x.$$

Note that $x = FW_N \hat{x} = FW_N \cdot W_N^H x$, so that $W_N W_N^H = (1/F)I = N \cdot I$.

Suppose in particular that we wish to find the Fourier series expansion of the signal x given by

$$x(0) = x(1) = 1, \qquad x(2) = x(3) = 0,$$

as shown in Fig. 6.3. For this signal we have $N = 4$,

$$W_4 = \begin{bmatrix} 1 & 1 & 1 & 1 \\ 1 & j & -1 & -j \\ 1 & -1 & 1 & -1 \\ 1 & -j & -1 & j \end{bmatrix}, \qquad W_4^H = \begin{bmatrix} 1 & 1 & 1 & 1 \\ 1 & -j & -1 & j \\ 1 & -1 & -1 & -1 \\ 1 & j & -1 & -j \end{bmatrix},$$

and $x = \mathrm{col}(1, 1, 0, 0)$. It follows that

$$\hat{x} = W_4^H x = W_4^H \begin{bmatrix} 1 \\ 1 \\ 0 \\ 0 \end{bmatrix} = \begin{bmatrix} 2 \\ 1 - j \\ 0 \\ 1 + j \end{bmatrix}.$$

The finite Fourier series expansion of x thus is

$$x(n) = F[\hat{x}(0)\eta_0(n) + \hat{x}(F)\eta_F(n) + \hat{x}(2F)\eta_{2F}(n) + \hat{x}(3F)\eta_{3F}(n)]$$
$$= \tfrac{1}{4}[2 + (1 - j)e^{j2\pi n/4} + 0 \cdot e^{j2\pi 2n/4} + (1 + j)e^{j2\pi 3n/4}], \qquad n \in \underline{4}.$$

(b) *Fourier series expansion of a continuous-time rectangular pulse.* As a second example we consider a rectangular continous-time pulse of width a and height 1 as in Fig. 6.4, given by

$$x(t) = \begin{cases} 1 & \text{for } 0 \le t < a, \\ 0 & \text{for } a \le t < P, \end{cases} \qquad t \in [0, P),$$

with $0 \le a \le P$. The expansion coefficients of the pulse are given by

$$\hat{x}(f) = \int_0^a e^{-j2\pi ft}\, dt = \begin{cases} a & \text{for } f = 0, \\ ae^{-j\pi fa}\dfrac{\sin(\pi fa)}{\pi fa} & \text{for } f \ne 0, \end{cases} \qquad f \in \mathbb{Z}(F),$$

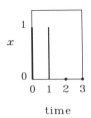

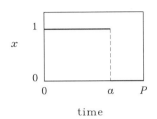

Figure 6.3. A discrete-time signal.

Figure 6.4. A rectangular continuous-time pulse.

where $F = 1/P$. In particular, for $a = P/2$ it follows that

$$\hat{x}(kF) = \begin{cases} P/2 & \text{for } k = 0, \\ 0 & \text{for } k \neq 0,\ k \text{ even,} \\ -jP/k\pi & \text{for } k \text{ odd,} \end{cases} \qquad k \in \mathbb{Z}.$$

In Fig. 6.5 plots are shown of the *truncated sum signals* x_N, given by

$$x_N(t) = F \sum_{k=-N}^{N} \hat{x}(kF)\, e^{j2\pi kFt}, \qquad t \in [0, P),$$

for $N = 0, 1, 3$, and 5, with $a = P/2$ and $P = 1$. It is seen that the approximation of the original waveform improves as N increases.

The "frequency content" of the rectangular pulse may be assessed by considering the magnitude $|\hat{x}(f)|, f \in \mathbb{Z}(F)$, of the Fourier coefficients, given by

$$|\hat{x}(f)| = |a| \cdot \left| \frac{\sin(\pi fa)}{\pi fa} \right|, \qquad f \in \mathbb{Z}(F).$$

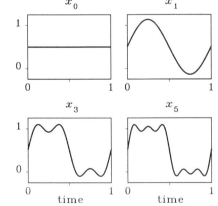

Figure 6.5. Truncated sum signals x_N for the rectangular pulse for $N = 0, 1, 3$, and 5.

Figure 6.6 illustrates that if $P = 1$ and a increases from 0 to 1, then the pattern formed by the Fourier coefficients becomes more and concentrated at low frequencies. Evidently, if the pulse is long it "contains" less high frequencies than if it is short.   ∎

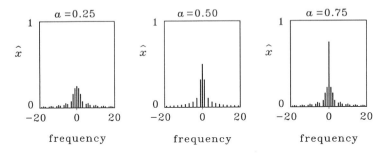

Figure 6.6. Dependence of the Fourier coefficients of the rectangular pulse on its width a.

Identities of Plancherel and Parseval

The *energy* of a discrete-time signal x defined on the time axis $\underline{N}$ is

$$\| x \|_2^2 = \langle x, x \rangle = \sum_{n=0}^{N-1} | x(n) |^2.$$

The energy may be obtained directly from the Fourier coefficients of x and, in fact, equals the *energy* of the Fourier coefficients themselves, if properly defined. This result is known as *Parseval's identity* and also applies to the continuous-time case. We present it in a generalized form, which is sometimes referred to as *Plancherel's identity*.

6.4.7.   Identities of Plancherel and Parseval.

Define the inner product on the signal space $\ell_{\underline{N}}$ as

$$\langle x, y \rangle_{\ell_{\underline{N}}} = \sum_{n=0}^{N-1} x(n) \overline{y(n)},$$

and on the frequency signal space $\ell_{\underline{N}}(\mathrm{F})$, with $F = 1/N$, as

Define the inner product on the signal space $\mathscr{L}_2[0, P)$ as

$$\langle x, y \rangle_{\mathscr{L}_2[0, P)} = \int_0^P x(t) \overline{y(t)} \, dt,$$

and on the frequency signal space $\ell_2(F)$, with $F = 1/P$, as

$$\langle \hat{x}, \hat{y} \rangle_{\ell_N(F)} = F \sum_{n=0}^{N-1} \hat{x}(nF)\overline{\hat{y}(nF)}. \qquad\qquad \langle \hat{x}, \hat{y} \rangle_{\ell_2(F)} = F \sum_{n=-\infty}^{\infty} \hat{x}(nF)\overline{\hat{y}(nF)}.$$

Let $\hat{x} \in \ell_N(F)$ be the Fourier coefficients of the time signal $x \in \ell_N$ and $\hat{y}$ those of y, as defined in 6.4.3.

    (a) *Plancherel's identity:* The inner product of x and y and that of $\hat{x}$ and $\hat{y}$ are related by

$$\langle x, y \rangle_{\ell_N} = \langle \hat{x}, \hat{y} \rangle_{\ell_N(F)}.$$

    (b) *Parseval's identity:* In particular, the time signal x and its Fourier coefficients $\hat{x}$ have equal energy, that is,

$$\|x\|_2^2 = \|\hat{x}\|_2^2,$$

with the norms the natural norms.

Let $\hat{x} \in \ell_2(F)$ be the Fourier coefficients of the time signal $x \in \mathcal{L}_2[0, P)$ and $\hat{y}$ those of y, as defined in 6.4.3.

    (a') *Plancherel's identity:* The inner product of x and y and that of $\hat{x}$ and $\hat{y}$ are related by

$$\langle x, y \rangle_{\mathcal{L}_2[0, P)} = \langle \hat{x}, \hat{y} \rangle_{\ell_2(F)}.$$

    (b') *Parseval's identity:* In particular, the time signal x and its Fourier coefficients $\hat{x}$ have equal energy, that is,

$$\|x\|_2^2 = \|\hat{x}\|_2^2,$$

with the norms the natural norms.   ■

Michel Plancherel (1885–1967) was a Swiss mathematician who taught at the Eidgenössische Hochschule at Zürich. Marc-Antoine Parseval (?–1836) was a French astronomer and mathematician who published the theorem named after him in 1805.

6.4.8.   Proof.

    (a) The proof of the identity of Plancherel follows from 6.2.9(b) by setting $v_i = F\eta_{iF}$, $w_i = \eta_{iF}$ and $\mathbb{I} = \{0, 1, \cdots, N - 1\}$. As a result,

$$\langle x, w_i \rangle_{\ell_N} = \langle x, \eta_{iF} \rangle_{\ell_N} = \hat{x}(iF),$$
$$\langle v_i, y \rangle_{\ell_N} = \overline{\langle y, v_i \rangle_{\ell_N}} = \overline{\langle y, F\eta_{iF} \rangle_{\ell_N}} = F\hat{y}(iF).$$

It follows from 6.2.9(b) that

$$\langle x, y \rangle_{\ell_N} = F \sum_{i=0}^{N-1} \hat{x}(iF)\overline{\hat{y}(iF)} = \langle \hat{x}, \hat{y} \rangle_{\ell_N(F)},$$

which proves the identity.

    (b) Parseval's equality immediately follows from Plancherel's identity by setting $y = x$.

(a′), (b′) The proof for the continuous-time case is similar to that for the discrete-time case.   ∎

Parseval's identity may also be written in the well-known form

$$\sum_{n=0}^{N-1} |x(n)|^2 = F \sum_{k=0}^{N-1} |\hat{x}(kF)|^2$$

in the discrete-time case and

$$\int_0^P |x(t)|^2 \, dt = F \sum_{k=-\infty}^{\infty} |\hat{x}(kF)|^2$$

in the continuous-time case.

6.4.9.   Example: Parseval's identity.   In Example 6.4.6(b) we found that the Fourier coefficients of the rectangular pulse

$$x(t) = \begin{cases} 1 & \text{for } 0 \le t < P/2, \\ 0 & \text{otherwise,} \end{cases} \qquad t \in [0, P),$$

are given by

$$\hat{x}(kF) = \begin{cases} P/2 & \text{for } k = 0, \\ 0 & \text{for } k \ne 0, \, k \text{ even}, \\ -jP/k\pi & \text{for } k \text{ odd}, \end{cases} \qquad k \in \mathbb{Z}.$$

On the one hand,

$$\int_0^P |x(t)|^2 \, dt = \int_0^{P/2} dt = \frac{P}{2},$$

while, on the other,

$$F \sum_{k=-\infty}^{\infty} |\hat{x}(kF)|^2 = F\left[\frac{P^2}{4} + 2 \sum_{\substack{k=1, \\ k \text{ odd}}}^{\infty} \frac{P^2}{k^2 \pi^2} \right].$$

By Parseval's equality the two expressions are equal so that

$$\frac{P}{2} = F\left[\frac{P^2}{4} + 2 \sum_{\substack{k=1, \\ k \text{ odd}}}^{\infty} \frac{P^2}{k^2 \pi^2} \right].$$

Since $F = 1/P$ this may be rearranged to the standard infinite sum

$$\sum_{\substack{k=1, \\ k \text{ odd}}}^{\infty} \frac{1}{k^2} = \frac{\pi^2}{8}.$$ ∎

Fourier Series Expansion of Periodic Signals

The finite Fourier series expansion of 6.4.3 deals with *finite-time* signals x defined on the discrete time axis $\underline{N} = \{0, 1, 2, \cdots, N-1\}$. The expansion is in terms of the harmonics

$$e^{j2\pi kFn}, \qquad n \in N, \tag{3}$$

with $k = 0, 1, \cdots, N-1$ and $F = 1/N$. The harmonic (3) has frequency $kF = k/N$. If n ranges over the *infinite* time axis $\mathbb{Z}$ and $k \neq 0$, by 2.2.12 the harmonic (3) is *periodic* with period N. As a result, the signal

$$F \sum_{f \in \underline{N}(F)} \hat{x}(f) e^{j2\pi fn}, \qquad n \in \mathbb{Z},$$

is also periodic with period N. This means that the finite Fourier series expansion of 6.4.3, taken on the infinite time axis $\mathbb{Z}$, is the *periodic continuation* of the signal x as defined on the finite time axis $\underline{N}$. Figure 6.7 illustrates the periodic continuation.

finite–time
signal

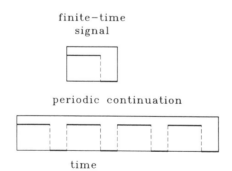

periodic continuation

time

Figure 6.7. Top: a finite-time signal. Bottom: its periodic continuation.

Note the difference between periodic *continuation* (which applies to a signal defined on a *finite* signal axis) and periodic *extension* as defined in 3.8.1. Also note that if the finite-time signal x repeats itself within the finite time interval where it is defined, the period of its periodic continuation in the strict sense of 2.2.8(b) is not N. In the context of periodic continuation and extension it is convenient to abuse terminology, however, and refer to N as the period.

Similar to the finite Fourier series expansion, the infinite Fourier series expansion consists of harmonics of the form

$$e^{j2\pi kFt}, \qquad t \in [0, P),$$

with $F = 1/P$. Taken on the infinite time axis $\mathbb{R}$ the harmonics are periodic with period $|1/kF| = |P/k|$. The largest of these periods is P and, hence, the infinite Fourier series expansion

$$F \sum_{f \in \mathbb{Z}(F)} \hat{x}(f) e^{j2\pi ft}, \qquad t \in \mathbb{R},$$

taken on the infinite time axis $\mathbb{R}$, is the periodic continuation of the finite-time signal x as defined on $[0, P)$.

Thus, the Fourier series expansion of 6.4.3 may equivalently be viewed as the expansion of *finite-time* signals and that of *periodic* signals. The Fourier series expansion of a periodic signal is simply obtained by expanding a single period according to 6.4.3, and extending the expansion to the infinite time axis $\mathbb{Z}$ or $\mathbb{R}$.

We define a few useful notions connected with the Fourier series expansion.

6.4.10.   Definitions: Fundamental frequency and fundamental period, constant and fundamental components, higher harmonics.
(a) *Fundamental frequency and period.* The constant F, both for the finite and the infinite Fourier series, is called the *fundamental frequency* of the expansion, and $1/F$ its *fundamental period.*

(b) *Constant component.* The term in the finite and infinite Fourier series expansion corresponding to $f = 0$ is called the *constant component.*

(c) *Fundamental component and higher harmonics.* In the *infinite* Fourier series expansion, the terms corresponding to $f = F$ and $f = -F$, added together, are called the *fundamental component* of the expansion. Its period equals the fundamental period. All other terms—the constant component excluded—are called the *higher harmonics.* The terms corresponding to $f = kF$ and $f = -kF$, with $k = 2$, $3, \cdots$, added together constitute the kth *harmonic.*

In the *finite* Fourier series expansion, the terms corresponding to $f = F$ and $f = (N - 1)F$ by aliasing have the same frequency. Added together they form the *fundamental component.* The kth *harmonic,* with $k = 2, 3, \cdots$, and $k \leq N/2$, consists of the term corresponding to $f = kF$ added to that for $f = (N - k)F$.   ∎

We consider the following example.

6.4.11.   Example: Fourier series expansion of a periodic rectangular pulse.   In 6.4.6(b) we obtained the Fourier series expansion of a finite-time continuous-time rectangular pulse as

$$x(t) = F \sum_{f \in \mathbb{Z}(F)} \hat{x}(f) e^{j2\pi ft}, \qquad t \in [0, P),$$

with the coefficients $\hat{x}$ given by

$$\hat{x}(f) = \begin{cases} a & \text{for } f = 0, \\ ae^{-j\pi fa} \dfrac{\sin(\pi fa)}{\pi fa} & \text{for } f \neq 0, \end{cases} \qquad f \in \mathbb{Z}(F),$$

where $F = 1/P$. By taking the time axis of the expansion as $\mathbb{R}$ we obtain the expansion of the periodic continuation of the pulse. Several periods of the periodic pulse and the truncated sum signal x_5, obtained by summing the harmonics with frequencies from $-5F$ to $5F$, are shown in Fig. 6.8 for $a = P/2$ and $P = 1$.

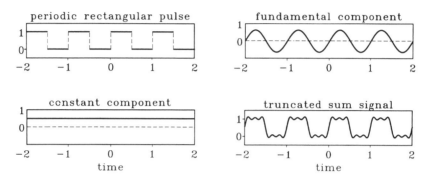

Figure 6.8. Periodic rectangular pulse, its constant and fundamental components and its truncated sum signal with five terms.

The fundamental period of the expansion is P, and the fundamental frequency $1/P$. The constant component is

$$x_{\text{constant}}(t) = F\hat{x}(0) = Fa = \frac{a}{P}, \qquad t \in \mathbb{R}.$$

The fundamental component is

$$\begin{aligned} x_{\text{fundamental}}(t) &= F[\hat{x}(F)\, e^{j2\pi Ft} + \hat{x}(-F)\, e^{-j2\pi Ft}] \\ &= aF\left[e^{-j\pi Fa} \frac{\sin(\pi Fa)}{\pi Fa} e^{j2\pi Ft} + e^{j\pi Fa} \frac{\sin(\pi Fa)}{\pi Fa} e^{-j2\pi Ft} \right] \\ &= \frac{2a}{P} \frac{\sin(\pi Fa)}{\pi Fa} \cos\left[2\pi F\left(t - \frac{a}{2} \right) \right], \qquad t \in \mathbb{R}. \end{aligned}$$

For $a = P/2$ this reduces to

$$x_{\text{fundamental}}(t) = \frac{2}{\pi} \sin(2\pi Ft), \qquad t \in \mathbb{R}.$$

Plots of the constant and fundamental components may be found in Fig. 6.8 for $P = 1$. ∎

Convergence Properties of the Infinite Fourier Series Expansion

The infinite Fourier series expansion of 6.4.3 involves an infinite sum for the signal x. This sum does not necessarily converge *pointwise* to x, however. In what follows we collect a number of facts about the convergence of the infinite Fourier series expansion.

6.4.12.   Summary: Convergence of the infinite Fourier series expansion.

(a) $\mathcal{L}_2$-*convergence*. The infinite Fourier series expansion

$$x = F \sum_{f \in \mathbb{Z}(F)} \hat{x}(f) \eta_f$$

is to be understood in the $\mathcal{L}_2$ sense, that is, if we define the *truncated sum signal*

$$x_N = F \sum_{k=-N}^{N} \hat{x}(kF)\, \eta_{kF}, \qquad N \in \mathbb{Z}_t,$$

then

$$\lim_{N \to \infty} \| x_N - x \|_2 = 0,$$

with the norm in the $\mathcal{L}_2[0, P)$ sense.

(b) *Conditions for pointwise convergence.* If x is a signal of *bounded variation* on $[0, P)$ (see below), the infinite Fourier series expansion represents the periodic continuation of x at every $t \in \mathbb{R}$ in the sense that

$$F \sum_{f \in \mathbb{Z}(F)} \hat{x}(f) e^{j2\pi f t} = \frac{x(t^+) + x(t^-)}{2}, \qquad t \in \mathbb{R},$$

where

$$x(t^+) = \lim_{\epsilon \downarrow 0} x(t + \epsilon), \qquad x(t^-) = \lim_{\epsilon \downarrow 0} x(t - \epsilon).$$

This means that the sum converges *pointwise* to $x(t)$ at every t where x is *continuous,* and to the *midpoint* of the left and right limits at every t where x has a jump.

The signal $x \in \mathcal{L}_2[0, P)$ has *bounded variation* if there exists a real constant α such that

$$\sum_{i=1}^{N} | x(t_i) - x(t_{i-1}) | \leq \alpha$$

for all $N \in \mathbb{N}$ and for all $0 \leq t_0 \leq t_1 \leq \cdots \leq t_N < P$. A set of *sufficient* conditions for x to be of bounded variation is that x be *bounded* and have a *finite* number of maxima, minima and jumps on $[0, P)$. Another set of sufficient conditions, known as *Dirichlet's conditions,* is that x have a *finite* derivative everywhere on $[0, P)$ except at a finite number of points.

Peter Gustav Lejeune Dirichlet (1805–1859), born of a French family in Germany, lived in Germany most of his life. He contributed to many aspects of mathematics.

(c) *The convergence is not uniform.* In general, if x_N is the truncated sum signal of (a), the sequence x_N does *not* converge uniformly to x, that is,

$$\lim_{N \to \infty} \| x_N - x \|_\infty \neq 0.$$

In other words,

$$\sup_{0 \leq t < P} \left| x_N(t) - x(t) \right|$$

does not converge to 0 as $N \to \infty$. If the derivative of x exists and is finite everywhere, uniform convergence is assured, however. The fact that x_N does not converge uniformly to x is known as *Gibbs's phenomenon,* and is illustrated in Example 6.4.13.

Josiah W. Gibbs (1839–1903) was an American physicist. He is best known for his work on the phase rule.

(d) *Best approximation.* In view of 6.2.20, the truncated sum signal x_N is the *best truncated approximation* to x in the subspace spanned by η_{kF}, $k \in \{-N, -N + 1, \cdots, 0, 1, \cdots, N\}$, in the sense of the $\mathscr{L}_2[0, P)$ norm.   ∎

6.4.13.   Example: Convergence of the infinite Fourier series expansion of the rectangular pulse.   In Figure 6.5 the truncated sum signals x_N for the rectangular pulse of Example 6.4.6(b) are shown for $N = 0, 1, 3$, and 5, with $a = P/2$ and $P = 1$. The approximation to the original waveform improves as N increases. It follows from 6.4.12(b) that the Fourier series converges to the "midpoint" value $1/2$ at $t = 0$ and $t = 1/2$. As pointed out in 6.4.12(c), the infinite sum does not necessarily converge *uniformly.* In fact, if x has a *jump,* the truncated sum signal x_N "overshoots" and "undershoots" the jump by about 9% for every N, no matter how large N becomes. This is known as *Gibbs's phenomenon* and is illustrated for the rectangular pulse in Fig. 6.9. Even if 25 or 50 terms are included in the truncated

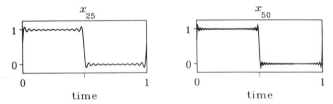

Figure 6.9. Gibbs's phenomenon for the rectangular pulse. Left: partial sum signal for $N = 25$. Right: partial sum signal for $N = 50$.

sum, the Fourier series for the rectangular pulse still exhibits overshoot and undershoot at the jumps. An explanation of Gibbs's phenomenon is offered in 7.3.11. ∎

Trigonometric Form of the Infinite Fourier Series Expansion

The infinite Fourier series expansion is based on the *complex* harmonic functions $e^{j2\pi ft}$, $t \in \mathbb{R}$, $f \in \mathbb{Z}(F)$. Sometimes it is more convenient to expand the signal in the *real* harmonic (trigonometric) functions

$$1, \qquad\qquad t \in \mathbb{R},$$

$$\cos(2\pi kt/P), \qquad t \in \mathbb{R}, k \in \mathbb{N},$$

$$\sin(2\pi kt/P), \qquad t \in \mathbb{R}, k \in \mathbb{N},$$

which together form an orthogonal set and are a basis for $\mathcal{L}_2[0, P)$.

6.4.14.   Summary: Trigonometric form of the Fourier series. Any $x \in \mathcal{L}_2[0, P)$ may be expanded as

$$x(t) = a_0 + \sum_{k=1}^{\infty} a_k \cos(2\pi kt/P) + \sum_{k=1}^{\infty} b_k \sin(2\pi kt/P), \qquad t \in \mathbb{R},$$

where

$$a_0 = \frac{1}{P} \int_0^P x(t)\, dt,$$

$$a_k = \frac{2}{P} \int_0^P x(t) \cos(2\pi kt/P)\, dt, \qquad k \in \mathbb{N},$$

$$b_k = \frac{2}{P} \int_0^P x(t) \sin(2\pi kt/P)\, dt, \qquad k \in \mathbb{N}. \qquad\qquad ∎$$

The proof is straightforward.

6.4.15.   Exercise: Relation between the complex exponential and the trigonometric form of the Fourier series.   Prove that the coefficients $\hat{x}(kF)$, $k \in \mathbb{Z}$, of the complex exponential form of 6.4.3 and the coefficients a_0, a_k, and b_k, $k \in \mathbb{N}$, of the trigonometric form of 6.4.14 of the infinite Fourier series are related as follows:

$$\begin{cases} F\hat{x}(0) = a_0, \\ F\hat{x}(kF) = \frac{1}{2}(a_k - jb_k), & k \in \mathbb{N}, \\ F\hat{x}(-kF) = \frac{1}{2}(a_k + jb_k), & k \in \mathbb{N}, \end{cases}$$

$$\begin{cases} a_0 = F\hat{x}(0), \\ a_k = F[\hat{x}(kF) + \hat{x}(-kF)], & k \in \mathbb{N}, \\ b_k = jF[\hat{x}(kF) - \hat{x}(-kF)], & k \in \mathbb{N}. \end{cases}$$ ∎

6.4.16.   Example: Trigonometric form of the expansion of the rectangular pulse.   In 6.4.6(b) we found that if $a = P/2$ the Fourier coefficients of the rectangular pulse are given by

$$\hat{x}(kF) = \begin{cases} P/2 & \text{for } k = 0, \\ 0 & \text{for } k \neq 0,\ k \text{ even}, \\ -jP/k\pi & \text{for } k \text{ odd}, \end{cases} \qquad k \in \mathbb{Z}.$$

It follows by direct application of 6.4.15 that the coefficients of the trigonometric form of the Fourier series are

$$a_0 = 1/2,$$

$$a_k = 0 \text{ for } k \in \mathbb{N},$$

$$b_k = \begin{cases} 0 & \text{for } k \text{ even}, \\ 2k\pi & \text{for } k \text{ odd}, \end{cases} \qquad k \in \mathbb{N}.$$ ∎

Symmetry Properties of the Fourier Coefficients

Sometimes it is helpful to be aware of the following symmetry properties of the Fourier coefficients of the complex exponential and trigonometric forms of the infinite Fourier series. We state the properties for the expansion of a *periodic* signal x defined on the infinite time axis $\mathbb{R}$.

6.4.17.   Summary: Symmetry properties of the expansion coefficients of the infinite Fourier series.
(a) Each of the three following conditions is necessary and sufficient for the periodic signal x to be *real-valued*:

(i) The coefficients a_0, a_k, and b_k, $k \in \mathbb{N}$, of the trigonometric expansion of x are all real.

(ii) The Fourier coefficients $\hat{x}$ of x are *conjugate symmetric*, that is,

$$\hat{x}(-f) = \overline{\hat{x}(f)} \qquad \text{for all } f \in \mathbb{Z}(F).$$

(iii) The magnitude $|\hat{x}|$ of the Fourier coefficients $\hat{x}$ of x is an *even* and the phase arg $(\hat{x})$ is an *odd* function of frequency.

(b) Each of the two following conditions is necessary and sufficient for the periodic signal x to be *even:*

(i) The coefficients b_k, $k \in \mathbb{N}$, of the sine terms in the trigonometric expansion of x are all zero.

(ii) The Fourier coefficients $\hat{x}$ of x are an *even* function of frequency, that is,

$$\hat{x}(-f) = \hat{x}(f) \qquad \text{for all } f \in \mathbb{Z}(F).$$

(c) Each of the two following conditions is necessary and sufficient for the periodic signal x to be *odd:*

(i) The coefficients a_0 and a_k, $k \in \mathbb{N}$, of the cosine terms in the trigonometric expansion of x are all zero.

(ii) The Fourier coefficients $\hat{x}$ of x are an *odd* function of frequency, that is,

$$\hat{x}(f) = -\hat{x}(-f) \qquad \text{for all } f \in \mathbb{Z}(F). \qquad \blacksquare$$

The proof is left as an exercise.

6.4.18.   Exercise. Necessary and sufficient conditions for the Fourier coefficients to be real.   Prove that the Fourier coefficients $\hat{x}$ of x are real if and only if x is conjugate symmetric, that is, $x(-t) = \overline{x(t)}$ for all $t \in \mathbb{R}$. $\blacksquare$

6.4.19.   Example: Symmetry properties of the Fourier coefficients of the periodic rectangular pulse.   From Example 6.4.16 we see that for $a = P/2$ the coefficients of the trigonometric expansion of the periodic rectangular pulse, which is a real-valued signal, are all real, which agrees with 6.4.17(a)(i). From Example 6.4.6(b) we observe that for any a the Fourier coefficients $\hat{x}$ of x are conjugate symmetric, which agrees with 6.4.17(a)(ii), and that their magnitude is an even function of frequency, which agrees with 6.4.17(a)(iii). The periodic rectangular pulse is neither even nor odd, so that 6.4.17(b) and (c) do not apply. $\blacksquare$

Generalized Infinite Fourier Series

So far we dealt with the infinite Fourier series expansion of *regular* continuous-time signals. In the remainder of the section we extend the theory to *singular* signals, in particular signals that include delta functions.

Earlier in this section we found that continuous-time signals in $\mathscr{L}_2[0, P)$, or, equivalently, periodic continuous-time signals that have finite energy on any finite interval, may be expanded in an infinite Fourier series, whose coefficients are in $\ell_2(F)$. According to the result that follows, also periodic *singular* signals may be expanded in an infinite Fourier series, except that the coefficients no longer belong to $\ell_2(F)$. Instead, the coefficients are sequences of *polynomial growth,* a notion that is defined below.

6.4.20.   Summary: Infinite Fourier series of periodic generalized signals.   Let x be a periodic generalized signal with period P.

(a) The infinite Fourier series expansion of x is

$$x(t) = F \sum_{f \in \mathbb{Z}(F)} \hat{x}(f)\, e^{j2\pi ft}, \qquad t \in \mathbb{R},$$

where $F = 1/P$ and the Fourier coefficients $\hat{x}$ are given by

$$\hat{x}(f) = \int_{\alpha}^{\alpha+P} x(t) e^{-j2\pi ft}\, dt, \qquad f \in \mathbb{Z}(F), \tag{4}$$

where α is any real number such that x is regular at α.

(b) The sequence $\hat{x}(f)$, $f \in \mathbb{Z}(F)$, is of *polynomial growth,* that is, there exist real numbers β and γ and a nonnegative integer k such that

$$|\hat{x}(f)| \le \beta |f|^k + \gamma, \qquad f \in \mathbb{Z}(F).$$

(c) Every infinite Fourier series expansion

$$F \sum_{f \in \mathbb{Z}(F)} \hat{x}(f)\, e^{j2\pi ft}, \qquad t \in \mathbb{R},$$

whose coefficients $\hat{x}$ are of polynomial growth is a periodic generalized function. ∎

The proof of this result is well beyond the scope of this book. We content ourselves with pointing out that if x is a regular signal, then by the periodicity of x the expression (4) for the Fourier coefficients yields the usual result for any α.

By way of example we consider the Fourier series expansion of the infinite comb.

6.4.21.   Example: Infinite Fourier series expansion of the infinite comb.   The infinite comb w_P (compare C.31) is the singular periodic signal given by

$$w_P(t) = \sum_{n=-\infty}^{\infty} \delta(t + nP), \qquad t \in \mathbb{R}.$$

Figure 6.10 shows a graphical representation of the comb. To find the Fourier coefficients of w_P we choose a value for α anywhere in the interval $(0, P)$ but not equal to 0 or P. It follows that

$$\hat{w}_P(f) = \int_{\alpha}^{\alpha+P} \delta(t - P)e^{-j2\pi ft}\, dt = e^{-j2\pi fP} = 1, \qquad f \in \mathbb{Z}(F).$$

Note that the coefficients $\hat{w}_P$ do not belong to $\ell_2(F)$ but are of polynomial growth. The generalized Fourier series expansion for the infinite comb thus is

$$w_P(t) = F \sum_{f \in \mathbb{Z}(F)} e^{j2\pi ft}, \qquad f \in \mathbb{Z}(F).$$

This may be rewritten in the form of the remarkable equality

$$\sum_{n=-\infty}^{\infty} \delta(t + nP) = \frac{1}{P} \sum_{k=-\infty}^{\infty} e^{j2\pi kt/P}, \qquad t \in \mathbb{R},$$

which is of course to be understood in the distribution sense.   ∎

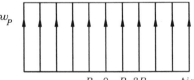

w_P

$-P \quad 0 \quad P \ 2P \qquad$ time   **Figure 6.10.**   The infinite comb.

6.4.22.   Exercise: Fourier series expansion of an infinite comb of derivatives of delta functions.   Prove that the infinite comb of derivatives of delta functions $w_P^{(k)}$, defined by

$$w_P^{(k)}(t) = \sum_{n=-\infty}^{\infty} \delta^{(k)}(t + nP), \qquad t \in \mathbb{R},$$

with $k \geq 1$, has the infinite Fourier series expansion

$$w_P^{(k)}(t) = F \sum_{f \in \mathbb{Z}(F)} (j2\pi f)^k e^{j2\pi ft}, \qquad t \in \mathbb{R}.$$

Note that the coefficients are of polynomial growth. *Hint:* According to C.32(b), infinite sums that define singular functions may be differentiated term by term.   ∎

6.4.23.   Review: Fourier series expansion of sampled signals.   The Fourier series expansion of discrete-time signals may be extended to the space $\ell_{\underline{N}}(T)$ of complex-valued sampled signals defined on the finite time axis

$$\underline{N}(T) = \{0, T, \cdots, (N-1)T\}.$$

This space has the orthogonal finite harmonic basis $\{F\eta_f, f \in \underline{N}(F)\}$, with $F = 1/NT$. Any signal x in the space $\ell_{\underline{N}}(T)$ may be expanded as

$$x(t) = F \sum_{f \in \underline{N}(F)} \hat{x}(f) e^{j2\pi ft}, \qquad t \in \underline{N}(T),$$

where the Fourier coefficients $\hat{x} \in \ell_{\underline{N}}(F)$ are given by

$$\hat{x}(f) = T \sum_{t \in \underline{N}(T)} x(t) e^{-j2\pi ft}, \qquad f \in \underline{N}(F).$$

Defining the inner product on $\ell_{\underline{N}}(T)$ as

$$\langle x, y \rangle_{\ell_{\underline{N}}(T)} = T \sum_{t \in \underline{N}(T)} x(t) \overline{y(t)},$$

Plancherel's equality takes the form

$$\langle x, y \rangle_{\ell_{\underline{N}}(T)} = \langle \hat{x}, \hat{y} \rangle_{\ell_{\underline{N}}(F)}.$$

Parseval's identity $\|x\|_2^2 = \|\hat{x}\|_2^2$ may explicitly be written as

$$T \sum_{t \in \underline{N}(T)} |x(t)|^2 = F \sum_{f \in \underline{N}(F)} |\hat{x}(f)|^2.$$

The finite Fourier series expansion for sampled signals may of course also be used to expand *periodic* signals defined on the time axis $\mathbb{Z}(T)$ with period NT.   ∎

6.5 LINEAR TIME-INVARIANT SYSTEMS WITH PERIODIC INPUTS

In the preceding section we saw how finite-time and periodic signals may be expanded in a harmonic basis. In the present section we analyze the response of *infinite-time* linear time-invariant systems to *periodic* inputs. The Fourier series expansion is a *spectral expansion* for such systems, which much simplifies the analysis.

We also show that harmonics form a spectral basis for *cyclical* convolution maps. Because as seen in Section 3.8 regular convolution systems with periodic inputs may be reduced to cyclical convolution systems, this offers an alternative view of the frequency domain analysis of the response of convolution systems to periodic inputs.

Frequency Domain Solution of the Response of Convolution Systems to Periodic Inputs

We consider the response of convolution systems to periodic inputs. The IO map of convolution systems may be expressed as

$$y = h * u,$$

with $*$ the discrete- or continuous-time convolution and h the impulse response of the system.

In Section 3.7 we found that a convolution system passes the harmonic signal η_f through as a harmonic with the same frequency. The harmonic signal η_f hence is an *eigensignal* of the convolution system. The corresponding eigenvalue is $\hat{h}(f)$, with $\hat{h}$ the frequency response function of the system.

In Section 6.4 we showed that discrete-time periodic signals with period N may be expanded in the harmonic basis $\{F\eta_{nF}, n \in \underline{N}\}$, with $F = 1/N$ the fundamental frequency. Similarly, continuous-time signals with period P may be expanded in the harmonic basis $\{F\eta_{nF}, n \in \mathbb{Z}\}$, with $F = 1/P$. Because harmonics are eigensignals for convolution systems, the basis $\{F\eta_{nF}, n \in \underline{N}\}$, $F = 1/N$, is a *spectral basis* for discrete-time convolution systems with periodic inputs with period N. By the same token, $\{F\eta_{nF}, n \in \mathbb{Z}\}$, $F = 1/P$, is a spectral basis for continuous-time convolution systems with periodic inputs with period P.

This makes it simple to obtain the response of convolution systems to periodic inputs.

6.5.1.   Summary: Frequency domain analysis of the response of convolution systems to a periodic input.

(a) The finite Fourier expansion

$$x = F \sum_{k=0}^{N-1} \hat{x}(kF)\eta_{kF},$$

(a') The infinite Fourier expansion

$$x = F \sum_{k \in \mathbb{Z}} \hat{x}(kF)\eta_{kF},$$

$F = 1/N$, is a spectral expansion for the discrete-time convolution system defined by

$$y = h * u$$

with periodic inputs with period N.

(b) The response of the convolution system to the periodic input

$$u = F \sum_{k=0}^{N-1} \hat{u}(kF)\eta_{kF}$$

exists if h has finite action and is given by

$$y = F \sum_{k=0}^{N-1} \hat{h}(kF)\hat{u}(kF)\eta_{kF},$$

where $\hat{h}$ is the frequency response function

$$\hat{h}(f) = \sum_{n=-\infty}^{\infty} h(n)e^{-j2\pi fn}, \qquad f \in \mathbb{R}.$$

$F = 1/P$, is a spectral expansion for the continuous-time convolution system defined by

$$y = h * u$$

with periodic inputs with period P.

(b') The response of the convolution system to the periodic input

$$u = F \sum_{k \in \mathbb{Z}} \hat{u}(kF)\eta_{kF}$$

exists if h has finite action and is given by

$$y = F \sum_{k \in \mathbb{Z}} \hat{h}(kF)\hat{u}(kF)\eta_{kF},$$

where $\hat{h}$ is the frequency response function

$$\hat{h}(f) = \int_{-\infty}^{\infty} h(t)e^{-j2\pi ft}\, dt, \qquad f \in \mathbb{R}. \quad \blacksquare$$

The central result is that the expansion coefficient $\hat{y}(f)$ of the output may simply be obtained by multiplying the corresponding expansion coefficient $\hat{u}(f)$ of the input by the frequency response function $\hat{h}$ of the system at the frequency f.

The proof of 6.5.1 relies of course on the superposition property of linear IOM systems together with the fact that the response of the system to a single component $\hat{u}(f)\eta_f$ of the input is $\hat{h}(f)\hat{u}(f)\eta_f$.

6.5.2.   Examples: Periodic inputs.

(a) *Exponential smoother*. In Example 3.7.5(a) we found that the frequency response function of the exponential smoother exists if the parameter a satisfies $|a| < 1$. The frequency response function then is given by

$$\hat{h}(f) = \frac{1-a}{1-ae^{-j2\pi f}}, \qquad f \in \mathbb{R}.$$

Assuming that $a = 1/2$, we determine the response of the system to the periodic input u with period $N = 4$, one period of which is given by

$$u(0) = u(1) = 1, \qquad u(2) = u(3) = 0.$$

This is the signal depicted in Fig. 6.3. In Example 6.4.6(a) it was found that the four coefficients of its finite Fourier series expansion are

$$\hat{u}(0) = 2, \qquad \hat{u}(F) = 1 - j, \qquad \hat{u}(2F) = 0, \qquad \hat{u}(3F) = 1 + j,$$

where $F = 1/4$. By 6.5.1 the expansion coefficients of the corresponding output are given by

$$\hat{y}(f) = \hat{h}(f)\hat{u}(f), \qquad f \in \{0, F, 2F, 3F\}.$$

Evaluation of $\hat{h}(f)$ at the frequencies 0, F, $2F$, and $3F$ with $F = 1/4$ and of the coefficients $\hat{y}(f)$ results in the following little table:

f	$\hat{u}(f)$	$\hat{h}(f)$	$\hat{y}(f)$
0	2	1	2
F	$1 - j$	$\dfrac{2 - j}{5}$	$\dfrac{1 - 3j}{5}$
$2F$	0	$\dfrac{1}{3}$	0
$3F$	$1 + j$	$\dfrac{2 + j}{5}$	$\dfrac{1 + 3j}{5}$

Consequently, the output is given by

$$y(n) = \tfrac{1}{4}\left(2 + \frac{1 - 3j}{5}e^{j2\pi n/4} + 0 \cdot e^{j2\pi n/2} + \frac{1 + 3j}{5}e^{j2\pi n3/4}\right), \qquad n \in \mathbb{Z},$$

Successive evaluation for $n = 0, 1, 2, 3$ shows that one period of the output is given by

$$y(0) = \frac{3}{5}, \qquad y(1) = \frac{4}{5}, \qquad y(2) = \frac{2}{5}, \qquad y(3) = \frac{1}{5}.$$

This result agrees with that of Example 3.8.10, where we solved the same problem by cyclical convolution.

(b) *RC network.* As we found in Example 3.7.5(b), the RC network with the voltage across the capacitor as output has the frequency response function

$$\hat{h}(f) = \frac{1}{1 + RCj\,2\pi f}, \qquad f \in \mathbb{R}.$$

Suppose that the input u to the network is the periodic rectangular pulse of Example 6.4.11. If $a = P/2$, by 6.4.6(b) its Fourier coefficients are given by

$$\hat{u}(kF) = \begin{cases} P/2 & \text{for } k = 0, \\ 0 & \text{for } k \neq 0,\ k \text{ even}, \qquad k \in \mathbb{Z}, \\ -jP/k\pi & \text{for } k \text{ odd}, \end{cases}$$

with $F = 1/P$. As a result, the expansion coefficients of the output are

$$\hat{y}(kF) = \begin{cases} P/2 & \text{for } k = 0, \\ 0 & \text{for } k \neq 0,\ k \text{ even}, \qquad k \in \mathbb{Z}. \\ \dfrac{1}{1+RCj\,2\pi kF}\ \dfrac{-jP}{k\pi} & \text{for } k \text{ odd}, \end{cases}$$

As the frequency $f = kF$ increases, the expansion coefficients $\hat{y}(kF)$ of the output decrease as $1/k^2$ rather than as $1/k$ like for the input. The result is that high frequencies are less strongly represented in the output than in the input, so that the output is smoother than the input.

Because the frequency response function of the network only decreases as $1/k$, the smoothing effect is limited, as illustrated in Fig. 6.11. The figure gives the response of the network to the periodic rectangular pulse for $RC = 1$ and $P = 1$. ∎

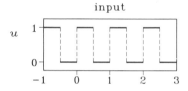

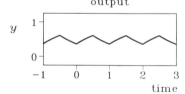

Figure 6.11.   Response of the RC network to the periodic rectangular pulse.

6.5.3.   Review: Response of sampled convolution systems to periodic inputs.

We restate the most important results of this section for sampled systems. The frequency response function of a sampled convolution system defined on the time axis $\mathbb{Z}(T)$ with impulse response h is given by

$$\hat{h}(f) = T \sum_{n=-\infty}^{\infty} h(nT)e^{-j2\pi nTf}, \qquad f \in \mathbb{R}.$$

The finite Fourier series expansion of 6.4.23 is the spectral expansion for sampled convolution systems with periodic inputs with period NT. If the Fourier coefficients of the input are $\hat{u}(f),\ f \in \underline{N}(F)$, with $F = 1/NT$, those of the output are given by

$$\hat{y}(f) = \hat{h}(f)\hat{u}(f), \qquad f \in \underline{N}(F).$$ ■

Frequency Domain Solution of Cyclical Convolution Systems

Cyclical convolution systems are described by a discrete- or continuous-time cyclical convolution IO map of the form

$$Y = H \odot U. \tag{1}$$

As seen in Section 3.8, cyclical convolution systems are useful for the analysis of the response of regular convolution systems to periodic inputs. In this case, Y is one period of the output, U one period of the input, while H is one period of the periodic extension of the impulse response h of the regular convolution system.

It follows from 3.8.11 that in the discrete-time case, where the time axis is $\underline{N}$, the harmonics $\eta_{k/N}$, $k \in \underline{N}$, are eigensignals of the cyclical convolution system (1). The eigenvalue corresponding to the eigensignal $\eta_{k/N}$ is $\hat{H}(k/N)$, with $\hat{H}$ the frequency response function of the cyclical convolution system, given by

$$\hat{H}(f) = \sum_{n=0}^{N-1} H(n)e^{-j2\pi fn}, \qquad f = k/N, \qquad k \in \underline{N}.$$

The set $\{\eta_{k/N}\}$, $k \in \underline{N}$, is the basis for the finite Fourier series expansion. As a result, the finite Fourier transform is a *spectral expansion* for the discrete-time cyclical convolution system (1).

In the continuous-time case, if (1) represents a cyclical convolution system on the time axis $[0, P)$, the harmonic $\eta_{k/P}$, $k \in \mathbb{Z}$, is an eigensignal, with eigenvalue $\hat{H}(k/P)$. The frequency response function is now given by

$$\hat{H}(f) = \int_0^P H(t)e^{-j2\pi ft}\, dt, \qquad f = k/P, \qquad k \in \mathbb{Z}.$$

It follows that the infinite Fourier series expansion, which uses the basis $\{\eta_{k/P}, k \in \mathbb{Z}\}$, is a spectral expansion for the continuous-time cyclical convolution system (1).

We summarize as follows.

6.5.4.   Summary: Frequency domain analysis of cyclical convolution systems.

(a) The finite Fourier expansion	(a') The infinite Fourier expansion
$$x = F \sum_{k=0}^{N-1} \hat{x}(kF)\eta_{kF},$$	$$x = F \sum_{k \in \mathbb{Z}} \hat{x}(kF)\eta_{kF},$$

$F = 1/N$, is a spectral expansion for the discrete-time cyclical convolution system $Y = H \odot U$ defined on the time axis $\underline{N}$.

(b) The response of the cyclical convolution system to the input

$$U = F \sum_{k=0}^{N-1} \hat{U}(kF)\eta_{kF}$$

is

$$Y = F \sum_{k=0}^{N-1} \hat{H}(kF)\hat{U}(kF)\eta_{kF},$$

where

$$\hat{H}(f) = \sum_{n=0}^{N-1} H(n)e^{-j2\pi fn}, \qquad f \in \underline{N}(F).$$

$F = 1/P$, is a spectral expansion for the continuous-time cyclical convolution system $Y = H \odot U$ defined on the time axis $[0, P)$.

(b′) The response of the cyclical convolution system to the input

$$U = F \sum_{k \in \mathbb{Z}} \hat{U}(kF)\eta_{kF}$$

is

$$Y = F \sum_{k \in \mathbb{Z}} \hat{H}(kF)\hat{U}(kF)\eta_{kF},$$

where

$$\hat{H}(f) = \int_{0}^{P} H(t)e^{-j2\pi ft}\, dt, \qquad f \in \mathbb{Z}(F).$$

■

It remains to establish the connection between the direct frequency domain analysis of the response of convolution systems to periodic inputs and that via cyclical convolution.

The direct approach is to find the Fourier coefficients $\hat{u}$ of the periodic input u, and to determine the Fourier coefficients of the output as

$$\hat{y}(f) = \hat{h}(f)\hat{u}(f),$$

where f ranges over $\underline{N}(F)$ in the discrete-time case and over $\mathbb{Z}(F)$ in the continuous-time case, and $\hat{h}$ is the frequency response function of the system.

The approach via cyclical convolution is to find the Fourier coefficients $\hat{U}$ of the one-period restriction U of the input u, and to determine the Fourier coefficients $\hat{Y}$ of the one-period restriction Y of the output y as

$$\hat{Y}(f) = \hat{H}(f)\hat{U}(f),$$

with f ranging over the same sets as in the direct approach, and $\hat{H}$ given as in 6.5.4.

Because the Fourier coefficients of u and y are the same as the Fourier coefficients of their one-period restrictions we immediately have $\hat{U} = \hat{u}$ and $\hat{Y} = \hat{y}$. It follows that necessarily

$$\hat{H}(f) = \hat{h}(f),$$

with $f \in \underline{N}(F)$ in the discrete-time case and $f \in \mathbb{Z}(F)$ in the continuous-time case. Indeed, this was already established in 3.8.12.

6.6 PROBLEMS

The first problems in this section deal with bases, orthonormalization, the projection theorem, and spectral expansion.

6.6.1. Reciprocal basis.

(a) *Basis and reciprocal basis for* $\mathbb{C}^3$. Find the reciprocal basis to the basis $a_1 = \text{col}(1, 0, 0)$, $a_2 = \text{col}(1, 1, 0)$ and $a_3 = \text{col}(1, 1, 1)$ for $\mathbb{C}^3$. Use the reciprocal basis to expand the vector $\text{col}(1, 2, 3)$ in the basis $\{a_1, a_2, a_3\}$.

(b) *Basis and reciprocal basis for a space of polynomials.* The set of all polynomials of degree two with complex-valued coefficients defined on $[0, 1)$ forms an inner product space with the usual inner product

$$\langle x, y \rangle = \int_0^1 x(t)\overline{y(t)}\, dt.$$

Find the reciprocal to the basis formed by the polynomials p_1, p_2 and p_3 given by

$$p_1(t) = 1, \qquad p_2(t) = t, \qquad p_3(t) = t^2, \qquad t \in [0, 1).$$

Hint: Suppose that the reciprocal basis is given by the polynomials q_1, q_2 and q_3, and write $q_i = a_{i1} p_1 + a_{i2} p_2 + a_{i3} p_3$, $i = 1, 2, 3$, with the coefficients a_{ik} to be determined.

6.6.2. Gram-Schmidt orthonormalization procedure.
What follows is known as the *orthonormalization procedure of Gram-Schmidt*. Suppose that a linearly independent countable set of elements $\{v_1, v_2, v_3, \cdots\}$ of some inner product space X is given. Form from this another countable set $\{u_1, u_2, u_3, \cdots\}$ as follows. First let

$$u_1 := v_1/\|v_1\|,$$

where $\|\cdot\|$ is the natural norm, i.e., $\|x\| = \sqrt{\langle x, x \rangle}$. Then for $i = 2, 3, \cdots$, first take

$$w_i := v_i - \sum_{k=1}^{i-1} \langle v_i, u_k \rangle u_k,$$

and then normalize as

$$u_i := w_i/\|w_i\|.$$

(a) Prove that for each N the set $\{u_1, u_2, \cdots, u_N\}$ spans the same linear subspace as $\{v_1, v_2, \cdots, v_N\}$.

(b) Prove that the set $\{u_1, u_2, u_3, \cdots\}$ is orthonormal.

(c) The set of all polynomials of degree two with complex-valued coefficients defined on $[0, 1)$ forms an inner product space with the usual inner product as in

6.6.1(b). Use the Gram-Schmidt procedure to determine an orthonormal basis for this subspace from the polynomials p_1, p_2 and p_3 given by

$$p_1(t) = 1, \qquad p_2(t) = t, \qquad p_3(t) = t^2, \qquad t \in [0, 1).$$

> Jorgen Pederen Gram (1850–1916) was a Danish insurance mathematician who occupied himself with the Γ-function and orthogonal systems. The mathematician Erhard Schmidt (1876–1959) was born in the Baltic countries but studied in Germany and taught at the University of Berlin.

6.6.3. Projection theorem.

(a) *Projection in $\mathbb{C}^3$.* Use the projection theorem to determine the point in the plane spanned by the vectors col(0, 1, 0) and col(1, 0, 1) in $\mathbb{C}^3$ that is closest to the point (1, 2, 3).

(b) *Projection in a space of polynomials.* Consider the space of polynomials of degree two and the polynomials p_1, p_2 and p_3 as introduced in 6.6.1(b) and 6.6.2(c). Use the projection theorem to determine the point in the subspace spanned by the polynomials of degree *one* (i.e., the subspace spanned by the polynomials p_1 and p_2) that is closest to the polynomial p_3. Plot p_3 and its projection, which is also the best approximation of p_3 by a polynomial of degree one.

6.6.4. Spectral expansion.

(a) Consider the map $\mathbb{C}^N \to \mathbb{C}^N$ given by $y = Ku$, with $N = 2$ and K the 2×2 matrix

$$K = \begin{bmatrix} 2 & 1 \\ 1 & 2 \end{bmatrix}.$$

(a.1) Compute the eigenvalues and eigenvectors of K.

(a.2) Show that the eigenvectors form a basis for $\mathbb{C}^N$, and hence provide a *spectral* expansion for the map.

(a.3) Determine the map from the expansion coefficients $\hat{u}_i$, $i = 1, 2, \cdots, N$, of the input u in the spectral basis to the expansion coefficients $\hat{y}_i$, $i = 1, 2, \cdots, N$, of the output y.

(a.4) Show that the matrix K is normal, and correspondingly the spectral basis is orthogonal.

(b) Repeat (a) for $N = 3$ and K the 3×3 matrix

$$K = \begin{bmatrix} 1 & 0 & 2 \\ 2 & 1 & 0 \\ 0 & 2 & 1 \end{bmatrix}.$$

The following problems concern the finite and infinite Fourier series expansions.

6.6.5. Finite Fourier series. Determine the finite Fourier series of the following discrete-time signals x defined on the time axis $\underline{N} = \{0, 1, \cdots, N - 1\}$:

(a) $N = 4$, and

$$x(n) = \begin{cases} 1 & \text{for } n = 0 \text{ and } n = 2, \\ 0 & \text{for } n = 1 \text{ and } n = 3, \end{cases} \qquad n \in \underline{4}.$$

(b) $N = 8$, and

$$x(n) = \begin{cases} 1 & \text{for } n = 0 \text{ and } n = 4, \\ 0 & \text{otherwise}, \end{cases} \qquad n \in \underline{8}.$$

Determine the discrete-time signal $x \in \ell_N$ whose finite Fourier series has the following coefficients:

(c) $\hat{x}(F) = \hat{x}((N - 1)F) = 1/2F,$

  $\hat{x}(kF) = 0, \qquad k = 0, 2, 3, \cdots, N - 2,$

  with $F = 1/N$.

(d) $\hat{x}(kF) = 1, \qquad k = 0, 1, \cdots, N - 1,$

  with $F = 1/N$.

6.6.6. Infinite Fourier series. Determine the infinite Fourier series, both in complex exponential and in trigonometric form, of the following continuous-time signals. In each case, plot the signal and the magnitude of its Fourier coefficients. Check which of the various symmetry properties of 6.4.17 apply.

(a) The periodic triangular singal x, one period of which is given by

$$x(t) = 1 - \frac{|t|}{P}, \qquad -P/2 \le t < P/2.$$

(b) The half-wave rectified sine x, one period of which is given by

$$x(t) = \begin{cases} \sin\left(\dfrac{2\pi t}{P}\right) & \text{for } 0 \le t < P/2, \\ 0 & \text{for } P/2 \le t < P, \end{cases} \qquad t \in [0, P).$$

(c) The full-wave rectified sine, which is the signal x given by

$$x(t) = \left| \sin\left(\dfrac{2\pi t}{P}\right) \right|, \qquad t \in \mathbb{R}.$$

6.6.7. Parseval's identity. Apply Parseval's identity to each of the infinite Fourier series expansions of 6.6.6, and find the resulting infinite sums.

6.6.8. Trigonometric form of the infinite Fourier series.

(a) Prove that the real harmonic functions

$$1, \qquad\qquad\qquad t \in \mathbb{R},$$

$$\cos{(2\pi kt/P)}, \qquad t \in \mathbb{R}, k \in \mathbb{N},$$

$$\sin{(2\pi kt/P)}, \qquad t \in \mathbb{R}, k \in \mathbb{N},$$

form an orthogonal basis for $\mathscr{L}_2[0, P)$. *Hint:* Show that the functions are orthogonal, and that they span the same space as the complex harmonic functions $e^{j2\pi kt/P}, t \in \mathbb{R}, k \in \mathbb{Z}$.

(b) Derive the trigonometric form of the infinite Fourier series as given in 6.4.14.

(c) Derive the relations of 6.4.15 that connect the coefficients of the exponential and the trigonometric form of the infinite Fourier series.

(d) Prove that in terms of the trigonometric infinite Fourier series Parseval's equality takes the form

$$\int_0^P |x(t)|^2\, dt = P\left[a_0^2 + \tfrac{1}{2} \sum_{k=1}^{\infty} (a_k^2 + b_k^2)\right].$$

6.6.9. Symmetry properties of the Fourier coefficients. Prove the symmetry properties of the coefficients of the infinite Fourier series as summarized in 6.4.17.

6.6.10. Fourier series expansion of a real-valued signal. Suppose that x is a continuous-time periodic signal with period P that has the infinite Fourier series expansion

$$x(t) = F \sum_{f \in \mathbb{Z}(F)} \hat{x}(f) e^{j2\pi ft}, \qquad t \in \mathbb{R}.$$

(a) Show that if x is real-valued it may be expressed as

$$x(t) = F \sum_{f \in \mathbb{Z}(F)} |\hat{x}(f)| \cos{[2\pi ft + \phi(f)]}, \qquad t \in \mathbb{R},$$

where $\phi(f) = \arg{[\hat{x}(f)]}, f \in \mathbb{Z}(F)$. *Hint:* Use the conjugate symmetry of the Fourier coefficients $\hat{x}$.

(b) Prove that the expansion of (a) in turn may be written as

$$x(t) = F\hat{x}(0) + 2F \sum_{k=1}^{\infty} |\hat{x}(kF)| \cos{[2\pi kFt + \phi(kF)]}, \qquad t \in \mathbb{R}.$$

This shows that the periodic real signal x may be expanded as a linear combination of a constant and a countable set of real harmonics whose frequencies are multiples of the fundamental frequency F, such that the amplitude of the kth harmonic is $2F|\hat{x}(kF)|$ and its phase $\arg[\hat{x}(kF)]$.

(c) Expand the rectangular pulse of 6.4.6(b) as in (b).

6.6.11. Generalized Fourier series.

(a) Determine the generalized Fourier series expansion of the periodic signal

$$x(t) = \sum_{k=-\infty}^{\infty} (-1)^k \delta\left(t - \frac{kP}{2}\right), \qquad t \in \mathbb{R}.$$

Plot the signal and the magnitude of its Fourier coefficients.

(b) Determine the Fourier series expansion of the periodic sawtooth signal x, one period of which is given by

$$x(t) = t/P \qquad \text{for} \quad 0 \le t < P.$$

Express x in terms of its infinite Fourier series expansion and use term-by-term differentiation to derive the Fourier series expansion of the infinite comb w_P.

6.6.12. Filtering a periodic rectangular pulse. One period of the periodic rectangular pulse u is given by

$$u(t) = \begin{cases} 1 & \text{for } 0 \le t < P/2, \\ 0 & \text{for } P/2 \le t < P, \end{cases} \qquad t \in \mathbb{R}.$$

The signal u, which has the fundamental frequency $f_o = 1/P$, is passed through an ideal low-pass filter with frequency response function

$$\hat{h}(f) = \begin{cases} 1 & \text{for } |f| \le f_1, \\ 0 & \text{otherwise}, \end{cases} \qquad f \in \mathbb{R}.$$

Determine the response y of the filter to this input in the following two cases:

(a) $0 < f_1 < f_o$,
(b) $f_o < f_1 < 2f_o$.

6.7 COMPUTER EXERCISES

The first exercise in this section deals with the expansion of a signal in a polynomial basis.

6.7.1. Trend estimation. Many observed time series, in particular economic figures, show *trends,* which are systematic changes in a certain direction. It often is desirable to either estimate the trend, or to remove it. Trends may be represented as linear combinations of certain fixed time signals, for which often polynomials of low degree are taken. In this exercise we consider trend estimation as the projection of the observed signal x on the linear subspace spanned by the signals p_i, $i = 1, 2, \cdots, K$, given by

$$p_i(t) = t^{i-1}, \qquad t \in \mathbb{T},$$

with $\mathbb{T}$ the time axis. The projection is the *best approximation* of the signal x in the subspace.

The projection is obtained as follows. Let $\{u_i, \ i = 1, \ 2, \ \cdots \}$ denote the orthonormal basis that results by applying the Gram-Schmidt procedure (see 6.6.2) to the sequence $p_i, \ i = 1, \ 2, \ \cdots$. By 6.6.2(a), the subspace spanned by $u_i, \ i = 1, \ 2, \ \cdots, \ K$, coincides with the subspace spanned by $p_i, \ i = 1, \ 2, \ \cdots, \ K$. By 6.2.20 the best approximation of x in this subspace may be obtained by truncating the expansion of x in the orthonormal basis $\{u_i, \ i = 1, \ 2, \ \cdots \}$ after K terms.

(a) Generate a signal x with a constant and linear trend on the time axis $\mathbb{T} = \{0, T, 2T, \cdots, 1\}$, with $T = 0.01$, as

$$x(t) = 0.5 + t + n(t), \qquad t \in \mathbb{T},$$

where $n(t), \ t \in \mathbb{T}$, is a sequence of independent samples of a Gaussian random variable with zero mean and standard deviation 0.1. *Hint:* In SIGSYS, use the permanent signal `noiseplus`.

(b) We wish to fit a polynomial of order two, and hence take $K = 3$. Use the Gram-Schmidt orthonormalization procedure (see 6.6.2) to generate an orthonormal basis u_1, u_2, u_3 for the space spanned by the polynomials p_1, p_2, and p_3. Plot the signals u_1, u_2, and u_3. *Hint:* Generate the polynomials and use inner products to obtain the orthonormal basis.

(c) Compute the expansion coefficients $\hat{x}_i, \ i = 1, \ 2, \ 3$ of the signal x in the orthonormal basis $\{u_1, u_2, u_3\}$. Compute and plot the estimated trend

$$x^* = \sum_{i=1}^{3} \hat{x}_i u_i$$

together with x. Also compute and plot the error signal $e = x - x^*$, and determine the root mean square value of the error signal. Comment on the values of the expansion coefficients that are obtained and on the rms value of the error signal.

The next few exercises relate to the Fourier series expansion.

6.7.2. Gibbs's phenomenon. In 6.4.13 Gibbs's phenomenon is illustrated for the rectangular pulse.

(a) Create and plot one period of the rectangular pulse x with period $P = 1$ and $a = P/2$, using 256 sample points. Compute the expansion coefficients $\hat{x}$ of the infinite Fourier series of x. Plot their magnitude.

(b) Truncate the coefficients $\hat{x}$ by setting $\hat{x}(kF) = 0$ for $|k| > N$. Sum the resulting truncated Fourier series x_N for $N = 0, 1, 3, 5, 25$, and 50. Plot the results, and compare with the partial sum signals of Figs. 6.5 and 6.9.

(c) Use the same method to produce the partial sum signals y_N with $N = 10$ and 20 for the periodic sawtooth y, one period of which is given by

$$y(t) = t, \qquad 0 \le t < 1.$$

Think of a way to plot several periods of y and y_N.

6.7.3. Filtering of periodic signals. The response of a linear time-invariant system to a periodic input u may be determined as follows. First compute the Fourier coefficients $\hat{u}$ of u. Next multiply $\hat{u}$ by the frequency response function $\hat{h}$ of the system to obtain the Fourier coefficients $\hat{y} = \hat{h}\hat{u}$ of the output y. The output y itself follows by summing the Fourier series with coefficients $\hat{y}$.

Numerical computation of Fourier expansions.

In SIGSYS, the coefficients of the finite Fourier series may be found by using the function ddft. The resulting finite Fourier series may be summed by using the inverse function iddft. Because the computation relies on the fast Fourier transform (see Section 9.7) it is most effective to take the number of points on which the signal is defined as an integral power of 2.

The coefficients of the infinite Fourier series may be found with the SIGSYS function cdft. For the alternative form of 6.4.5 the function fc is available. Continuous-time signals whose infinite Fourier series is to be computed need be represented in sampled form. To make the sampling errors small, the resolution should be sufficiently high. Because also this computation relies on the FFT, again it is most effective to take the number of sample points as an integral power of 2. Infinite Fourier series may be summed by using the inverse functions icdft or fs, depending on which form of the infinite series is used. Note that infinite Fourier series numerically are always obtained as truncated sums. Sampling the original continuous-time signal automatically results in a finite, truncated Fourier series. Summation of this truncated series reconstructs the original continuous-time signal exactly *at* the sampling instants.

The Fourier coefficients of a real signal constitute a frequency signal that is generally complex-valued. Summing the Fourier series with these complex coefficients should restore the original real signal. If the calculations are performed numerically, round-off errors usually cause the result to have a very small imaginary part. The imaginary part constitutes "numerical noise." This phenomenon may be observed in some of the exercises. The imaginary part may be removed by replacing the signal with its real part.

In this exercise we first consider the response of a continuous-time system with the low-pass frequency response function

$$\hat{h}_K(f) = \frac{1}{(1 + jf/f_o)^K}, \qquad f \in \mathbb{R},$$

to the periodic sawtooth, one period of which is given by

$$u(t) = \frac{t}{P} - \tfrac{1}{2}, \qquad 0 \le t < P.$$

(a) Create and plot one period of the sawtooth with $P = 1$, using 128 sample points. Think of a way of plotting several periods. Compute the coefficients $\hat{u}$ of the infinite Fourier series expansion of u. Hint: In SIGSYS, set inc and num equal to the correct values, and use dotplus to generate the signal u.

(b) Let $f_o = 2.5$. Generate and plot the frequency response function $\hat{h}_K$ for $K = 1$, 2, and 3. Compute the Fourier coefficients $\hat{y}_K$ of the response of the system with the frequency response function $\hat{h}_K$ to the periodic input u for $K = 1$, 2, and 3. *Hint:* In SIGSYS, use the command set INC to define the correct frequency increment corresponding to inc and num, and generate $\hat{h}_K$ using the frequency axis DOT.

(c) Sum the Fourier series expansion to obtain one period of the output y_K for $K = 1$, 2, and 3. Plot several periods of the resulting outputs. As K increases, the low-pass character of the filter becomes more and more pronounced. Observe the effect on the output.

(d) Repeat (b) and (c) for the system with the *high-pass* frequency response function

$$\hat{g}(f) = \frac{(jf/f_o)^K}{(1 + jf/f_o)^K}, \qquad f \in \mathbb{R},$$

again for $K = 1$, 2, and 3. Comment on the outputs that are obtained.

6.7.4. Finite and infinite Fourier series and frequency response. In 6.5.2(a) it is shown how the response of the exponential smoother to a simple periodic input with period 4 may be obtained, and in 6.5.2(b) the response of the RC network to a periodic rectangular signal is found. In this exercise we repeat these examples for more complicated periodic inputs.

(a) Find the expansion coefficients $\hat{u}$ of the finite Fourier series of the discrete-time periodic signal u with period 64, one period of which is given by

$$u(n) = \text{trian}\left(\frac{n-8}{8}\right) - \text{trian}\left(\frac{n-24}{8}\right), \qquad n = 0, 1, 2, \cdots, 63.$$

Plot the signal and its Fourier coefficients. *Hint:* In SIGSYS, set inc and num equal to the correct values and use the time axis dotplus to generate one period of u.

(b) Find the expansion coefficients $\hat{y}$ of the response y of the exponential smoother to this periodic input. Let $a = 0.5$. *Hint:* Use the SIGSYS command set INC to define the correct frequency increment corresponding to num and inc, and use the frequency axis DOTPLUS to compute the frequency response function $\hat{h}$ of the smoother.

(c) Find and plot the response y itself. Think of a way to plot several periods of the input u and the output y.

(d) Repeat (b) and (c) for $a = 0.9$. Comment on the difference.

(e) In 6.5.2(b) it is explained how the response of the RC network to the periodic rectangular pulse may be obtained. Create one period of the periodic signal u, which is given by

$$u(t) = \text{trian}\left(\frac{t-1/8}{1/8}\right) - \text{trian}\left(\frac{t-3/8}{1/8}\right), \qquad 0 \le t < 1,$$

using 64 sample points. Determine the expansion coefficients $\hat{u}$. Next compute the expansion coefficients $\hat{y}$ of the response of the RC network to this input, with $RC = 1$. Finally find one period of the response y itself. Think of a way to plot several periods of y. Repeat the calculations for $RC = 0.1$, and comment on the effect of this change on the output y. *Hint:* In SIGSYS, set inc and num to the correct values and use the time axis dotplus to generate one period of u. Use the SIGSYS command set INC and the frequency axis DOT to compute the frequency response function $\hat{h}$ of the RC network.

7

Fourier Transforms

7.1 INTRODUCTION

In the previous chapter we discussed the frequency domain analysis of linear time-invariant systems with periodic inputs. The analysis is based on *signal expansion* in harmonic basis functions. In this chapter the notion of system analysis by means of *transforms* is introduced.

The principle of the various transform methods in system analysis is to find a different representation of the input signal, called its *transform,* and to study the effect of the system on the input in terms of this alternative representation. The strength of the transform method is that often the transformed system is easier to handle than the original system.

The evaluation of the output by using transform methods involves transformation of the input to the *transformed* input, transformation of the system IO map to the *transformed* IO map, computation of the transformed output from the transformed input by using the transformed IO map, and computation of the output by *inverse* transformation.

The transforms we consider in this chapter are the various *Fourier* transforms. They are closely related to signal expansions. The Fourier transform of a signal simply consists of the expansion coefficients of the signal in terms of harmonic basis signals.

Chapter 8 deals with the *z*- and *Laplace* transforms, the former for discrete-time signals and the latter for continuous-time signals. These transforms relate to the expansion of signals in *exponential* signals and are intimately connected with the Fourier transform.

In Section 7.2 transform theory is presented in an abstract setting. The notion of *expansion transform* establishes the connection between expansions and transforms.

Section 7.3 deals with the *discrete-to-discrete Fourier transform* (DDFT) and the *continuous-to-discrete Fourier transform* (CDFT). They are an alternative way of looking at the Fourier series expansion: The DDFT transforms a finite-time discrete-time signal to the coefficients of its finite Fourier series, while the CDFT transforms a finite-time continuous-time signal to the coefficients of its infinite Fourier series. Thus, Section 7.3 essentially recapitulates Sections 6.4 and 6.5.

Section 7.4 is devoted to the *discrete-to-continuous Fourier transform* (DCFT) and the *continuous-to-continuous Fourier transform* (CCFT). The DCFT applies to infinite-time discrete-time signals and the CCFT to infinite-time continuous-time signals. Infinite-time signals may be expanded in a *continuum* of harmonics, rather than in a finite or countable set as for finite-time signals. The DCFT and CCFT transform the time signal to expansion "coefficients" that now are not a *sequence* of coefficients but a *function* on a finite or infinite interval.

The chapter concludes in Section 7.5 with a discussion of the application of the various Fourier transforms to the analysis of linear time-invariant systems.

7.2 TRANSFORM THEORY

In this section we first present transform theory in a quite general context and then show how expansion theory fits in as a special case.

Transforms

The idea behind transform theory is simple. What follows serves as an illustration. Suppose that design calculations need be carried out on an engineering project, whose specifications are in practical units, such as the "foot-pound-second" system. Physical calculations are often most conveniently performed in scientific units, say, the "meter-kilogram-second" (SI) system. The designer might nevertheless proceed in the *direct* way and perform the design computations in practical units. The alternative is the *transform* way, which is first to convert the specifications from practical to scientific units, do the calculations in the scientific manner, and convert the final results back to practical units.

This illustration may help as a visualization of the abstract definition of the transform of a signal space and that of a system, whose formal definitions now follow.

7.2.1.   Definition: Transformation of a signal space and of a system.

(a) *Transformation of the signal space*. Let

$$\mathcal{T}: X \;\rightarrow\; \hat{X}$$

be a map from a signal space X to another signal space $\hat{X}$. If $\mathcal{T}$ is a bijection, that is, there exists an inverse map $\mathcal{T}^{-1}$ from $\hat{X}$ to X, then $\hat{X}$ is called the *transformation* of the signal space X under the *transform* $\mathcal{T}$.

(b) *Transformation of the system IO map*. Consider an IOM system whose input and output sets are both X and that has the IO map

$$\phi: X \;\rightarrow\; X.$$

Let $\hat{X}$ be the transformation of the signal space X under the transform $\mathcal{T}$. Then the *transformed IOM system* is the IOM system whose input and output sets are both $\hat{X}$ and that has the transformed IO map $\hat{\phi}: \hat{X} \rightarrow \hat{X}$ given by

$$\hat{\phi} = \mathcal{T} \circ \phi \circ \mathcal{T}^{-1}.$$

Here, $\circ$ denotes map composition.   ■

The transformed map $\hat{\phi}$ is equivalently defined by

$$\hat{\phi}(\hat{u}) = \mathcal{T}(\phi(\mathcal{T}^{-1}(\hat{u}))) \qquad \text{for all} \quad \hat{u} \in \hat{X}.$$

Conversely,

$$\phi(u) = \mathcal{T}^{-1}(\hat{\phi}(\mathcal{T}(u))) \qquad \text{for all} \quad u \in X. \tag{1}$$

For brevity, in this and later chapters we use the symbol $\hat{\ }$ to denote the transform operation, and occasionally $\check{\ }$ to indicate the inverse transform operation, whenever it is clear from the context which transform is meant. Thus, if $\mathcal{T}: X \rightarrow \hat{X}$ we write

$$\hat{x} := \mathcal{T}(x), \qquad \check{x} := \mathcal{T}^{-1}(x).$$

With this convention we may rewrite (1) as

$$\phi(u) = (\hat{\phi}(\hat{u}))^{\check{}} \qquad \text{for all} \quad u \in X,$$

which illustrates that $\hat{\ }$ and $\check{\ }$ may be used both diacritically and as a superscript. It also shows that there are two ways to determine the response $y = \phi(u)$ of an IOM system

(i) the *direct* way, using

$$y = \phi(u),$$

or

(ii) the *transform* way, using

$$y = (\hat{\phi}(\hat{u}))^{\vee}.$$

The transform way involves a three-step procedure, namely, (1) determine the transformed input $\hat{u}$ by transforming the input u, (2) determine the image $\hat{y} = \hat{\phi}(\hat{u})$ of the transformed input under the transformed map (this image is the transform of the output y), and (3) determine the output y by inverse transformation of $\hat{y}$.

The procedure is illustrated in Fig. 7.1. The following example demonstrates how the idea of transformation may be advantageously used when performing computations on a large sparse vector (i.e., a vector of high dimension most of whose elements are zero).

The DIRECT way

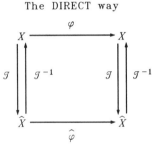

The TRANSFORM way

Figure 7.1. The direct and the transform way of determining the output of an IOM system.

7.2.2.   Example: Sparse computations.

Suppose that we need perform numerical computations on vectors $x = (x_1, x_2, \cdots, x_N) \in \mathbb{R}^N$ that have very large dimension N (say, 1,000,000 or more), but whose elements are mostly zero (say only about 100 elements are nonzero). Then much computation time can be saved by transforming the vector

$$x = (0, \cdots, 0, x_{n_1}, 0, \cdots, 0, x_{n_2}, 0, \cdots, 0, x_{n_k}, 0, \cdots),$$

where n_k is the coordinate number of the kth nonzero entry, to the double entry vector $\hat{x} \in (\mathbb{N} \times \mathbb{R})^N$ given by

$$\hat{x} = ((n_1, x_{n_1}), (n_2, x_{n_2}), \cdots, (n_K, x_{n_K}), (0, 0), \cdots, (0, 0)),$$

with K the number of nonzero entries of x. In practice one would of course only store the first K pairs, so that also much memory is saved. Given the transform $\hat{x}$ of x, the original vector x may easily be reconstructed.

Suppose that we wish to perform a simple operation on the vector x, such as taking the pointwise absolute value. This involves the map $\phi: \mathbb{R}^N \to \mathbb{R}^N$ given by

$$\phi(x) = (|x_1|, |x_2|, \cdots, |x_N|).$$

Obviously, the transformation $\hat{\phi}: (\mathbb{N} \times \mathbb{R})^N \to (\mathbb{N} \times \mathbb{R})^N$ of the map ϕ is defined by

$$\hat{\phi}(\hat{x}) = ((n_1, |x_{n_1}|), (n_2, |x_{n_2}|), \cdots, (n_K, |x_{n_K}|), (0, 0), \cdots, (0, 0)).$$

Clearly in transformed form the computation is much more efficient because only the first K elements of the transformed vector need be processed. ∎

Linear Transformation of Finite-Time Linear Discrete-Time Systems

Having introduced the idea of transformation, we continue by considering *linear* transforms for finite-time linear discrete-time systems. As in 6.3.1, we take the time axis as $\underline{N} = \{0, 1, 2, \cdots, N - 1\}$ and represent the system in matrix form as

$$y = Ku,$$

with $u \in \mathbb{C}^N$ and $y \in \mathbb{C}^N$ the input and output in vector form, and K an $N \times N$ matrix. Suppose that the input u is transformed to

$$\hat{u} = Uu,$$

with U an $N \times N$ nonsingular matrix. Then we may write for the transformed output

$$\hat{y} = Uy = UKu = UKU^{-1}Uu = UKU^{-1}\hat{u}.$$

This shows that the transformed map is $\hat{y} = \hat{K}\hat{u}$, where the $N \times N$ matrix $\hat{K}$ is given by $\hat{K} = UKU^{-1}$.

7.2.3.   Summary: Linear transformation of finite-time linear discrete-time systems.   Consider a finite-time linear discrete-time IOM system represented in matrix form as

$$y = Ku,$$

with $u \in \mathbb{C}^N$ and $y \in \mathbb{C}^N$ the input and output written in vector form and K an $N \times N$ matrix. Let the $N \times N$ nonsingular matrix U define a transform $\mathcal{T}: \mathbb{C}^N \to \mathbb{C}^N$ given by

$$\hat{x} = \mathcal{T}(x) = U(x), \qquad \check{x} = \mathcal{T}^{-1}(x) = U^{-1}x.$$

Then in matrix form the transformed system map is

$$\hat{y} = \hat{K}\hat{u},$$

where the $N \times N$ matrix $\hat{K}$ is given by

$$\hat{K} = UKU^{-1}.$$ ∎

Thus, for finite-time linear discrete-time systems the determination of the response y by the transform method amounts to the three-stage computation

$$y = (U^{-1} \cdot (\hat{K} \cdot (Uu))).$$

The idea is to choose the transformation matrix U such that multiplication by $\hat{K}$ is simpler than by K.

7.2.4.   Example: Linear transformation of a finite-time linear discrete-time system.   In Example 6.3.6 a linear discrete-time system with time axis $\underline{2} = \{0, 1\}$ is considered that in matrix form is represented by the 2×2 matrix

$$K = \begin{bmatrix} 3/2 & 1/2 \\ 1/2 & 3/2 \end{bmatrix}.$$

We choose the transformation matrix U and the resulting inverse as

$$U = \begin{bmatrix} 1/2 & 1/2 \\ -1/2 & 1/2 \end{bmatrix}, \qquad U^{-1} = \begin{bmatrix} 1 & -1 \\ 1 & 1 \end{bmatrix}.$$

The matrix $\hat{K}$ that represents the transformed system hence is

$$\hat{K} = UKU^{-1} = \begin{bmatrix} 1/2 & 1/2 \\ -1/2 & 1/2 \end{bmatrix} \begin{bmatrix} 3/2 & 1/2 \\ 1/2 & 3/2 \end{bmatrix} \begin{bmatrix} 1 & -1 \\ 1 & 1 \end{bmatrix} = \begin{bmatrix} 2 & 0 \\ 0 & 1 \end{bmatrix}.$$

The matrix U has been so selected that the matrix $\hat{K}$ that represents the transformed map is *diagonal*. This considerably simplifies the computation of the image $\hat{y}$ of the transformed input $\hat{u}$ under the transformed map, because multiplication of a vector by a diagonal matrix amounts to element-by-element multiplication of the vector. ∎

Expansion Transforms

We next make the simple but profound observation that expansion of a signal in a basis may be viewed as a transformation, called an *expansion transform*.

7.2.5.   Summary: Expansion transform.   Suppose that $\{v_i,\ i \in \mathbb{I}\}$, with $\mathbb{I}$ a finite or countable index set, is a basis for the linear space X. Then if $x \in X$ has the expansion

$$x = \sum_{i \in \mathbb{I}} v_i \hat{x}_i,$$

the map $\mathcal{T}: X \rightarrow \hat{X}$ that maps $x \in X$ into its expansion coefficients $\hat{x} = \{\hat{x}_i,\ i \in \mathbb{I}\} \in \hat{X}$ is a linear transform, called an *expansion transform*.   ∎

7.2.6.   Proof.   Since for any $x \in X$ the expansion coefficients $\hat{x} = \{\hat{x}_i,\ i \in \mathbb{I}\}$ are uniquely determined, $\mathcal{T}$ is a well-defined map. Because for every $\hat{x} \in \hat{X}$ the expansion defines a unique $x \in X$, the map $\mathcal{T}$ has an inverse and hence is a transform. It is easy to see that the transform is linear.   ∎

Given a basis, expanding a signal in that basis may be viewed as transforming the signal to its expansion coefficients. In what follows we apply 7.2.5 to the finite-dimensional space $\mathbb{C}^N$.

7.2.7.   Summary: Expansion transform for $\mathbb{C}^N$.   Suppose that $\{v_1, v_2, \cdots, v_N\}$ forms a basis for $\mathbb{C}^N$, and let V be the matrix

$$V = [v_1\ v_2\ \cdots\ v_N]$$

whose columns are the basis vectors. Define the linear transform $\mathcal{T}$ by

$$\mathcal{T}(x) = V^{-1}x, \qquad x \in \mathbb{C}^N.$$

Then the elements of

$$\hat{x} = \mathcal{T}(x) = V^{-1}x$$

are the coefficients of the expansion of x with respect to the basis, so that $\mathcal{T}$ is an expansion transform for $\mathbb{C}^N$.   ∎

7.2.8.   Proof.   By writing $x = VV^{-1}x = V\hat{x}$, with $\hat{x} = V^{-1}x$, we see that

$$x = V\hat{x} = [v_1\ v_2\ \cdots\ v_N]\begin{bmatrix} \hat{x}_1 \\ \hat{x}_2 \\ \cdots \\ \hat{x}_N \end{bmatrix} = \sum_{i=1}^{N} v_i \hat{x}_i,$$

where $\hat{x} = \text{col}(\hat{x}_1, \hat{x}_2, \cdots, \hat{x}_N)$. This shows that the elements of the transformed vector $\hat{x} = \mathcal{T}(x) = V^{-1}x$ form the coefficients of the expansion of x in the basis $\{v_1, v_2, \cdots, v_N\}$. ∎

We consider an example.

7.2.9.   Example: Expansion transformation of a finite-time linear discrete-time system.   In Example 7.2.4 the columns of the inverse U^{-1} of the transforming matrix U are $v_1 = \text{col}(1, 1)$ and $v_2 = \text{col}(-1, 1)$. Hence, the transformation of this example amounts to expanding in the basis formed by $\text{col}(1, 1)$ and $\text{col}(-1, 1)$. ∎

Spectral Transforms

In Section 6.3 it has become clear that when we study the response of a linear system it is helpful to expand the input in *eigensignals* of the system, provided they form a basis, of course. The expansion transform that corresponds to choosing eigensignals of the system as a basis is called the *spectral transform* for the system. It has the unique feature that it "diagonalizes" the system. The following result simply rephrases the spectral expansion of 6.3.5 in terms of transforms.

7.2.10.   Summary: Spectral transform of a map.   Let X be a linear space, $\phi: X \to X$ a linear IO map, and $\{v_i, i \in \mathbb{I}\}$ a finite or countable set of eigenvectors of ϕ. Suppose that $\{v_i, i \in \mathbb{I}\}$ forms a basis for X, so that every $x \in X$ may uniquely be expanded as

$$x = \sum_{i \in \mathbb{I}} v_i \hat{x}_i.$$

The transform $\mathcal{T}: X \to \hat{X}$ that maps each element $x \in X$ into its expansion coefficients $\hat{x} = \{\hat{x}_i, i \in \mathbb{I}\}$ is called the *spectral transform* for ϕ.

Let $\hat{\phi}: \hat{X} \to \hat{X}$ denote the transformed IO map under the spectral transform. Then $\hat{\phi}$ maps the transformed input

$$\hat{u} = \{\hat{u}_i, i \in \mathbb{I}\}$$

into the transformed output

$$\hat{y} = \{\hat{y}_i, i \in \mathbb{I}\},$$

such that

$$\hat{y}_i = \lambda_i \hat{u}_i, \qquad i \in \mathbb{I},$$

with λ_i the eigenvalue of ϕ corresponding to the eigenvector v_i. ∎

It follows that when using a spectral transform, application of the transformed IO map amounts to simply multiplying the transformed input $\hat{u}$ *element by element* by the eigenvalues of the map.

We consider in particular the spectral expansion of finite-time linear discrete-time systems.

7.2.11.   Summary: Spectral transform of finite-time linear discrete-time systems.   Consider a finite-time linear discrete-time IOM system represented in matrix form as

$$y = Ku,$$

with $u \in \mathbb{C}^N$ and $y \in \mathbb{C}^N$ the input and output written in vector form, and K an $N \times N$ matrix.

(a) There exists a spectral transform for the system if and only if the matrix K has N linearly independent eigenvectors.

(b) Suppose that K has the N linearly independent eigenvectors $v_1, v_2, \cdots,$ v_N, with corresponding eigenvalues $\lambda_1, \lambda_2, \cdots, \lambda_N$. Then, the spectral transform is given by

$$\hat{x} = \mathcal{T}(x) = V^{-1}x, \qquad x \in \mathbb{C}^N,$$

where $V = [v_1 \ v_2 \ \cdots, \ v_N]$ is the $N \times N$ matrix whose columns are the eigenvectors of K.

(c) The IO map of the transformed system is

$$\hat{y} = \hat{K}\hat{u},$$

where $\hat{K}$ is the $N \times N$ diagonal matrix

$$\hat{K} = V^{-1}KV = \begin{bmatrix} \lambda_1 & 0 & \cdots & 0 \\ 0 & \lambda_2 & \cdots & 0 \\ \cdots & \cdots & \cdots & \cdots \\ 0 & 0 & \cdots & \lambda_N \end{bmatrix}. \qquad\blacksquare$$

7.2.12.   Proof.
(a) If the matrix K does not have N linearly independent eigenvectors, no spectral transform exists.

(b) If K has N linearly independent eigenvectors the existence of a spectral transform and its form follow from 7.2.7.

(c) It follows from 7.2.10 that the transformed map is represented by the diagonal matrix $\hat{K}$.

7.2.13.   Example: Spectral transformation of a finite-time linear discrete-time system.   Consider the finite-time linear discrete-time system represented by the 2×2 matrix

$$K = \begin{bmatrix} 3/2 & 1/2 \\ 1/2 & 3/2 \end{bmatrix},$$

that was studied in Example 7.2.4. From Example 6.3.6 we know that K has the (unnormalized) eigenvectors $v_1 = \text{col}(1, 1)$ and $v_2 = \text{col}(-1, 1)$ with corresponding eigenvalues $\lambda_1 = 2$ and $\lambda_2 = 1$. As a result, the matrix V and its inverse that define the spectral transform are given by

$$V = [v_1 \ v_2] = \begin{bmatrix} 1 & -1 \\ 1 & 1 \end{bmatrix}, \qquad V^{-1} = \tfrac{1}{2}\begin{bmatrix} 1 & 1 \\ -1 & 1 \end{bmatrix},$$

while the transformed map is defined by the diagonal matrix

$$\hat{K} = \begin{bmatrix} \lambda_1 & 0 \\ 0 & \lambda_2 \end{bmatrix} = \begin{bmatrix} 2 & 0 \\ 0 & 1 \end{bmatrix}.$$

This is precisely the transform considered in Example 7.2.4.   ∎

Unitary Transforms

A special type of transform arises when the original space X and the transformed space $\hat{X}$ are both inner product spaces and the transform leaves the inner product of any two elements of X unchanged. Such transforms are called *unitary* transforms.

7.2.14.   Definition: Unitary transform.   Suppose that X and $\hat{X}$ are inner product spaces and $\mathcal{T}: X \rightarrow \hat{X}$ a transform. Denote by $\langle \cdot, \cdot \rangle_x$ the inner product in the space X and by $\langle \cdot, \cdot \rangle_{\hat{x}}$ that in $\hat{X}$. Then the transform $\mathcal{T}$ is *unitary* if

$$\langle x, y \rangle_x = \langle \hat{x}, \hat{y} \rangle_{\hat{x}}$$

for every x and y in X and their transforms $\hat{x}$ and $\hat{y}$.   ∎

If a transform is unitary, then it follows that

$$\langle x, x \rangle_X = \langle \hat{x}, \hat{x} \rangle_{\hat{X}},$$

which shows that a unitary transform preserves the (natural) norm. For this reason unitary transforms are said to be *norm preserving*.

7.2.15.   Example: Unitary transform on $\mathbb{C}^N$.   Let $\mathcal{T}: \mathbb{C}^N \to \mathbb{C}^N$ be the linear transform defined by the $N \times N$ matrix U, i.e., $\hat{x} = \mathcal{T}(x) = Ux$. Taking the inner product in $\mathbb{C}^N$ as

$$\langle x, y \rangle = \sum_{i=1}^{N} x_i \overline{y_i} = y^H x,$$

with the superscript ^{H} denoting the complex conjugate transpose, it follows that $\langle \hat{x}, \hat{y} \rangle = \langle Ux, Uy \rangle = (Uy)^H Ux = y^H U^H Ux$. For the transform to be unitary, the latter expression should equal $\langle x, y \rangle = y^H x$, which is the case if and only if $U^H U = I$, with I the $N \times N$ unit matrix. A matrix U such that

$$U^H U = I$$

is said to be a *unitary matrix*. Thus, the transform $\mathcal{T}$ defined by $\mathcal{T}(x) = Ux$ is unitary if and only if the matrix U is unitary.

The matrix U is unitary if and only if its columns form an orthonormal set. In Example 6.3.6 we noted that the vectors $\mathrm{col}(1, 1)/\sqrt{2}$ and $\mathrm{col}(1, -1)/\sqrt{2}$ constitute an orthonormal set in $\mathbb{C}^2$. It follows that the 2×2 matrix

$$U = \frac{1}{\sqrt{2}} \begin{bmatrix} 1 & 1 \\ 1 & -1 \end{bmatrix}$$

is unitary. Hence, the transform $\hat{x} = Ux$ defined by this matrix is unitary.   ∎

7.3 FOURIER TRANSFORMS: THE DDFT AND THE CDFT

In this section we discuss the expansion transforms corresponding to the Fourier series expansions. The *discrete-to-discrete Fourier transform* (DDFT) is the expansion transform originating from the finite Fourier series expansion, while the *continuous-to-discrete Fourier transform* (CDFT) follows from the infinite Fourier series expansion. The DDFT and CDFT are *spectral transforms* for infinite-time convolution systems with periodic inputs, or, equivalently, for cyclical convolution systems.

We first introduce the transforms and their inverses, show that they are unitary, list a number of useful properties, and finally demonstrate how they are applied in the analysis of the response of convolution systems to periodic inputs.

DDFT and CDFT

We begin by defining the DDFT and CDFT. The *DDFT* maps a *discrete*-time finite-time signal into its Fourier coefficients, while the *CDFT* maps a *continuous*-time finite-time signal into its Fourier coefficients. The Fourier coefficients in both cases are defined on a *discrete* frequency axis.

7.3.1.   Definition: DDFT and CDFT.

The transform

$$\mathscr{F}_{DD}: \ell_{\underline{N}} \;\rightarrow\; \ell_{\underline{N}}(F),$$

with $F = 1/N$, that maps the discrete-time signal x with time axis $\underline{N}$ into its Fourier coefficients $\hat{x} = \mathscr{F}_{DD}(x)$, given by

$$\hat{x}(f) = \sum_{n=0}^{N-1} x(n)e^{-j2\pi fn}, \qquad f \in \underline{N}(F),$$

is called the *discrete-to-discrete Fourier transform* (DDFT).

The transform

$$\mathscr{F}_{CD}: \mathscr{L}_2[0, P) \;\rightarrow\; \ell_2(F),$$

with $F = 1/P$, that maps the finite-energy continuous-time signal x with time axis $[0, P)$ into its Fourier coefficients $\hat{x} = \mathscr{F}_{CD}(x)$, given by

$$\hat{x}(f) = \int_0^P x(t)e^{-j2\pi ft}\,dt, \qquad f \in \mathbb{Z}(F),$$

is called the *continuous-to-discrete Fourier transform* (CDFT). ∎

The DDFT and CDFT are expansion transforms and hence their inverses are easily established.

7.3.2.   Summary: Inverses of the DDFT and CDFT.

The *inverse DDFT*

$$\mathscr{F}_{DD}^{-1}: \ell_{\underline{N}}(F) \;\rightarrow\; \ell_{\underline{N}}$$

maps the frequency signal $\hat{x}$ back into the time signal x according to

$$x(n) = F\sum_{f \in \underline{N}(F)} \hat{x}(f)e^{j2\pi fn}, \qquad n \in \underline{N}.$$

The *inverse CDFT*

$$\mathscr{F}_{CD}^{-1}: \ell_2(F) \;\rightarrow\; \mathscr{L}_2[0, P)$$

maps the frequency signal $\hat{x}$ back into the time signal x according to

$$x(t) = F\sum_{f \in \mathbb{Z}(F)} \hat{x}(f)e^{j2\pi ft}, \qquad t \in [0, P).$$
                                                                          ∎

Performing the inverse DDFT or CDFT amounts to summing the respective Fourier series.

   In 7.2.14 *unitary* transforms are defined as transforms that preserve the inner product. That the DDFT and CDFT are unitary is an immediate consequence of *Plancherel's identity* of 6.4.7.

7.3.3.   Summary: Unitariness of the DDFT and CDFT.

Define the inner product on the signal space $\ell_{\underline{N}}$ as

$$\langle x, y \rangle_{\ell_{\underline{N}}} = \sum_{n=0}^{N-1} x(n)\overline{y(n)},$$

and on the transformed space $\ell_{\underline{N}}(F)$ as

$$\langle \hat{x}, \hat{y} \rangle_{\ell_{\underline{N}}(F)} = F \sum_{f \in \underline{N}(F)} \hat{x}(f)\overline{\hat{y}(f)}.$$

(a) *Plancherel's identity.* The DDFT is a *unitary* transform, that is, for x and y time signals and $\hat{x}$ and $\hat{y}$ their transforms,

$$\langle x, y \rangle_{\ell_{\underline{N}}} = \langle \hat{x}, \hat{y} \rangle_{\ell_{\underline{N}}(F)}.$$

(b) *Parseval's identity.* In particular, a time signal x and its transform $\hat{x}$ have equal energies, that is,

$$\|x\|_2^2 = \|\hat{x}\|_2^2.$$

Define the inner product on the signal space $\mathcal{L}_2[0, P)$ as

$$\langle x, y \rangle_{\mathcal{L}_2[0, P)} = \int_0^P x(t)\overline{y(t)}\, dt,$$

and on the transformed space $\ell_2(F)$ as

$$\langle \hat{x}, \hat{y} \rangle_{\ell_2(F)} = F \sum_{f \in \mathbb{Z}(F)} \hat{x}(f)\overline{\hat{y}(f)}.$$

(a') *Plancherel's identity.* The CDFT is a *unitary* transform, that is, for x and y time signals and $\hat{x}$ and $\hat{y}$ their transforms,

$$\langle x, y \rangle_{\mathcal{L}_2[0, P)} = \langle \hat{x}, \hat{y} \rangle_{\ell_2(F)}.$$

(b') *Parseval's identity.* In particular, a time signal x and its transform $\hat{x}$ have equal energies, that is,

$$\|x\|_2^2 = \|\hat{x}\|_2^2.$$ ∎

7.3.4.   Remark: Discrete Fourier Transform.   The DDFT is based on expansion in the orthogonal basis $\{F\eta_f, f \in \underline{N}(F)\}$. An alternative is to choose the basis $\{\eta_f/\sqrt{N}, f \in \underline{N}(F)\}$, which is orthonormal under the inner product

$$\langle x, y \rangle_{\ell_{\underline{N}}} = \sum_{n=0}^{N-1} x(n)\overline{y(n)},$$

This basis results in the expansion

$$x(n) = \frac{1}{\sqrt{N}} \sum_{f \in \underline{N}(F)} \hat{x}(f)e^{j2\pi fn}, \qquad n \in \underline{N},$$

$$\hat{x}(f) = \frac{1}{\sqrt{N}} \sum_{n \in \underline{N}} x(n)e^{-j2\pi fn}, \qquad f \in \underline{N}(F).$$

Denoting $x(n) =: x_n$ and $\hat{x}(kF) =: \hat{x}_k$, this may be rewritten as

$$\hat{x}_k = \frac{1}{\sqrt{N}} \sum_{n=0}^{N-1} x_n e^{-j2\pi nk/N}, \qquad k = 0, 1, \cdots, N-1,$$

$$x_n = \frac{1}{\sqrt{N}} \sum_{k=0}^{N-1} \hat{x}_k e^{j2\pi nk/N}, \qquad n = 0, 1, \cdots, N-1.$$

The sequence $\hat{x}$ is often known as the *discrete Fourier transform* (DFT) of x. The DFT may be seen as a transform from $\mathbb{C}^N$ to $\mathbb{C}^N$, and is unitary under the usual inner product on $\mathbb{C}^N$.   ∎

Properties of the DDFT and the CDFT

In what follows a number of important properties of the DDFT and CDFT are presented. As before, ˆ denotes the transform and ˇ the inverse transform.

From their definitions, the DDFT and CDFT and their inverses are immediately recognized to be linear.

7.3.5.   Summary: Linearity of the DDFT and CDFT.   The DDFT and CDFT and their inverses are linear, that is, for every complex α and β

$$(\alpha x + \beta y)\hat{} = \alpha\hat{x} + \beta\hat{y}$$

for each x and y in the signal space and

$$(\alpha u + \beta v)\check{} = \alpha\check{u} + \beta\check{v}$$

for each u and v in the transformed space.   ∎

According to 6.5.4, the Fourier series expansions are spectral expansions for cyclical convolution systems. This immediately leads to the *convolution property* of the DDFT and CDFT, which implies that cyclical convolutions transform into products. The symmetry of the DDFT and CDFT and their inverses results in the *converse convolution property*, which implies that products transform into cyclical convolution for the DDFT and regular convolution for the CDFT. A direct proof of these properties is given in E.3 in Supplement E.

7.3.6.   Summary: Convolution and converse convolution properties of the DDFT and CDFT.

(a) *Convolution property.* Cyclical convolution transforms under the DDFT into multiplication, that is, if x transforms

(a′) *Convolution property.* Cyclical convolution transforms under the CDFT into multiplication, that is, if x transforms into $\hat{x}$ and y into $\hat{y}$ then

$$(x \odot y)^\wedge = \hat{x} \cdot \hat{y}.$$

(b) *Converse convolution property.* Vice-versa, multiplication transforms under the DDFT into cyclical convolution, that is,

$$(x \cdot y)^\wedge = \hat{x} \odot \hat{y}.$$

into $\hat{x}$ and y into $\hat{y}$, then

$$(x \odot y)^\wedge = \hat{x} \cdot \hat{y}.$$

(b') *Converse convolution property.* Vice-versa, multiplication transforms under the CDFT into convolution, that is,

$$(x \cdot y)^\wedge = \hat{x} * \hat{y}.$$ ∎

Note that for the DDFT cyclical convolution on the transformed space $\ell_{\underline{N}}(F)$ is defined as

$$(\hat{x} \odot \hat{y})(f) = F \sum_{\phi \in \underline{N}(F)} \hat{x}((f - \phi) \bmod NF)\hat{y}(\phi), \qquad f \in \underline{N}(F),$$

while for the CDFT convolution on the transformed space $\ell_2(F)$ is given by

$$(\hat{x} * \hat{y})(f) = F \sum_{\phi \in \mathbb{Z}(F)} \hat{x}(f - \phi)\hat{y}(\phi), \qquad f \in \mathbb{Z}(F).$$

Another useful pair of properties are the *shift property* and its converse. The proofs may be found in E.4 in Supplement E.

7.3.7.   Summary: Shift and converse shift properties of the DDFT and CDFT.

(a) *Shift property.* Cyclical shifting transforms under the DDFT into multiplication by a harmonic, that is, if x transforms into $\hat{x}$ then for every $m \in \mathbb{Z}$

$$x((n + m) \bmod N), \qquad n \in \underline{N},$$

transforms into

$$e^{j2\pi fm}\hat{x}(f), \qquad f \in \underline{N}(F).$$

(b) *Converse shift property.* Conversely, for every $\phi \in \mathbb{Z}(F)$

$$e^{j2\pi\phi n}x(n), \qquad n \in \underline{N},$$

under the DDFT transforms into

$$\hat{x}((f - \phi) \bmod NF), \qquad f \in \underline{N}(F).$$

(a') *Shift property.* Cyclical shifting transforms under the CDFT into multiplication by a harmonic, that is, if x transforms into $\hat{x}$ then for every real θ

$$x((t + \theta) \bmod P), \qquad t \in [0, P),$$

transforms into

$$e^{j2\pi f\theta}\hat{x}(f), \qquad f \in \mathbb{Z}(F).$$

(b') *Converse shift property.* Conversely, for every $\phi \in \mathbb{Z}(F)$

$$e^{j2\pi\phi t}x(t), \qquad t \in [0, P),$$

under the CDFT transforms into

$$\hat{x}(f - \phi), \qquad f \in \mathbb{Z}(F).$$ ∎

The DDFT transforms a time signal to a frequency signal defined on the finite axis $\underline{N}(F)$. When the DDFT is evaluated on the *infinite* frequency axis $\mathbb{Z}(F)$, it may easily be seen to be *periodic* with period NF. Similarly, when the *inverse* DDFT is evaluated on the infinite time axis $\mathbb{Z}$ rather than on $\underline{N}$, it is periodic with period N. Finally, evaluation of the inverse CDFT on the entire real axis $\mathbb{R}$ rather than the interval $[0, P)$ again leads to a periodic signal, this time with period P.

7.3.8.   Summary: Periodicity properties of the DDFT and CDFT.

(a) *Periodicity of the DDFT.* The DDFT $\hat{x}$ of x, defined on $\mathbb{Z}(F)$ rather than on $\underline{N}(F)$ by

$$\hat{x}(f) = \sum_{n \in \underline{N}} x(n)e^{-j2\pi fn}, \qquad f \in \mathbb{Z}(F),$$

is periodic with period NF.

(b) *Periodicity of the inverse DDFT.* The inverse DDFT x of $\hat{x}$, defined on $\mathbb{Z}$ rather than on $\underline{N}$ by

$$x(n) = F \sum_{f \in \underline{N}(F)} \hat{x}(f)e^{j2\pi fn}, \qquad n \in \mathbb{Z},$$

is periodic with period N.

(b') *Periodicity of the inverse CDFT.* The inverse CDFT x of $\hat{x}$, defined on $\mathbb{R}$ rather than on $[0, P)$ by

$$x(t) = F \sum_{f \in \mathbb{Z}(F)} \hat{x}(f)e^{j2\pi ft}, \qquad t \in \mathbb{R},$$

is periodic with period P.   ∎

For the CDFT, finally, there is a simple relation between the CDFT of a signal and that of its derivative, whose proof may be found in E.5 in Supplement E.

7.3.9.   Summary: Differentiation property of the CDFT.   Suppose that x is *cyclically continuous* on $[0, P)$, that is, x is continuous on $[0, P)$ and $x(0) = x(P^-)$. Then, if x is differentiable on $[0, P)$ with derivative Dx, the CDFT of Dx is

$$j2\pi f \cdot \hat{x}(f), \qquad f \in \mathbb{Z}(F),$$

with $\hat{x}$ the CDFT of x.   ∎

Examples

The various properties of the Fourier transform may often be exploited to facilitate the calculation of a transform or its inverse. We demonstrate this by an example.

7.3.10.   Example: CDFT of a triangular pulse.   By way of illustration we use two different methods to calculate the CDFT of the triangular pulse of Fig. 7.2, which is given by

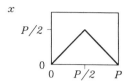

x

$P/2$

0

$0 \quad P/2 \quad P$

time   **Figure 7.2.**   A triangular pulse.

$$x(t) = \begin{cases} t & \text{for } 0 \le t < P/2, \\ P - t & \text{for } P/2 \le t < P, \end{cases} \quad t \in [0, P).$$

(a) One way of saving work when determining the CDFT of this signal is to note that it is the cyclical convolution of the rectangular pulse

$$y(t) = \begin{cases} 1 & \text{for } 0 \le t < P/2, \\ 0 & \text{for } P/2 \le t < P, \end{cases} \quad t \in [0, P),$$

as shown in Fig. 7.3, with itself. The signal y is easily found to have the CDFT

$$\hat{y}(f) = \begin{cases} 1/2F & \text{for } f = 0, \\ 1/j\pi f & \text{for } f/F \text{ odd}, \\ 0 & \text{for } f \ne 0, f/F \text{ even}, \end{cases} \quad f \in \mathbb{Z}(F).$$

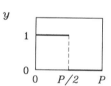

y

1

0

$0 \quad P/2 \quad P$

time   **Figure 7.3.**   A rectangular pulse.

Since $x = y \odot y$, by the convolution property $\hat{x} = \hat{y}^2$, so that

$$\hat{x}(f) = \begin{cases} 1/4F^2 & \text{for } f = 0, \\ -1/\pi^2 f^2 & \text{for } f/F \text{ odd}, \\ 0 & \text{for } f \ne 0, f/F \text{ even}, \end{cases} \quad f \in \mathbb{Z}(F).$$

A plot of $\hat{x}$, which happens to be real-valued, is given in Fig. 7.4. The signal has large constant and fundamental components, but small higher harmonics.

(b) A second way to determine the CDFT of the triangular pulse x other than by direct calculation is to observe that the derivative $z = Dx$ of x is given by

$$z(t) = \begin{cases} 1 & \text{for } 0 \le t < P/2, \\ -1 & \text{for } P/2 \le t < P, \end{cases} \quad t \in [0, P).$$

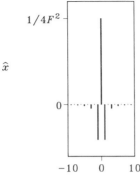

$1/4F^2$

$\hat{x}$

0

$-10 \quad 0 \quad 10$

normalized frequency f/F

Figure 7.4.   CDFT of the triangular pulse.

The CDFT of z is easily found to be

$$\hat{z}(f) = \begin{cases} 2/j\pi f & \text{for } f/F \text{ odd,} \\ 0 & \text{for } f/F \text{ even,} \end{cases} \qquad f \in \mathbb{Z}(F).$$

Since $z = Dx$, by the differentation property 7.3.9 we have $\hat{z}(f) = j2\pi f \cdot \hat{x}(f)$, $f \in \mathbb{Z}(F)$. It follows that

$$\hat{x}(f) = \frac{\hat{z}(f)}{j2\pi f}, \qquad f \neq 0, \qquad f \in \mathbb{Z}(F),$$

so that

$$\hat{x}(f) = \begin{cases} -1/\pi^2 f^2 & \text{for } f/F \text{ odd,} \\ 0 & \text{for } f/F \text{ even,} \end{cases} \qquad f \neq 0, \qquad f \in \mathbb{Z}(F).$$

This agrees with what we found before, but leaves $\hat{x}(0)$ undetermined. Direct calculation yields

$$\hat{x}(0) = \int_0^P x(t)\, dt = P^2/4 = 1/4F^2. \qquad\qquad \blacksquare$$

In Tables 7.1 and 7.2 lists are given of standard DDFT and CDFT pairs.

In the following we use the converse convolution property of the CDFT to explain Gibbs's phenomenon.

7.3.11.   Example: Explanation of Gibbs's phenomenon.   Let $\hat{x}$ be the CDFT of the time signal $x \in \mathcal{L}_2[0, P]$. Then Gibbs's phenomenon is the fact that the truncated sum signal x_N, defined by

TABLE 7.1   STANDARD DDFT PAIRS / FINITE FOURIER SERIES EXPANSIONS

Time signal $x(n),\, n \in \underline{N}$	DDFT, Fourier coefficients $\hat{x}(f),\, f \in \underline{N}(1/N)$	Conditions
$x(n)$	$\displaystyle\sum_{n=0}^{N-1} x(n)e^{-j2\pi fn}$	
$\displaystyle\sum_{f \in \underline{N}(1/N)} \hat{x}(f)e^{j2\pi fn}$	$\hat{x}(f)$	
$\begin{cases} 1, & n = 0 \\ 0, & \text{otherwise} \end{cases}$	1	
$\begin{cases} 1, n = 0, 1, \cdots, M-1 \\ 0, \text{otherwise} \end{cases}$	$e^{-j\pi f(M-1)}d_M(f)$	$M \in \underline{N}$
1	$\begin{cases} N, & f = 0 \\ 0, & \text{otherwise} \end{cases}$	
$\begin{cases} a^n, & n = 0, 1, \cdots, M-1 \\ 0, & \text{otherwise} \end{cases}$	$\dfrac{1 - a^M e^{-j2\pi fM}}{1 - ae^{j2\pi f}}$	$M \in \underline{N},\, a \in \mathbb{C}$
$\begin{cases} e^{j2\pi f_o n}, & n = 0, 1, \cdots, M-1 \\ 0, & \text{otherwise} \end{cases}$	$e^{-j\pi(f-f_o)(M-1)}d_M(f - f_o)$	$M \in \underline{N},\, f_o \in \mathbb{R}$
$e^{j2\pi f_o n}$	$\begin{cases} N, & f = f_o \\ 0, & \text{otherwise} \end{cases}$	$f_o \in \underline{N}(1/N)$

Notes:

(1) Dirichlet's kernel d_M is defined as $d_M(\xi) = M\dfrac{\text{sinc}\,(\pi\xi/M)}{\text{sinc}\,(\pi\xi)}$, $\xi \in \mathbb{R}$ (see 7.3.11).

(2) If the sequence x defined on $\underline{N}$ has the DDFT $\hat{x}(f),\, f \in \underline{N}(1/N)$, the sampled signal x^* defined on $\underline{N}(T)$ by

$$x^*(nT) = x(n), \qquad n \in \underline{N},$$

has the sampled DDFT $\hat{x}^*$ defined on $\underline{N}(1/NT)$ by

$$\hat{x}^*(f) = T\,\hat{x}(fT), \qquad f \in \underline{N}(1/NT).$$

$$x_N(t) = F \sum_{k=-N}^{N} \hat{x}(kF)e^{j2\pi kFt}, \qquad t \in [0, P), \tag{1}$$

does not necessarily converge uniformly to x as $N \to \infty$. In particular, if x exhibits a jump at some time t_o, the partial sum x_N *always* overshoots and undershoots x near t_o by an amount that does not decrease, no matter how large N is taken. Gibbs's phenomenon is illustrated in Fig. 6.9.

TABLE 7.2 STANDARD CDFT PAIRS/INFINITE FOURIER SERIES EXPANSIONS

Name[1]	Time signal $x(t), t \in [0, 1)$	CDFT, Fourier coefficients $\hat{x}(f), f \in \mathbb{Z}$
	$x(t)$	$\int_{0^-}^{1} x(t) e^{-j2\pi ft} \, dt$
	$\sum_{f \in \mathbb{Z}} \hat{x}(f) e^{j2\pi ft}$	$\hat{x}(f)$
Rectangular pulse $a \in [0, 1]$	$\begin{cases} 1, 0 \le t < a \\ 0, a \le t < 1 \end{cases}$	$\begin{cases} a, & f = 0 \\ ae^{-j\pi fa} \operatorname{sinc}(\pi fa), & \text{otherwise} \end{cases}$
Square wave	$\begin{cases} 1, 0 \le t < \frac{1}{2} \\ 0, \frac{1}{2} \le t < 1 \end{cases}$	$\begin{cases} \frac{1}{2}, & f = 0 \\ 0, & f \text{ even}, f \ne 0 \\ -j/f\pi, & f \text{ odd} \end{cases}$
Triangular pulse	$\begin{cases} t, & 0 \le t < \frac{1}{2} \\ 1-t, & \frac{1}{2} \le t < 1 \end{cases}$	$\begin{cases} \frac{1}{4}, & f = 0 \\ 0, & f \text{ even}, f \ne 0 \\ -1/\pi^2 f^2, & f \text{ odd} \end{cases}$
Sawtooth	$t, 0 \le t < 1$	$\begin{cases} \frac{1}{2}, & f = 0, \\ -1/j2\pi f, & \text{otherwise} \end{cases}$
Half-wave rectified sine	$\begin{cases} \sin(2\pi t), 0 \le t < \frac{1}{2} \\ 0, \quad \frac{1}{2} \le t < 1 \end{cases}$	$\begin{cases} \pm\frac{1}{4}j, & f = \pm 1 \\ 0, & f \text{ odd}, f \ne \pm 1 \\ -2/\pi(f^2-1), & f \text{ even} \end{cases}$
Full-wave rectified sine	$|\sin(2\pi t)|$	$\begin{cases} 0, & f \text{ odd} \\ -4/\pi(f^2-1), & f \text{ even} \end{cases}$
Infinite comb	$\sum_{k=-\infty}^{\infty} \delta(t-k)$	1

1. See Fig. 7.5 for plots of the various signals.

Note: If x is defined on $[0, 1)$ with CDFT $\hat{x}$, the CDFT $\hat{x}_p$ of the signal x_p defined on $[0, P)$ by

$$x_p(t) = x(t/P), \qquad t \in [0, P),$$

is given by

$$\hat{x}_p(f) = P\hat{x}(fP), \qquad f \in \mathbb{Z}(1/P).$$

To explain the phenomenon, let $\hat{w}_N$ be the "frequency window" defined by

$$\hat{w}_N(f) = \begin{cases} 1 & \text{for } -NF \le f \le NF, \\ 0 & \text{otherwise,} \end{cases} \qquad f \in \mathbb{Z}(F).$$

A plot of $\hat{w}_N$ is given in Fig. 7.6. The inverse CDFT w_N of $\hat{w}_N$ follows as

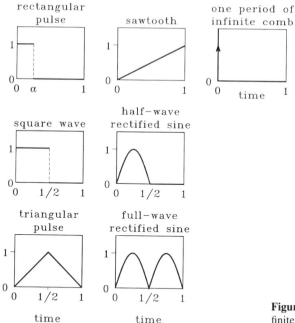

Figure 7.5. Plots of various standard finite-time signals.

$$w_N(t) = F \sum_{k=-N}^{N} e^{j2\pi kFt} = F\frac{e^{-j2\pi NFt} - e^{j2\pi(N+1)Ft}}{1 - e^{j2\pi Ft}}$$

$$= \frac{2N + 1}{P} \frac{\text{sinc}\,(\pi\,(2N + 1)t/P)}{\text{sinc}\,(\pi t/P)},$$

$$= \frac{1}{P} d_{2N+1}(t/P), \qquad t \in [0, P).$$

The function d_M, given by

$$d_M(t) = M\frac{\text{sinc}\,(\pi Mt)}{\text{sinc}\,(\pi t)}, \qquad t \in \mathbb{R},$$

with M an integer, is known as *Dirichlet's kernel*. Plots of d_M for several odd values of M are given in Fig. 7.7. For odd M the function d_M is periodic with period 1 and

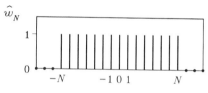

Figure 7.6. The rectangular window $\hat{w}_N$.

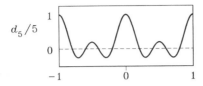

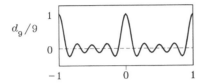

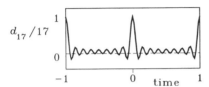

Figure 7.7. Dirichlet's kernel d_M for $M = 5, 9,$ and 17.

has unit area on any interval of length 1. As M approaches infinity with M odd the sequence d_M approaches an infinite comb of δ-functions. The function d_M has its zero crossings at those integral multiples of $1/M$ that are not integers.

To explain Gibbs's phenomenon we note that x_N as given by (1) is the inverse CDFT of

$$\hat{x}_N := \hat{x} \cdot \hat{w}_N.$$

By the converse convolution property it follows that

$$x_N = w_N \odot x.$$

Thus, x_N is obtained by convolving x cyclically with Dirichlet's kernel. If at a certain time instant t the signal x exhibits a jump, cyclical convolution with w_N causes a "ringing" effect with an associated undershoot and overshoot as observed in Example 6.4.13. As N increases, the ringing effect is compressed in time but otherwise does not change shape very much. ∎

Symmetry Properties of the DDFT and CDFT

The DDFT and the CDFT possess a number of symmetry properties. We first review some definitions.

7.3.12.   Definition: Conjugate symmetric, even and odd signals.

(a) *Infinite-axis signals.* Let x be a complex-valued signal on the infinite signal axis $\mathbb{T} = \mathbb{Z}$ or $\mathbb{T} = \mathbb{R}$. Then, x is *conjugate symmetric* if $x(-t) = \overline{x(t)}$ for all $t \in \mathbb{T}$,

x is *even* if $x(-t) = x(t)$ for all $t \in \mathbb{T}$, and

x is *odd* if $x(-t) = -x(t)$ for all $t \in \mathbb{T}$.

(b) *Finite-axis signals.* A complex-valued signal defined on a finite axis such as $\underline{N}$ or $[0, P)$ is conjugate symmetric, even or odd if its *periodic continuation* is conjugate symmetric, even or odd. ∎

Periodic continuation is defined in Section 6.4 (Fig. 6.7.) Figure 7.8 gives examples of even and odd infinite- and finite-time signals. The definitions may of course equally well be applied to time signals and frequency signals.

In what follows we observe that the symmetry properties of finite-axis signals may be interpreted cyclically.

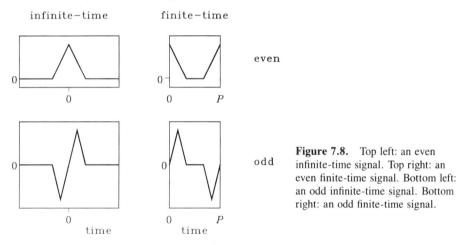

Figure 7.8. Top left: an even infinite-time signal. Top right: an even finite-time signal. Bottom left: an odd infinite-time signal. Bottom right: an odd finite-time signal.

7.3.13.   Exercise: Symmetry of finite-axis signals.   Suppose that x is a discrete-time signal on the finite time axis $\mathbb{T} = \underline{N} = \{0, 1, \cdots, N - 1\}$ or a continuous-time signal defined on $\mathbb{T} = [0, P)$. In the former case, let P denote N.

(a) Prove that x is conjugate symmetric if and only if it is *cyclically conjugate symmetric*, that is, $x(0)$ is real and

$$x(P - t) = \overline{x(t)} \qquad \text{for} \quad t \neq 0, t \in \mathbb{T}.$$

(b) Similarly, prove that x is even if and only if it is *cyclically even*, that is,

$$x(P - t) = x(t) \qquad \text{for} \quad t \neq 0, t \in \mathbb{T},$$

and odd if and only if it is *cyclically odd*, that is, $x(0) = 0$ and

$$x(P - t) = -x(t) \qquad \text{for} \quad t \neq 0, t \in \mathbb{T}.$$ ∎

The symmetry properties of the DDFT and CDFT may be summarized as follows. They are of course closely related to the symmetry properties 6.4.17 of the finite and infinite Fourier series expansions.

7.3.14.   Summary: Symmetry properties of the DDFT and CDFT.   Let x be a time signal defined on $\underline{N}$ or $[0, P)$ and $\hat{x}$ its DDFT or CDFT, respectively. Then

(a) x is real if and only if $\hat{x}$ is conjugate symmetric,

(b) x is conjugate symmetric if and only if $\hat{x}$ is real,

(c) x is even if and only if $\hat{x}$ is even, and

(d) x is odd if and only if $\hat{x}$ is odd.   ∎

The proof is not difficult and is left as an exercise.

7.3.15.   Example: Symmetry of the triangular pulse and its CDFT.
     (a) The triangular pulse of Example 7.3.10 is real. Hence, by 7.3.14(a) its CDFT is conjugate symmetric.

     (b) The triangular pulse is real and (cyclically) even, so that it is conjugate symmetric. It follows from 7.3.14(b) that its CDFT is real.

     (c) Because the triangular pulse is (cyclically) even, by 7.3.14(c) its CDFT is also even.   ∎

7.3.16.   Exercise: Time and frequency reversal.   If x is a signal defined on an infinite axis $\mathbb{T}$, the *time reversed* signal $\mu^{-1}x$ is the signal given by

$$(\mu^{-1}x)(t) = x(-t), \qquad t \in \mathbb{T}.$$

If x is defined on a finite time axis, time reversal consists of time reversal of the periodic continuation and restricting the result to the original time axis. Time reversal of finite-time signals is equivalent to *cyclical* time reversal. Frequency reversal is defined like time reversal.
     Prove that both for the DDFT and the CDFT time reversal transforms into frequency reversal, i.e., if x transforms into $\hat{x}$, then $\mu^{-1}x$ transforms into $\mu^{-1}\hat{x}$.   ∎

Generalized CDFT

According to 6.4.20 periodic generalized signals may be expanded in a generalized infinite Fourier series whose coefficients grow polynomially. It follows that the CDFT transforms one period of a periodic generalized signal into a frequency signal of polynomial growth. This is the *generalized CDFT*. To allow for singularities at time 0, it is useful to redefine the CDFT $\hat{x}$ of the signal x as

$$\hat{x}(f) = \int_{0^-}^{P} x(t)e^{-j2\pi ft}\, dt = \lim_{\epsilon \downarrow 0} \int_{-\epsilon}^{P} x(t)e^{-j2\pi ft}\, dt, \qquad f \in \mathbb{Z}(F),$$

where $F = 1/P$.

7.3.17.   Example: Generalized CDFT of a delta function. The signal x given by

$$x(t) = \delta(t), \qquad t \in [0^-, P),$$

may be viewed as one period of the infinite comb

$$w_P(t) = \sum_{k=-\infty}^{\infty} \delta(t - kP), \qquad t \in \mathbb{R}.$$

By 6.4.21 the coefficients of the infinite Fourier series expansion of the infinite comb are all 1. Hence, the CDFT of the signal x is

$$\hat{x}(f) = 1, \qquad f \in \mathbb{Z}(F),$$

with $F = 1/P$. ∎

Application of the DDFT and CDFT to Systems Analysis

The DDFT and CDFT may be used in the analysis of the response of convolution systems to *periodic* inputs. As seen in Section 3.8, the response of a discrete-time convolution system on the time axis $\mathbb{Z}$ to a bounded periodic input with period N exists and is periodic if the impulse response h of the system has finite action. One period Y of the output may be obtained by discrete-time cyclical convolution as

$$Y = H \odot U, \tag{2}$$

where U is one period of the input and H one period of the periodic extension h_{per} of the impulse response h. Because by 6.5.4 the finite Fourier series is a spectral expansion for the cyclical convolution, the DDFT is a *spectral transform* for the cyclical convolution system (2). By the convolution property of the DDFT, the transformed IO map of the system is

$$\hat{Y} = \hat{H}\hat{U}, \tag{3}$$

with $\hat{\ }$ denoting the DDFT.

Similarly, the response of a continuous-time convolution system to a periodic input may be described by continuous-time cyclical convolution. The CDFT is the spectral transform for the resulting continuous-time convolution system, again with a transformed IO map of the form (3).

7.3.18.   Summary: The DDFT and CDFT are spectral transforms for cyclical convolution systems.

The DDFT is a spectral transform for the discrete-time cyclical convolution system

$$Y = H \odot U,$$

defined on the time axis $\underline{N}$. The transformed IO map is

$$\hat{Y} = \hat{H}\hat{U}.$$

The CDFT is a spectral transform for the continuous-time cyclical convolution system

$$Y = H \odot U,$$

defined on the time axis $[0, P)$. The transformed IO map is

$$\hat{Y} = \hat{H}\hat{U}. \qquad\blacksquare$$

The response of a discrete- or continuous-time convolution system to a periodic input thus may be obtained as follows:

(i)   Find the DDFT or CDFT $\hat{U}$ of one period U of the input,

(ii)   multiply this by the DDFT or CDFT $\hat{H}$ of one period H of the periodic extension h_{per} of the impulse response of the system, and

(iii) determine one period Y of the output as the inverse DDFT or CDFT of $\hat{Y} = \hat{H}\hat{U}$.

By 3.8.2 the DDFT or CDFT $\hat{H}$ of one period H of the periodic extension h_{per} of the impulse response h of the system coincides with the frequency response function $\hat{h}$ of the system on the appropriate frequency axis. Hence,

$$\hat{Y}(f) = \hat{h}(f)\hat{U}(f), \qquad f \in \underline{N}(F) \quad \text{or} \quad f \in \mathbb{Z}(F).$$

Because $\hat{U}$ is nothing but the Fourier coefficients of the periodic input and $\hat{Y}$ those of the output this analysis is completely equivalent to the method described in 6.5 to obtain the response of convolution systems to periodic inputs by Fourier series expansion.

Figure 7.9 illustrates the role of the DDFT and CDFT in the analysis of the response of convolution systems to periodic inputs. Figure 7.10 is the specialization of this figure to the case where the input is the unit periodic input Δ_{per} or δ_{per} defined in Section 3.8.

We end this section with a brief review of the DDFT for sampled signals.

7.3.19.   Review: The DDFT of sampled signals. The sampled DDFT is the transform

$$\mathscr{F}_{\text{DD}} : \ell_{\underline{N}}(T) \to \ell_{\underline{N}}(F),$$

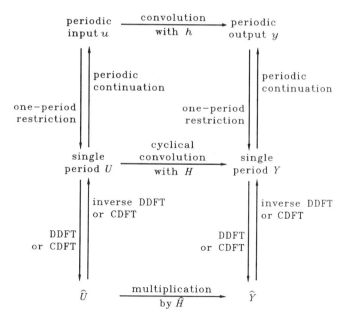

Figure 7.9.   The response of a convolution system to a periodic input may be obtained directly by regular or cyclical convolution, or by application of the DDFT or CDFT.

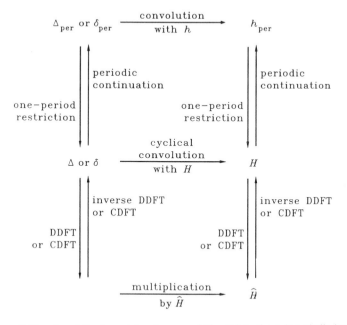

Figure 7.10.   Specialization of the diagram of Fig. 7.9 to the unit periodic input Δ_{per} or δ_{per}.

with $F = 1/NT$, that transforms the sampled signal x defined on the finite time axis $\underline{N}(T) = \{0, T, \cdots, (N-1)T\}$ to the frequency signal $\hat{x}$ given by

$$\hat{x}(f) = T \sum_{t \in N(T)} x(t)e^{-j2\pi ft}, \qquad f \in \underline{N}(F).$$

The formula for the inverse DDFT is almost the same as in 7.3.2, namely,

$$x(t) = F \sum_{f \in N(F)} \hat{x}(f)e^{j2\pi ft}, \qquad t \in \underline{N}(T).$$

Unitarity.   The sampled DDFT is unitary, that is, $\langle x, y \rangle_{\ell_N(T)} = \langle \hat{x}, \hat{y} \rangle_{\ell_N(F)}$, where the inner products on the signal spaces $\ell_{\underline{N}}(T)$ and $\ell_{\underline{N}}(F)$ are defined as

$$\langle x, y \rangle_{\ell_{\underline{N}}(T)} = T \sum_{t \in N(T)} x(t)\overline{y(t)},$$

$$\langle \hat{x}, \hat{y} \rangle_{\ell_{\underline{N}}(F)} = F \sum_{f \in N(F)} \hat{x}(f)\overline{\hat{y}(f)}.$$

Linearity.   The sampled DDFT and its inverse are linear.

Convolution property.   The sampled DDFT has the convolution property $(x \odot y)\hat{} = \hat{x} \cdot \hat{y}$ and the converse convolution property $(x \cdot y)\hat{} = \hat{x} \odot \hat{y}$, provided the circular convolution of two signals x and y on the time axis $\underline{N}(T)$ is defined as

$$(x \odot y)(t) = T \sum_{\tau \in N(T)} x((t - \tau) \bmod NT)y(\tau), \qquad t \in \underline{N}(T).$$

Shift Property.   The shift property holds in the sense that the shifted signal

$$x((t + \theta) \bmod NT), \qquad t \in \underline{N}(T),$$

for any $\theta \in \mathbb{Z}(T)$ transforms into

$$e^{j2\pi f\theta}\hat{x}(f), \qquad f \in \underline{N}(F).$$

Conversely, for every $\phi \in \mathbb{Z}(F)$, the signal

$$e^{j2\pi\phi t}x(t), \qquad t \in \underline{N}(T),$$

transforms into

$$\hat{x}((f - \phi) \bmod NF), \qquad f \in \underline{N}(F).$$

Periodicity Property.   When taken on the infinite frequency axis $\mathbb{Z}(F)$, the DDFT $\hat{x}$ of x is periodic with period NF. Conversely, when taken on the infinite time axis $\mathbb{Z}(T)$, the inverse DDFT x of $\hat{x}$ is periodic with period NT.

Symmetry.   The symmetry properties of 7.3.14 also apply to the sampled DDFT.   ∎

7.4 THE DCFT AND THE CCFT

In this section we explain how aperiodic *infinite-time* signals may be expanded in a *continuum* of harmonics. This leads to the *discrete-to-continuous Fourier transform* (DCFT) and the *continuous-to-continuous Fourier transform* (CCFT) as the corresponding expansion transforms. These transforms are *spectral transforms* for (regular) convolution systems.

The section is mainly devoted to the expansion and transformation of discrete- and continuous-time signals with *finite energy*. At the end we briefly consider the *generalized DCFT*, which extends the DCFT to signals of polynomial growth, and the generalized CCFT, which is the extension of the CCFT to signals of polynomial growth and generalized signals. A more elaborate discussion of the generalized CCFT is found in Supplement C.

Fourier Integral Expansion

By the infinite Fourier series the finite-energy continuous-time signal x defined on the time axis $[0, P)$ may be expanded as

$$x(t) = F \sum_{f \in \mathbb{Z}(F)} \hat{x}(f) e^{j2\pi ft}, \qquad t \in [0, P), \tag{1}$$

where

$$\hat{x}(f) = \int_0^P x(t) e^{-j2\pi ft}\, dt, \qquad f \in \mathbb{Z}(F). \tag{2}$$

This shows how by (1) a *finite-time* continuous-time signal may be expanded as a *countably infinite* linear combination of harmonics. Inspection reveals that conversely by (2) the frequency signal $\hat{x}$, defined on the *countably infinite* signal axis $\mathbb{Z}(F)$, is expanded as a linear combination of a *continuum* of harmonics with the "parameter" t ranging over the finite interval $[0, P)$.

By interchanging time and frequency, this observation allows expanding discrete-time signals defined on the infinite time axis $\mathbb{Z}$ as a linear combination of harmonics whose frequencies range over the interval $[0, 1)$. The details are given in

E.6 in Supplement E. The result is that the finite-energy discrete-time signal x defined on the time axis $\mathbb{Z}$ may be expanded as

$$x(n) = \int_0^1 \hat{x}(f)e^{j2\pi fn}\, df, \qquad n \in \mathbb{Z}, \tag{3}$$

where the "expansion coefficients" $\hat{x}$ may be obtained as

$$\hat{x}(f) = \sum_{n=-\infty}^{\infty} x(n)e^{-j2\pi fn}, \qquad f \in [0, 1).$$

The expansion (3) has the form of an integral, and is called a *Fourier integral expansion*.

Also finite-energy *continuous-time* signals defined on the infinite time-axis $\mathbb{R}$ may be expanded by a Fourier integral expansion. The proof is outside the scope of this book, but the result may easily be made plausible. By the infinite Fourier series expansion a finite-energy continuous-time signal x defined on the time axis $[0, P)$ may be expanded as

$$x(t) = F \sum_{f \in \mathbb{Z}(F)} \hat{x}(f)e^{j2\pi ft}, \qquad t \in [0, P),$$

where $F = 1/P$, and the expansion coefficients $\hat{x}$ are given by

$$\hat{x}(f) = \int_0^P x(t)e^{-j2\pi ft}\, dt, \qquad f \in \mathbb{Z}(F).$$

By the periodicity property of the CDFT, this expansion may equally well be used to expand a finite-energy continuous-time signal x defined on the finite time-axis $[-P/2, P/2)$ as

$$x(t) = F \sum_{f \in \mathbb{Z}(F)} \hat{x}(f)e^{j2\pi ft}, \qquad t \in [-P/2, P/2), \tag{4}$$

where $F = 1/P$ and

$$\hat{x}(f) = \int_{-P/2}^{P/2} x(t)e^{-j2\pi ft}\, dt, \qquad f \in \mathbb{Z}(F).$$

Suppose now that the finite-energy continuous-time signal x is actually defined on the infinite continuous time axis $\mathbb{R}$. Then (4) may be used to expand x on any finite interval $[-P/2, P/2)$. By letting P approach ∞ we may hope to obtain an expansion of x on the *entire* time axis $\mathbb{R}$. As P approaches ∞, its reciprocal F approaches 0, and it is plausible that the infinite sum (4) approaches the *integral*

$$x(t) = \int_{-\infty}^{\infty} \hat{x}(f)e^{j2\pi ft}\, df, \qquad t \in \mathbb{R},$$

where

$$\hat{x}(f) = \int_{-\infty}^{\infty} x(t)e^{-j2\pi ft}\, dt, \qquad f \in \mathbb{R}.$$

This result is indeed correct, and is known as the *Fourier integral*. It is the Fourier integral expansion for a finite-energy continuous-time signal defined on the time axis $\mathbb{R}$.

7.4.1.   Summary: Fourier integral expansions.

Discrete-time Fourier integral expansion: Let x be a finite-energy signal defined on the infinite discrete time axis $\mathbb{Z}$. Then the expansion of x in harmonics is of the form

$$x(n) = \int_{0}^{1} \hat{x}(f)e^{j2\pi fn}\, df, \qquad n \in \mathbb{Z},$$

where the finite-energy "expansion coefficients" $\hat{x}$ are given by

$$\hat{x}(f) = \sum_{n\in\mathbb{Z}} x(n)e^{-j2\pi fn}, \qquad f \in [0, 1).$$

Continuous-time Fourier integral expansion: Let x be a finite-energy signal defined on the infinite continuous time axis $\mathbb{R}$. Then the expansion of x in harmonics takes the form

$$x(t) = \int_{-\infty}^{\infty} \hat{x}(f)e^{j2\pi ft}\, df, \qquad t \in \mathbb{R},$$

where the finite-energy "expansion coefficients" $\hat{x}$ are given by

$$\hat{x}(f) = \int_{-\infty}^{\infty} x(t)e^{-j2\pi ft}\, dt, \qquad f \in \mathbb{R}. \blacksquare$$

In the integral expansions of x, both in the discrete- and the continuous-time case the harmonic $e^{j2\pi ft}$ with frequency f has the infinitesimal coefficient $\hat{x}(f)\, df$. Thus, $\hat{x}(f)$ may be thought of as the "density," per unit frequency, of the harmonics at the frequency f.

The DCFT and the CCFT

Before looking at some examples of Fourier integral expansions we introduce the *transforms* corresponding to the expansions of 7.4.1.

7.4.2.   Definition: The DCFT and the CCFT.

The transform

$$\mathscr{F}_{DC}\colon \ell_2 \;\rightarrow\; \mathscr{L}_2[0, 1),$$

that maps the finite-energy discrete-time signal x with time axis $\mathbb{Z}$ into its "Fourier coefficients" $\hat{x}$ given by

$$\hat{x}(f) = \sum_{n \in \mathbb{Z}} x(n)e^{-j2\pi fn}, \qquad f \in [0, 1),$$

is called the *discrete-to-continuous Fourier transform* (DCFT).

The transform

$$\mathscr{F}_{CC}\colon \mathscr{L}_2 \;\rightarrow\; \mathscr{L}_2$$

that maps the finite-energy continuous-time signal x with time axis $\mathbb{R}$ into its "Fourier coefficients" $\hat{x}$ given by

$$\hat{x}(f) = \int_{-\infty}^{\infty} x(t)e^{-j2\pi ft}\, dt, \qquad f \in \mathbb{R},$$

is called the *continuous-to-continuous Fourier transform* (CCFT). ∎

Examples

We consider the following examples of the DCFT and the CCFT.

7.4.3.   Examples: DCFT and CCFT.
(a) *DCFT of the unit pulse.* The unit pulse Δ is the discrete-time signal

$$\Delta(n) = \begin{cases} 1 & \text{for } n = 0, \\ 0 & \text{otherwise,} \end{cases} \qquad n \in \mathbb{Z}.$$

Its DCFT is immediately seen to be given by

$$\hat{\Delta}(f) = 1, \qquad f \in [0, 1).$$

Thus, the frequency content of the unit pulse is evenly distributed over all frequencies.

(b) *DCFT of a rectangular pulse.* Let x be the rectangular discrete-time pulse given by

$$x(n) = \begin{cases} 1 & \text{for } 0 \le n \le M - 1, \\ 0 & \text{otherwise,} \end{cases} \qquad n \in \mathbb{Z},$$

with M a natural number. A plot of x is given in Fig. 7.11. The DCFT of this signal is

$$\hat{x}(f) = \sum_{n=0}^{M-1} e^{-j2\pi fn} = \frac{1 - e^{-j2\pi fM}}{1 - e^{-j2\pi f}} = e^{-j\pi f(M-1)} M \frac{\text{sinc}\,(\pi fM)}{\text{sinc}\,(\pi f)}$$

$$= e^{-j\pi f(M-1)} d_M(f), \qquad 0 \le f < 1,$$

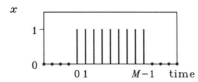

Figure 7.11. A discrete-time rectangular pulse of length M.

where d_M is Dirichlet's kernel as defined in 7.3.11. Plots of the magnitude of the DCFT $\hat{x}$ are given in Fig. 7.12 for three values of M. As M increases, the peaks contract near the ends of the frequency interval.

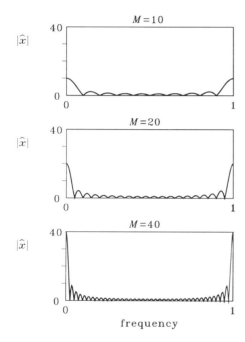

Figure 7.12. Magnitude of the DCFT $\hat{x}$ of the rectangular pulse for $M = 10, 20,$ and 40.

(c) *The CCFT of a one-sided exponential signal.* As a third example, consider the one-sided exponential continuous-time signal x given by

$$x(t) = e^{-t/\theta}\mathbb{1}(t), \qquad t \in \mathbb{R},$$

with $\theta > 0$. Its CCFT $\hat{x}$ is given by

$$\hat{x}(f) = \int_0^\infty e^{-t/\theta} e^{-j2\pi ft}\,dt = \frac{1}{\dfrac{1}{\theta} + j2\pi f}, \qquad f \in \mathbb{R}.$$

Plots of x and $\hat{x}$ are given in Figs. 7.13 and 7.14, respectively, where $\hat{x}$ is plotted as a function

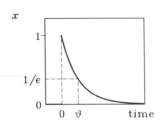

Figure 7.13.   A one-sided exponential signal.

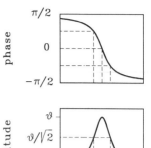

Figure 7.14.   CCFT of the one-sided exponential signal. Top: phase. Bottom: magnitude.

$$\hat{x}(\omega/2\pi) = \frac{1}{\dfrac{1}{\theta} + j\omega}, \qquad \omega \in \mathbb{R},$$

of the angular frequency $\omega = 2\pi f$. At the angular frequency $|\omega| = 1/\theta$ the magnitude

$$\left| \hat{x}(\omega/2\pi) \right| = \frac{1}{\sqrt{1/\theta^2 + \omega^2}}, \qquad \omega \in \mathbb{R},$$

of $\hat{x}$ assumes a value that is a fraction $1/\sqrt{2}$ of the peak value at $\omega = 0$. Thus, the shorter the time signal is (i.e., the smaller the time constant θ), the wider is its CCFT. This is also true for signals with a different shape.   ∎

Examples of well-known DCFT and CCFT pairs are listed in Tables 7.3 and 7.4, respectively.

Inverse and Unitariness of the DCFT and CCFT

The DCFT and CCFT are expansion transforms, so that their inverses may immediately be established.

7.4.4.   Summary: Inverses of the DCFT and CCFT.

The *inverse DCFT*   The *inverse CCFT*

$$\mathscr{F}_{\mathrm{DC}}^{-1}\colon \mathscr{L}_2[0,\,1) \;\to\; \ell_2 \qquad\qquad\qquad \mathscr{F}_{\mathrm{CC}}^{-1}\colon \mathscr{L}_2 \;\to\; \mathscr{L}_2$$

maps the frequency signal $\hat{x}$ back into the discrete-time signal $x = \mathcal{F}_{DC}^{-1}(\hat{x})$ according to

maps the frequency signal $\hat{x}$ back into the continuous-time signal $x = \mathcal{F}_{CC}^{-1}(\hat{x})$ according to

$$x(n) = \int_0^1 \hat{x}(f)e^{j2\pi fn}\,df, \qquad n \in \mathbb{Z}.$$

$$x(t) = \int_{-\infty}^{\infty} \hat{x}(f)e^{j2\pi ft}\,df, \qquad t \in \mathbb{R}. \quad \blacksquare$$

Like the DDFT and CDFT, the DCFT and CCFT are *unitary* transforms.

TABLE 7.3 STANDARD DCFT PAIRS

Time signal $x(n),\, n \in \mathbb{Z}$	DCFT $\hat{x}(f),\, f \in [0^-, 1)$ or $f \in \mathbb{R}$	Conditions
$x(n)$	$\displaystyle\sum_{n=-\infty}^{\infty} x(n)e^{-j2\pi fn}$	
$\displaystyle\int_{0^-}^{1} \hat{x}(f)e^{j2\pi fn}\,df$	$\hat{x}(f)$	
$\begin{cases} 1,\ n=0 \\ 0,\ \text{otherwise} \end{cases}$	1	
$\begin{cases} 1,\ n=0,1,\cdots,M-1 \\ 0,\ \text{otherwise} \end{cases}$	$e^{-j\pi f(M-1)}d_M(f)$	$M \in \mathbb{N}$
1	$\displaystyle\sum_{k=-\infty}^{\infty} \delta(f-k)$	
$\begin{cases} a^n,\ n=0,1,\cdots,M-1 \\ 0,\ \text{otherwise} \end{cases}$	$\dfrac{1-a^M e^{-j2\pi fM}}{1-ae^{-j2\pi f}}$	$a \in \mathbb{C},\, M \in \mathbb{N}$
$a^n 1(n)$	$\dfrac{1}{1-ae^{-j2\pi f}}$	$a \in \mathbb{C},\, \lvert a \rvert < 1$
$-a^n 1(-n-1)$	$\dfrac{1}{1-ae^{-j2\pi f}}$	$a \in \mathbb{C},\, \lvert a \rvert > 1$
$\begin{cases} e^{j2\pi f_o n},\ n=0,1,\cdots,M-1 \\ 0,\ \text{otherwise} \end{cases}$	$e^{-j\pi(f-f_o)(M-1)}\,d_M(f-f_o)$	$M \in \mathbb{N},\, f_o \in \mathbb{R}$
$e^{j2\pi f_o n}$	$\displaystyle\sum_{k=-\infty}^{\infty} \delta(f-f_o-k)$	$f_o \in \mathbb{R}$

Notes:

(1) Dirichlet's kernel d_M is defined as $d_M(\xi) = M\,\dfrac{\text{sinc}\,(\pi\xi/M)}{\text{sinc}\,(\pi\xi)}$, $\xi \in \mathbb{R}$ (see 7.3.11).

(2) If the sequence x defined on $\mathbb{Z}$ has the DCFT $\hat{x}$ defined on $[0, 1)$, the sampled signal x^* defined on $\mathbb{Z}(T)$ by

$$x^*(nT) = x(n), \qquad n \in \mathbb{Z},$$

has the sampled DCFT $\hat{x}^*$ defined on $[0, 1/T)$ by

$$\hat{x}^*(f) = T\,\hat{x}(fT), \qquad f \in [0, 1/T).$$

TABLE 7.4 STANDARD REGULAR CCFT PAIRS

Time signal $x(t), t \in \mathbb{R}$	CCFT $\hat{x}(f), f \in \mathbb{R}$	Conditions
$x(t)$	$\displaystyle\int_{-\infty}^{\infty} x(t)e^{-j2\pi ft}\,dt$	
$\displaystyle\int_{-\infty}^{\infty} \hat{x}(f)e^{j2\pi ft}\,df$	$\hat{x}(f)$	
$e^{-at}\mathbb{1}(t)$	$\dfrac{1}{j2\pi f + a}$	$a \in \mathbb{C}, \operatorname{Re}(a) > 0$
$-e^{-at}\mathbb{1}(-t)$	$\dfrac{1}{j2\pi f + a}$	$a \in \mathbb{C}, \operatorname{Re}(a) < 0$
$\dfrac{t^{k-1}e^{-at}}{(k-1)!}\mathbb{1}(t)$	$\dfrac{1}{(j2\pi f + a)^k}$	$k \in \mathbb{N}, a \in \mathbb{C}, \operatorname{Re}(a) > 0$
$-\dfrac{t^{k-1}e^{-at}}{(k-1)!}\mathbb{1}(-t)$	$\dfrac{1}{(j2\pi f + a)^k}$	$k \in \mathbb{N}, a \in \mathbb{C}, \operatorname{Re}(a) < 0$
$e^{-a|t|}$	$\dfrac{2a}{4\pi^2 f^2 + a^2}$	$a \in \mathbb{C}, \operatorname{Re}(a) > 0$
$\dfrac{1}{a^2 + t^2}$	$\dfrac{\pi}{a}e^{-2\pi a|f|}$	$a \in \mathbb{C}, \operatorname{Re}(a) > 0$
e^{-at^2}	$\sqrt{\dfrac{\pi}{a}}\,e^{-\pi^2 f^2/a}$	$a \in \mathbb{R}, a > 0$
$e^{-at}\cos(\omega t)\mathbb{1}(t)$	$\dfrac{j2\pi f + a}{(j2\pi f + a)^2 + \omega^2}$	$a, \omega \in \mathbb{R}, a > 0$
$e^{-at}\sin(\omega t)\mathbb{1}(t)$	$\dfrac{\omega}{(j2\pi f + a)^2 + \omega^2}$	$a, \omega \in \mathbb{R}, a > 0$
$\begin{cases} 1, & -a \leq t < a \\ 0, & \text{otherwise} \end{cases}$	$2a\,\operatorname{sinc}(2\pi fa)$	$a \in \mathbb{R}, a \geq 0$
$2a\,\operatorname{sinc}(2\pi at)$	$\begin{cases} 1, & -a \leq f < a \\ 0, & \text{otherwise} \end{cases}$	$a \in \mathbb{R}, a \geq 0$
$\begin{cases} 1 - |t|/a, & -a \leq t < a \\ 0, & \text{otherwise} \end{cases}$	$a\,\operatorname{sinc}^2(\pi fa)$	$a \in \mathbb{R}, a \geq 0$

7.4.5.   Summary: Unitarity of the DCFT and CCFT.

Define the inner product on the signal space ℓ_2 as

$$\langle x, y \rangle_{\ell_2} = \sum_{n \in \mathbb{Z}} x(n)\overline{y(n)},$$

and on the transformed space $\mathcal{L}_2[0, 1)$ as

Define the inner product on the signal and transformed space $\mathcal{L}_2$ as

$$\langle x, y \rangle_{\mathcal{L}_2} = \int_{-\infty}^{\infty} x(t)\overline{y(t)}\,dt,$$

and

$$\langle \hat{x}, \hat{y} \rangle_{\mathscr{L}_2[0,1)} = \int_0^1 \hat{x}(f) \overline{\hat{y}(f)} \, df. \qquad\qquad \langle \hat{x}, \hat{y} \rangle_{\mathscr{L}_2} = \int_{-\infty}^{\infty} \hat{x}(f) \overline{\hat{y}(f)} \, df.$$

Then, $\langle x, y \rangle_{\ell_2} = \langle \hat{x}, \hat{y} \rangle_{\mathscr{L}_2[0,1)}$, that is, the   Then, $\langle x, y \rangle_{\mathscr{L}_2} = \langle \hat{x}, \hat{y} \rangle_{\mathscr{L}_2}$, that is, the
DCFT is a unitary transform.   CCFT is a unitary transform.   ∎

Like in the case of the DDFT and the CDFT, the unitarity of the DCFT and CCFT
is known as *Plancherel's identity*. It implies norm preservation, also known as *Parseval's identity*. Specifically, we have

$$\sum_{n \in \mathbb{Z}} |x(n)|^2 = \int_0^1 |\hat{x}(f)|^2 \, df$$

for the DCFT and

$$\int_{-\infty}^{\infty} |x(t)|^2 \, dt = \int_{-\infty}^{\infty} |\hat{x}(f)|^2 \, df$$

for the CCFT.

7.4.6.   Example: Parseval's equality.   In Example 7.4.3(b) we found that the
DCFT of the discrete-time rectangular pulse

$$x(n) = \begin{cases} 1 & \text{for } 0 \leq n \leq M - 1, \\ 0 & \text{otherwise,} \end{cases} \qquad n \in \mathbb{Z},$$

is

$$\hat{x}(f) = e^{-j\pi f(M-1)} d_M(f), \qquad f \in [0, 1),$$

with d_M Dirichlet's kernel of order M. On the one hand we have

$$\sum_{n \in \mathbb{Z}} |\hat{x}(n)|^2 = M,$$

while on the other

$$\int_0^1 |\hat{x}(f)|^2 \, df = \int_0^1 |d_M(f)|^2 \, df.$$

Since by Parseval's equality the two expressions are equal, it follows that

$$\int_0^1 |d_M(f)|^2 \, df = M.$$

It is not easy to prove this directly.   ■

Properties of the DCFT and the CCFT

The properties of the DCFT and the CCFT are analogous to those of the DDFT and the CDFT. First of all, both transforms are linear.

7.4.7.   Summary: Linearity of the DCFT and CCFT. The DCFT and CCFT and their inverses are linear transforms.   ■

The DCFT and CCFT possess the convolution and converse convolution properties.

7.4.8.   Summary: Convolution and converse convolution properties of the DCFT and CCFT.

(a) *Convolution property.* Convolution transforms under the DCFT into multiplication, that is, if x transforms into $\hat{x}$ and y into $\hat{y}$, then

$$(x * y)\hat{} = \hat{x} \cdot \hat{y}.$$

(b) *Converse convolution property.* Conversely, products transform under the DCFT into convolutions, that is,

$$(x \cdot y)\hat{} = \hat{x} \odot \hat{y},$$

with $\odot$ the cyclical convolution for signals on $[0, 1)$.

(a′) *Convolution property.* Convolution transforms under the CCFT into multiplication, that is, if x transforms into $\hat{x}$ and y into $\hat{y}$, then

$$(x * y)\hat{} = \hat{x} \cdot \hat{y}.$$

(b′) *Converse convolution property.* Vice-versa, multiplication transforms under the CCFT into convolution, that is,

$$(x \cdot y)\hat{} = \hat{x} * \hat{y}. \qquad ■$$

Also the shift and converse shift properties are shared by the DCFT and CCFT.

7.4.9.   Summary: Shift and converse shift properties of the DCFT and CCFT.

(a) *Shift property.* Under the DCFT, time shifting transforms into multiplication by a harmonic, that is, if x transforms into $\hat{x}$, then for every $\theta \in \mathbb{Z}$

$$x(n + \theta), \qquad n \in \mathbb{Z},$$

transforms into

$$e^{j2\pi f\theta}\hat{x}(f), \qquad f \in [0, 1).$$

(a′) *Shift property.* Under the CCFT, time shifting transforms into multiplication by a harmonic, that is, if x transforms into $\hat{x}$, then for every $\theta \in \mathbb{R}$

$$x(t + \theta), \qquad t \in \mathbb{R},$$

transforms into

$$e^{j2\pi f\theta}\hat{x}(f), \qquad f \in \mathbb{R}.$$

(b) *Converse shift property*. On the other hand, for every $\phi \in \mathbb{R}$

$$e^{j2\pi\phi n}x(n), \qquad n \in \mathbb{Z},$$

transforms under the DCFT into

$$\hat{x}((f - \phi) \bmod 1), \qquad f \in [0, 1).$$

(b') *Converse shift property*. Conversely, for every $\phi \in \mathbb{R}$

$$e^{j2\pi\phi t}x(t), \qquad t \in \mathbb{R},$$

transforms under the CCFT into

$$\hat{x}(f - \phi), \qquad f \in \mathbb{R}. \qquad \blacksquare$$

The DCFT possesses a *periodicity* property.

7.4.10.   Summary: Periodicity property of the DCFT.
The DCFT $\hat{x}$ of x, defined on $\mathbb{R}$ rather than on $[0, 1)$ by

$$\hat{x}(f) = \sum_{n \in \mathbb{Z}} x(n)e^{-j2\pi fn}, \qquad f \in \mathbb{R},$$

is periodic with period 1. $\qquad\qquad\qquad\qquad\qquad\qquad\qquad\qquad\blacksquare$

Finally, both the DCFT and the CCFT enjoy certain differentiation properties.

7.4.11.   Summary: Differentiation properties of the DCFT and CCFT.

(a') *Differentiation property of the CCFT*. Under the CCFT, differentiation with respect to time transforms into multiplication by frequency. Suppose that x transforms to $\hat{x}$. Then, if x is differentiable, its derivative Dx transforms into

$$j2\pi f \cdot \hat{x}(f), \qquad f \in \mathbb{R}.$$

(b) *Converse differentiation property of the DCFT*. Under the DCFT, multiplication by time transforms into differentiation with respect to frequency. Suppose that x transforms into $\hat{x}$, and that $\hat{x}$ is cyclically continuous and differentiable on $[0, 1)$ with derivative $D\hat{x}$. Then

$$-j2\pi n \cdot x(n), \qquad n \in \mathbb{Z},$$

transforms into $D\hat{x}$.

(b') *Converse differentiation property of the CCFT*. Conversely, under the CCFT differentiation with respect to frequency transforms into multiplication by time. Suppose that x transforms into $\hat{x}$, and that $\hat{x}$ is continuous and differentiable with derivative $D\hat{x}$. Then

$$-j2\pi t \cdot x(t), \qquad t \in \mathbb{R},$$

transforms into $D\hat{x}$. $\qquad\qquad\blacksquare$

Cyclical continuity is defined in 7.3.9.

The proofs of the various properties are not very difficult. As for the DDFT and the CDFT, the properties may often be used to simplify the computation of Fourier transforms.

7.4.12.   Example: Application of the differentiation property.   In Example 7.4.3(c) we saw that the continuous-time signal

$$x(t) = e^{-t/\theta}\mathbb{1}(t), \qquad t \in \mathbb{R},$$

with $\theta > 0$, has the CCFT

$$\hat{x}(f) = \frac{1}{\dfrac{1}{\theta} + j2\pi f}, \qquad f \in \mathbb{R}.$$

The derivative $D\hat{x}$ of the CCFT is

$$D\hat{x}(f) = \frac{-j2\pi}{\left(\dfrac{1}{\theta} + j2\pi f\right)^2}, \qquad f \in \mathbb{R}.$$

By the converse differentiation property of the CCFT, this is the CCFT of the signal

$$-j2\pi t \cdot x(t) = -j2\pi t e^{-t/\theta}\mathbb{1}(t), \qquad t \in \mathbb{R}.$$

Dividing by $-j2\pi$, it follows that the signals

$$z(t) = te^{-t/\theta}\mathbb{1}(t), \qquad t \in \mathbb{R},$$

and

$$\hat{z}(f) = \frac{1}{\left(\dfrac{1}{\theta} + j2\pi f\right)^2}, \qquad f \in \mathbb{R},$$

form a CCFT pair.   ∎

The symmetry properties of the DCFT and CCFT are like those of the DDFT and CDFT.

7.4.13.   Summary: Symmetry properties of the DCFT and CCFT.   Let x be a finite-energy signal on the time axis $\mathbb{Z}$ or $\mathbb{R}$, and $\hat{x}$ its DCFT or CCFT, respectively. Then,

(a) x is real if and only if $\hat{x}$ is conjugate symmetric,

(b) x is conjugate symmetric if and only if $\hat{x}$ is real,

(c) x is even if and only if $\hat{x}$ is even, and

(d) x is odd if and only if $\hat{x}$ is odd.   ∎

Convergence Properties of the CCFT

In 6.4.12 we listed some facts about the convergence of the infinite Fourier series expansion. With slight rephrasing, these statements also apply to the convergence of the inverse CDFT. For the inverse CCFT, a number of corresponding statements hold.

7.4.14.   Remarks: Convergence of the inverse CDFT and CCFT.
    (a) *Inverse CDFT.* If x is a finite-energy continuous-time signal defined on the finite time-axis $[0, P)$, then it may be expressed as

$$x(t) = \lim_{N \to \infty} F \sum_{k=-N}^{N} \hat{x}(kF)e^{j2\pi kFt}, \qquad t \in [0, P), \tag{5}$$

with $\hat{x}$ the CDFT of x. The sum on the right-hand side of (5) converges in the $\mathcal{L}_2[0, P)$ sense (i.e., the 2-norm of the difference between x and the truncated sum converges to 0).

    If x has *bounded variation,* as defined in 6.4.12(b), the sum converges *pointwise* at those instants t where x is continuous. At those instants t where x is discontinuous the right-hand side of (5) converges to the midpoint value $[x(t^-) + x(t^+)]/2$.

    (b) *Inverse CCFT.* If x is a finite-energy continuous-time signal defined on the infinite time axis $\mathbb{R}$, by the inverse CCFT it may be represented as

$$x(t) = \lim_{B \to \infty} \int_{-B}^{B} \hat{x}(f)e^{j2\pi ft}\, df, \qquad t \in \mathbb{R},$$

with $\hat{x}$ the CCFT of x. Again, convergence is in the $\mathcal{L}_2$ sense, so that the 2-norm of the difference between the left-hand side and the truncated integral on the right-hand side converges to 0 as B approaches ∞.

    If x has *bounded variation* on $\mathbb{R}$, the truncated integral converges *pointwise* at those time instants where x is continuous. At discontinuities it converges to the midpoint value.

    By way of example, consider 7.4.3(c), where we found that the CCFT of the one-sided exponential

$$x(t) = e^{-t/\theta}\mathbb{1}(t), \qquad t \in \mathbb{R},$$

with $\theta > 0$, is

$$\hat{x}(f) = \frac{1}{\dfrac{1}{\theta} + j2\pi f}, \qquad f \in \mathbb{R}.$$

The signal x has a discontinuity at time 0, so we expect the inverse CCFT of $\hat{x}$ to assume the midpoint value $1/2$ at time 0. Indeed,

$$\int_{-B}^{B} \frac{1}{\dfrac{1}{\theta} + j2\pi f} \, df = \left. \frac{\log\left(\dfrac{1}{\theta} + j2\pi f\right)}{j2\pi} \right|_{-B}^{B} = \frac{1}{j2\pi} \log\left(\frac{\dfrac{1}{\theta} + j2\pi B}{\dfrac{1}{\theta} - j2\pi B}\right)$$

$$= \frac{1}{2\pi} \arg\left(\frac{\dfrac{1}{\theta} + j2\pi B}{\dfrac{1}{\theta} - j2\pi B}\right).$$

In the limit $B \to \infty$ the latter expression converges to $1/2$, as expected.   ∎

Summary of Fourier Transforms

Table 7.5 summarizes the formulas for various Fourier transforms. In Table 7.6 the main properties of the Fourier transform are reviewed.

7.4.15.   Review: Sampled version of the DCFT.   The sampled DCFT is the transform

$$\mathscr{F}_{\mathrm{DC}}\colon \ell_2(T) \quad \to \quad \mathscr{L}_2[0, B),$$

with $B = 1/T$, that transforms the finite-energy signal x defined on $\mathbb{Z}(T)$ into $\hat{x}$ given by

$$\hat{x}(f) = T \sum_{t \in \mathbb{Z}(T)} x(t) e^{-j2\pi ft}, \qquad f \in [0, B).$$

The inverse DCFT follows as

$$x(t) = \int_0^B \hat{x}(f)\, e^{j2\pi ft}\, df, \qquad t \in \mathbb{Z}(T).$$

   Unitarity.   The sampled DCFT is unitary, that is, $\langle x, y \rangle_{\ell_2(T)} = \langle \hat{x}, \hat{y} \rangle_{\mathscr{L}_2[0, P)}$, provided the inner product on the signal space $\ell_2(T)$ is defined as

TABLE 7.5 SUMMARY OF FOURIER TRANSFORMS

Type of transform	Time axis	Frequency axis	Type of time-signal	Transform	Inverse Transform	Relation between parameters
DDFT	$\underline{N}$	$\underline{N}(F)$	finite-time[1] sequence	$\hat{x}(f) = \sum_{n\in\underline{N}} x(n)e^{-j2\pi fn}$	$x(n) = F\sum_{f\in\underline{N}(F)} \hat{x}(f)e^{j2\pi fn}$	$NF = 1$
DCFT	$\mathbb{Z}$	$[0,1)$	infinite-time sequence	$\hat{x}(f) = \sum_{n\in\mathbb{Z}} x(n)e^{-j2\pi fn}$	$x(n) = \int_0^1 \hat{x}(f)e^{j2\pi fn}\,df$	
DDFT sampled	$\underline{N}(T)$	$\underline{N}(F)$	finite-time[2] sampled signal	$\hat{x}(f) = T\sum_{t\in\underline{N}(T)} x(t)e^{-j2\pi ft}$	$x(t) = F\sum_{f\in\underline{N}(F)} \hat{x}(f)e^{j2\pi ft}$	$NFT = 1$
DCFT sampled	$\mathbb{Z}(T)$	$[0,B)$	infinite-time sampled signal	$\hat{x}(f) = T\sum_{t\in\mathbb{Z}(T)} x(t)e^{-j2\pi ft}$	$x(t) = \int_{0^-}^B \hat{x}(f)e^{j2\pi ft}\,df$	$BT = 1$
CDFT	$[0,P)$	$\mathbb{Z}(F)$	finite-time[3] continuous-time	$\hat{x}(f) = \int_{0^-}^P x(t)e^{-j2\pi ft}\,dt$	$x(t) = F\sum_{f\in\mathbb{Z}(F)} \hat{x}(f)e^{j2\pi ft}$	$PF = 1$
CCFT	$\mathbb{R}$	$\mathbb{R}$	infinite-time continuous-time	$\hat{x}(f) = \int_{-\infty}^\infty x(t)e^{-j2\pi ft}\,dt$	$x(t) = \int_{-\infty}^\infty \hat{x}(f)e^{j2\pi ft}\,df$	

1. Also applies to a single period of a periodic signal x defined on the time axis $\mathbb{Z}$ with period N.
2. Also applies to a single period of a periodic signal x defined on the time axis $\mathbb{Z}(T)$ with period NT.
3. Also applies to a single period of a periodic signal x defined on the time axis $\mathbb{R}$ with period P.

TABLE 7.6 PROPERTIES OF THE FOURIER TRANSFORM

Property	Transform	Time signal	Transformed signal
Linearity	all	$\alpha x + \beta y$	$\alpha \hat{x} + \beta \hat{y}$
Convolution	DDFT, CDFT DCFT, CCFT	$x \odot y$ $x * y$	$\hat{x}\hat{y}$ $\hat{x}\hat{y}$
	DDFT, DCFT CDFT, CCFT	xy xy	$\hat{x} \odot \hat{y}$ $\hat{x} * \hat{y}$
Shift	all; $\theta \in \mathbb{T}$	$x(t + \theta), t \in \mathbb{T}$	$e^{j2\pi f\theta}\hat{x}(f), f \in \mathbb{F}$
	all; $\phi \in \mathbb{F}$	$e^{j2\pi\phi t} x(t), t \in \mathbb{T}$	$\hat{x}(f - \phi), f \in \mathbb{F}$
Periodicity	DDFT (sampled) CDFT DCFT	period NT period P not periodic	period NF not periodic period B
Differentiation	CDFT, CCFT DCFT, CCFT	$(Dx)(t), t \in \mathbb{T}$ $(-j2\pi t)x(t), t \in \mathbb{T}$	$(j2\pi f)\hat{x}(f), f \in \mathbb{F}$ $D\hat{x}(f), f \in \mathbb{F}$
Symmetry	all	real conjugate symmetric even odd real and even real and odd	conjugate symmetric real even odd real and even imaginary and odd

Notes:
(1) $\mathbb{T}$ is the time axis and $\mathbb{F}$ the corresponding frequency axis.
(2) For finite time and frequency axes the addition and subtraction in the shift property are modulo the length of the time or frequency axis.

$$\langle x, y \rangle_{\ell_2(T)} = T \sum_{t \in \mathbb{Z}(T)} x(t)\overline{y(t)}.$$

Linearity.   The sampled DCFT and its inverse are linear.

Convolution property.   The sampled DCFT has the convolution property $(x * y)\hat{} = \hat{x} \cdot \hat{y}$ and the converse convolution property $(x \cdot y)\hat{} = \hat{x} \odot \hat{y}$, provided the convolution of signals in $\ell_2(T)$ is defined as

$$(x * y)(t) = T \sum_{\tau \in \mathbb{Z}(T)} x(t - \tau)y(\tau), \qquad t \in \mathbb{Z}(T).$$

Shift property.   The shift property holds in the sense that for any $\theta \in \mathbb{Z}(T)$ the shifted signal

$$x(t + \theta), \qquad t \in \mathbb{Z}(T),$$

transforms into

$$e^{j2\pi f\theta}\hat{x}(f), \qquad f \in [0, B).$$

Conversely, for every $\phi \in \mathbb{R}$ the signal

$$e^{j2\pi\phi t}x(t), \qquad t \in \mathbb{Z}(T),$$

has the DCFT

$$\hat{x}((f - \phi) \bmod B), \qquad f \in [0, B).$$

Periodicity property.   Taken on $\mathbb{R}$ rather than $[0, B)$ the DCFT of a sampled signal is periodic with period B.

Converse differentiation property.   Suppose that x has a DCFT $\hat{x}$ that is cyclically continuous on $[0, B)$ and differentiable with derivative $D\hat{x}$. Then the signal

$$-j2\pi t \cdot x(t), \qquad t \in \mathbb{Z}(T),$$

has $D\hat{x}$ as its DCFT.

Symmetry properties.   The symmetry properties of 7.4.13 also apply to the sampled DCFT.   ∎

Generalized DCFT

As seen in Section 7.2, the generalized CDFT transforms generalized time signals into frequency signals of polynomial growth. Passing from the CDFT to the DCFT amounts to reversing the roles of time and frequency, and, hence, it is no surprise that the *generalized DCFT* transforms time signals of polynomial growth into singular frequency signals. We use the inverse transform to establish several generalized DCFT pairs. To allow for singularities at zero frequency, it is useful to redefine the inverse DCFT as

$$x(n) = \int_{0^-}^{1} \hat{x}(f)e^{j2\pi fn}, \qquad n \in \mathbb{Z}.$$

7.4.16.   Example: Generalized DCFT.

(a) *DCFT of constant and polynomial time signals.* Application of the inverse (generalized) DCFT to the singular frequency signal $\hat{x}(f) = \delta^{(k)}(f), f \in [0^-, 1)$, with k a nonnegative integer, results in the time signal

$$x(n) = \int_{0^-}^{1} \delta^{(k)}(f)e^{j2\pi fn}\, dt = (-j2\pi n)^k, \qquad n \in \mathbb{Z}.$$

It follows for $k = 0$ that

$$1, \quad n \in \mathbb{Z}, \qquad \delta(f), \quad f \in [0^-, 1),$$

form a DCFT pair, while for $k \in \mathbb{N}$

$$(-j2\pi n)^k, \quad n \in \mathbb{Z}, \qquad \delta^{(k)}(f), \quad f \in [0^-, 1),$$

are a DCFT pair.

(b) *DCFT of a harmonic.* Again by application of the inverse DCFT it follows that for $f_o \in [0, 1)$ the signals

$$e^{j2\pi f_o n}, \quad n \in \mathbb{Z}, \qquad \delta(f - f_o), \quad f \in [0^-, 1),$$

form a DCFT pair.

(c) *DCFT of a periodic discrete-time signal.* If x is a periodic discrete-time signal on $\mathbb{Z}$ with period N, it may be expanded in the finite Fourier series

$$x(n) = F \sum_{k=0}^{N-1} \hat{X}(kF)e^{j2\pi nkF}, \qquad n \in \mathbb{Z},$$

$F = 1/N$, where the Fourier coefficients $\hat{X}$ are given by

$$\hat{X}(kF) = \sum_{n=0}^{N-1} x(n)e^{-j2\pi nkF}, \qquad k \in \underline{N}.$$

The coefficients $\hat{X}$ actually are the *DDFT* of a single period X of x. By using the linearity of the DCFT and (b), it follows that the (generalized) DCFT of the periodic signal x is

$$\hat{x}(f) = F \sum_{k=0}^{N-1} \hat{X}(kF)\delta(f - kF), \qquad f \in [0^-, 1).$$

Hence, the DCFT of a periodic discrete-time signal is a finite comb of delta functions, whose coefficients coincide with the coefficients of the DDFT of a single period of the signal. Figure 7.15 illustrates this. ∎

Generalized CCFT

So far we considered the CCFT of regular continuous-time signals with finite energy. In later chapters the generalization of the CCFT to singular signals on the one hand and signals that have a singular CCFT on the other turns out to be a useful tool.

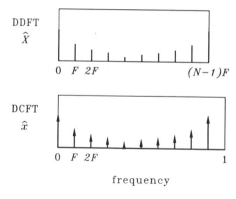

Figure 7.15. Top: the Fourier coefficients $\hat{X}$ of a periodic discrete-time signal x. Bottom: the DCFT $\hat{x}$ of the periodic signal. An arrow represents a δ-function, with its height indicating the coefficient of the δ-function.

An outline of the fundamentals of the generalized CCFT is given in Supplement C. At this point we content ourselves with a few rules of thumb that usually lead to correct answers.

For the evaluation of the CCFT of *singular* signals the properties of the δ-function and its derivatives as listed in Table 2.1 may often successfully be used.

7.4.17.   Example: CCFT of the δ-function and its derivatives.

(a) *CCFT of the δ-function.* Direct application of property (1) of Table 2.1 shows that the CCFT of the δ-function $x = \delta$ is

$$\hat{x}(f) = \int_{-\infty}^{\infty} \delta(t)e^{-j2\pi ft}\,dt = 1, \qquad f \in \mathbb{R}.$$

It follows that the *inverse* CCFT of the constant 1, which only exists in the sense of the generalized CCFT, is the δ-function.

(b) *CCFT of derivatives of the δ-function.* By using property (7) of Table 2.1 we find that the CCFT of the kth derivative $x = \delta^{(k)}$ of the δ-function is

$$\hat{x}(f) = \int_{-\infty}^{\infty} \delta^{(k)}(t)e^{-j2\pi ft}\,dt = (j2\pi f)^k, \qquad f \in \mathbb{R}.$$

Conversely, the (generalized) inverse CCFT of the frequency signal $(j2\pi f)^k$, $f \in \mathbb{R}$, which in the usual sense does not exist, is the kth derivative delta-function $\delta^{(k)}$.

It follows that the signals

$$\delta^{(k)}(t), \quad t \in \mathbb{R}, \qquad (j2\pi f)^k, \quad f \in \mathbb{R},$$

form a (generalized) CCFT pair for $k = 0, 1, 2, \cdots$. Note that the CCFTs have infinite energy but are of *polynomial growth*. ∎

In the same spirit, generalized CCFT pairs where the CCFT is singular may be obtained by formal application of the *inverse* CCFT.

7.4.18.   **Example: Inverse CCFT of singular frequency signals.**
    (a) *Inverse CCFT of* $\delta(f), f \in \mathbb{R}$. Again, using property (1) of Table 2.1 it follows that the inverse CCFT of $\hat{x} = \delta$ is

$$x(t) = \int_{-\infty}^{\infty} \delta(f) e^{j2\pi ft} \, df = 1.$$

Consequently, the (generalized) CCFT of the constant signal $x(t) = 1, t \in \mathbb{R}$, is the frequency signal $\hat{x}(f) = \delta(f), f \in \mathbb{R}$.

    (b) *Inverse CCFT of* $\delta^{(k)}(f), f \in \mathbb{R}$. From property (7) of Table 2.1 we obtain

$$\int_{-\infty}^{\infty} \delta^{(k)}(f) e^{j2\pi ft} \, df = (-1)^k (j2\pi t)^k, \qquad t \in \mathbb{R}.$$

It follows that the signals

$$(-j2\pi t)^k, \quad t \in \mathbb{R}, \qquad \delta^{(k)}(f), \quad f \in \mathbb{R},$$

form a (generalized) CCFT pair for $k = 0, 1, 2, \cdots$. Note that although the time signals have infinite energy they are of polynomial growth.   ∎

By using these naive techniques we may easily obtain the CCFT of *harmonic* and *periodic* signals.

7.4.19.   **Example: Generalized CCFT of harmonic and periodic signals.**
    (a) *CCFT of harmonic signals.* Application of the converse shift property of the CCFT to the CCFT pair

$$x(t) = 1, \qquad t \in \mathbb{R},$$
$$\hat{x}(f) = \delta(f), \qquad f \in \mathbb{R},$$

shows that also

$$z(t) = e^{j2\pi f_o t}, \qquad t \in \mathbb{R},$$
$$\hat{z}(f) = \delta(f - f_o), \qquad f \in \mathbb{R},$$

form a CCFT pair for any real number f_o. Thus, the CCFT of a (complex) harmonic with frequency f_o is a δ-function located at the frequency f_o, as illustrated in Fig. 7.16.

(b) *CCFT of a periodic signal.* Suppose that the continuous-time signal x is periodic with period P. Then it may be expanded according to the infinite Fourier series expansion as

$$x(t) = F \sum_{k=-\infty}^{\infty} \hat{X}(kF)e^{j2\pi tkF}, \qquad t \in \mathbb{R}, \tag{6}$$

$F = 1/P$, where the Fourier coefficients $\hat{X}$ are given by

$$\hat{X}(f) = \int_0^P x(t)e^{-j2\pi ft}\,dt, \qquad f \in \mathbb{Z}(F).$$

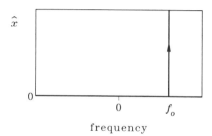

Figure 7.16. CCFT of the complex harmonic $e^{j2\pi f_o t}$, $t \in \mathbb{R}$.

$\hat{X}$ is actually the CDFT of one period X of x. By (6), the periodic signal x is a linear combination of the harmonic signals $e^{j2\pi tkF}$, $k \in \mathbb{Z}$, each of which according to (a) has the CCFT $\delta(f - kF)$, $f \in \mathbb{R}$. It follows by the linearity of the CCFT that the periodic signal x has the (generalized) CCFT

$$\hat{x}(f) = F \sum_{k=-\infty}^{\infty} \hat{X}(kF)\delta(f - kF), \qquad f \in \mathbb{R}.$$

Thus, the CCFT of a periodic signal consists of an infinite comb of δ-functions spaced with interval F, as illustrated in Fig. 7.17.

(c) *CCFT of the infinite comb.* The infinite comb with period P is the periodic signal given by

$$w_P(t) = \sum_{n=-\infty}^{\infty} \delta(t + nP), \qquad t \in \mathbb{R}.$$

CDFT
$\hat{X}$

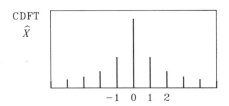

CCFT
$\hat{x}$

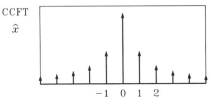

normalized frequency f/F

Figure 7.17. Top: the CDFT $\hat{X}$ of a periodic signal x. Bottom: the CCFT $\hat{x}$ of the periodic signal. An arrow represents a δ-function, with its height indicating the coefficient of the δ-function.

In 6.4.21 we found that the comb has a Fourier series expansion with coefficients $c_x(f) = 1$ for $f \in \mathbb{Z}(F)$. Thus, according to (b) the infinite comb has the CCFT

$$\hat{w}_P(f) = F \sum_{\phi \in \mathbb{Z}(F)} \delta(f - \phi) = F \sum_{k=-\infty}^{\infty} \delta(f - kF), \qquad f \in \mathbb{R},$$

with $F = 1/P$. This is again an infinite comb.   ∎

Rather more difficult is the question what the CCFT of the *unit step* is.

7.4.20.   Example: CCFT of the unit step.   What follows is a thumbnail derivation of the CCFT of the unit step. We may write the unit step as

$$x(t) = \mathbb{1}(t) = \tfrac{1}{2} + \tfrac{1}{2} \operatorname{sign}(t), \qquad t \in \mathbb{R}.$$

The CCFT of the "even part" of the unit step,

$$x_e(t) = \tfrac{1}{2}, \qquad t \in \mathbb{R},$$

by 7.4.18(a) is

$$\hat{x}_e(f) = \tfrac{1}{2}\delta(f), \qquad f \in \mathbb{R}.$$

By differentiating the "odd part"

$$x_o(t) = \tfrac{1}{2}\,\text{sign}\,(t), \qquad t \in \mathbb{R},$$

with respect to time it follows that

$$Dx_o(t) = \delta(t), \qquad t \in \mathbb{R}.$$

Application of the differentiation property of the CCFT results in

$$j2\pi f \cdot \hat{x}_o(f) = 1, \qquad f \in \mathbb{R},$$

so that

$$\hat{x}_o(f) = \frac{1}{j2\pi f}, \qquad f \neq 0, \qquad f \in \mathbb{R}.$$

This leaves $\hat{x}_o(0)$ undetermined. Since x_o is odd, so is $\hat{x}_o$, and, hence, $\hat{x}_o(0) = 0$. Thus, finally, we find that the CCFT of the unit step is

$$\hat{x}(f) = \hat{x}_e(f) + \hat{x}_o(f) = \tfrac{1}{2}\delta(f) + \frac{1}{j2\pi f}, \qquad f \in \mathbb{R}.$$

An impression of the CCFT of the unit step is given in Fig. 7.18. It has a δ-function at zero frequency, while at the other frequencies its magnitude behaves as $1/f$. It is clear that the low-frequency content of the step is large. ∎

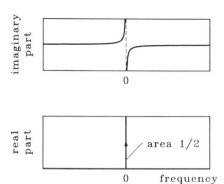

Figure 7.18. The CCFT of the unit step.

Table 7.7 lists several well-known generalized CCFT pairs.

TABLE 7.7 STANDARD GENERALIZED CCFT PAIRS

Time signal $x(t), t \in \mathbb{R}$	CCFT $\hat{x}(f), f \in \mathbb{R}$	Conditions
$\delta(t)$ $\delta^{(k)}(t)$	1 $(j2\pi f)^k$	$k \in \mathbb{N}$
1 $(-j2\pi t)^k$	$\delta(f)$ $\delta^{(k)}(f)$	$k \in \mathbb{N}$
$\delta(t + \theta)$	$e^{j2\pi f\theta}$	$\theta \in \mathbb{R}$
$e^{j2\pi f_o t}$	$\delta(f - f_o)$	$f_o \in \mathbb{R}$
$\sin(2\pi f_o t)$	$-\dfrac{j}{2}[\delta(f - f_o) - \delta(f + f_o)]$	$f_o \in \mathbb{R}$
$\cos(2\pi f_o t)$	$\frac{1}{2}[\delta(f - f_o) + \delta(f + f_o)]$	$f_o \in \mathbb{R}$
$1(t)$	$\frac{1}{2}\delta(f) + \dfrac{1}{j2\pi f}$	
$\text{sign}(t)$	$\dfrac{1}{j\pi f}$	
$t1(t)$	$\dfrac{j}{4\pi}\delta^{(1)}(f) - \dfrac{1}{4\pi^2 f^2}$	
$|t|$	$\dfrac{-1}{2\pi^2 f^2}$	
$\displaystyle\sum_{n=-\infty}^{\infty} a_n\delta(t - nT)$	$\displaystyle\sum_{k=-\infty}^{\infty} a_k e^{-j2\pi kfT}$	$\{a_k\}$ of polynomial growth, $T \in \mathbb{R}, T \neq 0$
$\displaystyle\sum_{n=-\infty}^{\infty} a_n e^{j2\pi nf_o t}$	$\displaystyle\sum_{k=-\infty}^{\infty} a_k\delta(f - kf_o)$	$\{a_k\}$ of polynomial growth, $f_o \in \mathbb{R}, f_o \neq 0$
$\displaystyle\sum_{n=-\infty}^{\infty} \delta(t + nT)$	$\dfrac{1}{T}\displaystyle\sum_{k=-\infty}^{\infty} \delta\left(f + \dfrac{k}{T}\right)$	$T \in \mathbb{R}, T \neq 0$
$\displaystyle\sum_{n=-\infty}^{\infty} e^{j2\pi nf_o t}$	$f_o\displaystyle\sum_{k=-\infty}^{\infty} e^{j2\pi kf/f_o}$	$f_o \in \mathbb{R}, f_o \neq 0$

7.5 FREQUENCY DOMAIN ANALYSIS OF LINEAR TIME-INVARIANT SYSTEMS

By the various Fourier transforms, inputs to linear time-invariant systems may be expanded in harmonics. As seen in Chapter 3, harmonics are *eigensignals* for linear time-invariant systems. Consequently, the Fourier expansions are *spectral* expansions for such systems, and correspondingly the various Fourier transforms are *spectral transforms*.

In terms of Fourier transforms, frequency domain analysis of linear time-invariant systems is a direct application of the convolution property. A linear time-invariant IOM system defined on the infinite time axis $\mathbb{Z}$ or $\mathbb{R}$ is a convolution system, whose IO map is characterized by

$$y = h * u,$$

with h the impulse response of the system. By the convolution property of the Fourier transform it follows that

$$\hat{y} = \hat{h}\hat{u},$$

where $\hat{}$ indicates the DCFT in the discrete-time case and the CCFT in the continuous-time case. This represents the transformed IO map. Figure 7.19 shows the *direct* way (by convolution) and the *transform* way to obtain the output.

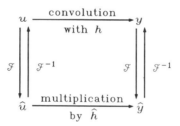

Figure 7.19.

The DIRECT way

The TRANSFORM way

Figure 7.19.   The direct and the Fourier transform way of characterizing the output of a convolution system.

Impulse Response and Frequency Response Function

In the discrete-time case, the DCFT $\hat{h}$ of the impulse response h of a convolution system is given by

$$\hat{h}(f) = \sum_{n=-\infty}^{\infty} h(n)e^{-j2\pi fn}, \qquad f \in [0, 1).$$

But this is precisely what we defined in 3.7.1 as the *frequency response function* of the convolution system. Similarly, in the continuous-time case the CCFT $\hat{h}$ of the impulse response h of a convolution system is

$$\hat{h}(f) = \int_{-\infty}^{\infty} h(t)e^{-j2\pi ft} \, dt, \qquad f \in \mathbb{R},$$

which by 3.7.1 is again the frequency response function. We conclude the following.

7.5.1.   Summary: Impulse response and frequency response function.   The
impulse response h and the frequency response function $\hat{h}$ of a discrete- or continuous-time convolution system form a Fourier transform pair.   ∎

The eminent importance of this result is that it allows to determine the impulse response from the frequency response function, namely, by *inverse Fourier transformation*.
     Figure 7.20 illustrates these results. It is a specialization of Fig. 7.19 to the sit-
uation where the input is a unit pulse Δ (in the discrete-time case) or a δ-function (in
the continuous-time case). Following the "direct route" confirms the well-known
fact that the impulse response of the system is the response to the unit pulse in the
discrete-time case or the δ-function in the continuous-time case. Following the
"transform route," we first observe that both the unit pulse and the δ-function transform to the constant 1, the former under the DCFT (Example 7.4.3(a)), the latter
under the CCFT (Example 7.4.16(a)). As a result, the frequency-domain response
of the system is the frequency response function $\hat{h}$ which after inverse transformation
results in the impulse response h.

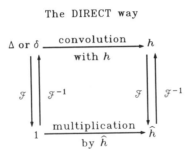

Figure 7.20.   Impulse response and frequency response function.

Frequency Content of Discrete-Time Signals

In the last part of this section we illustrate the application of frequency domain analysis by several examples. Before doing this we discuss what is meant by the
"frequency content" of discrete-time signals defined on the time axis $\mathbb{Z}$.
     Recall that by the aliasing effect, as discussed in 2.2.10, discrete-time har-
monic signals defined on $\mathbb{Z}$ whose frequencies differ by an integral multiple of 1 are
identical. This is the reason why in the DCFT only harmonics within the frequency
interval $[0, 1)$ enter. By the periodicity property of the DCFT we may as well consider any other frequency interval of length 1. To judge the "frequency content" of a
signal it makes most sense to choose the interval $[-\frac{1}{2}, \frac{1}{2})$.
     The reason is the following. As noted at the end of Section 2.2 and in Problem
2.6.9, a discrete-time harmonic defined on the time axis $\mathbb{Z}$ with frequency f in the
interval $[-\frac{1}{2}, \frac{1}{2})$ may not be exactly periodic, but repeats itself *approximately* with
repetition rate $|f|$. Repeating itself "approximately" means that the signal has $|f|$

maxima and minima per unit time and crosses $2|f|$ times through zero per unit time. If the frequency f lies outside the interval $[-\frac{1}{2}, \frac{1}{2})$, then it may be shifted into this interval by the modulo operation. Evidently, of all the frequencies that by aliasing might replace f, the one in the interval $[-\frac{1}{2}, \frac{1}{2})$ is indicative for the periodicity properties of the harmonic, and thus may be considered the "genuine" frequency.

In Section 3.7 we associated adjectives such as *low-pass, high-pass, band-pass,* and *band-stop* with frequency response functions, depending on their shape. We use the same terminology for *signals,* depending on the shape of their Fourier transforms. For *discrete-time* signals, the shape of their Fourier transforms is judged on the basis of their behavior on the frequency interval $[-\frac{1}{2}, \frac{1}{2})$.

If a signal x is *real,* then by the symmetry properties of the Fourier transform its Fourier transform $\hat{x}$ is conjugate symmetric. A consequence of this is that if $\hat{x}$ is known for positive frequencies, then its values for negative frequencies follow by complex conjugation. If $\hat{x}$ is conjugate symmetric, then its magnitude $|\hat{x}|$ is an *even* function of frequency and its phase arg $(\hat{x})$ an *odd* function of frequency. It, therefore, is customary to plot the Fourier transform $\hat{x}$ of real signals x for *positive* frequencies only. For discrete-time real-valued signals with time axis $\mathbb{Z}$ the appropriate frequency axis is $[0, \frac{1}{2}]$, while for sampled signals with time axis $\mathbb{Z}(T)$ the corresponding frequency axis is $[0, 1/2T]$.

7.5.2.   Example: Frequency content of the rectangular pulse.   By way of example, consider the discrete-time rectangular pulse

$$x(n) = \begin{cases} 1 & \text{for } 0 \le n \le M-1, \\ 0 & \text{otherwise,} \end{cases} \qquad n \in \mathbb{Z},$$

of 7.4.3(b). Its DCFT is

$$\hat{x}(f) = e^{-j\pi f(M-1)} d_M(f), \qquad f \in [0, 1).$$

Plots of the magnitude of $\hat{x}$ for several values of M are given in Fig. 7.12 on the frequency interval $[0, 1)$. Replotting them on $[-\frac{1}{2}, \frac{1}{2})$, which is a simple matter of cyclical shifting, yields Fig. 7.21. The figure shows that the rectangular pulse is a low-pass signal. As M increases its low-pass character becomes more pronounced.   ∎

Examples of Frequency Domain Analysis

We consider several applications of frequency domain analysis to systems. The first concerns a discrete-time system.

7.5.3.   Example: An ideal discrete-time low-pass filter.   A low-pass filter is a system that transmits low-frequency signals more or less unchanged and attenuates high-frequency signals. An *ideal* low-pass filter transmits low-frequency signals intact up to a specified *cut-off* frequency and completely suppresses signals with higher frequencies.

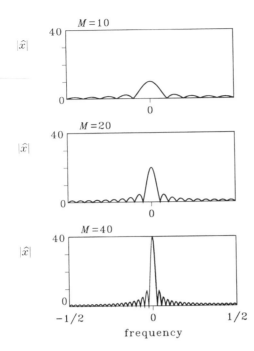

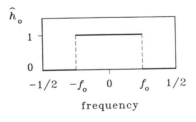

Figure 7.21. Frequency content of the rectangular pulse for $M = 10, 20,$ and 40.

The frequency response function of a discrete-time ideal low-pass filter has a rectangular shape as sketched in Fig. 7.22, and is given by

$$\hat{h}_o(f) = \begin{cases} 1 & \text{for } -f_o \le f < f_o, \\ 0 & \text{otherwise,} \end{cases} \qquad f \in [-\tfrac{1}{2}, \tfrac{1}{2}),$$

with f_o a fixed cut-off frequency such that $0 \le f_o < \tfrac{1}{2}$.

Figure 7.22. Frequency response function of a discrete-time ideal low-pass filter.

The impulse response h_o of the ideal low-pass filter follows by taking the inverse DCFT of the frequency response function $\hat{h}_o$ and, hence, is given by

$$h_o(n) = \int_{-f_o}^{f_o} e^{j2\pi f n}\, df = 2f_o \operatorname{sinc}(2\pi f_o n), \qquad n \in \mathbb{Z}.$$

A plot of the impulse response h_o is shown in Fig. 7.23 for $f_o = 0.25$. The plot shows what the effect is of removing all frequencies above f_o from the unit pulse, whose frequency content as we know from Example 7.4.3(a) is evenly distributed over all frequencies.

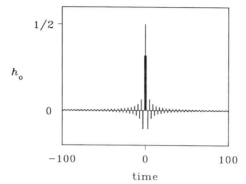

Figure 7.23.   Impulse response of the ideal low-pass filter.

The plot of the impulse response also shows that it is nonzero for negative times, so that the ideal low-pass filter is *anticipating*. As a result, ideal low-pass filtering is not possible in real time. One might try to overcome this difficulty by modifying the impulse response to

$$h(n) = \begin{cases} h_o(n - n_o) & \text{for } n \geq 0, \\ 0 & \text{otherwise,} \end{cases} \quad n \in \mathbb{Z},$$

as shown in Fig. 7.24 for $n_o = 10$. The impulse response has been *shifted*, as well as *truncated* (i.e., set equal to zero for negative times). If the impulse response were shifted by n_o but not truncated, then the output of the modified filter would simply be delayed by n_o compared to that of the ideal filter. The truncation is needed to make the filter non-anticipating. One may hope that if n_o is large enough, then truncation does not affect the ideal low-pass filtering effect much.

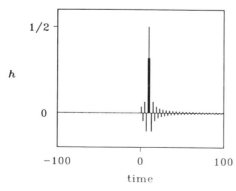

Figure 7.24.   Shifted and truncated impulse response of the ideal low-pass filter.

The magnitude of the frequency response function $\hat{h}$ corresponding to the shifted and truncated impulse response h may be computed numerically and is plotted in Fig. 7.25 for $n_o = 10$. The plot shows some deviation from the ideal low-pass behavior.

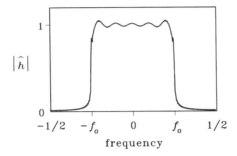

Figure 7.25. Magnitude of the frequency response function corresponding to the shifted and truncated impulse response.   ∎

For a second application of the use of frequency domain analysis we look at a continuous-time system.

7.5.4.   Example: Low-frequency differentiator.   A *differentiator* is a continuous-time system whose input and output are related as

$$y(t) = \frac{du(t)}{dt}, \qquad t \in \mathbb{R}.$$

In Example 3.7.5(d) we found that the differentiator has the frequency response function

$$\hat{h}(f) = j2\pi f, \qquad f \in \mathbb{R}.$$

Plots of the magnitude and phase of $\hat{h}$ are shown in Fig. 3.27. As frequency increases, the magnitude of $\hat{h}$ keeps increasing. This makes the differentiator very sensitive to imperfections in the input, which usually have a large high-frequency content. We therefore consider a system that acts as a differentiator for *low* frequencies, but whose frequency response function stops increasing and actually decays to zero for high frequencies. The frequency response function of such an *approximate differentiator* could for instance be chosen as

$$\hat{h}(f) = \frac{j2\pi f}{(1 + j2\pi f\theta)^2}, \qquad f \in \mathbb{R}, \tag{1}$$

with θ a positive number. A double-logarithmic plot of the magnitude and a semi-logarithmic plot of the phase of the frequency response function (1) as a function of the angular frequency $\omega = 2\pi f$ are given in Fig. 7.26. The plots follow the magnitude and phase plots of the ideal differentiator up to about the angular frequency

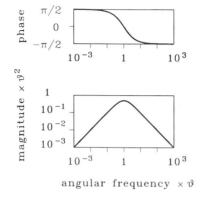

Figure 7.26. Amplitude and phase plots of an approximate differentiator with logarithmic scales for the amplitude and frequency.

$1/\theta$; above this frequency the magnitude drops off to zero and the phase changes over from $\pi/2$ to $-\pi/2$.

To find the impulse response of the approximate differentiator we need determine the inverse CCFT of $\hat{h}$. We do this indirectly by using a previous result and the properties of the CCFT. In Example 7.4.12 we found the CCFT pair

$$z(t) = te^{-t/\theta}\mathbb{1}(t), \qquad t \in \mathbb{R}, \qquad \hat{z}(f) = \frac{\theta^2}{(1 + j2\pi f\theta)^2}, \qquad f \in \mathbb{R}.$$

According to the differentiation property of the CCFT, multiplication of $\hat{z}$ by $j2\pi f$ is equivalent to differentiation of z. It follows that

$$\frac{dz(t)}{dt}, \qquad t \in \mathbb{R}, \qquad j2\pi f \cdot \hat{z}(f) = \frac{j2\pi f\theta^2}{(1 + j2\pi f\theta)^2}, \qquad f \in \mathbb{R},$$

form another CCFT pair. Comparing with $\hat{h}$ as given by (1) we conclude that the inverse CCFT h of $\hat{h}$ is given by

$$h(t) = \frac{1}{\theta^2}\frac{dz(t)}{dt} = \frac{1 - \dfrac{t}{\theta}}{\theta^2}e^{-t/\theta}\mathbb{1}(t), \qquad t \in \mathbb{R}.$$

A plot of the impulse response h is given in Fig. 7.27. It is non-anticipating, and may be viewed as an approximation to the impulse response δ' of the ideal differentiator. The same figure shows the response of the approximate differentiator both to a step and to a ramp. The responses of the *ideal* differentiator to these inputs are the delta function δ and the unit step $\mathbb{1}$, respectively. The plots show how the approximate differentiator "blurs" the ideal responses.   ∎

The final example illustrates the application of the generalized CCFT to the analysis of systems.

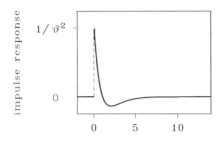

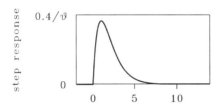

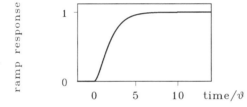

Figure 7.27. Time response of the approximate differentiator. Top: impulse response. Middle: step response. Bottom: response to a ramp signal.

7.5.5.   Example: Application of the generalized CCFT.

(a) *The frequency response function of an integrator.* An integrator is an IOM system with IO map

$$y(t) = \int_{-\infty}^{t} u(\tau)\, d\tau, \qquad t \in \mathbb{R}.$$

Its impulse response is

$$h(t) = \int_{-\infty}^{t} \delta(\tau)\, d\tau = \mathbb{1}(t), \qquad t \in \mathbb{R}.$$

It follows that the frequency response function $\hat{h}$ of the integrator is the CCFT of the unit step, so that by 7.4.20

$$\hat{h}(f) = \frac{1}{j2\pi f} + \tfrac{1}{2}\delta(f), \qquad f \in \mathbb{R}.$$

The integrator has infinite gain at zero frequency, which is why its response to a constant input does not exist.

(b) *The response of the RC network to a linearly increasing input.* According to 3.7.5(b) the RC network has the frequency response function

$$\hat{h}(f) = \frac{1}{1 + j2\pi fRC}, \qquad f \in \mathbb{R}.$$

Suppose that the input u to the system is the linearly increasing signal

$$u(t) = t, \qquad t \in \mathbb{R}.$$

By 7.4.18(b), the time signal

$$-j2\pi t, \qquad t \in \mathbb{R},$$

has the generalized CCFT

$$\delta^{(1)}(f), \qquad f \in \mathbb{R}.$$

This shows that the (generalized) CCFT of the input u is given by

$$\hat{u}(f) = -\frac{1}{j2\pi}\delta^{(1)}(f), \qquad f \in \mathbb{R}.$$

As a consequence, the CCFT of the output y is

$$\hat{y}(f) = \hat{h}(f)\hat{u}(f) = -\frac{1}{j2\pi}\frac{1}{1 + j2\pi fRC}\delta^{(1)}(f), \qquad f \in \mathbb{R}.$$

From property (8) of Table 2.1 it follows with $n = 1$ that

$$f(t)\delta^{(1)}(t) = f(0)\delta^{(1)}(t) - f^{(1)}(0)\delta(t), \qquad t \in \mathbb{R},$$

for any differentiable f. As a result,

$$\hat{y}(f) = -\frac{1}{j2\pi}(\delta^{(1)}(f) + j2\pi RC\delta(f)) = -\frac{\delta^{(1)}(f)}{j2\pi} - RC\delta(f), \qquad f \in \mathbb{R}.$$

By using the fact that $-\delta^{(1)}/j2\pi$ is the CCFT of $t, t \in \mathbb{R}$, and δ that of 1 we obtain

$$y(t) = t - RC, \qquad t \in \mathbb{R}.$$

Plots of the input and output are shown in Fig. 7.28. The input and the output both increase linearly with time, but the output lags behind by a time interval RC. The same result may of course be computed directly in the time domain.

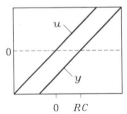

Figure 7.28. The response of the RC network to a linearly increasing input.

∎

Response to Periodic Inputs

In conclusion of this chapter, we reconsider the response of convolution systems to periodic inputs.

In Section 3.8 we found that the response of a discrete- or continuous-time regular convolution system to a bounded periodic input exists if the impulse response h of the system has finite action. One period of the response may be obtained by cyclical convolution of one period of the input with one period H of the periodic extension h_{per} of the impulse response h. The diagram of Fig. 3.37 demonstrates this.

In Section 7.3 it is shown how the DDFT (in the discrete-time case) or the CDFT (in the continuous-time case) may be used to transform cyclical convolution with H to multiplication by the frequency response function of the system. This is clarified in the diagram of Fig. 7.9.

We now consider how the response to periodic inputs may be found by using the DCFT or CCFT. According to 7.4.16(c), a periodic discrete-time signal on the time axis $\mathbb{Z}$, with period N, has a (generalized) DCFT that is a finite comb of delta functions, spaced with distance $F = 1/N$ on the interval $[0, 1)$, whose coefficients are obtained from the DDFT of one period of the signal. Similarly, as stated in 7.4.19(b), a periodic continuous-time signal with period P has a (generalized) CCFT that is an infinite comb of delta functions, spaced with distance $F = 1/P$ on $\mathbb{R}$, whose coefficients are obtained from the CDFT of one period of the signal.

Because periodic signals have a (generalized) DCFT or CCFT, we may apply the corresponding frequency domain analysis to obtain the response of convolution systems to periodic inputs. Suppose that u is a periodic input with period N to a discrete-time convolution system. From 7.4.16(c) its DCFT may be written as

$$\hat{u}(f) = F \sum_{k=0}^{N-1} \hat{U}(kF)\delta(f - kF), \qquad f \in [0^-, 1),$$

$F = 1/N$, where $\hat{U}$ is the DDFT of a single period U of u. If the system has the frequency response function $\hat{h}$, the DCFT of the output follows as

$$\hat{y}(f) = \hat{h}(f)\hat{u}(f) = \hat{h}(f)F \sum_{k=0}^{N-1} \hat{U}(kF)\delta(t - kF)$$

$$= F \sum_{k=0}^{N-1} \hat{U}(kF)\hat{h}(f)\delta(f - kF)$$

$$= F \sum_{k=0}^{N-1} \hat{U}(kF)\hat{h}(kF)\delta(f - kF), \qquad f \in [0^-, 1),$$

where we used property (4) of Table 2.1. This shows that the CDFT $\hat{Y}$ of one period Y of the output y is given by

$$\hat{Y}(kF) = \hat{h}(kF)\hat{U}(kF), \qquad k \in \underline{N}.$$

But this is simply a restatement of the fact, known from 7.3.18, that

$$\hat{Y} = \hat{H}\hat{U},$$

with $\hat{H}$ the DDFT of one period H of the periodic extension h_{per} of the impulse response h. Recall that according to 3.8.2 the frequency response function $\hat{h}$ and $\hat{H}$ coincide on the frequency axis $\underline{N}(F)$.

Thus, not surprisingly, frequency domain analysis of the response of discrete-time convolution systems to periodic inputs by application of the DCFT to the infinite-time signal leads to the same result as application of the DDFT to single periods of the input and output.

Similarly, if the CCFT of a periodic input u with period P to a continuous-time convolution system with frequency response function $\hat{h}$ is written as

$$\hat{u}(f) = F \sum_{k=-\infty}^{\infty} \hat{U}(kF)\delta(f - kF), \qquad f \in \mathbb{R},$$

$F = 1/P$, with $\hat{U}$ the CDFT of one period U of u, the CCFT of the output is

$$\hat{y}(f) = F \sum_{k=-\infty}^{\infty} \hat{Y}(kF)\delta(f - kF), \qquad f \in \mathbb{R},$$

with the CDFT $\hat{Y}$ of one period Y of y given by

$$\hat{Y}(kF) = \hat{h}(kF)\hat{U}(kF), \qquad k \in \mathbb{Z}.$$

The diagram of Fig. 7.29 illustrates the various ways the response of convolution systems to periodic inputs may be obtained.

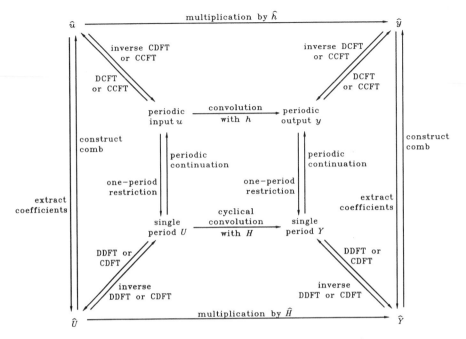

Figure 7.29. Various ways of obtaining the response of a convolution system to a periodic input.

7.5.6.   Exercise: Specialization of Fig. 7.29.   Specialize the diagram of Fig. 7.29 to the following cases:

(a) The periodic input u is the unit periodic input Δ_{per} or δ_{per}.
(b) The periodic input u is a periodic harmonic η_f.

Hint: Compare the diagrams of Figs. 3.38, 3.41, 7.10, and 7.20.   ∎

7.6 PROBLEMS

The first problems for this chapter deal with transforms.

7.6.1. "Slide rule" transform. A slide rule uses a logarithmic transformation to convert multiplications and divisions into additions and subtractions. Let $\phi: \mathbb{R}_+ \times \mathbb{R}_+ \to \mathbb{R}_+ \times \mathbb{R}_+$ be the map given by

$$(x, y) \quad \xrightarrow{\phi} \quad (xy, x/y).$$

Define the transform $\mathcal{T}: \mathbb{R}_+ \times \mathbb{R}_+ \to \mathbb{R} \times \mathbb{R}$ by

$$\mathcal{T}(x, y) = (\log (x), \log (y)).$$

(a) Determine the inverse transform $\mathcal{T}^{-1}$.
(b) Find the transformed map $\hat{\phi}: \mathbb{R} \times \mathbb{R} \to \mathbb{R} \times \mathbb{R}$.

7.6.2. Transform on $\mathbb{C}^3$. Consider the linear map $\phi: \mathbb{C}^3 \to \mathbb{C}^3$ defined by $\phi(x) = Kx$ with K the 3×3 matrix

$$K = \begin{bmatrix} 1 & 2 & 3 \\ 4 & 5 & 6 \\ 7 & 8 & 9 \end{bmatrix}.$$

Let $\mathcal{T}$ be the transform $\mathcal{T}: \mathbb{C}^3 \to \mathbb{C}^3$ defined by $\hat{x} = Ux$ where U is the 3×3 matrix

$$U = \begin{bmatrix} 1 & 1 & 1 \\ 0 & 1 & 1 \\ 0 & 0 & 1 \end{bmatrix}.$$

(a) Find the inverse transform $\mathcal{T}^{-1}$.
(b) Determine the transformed map $\hat{\phi}$.

7.6.3. Unitariness.

(a) *Linear transforms may always be made unitary.* Suppose that X is an inner product space with inner product $\langle \cdot, \cdot \rangle_X$, and $\mathcal{T}: X \to \hat{X}$ a linear transform. Prove that $\mathcal{T}$ is unitary if and only if the inner product on $\hat{X}$ is defined as

$$\langle \hat{x}, \hat{y} \rangle_{\hat{x}} = \langle \mathcal{T}^{-1}\hat{x}, \mathcal{T}^{-1}\hat{y} \rangle_X.$$

Hint: It is necessary to prove that $\langle \cdot, \cdot \rangle_{\hat{x}}$ is indeed an inner product.

(b) *Unitary orthogonal expansion transforms.* Suppose that the inner product space X has the countable orthogonal basis $\{v_i, i \in \mathbb{I}\}$, and let $\mathcal{T}$ be the corresponding expansion transform. This means that any element x of X is transformed to its expansion coefficients $\hat{x} = \{\hat{x}_i, i \in \mathbb{I}\}$, with respect to the orthogonal basis.

Prove that $\mathcal{T}$ is a unitary transform if and only if the inner product on the transformed space $\hat{X}$ is defined as

$$\langle \hat{x}, \hat{y} \rangle_{\hat{x}} = \sum_{i \in \mathbb{I}} \alpha_i \hat{x}_i \overline{\hat{y}}_i,$$

where $\alpha_i = \langle v_i, v_i \rangle_X$ for $i \in \mathbb{I}$. *Hint:* Use (a).

(c) *A unitary matrix.* For any natural number N, let

$$w_N = e^{j2\pi/N},$$

and W_N the $N \times N$ matrix whose (i, j)th element is

$$\frac{1}{\sqrt{N}} w_N^{(i-1)(j-1)}, \qquad i, j = 1, 2, \cdots, N.$$

Prove that W_N is unitary. *Hint:* Compare 6.4.6(a), but prove unitariness directly.

The next two problems are related to the CDFT.

7.6.4. Inverse CDFT. Let $F = 1/P$, with P a positive number, and determine the inverse CDFT of the following frequency signals:

(a)

$$\hat{x}_N(f) = \begin{cases} \dfrac{NF - |f|}{NF} & \text{for } |f| \le NF, \qquad f \in \mathbb{Z}(F), \\ 0 & \text{otherwise}, \end{cases}$$

with N an odd positive integer. *Hint:* Note that $\hat{x}_N$ is the convolution of a suitable rectangular frequency signal with itself.

(b)

$$\hat{x}(f) = \begin{cases} 0 & \text{for } f = 0, \\ \dfrac{1}{j 2\pi f} & \text{for } f \ne 0, \end{cases} \qquad f \in \mathbb{Z}(F).$$

Hint: Use the differentiation property of the CDFT.

7.6.5. Symmetry. The signal x is defined on the finite discrete time axis $\underline{N} = \{0, 1, \cdots, N - 1\}$ with $N = 8$, and partially given by

$$x(n) = \begin{cases} n & \text{for } n = 0, 1, 2, \\ 0 & \text{for } n = 3. \end{cases}$$

Complete the signal such that it is

(a) cyclically even,
(b) cyclically odd, or
(c) cyclically conjugate symmetric.

The following two problems concern the DCFT.

7.6.6. DCFT.

(a) Determine the DCFT of the discrete-time signal x given by

$$x(n) = a^n 1(n), \qquad n \in \mathbb{Z},$$

with $|a| < 1$. Sketch a plot of the time signal.

(b) Determine the DCFT of the discrete-time signal x given by

$$x(n) = -a^n 1(-n - 1), \qquad n \in \mathbb{Z},$$

with $|a| > 1$. Plot the time signal. Compare the DCFT with that of (a).
(c) Determine the inverse DCFT of the frequency signal

$$\hat{x}(f) = \frac{1}{1 - ae^{-j2\pi f}}, \qquad f \in [0, 1),$$

for (c.1) $|a| < 1$, and (c.2) $|a| > 1$.

7.6.7. Converse differentiation property of the DCFT. Prove the converse differentiation property 7.4.11(b) of the DCFT.

Next follows a series of problems concerning the CCFT.

7.6.8. CCFT. Determine the CCFT of the following signals. Plot the time-signals as well as the magnitude and phase of their transforms.

(a)
$$x(t) = \begin{cases} 1 - |t| & \text{for } |t| \le 1, \\ 0 & \text{otherwise,} \end{cases} \qquad t \in \mathbb{R}.$$

(b)
$$x(t) = \begin{cases} 1 - t & \text{for } 0 \le t < 1, \\ 0 & \text{otherwise,} \end{cases} \qquad t \in \mathbb{R}.$$

(c)
$$x(t) = \begin{cases} \cos(2\pi f_o t) & \text{for } -1/2f_o \le t < 1/2f_o, \\ 0 & \text{otherwise,} \end{cases} \qquad t \in \mathbb{R},$$

with f_o a positive real number.
(d)
$$x(t) = e^{at} 1(t), \qquad t \in \mathbb{R},$$

where Re $(a) < 0$.
(e)
$$x(t) = -e^{at} 1(-t), \qquad t \in \mathbb{R},$$

where Re $(a) > 0$. Compare with (d).
(f)
$$x(t) = e^{-\sigma t} \cos(\omega_o t) 1(t), \qquad t \in \mathbb{R},$$

with σ a positive and ω_o a nonnegative real number. *Hint:* Set successively $a = -\sigma + j\omega_o$ and $a = -\sigma - j\omega_o$ in (d), and add the results.

7.6.9. CCFT of a modulated signal. Suppose that x is a real continuous-time signal with CCFT $\hat{x}$. Determine the CCFT $\hat{x}_m$ of the *amplitude modulated* signal x_m given by

$$x_m(t) = x(t) \cdot \cos(2\pi f_o t), \qquad t \in \mathbb{R},$$

with f_o a nonnegative real number. Plot $|\hat{x}_m|$ assuming that $\hat{x}$ is low-pass and f_o is large. What does "large" mean in this context? *Hint:* Write

$$\cos(2\pi f_o t) = \tfrac{1}{2}(e^{j2\pi f_o t} + e^{-j2\pi f_o t})$$

and use the converse shift property of the CCFT.

7.6.10. Symmetries of CCFTs. Consider the following frequency signals and establish whether their inverse CCFT is even, odd, conjugate symmetric, real or several of these.

(a)

$$\hat{x}(f) = \frac{1}{1 + j2\pi f}, \qquad f \in \mathbb{R}.$$

(b)

$$\hat{x}(f) = e^{-2\pi^2 f^2}, \qquad f \in \mathbb{R}.$$

(c)

$$\hat{x}(f) = j2\pi f, \qquad f \in \mathbb{R}.$$

7.6.11. Inverse CCFT. Let $\hat{x}_n$, with $n \in \mathbb{N}$, be the frequency signal given by

$$\hat{x}_n(f) = \frac{1}{(a + j2\pi f)^n}, \qquad f \in \mathbb{R},$$

with Re $(a) > 0$. Determine the inverse CCFT x_n of $\hat{x}_n$. *Hint:* Apply the converse differentiation property of the CCFT to the transform pair x_1, $\hat{x}_1$ given by

$$x_1(t) = e^{-at}\mathbb{1}(t), \qquad t \in \mathbb{R},$$

$$\hat{x}_1(f) = \frac{1}{a + j2\pi f}, \qquad f \in \mathbb{R}.$$

7.6.12. CCFT of the Gaussian bell. The "Gaussian bell" with parameter 1 is the function $\phi \in \mathcal{L}$ given by

$$\phi(t) = \frac{1}{\sqrt{2\pi}}e^{-t^2/2}, \qquad t \in \mathbb{R}.$$

Prove that its CCFT is given by

$$\hat{\phi}(f) = e^{-2\pi^2 f^2}, \qquad f \in \mathbb{R}.$$

Note that ϕ and $\hat{\phi}$ are both of rapid decay. *Hint:* Compare C.44 and C.45 of Supplement C. Write $\hat{\phi}$ as the CCFT of ϕ, differentiate it with respect to frequency and use partial integration to show that $\hat{\phi}$ satisfies the differential equation $d\hat{\phi}(f)/df = -4\pi^2 f\hat{\phi}(f), f \in \mathbb{R}$. This equation may be solved by separation of variables, and the integration constant follows from the standard definite integral

$$\frac{1}{\sqrt{2\pi}} \int_{-\infty}^{\infty} e^{-t^2/2} \, dt = 1.$$

7.6.13. Generalized CCFTs. Find the generalized CCFTs of the following signals.

(a)
$$x(t) = t^k 1(t), \qquad t \in \mathbb{R},$$

with $k \in \mathbb{Z}_+$. *Hint:* Differentiate x sufficiently often and use the differentiation property of the CCFT.

(b)
$$x(t) = |t|^k, \qquad t \in \mathbb{R},$$

with $k \in \mathbb{Z}_+$.

(c)
$$x(t) = \text{sign}(t), \qquad t \in \mathbb{R}.$$

7.6.14. CCFT of periodic signals. Determine the generalized CCFT of the following periodic continuous-time signals:

(a) The full-wave rectified harmonic signal

$$x(t) = |\sin(2\pi f_o t)|, \qquad t \in \mathbb{R},$$

with f_o a positive real number.

(b) The singular periodic signal

$$x(t) = \sum_{k=-\infty}^{\infty} (-1)^k \delta\left(t - \frac{kP}{2}\right), \qquad t \in \mathbb{R},$$

with P a nonnegative real number.

7.6.15. Scaling property of the CCFT. Prove that if $\hat{x}$ is the CCFT of x, the CCFT of the time scaled signal

$$x(\alpha t), \qquad t \in \mathbb{R},$$

with α a nonzero real constant, is

$$\frac{\hat{x}(f/\alpha)}{|\alpha|}, \qquad f \in \mathbb{R}.$$

In particular, the CCFT of the time-reversed signal $x(-t)$, $t \in \mathbb{R}$, is the frequency reversed CCFT $\hat{x}(-f), f \in \mathbb{R}$.

7.6.16. Interchange of time and frequency for the CCFT.

(a) Suppose that x and $\hat{x}$ form a (generalized) CCFT pair. Prove that the CCFT of $\hat{x}(t), t \in \mathbb{R}$, is $x(-f), f \in \mathbb{R}$.

(b) Use (a) to determine the CCFT of the time signal

$$x(t) = \frac{1}{1+t}, \qquad t \in \mathbb{R}.$$

Hint: Compare 7.4.3(c).

7.6.17. CCFT of the complex conjugate. Prove that if $\hat{x}$ is the CCFT of x, then the CCFT of the complex conjugate $\overline{x}$ of x is $\hat{\overline{x}}(-f), f \in \mathbb{R}$.

We finish this section with some applications of Fourier transforms to systems.

7.6.18. Inverse of the exponential smoother. The exponential smoother has the frequency response function

$$\hat{h}(f) = \frac{1-a}{1-ae^{-j2\pi f}}, \qquad f \in [0, 1),$$

provided $|a| < 1$. The inverse of this frequency response function is

$$\hat{g}(f) = \frac{1-ae^{-j2\pi f}}{1-a}, \qquad f \in [0, 1).$$

(a) Determine the impulse response g of the inverse of the exponential smoother by computing the inverse DCFT of $\hat{g}$.
(b) Determine the frequency response function and the impulse response of the series connection of the exponential smoother and its inverse.
(c) Determine the impulse response of the inverse of the exponential smoother directly from the difference equation

$$y(n+1) = ay(n) + (1-a)u(n+1), \qquad n \in \mathbb{Z},$$

that describes the smoother.

7.6.19. Padé approximation of a delay. The frequency response function of a pure delay θ is given by

$$\hat{h}(f) = e^{-j2\pi f\theta}, \qquad f \in \mathbb{R}.$$

A pure delay cannot be realized with electrical networks containing resistors, capacitors, inductors, and amplifiers, because its frequency response function is not rational. It is therefore of interest to study how a pure delay may be *approximated* by a rational frequency response function. A *first-order Padé approximation* to a delay is a system with a frequency response function of the form

$$\hat{h}_1(f) = \frac{1+j2\pi fa}{1+j2\pi fb}, \qquad f \in \mathbb{R}.$$

The real constants a and b are chosen such that as many as possible of the coefficients of the Taylor series expansion of $\hat{h}_1$ in the frequency f equal those of the expansion of $\hat{h}$. This ensures that $\hat{h}_1$ is a good approximation of $\hat{h}$ at *low* frequencies. Because $\hat{h}_1(0) = \hat{h}(0)$, independently of a and b, the two available parameters may be used to match the coefficients of the terms in the expansions corresponding to f and f^2.

(a) Show that the first-order Padé approximation is given by

$$\hat{h}_1(f) = \frac{1 - j2\pi f\dfrac{\theta}{2}}{1 + j2\pi f\dfrac{\theta}{2}}, \qquad f \in \mathbb{R}.$$

(b) Sketch the magnitude and phase of $\hat{h}_1$ as a function of frequency and compare with the magnitude and phase of $\hat{h}$.

(c) Determine and sketch the impulse response h_1 of the first-order Padé approximation.

(d) Calculate the step response by integrating the impulse response. Sketch the step response, and compare it to the step response of the delay. The step response of the Padé filter appears to be a rather poor approximation of that of the delay. The cause is the relatively large high-frequency content of the unit step.

7.6.20. Removal of periodic component. Consider a continuous-time linear time-invariant system whose IO map is described by

$$y(t) = \frac{1}{a^2}\left(\int_t^{t+a} u(\tau)\, d\tau - \int_{t-a}^t u(\tau)\, d\tau\right), \qquad t \in \mathbb{R},$$

with a a positive real number.

(a) Determine the impulse response of the system.

(b) Find the frequency response function of the system.

(c) Show that the system filters out any periodic signal whose frequency is $1/a$ or an integral multiple of $1/a$. Explain this directly from the IO map.

7.6.21. Response of a discrete-time system to a periodic input. Consider a discrete-time convolution system with impulse response h that is zero for negative times and from time 0 onwards is given by the sequence $\{1, 1, 0, 1/2, 1/2, 0, 1/4, 1/4, 0, 1/8, 1/8, 0, \cdots\}$. Determine the response of this system to a periodic input with period 3 such that $u(0) = a$, $u(1) = b$, $u(2) = c$, with a, b and c constants, as follows:

(a) by regular convolution,

(b) by cyclical convolution,

(c) by application of the DCFT, and

(d) by application of the DDFT to a cyclical convolution.

7.7 COMPUTER EXERCISES

More numerical exercises involving Fourier transformation may be found in the computer exercises sections of Chapters 9–11. The numerical computation of Fourier transforms is discussed in Section 9.7. See also the box in Section 6.7.

7.7.1. Various Fourier transform pairs. The purpose of this exercise is to study the connection between time and frequency domain characteristics of signals.

(a) *Time duration and bandwidth.* In general, the shorter a signal is, the larger is its bandwidth, i.e., the "wider" its Fourier transform. Investigate this by computing the CCFT of the continuous-time signal x defined by

$$x(t) = e^{-|t|/P}, \qquad t \in \mathbb{R},$$

for different positive values of P (e.g., $P = 0.1, 0.3$, and 1). The larger P, the longer the signal. In each case, compute the fraction of the energy of the CCFT of the signal in the frequency interval $[-1, 1]$, and show that this fraction increases as P increases. Why is the CCFT of x real?

(b) *Smoothness and high-frequency content.* In general, the smoother a signal is, the smaller is its high-frequency content. Investigate this by comparing the CCFTs of the signals x and y, given by

$$x(t) = \begin{cases} 1 & \text{for } |t| \le 1, \\ 0 & \text{otherwise,} \end{cases} \qquad t \in \mathbb{R},$$

$$y(t) = \text{trian}\,(t), \qquad t \in \mathbb{R}.$$

These signals have the same time duration, but y is smoother than x. Normalize the two CCFTs so that their total energies are both 1. For the normalized CCFTs, compute the ratio

$$\frac{\int_F^\infty |\hat{y}(f)|^2\, df}{\int_F^\infty |\hat{x}(f)|^2\, df}, \qquad F \ge 0,$$

as a function of F to confirm that the high-frequency content of $\hat{y}$ is less than that of $\hat{x}$. Comment on the result.

(c) *Narrow-band signals are close to harmonic.* Band-pass signals with a small bandwidth behave as a harmonic signal with time-varying amplitude and phase. Investigate this by considering the signal x whose CCFT is given by

$$\hat{x}(f) = e^{-\frac{(f-f_o)^2}{b^2}} + e^{-\frac{(f+f_o)^2}{b^2}}, \qquad f \in \mathbb{R}.$$

Choose for instance $f_o = 1$ and $b = 0.1$. Define $\hat{x}$ on a well-chosen frequency axis with sufficient resolution and compute its inverse CCFT. Try also other values of the bandwidth b.

(d) *The CCFT of a periodic signal is an infinite comb.* Consider the periodic sawtooth x, one period of which is given by

$$x(t) = \frac{t}{P}, \qquad -P/2 \le t < P/2.$$

Take $P = 1$. Let the signal x_K consist of K periods of the sawtooth, symmetrically located with respect to the origin, and be zero otherwise. Define x_K for $K = 1, 3, 5,$ and 11 on a common time axis with sufficient resolution, and compute the CCFTs of these four signals. Observe that as K increases the CCFT becomes more and more spiked.

7.7.2. Response of convolution systems to aperiodic inputs. Determine the response of the continuous-time systems indicated in (a)–(c) below to each of the following inputs:

   (i) The δ-function $u_1 = \delta$; this yields the impulse response of the system.

  (ii) The triangular spike given by

$$u_2(t) = \begin{cases} t & \text{for } 0 \le t < 1, \\ 0 & \text{otherwise,} \end{cases} \qquad t \in \mathbb{R}.$$

 (iii) The noise burst given by

$$u_3(t) = \begin{cases} n(t)\, \text{trian}\,(t), & \text{for } -1 \le t < 1, \\ 0 & \text{otherwise,} \end{cases} \qquad t \in \mathbb{R},$$

where n is a white noise signal. *Hint:* In SIGSYS, generate n using the standard signal noise.

Define the time signals u_2 and u_3 on the common time axis $[-8, 8)$ with sufficient resolution (for instance on 256 sample points). *Hint:* In SIGSYS, use the command set INC to set the frequency increment equal to the correct value corresponding to inc and num, and use the frequency axis DOT to generate the following frequency response functions.

(a) *Low-pass filter.* The second-order low-pass filter with frequency response function

$$\hat{h}_1(f) = \frac{1}{(1 + jf/f_o)^2}, \qquad f \in \mathbb{R},$$

with $f_o = 1$.

(b) *Ideal low-pass filter.* The ideal low-pass filter with frequency response function

$$\hat{h}_2(f) = \begin{cases} 1 & \text{for } |f| \le f_o, \\ 0 & \text{otherwise,} \end{cases} \qquad f \in \mathbb{R},$$

with $f_o = 1$. Note that this filter is anticipating.

(c) *Band filter.* The band filter with frequency response function

$$\hat{h}_3(f) = e^{-\frac{(f-f_o)^2}{b^2}} + e^{-\frac{(f+f_o)^2}{b^2}}, \qquad f \in \mathbb{R},$$

with $f_o = 1$ and $b = 0.2$. Also this filter is anticipating.

Comment on the shapes of the outputs that you find.

7.7.3. Fast convolution. Given two discrete-time signals x and y, both defined on $\mathbb{Z}$, with finite supports of lengths N and M, respectively, the DDFT may be used to compute their convolution as follows.

    (i) Time shift the signals x and y so that they both start at time 0.

    (ii) Redefine the time axis of both x and y as $\{0, 1, \cdots, K - 1\}$, where K is the next largest integral power of 2 from $N + M$. This time axis includes the support of the convolution of the shifted signals x and y. The reason that K is chosen as an integral power of 2 is that this is required by the fast Fourier transform (FFT), which is a very efficient algorithm for performing the DDFT.

    (iii) Compute the DDFTs of the shifted signals x and y on the time axis $\{0, 1, \cdots, K - 1\}$.

    (iv) Multiply the DDFTs and apply the inverse DDFT. The result is the *cyclical* convolution of the shifted signals x and y. By 3.8.7(b) the cyclical and regular convolution coincide.

    (v) Shift the convolution to the correct interval by using the shift property of the convolution.

This is the algorithm that SIGSYS uses for fast convolution. We follow the steps explicitly to convolve the signals x and y given by

$$x(n) = \begin{cases} 1 & \text{for } n = 0, 1, \cdots, 10, \\ 0 & \text{otherwise}, \end{cases} \quad n \in \mathbb{Z},$$

$$y(n) = \begin{cases} 1 & \text{for } n = -5, -4, \cdots, 4, 5, \\ 0 & \text{otherwise}, \end{cases} \quad n \in \mathbb{Z}.$$

(a) Create and plot the signals x and y and compute and plot their convolution $z = x * y$. Verify analytically that the result is correct.

(b) Shift x and y so that they both start at time 0. Redefine their time axis as $\{0, 1, \cdots, K - 1\}$, with K the smallest integral power of 2 such that the time axis includes the support of the convolution of the shifted signals. *Hint:* Use 3.5.5.

(c) Compute the DDFTs of the shifted signals, multiply them, and determine the inverse DDFT of the product.

(d) Shift the convolution to the correct starting time. Compare with the result of (a).

(e) In step (b), take the incorrect value $K = 16$. Repeat step (c) and explain why the result is wrong. *Hint:* By 3.8.8, the result of steps (b)–(d) is one period of the periodic extension of z with period K.

8

The z- and Laplace Transforms

8.1 INTRODUCTION

In Chapter 4 we discussed difference and differential systems and showed how to find the response of such systems in the *time domain*. In the present chapter we introduce the *z*- and *Laplace transforms*. These provide us with *transform* methods to analyze the response of linear systems, which are often more efficient than time-domain methods. Moreover, they further enhance intuition concerning the behavior of difference and differential systems. The need for further transforms arises because the Fourier transforms we studied in Chapter 7 have two severe restrictions.

First of all, Fourier transforms cannot deal with signals of exponential growth. This means in particular that linear time-invariant systems that have inputs or impulse responses of exponential growth, or both, cannot be analyzed using Fourier transforms. To overcome this restriction, we *modify* the Fourier transform such that signals of exponential growth may be accommodated. The result is the *z-transform* in the discrete-time case and the *Laplace transform* in the continuous-time case.

Second, initial value problems for difference and differential systems cannot be handled. The reason is that the DCFT and CCFT are essentially *two*-sided. This difficulty may be resolved by the introduction of *one*-sided transforms, in particular the *one-sided z-transform* in the discrete-time case, and the *one-sided Laplace transform* in the continuous-time case.

We illustrate these points for a discrete-time system.

8.1.1.   Example: The exponential smoother.   Consider the discrete-time system

$$y(n + 1) = ay(n) + (1 - a)u(n + 1), \qquad n \in \mathbb{Z}.$$

When initially at rest, this is a convolution system with IO map $y = h * u$, where

$$h(n) = (1 - a)a^n 1(n), \qquad n \in \mathbb{Z}.$$

Suppose that $|a| > 1$, so that h grows exponentially. Then the DCFT cannot be used to analyze the convolution system, because the impulse response h has no DCFT. Nevertheless, for many inputs, in particular right one-sided inputs, the output is well-defined. Similarly, even if $|a| < 1$, so that h has a DCFT, the application of the DCFT breaks down when the input u is of exponential growth.

Next, consider initial value problems. Given $u(n)$, $n \in \mathbb{Z}_+$, as found in Section 4.5, the solution of the difference equation on $\mathbb{Z}_+$ is

$$y(n) = (h * u_+)(n) + \alpha a^n, \qquad n \in \mathbb{Z}_+,$$

with u_+ the input u truncated to $\mathbb{Z}_+$ and α a constant determined by the initial condition $y(0)$. Written out in full we have

$$y(n) = \sum_{k=0}^{n} h(n - k)u(k) + \alpha a^n, \qquad n \in \mathbb{Z}_+.$$

This solution cannot be obtained by using the DCFT, even if $|a| < 1$.   ∎

To make exponentially increasing signals amenable to Fourier transformation, we multiply them by an exponential signal, called *convergence factor,* that has the form

$$\rho^{-n}, \qquad n \in \mathbb{Z},$$

in the discrete-time case and

$$e^{-\sigma t}, \qquad t \in \mathbb{R},$$

in the continuous-time case. For a suitable choice of the real numbers ρ or σ this may result in signals that have a Fourier transform. For instance, the signal

$$u(t) = e^{at} 1(t), \qquad t \in \mathbb{R},$$

with a real, has no CCFT if $a > 0$, while the signal

$$v(t) = e^{-\sigma t} e^{at} 1(t), \qquad t \in \mathbb{R},$$

has a CCFT if $\sigma > a$.

The device of multiplying by a suitable converging factor leads to a modification of the Fourier transform. The transform thus obtained is the z-transform in the discrete-time case and the Laplace transform in the continuous-time case.

To deal with initial value problems, *one-sided* versions of the z- and Laplace transforms are introduced. They apply to signals that are defined on the time axis $\mathbb{Z}_+$ in the discrete-time case or $\mathbb{R}_+$ in the continuous-time case. Since *any* one-sided exponentially increasing signal may be made Fourier transformable by multiplying it by a signal of suitably fast exponential decay, the one-sided z- and Laplace transforms may be used to solve initial-value problems for difference and differential systems.

The Fourier transform is based on expanding signals in purely harmonic signals. The z- and Laplace transforms amount to the expansion in *exponentially weighted* harmonics of the form

$$\rho^n e^{j2\pi fn}, \qquad n \in \mathbb{Z},$$

in the discrete-time case or

$$e^{\sigma t} e^{j2\pi ft}, \qquad t \in \mathbb{R},$$

in the continuous-time case, with ρ or σ suitable real numbers.

The organization of the chapter is as follows. An important technique in obtaining the inverses of *rational* Fourier, z- and Laplace transforms, which are common when dealing with difference and differential systems, is *partial fraction expansion* of rational functions. In Section 8.2 a self-contained exposition of this technique is offered. In Section 8.3 the z- and Laplace transforms and their properties are defined. Section 8.4 is devoted to an account of the properties of the transforms, and a collection of standard transforms is presented. In Section 8.5 inverse z- and Laplace transformation is discussed. In Section 8.6 the z- and Laplace transforms are applied to convolution systems, leading to the notion of the *transfer function* of a system. In Section 8.7 is it shown how the z- and Laplace transform may be used for the analysis of linear time-invariant difference and differential systems and in Section 8.8 for that of linear time-invariant state systems. Both initially-at-rest systems and initial value problems are considered in these sections.

8.2 PARTIAL FRACTION EXPANSION

The technique of expanding a rational function in *partial fractions* is extensively used for finding inverse Fourier, z-, and Laplace transforms. This section is devoted to a presentation of the partial fraction expansion technique. We consider rational functions R of the form

$$R(\lambda) = \frac{P(\lambda)}{Q(\lambda)} = \frac{p_0 + p_1\lambda + \cdots + p_M\lambda^M}{q_0 + q_1\lambda + \cdots + q_N\lambda^N},$$

with λ an indeterminate variable, and $p_0, p_1, \cdots, p_M$ and $q_0, q_1, \cdots, q_N$ real or complex coefficients such that the *leading coefficients* p_M and q_N are nonzero. The rational function R is said to be *proper* if $M \leq N$, *strictly proper* if $M < N$, and *improper* if $M > N$.

Division Theorem

There are two steps in the partial fraction expansion of rational functions. The first is to reduce the expansion to that of a strictly proper rational function.

8.2.1.   Summary: Division theorem.   Suppose that the rational function $R = P/Q$, with P and Q polynomials, is not strictly proper. Then R may be expressed as

$$R = T + \frac{S}{Q},$$

where T is a polynomial whose degree is degree(P) $-$ degree(Q) and S a polynomial such that S/Q is strictly proper.   ∎

The theorem is a consequence of the well-known fact from algebra that division of the polynomial P by the polynomial Q results in a *quotient* polynomial T whose degree is degree(P) $-$ degree(Q) and a *remainder* polynomial S whose degree is degree(Q) $-$ 1 or less. The polynomials T and S may for instance be found by *long division*.

8.2.2.   Example: Long polynomial division.   We use long division to express the improper rational function

$$R(\lambda) = \frac{\lambda^4 + 2\lambda^3 + \lambda + 2}{\lambda^2 + 1}$$

as the sum of a polynomial and a strictly proper rational function. We have

$$
\begin{array}{r}
\lambda^2 + 2\lambda \; - 1 \\
\lambda^2 + 1 \,\overline{)\, \lambda^4 + 2\lambda^3 + \lambda + 2} \\
\underline{\lambda^4 + \lambda^2 } \\
2\lambda^3 - \lambda^2 + \lambda + 2 \\
\underline{2\lambda^3 + 2\lambda } \\
- \lambda^2 - \lambda + 2 \\
\underline{- \lambda^2 - 1} \\
-\lambda + 3
\end{array}
$$

As a result $T = \lambda^2 + 2\lambda - 1$ and $S = -\lambda + 3$, so that

$$R(\lambda) = \lambda^2 + 2\lambda - 1 + \frac{-\lambda + 3}{\lambda^2 + 1}.$$ ∎

Partial Fractions

The next step is the actual partial fraction expansion, which is performed on the strictly proper rational function S/Q. If S and Q have any common polynomial factors, it saves effort to cancel them at this point.

8.2.3.   Summary: Partial fraction expansion.   Suppose that $R = S/Q$, with S and Q polynomials, is a strictly proper rational function. Suppose also that the denominator polynomial Q has K different real or complex roots $\lambda_1, \lambda_2, \cdots, \lambda_K$, and that the multiplicity of the root λ_i is m_i, $i = 1, 2, \cdots, K$. Then R may be expanded in *partial fractions* as

$$R(\lambda) = \sum_{i=1}^{K} \left(\sum_{k=1}^{m_i} \frac{A_{ik}}{(\lambda - \lambda_i)^k} \right).$$

For $i = 1, 2, \cdots, K$, the real or complex coefficients A_{ik} are given by

$$A_{ik} = \frac{1}{(m_i - k)!} \frac{d^{m_i-k}}{d\lambda^{m_i-k}} [(\lambda - \lambda_i)^{m_i} R(\lambda)] \bigg|_{\lambda = \lambda_k}, \qquad k = 1, 2, \cdots, m_i. \qquad (1)$$

∎

In other words, corresponding to a root λ_i, of multiplicity m_i of the polynomial Q the partial fraction expansion of R has m_i terms as follows:

$$R(\lambda) = \cdots + \frac{A_{i1}}{\lambda - \lambda_i} + \frac{A_{i2}}{(\lambda - \lambda_i)^2} + \cdots + \frac{A_{im_i}}{(\lambda - \lambda_i)^{m_i}} + \cdots.$$

The formula (1) provides one way of obtaining the coefficients A_{ik}. An alternative way, demonstrated in the examples that follow, is to multiply the right-hand side of the partial fraction expansion by the common denominator Q, equate the result to S, and obtain a set of linear equations for the coefficients A_{ik} that may easily be solved.

   We illustrate partial fraction expansion by two examples.

8.2.4.   Example: Partial fraction expansion.
   (a) *Single roots only*. Consider the partial fraction expansion of the strictly proper rational function

$$R(\lambda) = \frac{1}{(\lambda + 1)(\lambda + 2)}.$$

The denominator has two different roots $\lambda_1 = -1$ and $\lambda_2 = -2$, both of which have multiplicity 1. As a result, by 8.2.3 the partial fraction expansion is of the form

$$R(\lambda) = \frac{1}{(\lambda + 1)(\lambda + 2)} = \frac{A_{11}}{\lambda + 1} + \frac{A_{21}}{\lambda + 2}. \tag{2}$$

Evaluation of the coefficients according to (1) yields

$$A_{11} = \frac{1}{0!}(\lambda + 1) \cdot \frac{1}{(\lambda + 1)(\lambda + 2)}\bigg|_{\lambda = -1} = 1,$$

$$A_{21} = \frac{1}{0!}(\lambda + 2) \cdot \frac{1}{(\lambda + 1)(\lambda + 2)}\bigg|_{\lambda = -2} = -1.$$

The alternative method to find the coefficients A_{11} and A_{21} is to multiply (2) by the common denominator $(\lambda + 1)(\lambda + 2)$, so that we obtain

$$1 = A_{11}(\lambda + 2) + A_{21}(\lambda + 1),$$

or

$$1 = (A_{11} + A_{21})\lambda + (2A_{11} + A_{21}).$$

Identification of the coefficients of terms with like powers of λ leads to the set of linear equations

$$A_{11} + A_{21} = 0,$$
$$2A_{11} + A_{21} = 1,$$

whose solution is easily found to be $A_{11} = 1$, $A_{21} = -1$.

   (b) *Repeated roots.* As a further example, consider the partial fraction expansion of

$$R(\lambda) = \frac{\lambda - 1}{(\lambda + 1)^2(\lambda - 2)}.$$

Since the denominator has the double root -1 and the single root 2, according to 8.2.3 the partial fraction expansion is of the form

$$R(\lambda) = \frac{\lambda - 1}{(\lambda + 1)^2(\lambda - 2)} = \frac{A_{11}}{\lambda + 1} + \frac{A_{12}}{(\lambda + 1)^2} + \frac{A_{21}}{\lambda - 2}. \tag{3}$$

The coefficients follow successively as

$$A_{11} = \frac{1}{1!} \frac{d}{d\lambda} \left[(\lambda + 1)^2 \cdot \frac{\lambda - 1}{(\lambda + 1)^2 (\lambda - 2)} \right] \Bigg|_{\lambda = -1} = \frac{-1}{(\lambda - 2)^2} \Bigg|_{\lambda = -1} = -\frac{1}{9},$$

$$A_{12} = \frac{1}{0!} (\lambda + 1)^2 \cdot \frac{\lambda - 1}{(\lambda + 1)^2 (\lambda - 2)} \Bigg|_{\lambda = -1} = \frac{2}{3},$$

$$A_{21} = \frac{1}{0!} (\lambda - 1) \cdot \frac{\lambda - 1}{(\lambda + 1)^2 (\lambda - 2)} \Bigg|_{\lambda = 2} = \frac{1}{9}.$$

As a result, the partial fraction expansion is

$$\frac{\lambda - 1}{(\lambda + 1)^2 (\lambda - 2)} = -\frac{1/9}{\lambda + 1} + \frac{2/3}{(\lambda + 1)^2} + \frac{1/9}{\lambda - 2}.$$

Alternatively, the coefficients A_{11}, A_{12} and A_{21} may be found by multiplying (3) by the common denominator $(\lambda + 1)^2 (\lambda - 2)$, equating the coefficients of like powers of λ and solving the resulting linear equations for the unknown coefficients A_{11}, A_{12}, and A_{21}.   ∎

Real Partial Fractions

When the polynomial Q has complex roots, the partial fraction expansion of 8.2.3 leads to complex arithmetic, even if all the coefficients of the rational function R are real. If Q has real coefficients, corresponding to every complex root λ_i the polynomial Q also has its complex conjugate $\bar{\lambda}_i$ as root. The terms in the partial fraction expansion corresponding to the complex conjugate pair of roots λ_i, $\bar{\lambda}_i$ may then be combined such that only real arithmetic is needed if the coefficients of R are all real.

8.2.5   Summary: Partial fraction expansion with complex conjugate roots.   Let $R = S/Q$ be a strictly proper rational function with real coefficients, and suppose that the polynomial Q has a complex conjugate pair of roots λ_i, $\bar{\lambda}_i$ of multiplicity m_i. Write

$$(\lambda - \lambda_i)(\lambda - \bar{\lambda}_i) = \lambda^2 + a_i \lambda + b_i,$$

with a_i and b_i real coefficients. Then corresponding to this pair of roots the partial fraction expansion of R has m_i terms as follows:

$$R(\lambda) = \cdots + \frac{B_{i1}\lambda + A_{i1}}{\lambda^2 + a_i\lambda + b_i} + \frac{B_{i2}\lambda + A_{i2}}{(\lambda^2 + a_i\lambda + b_i)^2} + \cdots$$

$$+ \frac{B_{im_i}\lambda + A_{im_i}}{(\lambda^2 + a_i\lambda + b_i)^{m_i}} + \cdots,$$

with A_{ik} and B_{ik}, $k = 1, 2, \cdots, m_i$, real constants.   ∎

The following example shows how to find the coefficients.

8.2.6.   Example: Partial fraction expansion with complex conjugate roots.   By way of example we consider the partial fraction expansion of the rational function

$$R(\lambda) = \frac{\lambda + 3}{(\lambda^2 + 2\lambda + 5)(\lambda + 1)}.$$

The factor $\lambda^2 + 2\lambda + 5$ has the complex conjugate pair of roots $-1 \pm 2j$. Rather than following 8.2.3 and working with complex roots we prefer a partial fraction expansion with a quadratic term. According to 8.2.5 this expansion is of the form

$$R(\lambda) = \frac{\lambda + 3}{(\lambda^2 + 2\lambda + 5)(\lambda + 1)} = \frac{B\lambda + A}{\lambda^2 + 2\lambda + 5} + \frac{C}{\lambda + 1},$$

with A, B, and C constants to be determined. Multiplication of the second and third members of the equality by $(\lambda^2 + 2\lambda + 5)(\lambda + 1)$ result in

$$\lambda + 3 = (B\lambda + A)(\lambda + 1) + C(\lambda^2 + 2\lambda + 5),$$

or

$$\lambda + 3 = (B + C)\lambda^2 + (A + B + 2C)\lambda + (A + 5C).$$

Equating the coefficients of like powers of λ results in the three linear equations

$$0 = B + C,$$
$$1 = A + B + 2C,$$
$$3 = A + 5C,$$

whose solution is easily found to be $A = 1/2$, $B = -1/2$, and $C = 1/2$.   ■

Application to Inverse Fourier Transformation

In system analysis, Fourier transforms that are rational functions are often encountered. By partial fraction expansion these rational functions may be reduced to simple fractions whose inverses are known.

8.2.7.   Example: Impulse response of the RCL network.   In Example 4.7.3(b) we found that the RCL network of Example 4.2.4(c) has the frequency response function

$$\hat{h}(f) = \frac{P(j2\pi f)}{Q(j2\pi f)}, \qquad f \in \mathbb{R},$$

where the polynomials P and Q are given by

$$P(\lambda) = \lambda^2, \qquad Q(\lambda) = \lambda^2 + \frac{R}{L}\lambda + \frac{1}{LC}.$$

We consider the partial fraction expansion of

$$\frac{P(\lambda)}{Q(\lambda)} = \frac{\lambda^2}{\lambda^2 + \frac{R}{L}\lambda + \frac{1}{LC}}$$

$$= \frac{\left(\lambda^2 + \frac{R}{L}\lambda + \frac{1}{LC}\right) - \left(\frac{R}{L}\lambda + \frac{1}{LC}\right)}{\lambda^2 + \frac{R}{L}\lambda + \frac{1}{LC}} = 1 - \frac{\frac{R}{L}\lambda + \frac{1}{LC}}{\lambda^2 + \frac{R}{L}\lambda + \frac{1}{LC}}.$$

Assuming the numerical values $R = 3$, $C = 1/200$, and $L = 1/100$, it follows that

$$\frac{P(\lambda)}{Q(\lambda)} = 1 - \frac{300\lambda + 20000}{\lambda^2 + 300\lambda + 20000}.$$

The denominator of the second term on the right-hand side may be factored as $\lambda^2 + 300\lambda + 20000 = (\lambda + 100)(\lambda + 200)$, so that this term may be expanded in partial fractions as

$$\frac{300\lambda + 20000}{\lambda^2 + 300\lambda + 20000} = \frac{A_1}{\lambda + 100} + \frac{A_2}{\lambda + 200}.$$

It is easily found that $A_1 = -100$ and $A_2 = 400$. It follows that the frequency response function of the RCL network may be expressed in partial fraction form as

$$\hat{h}(f) = \frac{P(j2\pi f)}{Q(j2\pi f)} = 1 + \frac{100}{j2\pi f + 100} - \frac{400}{j2\pi f + 200}, \qquad f \in \mathbb{R}.$$

Since the constant 1 is the CCFT of the δ-function, and from Table 7.4

$$x(t) = e^{-at}\mathbb{1}(t), \qquad t \in \mathbb{R}, \qquad \hat{x}(f) = \frac{1}{j2\pi f + a}, \qquad f \in \mathbb{R},$$

with a positive, form a CCFT pair, it follows that the inverse CCFT h of $\hat{h}$ is given by

$$h(t) = \delta(t) + (100e^{-100t} - 400e^{-200t})1(t), \qquad t \in \mathbb{R}.$$

This is the impulse response of the RCL network.   ∎

8.3 THE z- TRANSFORM AND THE LAPLACE TRANSFORM

In this section we introduce and explore the z- and Laplace transforms. The z- transform applies to discrete-time signals, the Laplace transform to continuous-time signals. Both the z- and the Laplace transform come in two versions: the *two*-sided transform and the *one*-sided transform. They are applied in Sections 8.6, 8.7, and 8.8 to systems defined on two- and one-sided time axes, respectively.

The Two-Sided z- and Laplace Transforms

The reason for introducing the two-sided z- and Laplace transforms is that the DCFT and CCFT cannot deal with signals that increase exponentially. A signal x defined on the time axis $\mathbb{T}$ is said to be of *exponential growth* if there exist real constants α and β such that

$$|x(t)| \le \beta \alpha^t, \qquad t \in \mathbb{T}.$$

The two-sided z- and Laplace transforms are *generalizations* of the DCFT and CCFT in the sense that they apply to all signals that have DCFTs and CCFTs, and additionally to signals that increase exponentially, provided the increase is in one direction only. The idea is to multiply the original signal by a *convergence factor*, which is an exponential signal of the form

$$\rho^{-n}, \qquad n \in \mathbb{Z},$$

in the discrete-time case or

$$e^{-\sigma t}, \qquad t \in \mathbb{R},$$

in the continuous-time case, with ρ and σ suitably chosen real numbers.

Thus, in the discrete-time case we take the DCFT of the signal $\rho^{-n}x(n)$, $n \in \mathbb{Z}$, and hence obtain

$$X(z) = \sum_{n=-\infty}^{\infty} x(n)\rho^{-n}e^{-j2\pi fn} = \sum_{n=-\infty}^{\infty} z^{-n}x(n),$$

where $z = \rho e^{j2\pi f}$. We consider X for all complex z for which the infinite sum converges, and call it the *z-transform* of the discrete-time signal x.

Likewise, in the continuous-time case we take the CCFT of the signal $x(t)e^{-\sigma t}$, $t \in \mathbb{R}$, and thus obtain

$$X(s) = \int_{-\infty}^{\infty} x(t)e^{-\sigma t}e^{-j2\pi ft} \, dt = \int_{-\infty}^{\infty} x(t)e^{-st} \, dt,$$

where $s = \sigma + j2\pi f$. Again we consider X for all complex s for which the integral exists and call it the *Laplace transform* of the continuous-time signal x.

8.3.1.   Definition: The two-sided z- and Laplace transforms.

The z-transform is the map $\mathscr{Z}$ that transforms the discrete-time signal x defined on the time axis $\mathbb{Z}$ to the complex function $X = \mathscr{Z}(x)$ of a complex variable given by

$$X(z) = \sum_{n=-\infty}^{\infty} x(n)z^{-n}, \qquad z \in \mathscr{E}.$$

The set $\mathscr{E} \subset \mathbb{C}$ consists of all complex numbers z for which the sum converges and is called the *existence region* of X.

The Laplace-transform is the map $\mathscr{L}$ that transforms the continuous-time regular or singular signal x defined on the time axis $\mathbb{R}$ to the complex function $X = \mathscr{L}(x)$ of a complex variable given by

$$X(s) = \int_{-\infty}^{\infty} x(t)e^{-st} \, dt, \qquad s \in \mathscr{E}.$$

The set $\mathscr{E} \subset \mathbb{C}$ consists of all complex numbers s for which the integral converges and is called the *existence region* of X. ∎

We usually indicate the z- or Laplace transform of a signal by a capital letter corresponding to the lower case letter that names the signal. The z- and Laplace transforms of 8.3.1 are sometimes called the *two-sided z-* and Laplace transforms, to distinguish then from the one-sided z- and Laplace transforms that follow.

Examples of z-Transforms

We present several examples of z-transforms of one- and two-sided signals.

> The name z-transform derives from the letter z that often denotes the independent variable of the transform. It is a nondescriptive name. In some branches of mathematics, for instance probability theory, a closely related concept is known under the name *generating function*. The z-transform has much to do with the theory of *power series*. The reason is that the z-transform of a discrete-time signal simply is a power series in z^{-1} with the signal values as coefficients.
>
> Pierre S. Laplace (1749–1827) was a French mathematician. The impetus for the application of the Laplace transform to electrical circuit theory was actually given by the work of the British self-educated physicist Oliver Heaviside (1850–1925). Heaviside developed what is called the *operational calculus* for the analysis of differential equations, whose mathematical justification is the Laplace transform. Heaviside's work was not appreciated in his time, and he died in neglect.

8.3.2.   Examples: *z*-transforms.

(a) *Right one-sided exponential.* Consider the right one-sided exponential discrete-time signal

$$x(n) = a^n 1(n), \qquad n \in \mathbb{Z},$$

as sketched in Fig. 8.1(a). Its *z*-transform $X = \mathscr{Z}(x)$ is given by

$$X(z) = \sum_{n=-\infty}^{\infty} a^n 1(n) z^{-n} = \sum_{n=0}^{\infty} a^n z^{-n} = \sum_{n=0}^{\infty} \left(\frac{a}{z}\right)^n = \frac{1}{1 - \dfrac{a}{z}} = \frac{z}{z-a}.$$

Since the infinite sum converges if and only if $|a/z| < 1$, the existence region of the *z*-transform is $|z| > |a|$.

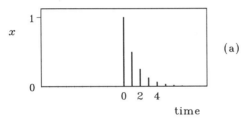

x

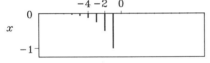

x

(a)

(b)   **Figure 8.1.**   Top: a right one-sided exponential signal. Bottom: a left one-sided exponential signal.

(b) *A left one-sided exponential.* The left one-sided exponential signal

$$x(n) = -a^n 1(-n - 1), \qquad n \in \mathbb{Z},$$

as in Fig. 8.1(b), has the *z*-transform

$$X(z) = -\sum_{n=-\infty}^{-1} a^n z^{-n} = -\sum_{k=1}^{\infty} \left(\frac{z}{a}\right)^k = -\frac{\dfrac{z}{a}}{1 - \dfrac{z}{a}} = \frac{z}{z-a}.$$

This *z*-transform looks the same as that of (a), but the infinite sum now converges provided $|z/a| < 1$. It follows that the existence region of this *z*-transform is $|z| < |a|$, the complement (excluding the line $|z| = |a|$) of the convergence region of the preceding example.

(c) *A two-sided exponential.* The two-sided exponential signal defined by

$$x(n) = a^n, \qquad n \in \mathbb{Z},$$

has no two-sided *z*-transform, because the infinite sum that defines it diverges for every $z \in \mathbb{C}$.

(d) *Shifted unit pulse*. It is easily verified that for any $k \in \mathbb{Z}$ the shifted unit pulse

$$x(n) = \Delta(n - k), \qquad n \in \mathbb{Z},$$

has the *z*-transform

$$X(z) = z^{-k} \quad \begin{cases} \text{for all } z \neq 0 \text{ if } k > 0, \\ \text{for all } z \in \mathbb{C} \text{ if } k \leq 0. \end{cases}$$

■

Existence Region of Two-Sided *z*-Transforms

Examples 8.3.2(a) and (b) show that it is important to know the existence region of a *z*-transform. The example moreover shows, and indeed it may be proved in general, that *right one-sided* signals, which may increase exponentially but not faster, have an existence region of the form

$$\mathcal{E} = \{z \in \mathbb{C} \mid |z| > \rho\}$$

for some real ρ that is large enough. This is a consequence of the fact that if the *z*-transform of a right one-sided signal exists for some $z_o \in \mathbb{C}$, it exists for all *z* such that $|z| > |z_o|$. The general form of the existence region is shown in Fig. 8.2(a).

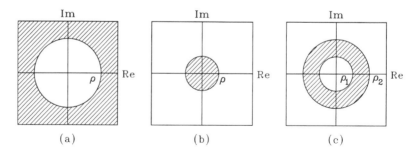

Figure 8.2.   Left: existence region of the *z*-transform of a right one-sided signal. Middle: existence region of a left one-sided signal. Right: annular existence region of a two-sided signal.

Left one-sided signals, which may increase exponentially in the negative time direction, have an existence region of the form

$$\mathcal{E} = \{z \in \mathbb{C} \mid |z| < \rho\}$$

for some real ρ that is small enough. Possibly ρ may be replaced with ∞. This type of existence region is shown in Fig. 8.2(b).

As 8.3.2(c) illustrates, *two*-sided exponentially increasing signals need not have a *z*-transform. If they do, their existence region is of the form

$$\mathscr{E} = \{z \in \mathbb{C} \mid \rho_1 < |z| < \rho_2\}$$

where ρ_1 and ρ_2 are real numbers. Possibly ρ_2 may be replaced with ∞. Figure 8.2(c) illustrates an existence region of this type, which is said to have an *annular* shape.

The next example illustrates such an annular existence region.

8.3.3.   **Example: Further *z*-transforms.**

(a) *A two-sided signal*. Consider the discrete-time signal defined by

$$x(n) = a^n \mathbb{1}(n) + b^n \mathbb{1}(-n - 1), \qquad n \in \mathbb{Z},$$

as sketched in Fig. 8.3(a) for $a = 1/2$ and $b = 3$. From Example 8.3.2 we know that the *z*-transform of the first term on the right-hand side exists provided $|z| > |a|$ while that of the second exists if $|z| < |b|$. It follows that the *z*-transform of x only exists if $|a| < |b|$, and its existence region then is given by

$$\mathscr{E} = \{z \in \mathbb{C} \mid |a| < |z| < |b|\}.$$

The *z*-transform itself is

$$X(z) = \frac{z}{z - a} - \frac{z}{z - b} = \frac{(a - b)z}{(z - a)(z - b)}, \qquad |a| < |z| < |b|.$$

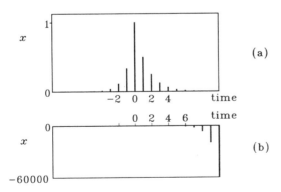

(a)

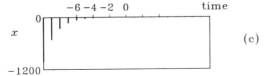

(b)

(c)

Figure 8.3.   Three signals that have the same *z*-transforms but different existence regions. Top: a two-sided signal. Middle: a right one-sided signal. Bottom: a left one-sided signal.

(b) *Signals with the same z-transforms but different existence regions*. To continue this example, suppose that we are given the same *z*-transform, but with a different existence region, namely

$$X(z) = \frac{(a - b)z}{(z - a)(z - b)}, \qquad |z| > \max(|a|, |b|).$$

It follows that

$$X(z) = \frac{z}{z - a} - \frac{z}{z - b}, \qquad |z| > \max(|a|, |b|),$$

which by Example 8.3.2(a) evidently is the *z*-transform of the right one-sided time signal

$$x(n) = [a^n - b^n]1(n), \qquad n \in \mathbb{Z},$$

of Fig. 8.3(b). If, on the other hand,

$$X(z) = \frac{z}{z - a} - \frac{z}{z - b}, \qquad |z| < \min(|a|, |b|),$$

by Example 8.3.2(b) we are clearly confronted with the *z*-transform of the left one-sided signal

$$x(n) = [-a^n + b^n]1(-n - 1), \qquad n \in \mathbb{Z},$$

of Fig. 8.3(c). ∎

Examples of Laplace Transforms

We continue with some examples of Laplace transforms.

8.3.4.   Examples: Laplace transforms.
(a) *Right one-sided exponential*. The Laplace transform of the right one-sided signal

$$x(t) = e^{at}1(t), \qquad t \in \mathbb{R},$$

is

$$X(s) = \int_{-\infty}^{\infty} e^{at}1(t)e^{-st}\, dt = \int_{0}^{\infty} e^{-(s-a)t}\, dt = \frac{e^{-(s-a)t}}{-(s - a)}\bigg|_{0}^{\infty}$$

$$= \frac{1}{s - a}, \qquad \text{Re}(s) > \text{Re}(a),$$

where the existence region Re $(s) > $ Re (a) is imposed by the requirement that the integral converge.

(b) *Left one-sided exponential.* The Laplace transform of the left one-sided signal

$$x(t) = -e^{at}1(-t), \qquad t \in \mathbb{R},$$

is

$$X(s) = -\int_{-\infty}^{0} e^{at}e^{-st}\, dt = \frac{1}{s-a}, \qquad \text{Re } (s) < \text{Re } (a).$$

The Laplace transform has the same form as that of the right one-sided signal of (a), but is defined on a different region.

(c) *A two-sided signal.* The two-sided signal x given by

$$x(t) = e^{at}1(t) + e^{bt}1(-t), \qquad t \in \mathbb{R},$$

has a two-sided Laplace transform if and only if Re $(a) < $ Re (b). If the Laplace transform exists, then it is given by

$$X(s) = \frac{1}{s-a} - \frac{1}{s-b}, \qquad \text{Re } (a) < \text{Re } (s) < \text{Re } (b).$$

In particular, setting $b = -a$ it follows that the two-sided Laplace transform of the signal z given by

$$z(t) = e^{a|t|}, \qquad t \in \mathbb{R},$$

exists provided Re $(a) < 0$, and then is given by

$$Z(s) = \frac{1}{s-a} - \frac{1}{s+a} = \frac{2a}{s^2 - a^2}, \qquad \text{Re } (a) < \text{Re } (s) < -\text{Re } (a).$$

(d) *A two-sided exponential.* The two-sided exponential signal

$$x(t) = e^{at}, \qquad t \in \mathbb{R},$$

does not have a two-sided Laplace transform for any $a \in \mathbb{C}$. Similarly, the two-sided polynomial signal

$$x(t) = t^k, \qquad t \in \mathbb{R},$$

does not have a two-sided Laplace transform for any $k \in \mathbb{Z}_+$.

(e) *Delta function*. The Laplace transform of the singular signal $x = \delta$ is

$$X(s) = \int_{-\infty}^{\infty} \delta(t)e^{-st}\, dt = 1 \qquad \text{for all} \quad s \in \mathbb{C}. \qquad \blacksquare$$

Existence Region of Two-Sided Laplace Transforms

We briefly review which signals have two-sided Laplace transforms. First of all, every *one*-sided signal of exponential growth has a two-sided Laplace transform. Right one-sided signals of exponential growth have an existence region of the form

$$\mathscr{E} = \{s \in \mathbb{C} \mid \text{Re}(s) > \sigma\}$$

for some real σ, possibly replaced with $-\infty$. The existence region of left one-sided signals of exponential growth is of the form

$$\mathscr{E} = \{s \in \mathbb{C} \mid \text{Re}(s) < \sigma\}$$

for some real σ, possibly replaced with ∞. Figures 8.4(a) and (b) show these existence regions.

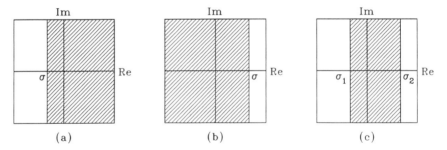

Figure 8.4.   Left: existence region of the Laplace transform of a right one-sided signal. Middle: existence region of a left one-sided signal. Right: existence strip of a two-sided signal.

Two-sided signals need not have a Laplace transform, as Example 8.3.4(d) illustrates. If they do, their existence region is of the form

$$\mathscr{E} = \{s \in \mathbb{C} \mid \sigma_1 < \text{Re}(s) < \sigma_2\},$$

with σ_1 and σ_2 real numbers. Possibly σ_1 may be replaced with $-\infty$ and σ_2 with ∞. Figure 8.4(c) shows that the existence region has the shape of a *strip* in the complex plane.

A comparison of Examples 8.3.4(a) and (b) reveals that it is important to keep track of the existence region of the Laplace transform, because the Laplace transforms of different time signals may be represented by identical formulas but have different existence regions.

Relation of the Two-Sided *z*- and Laplace Transforms With the DCFT and CCFT

The *z*- and Laplace transforms of a signal were introduced earlier in this section as the DCFT or CCFT, respectively, of the signal multiplied by an exponential signal. This results in the following relationship between the *z*-transform and DCFT of a discrete-time signal and the Laplace transform and CCFT of a continuous-time signal.

8.3.5.   Summary: Relationship between the *z*-transform and the DCFT and between the Laplace transform and the CCFT.

(a) Let X be the *z*-transform of the discrete-time signal x, with existence region $\mathscr{E}$. Then, if ρ is a positive real number such that the circle

$$\rho e^{j2\pi f}, \qquad f \in [0, 1),$$

is entirely contained in $\mathscr{E}$, the frequency signal

$$\hat{x}_\rho(f) := X(\rho e^{j2\pi f}), \qquad f \in [0, 1),$$

obtained by substituting $z = \rho e^{j2\pi f}$ in $X(z)$, is the DCFT of the signal x_ρ given by

$$x_\rho(n) = x(n)\rho^{-n}, \qquad n \in \mathbb{Z}.$$

(b) In particular, if the existence region $\mathscr{E}$ includes the unit circle, that is, $X(z)$ exists for $|z| = 1$, the signal x has a DCFT $\hat{x}$, and

$$\hat{x}(f) = X(e^{j2\pi f}), \qquad f \in [0, 1).$$

(a′) Let X be the Laplace transform of the continuous-time signal x with existence region $\mathscr{E}$. Then, if σ is a real number such that the vertical line

$$\sigma + j2\pi f, \qquad f \in \mathbb{R},$$

is entirely contained in $\mathscr{E}$, the frequency signal

$$\hat{x}_\sigma(f) = X(\sigma + j2\pi f), \qquad f \in \mathbb{R},$$

obtained by substituting $s = \sigma + j2\pi f$ in $X(s)$, is the CCFT of the signal x_σ given by

$$x_\sigma(t) = x(t)e^{-\sigma t}, \qquad t \in \mathbb{R}.$$

(b′) In particular, if the existence region $\mathscr{E}$ includes the imaginary axis, that is, $X(s)$ exists for Re $(s) = 0$, the signal x has a CCFT $\hat{x}$, and

$$\hat{x}(f) = X(j2\pi f), \qquad f \in \mathbb{R}. \qquad \blacksquare$$

Figure 8.5 illustrates the circle $\rho e^{j2\pi f}$, $0 \leq f < 1$, and the vertical line $\sigma + j2\pi f$, $f \in \mathbb{R}$. Both are pictured inside the existence region.

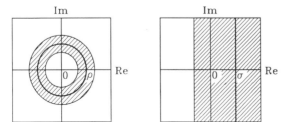

Figure 8.5. Left: the circle $\rho e^{j2\pi f}$, $0 \leq f < 1$. Right: the vertical line $\sigma + j2\pi f, f \in \mathbb{R}$.

8.3.6.   Examples: Relation of z-transform and DCFT, Laplace transform and CCFT.

(a) *Discrete-time exponential signal.* The right one-sided exponential signal

$$x(n) = a^n 1(n), \qquad n \in \mathbb{Z},$$

according to 8.3.2(a) has the two-sided z-transform

$$X(z) = \frac{z}{z - a}, \qquad |z| > |a|.$$

If $|a| < 1$, the existence region includes the unit circle, and the DCFT $\hat{x}$ of x is given by

$$\hat{x}(f) = X(e^{j2\pi f}) = \frac{e^{j2\pi f}}{e^{j2\pi f} - a}, \qquad f \in [0, 1).$$

For $|a| > 1$ the z-transform of x is well-defined, but the DCFT of x does not exist.

(b) *Continuous-time exponential signal.* The right one-sided exponential signal

$$x(t) = e^{at} 1(t), \qquad t \in \mathbb{R},$$

according to 8.3.4(a) has the two-sided Laplace transform

$$X(s) = \frac{1}{s - a}, \qquad \text{Re}(s) > \text{Re}(a).$$

If $\text{Re}(a) < 0$, the existence region includes the imaginary axis, and the CCFT $\hat{x}$ of x is given by

$$\hat{x}(f) = X(j2\pi f) = \frac{1}{j2\pi f - a}, \qquad f \in \mathbb{R}.$$

For $\text{Re}(a) > 0$ the Laplace transform of x is well-defined, but the CCFT of x does not exist.   ∎

The One-Sided z- and Laplace Transforms

The two-sided z- and Laplace transforms may be applied to the analysis of signals defined on two-sided infinite time axes. For signals defined on the one-sided infinite time axes $\mathbb{Z}_+$ or $\mathbb{R}_+$ we use the *one-sided* z- or Laplace transform, respectively. The one-sided transforms are particularly convenient for handling difference and differential systems that are not initially at rest.

The one-sided z- or Laplace transform of a signal x defined on a two-sided infinite time axis simply is the two-sided transform of the signal

$$x \cdot \mathbb{1},$$

with $\mathbb{1}$ the unit step. This multiplication by the unit step turns the signal into a one-sided signal.

8.3.7.   Definition: One-sided z- and Laplace transforms.

The one-sided z-transform is the map $\mathscr{Z}_+$ that transforms the discrete-time signal x defined on the time axis $\mathbb{Z}$ to the complex function $X_+ = \mathscr{Z}_+(x)$ of a complex variable given by

The one-sided Laplace transform is the map $\mathscr{L}_+$ that transforms the regular or singular continuous-time signal x defined on the time axis $\mathbb{R}$ to the complex function $X_+ = \mathscr{L}_+(x)$ of a complex variable given by

$$X_+(z) = \sum_{n=0}^{\infty} x(n)z^{-n}, \qquad z \in \mathscr{E}_+.$$

$$X_+(s) = \int_0^{\infty} x(t)e^{-st}\, dt, \qquad s \in \mathscr{E}_+.$$

The set $\mathscr{E}_+ \subset \mathbb{C}$ consists of all complex numbers z for which the sum converges and is called the *existence region* of X_+.

The set $\mathscr{E}_+ \subset \mathbb{C}$ consists of all complex numbers s for which the integral converges and is called the *existence region of X_+*. ∎

We usually denote the one-sided z- or Laplace transform of a signal by a capital letter with subscript $+$, such as X_+. If no confusion is possible, the subscript $+$ is sometimes omitted.

There is a close relation between the one- and two-sided z- and Laplace transforms, because

$$\mathscr{Z}_+(x) = \mathscr{Z}(x \cdot \mathbb{1}),$$
$$\mathscr{L}_+(x) = \mathscr{L}(x \cdot \mathbb{1}).$$

In particular, if x is a discrete- or continuous-time signal such that

$$x(t) = 0 \qquad \text{for} \quad t < 0,$$

the one- and two-sided z- or Laplace transforms are *identical,* with identical existence regions. To compute the one-sided transform of a time signal x its values at negative times are not needed; in fact, the signal need not even be defined for negative times.

Because the one-sided z-transform is the two-sided z-transform of a right one-sided signal, the existence region of one-sided z-transforms is of the form

$$\mathscr{E}_+ = \{z \in \mathbb{C} \mid |z| > \rho\},$$

for some real number ρ. For the same reason, the existence region of one-sided Laplace transforms has the appearance

$$\mathcal{E}_+ = \{s \in \mathbb{C} \mid \mathrm{Re}\,(s) > \sigma\},$$

with σ some real number. Possibly σ may be replaced with $-\infty$.

The one sided Laplace transform is defined in 8.3.7 as an integral from 0 to ∞ under the assumption that x has *no* singularity at time 0. If this assumption is not satisfied, such as when x has a δ-function at time 0, then we integrate from 0^-, that is, we define

$$X_+(s) = \int_{0^-}^{\infty} x(t)e^{-st}\,dt = \lim_{\epsilon \downarrow 0} \int_{-\epsilon}^{\infty} x(t)e^{-st}\,dt, \qquad s \in \mathcal{E}_+,$$

so that the singularity is included in the integration interval.

8.3.8.   Examples: One-sided z-transforms.

(a) *A two-sided signal.* Let the two-sided discrete-time signal x be given by

$$x(n) = \begin{cases} 0 & \text{for } n \leq -3, \\ -5 & \text{for } n = -2, \\ 2 & \text{for } n = -1, \\ 3 & \text{for } n = 0, \\ 4 & \text{for } n = 1, \\ 7 & \text{for } n = 2, \\ 1 & \text{for } n = 3, \\ 0 & \text{for } n \geq 4, \end{cases} \qquad n \in \mathbb{Z}.$$

Then its two-sided z-transform is

$$X(z) = -5z^{-2} + 2z^{-1} + 3 + 4z + 7z^2 + z^3, \qquad |z| > 0,$$

while its one-sided z-transform is

$$X_+(z) = 3 + 4z + 7z^2 + z^3, \qquad z \in \mathbb{C}.$$

(b) *Exponential signal and unit step.* Consider the discrete-time signal x given by

$$x(n) = \begin{cases} a^n & \text{for } n \geq 0, \\ \text{arbitrary} & \text{for } n < 0, \end{cases} \qquad n \in \mathbb{Z}.$$

By 8.3.2(a), x has the one-sided z-transform

$$X_+(z) = \frac{z}{z - a}, \qquad |z| > |a|.$$

In particular, this is the one-sided z-transform of x given by $x(n) = a^n$, $n \in \mathbb{Z}$, which by 8.3.2(c) has no two-sided z-transform.

By setting $a = 1$ it follows that the constant 1 and the unit step $\mathbb{1}$ have the same one-sided z-transform

$$X_+(z) = \frac{z}{z - 1}, \qquad |z| > 1.$$

(c) *Shifted unit pulse.* Let x be the shifted unit pulse given by

$$x(n) = \Delta(n - k), \qquad n \in \mathbb{Z},$$

with k an integer. For $k \geq 0$ the two- and one-sided z-transforms of x are both given by

$$X(z) = X_+(z) = z^{-k}, \qquad |z| > 0.$$

For $k < 0$, however,

$$X(z) = z^{-k}, \qquad z \in \mathbb{C},$$

while

$$X_+(z) = 0, \qquad z \in \mathbb{C}. \qquad \blacksquare$$

8.3.9.   Examples: One-sided Laplace transforms.

(a) *Exponential signal and unit step.* The one-sided Laplace transform of the signal x given by

$$x(t) = \begin{cases} e^{at} & \text{for } t \geq 0, \\ \text{arbitrary} & \text{for } t < 0, \end{cases} \qquad t \in \mathbb{R},$$

by Example 8.3.4(a) is

$$X_+(s) = \frac{1}{s - a}, \qquad \text{Re } (s) > \text{Re } (a).$$

In particular, by setting $a = 0$ it follows that the constant 1 and the unit step $\mathbb{1}$ have the *same* one-sided Laplace transform

$$X_+(s) = \frac{1}{s}, \qquad \text{Re } (s) > 0.$$

(b) *Shifted δ-function.* As a further example, consider the shifted δ-function

$$x(t) = \delta(t - \theta), \qquad t \in \mathbb{R},$$

with θ a real number. For $\theta \geq 0$ its one- and two-sided Laplace transforms are the same and given by

$$X(s) = X_+(s) = \int_{0^-}^{\infty} \delta(t - \theta)e^{-st}\, dt = e^{-s\theta}, \qquad s \in \mathbb{C}.$$

For $\theta < 0$ the two-sided Laplace transform of x is given by the same expression, but the one-sided Laplace transform is zero:

$$X(s) = e^{-s\theta}, \qquad s \in \mathbb{C},$$

$$X_+(s) = 0, \qquad s \in \mathbb{C}. \qquad\blacksquare$$

8.3.10.   Review: The sampled *z*-transform.   The two-sided sampled *z*-transform X of a sampled signal x defined on the time axis $\mathbb{Z}(T)$ is defined as

$$X(z) = T \sum_{t \in \mathbb{Z}(T)} x(t)z^{-t/T}, \qquad z \in \mathscr{E},$$

with $\mathscr{E}$ its existence region. Writing $z = \rho e^{j2\pi fT}$ it follows that

$$X(\rho e^{j2\pi fT}) = T \sum_{t \in \mathbb{Z}(T)} \rho^{-t/T} x(t)e^{-j2\pi ft},$$

which shows that for fixed ρ the *z*-transform is the DCFT of the signal $\rho^{-t/T}x(t)$, $t \in \mathbb{Z}(T)$.

The one-sided *z*-transform X_+ of x is similarly defined as

$$X_+(z) = T \sum_{t \in \mathbb{Z}_+(T)} x(t)z^{-t/T}, \qquad z \in \mathscr{E}_+.$$

It is not difficult to extend all the results that hold for the *z*-transform in sequence form to the sampled *z*-transform, but, since the sampled version of the *z*-transform is not common, we do not pursue it.   ■

8.4 PROPERTIES OF THE *z*- AND LAPLACE TRANSFORMS

We now turn to a review of some of the most important properties of the *z*- and Laplace transforms. Some apply to both the one- and two-sided versions, but several, notably those related to shifting and differentiation, differ.

8.4.1.   Summary: Linearity of the z- and Laplace transforms

Suppose that the one- or two-sided z-transforms X and Y of the discrete-time signals x and y exist on some common existence region $\mathscr{E}$. Then the one- or two-sided z-transform of any linear combination $\alpha x + \beta y$ of x and y also exists on $\mathscr{E}$ and is given by

$$\mathscr{Z}(\alpha x + \beta y) = \alpha X + \beta Y.$$

Suppose that the one- or two-sided Laplace transforms X and Y of the continuous-time signals x and y exist on some common existence region $\mathscr{E}$. Then the one- or two-sided Laplace transform of any linear combination $\alpha x + \beta y$ of x and y also exists on $\mathscr{E}$ and is given by

$$\mathscr{L}(\alpha x + \beta y) = \alpha X + \beta Y. \qquad \blacksquare$$

The linearity property is easily proved. The *largest* common existence region $\mathscr{E}$ referred to in 8.4.1 is the intersection $\mathscr{E}_x \cap \mathscr{E}_y$ of the individual existence regions $\mathscr{E}_x$ and $\mathscr{E}_y$ of X and Y, respectively.

An important tool for the analysis of linear time-invariant systems is the *convolution* property.

8.4.2.   Summary: Convolution property of the z- and Laplace transforms.

(a) *Convolution property of the two-sided z-transform.* Suppose that the two-sided z-transforms X and Y of the discrete-time signals x and y exist on some common existence region $\mathscr{E}$. Then the two-sided z-transform of the convolution $x * y$ of x and y also exists on $\mathscr{E}$ and is given by

$$\mathscr{Z}(x * y) = X \cdot Y.$$

(a′) *Convolution property of the two-sided Laplace transform.* Suppose that the two-sided Laplace transforms X and Y of the continuous-time signals x and y exist on some common existence region $\mathscr{E}$. Then the two-sided Laplace transform of the convolution $x * y$ of x and y also exists on $\mathscr{E}$ and is given by

$$\mathscr{L}(x * y) = X \cdot Y.$$

(b) *Convolution property of the one-sided z-transform.* Let x and y be discrete-time signals. Then if the one-sided z-transform of x and that of y exist on some common existence region $\mathscr{E}_+$, the one-sided z-transform of $x\mathbb{1} * y\mathbb{1}$ also exists on $\mathscr{E}_+$ and is given by

$$\mathscr{Z}_+(x\mathbb{1} * y\mathbb{1}) = \mathscr{Z}_+(x) \cdot \mathscr{Z}_+(y).$$

(b′) *Convolution property of the one-sided Laplace transform.* Let x and y be continuous-time signals. Then if the one-sided Laplace transform of x and that of y exist on some common existence region $\mathscr{E}_+$, the one-sided Laplace transform of $x\mathbb{1} * y\mathbb{1}$ also exists on $\mathscr{E}_+$ and is given by

$$\mathscr{L}_+(x\mathbb{1} * y\mathbb{1}) = \mathscr{L}_+(x) \cdot \mathscr{L}_+(y). \qquad \blacksquare$$

The proof of the convolution property is straightforward and is found in E.7 of Supplement E.

We next consider the *shift* properties of the z- and Laplace transforms.

8.4.3.   Summary: Shift properties of the z- and Laplace transforms.

(a) *Shift property of the two-sided z-transform.* Let the discrete-time signal x have the two-sided z-transform X with existence region $\mathscr{E}$. Then for every $k \in \mathbb{Z}$ the backward shifted signal

$$x(n + k), \qquad n \in \mathbb{Z},$$

has a two-sided z-transform that exists on the same region and is given by

$$z^k X(z), \qquad z \in \mathscr{E}.$$

(b) *Shift property of the one-sided z-transform.* If x has the one-sided z-transform X_+ with existence region $\mathscr{E}_+$, the once back shifted signal σx and the once forward shifted signal $\sigma^{-1} x$ have one-sided z-transforms that exists on the same region. They are given by

$$\mathscr{Z}_+(\sigma x)(z) = zX_+(z) - zx(0),$$
$$\mathscr{Z}_+(\sigma^{-1}x)(z) = z^{-1}X_+(z) + x(-1).$$

(c) *Converse shift property.* If x has a one- or two-sided z-transform X with existence region $\mathscr{E}$, the signal

$$a^n x(n), \qquad n \in \mathbb{Z},$$

with a a fixed nonzero complex number, has a one- or two-sided z-transform, respectively, given by

$$X\!\left(\frac{z}{a}\right), \qquad \frac{z}{a} \in \mathscr{E}.$$

(a$'$) *Shift property of the two-sided Laplace transform.* Let the regular or singular continuous-time signal x have the two-sided Laplace transform X with existence region $\mathscr{E}$. Then for every $\theta \in \mathbb{R}$ the backward shifted signal

$$x(t + \theta), \qquad t \in \mathbb{R},$$

has a two-sided Laplace transform that exists on the same region and is given by

$$e^{s\theta}X(s), \qquad s \in \mathscr{E}.$$

(b$'$) *Shift property of the one-sided Laplace transform.* If x has the one-sided Laplace transform X_+ with existence region $\mathscr{E}_+$, for $\theta \geq 0$ the *truncated* and *forward* shifted signal

$$x(t - \theta)\mathbb{1}(t-\theta), \qquad t \in \mathbb{R},$$

has the one-sided Laplace transform

$$e^{-s\theta}X_+(s), \qquad s \in \mathscr{E}_+.$$

(c$'$) *Converse shift property.* If x has a one- or two-sided Laplace transform X with existence region $\mathscr{E}$, the signal

$$e^{at}x(t), \qquad t \in \mathbb{R},$$

with a a fixed complex number, has a one- or two-sided Laplace transform, respectively, given by

$$X(s - a), \qquad s - a \in \mathscr{E}. \qquad \blacksquare$$

Also the shift property is easily proved, as shown in E.8 of Supplement E. In (b), σ is the usual backshift operator defined by $\sigma^k x(n) = x(n + k)$, $n \in \mathbb{Z}$, with k any integer. Note that the shift property (b$'$) of the one-sided Laplace transform involves back shifting *and* truncation. Replacing z with z/a in the converse shift property (c) of the z-transform may be thought of as a "radial shift."

8.4.4.   Exercise: Multiple shifts for the z-transform. Show that for any $k \in \mathbb{N}$

$$\mathscr{Z}_+(\sigma^k x)(z) = z^k X_+(z) - \sum_{i=0}^{k-1} z^{k-i} x(i), \qquad z \in \mathscr{C}_+,$$

$$\mathscr{Z}_+(\sigma^{-k} x)(z) = z^{-k} X_+(z) + \sum_{i=1}^{k} z^{-(k-i)} x(-i), \qquad z \in \mathscr{C}_+,$$

where X_+ is the one-sided z-transform of x and $\mathscr{C}_+$ its existence region.   ■

The shift property of the one-sided z-transform has an important application in the solution of initial value problems for constant coefficient linear difference systems.
    We next state the *differentiation* properties of the Laplace transform.

8.4.5.   Summary: Differentiation properties of the Laplace transform.
    (a) *Differentiation property of the two-sided Laplace transform.* Suppose that the regular or singular continuous-time signal x has the two-sided Laplace transform X with existence region $\mathscr{E}$. Then the derivative Dx of x has the two-sided Laplace transform

$$\mathscr{L}(Dx)(s) = sX(s), \qquad s \in \mathscr{E}.$$

    (b) *Differentiation property of the one-sided Laplace transform.* If x has the one-sided Laplace transform X_+ with existence region $\mathscr{E}_+$, its derivative Dx has the one-sided Laplace transform

$$\mathscr{L}_+(Dx)(s) = sX_+(s) - x(0^-), \qquad s \in \mathscr{E}_+.$$ ■

The proof is given in E.9 of Supplement E. The differentiation property of the one-sided Laplace transform plays an important role in the application of the Laplace transform to the solution of initial value problems for constant coefficient linear differential systems.

8.4.6.   Exercise: Multiple differentiation. Prove that for any $k \in \mathbb{N}$

$$\mathscr{L}_+(D^k x)(s) = s^k X_+(s) - \sum_{i=0}^{k-1} s^{k-i-1} (D^i x)(0^-), \qquad s \in \mathscr{E}_+,$$

with X_+ the one-sided Laplace transform of x and E_+ its existence region.   ■

The converse differentiation property of the z- and Laplace transforms may be formulated as follows.

8.4.7.   Summary: Converse differentiation property of the *z*- and Laplace transforms.

Suppose that the one- or two-sided *z*-transform X of the discrete-time signal x is differentiable with respect to z in its existence region $\mathscr{E}$. Then the one- or two-sided *z*-transform, respectively, of the signal

$$-nx(n), \qquad n \in \mathbb{Z},$$

exists on $\mathscr{E}$ and is given by

$$z\frac{dX(z)}{dz}, \qquad z \in \mathscr{E}.$$

Suppose that the one- or two-sided Laplace transform X of the continuous-time signal x is differentiable with respect to s in its existence region $\mathscr{E}$. Then the one- or two-sided Laplace transform, respectively, of the signal

$$-tx(t), \qquad t \in \mathbb{R},$$

exists on $\mathscr{E}$ and is given by

$$\frac{dX(s)}{ds}, \qquad s \in \mathscr{E}. \qquad \blacksquare$$

Also the proof of the converse differentiation property is found in E.10 of Supplement E.

The differentiation properties of the Laplace transform are complemented by the *integration* properties.

8.4.8   Summary: Integration properties of the Laplace transform.

(a) *Integration property of the two-sided Laplace transform.* Suppose that the continuous-time signal x has the two-sided Laplace transform X with existence region $\mathscr{E}$. Then the *integrated signal y* given by

$$y(t) = \int_{-\infty}^{t} x(\tau) \, d\tau, \qquad t \in \mathbb{R},$$

exists if the set

$$\mathscr{E}_y = \{s \in \mathscr{E} \mid \operatorname{Re}(s) > 0\}$$

is nonempty, and its two-sided Laplace transform is

$$Y(s) = \frac{X(s)}{s}, \qquad s \in \mathscr{E}_y.$$

(b) *Integration property of the one-sided Laplace transform.* Suppose that the continuous-time signal x has the one-sided Laplace transform X_+ with existence region $\mathscr{E}_+$. Then the integrated signal y defined by

$$y(t) = \int_{0-}^{t} x(\tau) \, d\tau, \qquad t \geq 0, t \in \mathbb{R},$$

has the one-sided Laplace transform

$$Y_+(s) = \frac{X_+(s)}{s}, \qquad s \in \mathscr{E}_y,$$

with

$$\mathscr{E}_y = \{s \in \mathscr{E}_+ \mid \mathrm{Re}\,(s) > 0\}. \qquad\qquad\qquad \blacksquare$$

The proof of the integration properties may be found in E.11 of Supplement E.

We next consider the *initial* and *final value* properties. These are sometimes useful for establishing asymptotic values of time signals or their transforms.

8.4.9.   Summary: Initial and final value properties of the one-sided z- and Laplace transforms.

(a) *Initial value property of the one-sided z-transform.* Let the discrete-time signal x have the one-sided z-transform X_+. Then,

(a') *Initial value property of the one-sided Laplace transform.* Let the continuous-time signal x have the one-sided Laplace transform X_+. Then, if x is regular on some open interval that includes the origin,

$$\lim_{|z| \to \infty} X_+(z) = x(0).$$

$$\lim_{s \to \infty} sX_+(s) = x(0^+).$$

(b) *Final value property of the one-sided z-transform.* Suppose that X_+ exists for $|z| > 1$. Then if $\lim_{n \to \infty} x(n)$ exists,

(b') *Final value property of the one-sided Laplace transform.* Suppose that X_+ exists for $\mathrm{Re}\,(s) > 0$. Then if $\lim_{t \to \infty} x(t)$ exists,

$$\lim_{z \to 1} (z - 1) X_+(z) = \lim_{n \to \infty} x(n).$$

$$\lim_{s \to 0} sX_+(s) = \lim_{t \to \infty} x(t). \qquad \blacksquare$$

The initial and final value properties are proved in E.12 of Supplement E.

The assumption in the statement of the final value property that x actually has a limit as time goes to ∞ is not vacuous. Consider, for instance, the discrete-time signal x defined by

$$x(n) = (-1)^n, \qquad n \in \mathbb{Z}_+,$$

which has no limit as $n \to \infty$. The one-sided z-transform of x is easily found by setting $a = -1$ in 8.3.8(b) and is given by

$$X_+(z) = \frac{z}{z+1}, \qquad |z| > 1.$$

Hence, the formula for the final value would yield

$$\lim_{z \to 1} (z-1) \cdot \frac{z}{z+1} = 0,$$

which is not $\lim_{n \to \infty} x(n)$.

The last property to be stated is the *scaling* property of the Laplace transform.

8.4.10.   Summary: Scaling property of the Laplace transform.

(a) *Two-sided Laplace transform.* Suppose that the continuous-time signal x has the two-sided Laplace transform X with existence region $\mathscr{E}$. Then the time scaled signal

$$x(\alpha t), \qquad t \in \mathbb{R},$$

with α real and $\alpha \neq 0$, has the two-sided Laplace transform

$$\frac{1}{|\alpha|} X \left(\frac{s}{\alpha} \right), \qquad s/|\alpha| \in \mathscr{E}.$$

(b) *One-sided Laplace transform.* If $\alpha > 0$, the time-scaled signal has the one-sided Laplace transform

$$\frac{1}{\alpha} X_+ \left(\frac{s}{\alpha} \right), \qquad s/\alpha \in \mathscr{E}_+,$$

with X_+ the one-sided transform of x and $\mathscr{E}_+$ its existence region.   ∎

The proof of the scaling property is easy.

In Tables 8.1 and 8.2 the various properties of the z- and Laplace transforms are summarized.

Application of the Properties of the z- and Laplace Transforms

In the remainder of this section we show how the various properties of the z- and Laplace transforms may be used to derive a number of useful standard transform pairs.

TABLE 8.1 PROPERTIES OF THE z-TRANSFORM

Property	Time signal $n \in \mathbb{Z}$	Two-sided z-transform $z \in \mathscr{E}$	Existence region $\mathscr{E}$	One-sided z-transform $z \in \mathscr{E}_+$	Existence region $\mathscr{E}_+$
Linearity	$(\alpha x + \beta y)(n)$, $\alpha, \beta \in \mathbb{C}$	$\alpha X(z) + \beta Y(z)$	$\mathscr{E}_x \cap \mathscr{E}_y$	$\alpha X_+(z) + \beta Y_+(z)$	$\mathscr{E}_{x,+} \cap \mathscr{E}_{y,+}$
Convolution	$(x * y)(n)$	$X(z)Y(z)$	$\mathscr{E}_x \cap \mathscr{E}_y$	$X_+(z)Y_+(z)$	$\mathscr{E}_{x,+} \cap \mathscr{E}_{y,+}$
	$(x \mid * y \mid)(n)$	$X_+(z)Y_+(z)$	$\mathscr{E}_{x,+} \cap \mathscr{E}_{y,+}$		
Shift	$x(n+k), k \in \mathbb{Z}$	$z^k X(z)$	$\mathscr{E}_x$	See 8.4.4	
	$x(n+1)$	$zX(z)$	$\mathscr{E}_x$	$zX_+(z) - zx(0)$	$\mathscr{E}_{x,+}$
	$x(n-1)$	$z^{-1}X(z)$	$\mathscr{E}_x$	$z^{-1}X_+(z) + x(-1)$	$\mathscr{E}_{x,+}$
Converse shift	$a^n x(n), a \in \mathbb{C}$	$X(z/a)$	$\{z \in \mathbb{C} \mid z/a \in \mathscr{E}_x\}$	$X_+(z/a)$	$\{z \in \mathbb{C} \mid z/a \in \mathscr{E}_{x,+}\}$
Converse differentiation	$-nx(n)$	$z\dfrac{dX(z)}{dz}$	$\mathscr{E}_x$	$z\dfrac{dX_+(z)}{dz}$	$\mathscr{E}_{x,+}$
Initial value				$x(0) = \lim\limits_{|z|\to\infty} X_+(z)$	
Final value				$\lim\limits_{n\to\infty} x(n) = \lim\limits_{z\to 1}(z-1)X_+(z)$ provided $\lim\limits_{n\to\infty} x(n)$ exists	

TABLE 8.2 PROPERTIES OF THE LAPLACE TRANSFORM

Property	Time signal $t \in \mathbb{R}$	Two-sided Laplace transform $s \in \mathcal{E}$	Existence region $\mathcal{E}$	One-sided Laplace transform $s \in \mathcal{E}_+$	Existence region $\mathcal{E}_+$
Linearity	$(\alpha x + \beta y)(t)$, $\alpha, \beta \in \mathbb{C}$	$\alpha X(s) + \beta Y(s)$	$\mathcal{E}_x \cap \mathcal{E}_y$	$\alpha X_+(s) + \beta Y_+(s)$	$\mathcal{E}_{x,+} \cap \mathcal{E}_{y,+}$
Convolution	$(x * y)(t)$	$X(s)Y(s)$	$\mathcal{E}_x \cap \mathcal{E}_y$		
	$(x\,\boxed{1} * y\,\boxed{1})(t)$	$X_+(s)Y_+(s)$	$\mathcal{E}_{x,+} \cap \mathcal{E}_{y,+}$	$X_+(s)Y_+(s)$	$\mathcal{E}_{x,+} \cap \mathcal{E}_{y,+}$
Shift	$x(t + \theta)$, $\theta \in \mathbb{R}$	$e^{s\theta}X(s)$	$\mathcal{E}_x$		
	$x(t - \theta)\,\mathbb{1}(t - \theta)$, $\theta \in \mathbb{R},\ \theta \geq 0$	$e^{-s\theta}X_+(s)$	$\mathcal{E}_{x,+}$	$e^{-s\theta}X_+(s)$	$\mathcal{E}_{x,+}$
Converse shift	$e^{at}x(t)$, $a \in \mathbb{C}$	$X(s - a)$	$\{s \in \mathbb{C} \mid s - a \in \mathcal{E}_x\}$	$X_+(s - a)$	$\{s \in \mathbb{C} \mid s - a \in \mathcal{E}_{x,+}\}$
Differentiation	$Dx(t)$	$sX(s)$	$\mathcal{E}_x$	$sX_+(s) - x(0^-)$	$\mathcal{E}_{x,+}$
Integration	$\displaystyle\int_{-\infty}^{t} x(\tau)\, d\tau$	$\dfrac{X(s)}{s}$	$\{s \in \mathcal{E}_x \mid \operatorname{Re}(s) > 0\}$		
	$\displaystyle\int_{0^-}^{t} x(\tau)\, d\tau$			$\dfrac{X_+(s)}{s}$	$\{s \in \mathcal{E}_{x,+} \mid \operatorname{Re}(s) > 0\}$
Converse differentiation	$-tx(t)$	$\dfrac{dX(s)}{ds}$	$\mathcal{E}_x$	$\dfrac{dX_+(s)}{ds}$	$\mathcal{E}_{x,+}$
Scaling	$x(\alpha t)$, $\alpha \in \mathbb{R},\ \alpha \neq 0$	$\dfrac{1}{|\alpha|}X\left(\dfrac{s}{\alpha}\right)$	$\left\{s \in \mathbb{C} \mid \dfrac{s}{|\alpha|} \in \mathcal{E}_x\right\}$		
	$x(\alpha t)$, $\alpha \in \mathbb{R},\ \alpha > 0$	$\dfrac{1}{\alpha}X\left(\dfrac{s}{\alpha}\right)$	$\left\{s \in \mathbb{C} \mid \dfrac{s}{\alpha} \in \mathcal{E}_x\right\}$	$\dfrac{1}{\alpha}X_+\left(\dfrac{s}{\alpha}\right)$	$\left\{s \in \mathbb{C} \mid \dfrac{s}{\alpha} \in \mathcal{E}_{x,+}\right\}$
Initial value				$x(0^+) = \lim_{s\to\infty} sX_+(s)$ provided x is regular at 0	
Final value				$\lim_{t\to\infty} x(t) = \lim_{s\to 0} sX_+(s)$ provided $\lim_{t\to\infty} x(t)$ exists	

8.4.11.   Examples: Some standard *z*-transform pairs.

(a) *Application of the shift property.* In 8.3.2(a) we found that the right one-sided exponential signal

$$x(n) = a^n 1(n), \qquad n \in \mathbb{Z},$$

has the two-sided *z*-transform, which also happens to be its one-sided *z*-transform, given by

$$X(z) = \frac{z}{z - a}, \qquad |z| > |a|.$$

Application of the shift property for the two-sided *z*-transform with $k = -1$ shows that the once forward shifted signal

$$\sigma^{-1} x(n) = x(n-1) = a^{n-1} 1(n - 1), \qquad n \in \mathbb{Z},$$

has the two-sided *z*-transform

$$z^{-1} X(z) = z^{-1} \cdot \frac{z}{z - a} = \frac{1}{z - a}, \qquad |z| > |a|.$$

Multiplication by *a* shows (by linearity) that

$$a^n 1(n - 1), \qquad n \in \mathbb{Z}, \qquad \text{and} \qquad \frac{a}{z - a}, \qquad |z| > |a|, \tag{1}$$

form a two-sided *z*-transform pair. Because the time signal is zero for negative times, it is also a *one-sided z*-transform pair.

(b) *Application of the converse differentiation property.* By using the converse differentiation property on the two-sided *z*-transform pair (1), we find that

$$-nx(n) = -na^n 1(n - 1), \qquad n \in \mathbb{Z},$$

and

$$z \frac{d}{dz} X(z) = -\frac{az}{(z - a)^2}, \qquad |z| > |a|,$$

form a two-sided *z*-transform pair. Application of the shift property for $k = -1$ and multiplication by $-a$ shows that

$$(n - 1)a^n 1(n - 2), \qquad n \in \mathbb{Z}, \qquad \text{and} \qquad \frac{a^2}{(z - a)^2}, \qquad |z| > |a|,$$

form a two-sided z-transform pair. By repeated application of the converse differentiation property it may be proved in this way that for $k \geq 1$

$$\binom{n - 1}{k - 1} a^n 1(n - k), \qquad n \in \mathbb{Z},$$

and

$$\frac{a^k}{(z - a)^k}, \qquad |z| > |a|,$$

form a two-sided z-transform pair. Because the time signal is zero for negative times, this is also a one-sided z-transform pair. Here we use the binomial coefficient notation

$$\binom{n}{k} = \frac{n!}{(n-k)!k!},$$

with n and k nonnegative integers such that $k \leq n$.   ∎

The transform pairs found in 8.4.11 together with several other standard pairs are collected in Table 8.3.

8.4.12.   Examples: Some standard Laplace transform pairs.
(a) *Application of linearity*. In 8.3.4(a) we found that the continuous-time signal

$$x(t) = e^{at} 1(t), \qquad t \in \mathbb{R},$$

has the two-sided Laplace transform

$$X(s) = \frac{1}{s-a}, \qquad \text{Re}(s) > \text{Re}(a).$$

This is at the same time the one-sided Laplace transform X_+ of x. Setting $a = \pm j\omega$, with ω real, we see that

TABLE 8.3   STANDARD z-TRANSFORMS

Time signal $n \in \mathbb{Z}$	Two-sided z-transform		One-sided z-transform		Conditions
$a^n 1(n)$	$\frac{z}{z-a}$	$|z|>|a|$	$\frac{z}{z-a}$	$|z|>|a|$	$a \in \mathbb{C}$
$-a^n 1(-n-1)$	$\frac{z}{z-a}$	$|z|<|a|$	0	$z \in \mathbb{C}$	$a \in \mathbb{C}$
a^n	—	—	$\frac{z}{z-a}$	$|z|>|a|$	$a \in \mathbb{C}$
$1(n)$	$\frac{z}{z-1}$	$|z|>1$	$\frac{z}{z-1}$	$|z|>1$	
$\Delta(n)$	1	$z \in \mathbb{C}$	1	$z \in \mathbb{C}$	
$\Delta(n-k)$	z^{-k}	$z \neq 0$	z^{-k}	$z \neq 0$	$k \in \mathbb{N}$
$\Delta(n+k)$	z^{k}	$z \in \mathbb{C}$	0	$z \in \mathbb{C}$	$k \in \mathbb{N}$
$a^n 1(n-1)$	$\frac{a}{z-a}$	$|z|>|a|$	$\frac{a}{z-a}$	$|z|>|a|$	$a \in \mathbb{C}$
$na^n 1(n-1)$	$\frac{az}{(z-a)^2}$	$|z|>|a|$	$\frac{az}{(z-a)^2}$	$|z|>|a|$	$a \in \mathbb{C}$
$\binom{n-1}{k-1}a^n 1(n-k)$	$\frac{a^k}{(z-a)^k}$	$|z|>|a|$	$\frac{a^k}{(z-a)^k}$	$|z|>|a|$	$a \in \mathbb{C}, k \in \mathbb{N}$
$-a^n 1(-n)$	$\frac{a}{z-a}$	$|z|<|a|$	-1	$z \in \mathbb{C}$	$a \in \mathbb{C}$
$(-1)^k\binom{-n+k-1}{k-1}a^n 1(-n)$	$\frac{a^k}{(z-a)^k}$	$|z|<|a|$	$(-1)^k$	$z \in \mathbb{C}$	$a \in \mathbb{C}, k \in \mathbb{N}$
$\rho^n\cos(\omega n + \phi)1(n-1)$	$\frac{z\rho\cos(\omega+\phi)-\rho^2\cos(\phi)}{z^2-2\rho z\cos(\omega)+\rho^2}$		$|z|>\rho$		$\rho, \omega, \phi \in \mathbb{R}.$
$\frac{a^n}{n!}1(n)$	$e^{a/z}$		$|z|>|a|$		$a \in \mathbb{C}$
$\frac{1}{n}1(n-1)$	$\ln\left(\frac{z}{z-1}\right)$		$|z|>1$		

$$e^{j\omega t}1(t), \qquad t \in \mathbb{R}, \qquad \text{and} \qquad \frac{1}{s-j\omega}, \qquad \text{Re}\,(s) > 0,$$

form a two-sided Laplace transform pair, as well as

$$e^{-j\omega t}1(t), \qquad t \in \mathbb{R}, \qquad \text{and} \qquad \frac{1}{s+j\omega}, \qquad \text{Re}\,(s) > 0.$$

Adding the signals and dividing by 2 we find by linearity that

$$\cos(\omega t)1(t), \qquad t \in \mathbb{R}, \qquad \text{and} \qquad \frac{s}{s^2+\omega^2}, \qquad \text{Re}\,(s) > 0, \tag{2}$$

form a two-sided pair. Because the time signal is zero for negative times, it is also a one-sided pair. Similarly, by subtracting and dividing by $2j$ we find that

$$\sin{(\omega t)}1(t), \qquad t \in \mathbb{R}, \qquad \text{and} \qquad \frac{\omega}{s^2 + \omega^2}, \qquad \text{Re}\,(s) > 0,$$

form both a two- and a one-sided Laplace transform pair.

TABLE 8.4 STANDARD LAPLACE TRANSFORMS

Time signal $t \in \mathbb{R}$	Two-sided Laplace transform		One-sided Laplace transform		Conditions
$e^{at}1(t)$	$\dfrac{1}{s-a}$	$\text{Re}\,(s) > \text{Re}\,(a)$	$\dfrac{1}{s-a}$	$\text{Re}\,(s) > \text{Re}\,(a)$	$a \in \mathbb{C}$
$-e^{at}1(-t)$	$\dfrac{1}{s-a}$	$\text{Re}\,(s) < \text{Re}\,(a)$	0	$s \in \mathbb{C}$	$a \in \mathbb{C}$
e^{at}	—	—	$\dfrac{1}{s-a}$	$\text{Re}\,(s) > \text{Re}\,(a)$	$a \in \mathbb{C}$
1	—	—	$\dfrac{1}{s}$	$\text{Re}\,(s) > 0$	
$1(t)$	$\dfrac{1}{s}$	$\text{Re}\,(s) > 0$	$\dfrac{1}{s}$	$\text{Re}\,(s) > 0$	
$\delta(t)$	1	$s \in \mathbb{C}$	1	$s \in \mathbb{C}$	
$\delta^{(k)}(t)$	s^k	$s \in \mathbb{C}$	s^k	$s \in \mathbb{C}$	$k \in \mathbb{N}$
$\dfrac{t^{k-1}}{(k-1)!}1(t)$	$\dfrac{1}{s^k}$	$\text{Re}\,(s) > 0$	$\dfrac{1}{s^k}$	$\text{Re}\,(s) > 0$	$k \in \mathbb{N}$
$\dfrac{t^{k-1}e^{at}}{(k-1)!}1(t)$	$\dfrac{1}{(s-a)^k}$	$\text{Re}\,(s) > \text{Re}\,(a)$	$\dfrac{1}{(s-a)^k}$	$\text{Re}\,(s) > \text{Re}\,(a)$	$a \in \mathbb{C}, k \in \mathbb{N}$
$-\dfrac{t^{k-1}e^{at}}{(k-1)!}1(-t)$	$\dfrac{1}{(s-a)^k}$	$\text{Re}\,(s) < \text{Re}\,(a)$	0	$s \in \mathbb{C}$	$a \in \mathbb{C}, k \in \mathbb{N}$
$\cos{(\omega t)}1(t)$	$\dfrac{s}{s^2 + \omega^2}$	$\text{Re}\,(s) > 0$	$\dfrac{s}{s^2 + \omega^2}$	$\text{Re}\,(s) > 0$	$\omega \in \mathbb{R}$
$\sin{(\omega t)}1(t)$	$\dfrac{\omega}{s^2 + \omega^2}$	$\text{Re}\,(s) > 0$	$\dfrac{\omega}{s^2 + \omega^2}$	$\text{Re}\,(s) > 0$	$\omega \in \mathbb{R}$
$e^{at}\cos{(\omega t)}1(t)$	$\dfrac{s-a}{(s-a)^2 + \omega^2}$	$\text{Re}\,(s) > a$	$\dfrac{s-a}{(s-a)^2 + \omega^2}$	$\text{Re}\,(s) > a$	$a, \omega \in \mathbb{R}$
$e^{at}\sin{(\omega t)}1(t)$	$\dfrac{\omega}{(s-a)^2 + \omega^2}$	$\text{Re}\,(s) > a$	$\dfrac{\omega}{(s-a)^2 + \omega^2}$	$\text{Re}\,(s) > a$	$a, \omega \in \mathbb{R}$

(b) *Application of the converse shift property*. Let *a* be a real number. Then application of the converse shift property to the Laplace transform pair (2) shows that

$$e^{at}\cos(\omega t)\mathbb{1}(t), \qquad t \in \mathbb{R}, \qquad \text{and} \qquad \frac{(s-a)}{(s-a)^2 + \omega^2}, \qquad \text{Re } (s) > a,$$

form both a two- and a one-sided pair. Similarly, also

$$e^{at}\sin(\omega t)\mathbb{1}(t), \qquad t \in \mathbb{R}, \qquad \text{and} \qquad \frac{\omega}{(s-a)^2 + \omega^2}, \qquad \text{Re } (s) > a,$$

form a two- and one-sided pair.   ∎

Table 8.4 displays the Laplace transform pairs found in 8.4.12 together with a collection of other well-known pairs. Many more may be found in for instance M. Abramowitz and I. A. Stegun (Eds.), *Handbook of Mathematical Functions*, Dover, New York, 1965.

8.5 INVERSE *z*- AND LAPLACE TRANSFORMATION

In Section 8.2 we introduced the *z*- or Laplace transform of the signal *x* as the DCFT or CCFT of a modification of *x*, obtained by multiplying it by a suitable exponential signal. As a result, the modified signal may be retrieved by the *inverse* DCFT or CCFT, which in turn makes it possible to recover the original signal *x*.

Although there are two kinds of transforms—the two-sided and the one-sided—there is only *one* inverse transform for both of them, which is the inverse two-sided transform. When applied to a one-sided transform, the inverse transform reconstructs the time signal for *nonnegative* times only.

Complex Inversion Formulas

We first consider the inverse *z*-transform. Suppose that *X* is the two-sided *z*-transform of the time signal *x*. Then if ρ is a positive real number such that the circle with radius ρ and center at the origin is entirely contained in the existence region $\mathscr{E}$ of *X*, according to 8.3.5(a) the function $X(\rho e^{j2\pi f})$, $f \in [0, 1)$, is the DCFT of the time signal $\rho^{-n}x(n)$, $n \in \mathbb{Z}$. By application of the inverse DCFT it follows that

$$\rho^{-n}x(n) = \int_0^1 X(\rho e^{j2\pi f})e^{j2\pi fn}\, df, \qquad n \in \mathbb{Z},$$

so that

$$x(n) = \rho^n \int_0^1 X(\rho e^{j2\pi f}) e^{j2\pi fn} \, df, \qquad n \in \mathbb{Z}.$$

This is an explicit formula that actually lends itself to numerical computation. The result may be put into a more compact form by rewriting it first as

$$x(n) = \int_0^1 X(\rho e^{j2\pi f})(\rho e^{j2\pi f})^n \, df, \qquad n \in \mathbb{Z},$$

and replacing the integration variable f with the complex variable $z = \rho e^{j2\pi f}$, where z varies along a circle in the complex plane with radius ρ and center at the origin. It follows that $dz = \rho e^{j2\pi f} j2\pi \, df$, so that $j2\pi \, df = dz/z$. Substitution of z for $\rho e^{j2\pi f}$ yields

$$x(n) = \frac{1}{j2\pi} \oint X(z) z^{n-1} \, dz, \qquad n \in \mathbb{Z},$$

where the integral is taken along a circle in the complex plane with center at the origin and radius ρ. This result is known as the *complex inversion formula* for the z-transform.

By a similar argument, it may be shown that if X is the two-sided Laplace transform of a signal x, whose existence region $\mathscr{E}$ contains the vertical line $\sigma + j2\pi f, f \in \mathbb{R}$, with σ a fixed real number, the signal x may be retrieved as

$$x(t) = e^{\sigma t} \int_{-\infty}^{\infty} X(\sigma + j2\pi f) e^{j2\pi ft} \, df, \qquad t \in \mathbb{R}.$$

By the substitution $\sigma + j2\pi f = s$ this takes the form

$$x(t) = \frac{1}{j2\pi} \int_{\sigma - j\infty}^{\sigma + j\infty} X(s) e^{st} \, ds, \qquad t \in \mathbb{R},$$

which is the *complex inversion formula* for the Laplace transform.

The one-sided z- or Laplace transform X_+ of a signal x is the two-sided transform of $x \cdot \mathbb{1}$. Hence, application of the inverse two-sided inverse transform to X_+ results in $x \cdot \mathbb{1}$. There is no way to reconstruct the signal x for negative times.

8.5.1.   Summary: Complex inversion formulas for the z- and Laplace transforms.

(a) *Inverse two-sided z-transform.* Given the two-sided z-transform X of a discrete-time signal x with existence region $\mathscr{E}$, the signal x may be retrieved from X by the inverse z-transform $x = \mathscr{Z}^{-1}(X)$ as

(a′) *Inverse two-sided Laplace transform.* Given the two-sided Laplace transform X of a continuous-time signal x with existence region $\mathscr{E}$, the signal x may be retrieved from X by the inverse Laplace transform $x = \mathscr{L}^{-1}(X)$ as

$$x(n) = \frac{1}{j2\pi} \oint X(z) z^{n-1}\, dz, \qquad n \in \mathbb{Z},$$

$$x(t) = \frac{1}{j2\pi} \int_{\sigma - j\infty}^{\sigma + j\infty} X(s) e^{st}\, ds, \qquad t \in \mathbb{R},$$

where the integral is taken along a circle in the complex plane with center at the origin that is entirely contained in the existence region $\mathscr{E}$.

where the integral is taken along a vertical line with abscissa σ that is entirely contained in the existence region $\mathscr{E}$.

(b) *Inverse one-sided z-transform.* Application of the inverse z-transform to the one-sided z-transform X_+ of the discrete-time signal x results in $x \cdot \mathbb{1}$, that is,

(b′) *Inverse one-sided Laplace transform.* Application of the inverse Laplace transform to the one-sided Laplace transform X_+ of the continuous-time signal x results in $x \cdot \mathbb{1}$, that is,

$$\mathscr{Z}^{-1}(X_+)(n) = \begin{cases} x(n) & \text{for } n \geq 0, \\ 0 & \text{for } n < 0, \end{cases}$$

$$\mathscr{L}^{-1}(X_+)(t) = \begin{cases} x(t) & \text{for } t \geq 0, \\ 0 & \text{for } t < 0, \end{cases}$$

for $n \in \mathbb{Z}$.

for $t \in \mathbb{R}$.   ∎

The complex inversion formula for the z-transform shows that by the z-transform the discrete-time signal x is expanded in a continuum of *exponential* signals of the form

$$z^{n-1}, \qquad n \in \mathbb{Z}.$$

Similarly, the complex inversion formula for the Laplace transform exhibits how the continuous-time signal x is expanded in a continuum of exponential signals of the form

$$e^{st}, \qquad t \in \mathbb{R}.$$

Inversion by Reduction

Effective use of the complex inversion formulas requires knowledge of complex function theory, which is outside the scope of this book. Often the formulas may be avoided by using *reduction* to obtain the inverse transform. Reduction consists of exploiting the properties of the z- and Laplace transforms and methods such as partial fraction and power series expansion to reduce the inverse transformation to that of elementary functions whose inverse transform is known and may be looked up in a table such as Table 8.3 or 8.4.

We illustrate this technique with several examples.

8.5.2.   Examples: Inverse z-transformation by reduction.   Consider the z-transform given by

$$X(z) = \frac{1}{z^2 - 1}, \qquad |z| > 1.$$

(a) *Inverse transformation by partial fraction expansion.* We first use partial fraction expansion to find the inverse z-transform. Factoring the denominator of X as $z^2 - 1 = (z - 1)(z + 1)$ it is easily found that

$$X(z) = \frac{\frac{1}{2}}{z - 1} + \frac{-\frac{1}{2}}{z + 1}, \qquad |z| > 1.$$

It follows from Table 8.3 that this is the z-transform of

$$x(n) = \tfrac{1}{2}\mathbb{1}(n - 1) + \tfrac{1}{2}(-1)^n \mathbb{1}(n - 1)$$
$$= \begin{cases} 1 & \text{for } n \text{ even}, n \geq 2, \\ 0 & \text{otherwise}, \end{cases} \qquad n \in \mathbb{Z}.$$

Because x is zero for negative times, X is the two-sided as well as the one-sided z-transform of x.

(b) *Inverse transformation by the power series method.* To apply the power series method we need expand X in a power series that converges for $|z| > 1$. A first attempt is the expansion

$$X(z) = -\frac{1}{1 - z^2} = -(1 + z^2 + z^4 + \cdots),$$

but this converges for $|z| < 1$ rather than for $|z| > 1$. We therefore rewrite and re-expand X as

$$X(z) = \frac{1}{z^2 - 1} = \frac{z^{-2}}{1 - z^{-2}} = z^{-2}(1 + z^{-2} + z^{-4} + \cdots)$$
$$= z^{-2} + z^{-4} + z^{-6} + \cdots, \qquad |z| > 1.$$

Identification of the coefficients of this expansion with those of

$$X(z) = \sum_{n=-\infty}^{\infty} x(n)z^{-n}$$

leads to the conclusion that $x(n) = 1$ for n even and $n \geq 2$, and $x(n) = 0$ otherwise. This is precisely what we found in (a).   ∎

8.5.3.   Examples: Inverse Laplace transformation by reduction.

(a) *Inverse Laplace transformation by partial fraction expansion.* Consider determining the inverse Laplace transform of X given by

$$X(s) = \frac{s - 1}{(s + 1)^2(s - 2)},$$

where the existence region is left open for the time being. In Example 8.2.4(b) we found that this rational function has the partial fraction expansion

$$X(s) = -\frac{1/9}{s + 1} + \frac{2/3}{(s + 1)^2} + \frac{1/9}{s - 2}.$$

Looking at the fractions on the right-hand side we see that there are three possible existence regions, as shown in Fig. 8.6.

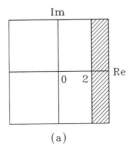

(a)

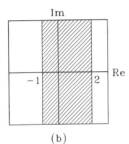

(b)

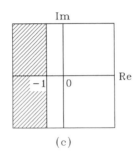

(c)

Figure 8.6. Possible existence regions for the Laplace transform of 8.5.3.

(i) Re $(s) > 2$. In this case, by Table 8.4 the two-sided inverse Laplace transform x of X is

$$x(t) = \left[-\frac{1}{9}e^{-t} + \frac{2}{3}te^{-t} + \frac{1}{9}e^{2t} \right] \mathbb{1}(t), \qquad t \in \mathbb{R},$$

as plotted in Fig. 8.7(a). Because this time signal is zero for $t < 0$, it is also the inverse *one-sided* Laplace transform of X.

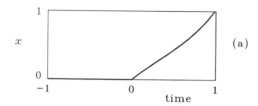

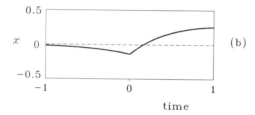

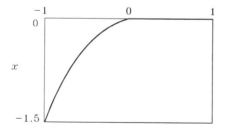

Figure 8.7. Three continuous-time signals that have the same two-sided Laplace transform but different existence regions. Top: a right one-sided time signal. Middle: a two-sided signal. Bottom: a left one-sided signal.

(ii) $-1 < $ Re $(s) < 2$. Again by Table 8.4 the inverse two-sided Laplace transform is

$$x(t) = \left[-\frac{1}{9}e^{-t} + \frac{2}{3}te^{-t} \right] \mathbb{1}(t) - \frac{1}{9}e^{2t}\mathbb{1}(-t), \qquad t \in \mathbb{R},$$

as plotted in Fig. 8.7(b). The *one-sided* Laplace transform of this two-sided signal is *not* X, but

$$X_+(s) = -\frac{1/9}{s+1} + \frac{2/3}{(s+1)^2} = \frac{-\frac{1}{9}s + \frac{5}{9}}{(s+1)^2}, \qquad \text{Re }(s) > -1.$$

(iii) Re $(s) < -1$. The inverse two-sided Laplace transform now is

$$x(t) = \left[\frac{1}{9}e^{-t} - \frac{2}{3}te^{-t} - \frac{1}{9}e^{2t}\right]\mathbb{1}(-t), \qquad t \in \mathbb{R},$$

as shown in Fig. 8.7(c). This is a left one-sided signal, and its one-sided Laplace transform is $X_+ = 0$.

(b) *Inverse Laplace transformation by power series expansion.* Occasionally power series expansion is also useful for inverse Laplace transformation. Consider finding the inverse Laplace transform of

$$X(s) = \frac{1}{1 - e^{-sP}}, \qquad \text{Re }(s) > 0,$$

with P a given positive number. Since it follows from Re $(s) > 0$ that $\left|e^{-sP}\right| = e^{-P\,\text{Re}\,(s)} < 1$, we may expand X in the power series

$$X(s) = 1 + e^{-sP} + e^{-2sP} + e^{-3sP} + \cdots, \qquad \text{Re }(s) > 0.$$

By Table 8.4, the first term on the right-hand side is the Laplace transform of the delta function δ. By the shift property, the second term is the Laplace transform of $\delta(t - P)$, $t \in \mathbb{R}$, the third that of $\delta(t - 2P)$, $t \in \mathbb{R}$, and so on. It follows that

$$x(t) = \delta(t) + \delta(t - P) + \delta(t - 2P) + \delta(t - 3P) + \cdots, \qquad t \in \mathbb{R},$$

which is the semi-infinite comb of Fig. 8.8. Because this signal is zero for $t < 0$, it is also the inverse one-sided Laplace transform of X. ∎

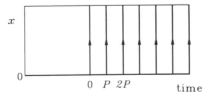

x

0

$0 \quad P \quad 2P$ time   **Figure 8.8.**   A semi-infinite comb.

Existence of the Inverse One-Sided z- and Laplace Transforms

The existence region of the two-sided z-transform of a right one-sided discrete-time signal is of the same form as the existence region of a one-sided z-transform, namely,

$$\{z \in \mathbb{C} \mid |z| > \rho\},$$

for some real ρ. However, by no means every two-sided z-transform that is well defined on such an existence region is also a one-sided z-transform. Consider for instance

$$X(z) = z^k, \qquad |z| > 0,$$

with k an integer. By 8.3.2(d) this is the two-sided z-transform of $x(n) = \Delta(n + k)$, $n \in \mathbb{Z}$. For $k > 0$, however, the one-sided z-transform of x is 0. It follows that X is *not* a one-sided z-transform for k positive.

By the same token, improper rational functions are not one-sided z-transforms. If X is nonproper rational in z, by the division theorem 8.2.1 it may be written as $X = T + R$, where T is polynomial in z and R strictly proper rational. Although R is a one-sided z-transform (see Exercise 8.5.4), T is not and, hence, also X is not a one-sided z-transform.

8.5.4.   Exercise: A proper rational function is a one-sided z-transform.   Suppose that X is a proper or strictly proper rational function in z. Prove that for $|z| > \rho$, with ρ a suitably large real number, X is a one-sided z-transform. *Hint:* Perform a partial fraction expansion in z.   ∎

The two-sided Laplace transform of a right one-sided continuous-time signal has an existence region of the form

$$\{s \in \mathbb{C} \mid \text{Re}\,(s) > \sigma\}, \tag{1}$$

for some real σ, the same as for a one-sided Laplace transform. However, not every two-sided Laplace transform that is well-defined for $\text{Re}\,(s) > \sigma$ is a one-sided Laplace transform. For $\theta > 0$ the function

$$X(s) = e^{s\theta}, \qquad s \in \mathbb{C},$$

for instance, is the two-sided Laplace transform of $\delta(t + \theta)$, $t \in \mathbb{R}$, whose one-sided Laplace transform is zero. Hence X is not a one-sided Laplace transform for $\theta > 0$.

Unlike for the *z*-transform, nonproper rational functions are one-sided Laplace transforms, provided their existence region is of the form (1) for some real σ. Writing $X = T + R$, with T a polynomial and R strictly proper rational, shows that the one-sided inverse Laplace transform is a linear combination of a number of δ-functions and derivative δ-functions placed at the origin and several exponentials.

8.6 TRANSFORM ANALYSIS OF CONVOLUTION SYSTEMS

In the final three sections of this chapter it is shown how *z*- and Laplace transformation may be applied to the analysis of linear time-invariant systems. We begin in this section with convolution systems, continue in Section 8.7 with difference and differential systems, and end in Section 8.8 with state systems.

Transfer Functions

The transform analysis of convolution system is based on the convolution property of the *z*- and Laplace transforms.

8.6.1.   Summary: Transform analysis of convolution systems.

Consider a discrete-time convolution system with impulse response h and IO map

$$y = h * u.$$

Suppose that h has a two-sided *z*-transform H that exists on $\mathscr{E}_h$ and u has a two-sided *z*-transform U that exists on $\mathscr{E}_u$.

   (a) A necessary and sufficient condition for the existence of the response y of the system is that $\mathscr{E}_h \cap \mathscr{E}_u$ be not empty.

   (b) If y exists its two-sided *z*-transform exists on $\mathscr{E}_h \cap \mathscr{E}_u$, and is given by

$$Y = HU,$$

so that

$$y = \mathscr{Z}^{-1}(HU).$$

Consider a continuous-time convolution system with impulse response h and IO map

$$y = h * u.$$

Suppose that h has a two-sided Laplace transform H that exists on $\mathscr{E}_h$ and u has a two-sided Laplace transform U that exists on $\mathscr{E}_u$.

   (a') A necessary and sufficient condition for the existence of the response y of the system is that $\mathscr{E}_h \cap \mathscr{E}_u$ be not empty.

   (b') If y exists its two-sided Laplace transform exists on $\mathscr{E}_h \cap \mathscr{E}_u$, and is given by

$$Y = HU,$$

so that

$$y = \mathscr{L}^{-1}(HU). \qquad \blacksquare$$

Note that if *both* the input *u and* the impulse response *h* are zero for negative times, then the two-sided *z*- and Laplace transforms may be replaced by one-sided transforms, so that

$$Y_+ = H_+ U_+.$$

The *z*-transform *H* of the impulse response *h* of a discrete-time system is given by

$$H(z) = \sum_{n=-\infty}^{\infty} h(n)z^{-n}, \qquad z \in \mathscr{E}_h, \tag{1}$$

and is known as the *transfer function* of the convolution system. Similarly, the Laplace transform

$$H(s) = \int_{-\infty}^{\infty} h(t)e^{-st}\, dt, \qquad s \in \mathscr{E}_h, \tag{2}$$

of the impulse response *h* of a continuous-time convolution system is the *transfer function* of the system.

We give a further interpretation of the transfer function of a convolution system, and discuss its connection with the frequency response function of the system. Suppose that the input to the discrete-time convolution system with impulse response *h* is the exponential signal

$$u(n) = z^n, \qquad n \in \mathbb{Z},$$

with *z* a complex number that lies in the existence region $\mathscr{E}_h$ of *H*. The response to this input is

$$y(n) = \sum_{k=-\infty}^{\infty} h(n-k)z^k = \sum_{m=-\infty}^{\infty} z^{n-m} h(m)$$

$$= \left(\sum_{m=-\infty}^{\infty} h(m)z^{-m} \right) z^n$$

$$= H(z)z^n, \qquad n \in \mathbb{Z}.$$

Thus, the response to the exponential input signal is again an exponential signal, with the same parameter *z*, premultiplied by the value of the transfer function $H(z)$ at the parameter *z*.

Similarly, it is easily confirmed that the response of a continuous-time convolution system with transfer function *H* to the exponential input

$$u(t) = e^{st}, \qquad t \in \mathbb{R},$$

with *s* a complex number in the existence region $\mathscr{E}_h$ of *H*, is given by

$$y(t) = H(s)e^{st}, \qquad t \in \mathbb{R}.$$

It follows that exponential signals are *eigensignals* of convolution signals, and, correspondingly, the *z*- and Laplace transforms are *spectral* transforms for discrete- and continuous-time convolution systems, respectively.

8.6.2.   Summary: Response of convolution systems to exponential inputs.

Suppose that the complex number *z* belongs to the existence region of the transfer function *H* of a discrete-time convolution system. Then, the response of the system to the exponential input

$$u(n) = z^n, \qquad n \in \mathbb{Z},$$

is

$$y(n) = H(z)z^n, \qquad n \in \mathbb{Z}.$$

Suppose that the complex number *s* belongs to the existence region of the transfer function *H* of a continuous-time convolution system. Then, the response of the system to the exponential input

$$u(t) = e^{st}, \qquad t \in \mathbb{R},$$

is

$$y(t) = H(s)e^{st}, \qquad t \in \mathbb{R}. \qquad \blacksquare$$

The transfer function is a generalization of the *frequency response function* of the system. The important difference is that the transfer function may exist even if the system has no frequency response function.

Conversely, if the frequency response function exists, then it is a special case of the transfer function. Suppose that the transfer function *H* of a discrete-time convolution system with impulse response *h* exists on the unit circle (i.e., the unit circle is included in the existence region). Then inspection of (1) shows that the frequency response function $\hat{h}$ of the convolution system is given by

$$\hat{h}(f) = H(e^{j2\pi f}), \qquad f \in [0, 1).$$

Similarly, if the existence region of the transfer function *H* of a continuous-time convolution system includes the imaginary axis, inspection of (2) shows that the frequency response function $\hat{h}$ of the system is

$$\hat{h}(f) = H(j2\pi f), \qquad f \in \mathbb{R}.$$

These results are a specialization of 8.3.5(b) and (b′) to the impulse response *h*. We summarize as follows.

8.6.3.   Summary: Transfer function and frequency response function.

Suppose that the existence region of the transfer function H of a discrete-time convolution system contains the unit circle. Then the system has a frequency response function $\hat{h}$, which is given by

$$\hat{h}(f) = H(e^{j2\pi f}), \qquad f \in [0, 1).$$

Suppose that the existence region of the transfer function H of a continuous-time convolution system contains the imaginary axis. Then the system has a frequency response function $\hat{h}$, which is given by

$$\hat{h}(f) = H(j2\pi f), \qquad f \in \mathbb{R}. \qquad \blacksquare$$

Examples

We consider several examples of the application of the z- and Laplace transform to the analysis of convolution systems.

8.6.4.   Examples: Transform analysis of discrete-time convolution systems.

(a) *Transfer function and step response of the exponential smoother*. The impulse response h of the exponential smoother according to 3.4.2(a) is given by

$$h(n) = (1 - a)a^n 1(n), \qquad n \in \mathbb{Z}.$$

From Table 8.3 we take that the z-transform of the signal a^n, $n \in \mathbb{Z}$, is $z/(z - a)$, $|z| > |a|$. It follows that the transfer function H of the smoother is

$$H(z) = \frac{(1 - a)z}{z - a}, \qquad |z| > |a|.$$

If $|a| < 1$, the existence region of H includes the unit circle, so that by 8.6.3 the system has the frequency response function

$$\hat{h}(f) = H(e^{j2\pi f}) = \frac{(1 - a)e^{j2\pi f}}{e^{j2\pi f} - a}$$

$$= \frac{1 - a}{1 - ae^{-j2\pi f}}, \qquad f \in \mathbb{R}.$$

This agrees with what we found in Example 3.7.5(a).

Suppose that the input to the smoother when initially at rest is the unit step $u = 1$. From Table 8.3 we see that the z-transform of this input is

$$U(z) = \frac{z}{z - 1}, \qquad |z| > 1.$$

It follows from 8.6.1 that the z-transform of the output is

$$Y(z) = H(z)U(z) = \frac{(1 - a)z^2}{(z - a)(z - 1)}, \qquad |z| > \max(1, |a|).$$

Expansion of Y in partial fractions results in

$$Y(z) = (1 - a) + \frac{1}{z - 1} - \frac{a^2}{z - a}, \qquad |z| > \max(1, |a|).$$

Inspection of Table 8.3 reveals that the inverse z-transform of Y is

$$y(n) = (1 - a)\Delta(n) + 1(n - 1) - a \cdot a^n 1(n - 1)$$
$$= (1 - a^{n+1})1(n), \qquad n \in \mathbb{Z}.$$

This is the step response that was already found in 3.4.6(a).

(b) *Response of the smoother to a two-sided input*. Suppose that the input to the smoother is the two-sided signal u given by

$$u(n) = b^n 1(n) + c^n 1(-n - 1), \qquad n \in \mathbb{Z},$$

as plotted in Fig. 8.9(a) for $|b| < 1$ and $|c| > 1$. From Table 8.3 it follows that this input has a two-sided z-transform provided $|b| < |c|$. Under this condition, the z-transform is given by

$$U(z) = \frac{z}{z - b} - \frac{z}{z - c}, \qquad |b| < |z| < |c|.$$

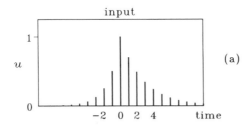

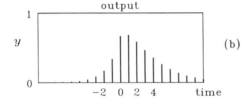

Figure 8.9.   Input and output to the exponential smoother.

Since the existence region of the transfer function of the smoother is $|z| > |a|$, by 8.6.1 the smoother has a well-defined response provided $|a| < |c|$, because this ensures that $\mathscr{E}_h \cap \mathscr{E}_u$ is nonempty. If this condition is satisfied, the z-transform Y of the output is

$$Y(z) = \frac{(1 - a)z}{z - a}\left(\frac{z}{z - b} - \frac{z}{z - c}\right), \qquad \max\,(|a|, |b|) < |z| < |c|.$$

By partial fraction expansion this may be rewritten as

$$Y(z) = \frac{a(1 - a)(b - c)}{(a - b)(a - c)} \cdot \frac{a}{z - a} - \frac{(1 - a)b}{a - b} \cdot \frac{b}{z - b} + \frac{(1 - a)c}{a - c} \cdot \frac{c}{z - c}.$$

Keeping an eye on the existence region, term-by-term inversion yields with Table 8.3

$$y(n) = \frac{a(1 - a)(b - c)}{(a - b)(a - c)} a^n 1(n - 1) - \frac{(1 - a)b}{a - b} b^n 1(n - 1) - \frac{(1 - a)c}{a - c} c^n 1(-n),$$

for $n \in \mathbb{Z}$. The output is plotted in Fig. 8.9(b) for $a = 0.5$, $b = 0.7$, and $c = 2$. ∎

8.6.5.   Example: Transform analysis of the sliding window averager. The sliding window averager is the continuous-time system with IO map

$$y(t) = \frac{1}{T_1 + T_2} \int_{t-T_1}^{t+T_2} u(\tau)\, d\tau, \qquad t \in \mathbb{R},$$

with T_1 and T_2 nonnegative real numbers such that $T_1 + T_2 \neq 0$. To find the transfer function of this system we consider the exponential input

$$u(t) = e^{st}, \qquad t \in \mathbb{R}.$$

The response to this input is

$$y(t) = \frac{1}{T_1 + T_2} \int_{t-T_1}^{t+T_2} e^{s\tau}\, d\tau = \frac{e^{sT_2} - e^{-sT_1}}{s(T_1 + T_2)} e^{st}, \qquad t \in \mathbb{R},$$

so that the system has the transfer function

$$H(s) = \frac{e^{sT_2} - e^{-sT_1}}{s(T_1 + T_2)}.$$

The transfer function is well-defined on the whole complex plane. (Note that $H(0) = 1$.) Rewriting H as

$$H(s) = \frac{1}{T_1 + T_2}\left(\frac{e^{sT_2}}{s} - \frac{e^{-sT_1}}{s}\right)$$

we see with Table 8.4 and the shift property that the inverse Laplace transform of H (i.e., the impulse response of the system) is given by

$$h(t) = \frac{1}{T_1 + T_2}\Big(1(t + T_2) - 1(t - T_1)\Big)$$

$$= \begin{cases} \dfrac{1}{T_1 + T_2} & \text{for } -T_2 \le t < T_1, \\ 0 & \text{otherwise}, \end{cases} \qquad t \in \mathbb{R}.$$

We now pose the question whether there exists an input u such that the corresponding output y is the ramp signal

$$y(t) = t \cdot 1(t), \qquad t \in \mathbb{R}.$$

From Table 8.4 we see that this output has the Laplace transform

$$Y(s) = \frac{1}{s^2}, \qquad \text{Re }(s) > 0.$$

Assuming that the input u we are looking for has a Laplace transform U, by 8.6.1 we have $Y = HU$, or

$$U(s) = \frac{Y(s)}{H(s)} = \frac{T_1 + T_2}{s(e^{sT_2} - e^{-sT_1})}, \qquad \text{Re }(s) > 0,$$

The inverse transform of U may be found by power series expansion as follows. First rewrite U as

$$U(s) = \frac{T}{se^{sT_2}(1 - e^{-sT})}, \qquad \text{Re }(s) > 0,$$

where $T = T_1 + T_2$. Because $\left|e^{-sT}\right| = e^{-T\,\text{Re}\,(s)} < 1$ for Re $(s) > 0$, we may expand in a converging power series as

$$U(s) = \frac{T}{s}e^{-sT_2}(1 + e^{-sT} + e^{-s2T} + e^{-s3T} + \cdots), \qquad \text{Re }(s) > 0.$$

By using the fact that $1/s$, Re $(s) > 0$, is the Laplace transform of the unit step, and exploiting the shift property, we find the inverse Laplace transform of U as

$$u(t) = T[\mathbb{1}(t - T_2) + \mathbb{1}(t - T - T_2) + \mathbb{1}(t - 2T - T_2) + \cdots], \qquad t \in \mathbb{R}.$$

A plot of this staircase-like input u is given in Fig. 8.10.   ■

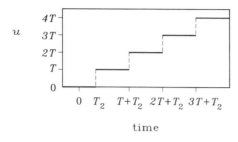

Figure 8.10.   The response of the sliding window averager to this input is a ramp.

8.6.6.   Review: Transfer function of sampled convolution systems.   A sampled convolution system defined on the time axis $\mathbb{Z}(T)$ is described by the IO map

$$y(t) = T \sum_{\tau \in \mathbb{Z}(T)} h(t - \tau)u(\tau), \qquad t \in \mathbb{Z}(T),$$

with h the impulse response of the convolution system. If the input is the exponential signal $u(t) = z^{t/T}$, $t \in \mathbb{Z}(T)$, then it may easily be found that the output is

$$y(t) = H(z)z^{t/T}, \qquad t \in \mathbb{Z}(T),$$

where H is the "sampled" transfer function

$$H(z) = T \sum_{t \in \mathbb{Z}(T)} h(t)z^{-t/T}$$

for those z for which it exists. Comparing with 8.3.10 we see that H is the sampled z-transform of the impulse response h.

As observed in 8.3.10, the sampled z-transform is not common. It is much more useful to describe sampled convolution systems in *sequence* form by the IO map

$$y(nT) = \sum_{m=-\infty}^{\infty} h*(n - m)u(mT), \qquad n \in \mathbb{Z}, \tag{3}$$

where the *pulse response* $h*$ is given by

$$h*(n) = Th(nT), \qquad n \in \mathbb{Z}.$$

Application of the z-transform in sequence form to (3) by the convolution property results in

$$Y^*(z) = H^*(z)U^*(z),$$

where

$$U^*(z) = \sum_{n=-\infty}^{\infty} u(nT)z^{-n}, \qquad Y^*(z) = \sum_{n=-\infty}^{\infty} y(nT)z^{-n},$$

and

$$H^*(z) = \sum_{n=-\infty}^{\infty} h^*(nT)z^{-n} = T\sum_{n=-\infty}^{\infty} h(nT)z^{-n} = H(z).$$

Thus, in this way we find the *same* transfer function as when using the sampled z-transform. The thing to keep in mind is that when analyzing the sampled system in sequence form the transfer function forms a z-transform pair with the *pulse* response, and not with the impulse response. ∎

8.7 TRANSFORM ANALYSIS OF DIFFERENCE AND DIFFERENTIAL SYSTEMS

In this section we study difference and differential systems in two situations: initially at rest on a two-sided infinite time axis and on a semi-infinite time axis with arbitrary initial conditions.

Transfer Functions of Initially-At-Rest Difference and Differential Systems

Initially-at-rest systems defined by constant coefficient difference and differential equations are convolution systems. In Section 8.6 we have seen that the z- or Laplace transform analysis of a convolution system relies on the *transfer function* of the system. The transfer function is the z- or Laplace transform of the impulse response of the system. Alternatively, the transfer function may be obtained by determining the response of the system to an exponential input. We use the former property to establish the *existence region* of the transfer function of a difference or differential system and the latter to find the transfer function itself.

Consider a difference system described by the constant coefficient linear difference equation

$$Q(\sigma)y = P(\sigma)u, \tag{1}$$

with u and y defined on the infinite time axis $\mathbb{Z}$, and Q and P given polynomials. From 4.5.1 we know that the impulse response h of the system is the sum of several shifted unit pulses and a linear combination of the basis solutions of the homogeneous equation multiplied by the unit step. The basis solutions that are involved are of the form

$$n^k \pi_i^n, \qquad n \in \mathbb{Z},$$

with k an integer and π_i a *pole* of the system. Recall that the poles of the system are those characteristic values of the system that do not cancel against roots of P in the rational function P/Q. As observed in Section 4.6, characteristic roots that cancel (i.e., that are not poles of the system) do not appear in h. We note that $n^k \pi_i^n z^{-n}$, $n \in \mathbb{Z}$, decays exponentially as n increases if and only if $|z| > |\pi_i|$. It follows that the transfer function

$$H(z) = \sum_{n=-\infty}^{\infty} h(n) z^{-n}$$

exists if and only if $|z| > |\pi_i|$ for each i. This establishes the existence region of H as

$$\mathcal{E}_h = \{ z \in \mathbb{C} \mid |z| > \max_i |\pi_i| \}.$$

To determine the actual form of the transfer function of the initially-at-rest system defined by the difference equation (1) we use the fact that if its input is exponential of the form $u(n) = z^n$, $n \in \mathbb{Z}$, with z in the existence region of H, then the corresponding output is $y(n) = H(z) z^n$, $n \in \mathbb{Z}$. Thus, we determine $H(z)$ by substituting

$$u(n) = z^n, \qquad y(n) = y_o z^n, \qquad n \in \mathbb{Z},$$

into the difference equation, and solving for y_o. Note that if u is exponential then

$$(\sigma^k u)(n) = u(n + k) = z^{n+k} = z^k \cdot z^n, \qquad n \in \mathbb{Z}.$$

By linearity, it follows that

$$P(\sigma)u(n) = (p_N \sigma^N + p_{N-1} \sigma^{N-1} + \cdots + p_0) z^n$$
$$= (p_N z^N + p_{N-1} z^{N-1} + \cdots + p_0) z^n = P(z) z^n, \qquad n \in \mathbb{Z}.$$

Similarly, $Q(\sigma)y(n) = Q(z) y_o z^n$, $n \in \mathbb{Z}$, so that

$$Q(z) y_o z^n = P(z) z^n, \qquad n \in \mathbb{Z}.$$

Cancellation of the common factor z^n and solution for y_o results in

$$y_o = H(z) = \frac{P(z)}{Q(z)}, \qquad z \in \mathscr{E}_h.$$

By an analogous argument we may establish the existence region and form of the transfer function of the initially-at-rest differential system defined by the constant coefficient differential equation

$$Q(D)y = P(D)u.$$

Its existence region now is given by

$$\mathscr{E}_h = \{s \in \mathbb{C} \mid \text{Re}\,(s) > \max_i \text{Re}\,(\pi_i)\},$$

with the π_i again the poles of the system, while the transfer function itself is

$$H(s) = \frac{P(s)}{Q(s)}, \qquad s \in \mathscr{E}_h.$$

Thus, both for difference and differential systems the transfer function H may immediately be obtained from the polynomials P and Q in the difference or differential equation. We summarize as follows.

8.7.1.   Summary: Transfer function of linear constant coefficient difference and differential systems.

The transfer function of the initially-at-rest difference system defined by the difference equation $Q(\sigma)y = P(\sigma)u$ is

$$H(z) = \frac{P(z)}{Q(z)}, \qquad z \in \mathscr{E}_h.$$

The existence region is given by

$$\mathscr{E}_h = \{z \in \mathbb{C} \mid |z| \geq \max_i |\pi_i|\},$$

where the complex numbers π_i are the poles of the system.

The transfer function of the initially-at-rest differential system defined by the differential equation $Q(D)y = P(D)u$ is

$$H(s) = \frac{P(s)}{Q(s)}, \qquad s \in \mathscr{E}_h.$$

The existence region is given by

$$\mathscr{E}_h = \{s \in \mathbb{C} \mid \text{Re}\,(s) > \max_i \text{Re}\,(\pi_i)\},$$

where the complex numbers π_i are the poles of the system.   ∎

We already encountered the transfer function H of a difference or differential system when defining poles and zeros of difference and differential systems in 4.6.1. After canceling all common factors of the numerator and denominator of $H = P/Q$ we may write H in the form

$$H(\lambda) = \frac{p_M \prod_i (\lambda - \zeta_i)}{q_N \prod_i (\lambda - \pi_i)},$$

where λ equals z in the discrete-time case and s in the continuous-time case. The complex numbers ζ_i are the zeros of the system, and the complex numbers π_i its poles, while p_M and q_N are constants. H vanishes at the zeros, and is infinite at the poles.

We consider a few examples of transfer functions.

8.7.2.   Examples: Transfer functions of difference and differential systems.

(a) *Exponential smoother*. The exponential smoother is described by the difference equation

$$y(n + 1) - ay(n) = (1 - a)u(n + 1), \qquad n \in \mathbb{Z}.$$

Inspection shows that according to 8.7.1 the transfer function of the system is

$$H(z) = \frac{(1 - a)z}{z - a}, \qquad z \in \mathcal{E}_h.$$

If $a \neq 0$ and $a \neq 1$, the system has a single pole a, so that by 8.7.1 the existence region is

$$\mathcal{E}_h = \{z \in \mathbb{C} \mid |z| > |a|\}.$$

This is what we found in Example 8.6.4(a), with much more effort. If $a = 0$ the pole cancels and the transfer function is $H(z) = 1$, $z \in \mathbb{C}$. If $a = 1$, then the transfer function trivially is $H(z) = 0$, $z \in \mathbb{C}$.

(b) *RCL network*. From Example 4.2.4(c), the RCL network with the voltage across the inductor as output is described by the differential equation

$$\frac{d^2y(t)}{dt^2} + \frac{R}{L} \frac{dy(t)}{dt} + \frac{1}{LC} y(t) = \frac{d^2u(t)}{dt^2}, \qquad t \in \mathbb{R}.$$

It follows that the transfer function is

$$H(s) = \frac{s^2}{s^2 + \dfrac{R}{L}s + \dfrac{1}{LC}}, \qquad s \in \mathcal{E}_h.$$

The poles of the transfer function are the characteristic roots

$$\lambda_{1,2} = \frac{-R \pm \sqrt{R^2 - \dfrac{4L}{C}}}{2L}.$$

If $R^2 \geq 4L/C$, then both roots are real, and the existence region of the transfer function is formed by those s such that $\text{Re}\,(s) > (\sqrt{R^2 - 4L/C} - R)/2L$. If $R^2 < 4L/C$, then the roots form a complex conjugate pair with real part $-R/2L$, so that the existence region is given by $\text{Re}\,(s) > -R/2L$. Figure 8.11 illustrates the existence region. ∎

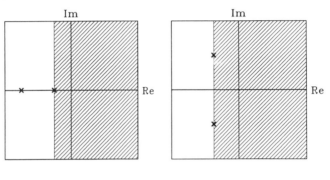

✕ pole

Figure 8.11. Existence region for the transfer function of the RCL network. Left: pair of real roots. Right: complex conjugate pair of roots.

Transform Analysis of Initially-At-Rest Difference and Differential Systems

Initially-at-rest difference and differential systems are convolution systems, whose transfer function H may be found as in 8.7.1. From 8.6.1 and 8.7.1 we immediately obtain the following.

8.7.3.   Summary: Transform analysis of initially-at-rest difference and differential systems.

Consider the initially-at-rest difference system described by the difference equation

$$Q(\sigma)y = P(\sigma)u$$

on the time axis $\mathbb{Z}$, and define

$$\mathscr{E}_h = \left\{ z \in \mathbb{C} \mid |z| > \max_i |\pi_i| \right\},$$

Consider the initially-at-rest differential system described by the differential equation

$$Q(D)y = P(D)u$$

on the time axis $\mathbb{R}$, and define

$$\mathscr{E}_h = \left\{ s \in \mathbb{C} \mid \text{Re}\,(s) > \max_i \text{Re}\,(\pi_i) \right\},$$

with the π_i the poles of the transfer function $H = P/Q$.

(a) A necessary and sufficient condition for the initially-at-rest system to have a well-defined response y is that $\mathscr{E}_h \cap \mathscr{E}_u$ be nonempty, where $\mathscr{E}_u$ is the existence region of the z-transform U of u.

(b) If the system has a well-defined response y, its z-transform is given by

$$Y = HU$$

with existence region $\mathscr{E}_h \cap \mathscr{E}_u$, so that

$$y = \mathscr{Z}^{-1}(HU).$$

(c) In particular, the impulse response h of the initially-at-rest system is given by

$$h = \mathscr{Z}^{-1}(H).$$

with the π_i the poles of the transfer function $H = P/Q$.

(a$'$) A necessary and sufficient condition for the initially-at-rest system to have a well-defined response y is that $\mathscr{E}_h \cap \mathscr{E}_u$ be nonempty, where $\mathscr{E}_u$ is the existence region of the Laplace transform U of u.

(b$'$) If the system has a well-defined response y, its Laplace transform is given by

$$Y = HU$$

with existence region $\mathscr{E}_h \cap \mathscr{E}_u$, so that

$$y = \mathscr{L}^{-1}(HU).$$

(c$'$) In particular, the impulse response h of the initially-at-rest system is given by

$$h = \mathscr{L}^{-1}(H). \qquad \blacksquare$$

The final results (c) and (c$'$) follow by taking $u = \Delta$ or δ. By (c) and (c$'$) we may obtain the impulse response of the difference or differential system by *inverse z- or Laplace transformation of the transfer function.* By using partial fraction expansion, this is often a much easier way of finding the impulse response than the time domain methods outlined in Section 4.5.

8.7.4.   Example: Impulse response of difference and differential systems.

(a) *Impulse response of a second-order difference system.* In 4.5.2(a) we considered the difference system

$$\frac{1}{6} y(n) - \frac{5}{6} y(n+1) + y(n+2) = u(n+3), \qquad n \in \mathbb{Z}.$$

Inspection shows that the transfer function of the system is

$$H(z) = \frac{z^3}{z^2 - \dfrac{5}{6} z + \dfrac{1}{6}}.$$

The denominator of H may be factored as $(z - 1/2)(z - 1/3)$. Neither of the roots $1/2$ and $1/3$ cancels, so that the existence region of H is

$$\mathscr{E}_h = \{z \in \mathbb{C} \mid |z| > 1/2\}.$$

It follows by partial fraction expansion that

$$G(z) = z^{-3}H(z) = \frac{1}{\left(z - \frac{1}{2}\right)\left(z - \frac{1}{3}\right)} = 12\frac{\frac{1}{2}}{z - \frac{1}{2}} - 18\frac{\frac{1}{3}}{z - \frac{1}{3}}.$$

By Table 8.3, on the given existence region the inverse z-transform g of G is given by

$$g(n) = 12\left(\frac{1}{2}\right)^n 1(n-1) - 18\left(\frac{1}{3}\right)^n 1(n-1), \qquad n \in \mathbb{Z}.$$

Since $H(z) = z^3 G(z)$, it follows by the shift property of the two-sided z transform that the inverse z-transform h of H is given by

$$h(n) = g(n+3) = \left[12\left(\frac{1}{2}\right)^{n+3} - 18\left(\frac{1}{3}\right)^{n+3}\right]1(n+2)$$

$$= \left[\frac{3}{2}\left(\frac{1}{2}\right)^n - \frac{2}{3}\left(\frac{1}{3}\right)^n\right]1(n+2), \qquad n \in \mathbb{Z}.$$

This is precisely the impulse response that we found in 4.5.2(a). Note that the transfer function H is improper, and that accordingly we obtain an anticipating impulse response.

(b) *Impulse response of the RCL network.* In the second part of this example we rework 4.5.2(b), where the impulse response was determined of the RCL network described by the differential equation

$$\frac{1}{LC}y + \frac{R}{L}\frac{dy(t)}{dt} + \frac{d^2y(t)}{dt^2} = \frac{d^2u(t)}{dt^2}, \qquad t \in \mathbb{R}.$$

The transfer function of the network is

$$H(s) = \frac{s^2}{s^2 + \frac{R}{L}s + \frac{1}{LC}}.$$

Assuming as in 4.5.2(b) that the network is critically damped (i.e., $R^2 = 4L/C$), it follows that

$$H(s) = \frac{s^2}{\left(s + \dfrac{1}{\sqrt{LC}}\right)^2},$$

which shows that the existence region is Re $(s) > -1/\sqrt{LC}$. By partial fraction expansion it follows that

$$H(s) = 1 - \frac{\dfrac{2}{\sqrt{LC}}}{s + \dfrac{1}{\sqrt{LC}}} + \frac{\dfrac{1}{LC}}{\left(s + \dfrac{1}{\sqrt{LC}}\right)^2}, \qquad \text{Re } (s) > -\frac{1}{\sqrt{LC}}.$$

With Table 8.4 we find that this is the Laplace transform of

$$h(t) = \delta(t) + \left[-\frac{2}{\sqrt{LC}} e^{-t/\sqrt{LC}} + \frac{t}{LC} e^{-t/\sqrt{LC}} \right] 1(t), \qquad t \in \mathbb{R}.$$

This agrees with the result of 4.5.2(b).   ∎

Let us consider more in general how the impulse response of a difference or differential system follows by inverse z- or Laplace transformation of the transfer function. According to Section 8.2 the transfer function $H = P/Q$ of a differential system may be expanded as

$$H(s) = B(s) + \sum_{i=1}^{K} \left(\sum_{k=1}^{m_i} \frac{A_{ik}}{(s - \pi_i)^k} \right). \tag{2}$$

B is a polynomial of degree $M - N$ that may be written as

$$B(s) = \sum_{i=0}^{M-N} \beta_i s^i,$$

with the β_i the coefficients of B, M the degree of P and N that of Q. The partial fraction expansion in (2) only contains terms involving the *poles* π_i of H. A_{ik} is the coefficient of the partial fraction and m_i the multiplicity of the pole π_i. Inverse transformation of (2) with the use of Tables 8.4 and 8.5 shows that the impulse response h is of the form

$$h(t) = \sum_{i=0}^{M-N} \beta_i \delta^{(i)}(t) + \sum_{i=1}^{K} \left(\sum_{k=1}^{m_i} A_{ik} \frac{t^{k-1} e^{\pi_k t}}{(k-1)!} 1(t) \right), \qquad t \in \mathbb{R}.$$

By 4.4.1, each of the terms of the double sum on the right-hand side of this expression is a basis solution of the homogeneous differential equation $Q(D)y = 0$. Hence, we may rewrite h as

$$h(t) = \sum_{i=0}^{M-N} \beta_i \delta^{(i)}(t) + \sum_{i=1}^{N} \alpha_i y_i(t) \mathbb{1}(t), \qquad t \in \mathbb{R},$$

where the y_i are basis solutions of the homogeneous equation, and the α_i suitable coefficients. This is the form in which the impulse response is given in 4.5.1. A similar derivation may be given for the discrete-time case.

The present derivation confirms that only the *poles* π_i of the system affect the impulse response of the difference or differential system, as proved in the time domain in E.1 in Supplement E. Those roots of Q that cancel play no role.

Transform Solution of Initial Value Problems

The *one*-sided z- and Laplace transforms are eminently suited for the solution of initial value problems for linear constant coefficient difference and differential systems. Consider for instance the difference equation

$$Q(\sigma)y = P(\sigma)u,$$

with $u(n)$ given for $n \geq 0$ and given initial conditions $y(0), y(1), \cdots, y(N - 1)$, with N the degree of Q. Application of the one-sided z-transform yields with the use of the shift property an *algebraic* equation for the one-sided z-transform Y_+ of y. After solving for Y_+, the time domain solution is recovered by inverse z-transformation.

Similarly, application of the one-sided Laplace transform to the differential equation

$$Q(D)y = P(D)u,$$

with $u(t)$ given for $t \geq 0^-$ and specified initial conditions $y(0^-), y^{(1)}(0^-), \cdots, y^{(N-1)}(0^-)$, results in an algebraic equation for the one-sided Laplace transform Y_+ of y. After solving for Y_+ the time domain solution follows by inverse Laplace transformation.

We illustrate the procedure with some examples.

8.7.5.   Examples: Transform solution of initial value problems.

(a) *The Fibonacci equation.* The Fibonacci equation is the homogeneous difference equation

$$y(n + 2) = y(n + 1) + y(n), \qquad n \in \mathbb{Z}_+, \tag{3}$$

with the initial conditions

$$y(0) = 0, \qquad y(1) = 1.$$

By the shift property, the one-sided z-transform of $y(n + 1)$, $n \in \mathbb{Z}$, is

$$zY_+(z) - zy(0) = zY_+(z),$$

while that of $y(n + 2)$, $n \in \mathbb{Z}$, is

$$z[zY_+(z) - zy(0)] - zy(1) = z^2Y_+(z) - z.$$

Thus, application of the one-sided z-transform to the difference equation (3) shows that

$$z^2Y_+(z) - z = zY_+(z) + Y_+(z).$$

As predicted, this is an algebraic equation, which may be solved for Y_+ as

$$Y_+(z) = \frac{z}{z^2 - z - 1}.$$

Because the poles of the rational function on the right-hand side are $\frac{1}{2} \pm \frac{1}{2}\sqrt{5}$, this is the z-transform of a right one-sided signal if $|z| > \frac{1}{2} + \frac{1}{2}\sqrt{5}$. By partial fraction expansion it follows that

$$Y_+(z) = \frac{\sqrt{5}}{5}\frac{\frac{1}{2} + \frac{1}{2}\sqrt{5}}{z - (\frac{1}{2} + \frac{1}{2}\sqrt{5})} - \frac{\sqrt{5}}{5}\frac{\frac{1}{2} - \frac{1}{2}\sqrt{5}}{z - (\frac{1}{2} - \frac{1}{2}\sqrt{5})}, \qquad |z| > \tfrac{1}{2} + \tfrac{1}{2}\sqrt{5}.$$

Consultation of Table 8.3 shows that this is the inverse z-transform of

$$y(n) = \frac{\sqrt{5}}{5}\left((\tfrac{1}{2} + \tfrac{1}{2}\sqrt{5})^n - (\tfrac{1}{2} - \tfrac{1}{2}\sqrt{5})^n\right)\mathbb{1}(n - 1), \qquad n \in \mathbb{Z},$$

which is the desired solution of the initial value problem.

(b) *RCL network*. As a second example we consider the solution of initial value problems for the RCL network with the voltage across the inductor as output, which by 4.2.4(c) is described by the differential equation

$$\frac{1}{LC}y(t) + \frac{R}{L}\frac{dy(t)}{dt} + \frac{d^2y(t)}{dt^2} = \frac{d^2u(t)}{dt^2}, \qquad t \in \mathbb{R}.$$

By the differentiation rule, the one-sided Laplace transform of the derivative Dy is

$$sY_+(s) - y(0^-), \qquad s \in \mathcal{E}_{y,+}.$$

Hence, the one-sided Laplace transform of the second derivative $D^2y = D(Dy)$ is

$$s[sY_+(s) - y(0^-)] - y'(0^-) = s^2Y_+(s) - sy(0^-) - y'(0^-), \qquad s \in \mathcal{E}_{y,+}.$$

For the Laplace transform of D^2u we find a similar result. Thus, application of the one-sided Laplace transform to the differential equation results in

$$\frac{1}{LC}Y_+(s) + \frac{R}{L}[sY_+(s) - y(0^-)] + [s^2Y_+(s) - sy(0^-) - y'(0^-)]$$

$$= s^2U_+(s) - su(0^-) - u'(0^-), \qquad s \in \mathcal{E}_{y,+} \cap \mathcal{E}_{u,+}.$$

This is an algebraic equation, which may be solved for Y_+ as

$$Y_+(s) = \frac{s^2}{s^2 + \dfrac{R}{L}s + \dfrac{1}{LC}}U_+(s)$$

$$+ \frac{s[y(0^-) - u(0^-)] + \left[\dfrac{R}{L}y(0^-) + y'(0^-) - u'(0^-)\right]}{s^2 + \dfrac{R}{L}s + \dfrac{1}{LC}}, \qquad s \in \mathcal{E}_{y,+}.$$

Inspection of the right-hand side shows that

$$\mathcal{E}_{y,+} = \mathcal{E}_{u,+} \cap \{s \in \mathbb{C} \mid \text{Re}\ (s) > \max\ [\text{Re}\ (\pi_1),\ \text{Re}\ (\pi_2)]\},$$

where π_1 and π_2 are the roots of the denominator polynomial $s^2 + (R/L)s + 1/LC$. For simplicity we assume the numerical values $R/L = 2$ and $LC = 1$, so that

$$Y_+(s) = \frac{s^2}{s^2 + 2s + 1}U_+(s)$$

$$+ \frac{s[y(0^-) - u(0^-)] + [2y(0^-) + y'(0^-) - u'(0^-)]}{s^2 + 2s + 1},$$

for $s \in \mathcal{E}_{y,+}$. Consider the two following special cases:

(i) $y(0^-) = 1$, $y'(0^-) = 0$, $u(t) = 1$ for $t \in \mathbb{R}$,
(ii) $y(0^-) = 1$, $y'(0^-) = 0$, $u(t) = \mathbb{1}(t)$ for $t \in \mathbb{R}$.

The initial conditions and the behavior of u for $t \geq 0$ are the same, but in case (i) u is continuous at time 0, and in case (ii) it has a jump. Thus, on $[0^-, \infty)$ the inputs differ. In the time domain the jump of u in case (ii) results in a derivative δ-function on the right-hand side of the differential equation, which affects the solution. In the transform domain in both cases $U_+(s) = 1/s$, Re $(s) > 0$, but because $u(0^-)$ is not the same, in case (i) we have

$$Y_+(s) = \frac{s^2}{s^2 + 2s + 1} \cdot \frac{1}{s} + \frac{(1 - 1)s + 2}{s^2 + 2s + 1} = \frac{s + 2}{s^2 + 2s + 1}$$

$$= \frac{1}{s + 1} + \frac{1}{(s + 1)^2}, \qquad \text{Re } (s) > -1,$$

while in case (ii) it follows that

$$Y_+(s) = \frac{s^2}{s^2 + 2s + 1} \cdot \frac{1}{s} + \frac{(1 - 0)s + 2}{s^2 + 2s + 1} = \frac{2(s + 1)}{s^2 + 2s + 1}$$

$$= \frac{2}{s + 1}, \qquad \text{Re } (s) > -1.$$

Because of the cancellation of the denominator s of U_+, the two Laplace transforms exist on Re $(s) > -1$. In case (i) the solution of the initial value problem is

$$y(t) = (1 + t)e^{-t}, \qquad t \geq 0,$$

and in case (ii)

$$y(t) = 2e^{-t}, \qquad t \geq 0. \qquad\blacksquare$$

8.7.6.   Example: Solution of initial value problems for sets of differential equations.   In this example we demonstrate the use of the one-sided Laplace transform for the solution of initial value problems for *sets* of differential and algebraic equations, which together describe a system. By way of example we consider again the RCL network. Rather than eliminating all intermediate system variables, as we did in 4.2.4(c) in order to arrive at the differential equation relating the input and output, we apply the Laplace transform directly to the various component equations.

The network diagram is repeated in Fig. 8.12. It follows by Kirchhoff's voltage law that

$$u = v_R + v_C + v_L.$$

For the three circuit components we have successively

$$v_L = \dot{\phi}_L, \qquad \phi = Li,$$

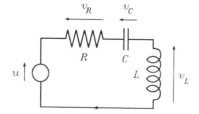

Figure 8.12. An RCL network.

with ϕ the flux contained by the inductor,

$$v_C = \frac{1}{C}q, \qquad \dot{q} = i,$$

with q the charge contained by the capacitor, and

$$v_R = Ri.$$

We apply the one-sided Laplace transform to each of these six equations, and obtain

$$U(s) = V_R(s) + V_C(s) + V_L(s),$$

$$V_L(s) = s\Phi(s) - \phi(0^-), \qquad \Phi(s) = LI(s),$$

$$V_C(s) = \frac{1}{C}Q(s), \qquad sQ(s) - q(0^-) = I(s),$$

$$V_R(s) = \frac{1}{R}I(s),$$

with U, V_R, V_C, V_L, Φ, Q and I denoting one-sided Laplace transforms, which we consider on some common existence region $\mathscr{E}$.

Suppose that we are interested in the current through the network. Straightforward algebraic elimination of V_R, V_C, V_L, Φ, and Q and solution for I yields

$$I(s) = \frac{\dfrac{s}{L}}{s^2 + \dfrac{R}{L}s + \dfrac{1}{LC}}U(s) + \frac{\dfrac{s}{L}\phi(0^-) - \dfrac{1}{LC}q(0^-)}{s^2 + \dfrac{R}{L}s + \dfrac{1}{LC}}.$$

The expression shows that the solution of the initial value problem is determined by the input u on $\mathbb{R}_+$, the initial charge $q(0^-)$ of the capacitor and the initial flux $\phi(0^-)$ of the inductor. Suppose that $u = 0$, $R = 2$, $C = \frac{1}{2}$, $L = 1$, $q(0^-) = 1$, and $\phi(0^-) = 1$. Then,

$$I(s) = \frac{s - 2}{s^2 + 2s + 2}.$$

The poles of I are $-1 \pm j$, so that the existence region is Re $(s) > -1$. Rewriting I as

$$I(s) = \frac{(s + 1) - 3}{(s + 1)^2 + 1} = \frac{s + 1}{(s + 1)^2 + 1} - 3\frac{1}{(s + 1)^2 + 1}, \qquad \text{Re } (s) > -1,$$

we see from Table 8.4 that in the time domain

$$i(t) = e^{-t}[\cos(t) - 3\sin(t)]\mathbb{1}(t), \qquad t \in \mathbb{R}.$$ ∎

8.7.7.   Review: Transfer function of sampled difference systems.   Sampled difference systems are described by a difference equation of the form

$$q_N y(t + NT) + q_{N-1} y(t + (N - 1)T) + \cdots + q_0 y(t)$$
$$= p_M u(t + MT) + p_{M-1} u(t + (M - 1)T) + \cdots + p_0 u(t), \qquad t \in \mathbb{Z}(T).$$

Defining the polynomials

$$Q(\lambda) = q_N \lambda^N + q_{N-1} \lambda^{N-1} + \cdots + q_0,$$
$$P(\lambda) = p_M \lambda^M + p_{M-1} \lambda^{M-1} + \cdots + p_0,$$

the difference equation may be written in the more compact form

$$Q(\sigma^T)y = P(\sigma^T)u.$$

The transfer function of the sampled difference system is as in 8.6.6 the "sampled" z-transform of the impulse response of the initially-at-rest system, and also the z-transform in sequential form of the pulse response. The transfer function may be obtained by determining the response of the initially-at-rest system to the exponential input $u(t) = z^{t/T}$, $t \in \mathbb{Z}(T)$, which easily leads to the conclusion that

$$H(z) = \frac{P(z)}{Q(z)}, \qquad z \in \mathscr{E}.$$

The existence region is $\mathscr{E} = \{z \in \mathbb{Z} \mid |z| > \max_i |\pi_i|\}$, where the π_i are the poles of H.

Initial value problems for sampled difference systems are most conveniently approached by applying the z-transform in sequential form to the equation

$$q_N y^*(n + N) + q_{N-1} y^*(n + N - 1) + \cdots + q_0 y^*(n)$$
$$= p_M u^*(n + M) + p_{M-1} u^*(n + M - 1) + \cdots + p_0 u^*(n), \qquad n \in \mathbb{Z}_+,$$

where $y^*(n) = y(nT)$ and $u^*(n) = u(nT)$, $n \in \mathbb{Z}_+$.   ∎

8.8 TRANSFORM ANALYSIS OF STATE SYSTEMS

In the final section of this chapter we briefly discuss the application of the z- and Laplace transforms, in particular the one-sided transforms, to the analysis of linear time-invariant state difference and differential systems.

Transformation of Vector-Valued Signals

Before discussing the transform analysis of state difference and differential systems we first need define the transform of *vector-valued* signals.

8.8.1.   Definition: The *z*- and Laplace transforms of vector-valued signals.   Suppose that x is a vector-valued discrete- or continuous-time signal with signal range $\mathbb{C}^N$ such that

$$x = \begin{bmatrix} x_1 \\ x_2 \\ \cdots \\ x_N \end{bmatrix}.$$

Then the two- or one sided *z*- or Laplace transform X of x is defined as

$$X = \begin{bmatrix} X_1 \\ X_2 \\ \cdots \\ X_N \end{bmatrix},$$

where X_i is the transform of the ith component x_i of x. The existence region $\mathscr{E}$ of X is the intersection of the existence regions $\mathscr{E}_i$ of the component transforms X_i, $i = 1$, $2, \cdots, N$.   ∎

We illustrate the definition with a simple example.

8.8.2   Example: Vector-valued Laplace transform.   By way of example, let x be the vector-valued continuous-time signal of dimension two given by

$$x(t) = \begin{bmatrix} \mathbb{1}(t) \\ e^{-t}\mathbb{1}(t) \end{bmatrix}, \qquad t \in \mathbb{R}.$$

Then the two-sided Laplace transform (which is also the one-sided transform) of x is

$$X(s) = \begin{bmatrix} \dfrac{1}{s} \\ \dfrac{1}{s+1} \end{bmatrix}, \qquad \text{Re}\,(s) > 0.$$

The existence region of X is the intersection

$$\{s \in \mathbb{C} \mid \text{Re}\,(s) > 0\} \cap \{s \in \mathbb{C} \mid \text{Re}\,(s) > -1\}$$

of the existence region of the first component and that of the second component.   ∎

The various properties of the z- and Laplace transforms may be generalized in reasonably obvious ways to the vector case. We need only linearity, the shift property of the one-sided z-transform and the differentiation property of the one-sided Laplace transform.

8.8.3.   Summary: Some properties of vector z- and Laplace transforms.

(a) *Linearity of the z-transform.* Suppose that x and y are vector-valued discrete-time signals of not necessarily the same dimensions and A and B constant matrices with real or complex elements of suitable dimensions such that Ax and By have the same dimensions.

Then, if X is the (one- or two-sided) z-transform of x and Y that of y, with existence regions $\mathscr{E}_x$ and $\mathscr{E}_y$, respectively, the z-transform of $Ax + By$ is $AX + BY$, with existence region $\mathscr{E}_x \cap \mathscr{E}_y$.

(b) *Shift property of the one-sided z-transform.* Let x be a vector-valued discrete-time signal with one-sided transform X_+ with existence region $\mathscr{E}_+$. Then the one-sided z-transform of the backward shifted signal $x(n + 1)$, $n \in \mathbb{Z}$, is $zX_+(z) - zx(0)$, $z \in \mathscr{E}_+$.

(a′) *Linearity of the Laplace transform.* Suppose that x and y are vector-valued continuous-time signals of not necessarily the same dimensions and A and B constant matrices with real or complex elements of suitable dimensions such that Ax and By have the same dimensions.

Then, if X is the (one- or two-sided) Laplace transform of x and Y that of y with existence regions $\mathscr{E}_x$ and $\mathscr{E}_y$, respectively, the Laplace transform of $Ax + By$ is $AX + BY$, with existence region $\mathscr{E}_x \cap \mathscr{E}_y$.

(b′) *Differentiation property of the Laplace transform.* Suppose that the regular or singular vector-valued continuous-time signal x has the one-sided Laplace transform X_+ with existence region $\mathscr{E}_+$. Then the (component-by-component) differentiated signal $\dot{x}$ has the one-sided Laplace transform $sX_+(s) - x(0^-)$, $s \in \mathscr{E}_+$.

■

The proof is straightforward.

Transform Analysis of State Difference and Differential Systems

We are now ready to apply the one-sided z-transform to the linear time-invariant state difference and output equations

$$x(n + 1) = Ax(n) + Bu(n), \tag{1a}$$

$$y(n) = Cx(n) + Du(n), \qquad n \in \mathbb{Z}_+, \tag{1b}$$

and to use the one-sided Laplace transform for the solution of the linear time-invariant state differential and output equations

$$\dot{x}(t) = Ax(t) + Bu(t), \tag{2a}$$

$$y(t) = Cx(t) + Du(t), \qquad t \in [0^-, \infty). \tag{2b}$$

One-sided z-transformation of the state difference and output equations (1) straight-forwardly yields with application of the shift property

$$zX(z) - zx(0) = AX(z) + BU(z),$$
$$Y(z) = CX(z) + DU(z), \qquad z \in \mathscr{E}.$$

X, U, and Y are the one-sided z-transforms of x, u and y. $\mathscr{E} = \{z \in \mathbb{C} \mid |z| > \rho\}$, with the real number ρ to be determined, is the common existence region of the three transforms. The first equation may be rearranged as

$$(zI - A)X(z) = zx(0) + BU(z),$$

where I is a unit matrix of the same dimensions as the matrix A. As long as z does not coincide with any of the eigenvalues of the matrix A the matrix $zI - A$ is non-singular, and we may solve for X as

$$X(z) = (zI - A)^{-1}zx(0) + (zI - A)^{-1}BU(z), \qquad z \in \mathscr{E}. \tag{3}$$

Substitution of this into the second equation results in

$$Y(z) = C(zI - A)^{-1}zx(0) + [C(zI - A)^{-1}B + D]U(z), \qquad z \in \mathscr{E}. \tag{4}$$

The equations (3) and (4) constitute an explicit solution for the state and output of the system in terms of z-transforms.

By using the differentiation property of the one-sided Laplace transform, it is easily established that the equivalent expressions for the one-sided Laplace transforms of the state and the output of the state differential system (2) are

$$X(s) = (sI - A)^{-1}x(0^-) + (sI - A)^{-1}BU(s),$$
$$Y(s) = C(sI - A)^{-1}x(0^-) + [C(sI - A)^{-1}B + D]U(s), \qquad s \in \mathscr{E},$$

with I a unit matrix with the same dimensions as A and $\mathscr{E}$ an existence region of the form $\mathscr{E} = \{s \in \mathbb{C} \mid \mathrm{Re}\,(s) > \sigma\}$ for some real σ.

We first establish a result for the z-transform of the transition matrix of a state difference system. To this end, suppose that the input u is zero, so that also $U = 0$. It follows from (3) that the z-transform of the zero-input response of the state is

$$X(z) = (zI - A)^{-1}zx(0), \qquad z \in \mathscr{E}_o,$$

with $\mathscr{E}_o$ the existence region of the transform, which is still to be determined. This expression is the transform equivalent of the time domain expression

$$x(n) = A^n x(0), \qquad n \in \mathbb{Z}_+.$$

Since this equivalence holds for *any* initial state $x(0)$, it follows that

$$A^n, \qquad n \in \mathbb{Z}_+, \qquad \text{and} \qquad (zI - A)^{-1}z, \qquad z \in \mathscr{E}_o,$$

form an (element-by-element) one-sided z-transform pair. Because as we know from Chapter 5 the elements of A^n are linear combinations of terms of the form $n^k\lambda_i^n$, with k an integer and λ_i an eigenvalue of the matrix A, we conclude that the existence region is

$$\mathscr{E}_o = \{z \in \mathbb{C} \mid |z| > \max_i |\lambda_i|\}. \tag{5}$$

By a parallel argument it follows for the continuous-time case that

$$e^{At}, \qquad t \in \mathbb{R}_+, \qquad \text{and} \qquad (sI - A)^{-1}, \, s \in \mathscr{E}_o,$$

form a one-sided Laplace transform pair, with existence region

$$\mathscr{E}_o = \{s \in \mathbb{C} \mid \text{Re}\,(s) > \max_i \text{Re}\,(\lambda_i)\}, \tag{6}$$

with the λ_i again the eigenvalues of the matrix A.

We summarize as follows.

8.8.4.   Summary: Transform of the transition matrix.

The element-by-element one- and two-sided z-transform of the transition matrix

$$A^n 1(n), \qquad n \in \mathbb{Z},$$

is the matrix

$$(zI - A)^{-1}z, \qquad z \in \mathscr{E}_o.$$

The existence region $\mathscr{E}_o$ is

$$\mathscr{E}_o = \{z \in \mathbb{C} \mid |z| > \max_i |\lambda_i|\},$$

where the λ_i are the eigenvalues of A.

The element-by-element one-sided Laplace transform of the transition matrix

$$e^{At} 1(t), \qquad t \in \mathbb{R},$$

is the matrix

$$(sI - A)^{-1}, \qquad s \in \mathscr{E}_o.$$

The existence region $\mathscr{E}_o$ is

$$\mathscr{E}_o = \{s \in \mathbb{C} \mid \text{Re}\,(s) > \max_i \text{Re}\,(\lambda_i)\},$$

where the λ_i are the eigenvalues of A.   ∎

This result provides a new method to compute the transition matrix of a time-invariant discrete- or continuous-time state system, namely, by inverse z- or Laplace transformation. We illustrate the procedure with two examples.

8.8.5.   Example: Computation of the transition matrix by z- or Laplace transformation.

(a) *Second-order smoother*. In Example 5.6.8(a) we considered the second-order smoother with state difference and output equations

$$x(n + 1) = Ax(n) + Bu(n),$$
$$y(n) = Cx(n) + Du(n), \qquad n \in \mathbb{Z},$$

with

$$A = \begin{bmatrix} \frac{1}{2} & 1 \\ 0 & 0 \end{bmatrix}, \qquad B = \begin{bmatrix} 1 \\ 0 \end{bmatrix},$$

$$C = [1 \quad 0], \qquad D = 0.$$

It follows that

$$(zI - A)^{-1} = \left(\begin{bmatrix} z - \frac{1}{2} & -1 \\ 0 & z \end{bmatrix} \right)^{-1} = \frac{1}{z(z - \frac{1}{2})} \begin{bmatrix} z & 1 \\ 0 & z - \frac{1}{2} \end{bmatrix}$$

$$= \begin{bmatrix} \dfrac{1}{z - \frac{1}{2}} & \dfrac{1}{z(z - \frac{1}{2})} \\ 0 & \dfrac{1}{z} \end{bmatrix}.$$

It follows that the one-sided z-transform of A^n, $n \in \mathbb{Z}_+$, is given by

$$(zI - A)^{-1} z = \begin{bmatrix} \dfrac{z}{z - \frac{1}{2}} & \dfrac{1}{z - \frac{1}{2}} \\ 0 & 1 \end{bmatrix}, \qquad |z| > \frac{1}{2}.$$

With Table 8.3 we obtain the inverse z-transform as

$$A^n 1(n) = \begin{bmatrix} (\frac{1}{2})^n 1(n) & 2(\frac{1}{2})^n 1(n - 1) \\ 0 & \Delta(n) \end{bmatrix}, \qquad n \in \mathbb{Z}.$$

This agrees with what we found in 5.6.8(a).

(b) *RCL network*. According to Example 5.2.1, the RCL network is described by the state differential equation $\dot{x} = Ax + Bu$, $y = Cx + Du$, where with the numerical values of 5.6.8(b) the matrix A is given by

$$A = \begin{bmatrix} 0 & 100 \\ -1000 & -1100 \end{bmatrix}.$$

It follows that

$$(sI - A)^{-1} = \left(\begin{bmatrix} s & -100 \\ 1000 & s + 1100 \end{bmatrix} \right)^{-1}$$

$$= \frac{1}{s^2 + 1100s + 100000} \begin{bmatrix} s + 1100 & 100 \\ -1000 & s \end{bmatrix}.$$

Since $s^2 + 1100s + 100000 = (s + 100)(s + 1000)$, this is the Laplace transform of $e^{At}1(t)$, $t \in \mathbb{R}$, on the existence region defined by Re $(s) > -100$. Element-by-element partial fraction expansion results in

$$
\begin{bmatrix}
\dfrac{1}{9}\left(\dfrac{10}{s + 100} + \dfrac{-1}{s + 1000}\right) & \dfrac{1}{9}\left(\dfrac{1}{s + 100} + \dfrac{-1}{s + 1000}\right) \\[3mm]
\dfrac{10}{9}\left(\dfrac{-1}{s + 100} + \dfrac{1}{s + 1000}\right) & \dfrac{1}{9}\left(\dfrac{-1}{s + 100} + \dfrac{10}{s + 1000}\right)
\end{bmatrix}, \qquad \text{Re } (s) > -100,
$$

whose inverse Laplace transform is

$$
e^{At}1(t) =
\begin{bmatrix}
\dfrac{10}{9}e^{-100t} - \dfrac{1}{9}e^{-1000t} & \dfrac{1}{9}e^{-100t} - \dfrac{1}{9}e^{-1000t} \\[3mm]
-\dfrac{10}{9}e^{-100t} + \dfrac{10}{9}e^{-1000t} & -\dfrac{1}{9}e^{-100t} + \dfrac{10}{9}e^{-1000t}
\end{bmatrix} 1(t), \qquad t \in \mathbb{R}.
$$

This corresponds to what we found in 5.6.8(b).   ∎

The Transfer Matrix of State Difference and Differential Systems

We next consider the *zero-state* response of the output of state difference and differential systems. By setting $x(0) = 0$ in (4), we see that the one-sided z-transform of the zero-state response of the state difference system (1) is

$$Y(z) = H(z)U(z), \qquad z \in \mathscr{C},$$

where

$$H(z) = C(zI - A)^{-1}B + D. \tag{7}$$

By the convolution property of the one-sided z-transform it follows that in the time domain

$$y = h1 * u1, \tag{8}$$

where h is the element-by-element inverse z-transform of the matrix H. As seen in Section 5.5 h is the *impulse response matrix* of the system, given by

$$h(n) = CA^{n-1}B1(n - 1) + D\Delta(n), \qquad n \in \mathbb{Z}. \tag{9}$$

In explicit form, the convolution (8) of the matrix h with the vector-valued signal u is to be read as

$$y_i = \sum_{j=1}^{k} h_{ij}\mathbb{1} * u_j\mathbb{1}, \qquad i = 1, 2, \cdots, M,$$

where the y_i are the components of y, the u_j those of u, and the h_{ij} the elements of the matrix function h. The expression (9) for the impulse response matrix also follows by direct inverse transformation of the transfer matrix H by using 8.8.4. Because h is zero for negative times, $h\mathbb{1} = h$.

The matrix H is called the *transfer matrix* of the system. Inspection of (7) shows that the existence region of H contains $\mathscr{E}_o$ as given in (5). The existence region $\mathscr{E}$ of Y hence may be taken as $\mathscr{E}_o \cap \mathscr{E}_u$, with $\mathscr{E}_u$ the existence region of U.

In the continuous-time case the one-sided Laplace transform of the zero-state response is given by

$$Y(s) = H(s)U(s), \qquad s \in \mathscr{E},$$

where the transfer matrix H now is given by

$$H(s) = C(sI - A)^{-1}B + D, \qquad s \in \mathscr{E}_o,$$

with $\mathscr{E}_o$ given by (6) and $\mathscr{E} = \mathscr{E}_o \cap \mathscr{E}_u$. H is the element-by-element (one- and two-sided) Laplace transform of the impulse response matrix

$$h(t) = Ce^{At}B\mathbb{1}(t) + D\delta(t), \qquad t \in \mathbb{R}.$$

We summarize as follows.

8.8.6.   Summary: Transfer matrix and impulse response matrix.

The impulse response matrix

$$h(n) = CA^{n-1}B\mathbb{1}(n-1)$$
$$+ D\Delta(n), \qquad n \in \mathbb{Z},$$

of the state difference system (1) is the inverse *z*-transform of the transfer matrix

$$H(z) = C(zI - A)^{-1}B + D,$$
$$z \in \mathscr{E}_o.$$

The impulse response matrix

$$h(t) = Ce^{At}B\mathbb{1}(t) + D\delta(t), \qquad t \in \mathbb{R},$$

of the state differential system (2) is the inverse Laplace transform of the transfer matrix

$$H(s) = C(sI - A)^{-1}B + D,$$
$$s \in \mathscr{E}_o. \quad \blacksquare$$

In what follows we consider the connection between the transfer matrix and the response to exponentials on the one hand and the frequency response matrix on the other.

8.8.7.   Exercise: Response to exponential inputs.

Consider the state difference system (1), with initial state $x(n_o) = 0$ and the exponential input

$$u(n) = u_o z^n, \qquad n \geq n_o, n \in \mathbb{Z},$$

where u_o is a constant vector and z a complex number in the set $\mathscr{E}_o$, with $\mathscr{E}_o$ given by (5). Prove that in the limit $n_o \to -\infty$ the response of the system is

$$y(n) = H(z)u_o z^n, \qquad n \in \mathbb{Z},$$

with H the transfer matrix of the system.

Consider the state differential system (2), with initial state $x(t_o) = 0$ and the exponential input

$$u(t) = u_o e^{st}, \qquad t \geq t_o, t \in \mathbb{R},$$

where u_o is a constant vector and s a complex number in the set $\mathscr{E}_o$, with $\mathscr{E}_o$ given by (6). Prove that in the limit $t_o \to -\infty$ the response of the system is

$$y(t) = H(s)u_o e^{st}, \qquad t \in \mathbb{R},$$

with H the transfer matrix of the system. ∎

8.8.8.   Exercise: Transfer matrix and frequency response matrix.
In Section 5.8 we introduced the *frequency response matrix* of a state difference or differential system.

Show that if the existence region $\mathscr{E}_o$ of the transfer matrix H of a state difference system includes the unit circle, the frequency response matrix $\hat{h}$ of the system is given by

$$\hat{h}(f) = H(e^{j2\pi f}), \qquad f \in \mathbb{R}.$$

Show that if the existence region $\mathscr{E}_o$ of the transfer matrix H of a state differential system includes the imaginary axis, the frequency response matrix $\hat{h}$ of the system is given by

$$\hat{h}(f) = H(j2\pi f), \qquad f \in \mathbb{R}. \qquad ∎$$

For multi-input multi-output (MIMO) systems with K-dimensional input and M-dimensional output, the transfer matrix H is an $M \times K$ matrix. The (i, j) element H_{ij} of this matrix may be interpreted as follows: If the jth component of the input is an exponential signal and the remaining components of the input are zero, then the response of the ith component of the output equals H_{ij} multiplied by the jth component of the input. For single-input single-output (SISO) systems H is 1×1 and called, as before, the *transfer function* of the system.

We use 8.8.6 to compute the transfer function or matrix and from this the impulse response function or matrix for two simple systems.

8.8.9.   Example: Transfer and impulse response functions and matrices.

(a) *Second-order smoother*. In Example 8.8.5 we found for the second-order smoother that

$$(zI - A)^{-1} = \begin{bmatrix} \dfrac{1}{z - \frac{1}{2}} & \dfrac{1}{z(z - \frac{1}{2})} \\ 0 & \dfrac{1}{z} \end{bmatrix}, \qquad |z| > \frac{1}{2}.$$

As a result, the transfer matrix of the system, which is 1×1 and, hence, a transfer function, is given by

$$H(z) = C(zI - A)^{-1}B + D = \begin{bmatrix} 1 & 0 \end{bmatrix} \begin{bmatrix} \dfrac{1}{z - \frac{1}{2}} & \dfrac{1}{z(z - \frac{1}{2})} \\ 0 & \dfrac{1}{z} \end{bmatrix} \begin{bmatrix} 1 \\ 0 \end{bmatrix}$$

$$= \frac{1}{z - \frac{1}{2}}, \qquad |z| > \frac{1}{2}.$$

The impulse response function h follows by inverse z-transformation according to Table 8.3 as

$$h(n) = 2(\tfrac{1}{2})^n 1(n - 1), \qquad n \in \mathbb{Z}.$$

This agrees with the result of Example 5.6.8(a).

(b) *Moon rocket.* As an example of a MIMO system we consider the moon rocket of Example 5.5.12, taking the elevation x_1 and the speed x_2 both as outputs. It follows that the state differential and output equations are of the form (2), with

$$A = \begin{bmatrix} 0 & 1 \\ 0 & 0 \end{bmatrix}, \qquad B = \begin{bmatrix} 0 \\ c/m \end{bmatrix}, \qquad C = \begin{bmatrix} 1 & 0 \\ 0 & 1 \end{bmatrix}, \qquad D = 0.$$

This system has the 2×1 transfer matrix

$$H(s) = C(sI - A)^{-1}B$$

$$= \begin{bmatrix} 1 & 0 \\ 0 & 1 \end{bmatrix} \left(\begin{bmatrix} s & -1 \\ 0 & s \end{bmatrix} \right)^{-1} \begin{bmatrix} 0 \\ c/m \end{bmatrix} = \begin{bmatrix} 1 & 0 \\ 0 & 1 \end{bmatrix} \frac{1}{s^2} \begin{bmatrix} s & 1 \\ 0 & s \end{bmatrix} \begin{bmatrix} 0 \\ c/m \end{bmatrix}$$

$$= \begin{bmatrix} \dfrac{c/m}{s^2} \\ \dfrac{c/m}{s} \end{bmatrix}, \qquad \text{Re}(s) > 0.$$

The existence region as given follows from the fact that the matrix A has a double eigenvalue 0.

The impulse response matrix h, which again is a 2×1 matrix, follows by element-by-element inverse Laplace transformation. With the help of Table 8.4 we find

$$h(t) = \begin{bmatrix} \dfrac{c}{m} t \mathbb{1}(t) \\[2mm] \dfrac{c}{m} \mathbb{1}(t) \end{bmatrix}, \qquad t \in \mathbb{R}.$$

This is what we found in 5.5.12.   ■

8.9 PROBLEMS

The first problem provides an opportunity to practice with partial fraction expansions.

8.9.1. Partial fraction expansion. Determine the partial fraction expansions, both in complex and in real form, of the following rational functions:

(a) $\dfrac{1}{\lambda^2 + 3\lambda + 2}$,   (b) $\dfrac{\lambda^2 + 1}{\lambda + 1}$,   (c) $\dfrac{\lambda + 2}{(\lambda + 1)^2}$,

(d) $\dfrac{2\lambda - 1}{\lambda^3 + 1}$,   (e) $\dfrac{(\lambda + 1)^2}{(\lambda^2 + 1)^2}$.

In the following problems, z- and Laplace transformation and their inversion are considered.

8.9.2. z-Transforms. Show that the z-transforms of x and y, with

$$x(n) = \cos(2\pi f n)\mathbb{1}(n), \qquad n \in \mathbb{Z},$$
$$y(n) = \sin(2\pi f n)\mathbb{1}(n), \qquad n \in \mathbb{Z},$$

and f a real number, are given by

$$X(z) = \frac{z[z - \cos(2\pi f)]}{z^2 - 2z \cos(2\pi f) + 1}, \qquad |z| > 1,$$

$$Y(z) = \frac{z \sin(2\pi f)}{z^2 - 2z \cos(2\pi f) + 1}, \qquad |z| > 1.$$

8.9.3. Laplace transforms. Determine the (two-sided) Laplace transforms of the following signals together with their existence regions.

(a) $x(t) = \mathbb{1}(t - \theta), \qquad t \in \mathbb{R},$

with θ a real number.

(b)
$$x(t) = \begin{cases} 1 & \text{for } nP \leq t < (n + \tfrac{1}{2})P,\, n \in \mathbb{Z}_+, \\ -1 & \text{for } (n + \tfrac{1}{2})P \leq t < (n + 1)P,\, n \in \mathbb{Z}_+, \\ 0 & \text{for } t < 0, \end{cases} \qquad t \in \mathbb{R},$$

with P a positive real number.

(c) $x(t) = t \cos{(\omega t)}1(t), \qquad t \in \mathbb{R},$

with ω a real number.

8.9.4. Convolution by the z-transform. Use the *z*-transform to determine the convolution of the signals x and y given by

$$x(n) = a^n 1(n), \qquad n \in \mathbb{Z},$$
$$y(n) = b^n 1(n), \qquad n \in \mathbb{Z},$$

with a and b complex numbers. Does the convolution exist for *every* a and b?

8.9.5. Convolution by Laplace transformation. Use the Laplace transform to compute the convolution $z = x * y$ of

$$x(t) = t^n 1(t), \qquad t \in \mathbb{R},$$
$$y(t) = t^m 1(t), \qquad t \in \mathbb{R},$$

with n and m nonnegative integers.

8.9.6. Initial and final value properties of the z-transform. Check whether the initial and final value properties of the one-sided *z*-transform apply to the following functions. If they do, determine the initial and final values.

(a) $\dfrac{1}{z^2 + 3z + 2}. \qquad |z| > \rho,$

(b) $\dfrac{z^2 + 1}{z + 1}, \qquad |z| > \rho,$

(c) $\dfrac{z + 2}{(z + 1)^2}, \qquad |z| > \rho,$

(d) $\dfrac{2z - 3/2}{(z - 1)(z - 1/2)}, \qquad |z| > \rho,$

with ρ a suitably large positive number.

8.9.7. Inverse z-transforms. Use partial fraction expansion to determine the inverses of the following two-sided *z*-transforms:

(a) $\dfrac{1}{z^2 + 3z + 2} \qquad |z| > \rho,$ (b) $\dfrac{z^2 + 1}{z + 1}, \qquad |z| > \rho,$

(c) $\dfrac{z + 2}{(z + 1)^2}, \qquad |z| > \rho,$ (d) $\dfrac{2z - 1}{z^3 + 1}, \qquad |z| > \rho,$

(e) $\dfrac{(z + 1)^2}{(z^2 + 1)^2}, \qquad |z| > \rho.$

In each case ρ is a sufficiently large positive number so that the z-transform is well-defined. In every case, determine ρ, and establish whether the transform is also a one-sided z-transform.

8.9.8. Inverse z-transformation by the power series method. Use the power series method to determine the inverse z-transforms of

(a) $X(z) = e^{-1/z^2}$, $|z| > 0$,

(b) $X(z) = \dfrac{(1 + z)^2}{z}$, $|z| > 0$.

8.9.9. Inverse Laplace transforms. Find the inverse Laplace transforms of the following functions:

(a) $\dfrac{s^2 - a^2}{(s^2 + a^2)^2}$,   Re $(s) > 0$,

with a real.

(b) $\dfrac{s^3}{(s - a)^3}$,   Re $(s) > a$,

with a real.

(c) $\dfrac{e^{-as}(1 - e^{-as})}{s}$, $s \in \mathbb{C}$,

with a a positive real number.

(d) $\dfrac{1}{s} \tanh \left(\dfrac{as}{2}\right)$,   Re $(s) > 0$,

with a a positive real number. *Hint:* $\tanh (x) = (e^x - e^{-x})/(e^x + e^{-x})$. Find a suitable power series expansion for $\tanh (as/2)$.

8.9.10. One-sidedness and finite action of inverse z- and Laplace transforms of rational functions. Let $R = P/Q$ be a rational function. *Hint:* See the discussion at the end of Section 8.5.

(a) *One-sidedness of the inverse z-transform.* Suppose that the existence region $\mathscr{E}$ of the two-sided z-transform $R(z)$, $z \in \mathscr{E}$, includes the unit circle.
   (a.1) Prove that the inverse z transform r of R is right one-sided if and only if R has all its poles strictly inside the unit circle.
   (a.2) Prove that $r(n) = 0$ for $n < 0$ if and only if R is proper and has all its poles strictly inside the unit circle.

(b) *One-sidedness of the inverse Laplace transform of a rational function.* Suppose that the existence region $\mathscr{E}$ of the two-sided Laplace transform $R(s)$, $s \in \mathscr{E}$, includes the imaginary axis. Prove that the inverse Laplace transform r is zero for negative times if and only if R has all its poles strictly inside the left-half complex plane.

(c) *Finite action of the inverse z-transform.* Suppose that the two-sided z-transform $R(z)$, $z \in \mathscr{E}$, exists on an existence region of the form $\mathscr{E} = \{z \in \mathbb{C} \mid |z| > \rho\}$. Prove that the inverse z-transform r of R has finite action if and only if R has all its poles strictly inside the unit circle.

(d) *Finite action of the inverse Laplace transform.* Suppose that the two-sided Laplace transform $R(s)$, $s \in \mathscr{E}$, exists on an existence region of the form $\mathscr{E} = \{s \in \mathbb{C} \mid \text{Re}\,(s) > \sigma\}$. Prove that the inverse Laplace transform r of R has finite action if and only if R is proper and has all its poles strictly in the left-half complex plane.

8.9.11. One-sided z- and Laplace transforms of periodic signals.

(a) Suppose that x is a discrete-time periodic signal with period N, and define the signal y as

$$y(n) = \begin{cases} x(n) & \text{for } n = 0, 1, \cdots N - 1, \\ 0 & \text{otherwise,} \end{cases} \qquad n \in \mathbb{Z}.$$

Prove that the one-sided z-transform of x is

$$X_+(z) = \frac{1}{1 - z^{-N}} Y(z), \qquad |z| > 1,$$

with Y the z-transform of y. *Hint:* Write $x\mathbb{1} = \Sigma_{k=0}^{\infty} \sigma^{-kN} y$ and use the linearity and shift property of the two-sided z-transform.

(b) Suppose that x is a continuous-time periodic signal with period P, and define the signal y as

$$y(t) = \begin{cases} x(t) & \text{for } 0 \leq t < P, \\ 0 & \text{otherwise,} \end{cases} \qquad t \in \mathbb{R}.$$

Prove that the one-sided z-transform of x is

$$X_+(s) = \frac{1}{1 - e^{-sP}} Y(s), \qquad \text{Re}\,(s) > 0,$$

with Y the Laplace transform of y.

We continue with a number of problems about various applications of the z- and Laplace transform to system analysis.

8.9.12. Computation of the impulse response via the z-transform. Determine the transfer functions and their existence regions of the following difference systems, and use inverse z-transformation to determine the impulse responses of the systems.

(a) The delay system

$$y(n) = u(n - M), \qquad n \in \mathbb{Z},$$

with M a nonnegative integer.

(b) The tapped delay line

$$y(n) = a_0 u(n) + a_1 u(n-1) + a_2 u(n-2)$$
$$+ \cdots + a_M u(n-M), \qquad n \in \mathbb{Z},$$

with $a_0, a_1, \cdots, a_M$ real constants and M a nonnegative integer.

(c) The auto-regressive system

$$y(n) = -\frac{1}{4} y(n-2) + u(n), \qquad n \in \mathbb{Z}.$$

(d) The auto-regressive system

$$y(n) = y(n-1) - \frac{5}{16} y(n-2) + u(n), \qquad n \in \mathbb{Z}.$$

(e) The ARMA system

$$y(n) = \frac{1}{4} y(n-2) + u(n) + u(n-1), \qquad n \in \mathbb{Z}.$$

8.9.13. Application of the z-transform to system analysis.

(a) Determine the two-sided z-transform of the signal x given by

$$x(n) = a^{|n|}, \qquad n \in \mathbb{Z},$$

with a real. Sketch the time signal. When does the z-transform exist and what is its existence region?

(b) The exponential smoother has the impulse response h given by

$$h(n) = (1-a)a^n 1(n), \qquad n \in \mathbb{Z},$$

with a real such that $0 < a < 1$. Find a nonzero input u to the smoother such that the output $y = h * u$ is an *even* time signal.

8.9.14. Response of difference systems to exponential and harmonic inputs. Consider the system described by the constant coefficient difference system $Q(\sigma)y = P(\sigma)u$, with transfer function $H = P/Q$. Let $\pi_1, \pi_2, \cdots, \pi_L$ be the poles of H and denote $\rho_o = \max_i |\pi_i|$.

(a) Show that if $|a| > \rho_o$, then the response of the initially-at-rest system to the *two-sided* exponential input $u(n) = a^n$, $n \in \mathbb{Z}$, is

$$y(n) = H(a)a^n, \qquad n \in \mathbb{Z}.$$

Prove in particular that if $\rho_o < 1$, the response of the system to the two-sided harmonic input $u(n) = e^{j2\pi fn}$, $n \in \mathbb{Z}$, is

$$y(n) = \hat{h}(f)e^{j2\pi fn}, \qquad n \in \mathbb{Z},$$

with $\hat{h}(f) = H(e^{j2\pi f})$, $f \in \mathbb{R}$, the frequency response function of the system.

(b) Next, use the *z*-transform to show that if the system is initially at rest its response to the *one-sided* exponential input u given by

$$u(n) = a^n 1(n), \qquad n \in \mathbb{Z},$$

again with $|a| > \rho_o$, is of the form

$$y(n) = H(a)a^n 1(n) + y_{\text{transient}}(n), \qquad n \in \mathbb{Z},$$

where

$$\frac{y_{\text{transient}}(n)}{H(a)a^n} \to 0 \text{ as } n \to \infty.$$

Prove in particular that if $\rho_o < 1$ the response of the system to the one-sided harmonic input

$$u(n) = e^{j2\pi fn} 1(n), \qquad t \in \mathbb{Z},$$

with f real, is of the form

$$y(n) = \hat{h}(f)e^{j2\pi fn} 1(n) + y_{\text{transient}}(n), \qquad n \in \mathbb{Z},$$

where $y_{\text{transient}}(n) \to 0$ as $n \to \infty$. The terms $H(a)a^n 1(n)$ and $\hat{h}(f)e^{j2\pi fn} 1(n)$, $n \in \mathbb{Z}$, respectively, are called the *steady-state* responses. *Hint:* Use partial fraction expansion.

(c) The exponential smoother is described by the difference equation

$$y(n + 1) = ay(n) + (1 - a)u(n), \qquad n \in \mathbb{Z}.$$

Determine the steady-state and transient response of the initially-at-rest exponential smoother to the input u given by

$$u(n) = \cos(2\pi fn) 1(n), \qquad n \in \mathbb{Z},$$

with $f = 1/4$ and $a = \sqrt{3}/3$. *Hint:* Determine the response to the given input as the real part of the response to the input $u(n) = e^{j2\pi fn} 1(n)$, $n \in \mathbb{Z}$.

8.9.15. Impulse response of the LRC network. The LRC network of 4.2.4(c) with the current through the network as output is described by the differential equation

$$\frac{d^2y(t)}{dt^2} + \frac{R}{L}\frac{dy(t)}{dt} + \frac{1}{LC}y(t) = \frac{1}{L}\frac{du(t)}{dt}, \qquad t \in \mathbb{R}.$$

(a) Show that the system has the transfer function

$$H(s) = \frac{sC}{1 + RCs + LCs^2}, \qquad \mathrm{Re}\,(s) > \sigma,$$

with σ a suitable negative constant (assuming that RC and LC are both positive).

(b) Assume $R = 0.001\ \Omega$, $L = 0.1$ H and $C = 0.001$ F, and determine the impulse response of the network.

8.9.16. Response of the RC network. The RC network has the transfer function

$$H(s) = \frac{1}{1 + RCs}, \qquad \mathrm{Re}\,(s) > -\frac{1}{RC}.$$

(a) *Response to a singular input.* Assume that the system is initially at rest, and use the Laplace transform to find its response to the input

$$u = \delta^{(1)}.$$

(b) *Steady-state and transient response to a ramp input.* Determine the steady-state and the transient response of the RC network to the input

$$u(t) = t \cdot \mathbb{1}(t), \qquad t \in \mathbb{R}.$$

Find suitable interpretations for the terms "steady-state" and "transient." *Hint:* The steady-state response is of the form $y_o \cdot (t - a) \cdot \mathbb{1}(t)$, $t \in \mathbb{R}$, with y_o and a constants.

The application of the one-sided z- and Laplace transforms to the solution of initial value problems is considered in the next two problems.

8.9.17. Solution of initial value problems with the one-sided z-transform. Solve the following initial value problems for difference equations using the one-sided z-transform.

(a) The exponential smoother

$$y(n) = ay(n-1) + (1-a)u(n), \qquad n \in \mathbb{Z}_+,$$

with $u(n) = \cos(2\pi fn)$, $n \in \mathbb{Z}_+$, and $y(-1) = 0$, where $f = 1/4$ and $a = \sqrt{3}/3$.

(b) The ARMA system

$$y(n) = \frac{5}{6}y(n-1) - \frac{1}{6}y(n-2) + u(n) + u(n-1), \qquad n \in \mathbb{Z}_+,$$

with $u(n) = \mathbb{1}(n)$, $n \in \mathbb{Z}$, and $y(n) = 0$ for $n < 0$, $n \in \mathbb{Z}$.

(c) The ARMA system

$$y(n) = \frac{1}{4}y(n-2) + u(n) + u(n-1), \qquad n \in \mathbb{Z}_+,$$

with $u(n) = 1$ for $n \in \mathbb{Z}$ and $y(-2) = y(-1) = 1$.

8.9.18. Solution of initial value problems by Laplace transformation.

(a) Use the one-sided Laplace transform to solve the differential equation

$$my''(t) = u(t), \qquad t \geq 0,$$

with $u(t) = 1$, $t \geq 0^-$, $y(0^-) = y_o$ and $y'(0^-) = y'_o$.

(b) Use the one-sided Laplace transform to solve the differential equation

$$y''(t) + y(t) = u'(t) + u(t), \qquad t \geq 0^-,$$

with $y(0^-) = 1$, $y'(0^-) = 0$ and

$$u(t) = 1 \qquad \text{for } t \geq 0^-.$$

(c) Repeat (b) but now with

$$u(t) = \mathbb{1}(t) \qquad \text{for } t \geq 0^-.$$

(d) The double spring-mass system of Problem 4.8.3(b) is described by the set of differential equations

$$m_1\ddot{z}_1 = -k_1(z_1 - u) + k_2(z_2 - z_1),$$
$$m_2\ddot{z}_2 = \qquad\quad - k_2(z_2 - z_1).$$

Use the one-sided Laplace transform to solve these equations for z_1 and z_2 on $t \geq 0$ when $u(t) = 0$ for $t \geq 0^-$, and the initial conditions are $z_1(0^-) = -1$, $z_2(0^-) = 1$, $\dot{z}_1(0^-) = \dot{z}_2(0^-) = 0$. Assume that $k_1/m_1 = 3/2$, $k_2/m_2 = 4/3$, and $m_2/m_1 = 1/8$.

The last problem for this chapter concerns the application of the z- and Laplace transform to the computation of state transition matrices.

8.9.19. State transition matrix.

(a) *Fibonacci system.* According to 5.9.13 the state representation of the Fibonacci system is $x(n + 1) = Ax(n)$, $n \in \mathbb{Z}_+$, with

$$A = \begin{bmatrix} 1 & 1 \\ 1 & 0 \end{bmatrix}.$$

Use z-transformation to determine the transition matrix A^n, $n \in \mathbb{Z}_+$, of the system.

(b) *Harmonic oscillator*. As seen in 5.9.14, the harmonic oscillator is described by the state differential equation $\dot{x} = Ax$, with

$$A = \begin{bmatrix} 0 & 1 \\ -1 & 0 \end{bmatrix}.$$

Use Laplace transformation to find the state transition matrix e^{At}, $t \geq 0$.

(c) *Double spring-mass system*. As shown in 5.9.15, the double spring mass system is described by the state differential equation $\dot{x} = Ax + Bu$, where

$$A = \begin{bmatrix} 0 & 1 & 0 & 0 \\ -\left(\omega_1^2 + \dfrac{m_2}{m_1}\omega_2^2\right) & 0 & \dfrac{m_2}{m_1}\omega_2^2 & 0 \\ 0 & 0 & 0 & 1 \\ \omega_2^2 & 0 & -\omega_2^2 & 0 \end{bmatrix}.$$

Assume the numerical values $\omega_1^2 = 3/2$, $\omega_2^2 = 4/3$, and $m_2/m_1 = 1/8$, and use Laplace transformation to compute the state transition matrix e^{At}, $t \geq 0$.

8.10 COMPUTER EXERCISES

The z- and Laplace transforms are important analytical tools. Their numerical computation is discussed in Section 9.7. Only a single computer exercise follows.

8.10.1.   Inverse z- and Laplace transformation.

(a) *Inverse z-transformation*. Consider the z-transform X of the signal x given by

$$X(z) = \frac{z^2 + z + 1}{z^3 + z^2 + z + 1}, \qquad z \in \mathscr{E},$$

with the existence region $\mathscr{E}$ to be determined.

(a.1) Compute the poles of X and determine the two possible existence regions $\mathscr{E}_1$ and $\mathscr{E}_2$ of X.

(a.2) For each existence region, choose the real number ρ such that the circle $\rho e^{j2\pi f}$, $f \in [0, 1)$, is in the existence region. Evaluate $\hat{x}_\rho(f) = X(\rho e^{j2\pi f})$, $f \in [0, 1)$.

(a.3) Compute the signal x_ρ, given by

$$x_\rho(n) = \rho^{-n}x(n), \qquad n \in \mathbb{Z},$$

by application of the inverse DCFT to $\hat{x}_\rho$. Determine x from x_ρ. Do this for both existence regions.

(b) *Inverse Laplace transformation.* Define Q as the polynomial with leading coefficient 1 whose roots are $-1 \pm 6j$ and -1. Let Y be the Laplace transform of the signal y given by

$$Y(s) = \frac{1}{Q(s)}, \qquad s \in \mathcal{E},$$

with the existence region $\mathcal{E}$ to be determined.

(b.1) Determine the two possible existence regions $\mathcal{E}_1$ and $\mathcal{E}_2$ of Y.

(b.2) For each existence region, choose the real number σ such that the vertical line $\sigma + j2\pi f$, $f \in \mathbb{R}$, is in the existence region. Evaluate $\hat{y}_\sigma(f) = Y(\sigma + j2\pi f)$, $f \in \mathbb{R}$,

(b.3) Compute the signal y_σ, given by

$$y_\sigma(t) = e^{-\sigma t} y(t), \qquad t \in \mathbb{R},$$

by application of the inverse CCFT to $\hat{y}_\sigma$. Determine y from y_σ. Do this for each existence region.

9

Applications to Signal Processing and Digital Filtering

9.1 INTRODUCTION

Signal processing plays an important role in diverse fields such as communication, control, electronic sound and image recording, image and speech processing, seismic exploration, navigation, medical diagnosis, economic analysis, and business forecasting.

Signal *processors* may be continuous- or discrete-time systems. They are nearly always implemented electronically. Continuous-time processors are often constructed by using resistors and capacitors together with operational amplifiers, and are usually referred to as *analog* signal processors. Discrete-time systems are put together by using digital hardware, including microprocessors, and are called *digital* signal processors.

The advantage of digital signal processors is that their accuracy is completely under control because it depends on the word length. By selecting a sufficiently large word length, for instance 16 or even 32 bits, great accuracy may be achieved at comparatively low cost. The accuracy of analog signal processors often cannot be ascertained precisely and high precision, say, better than 3%, requires considerable expense.

Besides precision, other advantages of digital hardware are reliability, reproducibility and programmability. On the other hand, digital hardware imposes an inherent processing delay, which may be a disadvantage. For on-line applications where very little delay is acceptable, analog implementation may be the only practicable solution.

Because of the advantages of digital signal processing, its importance is increasing over that of analog signal processing. By using samplers and interpolators, signal processing tasks for continuous-time signals may be implemented digitally as in Fig. 9.1. The effect of *sampling* continuous-time signals on their frequency content is the subject of Section 9.2. Conversely, the effect of *interpolation* on the frequency content of sampled signals is investigated. This leads to a celebrated result known as the *sampling theorem*. This theorem states that if a continuous-time signal is *band-limited*, it may be completely reconstructed after sampling, provided the sampling frequency is high enough.

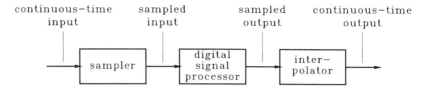

Figure 9.1.   Digital implementation of a continuous-time signal processing task.

Section 9.3 describes a number of signal processing tasks, and introduces the important notions of *on-line* versus *off-line* signal processing, as well as that of *processing delay*. In Section 9.4 we analyze the effect of *truncating* an infinite time or frequency signal to a finite signal. It will be seen how "soft" truncation, called *windowing*, may reduce the adverse effects of truncation.

In Sections 9.5 and 9.6 we study the problem how to design the digital signal processor in the configuration of Fig. 9.1. Section 9.5 deals with *FIR* (Finite Impulse Response) and Section 9.6 with *IIR* (Infinite Impulse Response) filters.

Section 9.7 describes the famous *Fast Fourier Transform* (FFT), which is a highly efficient algorithm for the numerical computation of the DDFT. Also, the numerical computation of the various other Fourier transforms, z-, and Laplace transforms and convolutions is briefly discussed.

9.2 SAMPLING, INTERPOLATION, AND THE SAMPLING THEOREM

In Section 2.3 sampling is introduced as a procedure to convert continuous-time signals to discrete-time signals. Sampling on the discrete time axis $\mathbb{Z}(T)$ reduces a continuous-time signal x to the sampled signal $x*$ given by

$$x*(t) = x(t), \qquad t \in \mathbb{Z}(T),$$

as illustrated in Fig. 9.2. Sampling is an important operation, because it allows manipulating signals that are originally continuous-time signals by digital processors

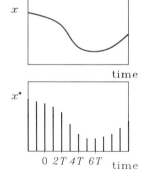

Figure 9.2. Sampling converts the continuous-time signal x to the sampled signal x^*.

and other inherently discrete-time devices. In this section we first study the effect of sampling on the *frequency content* of the signal and then see whether and how the original continuous-time signal may be reconstructed from the sampled signal.

Frequency Content of a Sampled Signal

Before stating the first result, we recall that the DCFT $\hat{x}^*$ of a sampled signal x^*, with sampling interval T, is defined by

$$\hat{x}^*(f) = T \sum_{n=-\infty}^{\infty} x^*(nT)e^{-j2\pi fnT}, \qquad f \in \mathbb{R}.$$

The DCFT as given here is defined on the entire frequency axis $\mathbb{R}$, but since it is periodic in the frequency f with period $F = 1/T$ it is enough to know the DCFT on *one* period only.

   The following summarizes the connection between the DCFT of a sampled signal and the CCFT of the original continuous-time signal.

9.2.1.   Summary: Frequency content of a sampled signal.   Suppose that the continuous-time signal x defined on the time axis $\mathbb{R}$ is continuous and has bounded variation. Sampling x on the time axis $\mathbb{Z}(T)$ results in the sampled signal x^* given by

$$x^*(t) = x(t), \qquad t \in \mathbb{Z}(T).$$

Then if x has the CCFT $\hat{x}$, the DCFT $\hat{x}^*$ of the sampled signal x^* is given by

$$\hat{x}^*(f) = \sum_{k=-\infty}^{\infty} \hat{x}(f - kF), \qquad f \in \mathbb{R},$$

that is, $\hat{x}^*$ is the periodic extension of $\hat{x}$ on the frequency axis $\mathbb{R}$, with period $F = 1/T$.   ∎

9.2.2.   **Proof.**   The DCFT of x^* may be expressed as

$$\hat{x}^*(f) = T \sum_{n=-\infty}^{\infty} x^*(nT)e^{-j2\pi fnT} = T \sum_{n=-\infty}^{\infty} x(nT)e^{-j2\pi fnT}$$

$$= T \int_{-\infty}^{\infty} x(t)w_T(t)e^{-j2\pi ft}\, dt, \qquad f \in \mathbb{R},$$

where w_T is the infinite comb

$$w_T(t) = \sum_{n=-\infty}^{\infty} \delta(t - nT), \qquad t \in \mathbb{R}.$$

It follows that the DCFT $\hat{x}^*$ of x^* is the CCFT of $Tx \cdot w_T$, so that by the converse convolution property of the CCFT, $\hat{x}^* = T\hat{x} * \hat{w}_T$. By 7.4.19(c) the CCFT $\hat{w}_T$ of the infinite comb is

$$\hat{w}_T(f) = F \sum_{k=-\infty}^{\infty} \delta(f - kF), \qquad f \in \mathbb{R},$$

where $F = 1/T$. As a result,

$$\hat{x}^*(f) = T(\hat{x} * \hat{w}_T)(f) = T \int_{-\infty}^{\infty} \hat{x}(f - \phi)\left(F \sum_{k=-\infty}^{\infty} \delta(\phi - kF) \right) d\phi$$

$$= \sum_{k=-\infty}^{\infty} \hat{x}(f - kF), \qquad f \in \mathbb{R}. \qquad\blacksquare$$

Figure 9.3 illustrates the effect of sampling the continuous-time signal x on its frequency content. Any frequency component of the continuous-time signal with frequency *outside* the frequency interval $[-F/2, F/2)$ is "folded back" into this interval. The cause is the *aliasing* effect described in Section 2.2: As a result of sampling, harmonics with frequencies outside $[-F/2, F/2)$ are mistaken for harmonics with frequencies inside this interval.

Bandwidth

Figure 9.4 shows that sampling does *not* affect the frequency content of continuous-time signals x whose frequency content $\hat{x}$ is *zero* outside the interval $[-F/2, F/2)$. For such signals the DCFT of the sampled signal and the CCFT of the continuous-time signal coincide on this interval, that is,

$$\hat{x}^*(f) = \hat{x}(f) \qquad \text{for} \quad -F/2 \leq f \leq F/2.$$

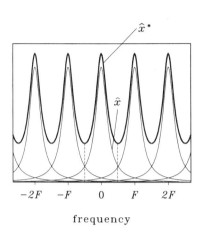

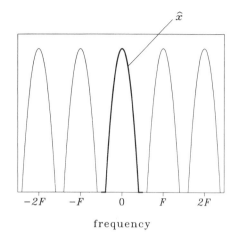

Figure 9.3. Sampling a continuous-time signal at the rate F folds its Fourier transform back into the interval $[-F/2, F/2)$.

Figure 9.4. The frequency content of signals that have a sufficiently small bandwidth is not folded.

This fact will enable us to reconstruct the continuous-time signal x *intact* from the sampled signal $x*$.

Continuous-time signals whose frequency content is zero outside a given interval are said to be *band-limited*. The *bandwidth* of band-limited signals is defined as follows.

9.2.3.   Definition: Bandwidth of a continuous-time signal.   Let x be a continuous-time signal with CCFT $\hat{x}$. Then the *bandwidth* B of x is the smallest $B \geq 0$ such that $\hat{x}(f) = 0$ for $|f| > B$.   ∎

Figure 9.5 illustrates the definition of the bandwidth. The figure shows that the bandwidth B is the highest frequency contained in the signal. The definition is most meaningful for signals that are low-pass.

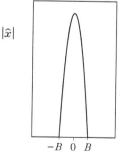

Figure 9.5. Fourier transform of a signal with bandwidth B.

Frequency Content of an Interpolated Signal

The converse of sampling is any method to construct a continuous-time signal from a sampled signal. As explained in Section 2.3, a particular method of converting a given sampled signal x^* defined on the discrete time axis $\mathbb{Z}(T)$ to a continuous-time signal x is *interpolation* with an *interpolating function* i according to

$$x(t) = \sum_{n=-\infty}^{\infty} x^*(nT)i\left(\frac{t-nT}{T}\right), \qquad t \in \mathbb{R}.$$

Normally the interpolating function i satisfies

$$i(t) = \begin{cases} 1 & \text{for } t = 0, \\ 0 & \text{for } t = k, \, k \neq 0, \, k \in \mathbb{Z}, \end{cases} \qquad t \in \mathbb{R},$$

so that

$$x(kT) = x^*(kT), \qquad k \in \mathbb{Z}.$$

Interpolating a sampled signal this way results in a continuous-time signal that equals the original continuous-time signal *at* the sampling instants, but not necessarily *between* sampling instants.

As described in Section 2.3, *step* interpolation is most common. Another well-known interpolation scheme is *linear* interpolation. An interpolation method that is important for theoretical considerations is δ-*interpolation*, that is, $i = \delta$. This converts the sampled signal x^* to the continuous-time signal $x^\dagger$ given by

$$
\begin{aligned}
x^\dagger(t) &= \sum_{n=-\infty}^{\infty} x^*(nT)\delta\left(\frac{t-nT}{T}\right) \\
&= T \sum_{n=-\infty}^{\infty} x^*(nT)\delta(t-nT), \qquad t \in \mathbb{R}.
\end{aligned}
\tag{1}
$$

The relation between the frequency content of a sampled signal x^* and that of the corresponding interpolated signal x may be summarized as follows.

9.2.4.   Summary: Frequency content of an interpolated signal. Suppose that x^* is a sampled signal defined on the time axis $\mathbb{Z}(T)$ with DCFT $\hat{x}^*$ and i an interpolating function with CCFT $\hat{i}$. Then the interpolated signal x given by

$$x(t) = \sum_{n=-\infty}^{\infty} x^*(nT)i\left(\frac{t-nT}{T}\right), \qquad t \in \mathbb{R},$$

has the CCFT

$$\hat{x}(f) = \hat{x}^*(f) \cdot \hat{\imath}(fT), \qquad f \in \mathbb{R}. \tag*{$\blacksquare$}$$

9.2.5.   Proof.   Define the continuous-time signal i_T as

$$i_T(t) = \frac{1}{T} i\!\left(\frac{t}{T}\right), \qquad t \in \mathbb{R}.$$

It is easily verified that the interpolated signal x may be written as the continuous-time convolution

$$x = x^\dagger * i_T,$$

where $x^\dagger$ is the δ-interpolated signal (1). By the convolution property, the CCFT $\hat{x}$ of x equals

$$\hat{x} = \hat{x}^\dagger \cdot \hat{\imath}_T,$$

with $\hat{x}^\dagger$ the CCFT of $x^\dagger$ and $\hat{\imath}_T$ that of i_T. The proof of the theorem is completed by noting that on the one hand

$$\hat{x}^\dagger(f) = \int_{-\infty}^{\infty} \left(T \sum_{n=-\infty}^{\infty} x^*(nT)\delta(t - nT)\right) e^{-j2\pi ft}\, dt = T \sum_{n=-\infty}^{\infty} x^*(nT) e^{-j2\pi fnT}$$

$$= \hat{x}^*(f), \qquad f \in \mathbb{R},$$

while, on the other,

$$\hat{\imath}_T(f) = \frac{1}{T} \int_{-\infty}^{\infty} i\!\left(\frac{t}{T}\right) e^{-j2\pi ft}\, dt = \int_{-\infty}^{\infty} i(\tau) e^{-j2\pi f\tau T}\, d\tau$$

$$= \hat{\imath}(fT), \qquad f \in \mathbb{R}. \tag*{$\blacksquare$}$$

Evidently, the effect of interpolating the sampled signal x^* with the interpolating function i is to multiply the frequency content $\hat{x}^*$ of the sampled signal by $\hat{\imath}(fT)$, $f \in \mathbb{R}$, where $\hat{\imath}$ is the CCFT of the interpolating function i.

The Sampling Theorem

We now consider the question in how far interpolation of *sampled* signals is suited to reconstruct the original continuous-time signal. Contemplation of 9.2.1 and 9.2.4 shows that given a continuous-time signal x that is continuous and has bounded variation, exact reconstruction of x from the sampled signal x^* using interpolation is possible under the following conditions:

(a) The bandwidth B of the signal x is less than $1/2T$.

(b) The CCFT $\hat{\imath}$ of the interpolating function i is chosen as

$$\hat{\imath}(f) = \begin{cases} 1 & \text{for } -\frac{1}{2} \leq f < \frac{1}{2}, \\ 0 & \text{otherwise}, \end{cases} \qquad f \in \mathbb{R}.$$

Condition (a) ensures that sampling does not affect the frequency content of the signal x. Choosing $\hat{\imath}$ as in (b) eliminates the periodically extended part of $\hat{x}$ and retains the relevant central part.

The interpolating function i corresponding to $\hat{\imath}$ as given under (b) follows by inverse transformation of $\hat{\imath}$ as

$$i(t) = \int_{-\frac{1}{2}}^{\frac{1}{2}} e^{j2\pi ft} \, dt = \text{sinc } (\pi t), \qquad t \in \mathbb{R}.$$

This is the sinc interpolating function of Section 2.3. A plot is given in Fig. 2.19.

We thus obtain the following celebrated result, which is known as the *sampling theorem*.

9.2.6. Summary: The sampling theorem. Suppose that the continuous-time signal x is continuous and of bounded variation, and moreover has finite bandwidth $B > 0$. Let x^* be obtained by sampling x on $\mathbb{Z}(T)$, that is,

$$x^*(t) = x(t), \qquad t \in \mathbb{Z}(T).$$

If $B < 1/2T$, then the continuous-time signal x may be *exactly* reconstructed from the sampled signal x^* as

$$x(t) = \sum_{n=-\infty}^{\infty} x^*(nT) \text{ sinc } [\pi (t - nT)/T], \qquad t \in \mathbb{R}. \qquad \blacksquare$$

The interpolation formula of the sampling theorem is sometimes known as the *cardinal series*. For a given continuous-time signal x with bandwidth B, the frequency $2B$ is called the *Nyquist rate*. According to the sampling theorem, the signal x may be reconstructed after sampling provided the sampling rate $1/T$ is *greater* than the Nyquist rate $2B$.

The physicist Harry Nyquist (1889–1976) was born in Sweden but moved to the United States at the age of 18. He worked with the Bell Telephone Laboratories and did important work on thermal noise, data transmission, and negative feedback.

9.2.7.   Remark: Aliasing due to undersampling.   Suppose that the continuous-time signal x has bandwidth B but is sampled at a sampling rate $f_s = 1/T$ that is *less* than the Nyquist rate $2B$. It follows from 9.2.1 that aliasing takes place, so that the content of the signal at frequencies greater than $1/2T$ spills over into the interval $[-1/2T, 1/2T)$. Interpolation of the sampled signal with the sinc function reconstructs the original continuous-time signal x exactly *at* the sampling instants but incorrectly *between* sampling instants. The closer the sampling rate is to the Nyquist rate, the smaller the interpolation error due to aliasing is.

To avoid aliasing errors caused by sampling, it is customary in engineering applications to precede any sampler with sampling rate $1/T$ by a *presampling filter*, or *conditioning filter*, as in Fig. 9.6, which removes all frequencies over $1/2T$ from the signal.   ∎

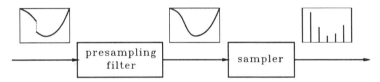

Figure 9.6.   Sampling with a presampling filter.

The application of the sampling theorem is confined to signals with finite bandwidth. In practice, signals that are exactly band-limited do not exist, but there are many signals that have by far the greater part of their energy concentrated in a limited band. Such signals may be sampled and subsequently interpolated without much loss of accuracy.

From an engineering point of view, interpolation of sampled signals with sinc functions is not a very practical proposition, because

(1) interpolation with sinc functions is an anticipating operation on the sampled signal x^*, and

(2) the sinc function decays quite slowly (inversely with time), so that accurate interpolation requires many past and future values of the sampled signal.

For this reason, usually other interpolation schemes are employed, such as step or linear interpolation, together with *oversampling* (i.e., sampling at a rate that is higher than the minimal rate required by the sampling theorem). The results of 9.2.1 and 9.2.4 may be used to assess the errors introduced by such schemes.

9.2.8.   Example: Sampling and reconstruction.   By way of example, consider the reconstruction of the continuous-time signal x of Example 2.3.11, which is given by

$$x(t) = \begin{cases} \cos(\pi t) & \text{for } -\tfrac{1}{2} \le t < \tfrac{1}{2}, \\ 0 & \text{otherwise}, \end{cases} \qquad t \in \mathbb{R}.$$

Plots of the signal x and the magnitude of its CCFT $\hat{x}$ of x are given in Fig. 9.7. The signal is not band-limited, but most of the energy of the signal is concentrated in the band, say, between -8 and 8. Thus, choosing the bandwidth as $B = 8$, the Nyquist sampling rate is $2B = 16$. Figure 9.8 shows the reconstruction of the signal with sinc interpolation at this rate and also with the rates 8 and 4. Because the bandwidth of the signal is infinite, the interpolation is exact for none of these rates, but the accuracy improves as the rate is taken higher. The step and linear interpolation of the signal are shown in Fig. 2.20. ∎

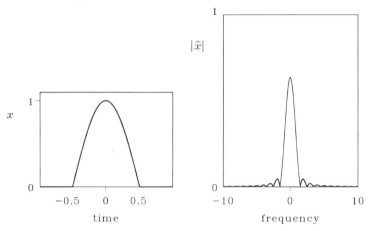

Figure 9.7.   Left: a signal x. Right: the magnitude of its CCFT $\hat{x}$.

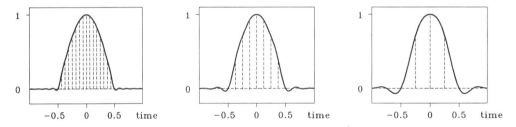

Figure 9.8.   The reconstruction of the signal x of Fig. 9.7 by sinc interpolation. Left: sampling rate 16. Middle: sampling rate 8. Right: sampling rate 4.

9.3 ON-LINE AND OFF-LINE SIGNAL PROCESSING

The *design problem* in signal processing is to find a system, called *signal processor*, whose input is a time signal and whose output is the desired information that is to be extracted from the signal. Often, but not always, the "desired information" is again a time signal. In this section we discuss *on-line* and *off-line* signal processing and consider a variety of engineering problems that call for signal processing.

On-line digital signal processing means that the input consists of a continuous flow of data, arriving at a certain *rate* (expressed in bits/second, for instance), and that the corresponding output is produced at the *same* rate. On-line processing is sometimes called *real-time processing*. Given the hardware, the algorithm that is implemented generally imposes a *maximal sampling rate* on the input and output signals.

Because of the computation time, in on-line signal processing the output is often *delayed* by a certain time interval, called the *processing delay*. If the signal processing task is *anticipating*, implementation involves an inherent additional time delay.

In *off-line* signal processing, sometimes referred to as *batch processing*, there is no match between the input and output rates. We illustrate the difference between on-line and off-line signal processing by several examples.

Examples of Off-Line and On-Line Signal Processing

9.3.1.   Examples of off-line signal processing tasks.
(a) *Fourier transformation*. A typical example of a signal processing task that is normally done off-line is Fourier transformation. The signal to be processed is a time signal x, while the output of the signal processor is a frequency signal $\hat{x}$. Fourier transformation requires the entire signal to be known, so that the signal cannot be Fourier transformed sample by sample as the samples arrive. The output is a signal that is not even defined on a time axis but on a frequency axis.

(b) *Off-line simulation*. Another example of off-line signal processing is (off-line) simulation, which is the task of computing responses of a system to various input signals and initial conditions. Although the input and output usually are time signals, their time axes are determined by the rate at which the signal processor does the calculations, and are not synchronized with the environment.

(c) *Image processing*. Another example of signal processing that is often done off-line is *image enhancing*. An image enhancer accepts the raw data that describe a static image, and processes it to remove blurs and improve contrast. The result is an improved image.   ∎

9.3.2.   Examples of on-line signal processing tasks.
(a) *Equalization*. An instance of a typical *on-line* signal processing task is *equalization*. An equalizer is needed when a continuous flow of data, such as speech, is received over a telephone line or other communication channel, which distorts the data because the frequency response function of the line is not flat. The task of the equalizer is to *compensate* for the distortion of the line. Figure 9.9 shows the configuration. If the telephone line is a linear time-invariant continuous- or discrete-time system with frequency response function $\hat{h}$, the Fourier transform of the received signal y is $\hat{y} = \hat{h}\hat{u}$, with $\hat{u}$ the Fourier transform of the transmitted data. The equalizer performs its task ideally if it is a linear time-invariant system with frequency response function $1/\hat{h}$. In this type of application, a small processing delay is acceptable. For a smooth two-way telephone conversation, the delay should not exceed, say, about 0.1 second.

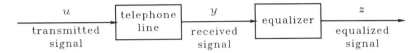

Figure 9.9.   Signal equalization.

(b) *Filtering.* Another common on-line processing task is *filtering*. The need for this arises for instance when a signal is sent by means of a channel, such as a radio link, from a transmitter to a receiver. The transmitter sends a time signal u, but as the result of environmental effects the received signal is $y = u + n$, with n representing an undesired signal referred to as *noise*. The task of the signal processor is to remove as much of the noise as possible without distorting the signal. This is called *filtering*. Figure 9.10 shows the arrangement. If the frequency content of the message signal u is confined to a certain frequency range, and that of the noise to another, the filtering task may be accomplished with a linear time-invariant system whose frequency response function is zero in the frequency range of the noise. If the frequency ranges of the message signal and that of the noise overlap, then the choice of the frequency response of the filter needs more sophistication.

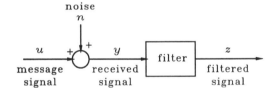

Figure 9.10.   Signal filtering.

(c) *On-line simulation.* As explained in 9.3.1(b), simulation is usually done off-line. There are instances, however, that call for on-line simulation. An example is a flight trainer, where the pilot occupies a cockpit without an airplane, and the motion of the plane is simulated and transmitted to the cockpit. Since the goal of the exercise is to train the pilot how to fly, it is necessary to compute the response of the airplane to the commands of the pilot in precise synchronization with clock time so that the cockpit and flight instruments respond without any delays. In this application processing delays that are as small as a few milliseconds may impair the viability of the system.

(d) *On-line image processing.* Application of image enhancing to TV requires *on-line* implementation.   ∎

Anticipating Signal Processing Tasks

In what follows, we describe a few anticipating signal processing tasks that introduce an inherent delay, in addition to the computational delay.

9.3.3.   Examples: Anticipating signal processing tasks.

(a) *Signal interpolation.* A signal processing task that cannot be carried out well on-line without introducing a delay is the interpolation of a sampled signal. If the sampled signal x^* is defined on the time axis $\mathbb{Z}(T)$, the interpolated continuous-time signal is

$$x(t) = \sum_{k=-\infty}^{\infty} x^*(kT) i\left(\frac{t - kT}{T}\right), \qquad t \in \mathbb{R},$$

with i the interpolating function. If i is anticipating (i.e., i is nonzero for negative times), then interpolation at time t requires knowledge of *future* values of the sampled signal x^*. Examples of interpolation with anticipating interpolating functions are linear and sinc interpolation. Linear interpolation requires *one* future value and thus may be implemented with a delay of one sampling interval with length T. Sinc interpolation needs infinitely many future values and, hence, cannot be realized in real time at all. *Approximate* on-line implementation of sinc interpolation may be achieved by introducing a delay of N sampling periods and truncating the interpolating function, so that

$$i(t) = \begin{cases} \text{sinc } (\pi (t - N)) & \text{for } 0 < t \leq 2N, \\ 0 & \text{otherwise,} \end{cases} \qquad t \in \mathbb{R}.$$

Only interpolation with the step interpolating function does not require future signal values and may be realized on-line.

(b) *Ideal low-pass filtering.* As another example of a filtering task that cannot be done on-line we consider ideal low-pass filtering. Applied to continuous-time signals, ideal low-pass filtering consists of sending the signal through a system with frequency response function

$$\hat{h}(f) = \begin{cases} 1 & \text{for } -f_o \leq f < f_o, \\ 0 & \text{otherwise,} \end{cases} \qquad f \in \mathbb{R},$$

where $f_o > 0$. The impulse response of the filter follows by application of the inverse CCFT and is given by

$$h(t) = \int_{-f_o}^{f_o} e^{j2\pi ft} \, dt = 2f_o \text{ sinc } (2\pi f_o t), \qquad t \in \mathbb{R}.$$

This is the impulse response of an anticipating system, which cannot be realized on-line. Approximate on-line realization may be achieved by delaying the impulse response by a time interval θ and truncating it so that it becomes non-anticipating. The output of the system with the resulting impulse response

$$h_d(t) = \begin{cases} h(t - \theta) & \text{for } t \geq 0, \\ 0 & \text{otherwise,} \end{cases}$$

$$= 2f_o \text{ sinc } (2\pi f_o(t - \theta)) \mathbb{1}(t - \theta), \qquad t \in \mathbb{R},$$

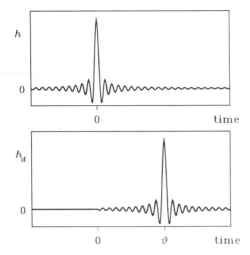

Figure 9.11.   Top: impulse response h for ideal low-pass filtering. Bottom: delayed approximation h_d.

is an approximation to the ideally filtered signal delayed by θ, with θ a positive number. Figure 9.11 shows the impulse responses h and h_d.   ■

9.4 WINDOWS AND WINDOWING

Sometimes it is necessary to *truncate* a time signal of infinite or very long duration to a signal with *finite* duration, that is, to set it equal to zero outside a given interval of finite length. This arises, for instance, when one wishes to compute the Fourier transform of the signal numerically by a digital computer, which, of course, can only handle signals of finite length.

Truncating a time signal x with infinite time axis $\mathbb{T}$ to an interval $\mathbb{I} \subset \mathbb{T}$ is equivalent to *multiplying* it by the *rectangular window* w_{rect} given by

$$w_{\text{rect}}(t) = \begin{cases} 1 & \text{for } t \in \mathbb{I}, \\ 0 & \text{otherwise,} \end{cases} \qquad t \in \mathbb{T}.$$

Thus, the truncated signal is $x \cdot w_{\text{rect}}$. Similarly, we may have to truncate a frequency signal, such as when computing the inverse Fourier transform.

Rectangular Windows

We explore the effect of multiplying a time signal x by a rectangular window w_{rect} by examining the Fourier transform of $x \cdot w_{\text{rect}}$. By the converse convolution property, this Fourier transform is given by

$$\hat{x} * \hat{w}_{\text{rect}},$$

where $*$ denotes cyclical or regular frequency convolution, depending on whether x is a discrete- or a continuous-time signal, respectively. Thus, a multiplicative window in the time domain is equivalent to a *convolutive window* in the frequency domain.

The effect of the convolutive window $\hat{w}_{rect}$ on $\hat{x}$ is that $\hat{x}$ is *locally averaged* at each frequency, with weights determined by the shape of the frequency window $\hat{w}_{rect}$. This local averaging procedure first causes *loss of resolution*. Second, because of peculiarities in the shape of the frequency window, there is additional distortion that is often called *leakage*.

To make the discussion more specific, we assume in the following that x is a discrete-time signal defined on the time axis $\mathbb{Z}$. Define

$$w_{rect,N}(n) = \begin{cases} 1 & \text{for } |n| \le N, \\ 0 & \text{otherwise,} \end{cases} \qquad n \in \mathbb{Z},$$

as in Fig. 9.12. The DCFT of $w_{rect,N}$ is

$$\hat{w}_{rect,N}(f) = d_{2N+1}(f), \qquad f \in [-1/2, 1/2).$$

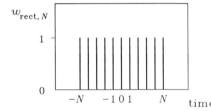

Figure 9.12.   The rectangular window $w_{rect,N}$.

Here d_M is *Dirichlet's kernel,* as introduced in 7.3.11 and given by

$$d_M(f) = M\frac{\text{sinc }(\pi M f)}{\text{sinc }(\pi f)}, \qquad f \in \mathbb{R}.$$

Figure 9.13 gives a plot of the frequency window $\hat{w}_{rect,N}$ for $N = 8$ on the interval $[-1/2, 1/2)$. The figure shows the *main lobe*, which has width $1/(N + \frac{1}{2})$, and the *side lobes*. The main lobe is the part of the frequency window located between the first zero crossings of the window to the left and the right of the zero frequency. The side lobes constitute the remainder of the window.

We may now see why convolution of $\hat{x}$ with the frequency window $\hat{w}_{rect,N}$ leads to loss of resolution and leakage. The convolutive frequency window $\hat{w}_{rect,N}$ has a main lobe of width $W := 1/(N + \frac{1}{2})$, which approximately equals $1/N$ for N large. If the side lobes were absent (i.e., if $\hat{w}_{rect,N}$ were zero outside $(-W/2, W/2)$), then convolving $\hat{x}$ with $\hat{w}_{rect,N}$ would result in local averaging of $\hat{x}$, causing loss of detail. A sharp peak in $\hat{x}$ of width less than W becomes lower and wider, while any other abrupt change in $\hat{x}$ over an interval of length less than W similarly becomes more gradual. This is called *loss of resolution.*

The side lobes result in additional distortion of $\hat{x}$. Because (a) the side lobes of the Dirichlet kernel are relatively large compared with the main lobe, (b) the side

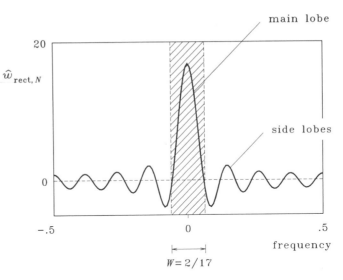

Figure 9.13. Main lobe and side lobes of the convolutive frequency window $\hat{w}_{\text{rect},N}$ for $N = 8$.

lobes decrease slowly, and (c) the side lobes have alternating signs, they have considerable effect, depending on the original shape of $\hat{x}$, of course. If $\hat{x}$ has any abrupt changes, the side lobes result in the *ringing* effect familiar from Gibbs's phenomenon. If the real or imaginary part of $\hat{x}$ is a nonnegative function of frequency, the Dirichlet window may make it negative at some frequencies, which is troublesome in certain applications. The effect of the side lobes is often called *leakage,* because frequency content belonging to one frequency "leaks" to other frequencies.

9.4.1.   Example: Effect of a rectangular window.   By way of example we consider Fourier transforming the time sequence

$$x(n) = a^{|n|} \cos (2\pi f_o n), \qquad n \in \mathbb{Z}.$$

A plot of the signal together with its DCFT for $f_o = 0.1$ and $a = 0.98$ is given in Fig. 9.14. The DCFT is nonnegative for all frequencies and has peaks at the frequencies $\pm f_o$.

Figure 9.15 shows the effect on the DCFT of truncating the time signal to the interval $[-N, N]$ with $N = 64$. As a result of the windowing effect, the peaks have become wider and lower, resulting in a notable loss of resolution. Moreover, there is a marked leakage effect: A strong ringing phenomenon occurs, and the computed Fourier transform assumes negative values for some frequencies, while the original signal has a nonnegative Fourier transform.   ∎

For *continuous-time* signals, symmetric truncation is equivalent to application of the symmetric rectangular multiplicative window of width $2a$, given by

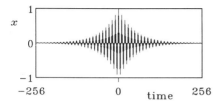

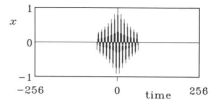

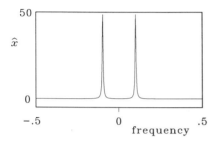

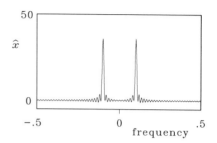

Figure 9.14. An exponentially decaying harmonic signal and its DCFT.

Figure 9.15. Truncated decaying harmonic signal and its DCFT.

$$w_{\text{rect},\,a}(t) = \begin{cases} 1 & \text{for } |t| \le a, \\ 0 & \text{otherwise,} \end{cases} \qquad t \in \mathbb{R}.$$

The corresponding convolutive frequency window is the CCFT of $w_{\text{rect},\,a}$, which is given by

$$\hat{w}_{\text{rect},\,a}(f) = 2a \ \text{sinc}\ (2\pi f a), \qquad f \in \mathbb{R}.$$

The width W of the main lobe of the window is $W = 1/a$. A plot of the window is shown in Fig. 9.16. The effect of the convolutive frequency window is similar to that for the discrete-time case.

Other Windows

As we have seen, truncation of time signals amounts to multiplication by a rectangular window, resulting in *loss of resolution* and *leakage* of the Fourier transform of the signal. *Modifying* the rectangular window to a different shape, while maintaining its width, changes the relation between loss of resolution and leakage. In particular, leakage may be drastically reduced, at the expense of a somewhat increased loss of resolution. Recall that the wider the main lobe is the greater the loss of resolution, and the smaller the side lobes are the less is the leakage.

In Table 9.1 we list some well-known *continuous-time* windows together with their CCFTs. Table 9.2 shows the width of the main lobe of each frequency window and the magnitude of the largest side lobe compared with that of the main lobe. The width of the time window is always $2a$; outside the interval $[-a, a]$ all time windows are zero. Plots of the various windows and their CCFTs are given in Fig. 9.16.

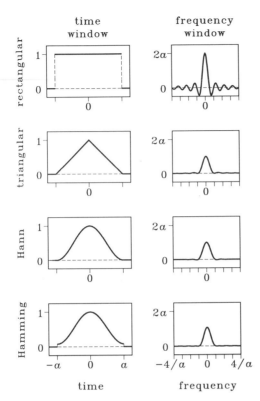

Figure 9.16. Four continuous-time windows and their CCFTs.

TABLE 9.1 CCFT WINDOW PAIRS

Window name	Time window on $[-a, a]$	Frequency window on $\mathbb{R}$
Rectangular	$w_{\text{rect},a}(t) = 1$	$\hat{w}_{\text{rect},a}(t) = 2a \operatorname{sinc}(2\pi f a)$
Triangular	$w_{\text{trian},a}(t) = 1 - \dfrac{|t|}{a}$	$\dfrac{1}{a}\hat{w}^2_{\text{rect},a/2}(f) = a \operatorname{sinc}^2(\pi f a)$
Hann	$w_{\text{Hann},a}(t) = \frac{1}{2}[1 + \cos(\pi t/a)]$	$\frac{1}{2}\hat{w}_{\text{rect},a}(f) + \frac{1}{4}\hat{w}_{\text{rect},a}\left(f + \dfrac{1}{2a}\right)$ $+ \frac{1}{4}\hat{w}_{\text{rect},a}\left(f - \dfrac{1}{2a}\right)$
Hamming	$w_{\text{Hamming},a}(t) = 0.54 + 0.46 \cos(\pi t/a)$	$0.54\hat{w}_{\text{rect},a}(f) + 0.23\hat{w}_{\text{rect},a}\left(f + \dfrac{1}{2a}\right)$ $+ 0.23\hat{w}_{\text{rect},a}\left(f - \dfrac{1}{2a}\right)$

TABLE 9.2   FREQUENCY WINDOW DATA

Window name	Width of main lobe	Relative magnitude of largest side lobe
Rectangular	$1/a$	21.7%
Triangular	$2/a$	4.7%
Hann	$2/a$	2.7%
Hamming	$2/a$	0.7%

The widths of the main lobes of the CCFTs of the triangular window, Hann's window, and Hamming's window all are $2/a$ (i.e., *double* that of the rectangular window). The CCFT of the triangular window is nonnegative, and its side lobes are considerably smaller than those for the rectangular window. The side lobes of the CCFT of Hann's and Hamming's windows are even smaller, but they assume negative values for some frequencies. The largest side lobe of Hamming's window is smaller than that of Hann's window, but the side lobes of Hann's window fall off more rapidly.

Hann's window is named after the Austrian meteorologist Julius von Hann. *Hamming's* window takes it name from the U.S. mathematician Richard W. Hamming.

Windowing of *time sequences* and *sampled* discrete-time signals is analogous to continuous-time windowing. Suppose that the sampled signal x is defined on the discrete time axis $\mathbb{Z}(T)$. Let w_a be a continuous-time window with width $2a$, such as any of the windows of Table 9.1. Then, we may define a corresponding discrete-time window w_{NT}^* of width $2NT$, with N a positive integer, by setting $a = NT$ and letting

$$w_{NT}^*(kT) = w_{NT}(kT), \qquad k \in \mathbb{Z}.$$

The DCFT of the windowed discrete-time signal $x \cdot w_{NT}^*$ is the cyclical convolution

$$\hat{x} \odot \hat{w}_{NT}^*,$$

where $\hat{x}$ is the DCFT of x and $\hat{w}_{NT}^*$ that of the discrete-time window w_{NT}^*. Both are defined on the frequency axis $[-1/2T, 1/2T]$. If the sampling interval T is small compared with the width $2NT$ of the time window (i.e., if N is large), then we may approximate $\hat{w}_{NT}^*$ as

$$\hat{w}_{NT}^*(f) \approx \hat{w}_{NT}(f), \qquad -1/2T \le f < 1/2T,$$

where $\hat{w}_a$ is the frequency window corresponding to the continuous-time window w_a. Cyclical convolutive windows have the same effects as regular convolutive windows: loss of resolution and leakage.

9.4.2.   Example: Windowing.   In Example 9.4.1 we showed the effect of a rectangular multiplicative time window on the Fourier transform of an exponentially decreasing harmonic time sequence. Figure 9.17 illustrates the result of modifying the rectangular window to a *triangular* window. The ringing phenomenon has all but disappeared, and the DCFT of the windowed signal assumes nonnegative values only. The price is that owing to the greater loss of resolution the height of the peaks is half the actual value and the peaks are wider. Application of Hann's or Hamming's window gives very similar results.   ∎

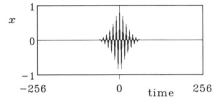

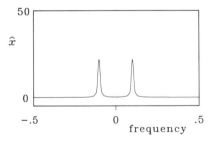

Figure 9.17.   Top: the truncated decaying harmonic signal after application of a triangular window. Bottom: the corresponding DCFT.

9.4.3.   Exercise: Smoothing "raw" frequency data.   If a time sequence x defined on the time axis $\mathbb{Z}$ is truncated to $[-N, N)$, then according to 9.7.7 it is sufficient to compute its DCFT on the set of frequencies

$$\{kW \mid k = -N, -N + 1, \cdots, N - 1\},$$

with $W = 1/2N$. The Fast Fourier Transform, discussed in Section 9.7, provides an efficient way of computing the DCFT on this set. The result is equivalent to applying a rectangular time window to x and then computing the DCFT on the given discrete set of frequencies. The Fourier transform found this way is sometimes called the *raw* frequency data corresponding to the given truncated time signal.

These raw data may be smoothed by taking for each frequency a weighted sum of the raw value at that frequency, with weight 0.5, and the raw values for the adjacent frequencies at either side, each with weight 0.25. Show that this procedure is equivalent to applying Hann's time window *before* Fourier transformation. How is the smoothing done for the data at the extreme frequencies $-NW$ and $(N - 1)W$?

Hint: Write $\frac{1}{2}[1 + \cos{(\pi n/N)}] = \frac{1}{2} + \frac{1}{4}e^{j\pi n/N} + \frac{1}{4}e^{-j\pi n/N}$, and use the converse shift property of the DCFT.   ∎

Frequency Windows

The presentation in this section so far centered on multiplicative time windows and the corresponding convolutive frequency windows. An entirely complementary discussion applies to multiplicative *frequency* windows and the resulting convolutive time windows. An application of multiplicative frequency windowing is encountered in the FIR filter design method of Section 9.5.

9.4.4.   Example: Multiplicative windowing in the frequency domain.   In Example 6.4.13 we showed how truncation of the infinite Fourier series of a rectangular pulse results in Gibbs's phenomenon. Truncating the Fourier series of a periodic signal with period P to $2N + 1$ terms is equivalent to truncating the CDFT of a single period of the signal to a frequency interval of the form $[-NF, NF]$, with $F = 1/P$ and N a nonnegative integer. This in turn is equivalent to application of a multiplicative rectangular *frequency* window. The multiplicative frequency window results in a convolutive time window with the shape of a Dirichlet kernel, which causes the ringing effect.

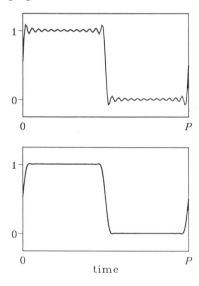

Figure 9.18. Top: the partial Fourier sum signal of the rectangular pulse with 25 terms. Bottom: the partial sum signal after application of Hann's window.

Using a different multiplicative frequency window, for instance, Hann's window, reduces the ringing effect. Figure 9.18 shows the partial sum Fourier series of the rectangular pulse for 25 terms. This is the same plot as in Fig. 6.9 and exhibits Gibbs's phenomenon very clearly. Figure 9.18 also shows the effect of again taking 25 terms, after multiplying the Fourier coefficients by Hann's window, with time replaced by frequency. The ringing effect is considerably less. Because the main lobe of the transform of Hann's window is twice as wide as that of the rectangular window, there is some loss of resolution, manifesting itself in a deterioration of the crispness of the transitions from $+1$ to -1 and vice-versa.   ∎

9.5 DESIGN OF FIR DIGITAL FILTERS

In this and the following section we present design procedures for digital signal processors. As a preamble we discuss how *continuous-time* signal processing tasks may be implemented digitally.

Digital Implementation of Analog Signal Processing Tasks

The configuration we consider is given in Fig. 9.19. It consists of the series connection of an *A/D converter*, which converts the continuous-time signal to a sampled and quantized signal, a *digital filter*, implemented by digital circuits, and a *D/A converter*, which converts the digital signal to a continuous-time signal. The effects of quantization will not be discussed, and the A/D converter is a *sampler*, which converts the continuous-time signal u to the sampled signal $u*$ with sampling interval T given by

$$u*(t) = u(t), \qquad t \in \mathbb{Z}(T).$$

According to 9.2.1, the frequency content $\hat{u}*$ of the sampled signal $u*$ (i.e., its DCFT), is given by

$$\hat{u}*(f) = \sum_{k=-\infty}^{\infty} \hat{u}(f - kf_s), \qquad f \in \mathbb{R},$$

where $\hat{u}$ is the frequency content of the continuous-time signal u (i.e., its CCFT), and $f_s = 1/T$ the sampling rate. Figure 9.3 shows that sampling folds the frequency content of u back into the interval $[-f_s/2, f_s/2)$. To prevent serious loss of information by sampling, the sampling rate should be so high that only a negligible fraction of the signal power lies outside the band $[-f_s/2, f_s/2)$.

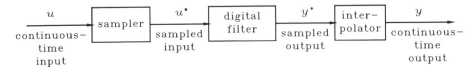

Figure 9.19.   Digital processing of a continuous-time signal.

On the output side, the D/A converter converts the sampled output of the digital signal processor back to a continuous-time signal. If the D/A converter consists of an interpolator, we know from 9.2.4 that the frequency content $\hat{y}$ of the output (i.e., the CCFT of y) is given by

$$\hat{y}(f) = \hat{i}(fT) \cdot \hat{y}*(f), \qquad f \in \mathbb{R},$$

where $\hat{i}$ is the CCFT of the interpolating function i, and $\hat{y}*$ the frequency content (i.e., the DCFT) of the sampled signal $y*$.

Figure 9.20 shows the magnitudes of the frequency response function $\hat{\imath}$ for the step and linear interpolating functions. The frequency response function of the linear interpolator has much smaller side lobes, so that the higher harmonics generated by sampling are more effectively suppressed than by the step interpolator. Passing the interpolated signal y through a low-pass filter improves the suppression of the higher harmonics.

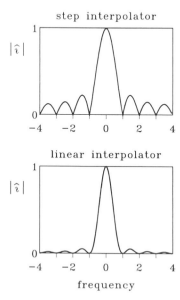

Figure 9.20. Magnitudes of the frequency responses of interpolators. Top: step interpolator. Bottom: linear interpolator.

Once the sampling rate f_s and the interpolator have been chosen, the frequency response function $\hat{h}*$ required for the digital signal processor is fully determined by the desired continuous-time frequency response. In what follows we concentrate on the question of how to implement a given frequency response function for the digital processor.

FIR Versus IIR Filters

An important decision in the design of the digital signal processor is whether to implement it as a *Finite Impulse Response* (FIR) or an *Infinite Impulse Response* (IIR) filter. FIR filters are described by an MA difference equation of the form

$$y(nT) = h*(0)u(nT) + h*(1)u((n-1)T)$$
$$+ \cdots + h*(M)u((n-M)T), \qquad n \in \mathbb{Z}, \tag{1}$$

where the finite sequence $h*$ is the pulse response of the system (see 8.6.6). The FIR filter may be implemented by the block diagram of Fig. 9.21, which contains M delays and $M + 1$ gains arranged in tapped delay line form. Because of its structure, such a system is also called a *transversal* filter.

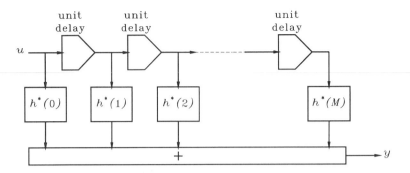

Figure 9.21.   Transversal filter.

IIR filters, as their name expresses, have an *infinite* impulse response. They are described by an ARMA difference equation of the form

$$q_N y((n + N)T) + q_{N-1} y((n + N - 1)T) + \cdots + q_0 y(nT)$$
$$= p_M u((n + M)T) + p_{M-1} u((n + M - 1)T)$$
$$+ \cdots + p_0 u(nT), \qquad n \in \mathbb{Z}. \qquad (2)$$

Systems of this type have an infinite impulse response unless $N = 0$.

We compare the main features of FIR and IIR filters. An important characteristic of FIR filters is that they are always BIBO stable. Implementation errors of FIR filters (in particular *quantization* errors caused by finite word length) therefore do not affect their stability. Moreover, the effect of such errors does not propagate indefinitely. On the other hand, some signal processing tasks may require a very long memory, resulting in a complex filter with many unit delays, and an associated long processing delay.

A distinct advantage of IIR filters is that for many signal processing tasks they require a much smaller complexity (i.e., fewer unit delays) than an FIR filter. On the other hand, IIR filters are less simple to design than FIR filters. Moreover, inaccuracies in their implementation (especially a too short word length) may strongly affect their operation and even make them unstable.

With the current state of technology, the main reason for preferring an IIR filter is that it has less processing delay. If processing delay is not crucial an FIR filter is the most obvious choice.

9.5.1.   Example: IIR and FIR implementation of the exponential smoother.
To compare IIR and FIR designs we consider the exponential smoother, which on the time axis $\mathbb{Z}(T)$ is described by the difference equation

$$y(nT) = a y((n - 1)T) + (1 - a) u(nT), \qquad n \in \mathbb{Z}.$$

Implementation as an IIR filter according to the block diagram of Fig. 9.22 requires a single unit delay and two coefficients. Because the pulse response

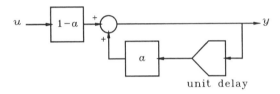

Figure 9.22. Implementation of the exponential smoother as an IIR filter.

$$h*(n) = (1 - a)a^n1(n), \qquad n \in \mathbb{Z},$$

of the smoother is infinite, implementation as an FIR filter as in Fig. 9.21 in principle is not possible. A simple way to *approximate* the system by an FIR filter is to *truncate* the impulse response. Unless a is very small, accurate approximation needs several or even many delays and coefficients. Suppose, for instance, that $a = 0.5$. Then we have successively

$$h*(0) = 0.5, \qquad h*(3) = 0.0625, \qquad h*(6) = 0.0078125,$$

$$h*(1) = 0.25, \qquad h*(4) = 0.03125, \qquad h*(7) = 0.0039063,$$

$$h*(2) = 0.125, \qquad h*(5) = 0.015625, \qquad h*(8) = 0.0019531.$$

All coefficients $h(n)$ with $n \geq 7$ are smaller than 1% of the largest coefficient. Neglecting these coefficients results in an FIR of length 7. The implementation of this FIR filter as in Fig. 9.21 requires 6 unit delays and 7 coefficients as compared with 1 delay and 2 coefficients for the IIR filter.   ∎

Filter Specification

The first step in designing a digital signal processor is the choice of the sampling rate $f_s = 1/T$. If the bandwidth of the input signal is B, by the sampling theorem the sampling rate f_s should be at least $2B$.

The next step consists of a specification of the desired frequency response function $\hat{h}_o$ of the sampled system on the frequency axis $[0, f_s/2]$. For a low-pass filter, the specification of the frequency response function might look as in Fig. 9.23. Up to the frequency f_{pass} the magnitude $|\hat{h}_o|$ is required to be equal to 1 within a *tolerance* δ_{pass}. Above the frequency f_{stop} the magnitude is specified to be less than a tolerance δ_{stop}. The interval $[0, f_{pass})$ is the *pass band*, while $[f_{stop}, f_s/2)$ is the *stop band*. The phase specification initially usually is of the form $\arg(\hat{h}) = 0$. As we shall see, however, to avoid anticipativity often a *delay* need be introduced, which causes a phase shift that is linear in frequency.

The Window Method for the Design of FIR Filters

There are many design procedures for FIR digital filters. We describe a simple method, called the *window* FIR design algorithm. It consists of the following steps:

(a) Choose a desired frequency response function $\hat{h}_o$ of the digital filter.

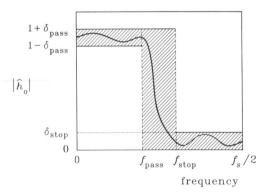

Figure 9.23. Specification of the frequency response function of a digital filter: The magnitude $|\hat{h}|$ is required to lie in the shaded area.

(b) Computation of the inverse DCFT results in the desired impulse response

$$h_o(nT) = \int_{-f_s/2}^{f_s/2} \hat{h}_o(f) e^{j2\pi fnT}\, df, \qquad n \in \mathbb{Z}.$$

(c) Usually, the impulse response h_o is infinite (i.e., has infinite support.) To make it finite, select a window w_{MT} such that

$$w_{MT}(nT) = 0 \qquad \text{for} \quad |n| > M,$$

and form the finite impulse response

$$h_w = h_o \cdot w_{MT}.$$

(d) Usually, the impulse response h_w is *anticipating*. To remedy this, shift the impulse response forward and form the finite non-anticipating impulse response

$$h_d(nT) = h_w((n - M)T), \qquad n \in \mathbb{Z}.$$

The pulse response of the FIR filter (1) finally follows as

$$h*(n) = T \cdot h_d(nT), \qquad n = 0, 1, 2, \cdots, M.$$

The engineering decisions to be made in the successive design steps follow.

(a) A desired frequency response function $\hat{h}_o$ need be chosen.

(c) A suitable window half-length M and window shape need be selected. The use of Hann's or Hamming's window ensures a small ripple. If the half-width of the time window is MT, the width of the corresponding Hann or Hamming frequency window is $2/MT$. This width determines the maximal resolution of the frequency response function after application of the window. Thus, for the low-pass frequency response function of Fig. 9.23, M has to be chosen so that $2/MT \approx f_{\text{stop}} - f_{\text{pass}}$, or

$$M \approx \frac{2f_s}{f_{\text{stop}} - f_{\text{pass}}}.$$

(d) Shifting the impulse response h_w to make it non-anticipating introduces a delay MT. If this delay is unacceptably large, the window length need be reduced. The specifications on the tolerances should be relaxed accordingly.

We illustrate the procedure with an example.

9.5.2.   Example: Design of a low-pass FIR filter.   Suppose that we wish to design a low-pass digital filter with a sampling rate of 180 kHz and a cut-off frequency of 60 kHz. Hence, the desired frequency response function is chosen as

$$\hat{h}_o(f) = \begin{cases} 1 & \text{for } |f| < f_o, \\ 0 & \text{for } f_o \leq |f| \leq f_s/2, \end{cases}$$

where $f_o = 60$ kHz and $f_s = 180$ kHz. The corresponding desired impulse response follows by the inverse DCFT and is given by

$$h_o(t) = 2f_o \text{ sinc } (2\pi f_o t), \qquad t \in \mathbb{Z}(T).$$

The desired frequency response function and the impulse response h_o are plotted in Fig. 9.24. The impulse response is infinite and anticipating.

Next, consider the choice of the window. Suppose that in the specification of Fig. 9.23 we take

$$f_{\text{pass}} = 57 \text{ kHz}, \qquad f_{\text{stop}} = 63 \text{ kHz},$$

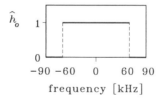

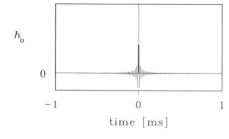

Figure 9.24.   Top: desired low-pass frequency response function. Bottom: the corresponding impulse response.

(i.e., 5% below and 5% above the cut-off frequency, respectively). The resulting time window half-length is $M = 2 \cdot 180/6 = 60$ samples. Figure 9.25 shows the impulse response h_w after application of Hann's window with a half-length of 60 samples, together with the resulting frequency response function $\hat{h}_w$. It turns out that the frequency response function conforms to the specifications of Fig. 9.23 with $f_{\text{pass}} = 57$ kHz, $f_{\text{stop}} = 63$ kHz, and tolerances $\delta_{\text{pass}} = \delta_{\text{stop}} = 0.006$.

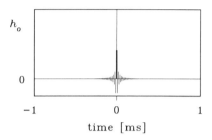

time [ms]

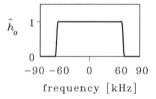

frequency [kHz]

Figure 9.25. Top: impulse response after application of Hann's window. Bottom: the resulting frequency response function.

To make the filter non-anticipating, the impulse response need be shifted by 60 samples, which introduces a delay of $MT = M/f_s = 1/3$ ms. The magnitude of the resulting frequency response function is as in Fig. 9.25, but the filter has an additional linear phase shift. The implementation of the transversal filter requires 119 unit delays and 120 coefficients.   ∎

9.6 DESIGN OF IIR DIGITAL FILTERS

When the FIR implementation of a digital signal processing task introduces too much *processing delay*, or the *processing rate* is too slow, implementation as an IIR filter may be necessary. IIR filters are discrete-time systems described by an ARMA difference equation of the form

$$q_N y((n + N)T) + q_{N-1} y((n - N + 1)T) + \cdots + q_0 y(nT)$$
$$= p_M u((n + M)T) + p_{M-1} u((n - M + 1)T) + \cdots + p_0 u(nT), \qquad n \in \mathbb{Z}.$$

By 8.7.7 the system has the transfer function

$$H(z) = \frac{P(z)}{Q(z)}, \qquad |z| > \rho,$$

where P and Q are the polynomials

$$P(\lambda) = p_M \lambda^M + p_{M-1} \lambda^{M-1} + \cdots + p_0,$$

$$Q(\lambda) = q_N \lambda^N + q_{N-1} \lambda^{N-1} + \cdots + q_0,$$

and $\rho = \max_i |\pi_i|$, with the π_i the poles of H. If $\rho < 1$ (i.e., all the poles π_i lie strictly inside the unit circle), then the system has the frequency response function

$$\hat{h}(f) = H(e^{j2\pi fT}), \qquad -f_s/2 \le f < f_s/2.$$

A common design procedure for IIR filters is as follows:

(a) Select a *continuous-time* filter with a desired frequency response function $\hat{h}(f)$, $f \in \mathbb{R}$.

(b) Find a *discrete-time* system with rational transfer function H^* such that the corresponding frequency response function

$$\hat{h}^*(f) = H^*(e^{j2\pi fT}), \qquad -f_s/2 \le f < f_s/2,$$

approximates the desired frequency response function as closely as possible.

Usually, the desired continuous-time frequency response function is also rational. For filtering applications the continuous-time frequency response function is often chosen from well-known filter families with rational transfer functions such as Butterworth filters, whose characteristics are widely known and well-documented.

In what follows we consider four methods to approximate a continuous-time rational transfer function by a discrete-time rational transfer function. Examples illustrate the methods.

The first step in determining an approximating discrete-time system is of course to choose the sampling rate $f_s = 1/T$. A simple rule of thumb is to choose f_s as at least twice the (approximate) bandwidth of the continuous-time filter.

Delta Equivalence Approximation

The first two methods to approximate continuous-time systems by discrete-time systems are *delta equivalence approximation* and *step equivalence approximation*. They require the intermediate computation of an impulse or step response.

The idea of delta approximation is the following. Suppose that we wish to approximate a continuous-time system with rational transfer function H. By application of the inverse Laplace transform we may determine the continuous-time impulse response h. Then at the sampling instants the output of the continuous-time system is given by

$$y(nT) = \int_{-\infty}^{\infty} h(nT - \tau) u(\tau) \, d\tau, \qquad n \in \mathbb{Z}. \tag{1}$$

We now replace the integral by an approximating sum as

$$y(nT) \approx \sum_{k=-\infty}^{\infty} T \cdot h(nT - kT) u(kT), \qquad n \in \mathbb{Z}.$$

This is equivalent to replacing the continuous-time input u in (1) by the "delta interpolated" input $u^{\dagger}$ given by

$$u^{\dagger}(t) = T \sum_{k \in \mathbb{Z}} u(t) \delta(t - kT), \qquad t \in \mathbb{R},$$

illustrated in Fig. 9.26. Delta approximation consists of replacing the continuous-time system with the discrete-time system

$$y(nT) = \sum_{k=-\infty}^{\infty} h^*(n - k) u(kT), \qquad n \in \mathbb{Z},$$

where the pulse response h^* is given by

$$h^*(k) = T \cdot h(kT), \qquad k \in \mathbb{Z}. \tag{2}$$

The transfer function of the sampled system follows as

$$H^*(z) = \sum_{k=-\infty}^{\infty} h^*(k) z^{-k},$$

$$= T \sum_{k=-\infty}^{\infty} h(kT) z^{-k}, \qquad |z| \geq \rho,$$

with ρ sufficiently large.

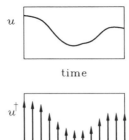

u

time

$u^{\dagger}$

time

Figure 9.26.   Top: a continuous-time signal. Bottom: the "delta interpolated" signal.

The key fact here is that if the continuous-time transfer function H is rational, the continuous-time impulse response h is a linear combination of signals that are exponential in time. Sampling h results in a pulse response h^* that is also a linear com bination of signals that are exponential in time. Application of the z-transform then yields a discrete-time transfer function H^* that is again rational, like the continuous-time transfer function H we started with.

We summarize the approximation procedure as follows.

9.6.1.   Summary: Delta equivalence approximation of a continuous-time system.

(a) Given the transfer function H of the continuous-time system, determine its impulse response h by inverse Laplace transformation.

(b) Sample the continuous-time impulse response h with sampling interval T to obtain the pulse response h^* as $h^*(n) = T \cdot h(nT)$, $n \in \mathbb{Z}$.

(c) Find the discrete-time transfer function H^* by z-transformation of h^*.   ∎

The delta equivalence approximation method applies both to systems that are BIBO stable and that are BIBO unstable. An example of delta equivalence approximation is given in 9.6.3.

Step Equivalence Approximation

In the *step equivalence approximation method* the continuous-time impulse response h is not approximated "pointwise" by the impulse response h^* as in (2) but by interval-wise integration as

$$h^*(n) = \int_0^T h(nT - \tau) \, d\tau, \qquad n \in \mathbb{Z}. \tag{3}$$

The name of this procedure is step equivalence approximation for the following reason. Given any continuous-time input u, let $u^\#$ be the step approximation given by

$$u^\#(t) = u(kT) \qquad \text{for} \quad kT \le t < (k + 1)T \text{ with } k \in \mathbb{Z}, t \in \mathbb{R},$$

as illustrated in Fig. 9.27. The signal $u^\#$ may be obtained by sampling u and step interpolating it. The response of the continuous-time system to the step approximation of the input at the sampling instants is

$$y(nT) = \int_{-\infty}^{\infty} h(t - \tau)u^\#(\tau) \, d\tau = \sum_{k=-\infty}^{\infty} \int_{kT}^{(k+1)T} h(t - \tau)u(kT) \, d\tau$$

$$= \sum_{k=-\infty}^{\infty} \left(\int_{kT}^{(k+1)T} h(t - \tau) \, d\tau \right) u(kT), \qquad n \in \mathbb{Z}.$$

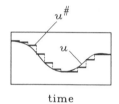

Figure 9.27. Step approximation of a continuous-time signal.

The change of variable $\tau = \theta - kT$ results in

$$y(nT) = \sum_{k=-\infty}^{\infty} \left(\int_0^T h((n-k)T - \theta)\, d\theta \right) u(kT)$$

$$= \sum_{k=-\infty}^{\infty} h*(n-k)\, u(kT), \qquad n \in \mathbb{Z}.$$

If the continuous-time system has a rational transfer function, so has the step equivalent discrete-time approximation. This may be seen as follows. The step response g of the continuous-time system is given by

$$g(t) = \int_{-\infty}^t h(\tau)\, d\tau, \qquad t \in \mathbb{R}.$$

Then $h*$ as given by (3) may be expressed as

$$h*(n) = g(nT) - g((n-1)T), \qquad n \in \mathbb{Z}. \tag{4}$$

Defining $G*$ as the z-transform (in sequence form) of $g(nT)$, $n \in \mathbb{Z}$, that is,

$$G*(z) = \sum_{n=-\infty}^{\infty} g(nT) z^{-n},$$

it follows from (4) by the shift property that the z-transform $H*$ of $h*$ is given by

$$H*(z) = [G*(z) - z^{-1}G*(z)] = \frac{z-1}{z} G*(z).$$

The step response g is the inverse Laplace transform of the rational function $H(s)/s$ and, hence, is a linear combination of exponential time functions. As a result, also $G*$ and $H*$ are rational.

We summarize the step equivalence approximation procedure as follows.

9.6.2.   Summary. Step equivalence approximation of continuous-time systems.

(a) Given the continuous-time transfer function H, determine the step response g by inverse Laplace transformation of $H(s)/s$.

(b) Sample the step response g with sampling interval T and compute the z-transform G^* of the sequence $g(nT)$, $n \in \mathbb{Z}$.

(c) Find the step equivalence approximating discrete-time transfer function H^* as

$$H^*(z) = \frac{z-1}{z} G^*(z).$$ ■

Like delta equivalence approximation, step equivalence approximation is suitable both for BIBO stable and BIBO unstable continuous-time systems.

We illustrate delta and step equivalence approximation by a simple example.

9.6.3.   Example: Delta and step equivalence approximation.   Consider the approximation of the continuous-time system with transfer function

$$H(s) = \frac{1}{s(1 + s\theta)}, \qquad \text{Re}\,(s) > 0,$$

with θ a positive real constant, by a discrete-time system using the delta and step equivalence approximations. This continuous-time system is BIBO unstable.

According to 9.6.1, delta equivalence approximation requires computation of the z-transform H^* of $Th(nT)$, $n \in \mathbb{Z}$, with h the impulse response of the continuous-time system. This is most easily done by first determining the partial fraction expansion of H, transforming each term back to the time domain, then z-transforming the individual terms, and finally adding the result together again after multiplication by T.

A useful result is that a term

$$\frac{1}{s+a}, \qquad \text{Re}\,(s) > -\text{Re}\,(a),$$

in the partial fraction expansion of H transforms to a term

$$e^{-at}\mathbb{1}(t), \qquad t \in \mathbb{R},$$

in the impulse response, which in turn after z-transformation in sequence form on $\mathbb{Z}(T)$ results in

$$\frac{z}{z - e^{-aT}}, \qquad |z| > |e^{-aT}|.$$

Partial fraction expansion of H leads to

$$H(s) = \frac{1}{s} - \frac{\theta}{1 + s\theta}, \qquad \text{Re}\,(s) > 0.$$

It follows that the *delta* equivalence approximation of H is

$$H^*(z) = T\left(\frac{z}{z-1} - \frac{z}{z-e^{-T/\theta}}\right) = \frac{Tz(1-e^{-T/\theta})}{(z-1)(z-e^{-T/\theta})}, \qquad |z| > 1.$$

To determine the *step* equivalence approximation of the system, we consider the partial fraction expansion

$$\frac{H(s)}{s} = \frac{1}{s^2(1+s\theta)} = \frac{1}{s^2} - \frac{\theta}{s} + \frac{\theta^2}{1+s\theta}, \qquad \text{Re}\ (s) > 0.$$

The inverse transform of $1/s^2$, Re $(s) > 0$, is the time signal $t\mathbb{1}(t)$, $t \in \mathbb{R}$, which —after sampling on the time axis $\mathbb{Z}(T)$—in sequence form has the z-transform $Tz/(z-1)^2$. Thus, the z-transform G^* as in 9.6.2 is

$$G^*(z) = \frac{Tz}{(z-1)^2} - \frac{z\theta}{z-1} + \frac{z\theta}{z-e^{-T/\theta}}, \qquad |z| > 1.$$

As a result, the step equivalence approximation is given by

$$H^*(z) = \frac{z-1}{z}G^*(z) = \frac{(T-\theta+\theta e^{-T/\theta})z - (Te^{-T/\theta}+\theta e^{-T/\theta}-\theta)}{(z-1)(z-e^{-T/\theta})},$$

for $|z| > 1$. In Fig. 9.28 plots are given of the step responses of the delta equivalence approximation and of the step equivalence approximation for $T = 1$ and $\theta = 2$. The response of the step equivalence approximation coincides with the step response of the continuous-time system. (This is because the step approximation of a step coincides with the step.) The step response of the delta equivalence approximation is not quite accurate. If the sampling interval T were smaller compared with the time constant θ the delta equivalence approximation would be better.   ∎

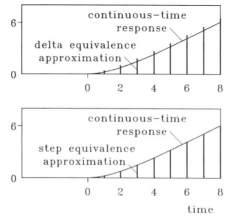

Figure 9.28. Step responses of the delta equivalence and the step equivalence approximations to a continuous-time system.

Staircase Approximation

Staircase and *trapezoidal* approximation of continuous-time systems with rational transfer functions are based on a rather different idea than delta and step equivalence approximation. We first explain staircase approximation.

Consider a realization of the continuous-time system with rational transfer function H as described in Section 5.3. The dynamic elements in this realization are *integrators*. Let the input to one of these integrators be the continuous-time signal v and its output the signal w. The input and output signals are related as $\dot{w} = v$, so that

$$w(t) = w(t - T) + \int_{t-T}^{t} v(\tau) \, d\tau, \qquad t \in \mathbb{R}. \tag{5}$$

A discrete-time approximation of the signal w at the sampling instants in terms of v at the sampling instants follows by replacing v with a (backward) staircase function as in Fig. 9.29. As a result, (5) takes the form

$$w(nT) = w((n - 1)T) + Tv(nT), \qquad n \in \mathbb{Z}.$$

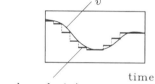

<table>
<tr><td>backward staircase
approximation</td><td>time</td><td>**Figure 9.29.** Backward staircase approximation of a continuous-time signal.</td></tr>
</table>

Application of the z-transform to this relation yields $W^*(z) = z^{-1}W^*(z) + TV^*(z)$, with W^* the z-transform of the sequence $w(nT)$, $n \in \mathbb{Z}$, and V^* similarly defined. It follows that

$$\frac{W^*(z)}{V^*(z)} = \frac{T}{1 - z^{-1}} = \frac{Tz}{z - 1}.$$

Replacement of all integrators in the realization of the continuous-time system by discrete-time blocks with transfer functions $Tz/(z - 1)$ results in a discrete-time system. Since each integrator has transfer function $1/s$, the transfer function $H^*(z)$ of this discrete-time system follows by replacing $1/s$ by $Tz/(z - 1)$ in the continuous-time transfer function $H(s)$. This amounts to the substitution $s = (z - 1)/zT$. We summarize as follows.

9.6.4.   Summary: Staircase approximation of continuous-time systems.
The *staircase* approximation of the continuous-time system with transfer function H is the discrete-time system with transfer function

$$H*(z) = H\left(\frac{z-1}{zT}\right).$$ ∎

If the continuous-time transfer function H is rational, so is the staircase approximation $H*$. Consider what happens to the frequency response function. Let $z = e^{j2\pi fT}$.
Then for f small compared with $1/2T$ we have

$$\frac{z-1}{zT} = \frac{1 - e^{-j2\pi fT}}{T} \approx j2\pi f,$$

so that the frequency response function of the staircase approximation satisfies

$$\hat{h}*(f) = H*(e^{j2\pi fT}) = H\left(\frac{1 - e^{-j2\pi fT}}{T}\right) \approx H(j2\pi f) = \hat{h}(f),$$

for $|f| << 1/2T$. Thus, the discrete-time frequency response function $\hat{h}*$ approxi-
mates the desired frequency response function $\hat{h}$ well at least for low frequencies.

Trapezoidal Approximation

Another better and therefore more commonly used approximation is to replace the
staircase approximation of Fig. 9.29 with a *trapezoidal* approximation as in Fig.
9.30. This leads to the discrete-time approximation

$$w(nT) = w((n-1)T) + \frac{T}{2}[v((n-1)T) + v(nT)], \qquad n \in \mathbb{Z}.$$

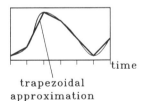

trapezoidal
approximation

Figure 9.30.   Trapezoidal approximation of a continuous-time signal.

After z-transformation of this expression we obtain

$$W*(z) = z^{-1}W*(z) + \frac{T}{2}[z^{-1}V*(z) + V*(z)],$$

or

$$\frac{W*(z)}{V*(z)} = \frac{T}{2}\frac{1 + z^{-1}}{1 - z^{-1}} = \frac{T}{2}\frac{z+1}{z-1}.$$ (6)

Replacing the integrators (whose transfer functions are $1/s$) with discrete-time blocks with the transfer function (6) leads to the substitution

$$s = \frac{2}{T} \frac{z - 1}{z + 1}.$$

This transformation from the z- to the s-domain is known as the *bilinear transformation*.

We summarize as follows.

9.6.5.   Summary: Trapezoidal approximation of continuous-time systems.
The *trapezoidal* approximation of the continuous-time system with transfer function H is the discrete-time system with transfer function

$$H^*(z) = H\!\left(\frac{2}{T} \frac{z - 1}{z + 1}\right). \qquad\qquad\blacksquare$$

Again, if H is rational, then so is the trapezoidal approximation H^*.

Let us look at the frequency response function. Substitution of $z = e^{j2\pi fT}$ results in

$$\frac{2}{T} \frac{z - 1}{z + 1} = \frac{2}{T} \frac{e^{j2\pi fT} - 1}{e^{j2\pi fT} + 1} = \frac{2}{T} \frac{e^{j\pi fT} - e^{-j\pi fT}}{e^{j\pi fT} + e^{-j\pi fT}}$$

$$= j2\pi \cdot \frac{\tan\,(\pi fT)}{\pi T}, \qquad |f| < 1/2T.$$

It follows that

$$\hat{h}^*(f) = H^*(e^{j2\pi fT}) = H\!\left(j2\pi \cdot \frac{\tan\,(\pi fT)}{\pi T}\right) = \hat{h}\!\left(\frac{\tan\,(\pi fT)}{\pi T}\right), \tag{7}$$

for $|f| < 1/2T$. For f small compared with $1/2T$, we may approximate $\tan\,(\pi fT)/\pi T \approx f$, so that at least for low frequencies $\hat{h}^*$ approximates $\hat{h}$ well.

From (7) we see that the trapezoidal approximation amounts to *nonlinear frequency scaling*: the value of the discrete-time frequency response function at the frequency f equals the continuous-time frequency response function at the frequency $\tan\,(\pi fT)/\pi T$. This is equivalent to the frequency transformation $f \to \phi(f)$, where ϕ is the function

$$\phi(f) = \frac{\tan\,(\pi fT)}{\pi T}, \qquad f \in \mathbb{R}.$$

A plot of ϕ is shown in Fig. 9.31. This frequency transformation is known as *warping*. The example that follows shows how to correct for it, which is called *pre-warping*.

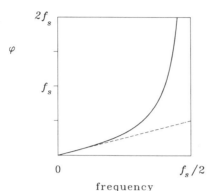

Figure 9.31. Frequency transformation for the trapezoidal approximation.

We end this section by showing the application of trapezoidal approximation to the design of an IIR filter.

9.6.6.   Example: Design of a Butterworth IIR filter.   In this example, like in 9.5.2, we consider the design of a low-pass digital filter with sampling rate $f_s = 1/T = 180$ kHz and a cut-off frequency of 60 kHz. This time we look for an IIR filter that approximates a *Butterworth* filter (compare 4.9.6.) A Butterworth filter of order N has a transfer function of the form

$$H_N(s) = \frac{1}{D_N(s)}, \qquad \text{Re } (s) \geq 0,$$

whose denominator D_N is a polynomial of degree N, with leading coefficient 1. The roots of D_N are evenly distributed along that part of the unit circle that lies in the left-half complex plane. For $N = 1, 2, 3,$ and 4, the Butterworth polynomials are given by

$$D_1(s) = s + 1,$$
$$D_2(s) = s^2 + s\sqrt{2} + 1,$$
$$D_3(s) = s^3 + 2s^2 + 2s + 1,$$
$$D_4(s) = s^4 + 2.6131259s^3 + 3.4142136s^2 + 2.6131259s + 1.$$

The locations of the roots are shown in Fig. 9.32. In general, the roots of the Nth-order Butterworth polynomial D_N are those N roots of $\lambda^{2N} + (-1)^N = 0$ that lie in the left-half complex plane.

The frequency response functions of Butterworth filters have a low-pass character. They cut off at the angular frequency 1, and the larger their order N is, the

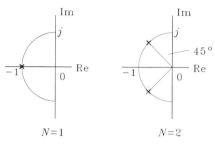

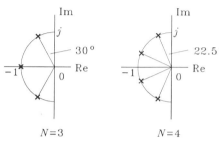

Figure 9.32. Pole locations of Butterworth filters of orders 1–4.

Figure 9.33. Magnitudes of the frequency response functions of Butterworth filters of orders 1–4.

steeper they roll off to zero for angular frequencies over 1. Figure 9.33 shows the magnitudes of the frequency response functions of several Butterworth filters.

Taking the transfer function as $H_N(s/\omega_c)$ shifts the cut-off angular frequency of the Nth order Butterworth filter from 1 to ω_c. The trapezoidal approximation of a fourth-order Butterworth approximation with cut-off angular frequency ω_c thus has the transfer function

$$H^*_{4,\text{trap}}(z) = H_4\left(\frac{2}{\omega_c T}\frac{z-1}{z+1}\right).$$

By warping, the relation between the cut-off frequency f_d of the digital filter and the cut-off frequency $f_c = \omega_c/2\pi$ of the Butterworth filter is $f_c = \phi(f_d) = \tan(\pi f_d T)/\pi T$. To correct for warping, we choose the cut-off frequency ω_c as

$$\omega_c T = 2\pi f_c T = 2\tan(\pi f_d/f_s) = 3.46410.$$

This correction is what is called prewarping. After substitution of the numerical values we obtain

$$H^*_{4,\text{trap}}(z) = \frac{9.00(z+1)^4}{38.3z^4 + 50.1z^3 + 39.5z^2 + 13.9z + 2.14}.$$

The implementation of the filter requires four unit delays. Figure 9.34 shows the magnitude of the resulting discrete-time frequency response function $\hat{h}_{4,\,\mathrm{trap}}^{*}(f)$ for $f \in [0, f_s/2)$. The frequency response function satisfies the specifications of Example 9.5.2 with tolerances $\delta_{\mathrm{pass}} = 0.15$ and $\delta_{\mathrm{stop}} = 0.52$. The large tolerances result from the fact that the frequency response function does not roll off very steeply at the cut-off frequency. Tighter tolerances may be achieved by increasing the filter order.   ∎

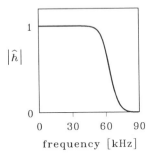

Figure 9.34. Magnitude of the frequency response function of a trapezoidal approximation to the fourth-order Butterworth filter.

9.6.7.   Exercise: Stability of the discrete-time approximations.

(a) Prove that all four methods (delta equivalence, step equivalence, staircase and trapezoidal approximation) to approximate continuous-time systems with rational transfer functions by discrete-time systems result in BIBO *stable* systems when the original continuous-time system is stable.

(b) Also show that for the staircase approximation replacing the *backward* by a *forward* approximation, that is, letting $y(nT) = y((n - 1)T) + Tu((n - 1)T)$, may result in unstable systems.   ∎

9.7 NUMERICAL COMPUTATION OF TRANSFORMS AND CONVOLUTIONS

In this section we discuss the digital computation of transforms and convolutions. Both *on-* and *off-line* computations are considered.

We begin with *off*-line computations. Off-line computations are commonly employed for the computation of the Fourier, z-, or Laplace transform of a signal. Infinite signals need be *windowed*, and continuous-time signals *sampled*, so that the computer faces a *finite* list of numbers. It will be seen that all Fourier transformations, and also the z- and Laplace transforms, may be reduced to the DDFT.

There exists a very efficient algorithm for computing the DDFT, called the *Fast Fourier Transform* (FFT). Thanks to the efficiency of the FFT the most economical way of computing convolutions is via Fourier transformation. This is called *fast convolution*.

For real-time filtering, *on-line* convolution is called for. This requires processing a *flow* of data, which arrives at a certain *rate*, and must be processed at this same rate. This may be achieved by partitioning the input into chunks, processing each chunk consecutively, and putting the chunks back together.

The organization of this section is as follows. First we introduce the FFT and next discuss the use of the FFT for evaluating the other Fourier transforms and the z- and Laplace transforms. The section ends with a discussion of off-and on-line fast convolution.

The Fast Fourier Transform

We present a simple treatment of the *Fast Fourier Transform,* abbreviated as FFT. The FFT is not yet another transform, but an algorithm for computing the DDFT numerically, arranged for maximal efficiency. Because of the similarity of the formula for the DDFT and that for the inverse DDFT, the algorithm serves for both. Moreover, by discretization any of the CD, DC, and CC transforms may be approximated as a DDFT, so that the FFT also serves for those other transforms.

Given a sampled discrete-time signal x defined on the finite time axis $\{0, T, \cdots, (N-1)T\}$, its DDFT $\hat{x}$ is defined as

$$\hat{x}(kF) = T \sum_{n=0}^{N-1} x(nT)e^{-j2\pi kn/N}, \qquad k \in \{0, 1, \cdots, N-1\}.$$

Denoting

$$\hat{x}(kF)/T =: \hat{x}_k, \qquad x(nT) =: x_n,$$

and

$$w_N := e^{-j2\pi/N},$$

we concentrate on the computation of

$$\hat{x}_k = \sum_{n=0}^{N-1} x_n(w_N)^{kn}, \qquad k = 0, 1, \cdots, N-1. \tag{1}$$

Redefining w_N as $e^{j2\pi/N}$ and replacing T with $F = 1/NT$ results in the *inverse* DDFT. With this modification the same algorithm applies both for the DDFT and its inverse.

9.7.1   Remark: Computational effort of the direct DDFT.   The direct computation of the DDFT according to (1) involves N (complex) multiplications and $N-1$ (complex) additions per frequency and, hence, all together requires N^2 multiplications and $N(N-1)$ additions. Since the multiplications are by far the most time-consuming, we only count the multiplications to estimate the computational effort.

Of course also the complex exponentials $(w_N)^{kn}$ need be computed. The complex numbers $(w_N)^m$, $m = 0, 1, \cdots, N - 1$, are evenly spaced along the unit circle, as illustrated in Fig. 9.35. For any integers k and n the exponential $(w_N)^{kn} = (w_N)^{kn \bmod N}$ is one of these complex numbers, which may be precomputed and tabulated, and are required for any algorithm for the DDFT.   ■

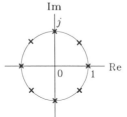

Figure 9.35. The complex numbers $(w_N)^m$, $m = 0, 1, \cdots, N - 1$, for $N = 8$.

The particular FFT we consider is the *Cooley-Tukey radix 2 FFT*, which is the most efficient and widely used algorithm. It involves, like all other FFTs, *multidimensional index mapping*, which converts a single numerically intensive computation to repeated easy computations. For radix 2 algorithms it is necessary that N be *even* so that N can be factored as $N = 2 \cdot (N/2)$. For the Cooley-Tukey radix 2 algorithm, multidimensional index mapping means replacing the *single* indices n and k in (1) with the *composite* indices

$$k = k_1 + 2k_2, \qquad k_1 \in \{0, 1\}, \qquad k_2 \in \{0, 1, \cdots, N/2 - 1\},$$
$$n = n_1 N/2 + n_2, \qquad n_1 \in \{0, 1\}, \qquad n_2 \in \{0, 1, \cdots, N/2 - 1\}.$$

We refer to this as the *radix 2 index map*.

9.7.2.   Summary: Radix 2 index mapping of the DDFT.
Under the radix 2 index map the DDFT (1) takes the form

$$\hat{x}_{k_1+2k_2} = \sum_{n_2=0}^{N/2-1} x_{n_2}^{k_1} (w_{N/2})^{k_2 n_2}$$

for $k_1 \in \{0, 1\}$ and $k_2 \in \{0, 1, \cdots, N/2 - 1\}$, where

$$x_{n_2}^0 = x_{n_2} + x_{n_2+N/2}, \qquad x_{n_2}^1 = (x_{n_2} - x_{n_2+N/2})(w_N)^{n_2}$$

for $n_2 \in \{0, 1, \cdots, N/2 - 1\}$.   ■

9.7.3.   Proof.
We consider radix R index mapping and specialize to $R = 2$ at the end of the proof. The general radix R Cooley-Tukey FFT algorithm requires N to be divisible by the integer R, and uses the index mappings

$$k = k_1 + k_2 R, \qquad k_1 \in \{0, 1, \cdots, R - 1\}, \quad k_2 \in \{0, 1, \cdots, N/R - 1\},$$

$$n = n_1 N/R + n_2, \quad n_1 \in \{0, 1, \cdots, R - 1\}, \quad n_2 \in \{0, 1, \cdots, N/R - 1\}.$$

Besides 2 the radix $R = 4$ is sometimes used; other values are not very common. Substitution into (1) yields

$$\hat{x}_{k_1 + k_2 R} = \sum_{n_1 = 0}^{R-1} \sum_{n_2 = 0}^{N/R-1} x_{n_1 N/R + n_2} (w_N)^{(k_1 + k_2 R)(n_1 N/R + n_2)}, \tag{2}$$

for $k_1 \in \{0, 1, \cdots, R - 1\}$ and $k_2 \in \{0, 1, \cdots, N/R - 1\}$. Expansion of the factor involving w_N results in

$$(w_N)^{(k_1 + k_2 R)(n_1 N/R + n_2)} = (w_N)^{k_1 n_1 N/R + k_1 n_2 + k_2 n_1 N + k_2 n_2 R}$$

$$= (w_R)^{k_1 n_1} (w_N)^{k_1 n_2} (w_{N/R})^{k_2 n_2},$$

where we use the facts that $(w_N)^{N/R} = w_R$, $(w_N)^N = 1$ and $(w_N)^R = w_{N/R}$. Substitution of this result into (2) and rearrangement of the order of summation leads to

$$\hat{x}_{k_1 + k_2 R} = \sum_{n_2 = 0}^{N/R-1} (w_{N/R})^{k_2 n_2} (w_N)^{k_1 n_2} \sum_{n_1 = 0}^{R-1} x_{n_1 N/R + n_2} (w_R)^{k_1 n_1}$$

$$= \sum_{n_2 = 0}^{N/R-1} x_{n_2}^{k_1} (w_{N/R})^{k_2 n_2},$$

for $k_1 \in \{0, 1, \cdots, R - 1\}$ and $k_2 \in \{0, 1, \cdots, N/R - 1\}$, where

$$x_{n_2}^{k_1} = (w_N)^{k_1 n_2} \sum_{n_1 = 0}^{R-1} x_{n_1 N/R + n_2} (w_R)^{k_1 n_1},$$

for $k_1 \in \{0, 1, \cdots, R - 1\}$ and $n_2 \in \{0, 1, \cdots, N/R - 1\}$. The formulas of 9.7.2 result by setting $R = 2$. ∎

By 9.7.2, the computation of the Fourier transform $\hat{x}$ amounts to the computation of the two sequences of numbers

$$\hat{x}_{2k_2} = \sum_{n_2 = 0}^{N/2-1} x_{n_2}^0 (w_{N/2})^{k_2 n_2}, \qquad k_2 = 0, 1, \cdots, N/2 - 1, \tag{3}$$

and

$$\hat{x}_{2k_2 + 1} = \sum_{n_2 = 0}^{N/2-1} x_{n_2}^1 (w_{N/2})^{k_2 n_2}, \qquad k_2 = 0, 1, \cdots, N/2 - 1, \tag{4}$$

where

$$x_{n_2}^0 = x_{n_2} + x_{n_2+N/2}, \qquad x_{n_2}^1 = (w_N)^{n_2}(x_{n_2} - x_{n_2+N/2}), \tag{5}$$

for $n_2 = 0, 1, 2, \cdots, N/2 - 1$. The computation of (3) and (4) amounts to calculating the DDFT of *two* sequences x^0 and x^1, each with *half* the length $N/2$ of the original sequence. These two DDFT's involve $2 \times (N/2)^2 = N^2/2$ multiplications, which is *half* the effort of the original problem. To arrive at this reduced problem, $N/2$ multiplications are needed, those required for the computation of the sequence x^1.

Stopping at this point would entail certain savings, but really impressive economies result by applying the reduction procedure of 9.7.2 *repeatedly*. If N is an integral power of 2, say, $N = 2^M$, then the procedure can be repeated M times, each time reducing N to half its value. In the final stage N equals 2, so that (3) and (4) simplify to two-term expressions of the form (5). This is the radix 2 Cooley-Tukey FFT.

9.7.4.   Remark: Computational effort of the radix 2 Cooley-Tukey FFT. Supposing that $N = 2^M$, the radix 2 Cooley-Tukey FFT requires the following number of complex multiplications:

Stage 1: $N/2$ multiplications for the computation of the subsequences x^0 and x^1 of x.

Stage 2: $(N/2)/2 = N/4$ multiplications for the computation of the subsequences x^{00} and x^{01} of x^0, and similarly $N/4$ multiplications for the computation of the subsequences x^{10} and x^{11} of x^1, so together $N/2$ multiplications.

Stage 3: $N/2$ multiplications.

$\cdots$ $\quad$ $\cdots$ $\cdots$ $\cdots$ $\cdots$ $\cdots$ $\cdots$

Stage M: $N/2$ multiplications.

All together, $NM/2$, that is, $\frac{1}{2}N \log_2 (N)$ multiplications are needed. Table 9.3 gives a comparison of this number with the N^2 multiplications needed for the direct DDFT.

■

TABLE 9.3 COMPARISON OF THE NUMBER OF COMPLEX MULTIPLICATIONS FOR THE DIRECT DDFT AND THE COOLEY-TURKEY RADIX 2 FFT

Signal length N	direct N^2	FFT $\frac{1}{2}N \log_2 (N)$	ratio FFT/direct
32	1024	80	.078
64	4096	192	.047
128	16384	448	.027
256	65536	1024	.016
512	262144	2304	.0088

It is seen from Table 9.3 that especially for large N the number of multiplications needed for the FFT is a small fraction of that for the direct DDFT. Some of the savings are lost because of additional bookkeeping required for the FFT. Moreover, if the signal whose DDFT is to be computed is not defined on a number of time instants N that is an integral power of 2, then it is necessary to extend the time axis and complement the signal with additional values (usually taken equal to 0), resulting in a loss of efficiency. Nevertheless, the advent of the FFT meant a significant advance in numerical signal processing.

9.7.5.   Remark: Code for the radix 2 Cooley-Tukey FFT, and in place calculations.   In Fig. 9.36 we show a SIGSYS code for the radix 2 FFT, which may easily be converted to Fortran, Pascal, BASIC, or other computer language. The code involves complex multiplications, which SIGSYS supports. For other languages conversion to real multiplication may be necessary.

```
function FFT(z)
    x=z; N=num(x)
    M=ceil(log(N)/log(2)); L=2^M      %Round length L up to power M of 2
    x=x(0:1:L-1); N=L                 %Pad with zeros. N is length of
                                      %subsequences
    for  i in 1:1:M                   %Step through M stages
        w=exp(-j*2*pi/N)
        for k in 0:N:L-1              %Step from one subsequence to the next
            for n in k:1:k+N/2-1 %Step through subsequence k
                xt=x(n)-x(n+N/2)
                x(n)=x(n)+x(n+N/2)
                x(n+N/2)=xt*w^n
            end
        end
        N=N/2
    end
    for  i in 1:1:L-1                 %Step through list for unscrambling
        ii=i; k=L/2; l=1; r=0        %Compute reversed bit index r from i
        while l<L
            if ii>=k ii=ii-k; r=r+1   end
            k=k/2; l=l*2
        end
        if r>i xt=x(r); x(r)=x(i); x(i)=xt  end
                                      %Interchange x(i) and x(r) if r>i
    end
    return x
end
```

Figure 9.36.   SIGSYS code for the FFT.

The code illustrates the use of "in place" calculations, which means that the complex numbers $x_{n_2}^0$, $n_2 = 0, 1, \cdots, N/2 - 1$ and $x_{n_2}^1$, $n_2 = 0, 1, \cdots, N/2 - 1$, in that order, are stored in the memory locations previously occupied by x_n, $n = 0, 1, \cdots, N - 1$. At the next stages of the computation, the same procedure is followed for each subsequence. This results in the following permutation of the data.

In the first stage of the computation, all even-indexed entries move to the first $N/2$ locations, while the odd-indexed entries go to the last $N/2$ locations. This means that if the indices are represented in *binary* form, then after the first stage the last bit of the index of each entry has become the *first* bit while all other bits have moved up one position to the right. Completing the second stage results in the removal of last bit to the *second* position. After completion of all M stages, the bits of the index of each entry have all *reversed* position compared with the original index.

> The FFT algorithm was published in 1965 by the American mathematician James W. Cooley and the American statistician John W. Tukey. The algorithm was anticipated by work of the German mathematician C. Runge (published in 1903 and 1905), which remained unnoticed for a long time. Only when the electronic digital computer was available did the method become important. See J. W. Cooley, P. A. W. Lewis, and P. Welch, "Historical Notes on the Fast Fourier Transform," *IEEE Trans. Audio Electroacoust.*, vol. AU-15, pp. 76–79, June 1967.

To restore the original indices, at the end of the code the reversed bit index corresponding to each index is computed, and the associated entries are interchanged. This is called *unscrambling*.

The efficiency of the code may be considerably improved by not computing the exponentials w^m when needed, but constructing at the beginning of the code a lookup table for $e^{-j2\pi m/N}$, $m = 0, 1, \cdots, N/2 - 1$. Various other improvements and complete Fortran codes may be found in the literature (see, for instance, C. S. Burrus and T. W. Parks, *DFT/FFT and Convolution Algorithms,* John Wiley, New York, 1985.)   ∎

9.7.6.   Example: FFT of a simple signal.   We illustrate the procedure for the signal represented by the sequence $x = (1, 1, 1, 1, 0, 0, 0, 0)$ of length 8. Application of 9.7.2 easily yields the two subsequences $x^0 = (1, 1, 1, 1)$ and $x^1 = (w_8^0, w_8^1, w_8^2, w_8^3)$. As $w_8 = e^{-j\pi/4} = (1 - j)/\sqrt{2}$, this works out to $x^1 = (1, (1 - j)/\sqrt{2}, -j, (-1 - j)/\sqrt{2})$.

By using 9.7.2 on the subsequence x^0 of length 4 two subsequences of length 2 result, given by $x^{00} = (2, 2)$ and $x^{01} = (0, 0)$, while from x^1 we obtain $x^{10} = (1 - j, -j\sqrt{2})$ and $x^{11} = (1 + j, -j\sqrt{2})$.

Application of 9.7.2 to these four subsequences results in the eight final entries $x^{000} = 4$, $x^{001} = 0$, $x^{010} = 0$, $x^{011} = 0$, $x^{100} = 1 + j(-1 - \sqrt{2})$, $x^{101} = 1 + j(-1 + \sqrt{2})$, $x^{110} = 1 + j(1 - \sqrt{2})$, and $x^{111} = 1 + j(1 + \sqrt{2})$. Figure 9.37 shows how the signal x at each stage branches out to the various subsequences. Table 9.4 lists the successive in place sequences.

The little table that follows shows how by bit reversal the original indices of the entries may be retrieved:

Final index	0	1	2	3	4	5	6	7
Binary form	000	001	010	011	100	101	110	111
Reversed binary form	000	100	010	110	001	101	011	111
Original index	0	4	2	6	1	5	3	7

As a result, the transformed sequence is given by

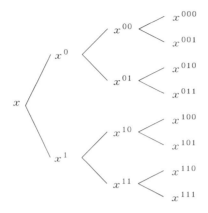

Figure 9.37.   At each stage of the radix 2 FFT each subsequence branches out into two subsequences.

TABLE 9.4   IN PLACE SEQUENCES

Stage 1	Stage 2	Stage 3	Stage 4	Unscrambled
1	1	2	4	4
1	1	2	0	$1 + j(-1 - \sqrt{2})$
1	1	0	0	0
1	1	0	0	$1 + j(1 - \sqrt{2})$
0	1	$1 - j$	$1 + j(-1 - \sqrt{2})$	0
0	$(1 - j)/\sqrt{2}$	$-j\sqrt{2}$	$1 + j(-1 + \sqrt{2})$	$1 + j(-1 + \sqrt{2})$
0	$-j$	$1 + j$	$1 + j(1 - \sqrt{2})$	0
0	$(-1 - j)/\sqrt{2}$	$-j\sqrt{2}$	$1 + j(1 + \sqrt{2})$	$1 + j(1 + \sqrt{2})$

$$\hat{x} = (4, 1 + j(-1 - \sqrt{2}), 0, 1 + j(1 - \sqrt{2}), 0,$$
$$1 + j(-1 + \sqrt{2}), 0, 1 + j(1 + \sqrt{2})). \qquad \blacksquare$$

Numerical Computation of Fourier Transforms

We next briefly discuss the numerical computation of the DCFT, CDFT, and CCFT.

The sampled DCFT transforms an *infinite* discrete-time signal with sampling interval T. If this is to be done numerically, then it is necessary to truncate the signal to a *finite* signal. Furthermore, the DCFT in principle is defined on the *continuous* frequency interval $[0, 1/T)$. Numerically, however, the Fourier transform may only be computed on a *finite* frequency grid. We show that there is a natural choice for the frequency grid. Suppose that the sampled signal x is defined on the time axis $\mathbb{Z}(T)$ and that it is symmetrically truncated to the interval $[-NT/2, (N/2 - 1)T]$, with N even and preferably an integral power of 2 if the FFT is used for Fourier transformation. The DCFT of the resulting truncated signal satisfies the following property.

9.7.7.   Summary: DCFT of a truncated signal.   Let the sampled signal x be defined on the time axis $\mathbb{Z}(T)$, and suppose that it is zero outside the interval $[-NT/2, (N/2 - 1)T]$. Then the DCFT $\hat{x}$ of x may be reconstructed by sinc interpolation of the *sampled* DCFT on the discrete frequency axis $\mathbb{Z}(F)$, where $F = 1/NT$, as

$$\hat{x}(f) = \sum_{n=-\infty}^{\infty} \hat{x}(nF) \, \text{sinc} \, [\pi (f - nF)/F], \qquad f \in \mathbb{R}. \qquad \blacksquare$$

This statement is a converse of the sampling theorem. The sampling theorem states that a time signal whose Fourier transform is zero outside the interval $[-B, B]$ may be sampled with sampling interval $1/2B$ and sinc interpolated without loss of information. By interchanging the roles of time and frequency it follows that a frequency signal such that the corresponding time signal is zero outside the interval $[-NT/2, (N/2 - 1)T]$ may be sampled with sampling interval $1/NT$ and sinc interpolated without loss of information.

By 9.7.7 it is sufficient to compute the DCFT of a truncated signal on the frequency axis $\{0, F, 2F, \cdots, (N - 1)F\}$ with $F = 1/NT$. If desired, the complete DCFT then follows by

(i)   periodically continuing the DCFT on the frequency axis $\mathbb{Z}(F)$ and

(ii)   sinc interpolation between the sampling frequencies.

The DCFT of the truncated signal, which is defined on the time axis $\{-NT/2, (-N/2 + 1)T, \cdots, (N/2 - 1)T\}$, may be computed on the frequency axis $\{0, F, 2F, \cdots, (N - 1)F\}$ by shifting the truncated signal *cyclically* to the time axis $\{0, T, 2T, \cdots, (N - 1)T\}$ and then calculating the DDFT by the fast Fourier transform.

Truncating the sampled signal x to the interval $[-NT/2, (N/2 - 1)T]$ is of course equivalent to application of a rectangular window and causes *loss of resolution* and *leakage* of the DCFT, as discussed in Section 9.4. Using a different window diminishes leakage. Loss of resolution may only be reduced by increasing the window length $2N$.

9.7.8.   Example: Numerical computation of the DCFT.   The DCFT of the time sequence

$$x(n) = a^{|n|} \cos (2\pi f_o n), \qquad n \in \mathbb{Z},$$

is shown in Fig. 9.38 for $f_o = 0.1$ and $a = 0.98$. The same figure shows the DCFT of the signal after truncation to $[-N/2, N/2 - 1]$ with $N = 128$, sampled with a frequency interval $1/N$. The plot clearly shows the loss of resolution and the effect of leakage due to truncation. $\blacksquare$

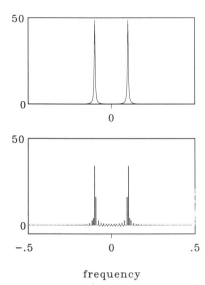

Figure 9.38. DCFT of a time sequence. Top: actual. Bottom: numerically computed after truncation.

frequency

We next consider the numerical computation of the CDFT. This transform involves a *continuous-time* signal, which for numerical transformation need be sampled, say with a sampling interval $T = P/N$, with P the period of the signal, and N an integer. Since normally the FFT is used for numerical calculations, it is most efficient to choose N as an integral power of 2. The CDFT may then be obtained by computing the DDFT by means of the FFT on the frequency axis $\{0, F, 2F, \cdots, (N-1)F\}$ with $F = 1/NT = 1/P$, and shifting the result cyclically to the frequency axis $\{-(N/2)F, -(N/2-1)F, \cdots, (N/2-1)F\}$. Because the continuous-time signal is sampled, the CDFT as computed suffers from *aliasing*. Since there is no truncation, loss of resolution and leakage play no role.

The numerical computation of the CCFT of an infinite continuous-time signal requires both truncation, say to $[-P, P)$, and sampling, say, with a sampling interval $T = P/N$ with N an integral power of 2. The approximate CCFT may then be computed as the DCFT of the resulting signal on a discrete frequency axis with frequency interval $F = 1/NT = 1/P$ at N frequencies symmetrically located with respect to zero frequency. The approximate CCFT suffers both from aliasing (due to sampling) and loss of resolution (due to truncation.)

The numerical computation of the various *inverse* Fourier transforms follows parallel lines. Sampling (for the inverse DCFT and CCFT) and truncation (for the inverse CDFT and CCFT) reduce the transform to the inverse DDFT which is carried out using the inverse FFT. Frequency sampling results in *time aliasing*, while frequency truncation causes *loss of resolution* and *leakage* in the *time* domain.

9.7.9   Example: Numerical inverse CCFT.   The CCFT

$$\hat{x}(f) = e^{-\frac{1}{1+j2\pi f}} + e^{-\frac{1}{1-j2\pi f}} - 2, \qquad f \in \mathbb{R},$$

does not look very tractable. The plot of $\hat{x}$, which is real, in Fig. 9.39(a) shows that the frequency signal has most of its energy in the frequency interval $[-2, 2)$. Sampling $\hat{x}$ on $N = 128$ points on this interval and inverse transformation by

(i)   shifting cyclically to the frequency axis $\{0, F, 2F, \cdots, (N-1)F\}$ with $F = 4/128$,

(ii)   application of the inverse DDFT using the FFT, resulting in a sampled signal with sampling interval $T = 1/NF = 1/4$ on the time axis $\{0, T, \cdots, (N-1)T\}$, and

(iii) cyclically shifting this to the time axis $\{-NT/2, \cdots, (N/2-1)T\}$, yields the signal of Fig. 9.39(b), which is probably a good approximation to the correct result.

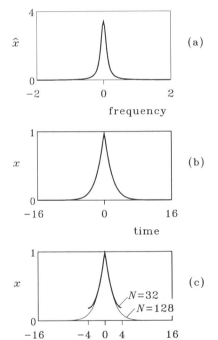

Figure 9.39.   Numerical inverse CCFT. Top: a CCFT. Middle: inverse CCFT on 128 points in $[-2, 2)$. Bottom: inverse CCFT on 32 points in $[-2, 2)$.

Reduction of the number of points N on which $\hat{x}$ is sampled to 32 (but keeping the frequency interval $[-2, 2)$) results in the signal of Fig. 9.39(c), which is defined on one fourth of the time axis of the previous signal and clearly shows the effect of time aliasing in the form of a deviation from the correct values near the ends of the time axis.   ■

Numerical z- and Laplace Transformation

Numerical computation of the z-transform relies on reduction to the DCFT. Suppose that the (two-sided) z-transform X of the discrete-time signal x with time axis $\mathbb{Z}$ exists in the complex plane on a circle $z = \rho e^{j2\pi f}, f \in [0, 1)$, with radius ρ. According to 8.3.5,

$$X(\rho e^{j2\pi f}) = \hat{x}_\rho(f), \qquad f \in [0, 1),$$

where $\hat{x}_\rho$ is the DCFT of the time signal x_ρ given by

$$x_\rho(n) = \rho^{-n} x(n), \qquad n \in \mathbb{Z}.$$

The DCFT $\hat{x}_\rho$ may be computed numerically as explained previously. In this way we find the z-transform X on a single circle with radius ρ.

   If on the other hand a z-transform X is *given* on a circle $z = \rho e^{j2\pi f}, f \in [0, 1)$, with radius ρ, the corresponding time signal x follows as

$$x(n) = \rho^n x_\rho(n), \qquad n \in \mathbb{Z},$$

where x_ρ is the inverse DCFT of the frequency signal $X(\rho e^{j2\pi f}), f \in [0, 1)$.

   Numerical computation of the Laplace transform is similarly accomplished by reduction to the CCFT. Let x be a continuous-time signal whose (two-sided) Laplace transform X exists on the vertical line $s = \sigma + j2\pi f, f \in \mathbb{R}$, in the complex plane, with σ a fixed real number. Then according to 8.3.5

$$X(\sigma + j2\pi f) = \hat{x}_\sigma(f), \qquad f \in \mathbb{R},$$

where $\hat{x}_\sigma$ is the CCFT of the time signal x_σ defined by

$$x_\sigma(t) = e^{-\sigma t} x(t), \qquad t \in \mathbb{R}.$$

The CCFT may be computed numerically as outlined before. In this way we find the Laplace transform X on a single fixed vertical line in the complex plane with abscissa σ.

   Conversely, given a Laplace transform X on a vertical line in the complex plane with abscissa σ, the corresponding time signal x follows as

$$x(t) = e^{\sigma t} x_\sigma(t), \qquad t \in \mathbb{R},$$

where x_σ is the inverse CCFT of the frequency signal $X(\sigma + j2\pi f), f \in \mathbb{R}$.

9.7.10.   Example: Numerical inverse z-transformation.   Consider the inverse z-transformation of the function X given by

$$X(z) = e^{-\frac{1}{1-2z}}, \qquad z \in \mathscr{E}.$$

We consider two existence regions, namely,

$$\mathscr{E}_1 = \{z \in \mathbb{C} \mid |z| < \tfrac{1}{2}\}$$

and

$$\mathscr{E}_2 = \{z \in \mathbb{C} \mid |z| > \tfrac{1}{2}\}.$$

To compute the inverse z-transform of X with existence region $\mathscr{E}_1$, we consider X on the circle $z = \rho e^{j2\pi f}$, $f \in [-\tfrac{1}{2}, \tfrac{1}{2})$, with radius $\rho = \tfrac{1}{4}$. Evaluation of $X(\rho e^{j2\pi f})$ on $[-\tfrac{1}{2}, \tfrac{1}{2})$ yields the DCFT $\hat{x}_{1/4}$ of the signal $x_{1/4}$. Figure 9.40 (a) shows a plot of the magnitude of $\hat{x}_{1/4}$. The signal $x_{1/4}$ itself follows by numerical inverse DC transformation as described before, and is displayed in Fig. 9.40(b). The signal x, finally, which is shown in Fig. 9.40(c), follows by multiplying $x_{1/4}$ by the signal $(\tfrac{1}{4})^n$, $n \in \mathbb{Z}$.

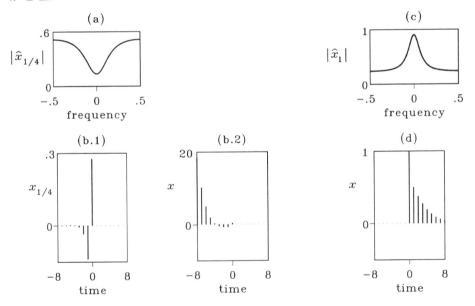

Figure 9.40. Numerical inverse z-transformation: (a) magnitude of a z-transform on a circle with radius $\tfrac{1}{4}$; (b.1) its inverse DCFT $x_{1/4}$, (b.2) the time signal x; (c) magnitude of the z-transform on a circle with radius 1; (d) its numerical inverse.

The inverse z-transform for the existence region $\mathscr{E}_2$ follows by the same procedure, except that ρ is chosen such that the circle $\rho e^{j2\pi f}$, $f \in [-\tfrac{1}{2}, \tfrac{1}{2})$ lies in $\mathscr{E}_2$. The choice $\rho = 1$ achieves this, and leads to the signal shown in Fig. 9.40(d). ∎

Fast Convolution

We end this section with a discussion of *fast convolution*. Fast convolution relies on the Fast Fourier Transform. The convolution $z = x * y$ of two sequences x and y is obtained by first computing the Fourier transforms $\hat{x}$ and $\hat{y}$ of x and y by the FFT, and then inversely Fourier transforming the product $\hat{z} = \hat{x} \cdot \hat{y}$. The computational savings over direct computation are considerable.

In digital filtering applications the output y of a filter with impulse response h is given by

$$y = h * u,$$

with h fixed and u the input to the filter. For off-line computations with very long inputs and for on-line computation the following procedure may be followed:

(i)   break up the input u into pieces of limited length,
(ii)   convolve each piece with the impulse response h using the FFT, and
(iii) patch the resulting pieces of output correctly together.

For *off-line* signal processing this procedure results both in a reduction of computation time and of storage requirements. The procedure allows *on-line* signal processing provided the processing time needed for step (ii) is less than the time each chunk of input takes to arrive at the processor. The processing delay equals precisely this time.

We briefly describe a simple way of breaking up the input u and patching the output back together that is known as the *overlap-add* method. Another procedure, called the *overlap-save* method, is discussed in 9.8.8. Consider the computation of the discrete-time convolution

$$y = h * u,$$

with h and u both defined on the discrete time axis $\mathbb{Z}$. We assume that the impulse response h is non-anticipating and of finite length M. Supposing the input u to have much longer or even infinite length, we break it up as in Fig. 9.41 into pieces u_k, $k \in \mathbb{Z}$, of length K, given by

$$u_k(n) = \begin{cases} u(n + kK) & \text{for } n = 0, 1, 2, \cdots, K - 1, \\ 0 & \text{otherwise,} \end{cases} \quad n \in \mathbb{Z}.$$

It follows that

$$u = \sum_{k \in \mathbb{Z}} \sigma^{-kK} u_k,$$

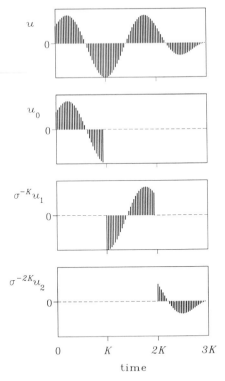

Figure 9.41. Breaking the signal u up into pieces u_k, $k \in \mathbb{Z}$.

with σ the backward shift operator. Using the linearity and shift properties of the convolution we have

$$y = h * u = h * \left(\sum_{k \in \mathbb{Z}} \sigma^{-kK} u_k \right) = \sum_{k \in \mathbb{Z}} h * (\sigma^{-kK} u_k) = \sum_{k \in \mathbb{Z}} \sigma^{-kK} (h * u_k)$$

$$= \sum_{k \in \mathbb{Z}} \sigma^{kK} y_k,$$

where

$$y_k := h * u_k, \qquad k \in \mathbb{Z}.$$

This shows that the output signal y may be written as the sum of the shifted signals y_k, $k \in \mathbb{Z}$.

By assumption, h has support $[0, M - 1]$, while that of u_k is $[0, K - 1]$. As a result, by 3.5.5 the support of y_k is contained in $[0, M + K - 2]$. If $M < K$, the positions of the shifted signals $\sigma^{kK} y_k$ are as in Fig. 9.42. At any time n the value of $y(n)$ follows by adding signal values from at most two shifted "pieces" y_k.

The convolutions $y_k = h * u_k$ may be computed using fast convolution as described earlier, based on the FFT. The DDFT of h of course need only be computed once.

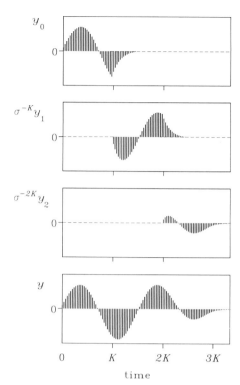

Figure 9.42. Reconstruction of y from the pieces y_k, $k \in \mathbb{Z}$.

9.8 PROBLEMS

The first two problems are connected with the sampling theorem.

9.8.1. Two integrals involving the sinc function. Compute the following two integrals:

(a) $\displaystyle\int_{-\infty}^{\infty} \operatorname{sinc}(x)\, dx,$

(b) $\displaystyle\int_{-\infty}^{\infty} |\operatorname{sinc}(x)|^2\, dx.$

Hint: The function $\operatorname{sinc}(\pi t)$, $t \in \mathbb{R}$, is the inverse CCFT of $\operatorname{rect}(f)$, $f \in \mathbb{R}$.

9.8.2. Band-limited input to a continuous-time convolution system. Suppose that the input u to a continuous-time convolution system with frequency response function $\hat{h}$ is band-limited with bandwidth B. It follows from the sampling theorem that u may be sampled with sampling interval T without loss of information provided $T \leq 1/2B$. Let the discrete-time signal u^*, defined on the time axis $\mathbb{Z}(T)$, be the sampled input given by

$$u^*(nT) = u(nT), \qquad n \in \mathbb{Z}.$$

Similarly, y^* is the sampled output y of the system, that is,

$$y*(nT) = y(nT), \qquad n \in \mathbb{Z}.$$

Prove that $y*$ may be obtained by discrete-time (sampled) convolution as

$$y* = h* * u*.$$

Here, $h*$ is the inverse DCFT of $\hat{h}*$, which on $[-1/2T, 1/2T)$ is given by

$$\hat{h}*(f) = \hat{h}(f), \qquad -1/2T \leq f \leq 1/2T,$$

and is periodically continued outside this interval. If also $\hat{h}$ is band-limited with bandwidth less than $1/2T$, $h*$ simply follows by sampling the impulse response h of the convolution system.

The next two problems concern windows and windowing.

9.8.3. Time and frequency windows. Derive the expressions for the frequency windows listed in Table 9.1.

9.8.4. Truncation of IIR filters. Given a discrete-time IIR filter whose impulse response has finite action, an approximate FIR realization of the filter may be obtained by simply *truncating* the infinite impulse response after the first N coefficients.

    (a) Show what the effect of truncation on the frequency response function of the filter is.

    (b) Discuss how the FIR filter could be modified to alter the effects of truncation.

The following two problems deal with the approximation of continuous-time systems by discrete-time systems.

9.8.5. State space versions of delta and step equivalence approximation. Suppose that we wish to find a discrete-time approximation to a continuous-time system described by the state differential and output equations

$$\dot{x}(t) = Ax(t) + Bu(t), \qquad y(t) = Cx(t), \qquad t \in \mathbb{R},$$

with A, B, and C constant matrices.

    (a) *Delta equivalence approximation.* Prove that the delta equivalence approximation to the continuous-time system is described by the state difference and output equations

$$\begin{aligned} x*(n + 1) &= A*x*(n) + B*u*(n), \\ y*(n) &= C*x*(n) + D*u*(n), \end{aligned} \qquad n \in \mathbb{Z}, \qquad (1)$$

where

$$u*(n) = u(nT), \qquad y*(n) = y(nT), \qquad n \in \mathbb{Z},$$

and

$$A^* = e^{AT}, \qquad B^* = e^{AT}BT, \qquad C^* = C, \qquad D^* = CBT.$$

(b) *Step equivalence approximation.* Similarly, prove that the step equivalence approximation is again of the form (1), but now with

$$A^* = e^{AT}, \qquad B^* = \int_0^T e^{A(T-\tau)}B \, d\tau, \qquad C^* = C, \qquad D^* = 0.$$

Find an infinite series expansion for B^*.

9.8.6. Discrete-time approximations of a continuous-time system. Find

(a) the delta equivalence approximation,
(b) the step equivalence approximation,
(c) the staircase approximation, and
(d) the trapezoidal approximation

of the continuous-time system with transfer function

$$H(s) = \frac{2}{(s+1)(s+2)}, \qquad \text{Re } (s) > -1.$$

In the following problem the converse sampling theorem is considered.

9.8.7. The converse sampling theorem. (Compare 9.7.7.)

(a) *The frequency sampling theorem.* Suppose that z is a continuous-time signal of finite duration, that is,

$$z(t) = 0 \text{ for } |t| > P,$$

with P a fixed positive number. Prove by interchanging time and frequency in the sampling theorem that the CCFT $\hat{z}$ may be recovered from its sampled values with frequency interval $F = 1/2P$ by sinc interpolation as

$$\hat{z}(f) = \sum_{k=-\infty}^{\infty} \hat{z}(kF) \text{ sinc} \left(\frac{\pi(f - kF)}{F} \right), \qquad f \in \mathbb{R}.$$

(b) *Converse sampling theorem for discrete-time signals.* Suppose that x is a discrete-time signal defined on the time axis $\mathbb{Z}(T)$ such that

$$x(t) = 0 \text{ for } |t| > P,$$

with P a fixed positive number. Prove that the DCFT $\hat{x}$ (taken on the infinite frequency axis $\mathbb{R}$) may be recovered from its sampled values with frequency interval $F = 1/2P$ by sinc interpolation as

$$\hat{x}(f) = \sum_{k=-\infty}^{\infty} \hat{x}(kF) \text{ sinc} \left(\frac{\pi(f - kF)}{F} \right), \qquad f \in \mathbb{R}.$$

Hint: Apply (a) to the "delta interpolated" continuous-time signal z given by

$$z(t) = T \sum_{n=-\infty}^{\infty} x(nT)\delta(t - nT), \qquad t \in \mathbb{R}.$$

(c) *Application to numerical Fourier transformation.* Suppose that x is a discrete-time signal defined on the time axis $\mathbb{Z}(T)$ such that x is zero outside the interval $[-N/2\,T, (N/2 - 1)T]$, with N even. By (b), computing the DCFT $\hat{x}$ of x on a discrete frequency axis with frequency interval $1/NT$ causes no loss of information. Show that this computation may be reduced to performing a DDFT, and, hence, may be done by using the FFT.

The final problem offers an alternative method for fast convolution.

9.8.8. Fast convolution by the overlap-save method. An alternative for the overlap-add method for fast convolution, discussed at the end of Section 9.7, is the *overlap-save* method. Again, we consider the discrete-time convolution $y = h * u$, where h has a finite and relatively small length M, and u may be a very long signal.

(a) Suppose that u_k is a segment of the signal u of length K, with $K > M$, that is, u_k equals u on an interval of length K, and is zero otherwise. Show that the convolution $h * u_k$ generally differs from $h * u$ on the first $M - 1$ time instants of the interval on which u_k is nonzero but agrees with it on the rest of the interval.

(b) Let u_{k-1} be the segment that *precedes* u_k, also defined on an interval of length K, which *overlaps* by M time instants with the interval on which u_k is nonzero. Show that $h * u_{k-1}$ coincides with $h * u$ on the first M time instants of the interval on which u_k is nonzero. Is there any constraint on K?

(c) Explain the overlap-save method for fast convolution.

9.9 COMPUTER EXERCISES

The first exercises in this section deal with various aspects of signal processing.

9.9.1. Interpolation of band-limited signals. According to the sampling theorem, continuous-time signals with bandwidth B may be sampled, with sampling interval T, and later interpolated with sinc functions without any loss of information, provided $T < 1/2B$. Signals that are not band-limited to $[-B, B]$ are not recovered exactly, but if the signal does not have much energy outside this band the error is small.

(a) *Interpolation of a band-limited signal.* Show that the signal x defined by

$$x(t) = [\text{sinc}(\pi t)]^4, \qquad t \in \mathbb{R},$$

has finite bandwidth B. What is B? Define the signal on a sufficiently long time axis. Sample the signal with sampling interval $T = 1/2B$, and interpolate it with sinc functions. Explain the interpolation error, if any. *Hint:* See 3.11.3 on how to interpolate.

(b) *Interpolation of a signal that does not have finite bandwidth.* The signal y, defined by

$$y(t) = e^{-\frac{t^2}{2}}, \qquad t \in \mathbb{R},$$

does not have finite bandwidth, but its CCFT decays very fast. Define the signal on a sufficiently long time axis. Choose $B = 0.5$, and check that the CCFT of y is small outside the frequency band $[-B, B]$. Sample the signal with the interval $T = 1/2B$, and interpolate the sampled signals with sinc functions. Comment on the accuracy. *Hint:* Compare 7.6.12.

9.9.2. Windowing. As explained in Section 9.4, windowing may soften the effect of truncation. We consider time and frequency windowing in this exercise.

(a) *Time windowing.* The CCFT of the continuous-time real harmonic signal

$$x(t) = \operatorname{sinc}(\pi t), \qquad t \in \mathbb{R},$$

is 1 between $-\frac{1}{2}$ and $\frac{1}{2}$, and 0 elsewhere. Because of the slow decay of x it is difficult to establish this numerically. Define the signal x with sufficient resolution on a two-sided time axis of length 16, say. Compute and plot the CCFT. Identify the effect of truncation. To mitigate this effect, apply a triangular multiplicative time window to the signal x, and compute the CCFT of the windowed signal. Discuss the effect of the window.

(b) *Frequency windowing.* In 6.7.2(c) Gibbs's phenomenon is investigated for the periodic sawtooth. Truncating the infinite Fourier series is equivalent to applying a rectangular window in the frequency domain. Study the effect of applying a triangular window to the truncated Fourier coefficients, again for $N = 10$ and 20 as in 6.7.2(c). *Hint:* Compare 9.4.4.

9.9.3. Zero padding. Suppose that the discrete-time signal x, defined on $\mathbb{Z}$, is zero outside some interval $[-N, N]$. Then by the converse sampling theorem (see 9.8.7) without loss of information the DCFT of x may be computed on a discrete frequency axis with frequency interval $F = 1/2N$. Values between the sampling points follow by sinc interpolation. The DCFT on the discrete frequency axis may be found by application of the DDFT to the restricted signal x after cyclical shifting to $[0, 2N - 1]$.

Suppose that x is *padded with zeros,* that is, the signal is considered on the interval $[-M, M]$, with $M > N$, and its DCFT is computed by cyclical shifting to $[0, 2M - 1]$ and application of the DDFT. Since the DCFT is obtained on $2M$ frequency points, the result of padding is that its resolution increases. The increase of resolution constitutes no gain of information, because the entire DCFT may again be reconstructed by sinc interpolation.

We consider the discrete-time rectangular pulse x defined by

$$x(n) = \begin{cases} 1 & |n| \le N, \\ 0 & \text{otherwise,} \end{cases} \qquad n \in \mathbb{Z},$$

with $N = 7$.

(a) Determine the DCFT of x analytically. *Hint:* Compare 7.3.11.

(b) Define x on the finite time axis $\{-8, -7, \cdots, 7\}$ and compute its DCFT. This yields 16 points of the DCFT on the frequency axis $[0, 1)$. Compare the values you find with the analytical result.

(c) Redefine x on the finite time axes $\{-M, -M + 1, \cdots, M - 1\}$, for $M = 16$, 32, 64, 128, and compute the DCFT's of these signals. Verify that each time the length of the time axis is doubled one new point of the DCFT is obtained between each two points of the previous computation. Check that these new values agree with the theoretical values.

The final two exercises concern the design of digital filters.

9.9.4. Design of an FIR digital filter. In this exercise we consider the design of an FIR filter with the triangular frequency response function

$$
\hat{h}_o(f) = \begin{cases} \dfrac{f - f_1}{f_2 - f_1} & \text{for } f_1 \le f < f_2, \\[2mm] \dfrac{f_3 - f}{f_3 - f_2} & \text{for } f_2 \le f < f_3, \\[2mm] 0 & \text{otherwise,} \end{cases} \qquad f \in [0, f_s/2),
$$

with $0 \le f_1 \le f_2 \le f_3 \le f_s/2$, where $f_s = 1/T$ is the sampling rate. A plot of the desired frequency response function is given in Fig. 9.43. A filter with this frequency response function might be one of the band-pass filters in a multi-band audio equalizer (see 3.9.6).

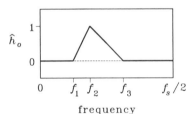

Figure 9.43. Desired frequency response function for an FIR digital filter.

The purpose of this exercise is to design a digital filter with this frequency response function by using the window method. Let $f_s = 50$ kHz, $f_1 = 1$ kHz, $f_2 = 2$ kHz and $f_3 = 4$ kHz. The filter tolerance specification is as follows. Let $\hat{h}_w$ be the frequency response function after windowing but *before* shifting (i.e., $\hat{h}_w$ is the DCFT of the windowed impulse response h_w of Section 9.5). Then we require that $|\hat{h}_w(f) - \hat{h}_o(f)| \le 0.1$ for all $f \in [0, f_s/2)$.

(a) Represent $\hat{h}_o$ on $[-f_s/2, f_s/2)$ with sufficient resolution, and compute its inverse DCFT h_o. *Hint:* Because $\hat{h}_o$ is represented on a "two-sided" frequency axis, use the function iccft in SIGSYS.

(b) To reduce the length of the filter, select a suitable time window w_{MT} with an appropriate width $2MT$. Hann's and Hamming's windows have small side lobes. Compute the resulting frequency response function $\hat{h}_w$. Check whether the tolerance specification is satisfied. If not, adapt the window width. Plot the magnitude of the frequency response function that is finally obtained.

(c) When the specifications are satisfied, shift the impulse response to obtain a real-izable filter and compute the pulse response $h*$. Plot $h*$. How much processing delay is incurred?

9.9.5. Design of an IIR digital filter. In this exercise we consider the design of an IIR bandpass filter by trapezoidal approximation of a bandpass Butterworth filter. A bandpass Butterworth filter is a filter with transfer function

$$H(s) = \frac{1}{D_N((s - j\omega_c)/\omega_o)} + \frac{1}{D_N((s + j\omega_c)/\omega_o)},$$

with ω_c the center angular frequency, ω_o the bandwidth (in terms of angular fre-quency), and D_N an Nth order Butterworth polynomial (see 9.6.7). Trapezoidal ap-proximation results in the discrete-time filter with transfer function

$$H*(z) = H\left(\frac{2}{T}\frac{z - 1}{z + 1}\right),$$

with T the sampling interval. The digital filter operates at the sampling frequency $f_s = 50$ kHz. Its frequency response function is required to have the center fre-quency $f_{cd} = 10$ kHz and a bandwidth $B_d = 2$ kHz.

(a) It is shown in Section 9.6 that the frequency response function of the trapezoidal approximation is given by

$$\hat{h}*(f) = \hat{h}\left(\frac{\tan(\pi f T)}{\pi T}\right), \qquad |f| < f_s/2,$$

where $\hat{h}$ is the frequency response function of the continuous-time system. This shows that by prewarping the center frequency $f_{cc} = \omega_c/2\pi$ of the continuous-time filter and the center frequency of the digital filter f_{cd} are related as

$$f_{cc} = \frac{\tan(\pi f_{cd} T)}{\pi T}.$$

Use this relation to compute the center angular frequency ω_c of the continuous-time filter. Let $B_c = \omega_o/2\pi$ be the bandwidth of the continuous-time filter. Then to obtain the correct bandwidth for the digital filter the frequencies $f_{cc} \pm B_c$ and $f_{cd} \pm B_d$ should be related as

$$f_{cc} \pm B_c = \frac{\tan[\pi(f_{cd} \pm B_d)T]}{\pi T}.$$

Because f_{cc} has already been fixed, only one of the two equalities may be satis-fied, say that for the $+$ sign. Use this to compute the bandwidth B_c of the continuous-time filter and from this ω_o.

(b) Given the parameters ω_c and ω_o, compute the frequency response function of the continuous-time Butterworth function. Plot its magnitude. *Hint:* Given the frequencies f_{cc} and B_c choose the SIGSYS parameters INC and num to define a suitable frequency axis DOTPLUS. Generate the polynomial D_4. Compute the arguments of the two denominators of H as (j*2*pi*DOTPLUS±j*omc)/omo. Finally compute $\hat{h}$ by substituting the arguments.

(c) Similarly, compute the frequency response function $\hat{h}*$ of the digital filter, and plot its magnitude. Comment on the shape. *Hint:* Choose the SIGSYS parameters INC and num such that DOTPLUS represents the discrete-time frequency axis $[0, f_s/2)$ with sufficient resolution. Compute the behavior of z on the unit circle as z=exp(j*2*pi*DOTPLUS*T). Next compute the argument of the trapezoidal approximation as s=2*(z−1)/(T*(z+1)). Finally, substitute s into the transfer function of the Butterworth filter after first computing the arguments (s±j*omc)/omo.

(d) If it is desired to compute the actual transfer function $H*$ of the digital filter it is useful to write first a simple SIGSYS procedure that substitutes the rational function $s = P(z)/Q(z)$ into the rational function $N(s)/D(s)$. *Hint:* If the polynomials N and D are given by

$$N(s) = n_0 + n_1 s + \cdots + n_K s^K,$$

$$D(s) = d_0 + d_1 s + \cdots + d_K s^K,$$

the substitution results in the rational function R/S, with

$$R = \sum_{i=0}^{K} n_i Q^{K-i} P^i, \qquad S = \sum_{i=0}^{K} d_i Q^{K-i} P^i.$$

10

Applications to Communication

10.1 INTRODUCTION

In this chapter we explain the principles of various electronic communication systems such as transmission by amplitude modulation (AM) and frequency modulation (FM). Low-frequency messages cannot be transmitted over long distances unless they are "mounted" on a high-frequency carrier. This mounting process is called *modulation*. The energy of modulated signals is concentrated about the carrier frequency. Signals with this property are called *narrow-band* signals. Section 10.2 presents tools for analyzing them. These tools are based on Fourier analysis and permit the description of narrow-band signals as harmonic signals with slowly varying amplitude and phase.

In Section 10.3 we discuss various practical modulation and demodulation schemes, including not only AM and FM but also single-side band (SSB) modulation and demodulation. Most of the material in this section may be understood without studying Section 10.2. We end the chapter by briefly describing in Section 10.4 *multiplexing* techniques, which allow the simultaneous transmission of *several* message signals along a single communication channel.

10.2 NARROW-BAND SIGNALS

In this section we introduce *narrow-band* signals and present techniques for handling such signals analytically and numerically. The most extreme example of a narrow-band signal is a real harmonic x given by

$$x(t) = \alpha \cos (2\pi f_o t + \phi), \qquad t \in \mathbb{R},$$

where the real nonnegative constant α is the *amplitude* and the real constant ϕ the *phase* of the harmonic. By 7.4.19 the continuous-to-continuous Fourier transform (CCFT) $\hat{x}$ of this signal is given by

$$\hat{x}(f) = \tfrac{1}{2}\alpha e^{j\phi}\delta(f - f_o) + \tfrac{1}{2}\alpha e^{-j\phi}\delta(f + f_o), \qquad f \in \mathbb{R},$$

which shows that x has *all* its energy concentrated at the frequencies $\pm f_o$.

Signals that do not have their energy concentrated exactly at a single frequency but in a narrow band about it look like harmonic signals with varying amplitude and phase. We illustrate this by two examples.

10.2.1. Examples: Narrow-band signals.

(a) Consider a signal z whose CCFT $\hat{z}$ is given by

$$\hat{z}(f) = \begin{cases} 1/B & \text{for } f_o - B \le |f| < f_o + B, \\ 0 & \text{otherwise,} \end{cases} \qquad f \in \mathbb{R},$$

as sketched in Fig. 10.1. The time signal z may easily be found by taking the inverse CCFT of $\hat{z}$, and is of the form

$$z(t) = \text{sinc}(2\pi Bt)\cos(2\pi f_o t), \qquad t \in \mathbb{R},$$

also plotted in Fig. 10.1. The signal z may be considered as the product $z = z_m z_o$ of the real harmonic signal $z_o(t) = \cos(2\pi f_o t)$, $t \in \mathbb{R}$, and the *time-varying amplitude*

$$z_m(t) = \text{sinc}(2\pi Bt), \qquad t \in \mathbb{R}.$$

If $B << f_o$, then the amplitude z_m varies *slowly* compared with the harmonic z_o.

(b) Another example of a narrow-band signal is the signal z given by

$$z(t) = \begin{cases} \cos(2\pi f_o t) & \text{for } |t| \le T/2, \\ 0 & \text{otherwise,} \end{cases} \qquad t \in \mathbb{R},$$

as sketched in Fig. 10.2. The CCFT of the signal is

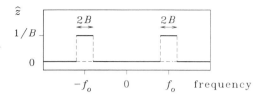

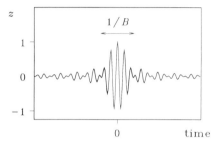

Figure 10.1. Narrow-band signal. Top: CCFT of a narrow-band signal. Bottom: the signal itself.

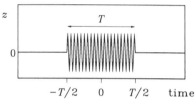

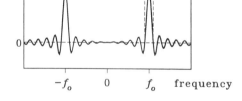

Figure 10.2. Another narrow-band signal. Top: a time signal. Bottom: its CCFT.

$$\hat{z}(f) = \frac{T}{2}(\text{sinc }(\pi(f_o - f)T) + \text{sinc }(\pi(f + f_o)T)), \qquad f \in \mathbb{R},$$

and is also shown in Fig. 10.2. If $f_o \gg 1/T$, the energy of z is concentrated in the proximity of the frequencies $\pm f_o$. ∎

Example 10.2.1(a) illustrates that if the CCFT of a real signal z is concentrated near the frequencies $\pm f_o$, the time signal is a real harmonic signal with frequency f_o and "slowly" varying amplitude and phase. Vice-versa, Example 10.2.1(b) suggests that if a time signal z is a real harmonic with frequency f_o whose amplitude and phase vary slowly, then its CCFT consists of two peaks, one near the frequency $-f_o$ and the other near f_o. In what follows these notions will be made precise by developing the concept of *complex envelope*.

Complex Envelope

The complex envelope may be seen as an extension of the notion of the *phasor* of a real harmonic signal, which we briefly review. The real harmonic signal x, given by

$$x(t) = \alpha_x \cos(2\pi f_o t + \phi_x), \qquad t \in \mathbb{R},$$

with amplitude $\alpha_x \geq 0$ and phase $\phi_x \in [0, 2\pi)$, may equivalently be written as

$$x(t) = \text{Re}\,(\alpha_x e^{j\phi_x} e^{2\pi f_o t}), \qquad t \in \mathbb{R}.$$

The complex number $\alpha_x e^{j\phi_x}$ is the *phasor* representing the harmonic signal. Its magnitude is the amplitude α_x, and its argument the phase ϕ_x of the harmonic signal.

The phasor $\alpha_x e^{j\phi_x}$ may be derived from the CCFT

$$\hat{x}(f) = \tfrac{1}{2}(\alpha_x e^{j\phi_x})\delta(f - f_o) + \tfrac{1}{2}(\alpha_x e^{-j\phi_x})\delta(f + f_o), \qquad f \in \mathbb{R},$$

of x as follows:

(i) Suppress the negative frequency part of $\hat{x}$ and multiply by two to obtain the frequency signal $\hat{x}_p$ given by

$$\hat{x}_p(f) = \alpha_x e^{j\phi_x}\delta(f - f_o), \qquad f \in \mathbb{R}.$$

(ii) Shift $\hat{x}_p$ to the left by f_o to obtain the frequency signal $\hat{x}_e$ given by

$$\hat{x}_e(f) = \alpha_x e^{j\phi_x}\delta(f), \qquad f \in \mathbb{R}.$$

(iii) Apply the inverse CCFT to $\hat{x}_e$ to obtain the constant time signal x_e given by

$$x_e(t) = \alpha_x e^{j\phi_x}, \qquad t \in \mathbb{R},$$

which is the phasor of x.

Application of this sequence of operations to *any* real continuous-time signal x results in a time signal x_e that is called the *complex envelope* of x.

10.2.2.   Definition: Pre-envelope and complex envelope. Let x be a real continuous-time signal whose CCFT $\hat{x}$ has no singularity (i.e., no delta function) at the frequency zero.

(a) The *pre-envelope* of x is the (complex-valued) signal x_p with CCFT

$$\hat{x}_p(f) = 2\hat{x}(f)\mathbb{1}(f), \qquad f \in \mathbb{R}.$$

(b) The *complex envelope of x with respect to the frequency* $f_o \in \mathbb{R}_+$ is the signal x_e with CCFT

$$\hat{x}_e(f) = \hat{x}_p(f + f_o) = 2\hat{x}(f + f_o)\mathbb{1}(f + f_o), \qquad f \in \mathbb{R}.$$ ∎

The definition specifies the pre-envelope x_p and complex envelope x_e of x in terms of their CCFTs. The time-domain relations between the signal x, its pre-envelope, and its complex envelope are less simple than the frequency-domain relations and are dealt with in 10.2.3. The pre-envelope x_p is sometimes called the *analytical signal* corresponding to x.

The results of this section are valid for *any* real signal x, but most useful if x is narrow-band. Let us define this notion. The real signal x is called *band-pass band-limited,* with *center frequency* f_c and *bandwidth B*, if the CCFT $\hat{x}$ of x satisfies

$$\hat{x}(f) = 0 \text{ for } \begin{cases} |f| < f_c - B, \\ |f| > f_c + B, \end{cases} \quad f \in \mathbb{R},$$

with B and f_c positive real numbers such that $B < f_c$. Figure 10.3 illustrates the CCFT $\hat{x}$ of a band-pass band-limited signal. Note that the center frequency f_c and the bandwidth are not uniquely defined, unless B is taken as the *smallest* real number such that the definition is satisfied. If the bandwidth B is much smaller than the center frequency f_c, then the signal is said to be *narrow-band*.

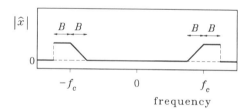

Figure 10.3. CCFT of a band-pass band-limited signal with center frequency f_c and bandwidth B.

Figure 10.4 shows the CCFTs of the pre-envelope and complex envelope when the signal x is a band-pass band-limited signal and f_o a frequency within the positive frequency band of x. It is seen that the complex envelope x_e contains low frequencies only. Note that because the frequency $f_o \in \mathbb{R}_+$ may be arbitrarily chosen the complex envelope is not uniquely defined.

Given the complex envelope x_e of a real signal x with respect to a given frequency f_o, the CCFTs of the pre-envelope and of the signal x itself may successively be retrieved as

$$\hat{x}_p(f) = \hat{x}_e(f - f_o), \qquad f \in \mathbb{R},$$
$$\hat{x}(f) = \tfrac{1}{2}[\hat{x}_p(f) + \overline{\hat{x}_p(-f)}] = \tfrac{1}{2}[\hat{x}_e(f - f_o) + \overline{\hat{x}_e(-f + f_o)}], \qquad f \in \mathbb{R}.$$

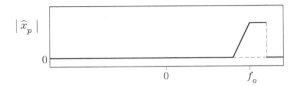

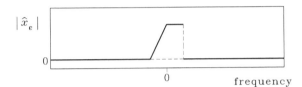

Figure 10.4. Top: the CCFT of a band-pass band-limited signal. Middle: the CCFT of its pre-envelope. Bottom: the CCFT of its complex envelope with respect to the frequency f_o.

This shows that determining the complex envelope of a real continuous-time signal is an invertible operation, called the *complex envelope transformation*.

The following theorem shows the time-domain relationships between the signal x, its pre-envelope x_p and its complex envelope x_e with respect to the frequency f_o.

10.2.3.   Summary: Pre-envelope and complex envelope.

(a) The pre-envelope x_p of a real continuous-time signal x is

$$x_p = x + j\overset{\circ}{x}$$

where the real signal $\overset{\circ}{x}$ is the *Hilbert transform* of x, given by

$$\overset{\circ}{x}(t) = \frac{1}{\pi} \int_{-\infty}^{\infty} \frac{x(\tau)}{t - \tau}\, d\tau, \qquad t \in \mathbb{R}.$$

The CCFT of the Hilbert transform $\overset{\circ}{x}$ of x is

$$\overset{\circ}{\hat{x}}(f) = -j \operatorname{sign}(f)\hat{x}(f), \qquad f \in \mathbb{R}. \tag{1}$$

(b) The complex envelope x_e of x with respect to the frequency $f_o \in \mathbb{R}_+$ is

$$x_e(t) = x_p(t)e^{-j2\pi f_o t} = [x(t) + j\overset{\circ}{x}(t)]e^{-j2\pi f_o t}, \qquad t \in \mathbb{R}.$$

(c) The signal x may be expressed in terms of its complex envelope x_e with respect to f_o as

$$x(t) = \text{Re } [x_e(t)e^{j2\pi f_o t}], \qquad t \in \mathbb{R}.$$ ∎

The German mathematician David Hilbert (1862–1943) was one of the greatest mathematicians of the first half of the 20th century. At the international congress of mathematicians in Paris in 1900 he put forth 23 problems as the targets for mathematics for the 20th century. Some of these problems have remained unsolved until today.

The integral in 10.2.3(a) denotes the principal value in the sense of Cauchy, that is,

$$\fint_{-\infty}^{\infty} = \lim_{\epsilon \downarrow 0} \left(\int_{-\infty}^{-\epsilon} + \int_{\epsilon}^{\infty} \right).$$

10.2.4.   Proof of 10.2.3.

(a) We have

$$\hat{x}_p(f) = 2\hat{x}(f)\mathbb{1}(f) = \hat{x}(f) + \text{sign } (f)\hat{x}(f)$$
$$= \hat{x}(f) + j\overset{\circ}{\hat{x}}(f), \qquad f \in \mathbb{R},$$

where

$$\overset{\circ}{\hat{x}}(f) = -j \text{ sign } (f)\hat{x}(f), \qquad f \in \mathbb{R}.$$

This proves (1). From 7.4.20 we have the CCFT pair

$$\text{sign } (t), \qquad t \in \mathbb{R}, \qquad\qquad 1/j\pi f, \qquad f \in \mathbb{R}.$$

By interchanging time and frequency we obtain by 7.6.16(a) the CCFT pair

$$1/j\pi t, \qquad t \in \mathbb{R}, \qquad\qquad -\text{sign } (f), \qquad f \in \mathbb{R}.$$

Hence, the inverse CCFT of $-j$ sign $(f), f \in \mathbb{R}$, is $1/\pi t, t \in \mathbb{R}$, and by the converse convolution property of the CCFT we have from (1)

$$\overset{\circ}{x}(t) = \frac{1}{\pi} \int_{-\infty}^{\infty} \frac{x(\tau)}{t - \tau} \, d\tau, \qquad t \in \mathbb{R}.$$

The singularity at zero of the function $1/\pi t, t \in \mathbb{R}$, necessitates taking the principal value of the integral.

(b) The second part of the theorem is an immediate consequence of the relation $\hat{x}_e(f) = \hat{x}_p(f + f_o), f \in \mathbb{R}$, and the converse shift property of the CCFT.

(c) Combining (a) and (b) it follows that

$$x(t) = \text{Re } [x_p(t)] = \text{Re } [x_e(t)e^{j2\pi f_o t}], \qquad t \in \mathbb{R}. \qquad \blacksquare$$

We illustrate the notions of pre-envelope and complex envelope by some examples.

10.2.5.   Examples: Pre-envelope, Hilbert transform, and complex envelope.

(a) *A pure harmonic.* The CCFT of the purely harmonic real signal

$$x(t) = \alpha \cos (2\pi f_1 t + \phi), \qquad t \in \mathbb{R},$$

with $\alpha \geq 0$ and $0 \leq \phi < 2\pi$, is

$$\hat{x}(f) = \tfrac{1}{2}\alpha e^{j\phi}\delta(f - f_1) + \tfrac{1}{2}\alpha e^{-j\phi}\delta(f + f_1), \qquad f \in \mathbb{R}.$$

As a result, for $f_1 > 0$ the CCFT $\hat{x}_p$ of the pre-envelope and the pre-envelope itself are given by

$$\hat{x}_p(f) = 2\hat{x}(f)\mathbb{1}(f) = \alpha e^{j\phi}\delta(f - f_1), \qquad f \in \mathbb{R},$$
$$x_p(t) = \alpha e^{j\phi}e^{j2\pi f_1 t}, \qquad t \in \mathbb{R}.$$

The CCFT of the Hilbert transform $\overset{\circ}{x}$ of x and the Hilbert transform itself are

$$\overset{\hat{\circ}}{x}(f) = -j \text{ sign } (f)\hat{x}(f) = \frac{\alpha}{2j}e^{j\phi}\delta(f - f_1) - \frac{\alpha}{2j}e^{-j\phi}\delta(f + f_1), \qquad f \in \mathbb{R},$$
$$\overset{\circ}{x}(t) = \alpha \sin (2\pi f_1 t + \phi), \qquad t \in \mathbb{R}.$$

The CCFT of the complex envelope x_e with respect to the frequency $f_o > 0$ and the envelope itself are

$$\hat{x}_e(f) = \hat{x}_p(f + f_o) = \alpha e^{j\phi}\delta(f - f_1 + f_o), \qquad f \in \mathbb{R},$$
$$x_e(t) = \alpha e^{j\phi}e^{j2\pi(f_1 - f_o)t}, \qquad t \in \mathbb{R}.$$

If $f_1 = f_o$, then the complex envelope is the phasor of the real harmonic signal x. If $f_1 \neq f_o$ such that $|f_1 - f_o| << f_o$, then the complex envelope x_e is not constant but varies slowly compared with $e^{j2\pi f_o t}, t \in \mathbb{R}$.

(b) *Amplitude modulated harmonic.* Let x be the signal given by

$$x(t) = g(t) \cos (2\pi f_o t + \phi), \qquad t \in \mathbb{R},$$

where g is real and low-pass with bandwidth B such that $B \leq f_o$. The CCFTs of the various signals associated with the complex envelope are qualitatively those of Fig. 10.4. The CCFT of x is

$$\hat{x}(f) = \tfrac{1}{2}[e^{j\phi}\hat{g}(f - f_o) + e^{-j\phi}\hat{g}(f + f_o)], \qquad f \in \mathbb{R}.$$

By using the converse shift property of the CCFT and the fact that $B \leq f_o$, we find that the pre-envelope of x is

$$\hat{x}_p(f) = e^{j\phi}\hat{g}(f - f_o), \qquad f \in \mathbb{R},$$

$$x_p(t) = g(t)e^{j(2\pi f_o t + \phi)}, \qquad t \in \mathbb{R}.$$

Again with the converse shift property and the fact that $B \leq f_o$ we obtain the Hilbert transform of x as

$$\overset{\circ}{\hat{x}}(f) = \frac{1}{2j}[e^{j\phi}\hat{g}(f - f_o) - e^{-j\phi}\hat{g}(f + f_o)], \qquad f \in \mathbb{R},$$

$$\overset{\circ}{x}(t) = g(t)\sin(2\pi f_o t + \phi), \qquad t \in \mathbb{R}.$$

The complex envelope of x with respect to f_o is

$$\hat{x}_e(f) = e^{j\phi}\hat{g}(f), \qquad f \in \mathbb{R},$$

$$x_e(t) = g(t)e^{j\phi}, \qquad t \in \mathbb{R}.$$

If the bandwidth B is small compared with f_o, then the signal is narrow-band and the complex envelope x_e varies slowly with respect to the harmonic $e^{j2\pi f_o t}$, $t \in \mathbb{R}$.

   (c) *Phase modulated harmonic.* As a final example, let x be the signal

$$x(t) = \cos[2\pi f_o t + \phi(t)], \qquad t \in \mathbb{R},$$

with ϕ a real signal such that the complex signal g given by

$$g(t) = e^{j\phi(t)}, \qquad t \in \mathbb{R},$$

is low-pass with bandwidth B such that $B \leq f_o$. Writing

$$x(t) = \tfrac{1}{2}[e^{j[2\pi f_o t + \phi(t)]} - e^{-j[2\pi f_o t + \phi(t)]}]$$

$$= \tfrac{1}{2}[g(t)e^{j2\pi f_o t} + \overline{g(t)}e^{-j2\pi f_o t}], \qquad t \in \mathbb{R},$$

and by using 7.6.17 (CCFT of the complex conjugate) and the converse shift property we find that the CCFT of x is

$$\hat{x}(f) = \tfrac{1}{2}[\hat{g}(f - f_o) + \overline{\hat{g}(-f - f_o)}], \qquad f \in \mathbb{R}.$$

Because by assumption the bandwidth B of g is less than f_o, it follows that the CCFT of the pre-envelope $\hat{x}_p$ and the pre-envelope x_p of x itself are

$$\hat{x}_p(f) = 2\hat{x}(f)\mathbb{1}(f) = \hat{g}(f - f_o), \qquad f \in \mathbb{R},$$

$$x_p(t) = g(t)e^{j2\pi f_o t} = e^{j[2\pi f_o t + \phi(t)]}, \qquad t \in \mathbb{R}.$$

As a result, the Hilbert transform $\overset{\circ}{x}$ of x, the CCFT of the complex envelope and the complex envelope x_e with respect to f_o are

$$\overset{\circ}{x}(t) = \text{Im}\,[x_p(t)] = \sin\,[2\pi f_o t + \phi(t)], \qquad t \in \mathbb{R},$$

$$\hat{x}_e(f) = \hat{x}_p(f + f_o) = \hat{g}(f), \qquad f \in \mathbb{R},$$

$$x_e(t) = g(t) = e^{j\phi(t)}, \qquad t \in \mathbb{R}.$$

If the bandwidth B of $e^{j\phi}$ is small compared with f_o, then the complex envelope x_e varies slowly with respect to $e^{j2\pi f_o t}$, $t \in \mathbb{R}$. $\blacksquare$

The Hilbert transform defined in 10.2.3 is usually most conveniently determined by means of the frequency domain relationship (1). In the following we summarize some of the properties of the Hilbert transform.

10.2.6.   Exercise: Hilbert transform.   Prove the following properties of the Hilbert transform.

(a) If it exists, then the Hilbert transform $\overset{\circ}{x}$ of the real continuous-time signal x is again real.

(b) Taken as a map $\mathcal{H}$ from real signals to real signals, the Hilbert transform is linear over the real numbers.

(c) Let $\mathcal{H}(x)$ denote the Hilbert transform of the real signal x. Then,

$$\mathcal{H}(\mathcal{H}(x)) = -x,$$

and, hence, the inverse Hilbert transform is given by

$$\mathcal{H}^{-1} = -\mathcal{H}.$$

(d) Let g and ϕ be real continuous-time signals such that the CCFT of $g(t)e^{j\phi(t)}$, $t \in \mathbb{R}$, is zero outside the frequency band $[-B, B]$, with $0 \le B < f_o$. Then,

$$x(t) = g(t)\cos\,[2\pi f_o t + \phi(t)], \qquad t \in \mathbb{R},$$

$$\overset{\circ}{x}(t) = g(t)\sin\,[2\pi f_o t + \phi(t)], \qquad t \in \mathbb{R},$$

form a Hilbert transform pair. In particular, the following signals form Hilbert transform pairs:

$$c(t) = \cos{(2\pi f_o t + \phi)}, \quad t \in \mathbb{R}, \qquad \overset{\circ}{c}(t) = \sin{(2\pi f_o t + \phi)}, \quad t \in \mathbb{R},$$

$$s(t) = \sin{(2\pi f_o t + \phi)}, \quad t \in \mathbb{R}, \qquad \overset{\circ}{s}(t) = -\cos{(2\pi f_o t + \phi)}, \quad t \in \mathbb{R},$$

with ϕ a constant.

(e) A real signal x and its Hilbert transform $\overset{\circ}{x}$ are orthogonal, that is,

$$\langle x, \overset{\circ}{x} \rangle = 0. \qquad \blacksquare$$

10.2.7.   Remark: Sampling theorem for band-pass band-limited signals.
What follows illustrates the advantage of dealing with the complex envelope of a narrow-band signal instead of the signal itself. As seen in 10.2.3(c), any real signal x may be reconstructed from its complex envelope x_e. Suppose that x is a real band-pass band-limited signal with CCFT as in Fig. 10.3, with center frequency f_c and bandwidth B. Then the complex envelope with respect to the frequency f_c is low-pass with bandwidth B, and by the sampling theorem 9.2.6 the envelope may be sampled with a sampling rate $2B$ without loss of information. This sampling rate is lower than that needed when sampling the narrow-band signal x itself, and even much lower if $B \ll f_c$. Note, however, that since the envelope is complex-valued, sampling involves storing and processing *two* real numbers per sample. $\blacksquare$

Envelope and Phase, In-Phase, and Quadrature Components

The relation

$$x(t) = \text{Re}\,[x_e(t)e^{j2\pi f_o t}], \qquad t \in \mathbb{R},$$

shows that the complex envelope x_e plays the role of *time-varying phasor* for the signal x. This is true for any signal x, but most useful if x is narrow-band and f_o some frequency in the positive frequency band of x. Then, the complex envelope x_e is low-pass and slowly-varying with respect to $e^{j2\pi f_o t}, t \in \mathbb{R}$.

Representing the complex envelope in *polar* and *Cartesian* forms, respectively, leads to useful interpretations.

10.2.8.   Summary: Envelope and phase, in-phase, and quadrature components.
(a) *Envelope and phase.* Let x_e be the complex envelope of the real continuous-time signal x with respect to the frequency f_o. Then the time signals

$$\alpha_x = |x_e|, \qquad \phi_x = \arg{(x_e)},$$

are called the *envelope* and *phase*, respectively, of x with respect to f_o, and one may write

$$x(t) = \alpha_x(t) \cos [2\pi f_o t + \phi_x(t)], \qquad t \in \mathbb{R}.$$

The *instantaneous frequency* f_x of x is the signal

$$f_x = f_o + \frac{1}{2\pi} \dot{\phi}_x,$$

with the overdot denoting the time derivative.

(b) *In-phase and quadrature components.* The signals

$$x_c = \text{Re }(x_e), \qquad x_s = \text{Im }(x_e)$$

are called the *in-phase* and *quadrature* components, respectively, of x with respect to f_o, and

$$x(t) = x_c(t) \cos (2\pi f_o t) - x_s(t) \sin (2\pi f_o t), \qquad t \in \mathbb{R}.$$

(c) *Relation between envelope, phase, in-phase, and quadrature components.* The envelope, phase, in-phase, and quadrature components are related as

$$\alpha_x = \sqrt{x_c^2 + x_s^2}, \qquad \phi_x = \arg (x_c + jx_s),$$
$$x_c = \alpha_x \cos (\phi_x), \qquad x_s = \alpha_x \sin (\phi_x) \qquad\qquad\blacksquare$$

The proof is simple. We continue Example 10.2.5 by determining the envelope, phase, in-phase, and quadrature components of the three signals that are considered.

10.2.9.   Examples: Envelope, phase, in-phase, and quadrature components.

(a) *Purely harmonic signal.* From the complex envelope

$$x_e(t) = \alpha e^{j\phi} e^{j2\pi(f_1 - f_o)t}, \qquad t \in \mathbb{R},$$

of the real harmonic signal

$$x(t) = \alpha \cos (2\pi f_1 t + \phi), \qquad t \in \mathbb{R},$$

with α and ϕ real constants such that $0 \le \phi < 2\pi$, which was obtained in Example 10.2.5(a), we find that the envelope and phase of the signal with respect to the frequency f_o are given by

$$\alpha_x(t) = |x_e(t)| = \alpha,$$

$$\phi_x(t) = \arg(x_e(t)) = \phi + 2\pi(f_1 - f_o)t, \qquad t \in \mathbb{R}.$$

The envelope is constant. If $f_1 = f_o$, also the phase is constant; otherwise, it varies linearly with time. The instantaneous frequency is given by

$$f_x(t) = f_o + \frac{1}{2\pi}\frac{d\phi_x(t)}{dt} = f_1, \qquad t \in \mathbb{R},$$

and is constant. The in-phase and quadrature components are

$$x_c(t) = \text{Re}[x_e(t)] = \alpha\cos[2\pi(f_1 - f_o)t + \phi],$$

$$x_s(t) = \text{Im}[x_e(t)] = \alpha\sin[2\pi(f_1 - f_o)t + \phi], \qquad t \in \mathbb{R}.$$

If $f_1 = f_o$, the in-phase and quadrature components are constant and equal to $\alpha\cos(\phi)$ and $\alpha\sin(\phi)$, respectively. If $f_1 \neq f_o$ such that $|f_1 - f_o| << f_o$, then the two components vary with time but slowly compared with x itself.

(b) *Amplitude modulated harmonic.* From the complex envelope

$$x_e(t) = g(t)e^{j\phi}, \qquad t \in \mathbb{R},$$

of the band-pass band-limited signal

$$x(t) = g(t)\cos(2\pi f_o t + \phi), \qquad t \in \mathbb{R},$$

as obtained in Example 10.2.5(b), we find what follows:

Envelope and phase of x with respect to f_o:

$$\alpha_x(t) = |g(t)|, \qquad t \in \mathbb{R},$$

$$\phi_x(t) = \begin{cases} \phi & \text{if } g(t) \geq 0, \\ \phi - \pi & \text{if } g(t) < 0, \end{cases} \qquad t \in \mathbb{R}.$$

In-phase and quadrature components:

$$x_c(t) = g(t)\cos(\phi), \qquad t \in \mathbb{R},$$

$$x_s(t) = g(t)\sin(\phi), \qquad t \in \mathbb{R}.$$

(c) *Phase modulated harmonic.* In Example 10.2.5(c) we determined the complex envelope with respect to f_o of the phase modulated harmonic

$$x(t) = \cos[2\pi f_o t + \phi(t)], \qquad t \in \mathbb{R},$$

as

$$x_e(t) = e^{j\phi(t)}, \qquad t \in \mathbb{R},$$

under the condition that x_e is a low-pass signal whose bandwidth B is less than f_o. It follows that the amplitude and phase of x with respect to f_o are

$$\alpha_x(t) = 1, \qquad \phi_x(t) = \phi(t), \qquad t \in \mathbb{R}.$$

The instantaneous frequency f_x of the signal x is

$$f_x(t) = f_o + \frac{1}{2\pi}\dot{\phi}(t), \qquad t \in \mathbb{R},$$

with the overdot denoting differentiation with respect to time. *Exercise:* Find the in-phase and quadrature components of x.   ∎

In-Phase and Quadrature Component Extraction

In Section 10.3 it is shown that modulation of a harmonic carrier with a message signal amounts to letting the amplitude, frequency, or phase of the carrier vary with the message signal, possibly simultaneously. *Demodulation* consists of retrieving the message signal. This may often be done by obtaining the complex envelope and associated signals.

Figure 10.5 shows a configuration that allows the extraction of the in-phase and quadrature components x_c and x_s of a signal x with respect to the frequency f_o. In this and later block diagrams, c and s are the real harmonic signals

$$c(t) = \cos(2\pi f_o t), \qquad t \in \mathbb{R},$$
$$s(t) = \sin(2\pi f_o t), \qquad t \in \mathbb{R}.$$

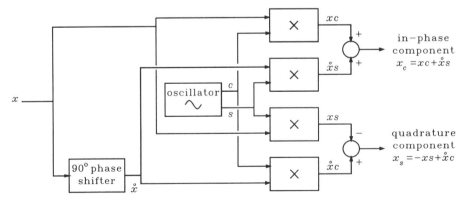

Figure 10.5.   In-phase and quadrature component extraction.

The configuration of Fig. 10.5 requires a component block whose input is the signal x and whose output is the Hilbert transform $\overset{\circ}{x}$. Such a block has the frequency response function

$$\hat{h}(f) = -j \operatorname{sign}(f), \qquad f \in \mathbb{R},$$

and is called a *90° phase shifter* or *quadrature filter*. Since the impulse response of the 90° phase shifter is

$$h(t) = \frac{1}{\pi t}, \qquad t \in \mathbb{R},$$

the phase shifter is anticipating, and only an approximation may be realized on-line. Fortunately the phase shifter usually only need operate over a limited frequency band; under this condition approximate on-line realization is feasible.

The block diagram of Fig. 10.5 may be complemented with the diagrams of Figs. 10.6 and 10.7 to obtain the envelope α_x, phase ϕ_x, and instantaneous frequency deviation

$$f_x - f_o = \frac{x_c \dot{x}_s - \dot{x}_c x_s}{x_c^2 + x_s^2}.$$

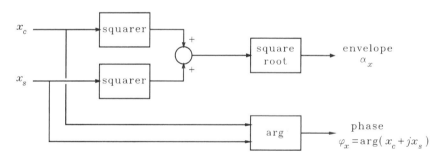

Figure 10.6.   Envelope and phase construction.

10.2.10.   Exercise: In-phase and quadrature component extraction.   Verify the in-phase and quadrature component extraction scheme of Fig. 10.5.   ∎

10.2.11.   Exercise: Alternative in-phase and quadrature extraction scheme.
Show that an alternative to the scheme of Fig. 10.5 for extracting the in-phase and quadrature components is that of Fig. 10.8, which employs two ideal low-pass filters rather than a 90° phase shifter. Assume that the complex envelope of x with respect to f_o is low-pass with bandwidth $B < f_o$. The frequency response function of the ideal low-pass filter is

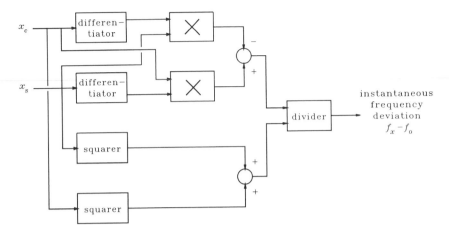

Figure 10.7. Instantaneous frequency extraction.

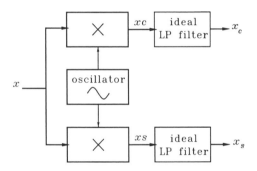

Figure 10.8. Alternative scheme for in-phase and quadrature component extraction.

$$\hat{h}(f) = \begin{cases} 1 & \text{for } |f| \le B, \\ 0 & \text{otherwise,} \end{cases} \qquad f \in \mathbb{R}.$$

Hint: Write x in terms of its in-phase and quadrature components and use simple trigonometric identities.   ∎

Response of Narrow-Band Filters to Narrow-Band Inputs

Envelope transformation facilitates the computation and interpretation of the response of narrow-band filters to narrow-band inputs. What follows shows that this response may be obtained by manipulating *envelopes* rather than the high-frequency signals themselves.

10.2.12.   Summary: Convolution of signals in terms of complex envelopes.
Let x and y be real continuous-time signals with complex envelopes x_e and y_e, respectively. Then, the CCFT $\hat{z}_e$ of the complex envelope z_e of the convolution $z = x * y$ equals

$$\hat{z}_e = \tfrac{1}{2}\hat{x}_e \cdot \hat{y}_e,$$

so that the complex envelope z_e itself is given by

$$z_e = \tfrac{1}{2}x_e * y_e. \qquad\qquad ■$$

The proof of the theorem is simple if Fourier transforms are used, and is left as an exercise. Note the factor $\tfrac{1}{2}$ that enters. We illustrate the application of the result with an example.

10.2.13.   Example: Response of a band-filter.   We consider the response of the LCR band-filter of Fig. 10.9 when the input voltage u is a signal with a step change in frequency given by

$$u(t) = \begin{cases} \cos\,(2\pi f_1 t) & \text{for } t < 0, \\ \cos\,(2\pi f_2 t) & \text{for } t \geq 0, \end{cases} \qquad t \in \mathbb{R}.$$

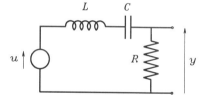

Figure 10.9.   LCR band-filter.

The frequencies f_1 and f_2 are both positive. A plot of the input u is given in Fig. 10.10. We use 10.2.12 to find the response to this input. It will be seen that some approximations are needed.

Considering the LCR network as a convolution system with impulse response h, application of 10.2.12 requires the complex envelopes of the input u and the impulse response to be known.

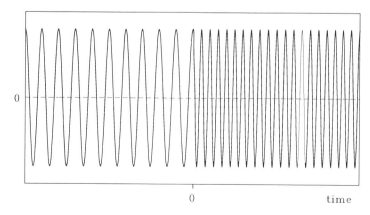

Figure 10.10.   A harmonic with a step change in frequency.

We first consider the complex envelope of the input. Writing u as

$$u(t) = \cos[2\pi f_o t + \phi(t)], \qquad t \in \mathbb{R},$$

with f_o some positive frequency and

$$\phi(t) = \begin{cases} 2\pi(f_1 - f_o)t & \text{for } t < 0, \\ 2\pi(f_2 - f_o)t & \text{for } t \geq 0, \end{cases} \qquad t \in \mathbb{R},$$

we see that u is a phase modulated harmonic as introduced in Example 10.2.5(c). From 10.2.9(c) the complex envelope u_e of u is

$$u_e(t) = e^{j\phi(t)} = \begin{cases} e^{j2\pi(f_1 - f_o)t} & \text{for } t < 0, \\ e^{j2\pi(f_2 - f_o)t} & \text{for } t \geq 0, \end{cases} \qquad t \in \mathbb{R}, \qquad (2)$$

provided u_e is band-pass band-limited with bandwidth B less than f_o. Writing u_e in the form

$$u_e(t) = e^{j2\pi(f_1 - f_o)t}\mathbb{1}(-t) + e^{j2\pi(f_2 - f_o)t}\mathbb{1}(t), \qquad t \in \mathbb{R},$$

we find by application of 7.4.20 (CCFT of the unit step), the converse shift property of the CCFT, and 7.6.15 (scaling property of the CCFT) that the CCFT of u_e is

$$\hat{u}_e(f) = \tfrac{1}{2}\delta(-f - f_o + f_1) + \frac{1}{j2\pi(-f - f_o + f_1)}$$

$$+ \tfrac{1}{2}\delta(f - f_2 + f_o) + \frac{1}{j2\pi(f - f_2 + f_o)}, \qquad f \in \mathbb{R}.$$

The difficulty is that this is the CCFT of a signal that is low-pass but *not* band-limited. Scrutiny of the CCFT, however, shows that most of the power of the signal u_e is concentrated in a band of width B such that $B \gg |f_2 - f_o|$ and $B \gg |f_1 - f_o|$. If we assume that f_1 and f_2 are not very far apart, that is, $|f_1 - f_2| \ll f_1$ and $|f_1 - f_2| \ll f_2$, then the condition $f_o < B$ is satisfied if f_o is chosen not much different from f_1 and f_2. With these assumptions we ignore the fact that u_e is not band-limited and take u_e as given by (2) as an approximation to the complex envelope of u.

We now turn to the determination of the complex envelope h_e of the impulse response h of the network. The transfer function of the LCR network from the input voltage u to the output voltage y is

$$H(s) = \frac{R}{sL + \dfrac{1}{sC} + R} = \frac{sRC}{1 + sRC + s^2LC},$$

so that the frequency response function of the filter is given by

$$\hat{h}(f) = H(j2\pi f) = \frac{j2\pi fRC}{1 + j2\pi fRC - 4\pi^2 f^2 LC}, \qquad f \in \mathbb{R}.$$

Defining the *resonance angular frequency* ω_r and the *quality factor q* of the network as

$$\omega_r := 1/\sqrt{LC}, \qquad q := \omega_r L/R,$$

the transfer function may be rewritten as

$$H(s) = \frac{s\omega_r/q}{s^2 + s\omega_r/q + \omega_r^2}.$$

The poles of the transfer function are $(-1 \pm j\sqrt{4q^2 - 1}) \cdot \omega_r/2q$. To determine the complex envelope of the impulse response of the LCR network, we first obtain the partial fraction expansion

$$H(s) = \frac{\omega_r}{2q}\left(\frac{1 + jK}{s - \frac{\omega_r}{2q}\left(-1 + \frac{j}{K}\right)} + \frac{1 - jK}{s - \frac{\omega_r}{2q}\left(-1 - \frac{j}{K}\right)} \right),$$

where

$$K := \frac{1}{\sqrt{4q^2 - 1}}.$$

Replacing s with $j\omega$, the magnitudes of the two denominators are

$$\left(\omega \pm \frac{\omega_r}{2qK}\right)^2 + \frac{\omega_r^2}{4q^2},$$

respectively, which assume their minimal values at the angular frequencies $\pm\omega_r/2qK$. As a result, the frequency response function $\hat{h}$ exhibits peaks at the angular frequencies $\pm\omega_r/2qK$. The larger the quality factor q is, the more pronounced the peaks are. For $q \gg 1$ the corresponding frequencies may be approximated as

$$\pm\frac{1}{2\pi}\frac{\omega_r}{2qK} \approx \pm\frac{\omega_r}{2\pi}.$$

Figure 10.11 illustrates the peaks. The peak at $f_r := \omega_r/2\pi$ originates from the first term in the partial fraction expansion of H, the peak at $-f_r$ from the second. If

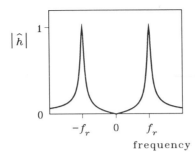

Figure 10.11. Magnitude of the frequency response function of the LCR filter for $q \gg 1$.

$q \gg 1$, then after substitution of $s = j2\pi f$ the first term contributes very little to the frequency response $\hat{h}$ for negative frequencies while the second term contributes very little for positive frequencies. Thus, for $q \gg 1$ we may obtain a good approximation to the CCFT of the pre-envelope $\hat{h}_p$ of the unit response h by simply omitting the second term and multiplying by 2. This results in

$$\hat{h}_p(f) \approx \frac{(1 + jK)\omega_r/q}{j2\pi f - \frac{\omega_r}{2q}(-1 + j/K)}, \qquad f \in \mathbb{R}.$$

Taking the complex envelope with respect to the frequency f_o defined by

$$f_o := \frac{1}{2\pi} \frac{\omega_r}{2qK} \approx \frac{\omega_r}{2\pi} = f_r$$

it follows that the CCFT of the complex envelope h_e of the unit response of the band-filter may be approximated as

$$\hat{h}_e(f) = \hat{h}_p(f - f_o) \approx \frac{(1 + jK)\omega_r/q}{j2\pi f + \frac{\omega_r}{2q}}, \qquad f \in \mathbb{R}.$$

As a result, the complex envelope h_e itself is approximately given by

$$h_e(t) \approx (1 + jK)\omega_r/q e^{-\omega_r t/2q} \mathbb{1}(t), \qquad t \in \mathbb{R}. \tag{3}$$

To find the response of the band-filter to the frequency modulated input u, we use 10.2.12 and hence compute the convolution of the complex envelope u_e of the input and the complex envelope h_e of the impulse response. As seen previously, the complex envelope u_e may be approximated by a complex harmonic signal, whose frequency is $f_1 - f_o$ until time zero and frequency $f_2 - f_o$ thereafter. The complex envelope h_e of the impulse response represents the impulse response of a first-order system with time constant $2q/\omega_r$. The response of this first-order system *before* time 0 to the complex envelope of the input is the steady-state response to the complex harmonic with frequency $f_1 - f_o$, so that by 10.2.12

$$y_e(t) \approx \tfrac{1}{2}\hat{h}_e(f_1 - f_o)e^{j2\pi(f_1 - f_o)t}, \qquad t < 0.$$

For $t \geq 0$, the response of the first-order system equals the steady-state response to the complex harmonic input with frequency $f_2 - f_o$ plus an exponential transient term with time constant $2q/\omega_r$, so that

$$y_e(t) \approx ae^{-\omega_r t/2q} + \tfrac{1}{2}\hat{h}_e(f_2 - f_o)e^{j2\pi(f_2 - f_o)t}, \qquad t \geq 0.$$

The constant a follows from the continuity condition $y_e(0^-) = y_e(0)$, so that $\tfrac{1}{2}\hat{h}_e(f_1 - f_o) = a + \tfrac{1}{2}\hat{h}_e(f_2 - f_o)$. The complete solution for y_e thus is given by

$$y_e(t) \approx \begin{cases} \tfrac{1}{2}\hat{h}_e(f_1 - f_o)e^{j2\pi(f_1 - f_o)t}, & t < 0, \\ \tfrac{1}{2}[\hat{h}_e(f_1 - f_o) - \hat{h}_e(f_2 - f_o)]e^{-\omega_r t/2q} + \tfrac{1}{2}\hat{h}_e(f_2 - f_o)e^{-j2\pi(f_2 - f_o)t}, & t \geq 0. \end{cases}$$

We assume the numerical values $R = 0.1\ \Omega$, $C = 10^{-6}$ F, and $L = 10^{-4}$ H. The resulting resonance angular frequency and quality factor are

$$\omega_r = 10^5 \text{ rad/s}, \qquad q = 100,$$

while $K = 0.0050001$. The large value of the quality factor justifies the approximation (3) for the complex envelope of the impulse response. The resulting frequency $f_o = \omega_r/4\pi qK$ is very close to the resonance frequency $f_r = \omega_r/2\pi = 15915.5$ Hz, and from (3) the complex envelope h_e numerically is given by

$$h_e(t) \approx (1 + 0.005j)e^{-500t}\mathbb{1}(t), \qquad t \in \mathbb{R}.$$

Suppose furthermore that $f_1 = 16000$ Hz and $f_2 = 17000$ Hz. The former frequency is close to the resonance frequency and the latter is a little higher. These values justify the approximation (2) for the complex envelope of the input.

Figure 10.12 shows plots of the envelope α_y and the phase ϕ_y of the output y of the band-pass filter with respect to the frequency f_o, obtained by computing the magnitude and argument of the complex envelope y_e. We note that the envelope after a transient settles at a new constant value. Both after and before the step change in frequency the phase varies linearly with time but at a different rate and with a transient in between. The sawtooth behavior of the phase plot results because the phase has been reduced modulo 2π to lie between $-\pi$ and π.

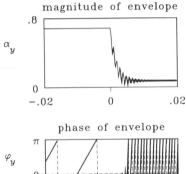

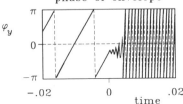

Figure 10.12. Magnitude and phase of the complex envelope of the response of the LCR band-filter to a harmonic input with a step change in frequency.

10.3 MODULATION AND DEMODULATION

In this section we present various modulation schemes and corresponding demodulation schemes. Because modulated signals usually consist of a carrier whose amplitude, frequency, or phase individually or simultaneously vary slowly with respect to the carrier itself, the theory of narrow-band signals developed in Section 10.2 directly applies. Most of the present section can be read without a thorough understanding of the material of Section 10.2, however.

Imagine that we wish to transmit audio voice signals, with frequencies between, say, 100 Hz and 3000 Hz, to some distant location. Since audio signals do not propagate over long distances in the atmosphere, it is necessary to "mount" the messages on a high-frequency electromagnetic carrier, whose frequency may lie anywhere between, say, 0.5 MHz and 3000 MHz. This process of mounting low-frequency "message" signals on a high-frequency carrier is called *modulation*. After receipt of the transmitted signal the low-frequency signal is extracted from the modulated signal by a process called *demodulation*.

We study four well-known modulation schemes: (1) double side-band suppressed carrier amplitude modulation (DSBSC AM), (2) (ordinary) amplitude modulation (AM), (3) single side-band amplitude modulation (SSB AM), and (4) frequency modulation (FM).

Double Side-Band Suppressed Carrier Amplitude Modulation

The simplest amplitude modulation scheme is double side-band suppressed carrier amplitude modulation (DSBSC AM).

10.3.1.   Definition: Double side-band suppressed carrier amplitude modulation.   Let m be a real continuous-time message signal and c a harmonic carrier given by

$$c(t) = \cos\,(2\pi f_c t), \qquad t \in \mathbb{R},$$

with positive carrier frequency f_c. DSBSC amplitude modulation consists of forming the modulated carrier

$$x_{\text{DSBSC}} = m \cdot c$$

by simply multiplying the carrier c by the message signal m.   ∎

Figure 10.13 illustrates DSBSC AM. The scheme is easily analyzed in the frequency domain. Suppose that the CCFT of the message signal m is $\hat{m}$. Then since

$$\begin{aligned} x_{\text{DSBSC}}(t) &= m(t)\cos\,(2\pi f_c t) \\ &= m(t)\tfrac{1}{2}(e^{j2\pi f_c t} + e^{-j2\pi f_c t}), \qquad t \in \mathbb{R}, \end{aligned}$$

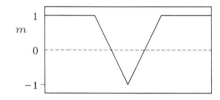

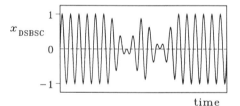

Figure 10.13. DSBSC amplitude modulation. Top: the message signal. Bottom: the modulated carrier.

by the converse shift property the CCFT of the modulated carrier is given by

$$\hat{x}_{\text{DSBSC}}(f) = \tfrac{1}{2}[\hat{m}(f - f_c) + \hat{m}(f + f_c)], \qquad f \in \mathbb{R},$$

as illustrated in Fig. 10.14. The figure shows that by the modulation process the CCFT $\hat{m}$ of the message signal is split into two equal parts, one of which is shifted to the right by f_c and the other to the left by the same amount.

Because the message signal m is real, the positive frequency part of the CCFT $\hat{x}_{\text{DSBSC}}$ of the modulated carrier is conjugate symmetric about the carrier frequency f_c, which is why the modulation scheme is called "double side band." It is moreover named "suppressed carrier" because the modulated carrier does not contain a distinct carrier component, as some other modulation schemes do.

DSBSC amplitude modulation may be implemented with the help of a harmonic oscillator and a multiplier as in Fig. 10.15. Demodulation, also indicated in

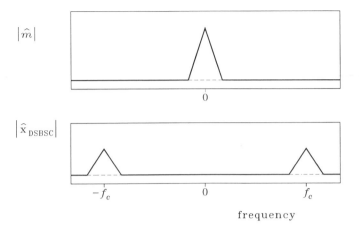

Figure 10.14. DSBSC amplitude modulation. Top: CCFT of the message signal. Bottom: CCFT of the DSBSC amplitude modulated carrier.

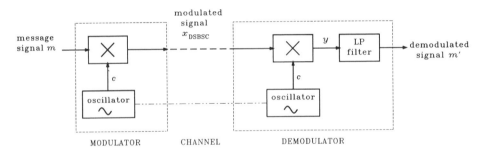

Figure 10.15. DSBSC amplitude modulation and demodulation.

Fig. 10.15, is achieved by multiplying the modulated signal x_{DSBSC} by the carrier c, which results in the signal y given by

$$y(t) = x_{\text{DSBSC}}(t) \cdot c(t) = m(t)[\cos{(2\pi f_c t)}]^2$$
$$= m(t)\tfrac{1}{2}[1 + \cos{(4\pi f_c t)}], \qquad t \in \mathbb{R}.$$

Assuming that the bandwidth B of the message signal is less than the carrier frequency f_c, the high-frequency component $m(t)\tfrac{1}{2}\cos{(4\pi f_c t)}$, $t \in \mathbb{R}$, may be removed by a low-pass filter with frequency response function

$$\hat{h}_{\text{LP}}(f) = \begin{cases} 1 & \text{for } |f| \le B, \\ 0 & \text{otherwise,} \end{cases} \qquad f \in \mathbb{R}.$$

Figure 10.16 indicates the CCFTs of y and the demodulated signal m', which equals m.

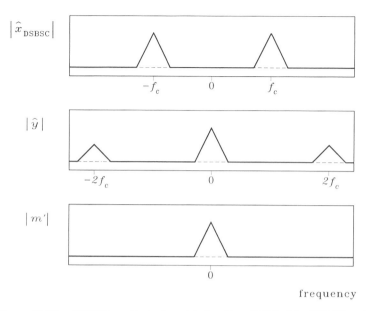

Figure 10.16.   DSBSC amplitude demodulation: Top: CCFT of the DSBSC modulated signal. Middle: CCFT of the signal y. Bottom: CCFT of the demodulated signal.

Note that in the terminology of Section 10.2 the demodulation scheme extracts the in-phase component of the modulated signal x_{DSBSC} with respect to the carrier frequency f_c (compare Fig. 10.8).

10.3.2.   Remarks: DSBCS amplitude modulation.

(a) Demodulation of a DSBSC amplitude modulated signal requires that the carrier frequency *and* phase (though not its amplitude) be exactly known at the receiver. This is called *synchronous* demodulation. One way of achieving this is to transmit the unmodulated carrier along another channel. When no other channel is available, a fraction α of the carrier may be *added* to the modulated signal as in Fig. 10.17. The receiver in turn is provided with a device that extracts the carrier from the received signal and makes it available for demodulation. The presence of the carrier in the received signal causes a constant (DC) component in the demodulated signal, which need be removed. The carrier extractor may consist of a band-filter that is sharply tuned to the carrier frequency f_c.

(b) The function of the low-pass filter in the DSBSC demodulator of Fig. 10.15 or 10.17 is to separate the low-pass message signal with frequency band $[-B, B]$ from the high-frequency component with frequency band $[2f_c - B, 2f_c + B]$ and a corresponding negative-frequency band. If $f_c \gg B$, then any reasonable low-pass filter with a suitable bandwidth is satisfactory. ∎

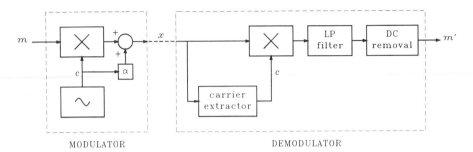

Figure 10.17.   DSBSC AM modulation and demodulation with added carrier.

10.3.3.   Exercise: DSBSC demodulation scheme.   Analyze the DSBSC demodulation scheme of Fig. 10.15 when the carrier used for demodulation has the
wrong frequency, amplitude, or phase.   ∎

Amplitude Modulation

Synchronous demodulation is facilitated or may even be eliminated if the carrier is
not suppressed. This is an important feature of ordinary amplitude modulation
(AM), which is the scheme employed in AM radio transmission.

10.3.4.   Definition: Amplitude modulation (AM).   Let m be a real continu-
ous-time message signal such that $|m(t)| \le m_o$, $t \in \mathbb{R}$, for some constant m_o, and c
a harmonic carrier given by

$$c(t) = \cos(2\pi f_c t), \qquad t \in \mathbb{R}.$$

Amplitude modulation with *modulation depth* α, where $0 \le \alpha \le 1$, consists of
forming the modulated carrier

$$x_{\mathrm{AM}} = \left(1 + \alpha \frac{m}{m_o}\right) \cdot c. \qquad\qquad ∎$$

Figure 10.18 illustrates this form of modulation. The CCFT of the amplitude modulated signal x_{AM} is given by

$$\hat{x}_{\mathrm{AM}}(f) = \tfrac{1}{2}[\delta(f - f_c) + \delta(f + f_c)] + \frac{\alpha}{2m_o}[\hat{m}(f - f_c) + \hat{m}(f + f_c)], \quad f \in \mathbb{R},$$

and is sketched in Fig. 10.19. The modulated signal includes the carrier as a distinct
component.

     The implementation of amplitude modulation is schematically indicated in Fig.
10.20, and is similar to that of DSBSC AM, except that the carrier is added. De-

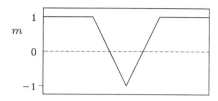

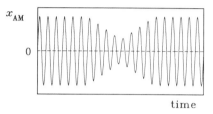

Figure 10.18. Amplitude modulation. Top: the message signal. Bottom: the modulated carrier with modulation depth $\alpha = 0.5$.

time

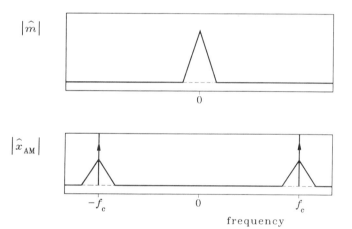

frequency

Figure 10.19. Amplitude modulation. Top: CCFT of the message signal. Bottom: CCFT of the modulated signal.

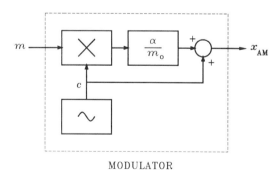

MODULATOR

Figure 10.20. Amplitude modulator.

modulation may be achieved by the same synchronous demodulation scheme as for DSBSC with added carrier given in Fig. 10.17.

A much more common and far less costly demodulation scheme for AM than synchronous demodulation as in Fig. 10.17 is *asynchronous* demodulation by a simple *envelope detector* as in Fig. 10.21. The scheme is feasible when the modulation depth is not greater than 1 and the carrier frequency f_c is much larger than the message signal bandwidth. Figure 10.22 shows the output y of the detector, which after low-pass filtering and removal of the constant component results in a demodulated signal.

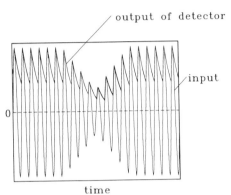

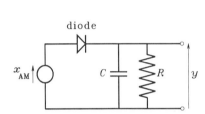

Figure 10.21.   Simple envelope detector.

Figure 10.22.   Input and output of the simple envelope detector.

10.3.5.   Remarks: Amplitude modulation.

(a) AM is less efficient than DSBSC AM because in the latter no or little signal power need be expended on transmitting the carrier. DSBSC AM requires synchronous demodulation, however.

(b) In AM radio transmission the message signal bandwidth is, say, 4000 Hz, and the carrier frequency (after heterodyning to IF) about 500 kHz. For this frequency separation the simple envelope detection of Fig. 10.19 works very well. This is what all AM radio receivers employ.   ∎

Single Side-Band Amplitude Modulation

We now turn to a transmission scheme that is efficient in bandwidth use, namely, *single side-band amplitude modulation* (SSB AM). In double side-band modulation the positive frequency part of the modulated carrier is conjugate symmetric with respect to the carrier frequency. By removing one of the side-bands (either the lower or the upper), no information is lost while the bandwidth is halved.

10.3.6.   Definition: Single side-band amplitude modulation (SSB AM).   Let m be a real continuous-time message signal and c a harmonic carrier given by

$$c(t) = \cos(2\pi f_c t), \qquad t \in \mathbb{R}.$$

Then (upper) single side-band amplitude modulation consists of passing the DSBSC amplitude modulated signal $x_{\text{DSBSC}} = m \cdot c$ through a filter with high-pass frequency response function

$$\hat{h}_{\text{HP}}(f) = \begin{cases} 2 & \text{for } |f| \geq f_c, \\ 0 & \text{for } |f| < f_c, \end{cases} \qquad f \in \mathbb{R}. \qquad \blacksquare$$

The high-pass filter includes a factor 2 so that no signal power is lost. Figure 10.23 indicates the theoretical modulation scheme, which is not always practical because of the sharp cut-off required for the high-pass filter. Figure 10.24 shows the CCFTs of the message signal and the SSB amplitude modulated signal. Figure 10.23 also shows how the SSB amplitude modulated signal may be demodulated by synchronous demodulation. The demodulation scheme is the same as that of Fig. 10.15 for DSBSC demodulation.

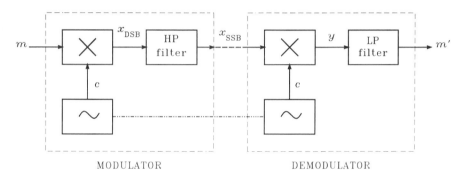

Figure 10.23.   SSB amplitude modulation and demodulation.

The need to transmit the carrier separately may be avoided by adding some of the carrier to the transmitted signal and employing carrier extraction at the receiver as in Fig. 10.17.

A common practical implementation of SSB amplitude modulation relies on the narrow-band manipulation techniques developed in Section 10.2.

10.3.7.   Summary: Single side-band amplitude modulation.   Suppose that the real message signal m is low-pass and band-limited with bandwidth B, where $0 \leq B < f_c$. Then the single side-band amplitude modulated signal x_{SSB} with carrier frequency f_c is given by

$$x_{\text{SSB}} = mc - \overset{\circ}{m}s,$$

where the real signal $\overset{\circ}{m}$ is the Hilbert transform of m, and c and s are the carrier and 90° phase shifted carrier given by

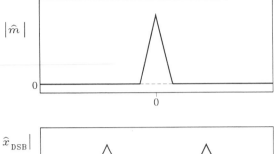

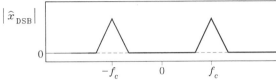

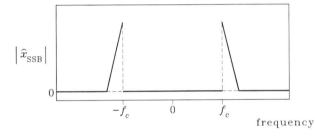

Figure 10.24. SSB AM. Top: CCFT of the message signal. Middle: CCFT of the DSB amplitude modulated signal. Bottom: CCFT of the SSB amplitude modulated signal.

frequency

$$c(t) = \cos(2\pi f_c t), \qquad t \in \mathbb{R},$$
$$s(t) = \sin(2\pi f_c t), \qquad t \in \mathbb{R}. \qquad\qquad \blacksquare$$

10.3.8.   Proof: Starting with the message signal m we first form its pre-envelope

$$m_p = m + j\overset{\circ}{m},$$

whose CCFT $\hat{m}_p$, given by $\hat{m}_p(f) = 2\hat{m}(f)\mathbb{1}(f)$, $f \in \mathbb{R}$, is indicated in Fig. 10.25(b). Multiplication by the complex harmonic $e^{j2\pi f_c t}$, $t \in \mathbb{R}$, yields the signal m_s such that

$$m_s(t) = m_p(t)e^{j2\pi f_c t} = [m(t) + j\overset{\circ}{m}(t)]e^{j2\pi f_c t}, \qquad t \in \mathbb{R}.$$

The CCFT of m_s is

$$\hat{m}_s(f) = \hat{m}_p(f - f_c) = 2\hat{m}(f - f_c)\mathbb{1}(f - f_c), \qquad f \in \mathbb{R},$$

as shown in Fig. 10.25(c). The real part r of m_s is given by

$$r(t) = \text{Re}\,[m_s(t)] = m(t)\cos(2\pi f_c t) - \overset{\circ}{m}(t)\sin(2\pi f_c t), \qquad t \in \mathbb{R}.$$

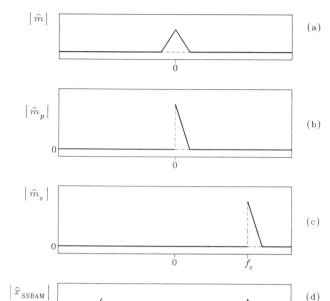

Figure 10.25. SSB amplitude modulation. (a) CCFT of the message signal. (b) CCFT of the pre-envelope of the message signal. (c) Frequency shifted pre-envelope. (d) CCFT of the SSB amplitude modulated signal.

Because by 7.6.17 the CCFT of the complex conjugate $\overline{m}_s$ of m_s is $\overline{\hat{m}_s(-f)}, f \in \mathbb{R}$, it follows from

$$r = \tfrac{1}{2}(m_s + \overline{m}_s)$$

that the CCFT $\hat{r}$ of r satisfies

$$
\begin{aligned}
\hat{r}(f) &= \tfrac{1}{2}[\hat{m}_s(f) + \overline{\hat{m}_s(-f)}] \\
&= \hat{m}(f - f_c)1(f - f_c) + \overline{\hat{m}(-f - f_c)1(-f - f_c)} \\
&= \hat{m}(f - f_c)1(f - f_c) + \hat{m}(f + f_c)1(-f + f_c), \qquad f \in \mathbb{R}.
\end{aligned}
$$

This is the CCFT of the desired single side-band amplitude modulated signal x_{SSB}, as shown in Fig. 10.25(d). ∎

A block diagram of the implementation of SSB amplitude modulation according to 10.3.7 is given in Fig. 10.26. As noted in Section 10.2, the "90° phase shifter" is a filter with frequency response function

$$\hat{h}(f) = -j \operatorname{sign}(f), \qquad f \in \mathbb{R},$$

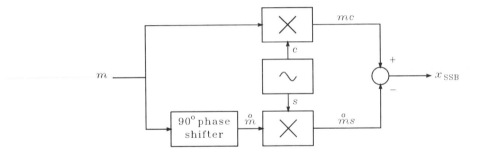

Figure 10.26. SSB amplitude modulation.

which may be approximately realized over the required frequency band with sufficient accuracy.

From 10.3.7 the time structure of the SSB amplitude modulated signal
$x_{\text{SSB}} = mc - \overset{\text{o}}{m}s$ may now be understood. The signal x_{SSB} is a narrow-band signal
whose in-phase component with respect to the carrier frequency f_c is the message
signal m and whose quadrature component is the Hilbert transform $\overset{\text{o}}{m}$ of m.

10.3.10.   Remarks: SSB AM.
(a) An alternative to the scheme of the right-hand side of Fig. 10.23 for de-
modulating SSB AM signals corresponds to the top half of the scheme for extracting
in-phase and quadrature components of Fig. 10.6 and is given in Fig. 10.27.

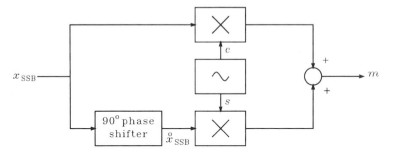

Figure 10.27.   Alternative scheme for SSB amplitude demodulation.

(b) The bandwidth of an SSB AM signal is half of that of the corresponding
AM signal. The price for this superior economy is the complexity of the modulator
and demodulator.   ∎

10.3.11.   Exercise: SSB duplexing.   Show that the SSB modulation and de-
modulation scheme of Fig. 10.28 allows the transmission of *two* message signals m_1
and m_2 by using a single carrier. Note that m_1 is the in-phase component and m_2 the
quadrature component of the modulated signal x with respect to the carrier frequency f_c.   ∎

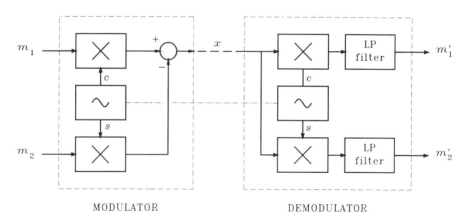

MODULATOR   DEMODULATOR

Figure 10.28. SSB duplexing.

Frequency Modulation

Frequency modulation (FM) provides far better immunity against noise than amplitude modulation, at the cost of a greater bandwidth.

10.3.12.   Definition: Frequency modulation.   Let m be a real continuous-time message signal such that $|m(t)| \leq m_o$, $t \in \mathbb{R}$, for some positive constant m_o, and c a harmonic carrier given by

$$c(t) = \cos(2\pi f_c t), \qquad t \in \mathbb{R}.$$

Frequency modulation with *frequency deviation* $\alpha > 0$ consists of forming the frequency modulated carrier

$$x_{\text{FM}}(t) = \cos(2\pi f_c t + \phi(t)), \qquad t \in \mathbb{R},$$

where

$$\phi(t) = \phi_o + 2\pi\alpha \int_{-\infty}^{t} \frac{m(\tau)}{m_o} \, d\tau, \qquad t \in \mathbb{R},$$

with ϕ_o an arbitrary constant.   ∎

Figure 10.29 shows an example of a frequency modulated signal. Note that while in DSBSC and AM the message information is imbedded in the envelope of the modulated signal, and in SSB both in the envelope and the phase, in FM the information is included in the *phase* only, while the envelope is constant. The constant envelope is the reason for the immunity of FM against noise: any noise and other deformation

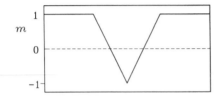

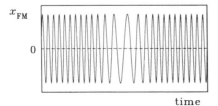

Figure 10.29. Frequency modulation. Top: a message signal. Bottom: the frequency modulated signal.

of the signal incurred during transmission that affect the *envelope* may be removed by preceding the demodulator by a limiter that restores the constant envelope.

The reason why this modulation scheme is called frequency modulation is that the *instantaneous frequency* of the modulated signal is

$$f_x(t) = \frac{1}{2\pi} \frac{d\phi(t)}{dt} = f_c + \alpha \frac{m(t)}{m_o}.$$

Thus, the instantaneous frequency deviation $f_x - f_c$ is proportional to the message signal. The maximal frequency deviation (achieved when the message equals its bound m_o) is α.

Figure 10.30 shows a frequency modulation scheme that uses sine and cosine function generators. The modulated signal is formed as

$$x_{\text{FM}}(t) = \cos(2\pi f_c t) \cos(\phi(t)) - \sin(2\pi f_c t) \sin(\phi(t))$$
$$= \cos(2\pi f_c t + \phi(t)), \quad t \in \mathbb{R}.$$

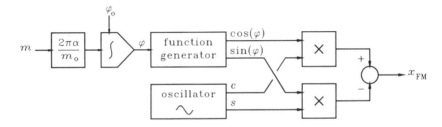

Figure 10.30. Frequency modulation using sine and cosine function generators.

Modulation schemes of this type, implemented with digital function generators, occur in state-of-the-art communication equipment. A more conventional scheme is to use a voltage controlled oscillator (VCO), whose instantaneous frequency is controlled by an external voltage, as shown in Fig. 10.31.

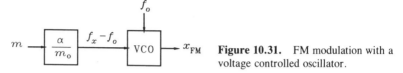

Figure 10.31.   FM modulation with a voltage controlled oscillator.

FM demodulation is achieved with a scheme as in Fig. 10.32. The received FM signal is first roughly band-filtered and shifted in frequency (this is called *heterodyning*; see 10.4.3) to an *intermediate frequency* (IF) signal, which is then more finely band-filtered to select the correct channel. The resulting signal is limited (i.e., sent through a hard or soft limiter), and again band-pass filtered to restore the harmonic shape and improve selectivity. Finally, the signal is differentiated and demodulated by means of envelope detection.

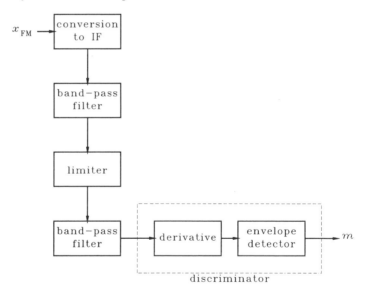

Figure 10.32.   FM demodulation.

The CCFT of the frequency modulated signal

$$x_{FM}(t) = \cos\left(2\pi f_c t + \phi(t)\right)$$
$$= \tfrac{1}{2}\left(e^{j[2\pi f_c t + \phi(t)]} + e^{-j[2\pi f_c t + \phi(t)]}\right), \qquad t \in \mathbb{R},$$

with

$$\phi(t) = \phi_o + 2\pi\alpha \int_{-\infty}^{t} \frac{m(\tau)}{m_o}\, d\tau, \qquad t \in \mathbb{R}.$$

is given by

$$\hat{x}_{\mathrm{FM}}(f) = \tfrac{1}{2}[\hat{p}(f - f_c) + \overline{\hat{p}(f + f_c)}], \qquad f \in \mathbb{R}.$$

Here $\hat{p}$ is the CCFT of the complex-valued signal

$$p(t) = e^{j\phi(t)} = \cos[\phi(t)] + j\sin[\phi(t)], \qquad t \in \mathbb{R},$$

which follows from the message signal by a nonlinear transformation. Thus, the frequency content of the FM signal $\hat{x}_{\mathrm{FM}}$ consists of the frequency content $\hat{p}$ of the signal p, after splitting and shifting to the frequencies $\pm f_c$. Because of the nonlinear transformation, the CCFT of p cannot be related easily to that of the message signal m. We therefore consider two special cases.

     First we study what is called *narrow band FM approximation,* which amounts to assuming that $\phi_o = 0$ and $|\phi(t)| << 1$ for all $t \in \mathbb{R}$. The latter holds if the frequency deviation α is sufficiently small. Then,

$$p(t) = e^{j\phi(t)} \approx 1 + j\phi(t),$$

so that by the integration property of the CCFT

$$\hat{p}(f) \approx \delta(f) + j\hat{\phi}(f)$$

$$= \delta(f) + \frac{\alpha}{f}\frac{\hat{m}(f)}{m_o}, \qquad f \in \mathbb{R}.$$

The resulting frequency content of the FM signal x_{FM} is sketched in Fig. 10.33. It has distinct carrier components. The bandwidth of the modulated signal about the carrier frequency is about the same as the bandwidth of the message signal.

     The second special case is to assume that the message signal m is *purely harmonic,* that is,

$$m(t) = a_m \cos(2\pi f_m t), \qquad t \in \mathbb{R},$$

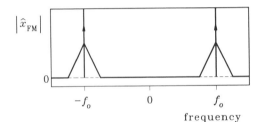

Figure 10.33. Frequency content of a narrow-band FM signal.

with a_m and f_m real constants. Correspondingly, let

$$\phi(t) = \frac{a_m}{m_o} \frac{\alpha}{f_m} \sin(2\pi f_m t) = \beta \sin(2\pi f_m t), \qquad t \in \mathbb{R},$$

where

$$\beta = \frac{a_m}{m_o} \frac{\alpha}{f_m}.$$

Consequently, the FM signal may be written as

$$\begin{aligned}
x_{\text{FM}}(t) &= \cos[2\pi f_c t + \beta \sin(2\pi f_m t)] \\
&= [\cos(2\pi f_c t) \cos(\beta \sin(2\pi f_m t)) \\
&\quad - \sin(2\pi f_c t) \sin(\beta \sin(2\pi f_m t))], \qquad t \in \mathbb{R}.
\end{aligned} \tag{1}$$

We use the well-known identities from the theory of special functions

$$\cos(z \sin(\theta)) = J_0(z) + 2 \sum_{k=1}^{\infty} J_{2k}(z) \cos(2k\theta),$$

$$\sin(z \sin(\theta)) = 2 \sum_{k=1}^{\infty} J_{2k+1}(z) \sin((2k+1)\theta),$$

where J_n is the *nth order Bessel function* (M. Abramowitz and I. A. Stegun, Eds., *Handbook of Mathematical Functions*, Dover, New York, 1965.) Substitution into (1) yields after some algebra

$$\begin{aligned}
x_{\text{FM}}(t) = J_0(\beta) \cos(2\pi f_c t) &+ \sum_{n=1}^{\infty} J_n(\beta) \cos[2\pi(f_c - nf_m)t] \\
&+ \sum_{n=1}^{\infty} (-1)^n J_n(\beta) \cos[2\pi(f_c + nf_m)t], \qquad t \in \mathbb{R}.
\end{aligned}$$

> Friedrich Wilhelm Bessel (1784–1846) was a German astronomer. He is well known as a correspondence partner of Gauss and introduced the function named after him in his work on planetary perturbations.

This shows that the frequency modulated signal x_{FM} besides the carrier frequency f_c also contains the frequencies $f_c \pm kf_m$, $k \in \mathbb{N}$. Figure 10.34 illustrates this. For fixed β and n large enough, the coefficients $J_n(\beta)$ decrease with increasing n. Asymptotically they behave as

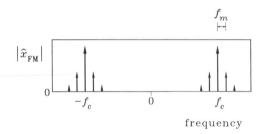

Figure 10.34. Frequency content of a frequency modulated harmonic with frequency f_m with carrier frequency f_c.

$$J_n(\beta) \approx \frac{1}{\sqrt{2\pi n}}\left(\frac{e\beta}{2n}\right)^n, \qquad n \gg 1.$$

If the infinite sums are truncated for n greater than once or twice β, then most of the signal energy is included. If β is less than one (which implies that the frequency of the harmonic f_m is less than the maximal frequency deviation α), then the higher harmonics play no great role. If β is greater than one, then the bandwidth of the modulated signal about the carrier frequency is several times greater than that of the message signal.

The final result shows that FM allows a transmission bandwidth that is *larger* than that of the message signal. This explains why FM is used for high-quality transmission. The larger the excess bandwidth, the less sensitive transmission is to distortion and noise.

10.4 MULTIPLEXING

The idea of multiplexing is to use a *single* carrier to transmit *several* messages at the same time, in such a way, of course, that the messages may be separated at the receiver. There are two distinct possibilities: *frequency* multiplexing and *time* multiplexing. Frequency multiplexing amounts to dividing the available *frequency band* among the various messages, taking care that the bands the messages occupy do not overlap. Time multiplexing involves allotting nonoverlapping *time slots* to the various messages. Figure 10.35 illustrates the ideas.

We briefly look at both multiplexig methods and show that the two methods are equivalent from the point of view of bandwidth utilization.

Frequency Multiplexing

We consider the problem of simultaneously transmitting N message signals m_1, m_2, $\cdots$, m_N, each of which has a finite bandwidth B. As a specific instance we may think of telephone conversations, with, say, $B = 3.2$ kHz. One way to achieve transmission is to use lower side-band SSB modulation for each of the message signals, as in Fig. 10.36, with carrier frequencies f_o, $2f_o$, $3f_o$, $\cdots$, Nf_o. The modulated signals are combined to the signal m. The frequency f_o is chosen slightly larger than B, for instance $f_o = 4$ kHz in the case of the telephone communication.

The next step is to use any modulation scheme (often FM) to transmit the com-

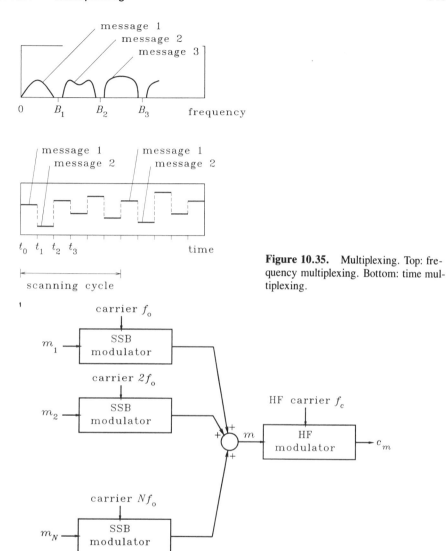

Figure 10.35. Multiplexing. Top: frequency multiplexing. Bottom: time multiplexing.

Figure 10.36. Frequency multiplexing.

bined message m by means of an ultra-high frequency carrier with frequency f_c. As long as the frequency bands of the message signals do not overlap, the signals may be demodulated at the receiver as in Fig. 10.37. The total bandwidth needed to transmit the combined message signal m is NB.

Time Multiplexing

Time multiplexing involves *sampling* and *scanning*. Again, let each of the N message signals m_1, m_2, $\cdots$, m_N have bandwidth B. Figure 10.38 schematically indicates the time multiplexer and modulator and the time-multiplexed combined signal

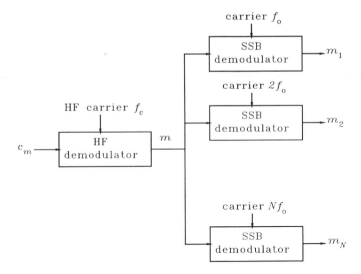

Figure 10.37. Demodulation of the frequency multiplexed signal of Fig. 10.36.

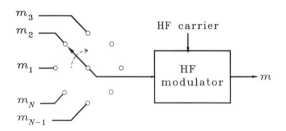

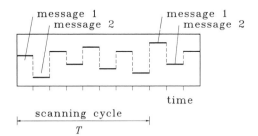

Figure 10.38. Time multiplexing.

m. To determine the scanning rate, we note that according to the sampling theorem each message signal need be sampled at a rate of $2B$. Thus, a scanning cycle of all N messages should be completed in a time slot of length $T = 1/2B$. This leaves individual time slots of duration $T/N = 1/2NB$. The transmission rate for the combined signal m is $2NB$ samples per unit of time. By the sampling theorem, the bandwidth of a signal sampled at a rate $2NB$ is NB. Hence, time multiplexing requires the same bandwidth as frequency multiplexing.

10.5 PROBLEMS

The first problem deals with various aspects of narrow-band signals.

10.5.1. Narrow-band signals. Compute the CCFT, pre-envelope, complex envelope with respect to f_o, Hilbert transform, in-phase and quadrature components, envelope, phase, and instantaneous frequency of the following modulated harmonics:

(a) The signal x given by

$$x(t) = \text{sinc}\,(2\pi Bt)\cos\,(2\pi f_o t), \qquad t \in \mathbb{R},$$

with f_o and B positive constants such that $B \leq f_o$. *Hint:* See 10.2.1(a).

(b) The signal x given by

$$x(t) = e^{-\frac{t^2}{2\sigma^2}}\cos\,(2\pi f_o t), \qquad t \in \mathbb{R},$$

with σ and f_o positive constants. Approximate the solution under the assumption that $\sigma \gg f_o$. *Hint:* See 7.6.12.

The second problem concerns a property of the frequency response function of non-anticipating real systems.

10.5.2. Frequency response function of a non-anticipating real system. Suppose that h is the impulse response of a non-anticipating real continuous-time system, that is, h is real and

$$h(t) = 0 \qquad \text{for}\quad t < 0.$$

Let the frequency response function $\hat{h}$ of the system be given by

$$\hat{h} = A + jB,$$

with A and B real.

(a) Write h as

$$h = a + b,$$

where the *even part* a and the *odd part* b of h are given by

$$a(t) = \tfrac{1}{2}[h(t) + h(-t)], \qquad t \in \mathbb{R},$$
$$b(t) = \tfrac{1}{2}[h(t) - h(-t)], \qquad t \in \mathbb{R}.$$

Prove that

$$\hat{a} = A, \qquad \hat{b} = jB,$$

with $\hat{a}$ the CCFT of a and $\hat{b}$ that of b. *Hint:* Use the symmetry properties of the CCFT.

(b) Prove that the real part A and the imaginary part B of $\hat{h}$ are related as

$$A = \mathcal{H}(B), \qquad B = -\mathcal{H}(A),$$

with $\mathcal{H}$ the Hilbert transform. *Hint:* Show that $b(t) = a(t)$ sign (t), $t \in \mathbb{R}$.

(c) Let the real part A of the frequency response function of a non-anticipating real system be given by

$$A(f) = \begin{cases} 1 & \text{for } |f| < f_o, \\ 0 & \text{otherwise}, \end{cases} \qquad f \in \mathbb{R},$$

with f_o a positive constant. Determine the impulse response h of the system.

The final four problems deal with various techniques in communication.

10.5.3. Heterodyning. Heterodyning is the operation of shifting the frequency range of a signal. It is often used in communication to select the correct signal from a frequency multiplexed signal. Figure 10.39 shows the arrangement. The frequency multiplexed signal x is multiplied by the real harmonic signal c given by

$$c(t) = \cos(2\pi f_o t), \qquad t \in \mathbb{R}.$$

The frequency f_o is fixed but adjustable. The resulting signal x_c is passed through a well-designed band-pass filter with a fixed frequency response. Its output x_{IF} is sent to the demodulator.

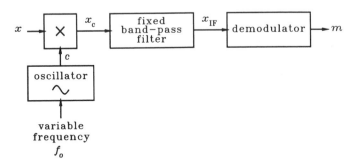

Figure 10.39.   Signal selection by heterodyning.

(a) Suppose that x is a narrow-band signal with center frequency f_c. Sketch the frequency content of the heterodyned signal x_c.

(b) Show that if the center frequency of the band-pass filter is f_{IF}, the arrangement of Fig. 10.39 may be used to select a band-pass signal whose center frequency is $f_c = f_{IF} + f_o$.

The scheme shifts the center frequency f_c to $f_{IF} = f_c - f_o$. The frequency f_{IF} is called the *intermediate frequency* (IF). By adjusting the oscillator frequency f_o, various signals may be tuned in. The band-pass filter and demodulator need only be designed for signals with a fixed band.

10.5.4. Synchronous DSBSC demodulation. In Fig. 10.15 a scheme is given for synchronous DSBSC demodulation. Explain how the cut-off frequency of the low-pass filter should be chosen. Indicate under what conditions the low-pass filter should be close to "ideal," and under what circumstances the requirements on the filter are less stringent.

10.5.5. Carrier extraction. In the description of Fig. 10.17 it is indicated that the "carrier extractor" usually involves a band-pass filter that is sharply tuned to the carrier frequency f_c. The detailed arrangement is shown in Fig. 10.40. For carrier extraction, the modulated signal x is first passed through a band-pass filter that is sharply tuned to the carrier frequency f_c. The output of the filter is limited, (i.e., sent through a hard or soft limiter) and once again band-pass filtered to restore the harmonic shape. Discuss the operation of this scheme.

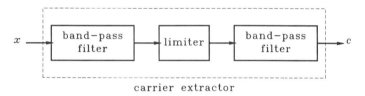

Figure 10.40. Carrier extraction.

10.5.6. FM demodulation. Discuss the FM demodulation scheme of Fig. 10.32 in detail. Explain the functions of the various subsystems. Sketch the shapes and frequency contents of the signals at the points between the subsystems.

10.6 COMPUTER EXERCISES

The first exercise in this section deals with narrow-band signals.

10.6.1. Narrow-band signals. For the following time signals, compute and plot their pre-envelope, Hilbert transform, complex envelope with respect to the frequency f_c, envelope, phase and instantaneous frequency, and the in-phase and quadrature components. In each case, m is the signal defined by

$$m(t) = \begin{cases} \frac{1}{2}[1 + \cos(\pi t)] & \text{for } |t| \le 1, \\ 0 & \text{otherwise,} \end{cases} \quad t \in \mathbb{R}.$$

Take $f_c = 10$ and define each time signal with a resolution of at least 256 points on the interval $[-1, 1)$.

(a) *Amplitude modulated signal:*

$$x(t) = m(t) \cdot \cos(2\pi f_c t), \qquad t \in \mathbb{R}.$$

(b) *Frequency modulated signal:*

$$x(t) = \cos\,[2\pi f_c t + \phi(t)], \qquad t \in \mathbb{R},$$

where

$$\phi(t) = \phi_o + 2\pi\alpha \int_{-\infty}^{t} \frac{m(\tau)}{m_o}\,d\tau, \qquad t \in \mathbb{R},$$

with $\phi_o = 0$, $m_o = 1$, $\alpha = f_c$.

(c) *Phase modulated signal:*

$$x(t) = \cos\,[2\pi f_c t + \psi(t)], \qquad t \in \mathbb{R},$$

where $\psi = \alpha m$, with $\alpha = 2\pi$.

The remaining two problems concern various modulation schemes.

10.6.2. DSBSC and amplitude modulation. Let m be the message signal defined by

$$m(t) = \begin{cases} \frac{1}{2}[1 + \cos\,(\pi t)] & \text{for } |t| \le 1, \\ 0 & \text{otherwise}, \end{cases} \qquad t \in \mathbb{R}.$$

Define the DSBSC amplitude modulated signal x_{DSBSC} and the amplitude modulated signal x_{AM} as

$$x_{\text{DSBSC}}(t) = m(t) \cdot \cos\,(2\pi f_c t), \qquad t \in \mathbb{R},$$

$$x_{\text{AM}}(t) = \left(1 + \alpha\frac{m(t)}{m_o}\right) \cdot \cos\,(2\pi f_c t), \qquad t \in \mathbb{R}.$$

Let $f_c = 10$, $m_o = 1$, $\alpha = \frac{1}{2}$, and define the time signals on a sufficiently long time axis with a resolution of at least 256 points on the interval $[-1, 1)$.

(a) Generate and plot the signals x_{DSBSC} and x_{AM} and their CCFTs.
(b) Demodulate the amplitude modulated signals by the scheme of Fig. 10.15. Plot the result.
(c) Test the effect of variations in the frequency, amplitude and phase of the carrier used for demodulation. Vary the conditions systematically and comment on the results.

10.6.3. Single side-band modulation. Let m be the message signal defined by

$$m(t) = \text{trian}\,(t) \cdot \cos\,(4\pi t), \qquad t \in \mathbb{R}.$$

Define x_{SSB} as the single side-band amplitude modulated signal as defined in 10.3.6 with carrier frequency $f_c = 10$. Study the signals on a sufficiently long time axis with at least 256 points on the interval $[-1, 1)$.

(a) Generate and plot x_{SSB} according to the scheme of Fig. 10.23.

(b) Generate and plot x_{SSB} according to the scheme of Fig. 10.26. Compare the result with that of (a).

(c) Demodulate the signal as in Fig. 10.23. Plot the result.

(d) Define the message signals m_1 and m_2 according to $m_1 = m$ and

$$m_2(t) = \text{trian } (t) \cdot \sin (3\pi t), \qquad t \in \mathbb{R}.$$

Generate and plot the duplexed SSB modulated signal x as in Fig. 10.28. Demodulate as indicated and plot the results.

11

Feedback and Applications to Automatic Control

11.1 INTRODUCTION

In this chapter we present a short introduction to *feedback* and its application in *automatic control*. Feedback is a powerful technique. It amounts to observing the effects of actions and correcting these actions according to the results.

The car cruise control system of Section 1.3 is an example of an automatic control system based on feedback. The difference between the actual car speed and the desired speed is measured and used to increase or decrease the throttle position as needed. Later in this section we describe the automatic cruise control system in more detail and illustrate central issues in control engineering such as *closed-loop response, robustness* to system changes, and *disturbance reduction*.

Feedback is the fundamental principle for the design of automatic control systems, but it has many other applications. Most audio amplifiers, for instance, feed a small portion of the output back to the input. This simple technique greatly improves the amplifier performance: The amplification factor is stabilized, linearity is improved, the bandwidth is increased, and the effect of noise is reduced.

Section 11.2 is devoted to a general analysis of the beneficial effects of feedback. It is shown in a quite general setting how feedback may improve system performance.

Stability is a major issue in feedback theory. Section 11.3 therefore presents several stability results for feedback systems, in particular the small gain theorem and the Nyquist stability criterion.

11.1.1.   Example: Automatic cruise control system.   The cruise control system has been introduced in Section 1.3. The block diagram is repeated in Fig. 11.1. The speed v of the car depends on the throttle opening u. The throttle opening is controlled by the cruise controller in such a way that the throttle opening is *increased* if the difference $v_r - v$ between the reference speed v_r and the actual speed in positive and *decreased* if the difference is negative.

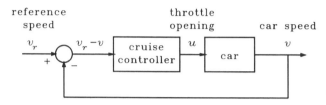

Figure 11.1.   Block diagram of the cruise control system.

As seen in 3.2.13, the motion of the car may be described by the differential equation

$$M\frac{dv(t)}{dt} = cu(t) - Bv^2(t), \qquad t \geq 0, \tag{1}$$

where M is the mass of the car, c a proportionality constant, and B the friction coefficient. The throttle opening ranges between 0 and 1, with 0 indicating a closed throttle and 1 a full opening. It is helpful to replace the speed v by the normalized speed $w = v/v_{max}$, with v_{max} the top speed of the car. The dimensionless quantity w varies between 0 and 1 and represents the speed as fraction of the top speed. The top speed v_{max} is the stationary solution of the differential equation (1) for $u(t) = 1$, $t \geq 0$, and is given by $v_{max} = \sqrt{c/B}$. As found in 4.8.1, in terms of w the car is described by the differential equation

$$\frac{dw(t)}{dt} = \alpha[u(t) - w^2(t)], \qquad t \geq 0, \tag{2}$$

with $\alpha = \sqrt{cB}/M$. Correspondingly replacing the reference speed v_r with the normalized reference speed w_r we obtain the feedback configuration of Fig. 11.2.

We assume that the controller is so constructed that the rate $du(t)/dt$ at which the throttle opening increases at time t is proportional to the difference $w_r(t) - w(t)$ of the reference speed and the actual speed. The mathematical model for the controller is

$$\frac{du(t)}{dt} = k[w_r(t) - w(t)], \qquad t \geq 0, \tag{3}$$

with k a parameter to be chosen by the designer of the control system. This controller is known as an *integral* controller, because (3) is equivalent to

$$u(t) = u(0) + k \int_0^t [w_r(\tau) - w(\tau)]\, d\tau, \qquad t \geq 0.$$

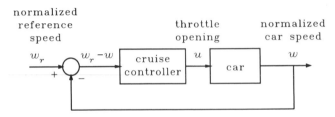

normalized
reference
speed

throttle
opening

normalized
car speed

Figure 11.2. Block diagram of the cruise control system with normalized speed.

The equations (2) and (3) together form the *state differential equation* of the cruise control system. The state of the system has the components w and u. The input to the overall system is the reference speed w_r.

We consider the *closed-loop response* of the cruise control system for different values of the design parameter k under the assumption that $\alpha = 0.1$ [s^{-1}]. In each case, initially the system is cruising at a constant speed $w_o = 0.5$, while at time 0 the reference speed w_r is changed from 0.5 to 0.6. The constant throttle opening u_o corresponding to a constant cruising speed w_o follows by setting $dw(t)/dt = 0$ in (2) and is given by $u_o = w_o^2$. Hence, the initial conditions for the state differential equation (2), (3) are $w(0) = w_o = 0.5$, $u(0) = u_o = w_o^2 = 0.25$.

Numerical integration of the state differential equation for the three parameter values

$$k = 0.02, \qquad k = 0.05, \qquad k = 0.1$$

results in the plots of Fig. 11.3. In each case the speed exactly reaches the reference value 0.6. For $k = 0.02$ the speed approaches the final value quite sluggishly, while for $k = 0.1$ it exhibits overshoot. For the intermediate value $k = 0.05$ the car speed achieves its final value most quickly.

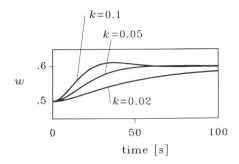

Figure 11.3. Closed-loop response of the normalized speed of the cruise control system for different values of the gain k.

A remarkable feature of the cruise control system is that the desired speed is maintained under widely varying conditions. Suppose for instance that the mass M of the car increases because the car is more heavily loaded. This results in a corresponding change in the constant $\alpha = \sqrt{cB/M}$. Figure 11.4 compares the response of the control system for the values $\alpha = 0.1$ and $\alpha = 0.08$, corresponding to a mass increase of about 20%. In both cases $k = 0.05$. The two responses differ little, and

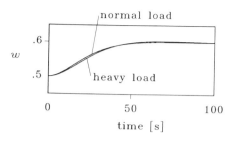

Figure 11.4.   Response of the cruise control system for two different loads.

the desired speed is reached exactly. The mass change affects the dynamic properties on the system. The example shows that the control system is *robust* with respect to such changes. This is one of the desirable properties that may be achieved by feedback.

The operation of the cruise control system may further be handicapped by difficulties such as head- or tailwind, or grades in the road. These effects, known in control engineering as *disturbances,* may be accounted for by modifying the car equation (2) to

$$\frac{dw(t)}{dt} = \alpha[u(t) - w^2(t)] + m(t), \qquad t \geq 0,$$

with m the disturbing force caused by the wind or grade. Figure 11.5 shows the response of the system under the same assumptions as before, for a constant disturbance $m(t) = m_o, t \geq 0$, which has been chosen such that the top speed of the car is reduced by 20%. The plot shows that because of the disturbance the car first slows down, but that soon the throttle position is adjusted to accommodate the changed conditions. Again the desired speed is reached exactly. The ability to counteract disturbances is an important characteristic of feedback systems.

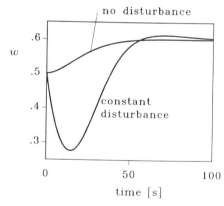

Figure 11.5.   Response of the cruise control system with and without constant disturbance.

The robustness and disturbance reduction properties of the cruise control system are impressive. The fact that the desired speed is always reached, in spite of system changes and disturbances, is a consequence of the integral control scheme: Inspection of (3) shows that if the reference speed w_r is constant, then the system can only be in a stationary condition with constant values for w and u if the error signal $w_r - w$ is zero. The concept of integral control is further pursued in 11.4.3.

A weak point of the control system design is that the speed of the car reaches the desired value quite slowly. This may be remedied by modifying the controller by adding what is called *proportional* control. This is explored in 11.5.1.   ∎

The cruise control system shows how feedback may be used to achieve robustness and disturbance reduction. The next section gives an overview of these and other important characteristics of feedback in a theoretical setting.

11.2 FEEDBACK THEORY

In this section, feedback theory is introduced, and it is shown how the simple concept of feedback has far-reaching technical implications. The cruise controller of 11.1.1 is an example of the application of feedback. Another instance is the use of feedback in electronic amplifiers.

11.2.1.   Example: A feedback audio amplifier.   An audio amplifier is an electronic device whose input is a low-power signal, such as the signal produced by a tuner or a CD or record player, and whose output has enough power to drive a loudspeaker. Such an amplifier commonly consists of a forward path in the form of a differential amplifier with several amplification stages, while a small part of the output is fed back to the input via a return path. Figure 11.6 shows the arrangement. Assuming that the forward amplifier has infinite input impedance and zero output impedance, in block diagram form the system may be represented as in Fig. 11.7.

The forward amplifier is easy to build, but suffers from such problems as distortion, limited bandwidth, variation of the amplification factor with environmental conditions and age, and noise. We shall see how feedback may be used to overcome these difficulties.   ∎

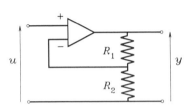

Figure 11.6.   Audio feedback amplifier.

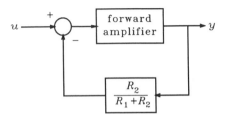

Figure 11.7.   Block diagram of the audio feedback amplifier.

Feedback Configurations

To understand and analyze feedback, we first consider the configuration of Fig.
11.8(a). The signal *r* is an external control input. The "plant" is a given system,
whose output is to be controlled. Often the function of this part of the feedback system is to provide power, and its dynamical properties are not always favorable. The
output *y* of the plant is fed back via the *return compensator* and subtracted from the
external input *r*. The difference *e* is called the *error signal* and is fed to the plant via
the *forward compensator*. The system of Fig. 11.8(b), in which the return compensator is a unit gain, is said to have *unit feedback*.

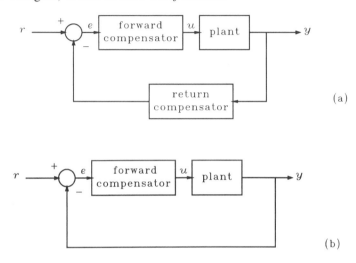

Figure 11.8.   Feedback configurations. Top: general. Bottom: unit feedback.

11.2.2.   Examples: Feedback configurations.

(a) *Cruise control system*. The cruise control system of Example 11.1.1 is a
unit feedback system. Its forward compensator is the cruise controller, and the plant
is the car.

(b) *Audio amplifier*. The audio feedback amplifier of Example 11.2.1 has a return compensator consisting of a gain $R_2/(R_1 + R_2)$. The forward compensator has
been absorbed into the plant.   ∎

For the purposes of this section we reduce the configuration of Fig. 11.8 to that of
Fig. 11.9, where the forward compensator has been absorbed into the plant. The
plant is represented as an IOM system with IO map ϕ, while the return compensator
has the IO map ψ. The control input *r*, the error signal *e*, and the output signal *y*
usually all are time signals. Correspondingly, ϕ and ψ are IO maps of dynamical
systems, mapping time signals to time signals.

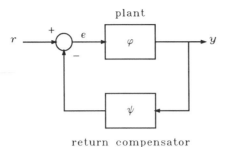

Figure 11.9. Feedback configuration with input-output maps.

The feedback system is represented by the equations

$$y = \phi(e), \qquad e = r - \psi(y).$$

These equations may or may not have a solution e and y for any given control input r. If a solution exists, the error signal e satisfies the equation $e = r - \psi(\phi(e))$, or

$$e + \gamma(e) = r. \tag{1}$$

Here,

$$\gamma = \psi \circ \phi$$

is the IO map of the series connection of the plant followed by the return compensator and is called the *loop IO map*. Equation (1) reduces the feedback system to a unit feedback system as in Fig. 11.10. Note that because γ maps time functions into time functions, (1) is a *functional* equation for the time signal e. We refer to it as the *feedback equation*.

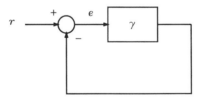

Figure 11.10. Equivalent unit feedback configuration.

11.2.3.   Example: Memoryless feedback loop.   To help intuition we consider the special case that the plant and return compensator are both memoryless and time-invariant. The plant and return compensator then may be described by memoryless IO maps $\phi: \mathbb{R} \to \mathbb{R}$ and $\psi: \mathbb{R} \to \mathbb{R}$. The memoryless loop IO map $\gamma: \mathbb{R} \to \mathbb{R}$ may be represented by its graph as in Fig. 11.11. Rewriting the feedback equation (1) as

$$r - e = \gamma(e),$$

with r and e assuming values in $\mathbb{R}$, we see that for any fixed value of r the solution of this equation may be found by intersecting the graph of the function $r - e$,

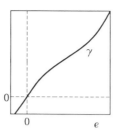

Figure 11.11.   Graph of the loop IO map.

$e \in \mathbb{R}$, with that of $\gamma(e)$, $e \in \mathbb{R}$. It is easy to think of a function γ such that the graphs do not intersect and, hence, no solution exists, as in Fig. 11.12(a). It is also easy to imagine functions γ such that there are several intersections, as in Fig. 11.12(b), and, hence, several solutions exist. There may also be a unique solution, as in Fig. 11.12(c).   ■

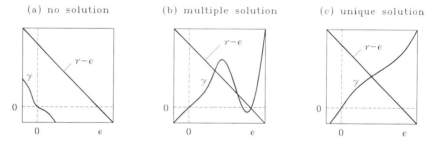

Figure 11.12.   Graphical solution of the feedback equation. Left: no solution exists. Middle: several solutions exist. Right: a unique solution exists.

High-Gain Feedback

Feedback is most effective if the loop IO map γ has a "large gain." One of the important consequences of this is that the map from the external input r to the output y is approximately the inverse ψ^{-1} of the IO map ψ of the return compensator. We first explain this loosely and then state the result more precisely.

Suppose that for a given external input r the feedback equation

$$e + \gamma(e) = r \tag{2}$$

has a solution e. Suppose also that the "gain" of the map γ is large, that is,

$$\|\gamma(e)\| >> \|e\| \qquad \text{for all} \quad e,$$

with $\|\cdot\|$ the norm on the signal space in which e is defined. Then in (2) we may neglect the first term on the left, so that

$$\gamma(e) \approx r.$$

Since by assumption $\|e\| << \|\gamma(e)\|$ it follows from this that

$$\|e\| << \|r\|.$$

In words: If the gain is large, the error e is small compared with the control input r. Going back to the configuration of Fig. 11.9 we see that this implies that

$$\psi(y) \approx r,$$

or

$$y \approx \psi^{-1}(r),$$

where ψ^{-1} is the *inverse* of the map ψ (assuming that it exists).

Before stating this result more precisely we look at an example.

11.2.4.   Example: Linear memoryless feedback system.   As an example where the approximations can be carried through explicitly we consider the system of Fig. 11.13, where the plant and the return compensator both are gains k_1 and k_2, respectively. The maps ϕ, ψ, and γ are memoryless, time-invariant, and linear and are given by

$$\phi(e) = k_1 e, \qquad \psi(y) = k_2 y, \qquad \gamma(e) = k_1 k_2 e.$$

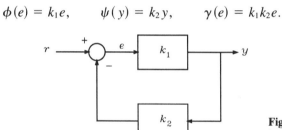

Figure 11.13.   Linear memoryless feedback system.

The "gain" of the map γ is large if $|k_1 k_2| >> 1$. The feedback equation (2) takes the form

$$e + k_1 k_2 e = r,$$

and has the explicit solution

$$e = \frac{1}{1 + k_1 k_2} r.$$

A necessary and sufficient condition for the existence of a solution e for all r is that $k_1 k_2 \neq -1$. The solution is unique. Given e, we find for the output y of the feedback system

$$y = k_1 e = \frac{k_1}{1 + k_1 k_2} r. \tag{3}$$

Suppose, for instance, that $k_1 k_2 = 100$. Then,

$$e = \frac{1}{101} r,$$

confirming that the error signal is small compared with the control input r. Rewriting (3) as

$$y = \frac{k_1 k_2}{1 + k_1 k_2} \frac{1}{k_2} r = \frac{100}{101} \frac{1}{k_2} r$$

confirms that the IO map of the feedback system is nearly equal to the inverse of the IO map of the return compensator. If $k_2 = 1$ (i.e., the system has *unit* feedback), then the IO map of the feedback system is given by

$$y = \frac{100}{101} r,$$

which is quite close to the unit map.   ∎

We now state our results in general for the case when the feedback equation defines a dynamical IO map from a normed signal space $\mathcal{Y}$ into itself. In what follows, the norm $\|\cdot\|$ on the space $\mathcal{Y}$ is always the amplitude $\|\cdot\|_\infty$. Hence, any signal in $\mathcal{Y}$ is *bounded*.

11.2.5.   Summary: High-gain feedback.   Suppose that the feedback equation

$$e + \gamma(e) = r$$

has a solution $e \in \mathcal{Y}$ for every $r \in \mathcal{Y}$, with $\mathcal{Y}$ a normed signal space. Assume that there exists a constant $k > 1$ such that

$$\|\gamma(e)\| \geq k \|e\| \qquad \text{for all} \quad e \in \mathcal{Y}. \tag{4}$$

The largest value of k for which this holds is called the *gain* of the map γ. Then the error signal e is bounded by

$$\|e\| \leq \frac{1}{k-1} \|r\|,$$

so that

$$\|r - \psi(y)\| \leq \frac{1}{k-1} \|r\|.$$

In particular, if $k \gg 1$, then

$$\|e\| \ll \|r\| \quad \text{and} \quad \|r - \psi(y)\| \ll \|r\|. \qquad \blacksquare$$

11.2.6.   Proof. By the triangle inequality, $\|r - e\| \leq \|r\| + \|e\|$. It follows from (4) with the feedback equation $r - e = \gamma(e)$ that

$$k\|e\| \leq \|\gamma(e)\| = \|r - e\| \leq \|r\| + \|e\|,$$

so that

$$\|e\| \leq \frac{1}{k-1}\|r\|. \qquad \blacksquare$$

The result 11.2.5 shows quite in general that *if* the feedback equation has a solution, then a large loop gain results in a small error signal. Note that it is assumed that the feedback equation has a bounded solution e for every bounded r. This is not necessarily always the case. If e is bounded for every bounded r, then the closed-loop system by definition is BIBO stable. Hence, the existence of solutions to the feedback equation is equivalent to the (BIBO) stability of the closed-loop system. The stability of feedback systems is discussed in the next section.

If the gain of the loop IO map is large, then the error signal is small, and $r \approx \psi(y)$, so that the IO map of the feedback system approximately equals the inverse ψ^{-1} of the IO map of the return compensator. When the return compensator is the identity map, its inverse map is also the identity map. It follows that the IO map of a high-gain unit feedback system approximately equals the identity map.

Robustness of Feedback Systems

The approximate identity

$$y \approx \psi^{-1}(r)$$

remains valid as long as the feedback equation has a bounded solution and the gain is large. The IO map ψ of the return compensator may often be implemented with good accuracy. This results in a matching accuracy for the IO map of the feedback system as long as the gain is large, even if the IO map of the plant is poorly defined or has unfavorable properties.

The fact that $y \approx \psi^{-1}(r)$ in spite of uncertainty about the plant dynamics is called *robustness* of the feedback system with respect to plant uncertainty. Mathematically, robustness amounts to the trivial observation that the inequality

$$\|r - \psi(y)\| \leq \frac{1}{k-1}\|r\|$$

remains valid as long as the loop IO map $\gamma = \phi \circ \psi$ satisfies

$$\| \gamma(e) \| \geq k \| e \|, \qquad e \in \mathcal{Y},$$

with $k > 1$, and as long as the feedback equation has a solution.

11.2.7.   Example: Robustness of a linear memoryless feedback system.   In the linear memoryless feedback system of Example 11.2.4 we found that if $k_1 k_2 = 100$, then the IO map of the feedback system is given by

$$y = \frac{100}{101} \frac{1}{k_2} r.$$

If the gain $k_1 k_2$ changes, say, from 100 to 200, with k_2 fixed, then the error signal and output become

$$e = \frac{1}{201} r, \qquad y = \frac{200}{201} \frac{1}{k_2} r.$$

This shows that the error signal remains small and the IO map of the feedback system remains approximately equal to the inverse of the IO map of the return compensator, in spite of the gain variation by a factor of two. Apparently, these properties are robust with respect to gain variations, provided the gain remains large.   ∎

11.2.8.   Example: Robustness of the cruise control system.   The robustness of the cruise control system of 11.1.1 with respect to a change in load is clearly demonstrated by Fig. 11.4. The reason for the robustness is that the integral controller provides a high loop gain.   ∎

Linearity and Bandwidth Improvement by Feedback

Besides robustness, several other favorable effects may be achieved by feedback. They include linearity improvement, bandwidth improvement, and disturbance reduction.

Linearity improvement is a consequence of the fact that if the loop gain is large enough, the IO map of the feedback system approximately equals the inverse ψ^{-1} of the IO map of the return compensator. If this IO map is linear, then so is the IO map of the feedback system, with good approximation, no matter how nonlinear the plant IO map ϕ is.

Also *bandwidth improvement* is a result of the high gain property. If the return compensator is a unit gain, then the IO map of the feedback system is close to unity over those frequencies for which the feedback gain is large.

Before explaining how feedback may reduce disturbances, we present two examples that illustrate the linearity and bandwidth improvement feedback may achieve.

11.2.9.   Example: Linearity improvement of a memoryless feedback system.   We consider a nonlinear memoryless feedback system with the configuration of Fig. 11.14. The plant is assumed to have an IO map ϕ given by

$$\phi(u) = u + \alpha u^3, \qquad u \in \mathbb{R},$$

where the nonnegative coefficient α is a measure for the nonlinearity of the plant. The forward compensator is chosen to consist of a constant gain $k > 0$, so that the loop gain map γ is given by

$$\gamma(e) = ke + \alpha k^3 e^3, \qquad e \in \mathbb{R}.$$

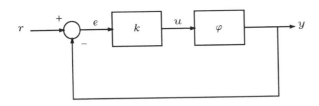

Figure 11.14.   Nonlinear memoryless feedback system.

Since

$$|\gamma(e)| = |ke(1 + \alpha k^2 e^2)| \geq k|e|, \qquad e \in \mathbb{R},$$

the loop gain is large if $k \gg 1$.

Suppose that $\alpha = 1$ and $k = 100$. Analytical solution of the feedback equation (2) is not feasible. Numerical solution of (2) for e for a range of values of r and substitution of the resulting values of e into $y = \gamma(e) = ke + \alpha k^3 e^3$, $e \in \mathbb{R}$, yields the graph of the IO map ϕ_{cl} of the closed-loop system given in Fig. 11.15. The map ϕ_{cl} gives the output $y = \phi_{cl}(r)$ of the feedback system for a given control input r. It is seen that compared with the IO map $\phi(u) = u + \alpha u^3$, $u \in \mathbb{R}$, of the plant linearity has considerably improved.   ∎

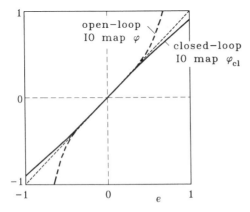

Figure 11.15.   Open- and closed-loop IO maps of the memoryless feedback system.

11.2.10.   Example: Bandwidth improvement of an audio amplifier.   To demonstrate how feedback may improve bandwidth, we consider the audio feedback amplifier of Example 11.2.1. Figure 11.7 is a feedback configuration of the type of Fig. 11.9, where the return compensator is a constant gain $\beta := R_2/(R_1 + R_2)$. Let us assume that the forward amplifier is a linear time-invariant system with first-order transfer function

$$H(s) = \frac{k}{1 + sT}, \qquad \text{Re } (s) > -1/T. \tag{5}$$

This system has a time constant $T > 0$ and a bandwidth roughly equal to $1/T$. The constant k is the zero frequency gain.

   We analyze this system in terms of Laplace transforms with the help of the block diagram of Fig. 11.16. Assuming that the external input r, the error signal e and the output y all have one-sided Laplace transforms given by R, E, and Y, respectively, and assuming zero initial conditions, the system is described by the equations

$$Y = HE, \qquad E = R - \beta Y.$$

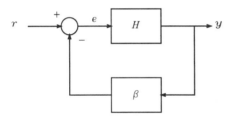

Figure 11.16.   Audio feedback amplifier.

Elimination of E yields $Y = H(R - \beta Y)$, or

$$Y = \frac{H}{1 + H\beta} R.$$

Thus, the closed-loop system is again linear time-invariant, with transfer function

$$H_{\text{cl}} = \frac{H}{1 + H\beta}.$$

H_{cl} is called the *closed-loop transfer function*. Substituting H as given by (5) we find

$$H_{\text{cl}}(s) = \frac{\dfrac{k}{1 + sT}}{1 + \dfrac{k\beta}{1 + sT}} = \frac{k}{(1 + k\beta) + sT} = \frac{\dfrac{k}{1 + k\beta}}{1 + s\dfrac{T}{1 + k\beta}}. \tag{6}$$

This is a well-defined one-sided Laplace transform for Re $(s) > -(1 + k\beta)/T$, and again represents a first-order system, with time constant $T/(1 + k\beta)$ and corresponding bandwidth $(1 + k\beta)/T$. The closed-loop system has zero frequency gain $k/(1 + k\beta)$. Figure 11.17 shows a double-logarithmic magnitude plot of the closed-loop frequency response function $\hat{h}_{cl}(f) = H_{cl}(j2\pi f)$, $f \in \mathbb{R}$, for $k\beta = 10$ and 100, with $k = 10^4$ and $T = 0.01$.

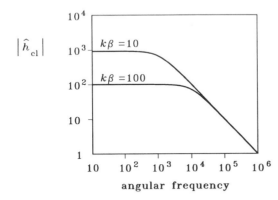

Figure 11.17. Magnitude plot of the closed-loop frequency response function of the audio amplifier.

We observe the following points:

(a) If $k\beta$ is large the zero frequency gain of the closed-loop system is approximately $1/\beta$.

(b) If $k\beta$ is large, the bandwidth $(1 + k\beta)/T \approx k\beta/T$ is much larger than the bandwidth $1/T$ of the forward amplifier.

This shows that the frequency response function of the closed-loop system approximately equals the inverse $1/\beta$ of the frequency response function of the return compensator over a frequency region with a bandwidth that is $k\beta$ times larger than that of the amplifier without feedback. The bandwidth improvement is obtained at the cost of a reduction of the zero frequency gain from k to $k/(1 + k\beta)$.   ∎

Disturbance Reduction

A further useful property of feedback is that the effect of (external) *disturbances* is reduced. It frequently happens that in the configuration of Fig. 11.8 external disturbances affect the output y. These disturbances are usually caused by environmental effects.

11.2.11.   Example: Disturbances in the cruise control system and amplifier.   External disturbances affecting the car speed in the cruise control system of Example 11.1.1 are head or tail winds and grades in the road. Disturbances that affect the output of the amplifier of Example 11.2.1 consist of electronic noise (which is internally generated in the amplifier circuits) and interference from other electrical or electronic equipment.   ∎

The effect of disturbances may often be modeled by adding a *disturbance signal d* at the output of the plant as in Fig. 11.18. For simplicity we study the effect of the disturbance in the *absence* of any external control input (i.e., we assume $r = 0$). The feedback system then is described by the equations

$$z = d + y, \qquad y = \phi(e), \qquad e = -\psi(z).$$

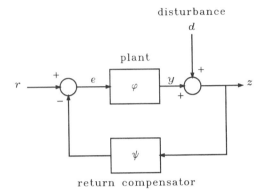

Figure 11.18.   Feedback system with disturbance.

Eliminating the output y and the error signal e we have $z = d + \phi(e) = d + \phi(-\psi(z))$, or

$$z = d - \delta(z), \tag{7}$$

where

$$\delta = (-\phi) \circ (-\psi).$$

The map δ is also called a *loop IO map*, but it is obtained by breaking the loop at a different point compared with when obtaining the loop IO map γ.

The equation (7) is a feedback equation for the configuration of Fig. 11.19. By analogy with the configuration of Fig. 11.10 it follows that if the gain is *large* in the sense that $\| \delta(z) \| >> \| z \|$ we have

$$\| z \| << \| d \|.$$

This means that the output z of the feedback system is small compared with the disturbance d, so that the effect of the disturbance is much reduced. All this holds *provided* the feedback equation (7) has at all a bounded solution z for any bounded d.

The precise result may be stated as follows.

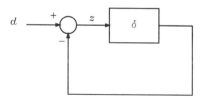

Figure 11.19.   Equivalent unit feedback configuration in the absence of the control input r.

11.2.12.   Summary: Disturbance reduction by feedback.   In the feedback configuration of Fig. 11.18, let $r = 0$, and suppose that the feedback equation

$$z + \delta(z) = d$$

with $\delta = (-\phi) \circ (-\psi)$, has a solution $z \in \mathcal{Y}$ for every $d \in \mathcal{Y}$, with $\mathcal{Y}$ a normed signal space. Assume that there exists a constant $k > 1$ such that

$$\|\delta(z)\| \geq k\|z\| \qquad \text{for all} \quad z \in \mathcal{Y}.$$

Then the system output is bounded by

$$\|z\| \leq \frac{1}{k-1}\|d\|.$$

In particular, if $k \gg 1$, then

$$\|z\| \ll \|d\|. \qquad\qquad\qquad\qquad\qquad\qquad \blacksquare$$

The proof is analogous to 11.2.6.

11.2.13.   Example: Cruise controller.   As was illustrated in Fig. 11.5, external disturbances in the form of wind or grades in the road have little effect on the speed of the car. This is caused by the high gain provided by the integral controller.

$\qquad$ ■

11.2.14.   Example: Noise reduction in the audio amplifier.   Electronic networks internally generate a spurious electrical signal called *noise*, which is caused by the thermal motion of electrons. In the feedback audio amplifier of Examples 11.2.1 and 11.2.10 we may represent the noise as an additive disturbance d at the output of the amplifier as in Fig. 11.20. In terms of Laplace transforms, the feedback system with noise is described by the equations

$$Z = D + HE, \qquad E = R - \beta Z,$$

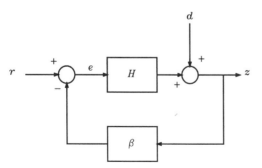

Figure 11.20.   Feedback amplifier with noise.

with Z, D, E, and R the one-sided Laplace transforms of the signals z, d, e, and r, respectively, and H the transfer function of the amplifier. Eliminating E we have $Z = D + H(R - \beta Z)$, so that

$$Z = \frac{H}{1 + H\beta}R + \frac{1}{1 + H\beta}D.$$

Denoting the closed-loop transfer function from the external input r to the output y as

$$H_{\text{cl}} = \frac{H}{1 + H\beta},$$

as in 11.2.10, and defining

$$S := \frac{1}{1 + H\beta}$$

as the *sensitivity function* of the closed-loop system, we have

$$Z = H_{\text{cl}}R + SD.$$

In 11.2.10 we analyzed the closed-loop transfer function H_{cl} assuming that $H(s) = k/(1 + sT)$, Re $(s) > -1/T$. For the sensitivity function we obtain

$$S(s) = \frac{1}{1 + \dfrac{k\beta}{1 + sT}} = \frac{1 + sT}{(1 + k\beta) + sT} = \frac{s + \dfrac{1}{T}}{s + \dfrac{1 + k\beta}{T}},$$

which is a valid one-sided Laplace transform for Re $(s) > -(1 + k\beta)/T$. Figure 11.21 shows a double-logarithmic magnitude plot of the frequency response function $S(j2\pi f)$, $f \in \mathbb{R}$. The asymptotic magnitude plot (compare 4.9.5) equals $1/(1 + k\beta)$ until the angular break frequency $1/T$, and then increases at a rate of 1

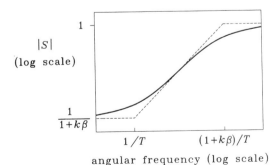

Figure 11.21. Double-logarithmic plot of the magnitude of the sensitivity function of the audio amplifier.

decade/decade until it reaches the value 1 at the break angular frequency $(1 + k\beta)/T$. If $k\beta$ is large, then the noise is attenuated by a factor $k\beta$ at low frequencies. For frequencies larger than $1/T$, the attenuation is less.   ∎

Pitfalls of Feedback

As we have seen, feedback may achieve very useful effects. Feedback also has pitfalls:

(a) Naively making the gain of the system large may easily result in an *unstable* feedback system. If the feedback system is unstable, then the feedback equation has no bounded solutions and the beneficial effects of feedback are nonexistent.

(b) Even if the feedback system is stable, high gain may result in overly large inputs to the plant that the plant cannot absorb. The result is reduction of the gain and an associated loss of performance.

(c) Feedback as in Fig. 11.8 usually involves measuring the output by means of an output sensor. The associated *measurement errors* and *measurement noise* may cause loss of accuracy.

Three examples illustrate these points.

11.2.15.   Example: Stability of the feedback audio amplifier.   In the arguments that were used in 11.2.10 and 11.2.14 to point out the favorable effects of feedback on the audio amplifier it was important that $k\beta$ be large, but its *sign* was irrelevant. From (6) we see by inverse Laplace transformation that the impulse response of the closed-loop system is

$$h_{\mathrm{cl}}(t) = k_c e^{-t/T_c} \mathbb{1}(t), \qquad t \in \mathbb{R},$$

with $k_c = k/(1 + k\beta)$ the closed-loop zero frequency gain and $T_c = T/(1 + k\beta)$ the closed-loop time constant. If $k\beta$ is large but *negative*, then the system has a negative closed-loop time constant T_c and, hence, is BIBO *unstable*. If the closed-loop system is BIBO unstable, then the error signal e and the output y are generally unbounded, so that 11.2.5 and 11.2.12 do not apply. An unstable feedback system is useless.

For the first-order amplifier, the stability problem is easily resolved by making the gain positive. In the next section we meet other examples where it is not so simple to guarantee the stability of the feedback system.   ∎

11.2.16.   Example: Saturation of the input in automatic control systems.   Automatic control systems such as the cruise control system of Example 11.1.1 are often designed as unit feedback systems with a forward compensator as in Fig. 11.8(b). The loop gain of such a feedback configuration may be made large by making the gain of the forward compensator large. Even if the error signal e

is small, a large gain of the forward compensator may easily result in plant inputs u that are large. Inputs that are too large may cause malfunction of the plant or even damage it. For this reason, the forward compensator usually has a limiting mechanism that prevents the input from exceeding certain bounds. This results in a reduction of the loop gain and adversely affects the performance of the feedback system.   ∎

11.2.17.   Example: Measurement noise in a unit feedback system.   Consider the unit feedback system of Fig. 11.22, where the combined plant and forward compensator form a linear time invariant system with transfer function H. The measurement errors and measurement noise are represented by a signal w that is added to the output y before it is fed back. In terms of Laplace transforms we write

$$Y = H(R - W - Y),$$

which may be solved for the Laplace transform Y of y in terms of the Laplace transforms R and W of r and w as

$$Y = \frac{1}{1 + H}(R - W).$$

This shows that within a minus sign the measurement noise w has the *same* contribution to the control system output y as the external input r. As a result, over the frequency band of the closed-loop system the measurement noise is transmitted to the output intact. The reason is, of course, that in the configuration of Fig. 11.22 the measured signal z is mistaken for the output y.

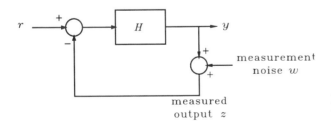

measured
output z

Figure 11.22.   Unit feedback system with measurement noise.

It is difficult to abate the effect of measurement noise (by reducing the loop gain, for instance) without sacrificing some of the advantages of feedback. It therefore is important to use good sensors that introduce as little measurement noise as possible.   ∎

11.3 STABILITY OF FEEDBACK SYSTEMS

In Section 11.2 we saw that making the loop gain of a feedback system large results in linearity and bandwidth improvement, reduction of disturbances, and robustness against plant variations. A basic assumption in the proofs of these properties is that the feedback equation has a bounded solution for any bounded input (i.e., that the closed-loop system is BIBO *stable*). Unfortunately, feedback systems, even those that are stable for small values of the loop gain, tend to become unstable if the loop gain is increased. Instability occurs if too much of the output is fed back and the timing (i.e., the phase) of the feedback signal is wrong. Two everyday examples illustrate this.

11.3.1.   Examples: Unstable feedback systems.

(a) *Acoustic feedback in a public address system*. A familiar example of a system rendered unstable by feedback is a public address system where the microphone is placed in front of the loudspeakers (Fig. 11.23(a)) and the gain of the amplifier is turned up too high. The output of the loudspeakers is directly picked up by the microphone and fed back so that the signal keeps increasing until the amplifier saturates.

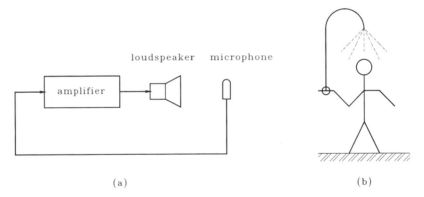

Figure 11.23. Two unstable feedback systems.

(b) *An impatient person taking a shower*. Another well-known instance of oscillation caused by feedback arises when an impatient person takes a shower, turning up the hot water tap too sharply when the water is too cold, and turning it back too abruptly when the water gets too hot (Fig. 11.23(b).) Because the water temperature responds with a delay, oscillations result.   ∎

In the remainder of this section the relation between the loop gain and the stability of a feedback system is studied. The results make it possible to develop methods to design a feedback system such that its loop gain is as large as possible while its stability is retained.

Our study of the stability of feedback systems is organized as follows. We first present a result, known as the *small gain theorem,* that is applicable to a quite general class of systems, including nonlinear systems. The price for this generality is that the small gain theorem only is a *sufficient* condition for stability. As the name suggests, its application is limited to feedback systems with a small loop gain, while we are usually interested in making the loop gain high. To find conditions for stability that are sufficient as well as necessary, we therefore subsequently restrict the discussion to linear time-invariant systems.

The Small Gain Theorem

Preparatory to formulating the small gain theorem we state a well-known fact from functional analysis.

11.3.2.   Summary: Fixed point theorem.   Let $\mathcal{Y}$ be a complete normed space (i.e., a normed space $\mathcal{Y}$ where every sequence x_n, $n \in \mathbb{N}$, such that $\lim_{n, m \to \infty}$ $\|x_n - x_m\| = 0$ has a limit in $\mathcal{Y}$), and $T: \mathcal{Y} \to \mathcal{Y}$ a map such that

$$\|T(u) - T(v)\| \le k\|u - v\| \qquad \text{for all} \quad u, v \in \mathcal{Y},$$

for some real constant $0 \le k < 1$. Such a map T is called a *contraction.* Then:

(a) There exists a *unique* solution $u^* \in \mathcal{Y}$ of the equation

$$u = T(u).$$

(b) The sequence u_n, $n = 0, 1, 2, \cdots$, defined by

$$u_{n+1} = T(u_n), \qquad n = 0, 1, 2, \cdots,$$

converges to u^* for every initial point $u_0 \in \mathcal{Y}$.   ∎

We apply the fixed point theorem to the feedback equation

$$e = r - \gamma(e),$$

which describes the feedback system of Fig. 11.9. The external input r is assumed to take its values in some (complete) normed signal space $\mathcal{Y}$. As before, the norm that is used is the amplitude $\| \cdot \|_\infty$. For any fixed r the feedback equation may be written in the form $e = T(e)$, with the map $T: \mathcal{Y} \to \mathcal{Y}$ given by

$$T(e) = r - \gamma(e).$$

It follows that $T(e_1) - T(e_2) = \gamma(e_2) - \gamma(e_1)$, so that T is a contraction if and only if γ is also a contraction, that is, if γ maps $\mathcal{Y}$ to $\mathcal{Y}$ and there exists a constant $0 \leq k < 1$ such that

$$\|\gamma(e_1) - \gamma(e_2)\| \leq k\|e_1 - e_2\| \qquad \text{for all} \quad e_1, e_2 \in \mathcal{Y}.$$

If T is a contraction, by the fixed point theorem the feedback equation has a unique bounded solution e for every bounded r. When the error signal e is taken as the output of the feedback system, this means that the feedback system is BIBO stable.

11.3.3.   Summary: Small-gain theorem.   Suppose that the loop IO map $\gamma: \mathcal{Y} \rightarrow \mathcal{Y}$ of the feedback system of Fig. 11.9 is a contraction, that is, $\mathcal{Y}$ is a complete normed space and there exists a constant $0 \leq k < 1$ such that

$$\|\gamma(e_1) - \gamma(e_2)\| \leq k\|e_1 - e_2\| \qquad \text{for all} \quad e_1, e_2 \in \mathcal{Y}.$$

Then the feedback equation

$$r = e + \gamma(e)$$

has a solution $e \in \mathcal{Y}$ for every $r \in \mathcal{Y}$.   ∎

The small gain theorem states, roughly, that if the loop gain is small enough, the closed-loop system is BIBO stable. Although this result applies to very general nonlinear and time-varying feedback systems, it only offers a *sufficient* condition for closed-loop stability. There are many stable feedback systems to which the small gain theorem cannot be applied. The reason is that the small gain theorem is based on amplitude considerations, and phase and timing information is ignored.
    The following examples illustrates the small gain theorem.

11.3.4.   Examples: Small gain theorem.
    (a) *Linear memoryless feedback system.* The loop gain map of the linear memoryless feedback system of Example 11.2.4 is given by

$$\gamma(e) = k_1 k_2 e.$$

It follows that

$$\|\gamma(e_1) - \gamma(e_2)\| = \|k_1 k_2(e_1 - e_2)\| = |k_1 k_2| \cdot \|e_1 - e_2\|,$$

so that γ is a contraction if and only if $|k_1 k_2| < 1$. By the small gain theorem, the latter is a sufficient condition for the existence of a unique solution of the feedback equation. From the direct analysis of Example 11.2.4, however, we know that a necessary as well as sufficient condition for the existence of the solution of the feedback equation is that $k_1 k_2 \neq -1$. Hence, the small gain theorem gives a quite con-servative sufficient condition for the existence of the solution.

(b) *Nonlinear memoryless feedback system.* In Example 11.2.9 we considered a nonlinear memoryless feedback system with memoryless loop IO map $\gamma \colon \mathbb{R} \to \mathbb{R}$, given by

$$\gamma(e) = ke + \alpha k^3 e^3.$$

If $\alpha > 0$ this is not a contraction for any positive value of k. Nevertheless, the feed-back equation has a unique solution for all positive values of α and k.   ∎

We next consider what the small gain theorem can tell us about the stability of a *lin-ear time-invariant* unit feedback system as in Fig. 11.24, where the forward com-pensator has been absorbed into the plant. The plant is assumed to have impulse re-sponse h, so that the loop IO map is the convolution map

$$\gamma(e) = h * e.$$

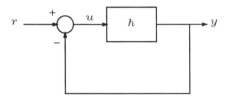

Figure 11.24.   Linear time-invariant
unit feedback system.

For the application of the small gain theorem the loop IO map must be BIBO stable. Hence, by 3.6.2(a) we need assume that the impulse h has finite action $\|h\|_1$. By 3.6.2(b) and linearity the loop IO map satisfies

$$\|\gamma(e_1) - \gamma(e_2)\|_\infty = \|h * (e_1 - e_2)\|_\infty$$
$$\leq \|h\|_1 \cdot \|e_1 - e_2\|_\infty \qquad \text{for any} \quad e_1, e_2 \in \mathcal{Y}.$$

By 3.6.2(b) equality may be assumed. Hence, the loop gain is a contraction if and only if $\|h\|_1 < 1$. It follows from the small gain theorem that this is a sufficient con-dition for BIBO stability of the closed-loop system.

We summarize as follows.

11.3.5.   Summary: Small gain theorem for linear feedback systems.   A sufficient condition for the discrete- or continuous-time linear time-invariant feed-back system of Fig. 11.24 to be BIBO stable is that the open-loop impulse response h satisfies

$$\|h\|_1 < 1.$$

The following example shows that also this result has limited applicability.

11.3.6.   Example: Application of the small gain theorem to the audio amplifier.   The loop gain transfer function of the audio amplifier of Example 11.2.10 is

$$L(s) = \frac{k\beta}{1 + sT}, \qquad \text{Re}\,(s) > -1/T.$$

The corresponding impulse response follows by inverse Laplace transformation as

$$h(t) = \frac{k\beta}{T}e^{-t/T}\mathbb{1}(t), \qquad t \in \mathbb{R}.$$

It easily follows that

$$\|h\|_1 = \frac{|k\beta|}{T}\int_0^\infty e^{-t/T}\,dt = |k\beta|.$$

Hence, by 11.3.5 a sufficient condition for the BIBO stability of the closed-loop system is that $|k\beta| < 1$, or

$$-1 < k\beta < 1.$$

In Example 11.2.10, however, we found that the transfer function of the closed-loop system has a single pole at $-(1 + k\beta)/T$. It follows that the closed-loop system is BIBO stable if and only if $(1 + k\beta)/T > 0$, or, equivalently, if and only if

$$k\beta > -1.$$

Again, the small gain theorem gives a conservative estimate of the conditions for stability.   ∎

Stability of Linear Time-Invariant Feedback Systems with Rational Transfer Functions

For linear time-invariant feedback systems we may derive much stronger stability conditions than that of 11.3.5, which are not only *sufficient* but also *necessary*. Consider the discrete- or continuous-time feedback system of Fig. 11.25, and assume that the plant, forward compensator, and return compensator have the rational transfer functions

$$H = \frac{P_p}{Q_p}, \qquad F = \frac{P_f}{Q_f}, \qquad G = \frac{P_r}{Q_r},$$

with the Ps and Qs polynomials. We suppose that the P and Q polynomials derive directly from the difference or differential equation describing each subsystem and that any common factors of the numerators and denominators have *not* been canceled.

Given H, G, and F we define the *loop transfer function L* as

$$L = HFG = \frac{P}{Q},$$

where

$$P := P_p P_f P_r, \qquad Q := Q_p Q_f Q_r.$$

The stability of the closed-loop system is fully determined by the loop gain transfer function L. To see this, we first write the equations that describe the closed-loop system in terms of one-sided Laplace transforms as

$$Y = HU, \qquad U = FE, \qquad E = R - GY.$$

Solution for E, U, and Y in terms of the control input R is straightforward and results in

$$E = \frac{1}{1 + L} R = \frac{1}{1 + \dfrac{P}{Q}} R = \frac{Q}{Q + P} R,$$

$$U = \frac{F}{1 + L} R = \frac{\dfrac{P_f}{Q_f}}{1 + \dfrac{P}{Q}} R = \frac{P_f Q_p Q_r}{Q + P} R, \qquad (1)$$

$$Y = \frac{HF}{1 + L} R = \frac{\dfrac{P_p P_f}{Q_p Q_f}}{1 + \dfrac{P}{Q}} R = \frac{P_p P_f Q_r}{Q + P} R.$$

Inspection shows that each of the transfer functions has the *same* denominator

$$Q_{\mathrm{cl}} := Q + P = Q_p Q_f Q_r + P_p P_f P_r.$$

The difference equations (in the discrete-time case) or differential equations (in the continuous-time case) describing the time behavior of the signals e, u, and y may be read off from the transfer functions as

$$Q_{cl}e = Qr,$$

$$Q_{cl}u = P_f Q_p Q_r r,$$

$$Q_{cl}y = P_p P_f Q_r r.$$

Each of these three equations has Q_{cl} as its characteristic polynomial, which is the reason why we refer to Q_{cl} as the *closed-loop characteristic polynomial*. The stability of the closed-loop system depends on the locations of the roots of Q_{cl}. We summarize as follows.

11.3.7.   Summary: Closed-loop characteristic polynomial and stability of a linear time-invariant feedback system.   Suppose that in the discrete- or continuous-time unit feedback configuration of Fig. 11.25 the transfer functions $H = P_p/Q_p$, $F = P_f/Q_f$, and $G = P_r/Q_r$ of the plant, the forward and the backward compensator, respectively, are rational. Then the characteristic polynomial of the closed-loop system is

$$Q_{cl} = Q_p Q_f Q_r + P_p P_f P_r.$$

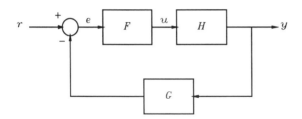

Figure 11.25. Linear time-invariant feedback system.

The closed-loop system is CICO stable if and only if the closed-loop characteristic roots (i.e., the roots of Q_{cl}) are all strictly inside the unit circle (in the discrete-time case) or strictly in the left-half complex plane (in the continuous-time case.)   ∎

Thus, the stability of the closed-loop system may be ascertained by determining the closed-loop characteristic polynomial Q_{cl} and verifying the location of its roots.

11.3.8.   Example: Audio amplifier.
    (a) *First-order plant*. In 11.2.10 we considered a feedback audio amplifier, consisting of a plant with transfer function

$$H(s) = \frac{k}{1 + sT}$$

and a return compensator with constant gain β. It follows that $F = 1$ and $G = \beta$, so that the loop gain transfer function of this feedback system is

$$L(s) = \frac{k\beta}{1 + sT}.$$

Hence, $Q(s) = 1 + sT$ and $P(s) = k\beta$. As a result, the closed-loop characteristic polynomial is

$$Q_{cl}(s) = Q(s) + P(s) = (1 + sT) + k\beta = sT + (1 + k\beta).$$

This polynomial has the single characteristic root

$$\lambda_1 = -\frac{1 + k\beta}{T}.$$

In Fig. 11.26(a) the locus of the root λ_1 in the complex plane is plotted as the gain $k\beta$ increases from 0 to $+\infty$. For $k\beta = 0$ (corresponding to no feedback at all) the locus starts at $-1/T$, which is the location of the (single) pole of the loop gain L. As $k\beta$ increases, the closed-loop root moves along the negative real axis toward $-\infty$. The root stays in the left-half plane for all positive $k\beta$, which means that stability is always ensured, even if the gain is made arbitrarily large.

(a)   (b)   (c)

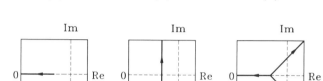

Figure 11.26. Loci of the closed-loop poles of the feedback audio amplifier. Left: first-order plant. Middle: second-order plant. Right: third-order plant.

The closed-loop transfer function, i.e., the transfer function from the control input r to the output y, is given by

$$H_{cl}(s) = \frac{H(s)F(s)}{1 + L(s)} = \frac{k}{(1 + sT) + k\beta},$$

as already obtained in 11.2.10. The corresponding closed-loop frequency response function $\hat{h}_{cl}(f) = H_{cl}(j2\pi f), f \in \mathbb{R}$, is plotted in Fig. 11.27(a) for several values of $k\beta$, where we have taken $T = 0.01$ and $k = 10^4$. In each case the frequency response function has been scaled by $\hat{h}_{cl}(0)$ before plotting. As $k\beta$ increases, the bandwidth increases. The closed-loop frequency response remains very well-behaved for all positive $k\beta$.

(b) *Second-order plant.* The situation is rather different when the plant transfer function is not first- but second-order of the form

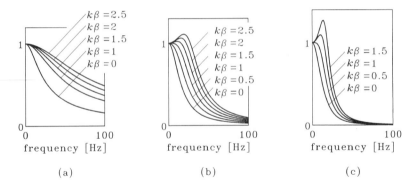

Figure 11.27. Magnitudes of the normalized closed-loop frequency response functions $\hat{h}_{cl}/\hat{h}_{cl}(0)$ of the feedback audio amplifier. Left: first-order plant. Middle: second-order plant. Right: third-order plant.

$$H(s) = \frac{k}{(1 + sT)^2}, \qquad \text{Re }(s) > -1/T,$$

which may well be a more accurate model for the dynamical properties of the forward amplifier. The loop gain transfer function now is

$$L(s) = \frac{k\beta}{(1 + sT)^2},$$

and the closed-loop characteristic polynomial, hence, takes the form

$$Q_{cl}(s) = (1 + sT)^2 + k\beta = s^2T^2 + 2sT + (1 + k\beta).$$

The characteristic polynomial now has degree two, and the two closed-loop characteristic values are

$$\lambda_{1,2} = -\frac{1}{T}(1 \pm j\sqrt{k\beta}).$$

For $k\beta = 0$ (no feedback) the closed-loop characteristic roots coincide at the point $-1/T$, which is the location of the two open-loop characteristic values. Figure 11.26(b) shows that as $k\beta$ increases the roots move on a vertical line in the complex plane that passes through the point $-1/T$. One root moves up, the other down. Both roots stay in the left-half plane, so that stability is ensured for every $k\beta \geq 0$, even as the gain goes to infinity.

For each $k\beta \geq 0$ the roots form a complex conjugate pair with real part $-1/T$ and imaginary part $\sqrt{k\beta}$. The real part is *constant*, which means that the transients decay to zero at the *same* rate as for the open-loop system. The imaginary part increases as $k\beta$ increases, which means that the response becomes more and more oscillatory, with the rate of decay constant.

The closed-loop transfer function is

$$H_{cl}(s) = \frac{H(s)F(s)}{1 + L(s)} = \frac{k}{(1 + sT)^2 + k\beta}.$$

The corresponding frequency response function is plotted in Fig. 11.27(b), again for $T = 0.01$ and $k = 10^4$. As $k\beta$ increases the bandwidth increases, but the frequency response function also exhibits a more and more pronounced resonance peak. The largest bandwidth without much peaking is achieved for $k\beta$ equal to about 1.5. The corresponding closed-loop characteristic values are located at about $100(-1 \pm j1.2)$.

(c) *Third-order plant*. We finally consider the case when the plant has the third-order transfer function

$$H(s) = \frac{k}{(1 + sT)^3}, \qquad \text{Re }(s) > -1/T,$$

which results in the loop gain transfer function

$$L(s) = \frac{k\beta}{(1 + sT)^3}.$$

The resulting closed-loop characteristic polynomial is

$$Q_{cl}(s) = (1 + sT)^3 + k\beta.$$

The three closed-loop characteristic values are

$$\lambda_1 = -\frac{1}{T}[1 + (k\beta)^{1/3}],$$

$$\lambda_{2,3} = \frac{1}{T}[-1 + \tfrac{1}{2}(1 \pm j\sqrt{3})(k\beta)^{1/3}].$$

For $k\beta = 0$ the three roots coincide at the location $-1/T$ of the three open-loop characteristic values. As $k\beta$ increases from 0 to $+\infty$, the closed-loop characteristic values move away from $-1/T$ as in Fig. 11.26(c). One root moves along the negative real axis toward $-\infty$, while the other two split into a complex conjugate pair. This pair crosses over into the right-half plane for $k\beta = 8$. For $k\beta$ larger than this value the closed-loop system is *unstable*.

The closed-loop transfer function is

$$H_{cl}(s) = \frac{H(s)F(s)}{1 + L(s)} = \frac{k}{(1 + sT)^3 + k\beta}.$$

The resulting closed-loop frequency response function is plotted in Fig. 11.27(c) for several values of $k\beta$. As $k\beta$ increases, the frequency response function becomes more and more peaked. The most suitable value of $k\beta$ appears to be about 0.5. The corresponding closed-loop characteristic values are located at about -270 and $-32 \pm j68$.   ∎

The example shows how the stability and response of the closed-loop system may be assessed by studying the loci of the closed-loop characteristic values in the complex plane. This is called *root locus* analysis of feedback systems and is extensively discussed in the control literature.

The Nyquist Stability Criterion

The Nyquist stability criterion makes it possible to check the stability of a linear time-invariant feedback system directly from the loop frequency response function $\hat{l}$. The derivation that we present in what follows applies to feedback systems with rational transfer functions, but the Nyquist criterion also holds for a class of systems with nonrational transfer functions, such as transfer functions that arise from time delays in continuous-time systems.

The Nyquist criterion is based on a result from complex function theory that is known as the *principle of the argument*.

11.3.9.   Summary: Principle of the argument.   Let R be a rational function, and $\mathscr{C}$ a closed contour in the complex plane as in Fig. 11.28. As the complex number λ traverses the contour $\mathscr{C}$ in clockwise direction its image $R(\lambda)$ under R traverses a closed contour that is denoted as $R(\mathscr{C})$, which is also shown in Fig. 11.28. Then as λ traverses the contour $\mathscr{C}$ exactly once in clockwise direction, we have

(the number of times $R(\lambda)$ encircles the origin in clockwise direction

as λ traverses $\mathscr{C}$)

$=$

(the number of zeros of R inside $\mathscr{C}$) $-$ (the number of poles of R inside $\mathscr{C}$).   ∎

We may apply the principle of the argument to verify the stability of a closed-loop system as follows. The loop gain transfer function of the feedback system of Fig. 11.25 is

$$L = HFG = \frac{P_p P_f P_r}{Q_p Q_f P_r} = \frac{P}{Q}.$$

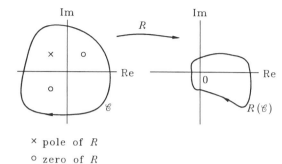

× pole of R

○ zero of R

Figure 11.28. Principle of the argument. Left: a closed contour $\mathscr{C}$ in the complex plane. Right: the image $R(\mathscr{C})$ of $\mathscr{C}$ under a rational function R.

The expression

$$J = 1 + L$$

appears in the denominators of each of the expressions (1), and is called the *return difference* of the closed-loop system. We have

$$J = 1 + L = 1 + \frac{P}{Q} = \frac{Q + P}{Q} = \frac{Q_{cl}}{Q},$$

with Q_{cl} the closed-loop characteristic polynomial of the feedback system. In applications the loop gain transfer function L is always *proper* (i.e., the degree of P is less than or at most equal to that of Q). Then the numerator polynomial Q_{cl} and the denominator polynomial Q of the rational function R have the same degree.

Suppose that the feedback system is continuous-time. Then the closed-loop system is CICO stable if and only if the denominator Q_{cl} has no roots in the right-half plane. It follows that the closed-loop system is stable if and only if

(the number of roots of Q_{cl} in the right-half plane)

−

(the number of roots of Q in the right-half plane)

=

−(the number of roots of Q in the right-half plane),

or, equivalently, if and only if

(the number of zeros of J in the right-half plane)

−

(the number of poles of J in the right half plane)

=

−(the number of roots of Q in the right-half plane).

By the principle of the argument we may determine the difference of the numbers of zeros and poles of J in the right-half plane by letting its variables s traverse a contour that encloses the right-half plane in clockwise direction and count the number of times that its image under J encircles the origin in clockwise direction.

Such a contour actually does not exist, but we may approximate it by a D- (so called after its shape) or *Nyquist* contour as in Fig. 11.29, consisting of a symmetric interval on the imaginary axis, closed by a semicircle in the right-half plane with radius ρ. As ρ increases to infinity, more and more of the right-half plane is included. Because the numerator and denominator of R have the *same* degree (under the assumption that L is proper), in the limit $\rho \rightarrow \infty$ the function J assumes the *same* value $J(j\infty)$ everywhere on the semicircle, and we obtain the image of the Nyquist contour in the limit $\rho \rightarrow \infty$ by simply plotting the image of the *imaginary axis* under J.

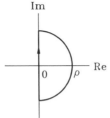

Figure 11.29. Nyquist contour.

Thus, by the principle of the argument the closed-loop system is stable if and only if the number of times $J(j\omega)$ encircles the origin in clockwise direction as ω increases from $-\infty$ to ∞ equals the negative of the number of right-half plane roots of the open-loop characteristic polynomial Q.

Our derivation of the Nyquist stability criterion is completed by noting that because $J = 1 + L$ the number of times the image of $J(j\omega)$ encircles the origin equals *the number of times the image of the loop frequency response function $L(j\omega)$ encircles the point* -1. We assume here that L has no poles on the imaginary axis (to prevent its image from going off to infinity) and that the image does not pass *through* the point -1 (to avoid difficulties in counting the number of encirclements.) These restrictive assumptions may easily be relaxed, for which we refer to the control literature. It is finally noted that plotting $L(j\omega)$, $-\infty \leq \omega \leq \infty$, is equivalent to plotting $\hat{l}(f) = L(j2\pi f)$ for $-\infty \leq f \leq \infty$. The plot of $\hat{l}$ is called the *Nyquist plot* of the feedback system.

11.3.10.   Summary: Nyquist stability criterion.   Suppose that the loop gain transfer function L of the feedback system of Fig. 11.25 is proper and has no poles on the imaginary axis, and that the Nyquist plot, that is, the plot of $\hat{l}$, defined by $\hat{l}(f) = L(j2\pi f), f \in \mathbb{R}$, does not pass through the point -1. Then

(the number of unstable closed-loop characteristic values)

=

(the number of times the Nyquist plot encircles the point -1)

−

(the number of unstable open-loop characteristic values).

Hence,

(a) The closed-loop system is CICO stable if and only if the number of encirclements equals the negative of the number of unstable open-loop characteristic values.

(b) In particular, if the open-loop system is CICO stable, the closed-loop system is CICO stable if and only if the number of encirclements is zero (i.e., the Nyquist plot does *not* encircle the point -1). ∎

We illustrate the use of the Nyquist criterion by an example.

11.3.11.   Example: Application of the Nyquist criterion to the feedback audio amplifier.   The loop frequency response function $\hat{l}$ of the feedback audio amplifier according to Example 11.3.8 is given by

$$\hat{l}_1(f) = \frac{k\beta}{1 + j2\pi fT}, \quad \hat{l}_2(f) = \frac{k\beta}{(1 + j2\pi fT)^2}, \quad \hat{l}_3(f) = \frac{k\beta}{(1 + j2\pi fT)^3}, \quad f \in \mathbb{R},$$

for the first-, second- and third-order case, respectively. In each case the open-loop system is stable, so a sufficient and necessary condition for the stability of the feedback system is that the Nyquist plot does *not* encircle the point -1 at all. In Fig. 11.30 the three Nyquist plots are given for $T = 0.01$ and $k\beta = 1$.

For this value of $k\beta$ none of the three plots encircles the point -1. Increasing $k\beta$ causes the plots to expand. Inspection shows that as $k\beta$ is increased, the Nyquist plot of $\hat{l}_1$ never encircles the point -1, or even approaches it. The plot of $\hat{l}_2$ also never encircles the critical point -1 but approaches it closely as $k\beta \to \infty$. This ac-

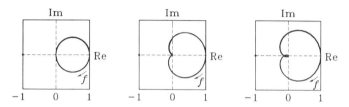

Figure 11.30.   Nyquist plots for the feedback audio amplifier. Left: first-order plant. Middle: second-order plant. Right: third-order plant.

counts for the increasingly oscillatory behavior of the closed-loop system. The plot of $\hat{l}_3$, finally, encircles the point -1 for any value of $k\beta$ larger than 8, so that the feedback system is unstable for these values of the return gain.

Exercise: Check that if $k\beta > 8$ the Nyquist plot of the third-order system encircles the point -1 twice so that the closed-loop system has *two* unstable poles. ∎

For feedback systems that are open-loop stable it is common to measure the degree of stability of the closed-loop system by establishing how far the Nyquist plot stays away from the point -1. This may be done by specifying two numbers. The first is the *gain margin*, which is the factor k_m by which the Nyquist plot should be multiplied so that it passes through the point -1. We have

$$ k_m = \frac{1}{|\hat{l}(f_1)|}, $$

where f_1 is the frequency for which the Nyquist plot intersects the negative real axis furthest from the origin (see Fig. 11.31). The second number that specifies the stability is the *phase margin*, which is the extra *phase* ϕ_m that must be added to make the Nyquist plot pass through the point -1. The phase margin ϕ_m is the angle between the negative real axis and $\hat{l}(f_2)$, where f_2 is the frequency where the Nyquist plot intersects the unit circle closest to the point -1 (see also Fig. 11.31).

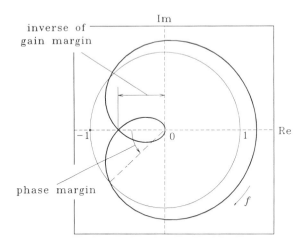

Figure 11.31.   Gain and phase margins.

11.3.12.   Example: Gain and phase margins of the audio feedback amplifier.

(a) *First-order amplifier.* Inspection of Fig. 11.30 shows that the gain margin for the first-order audio feedback amplifier is ∞ for any $k\beta > 0$. The phase margin is always greater than 90° but approaches this value as $k\beta$ increases to ∞.

(b) *Second-order amplifier.* In the second-order case the gain margin is always ∞. The phase margin approaches zero as $k\beta$ goes to infinity.

(c) *Third-order amplifier.* In the third-order case both the gain and the phase margin decrease to zero as $k\beta$ increases to 8.   ∎

11.4 PROBLEMS

The problems in this section relate to various aspects of feedback.

11.4.1. Linearity improvement by feedback. In the feedback configuration of Fig. 11.32, the forward compensator is a constant gain k, and the plant is memoryless with IO map ϕ given by

$$\phi(u) = \begin{cases} \frac{1}{2}(u - 1) & \text{for } u < -1, \\ u & \text{for } |u| \le 1, \\ \frac{1}{2}(u + 1) & \text{for } u > 1, \end{cases} \qquad u \in \mathbb{R}.$$

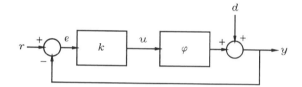

Figure 11.32.   A memoryless feedback system.

(a) Compute and plot the closed-loop IO map ϕ_{cl} from the external input r to the output y for different positive values of the gain k, assuming that the disturbance d is zero. Show that as k increases, the linearity of the closed-loop system improves.

(b) Assume that the external input r is zero, and compute the closed-loop map σ from the disturbance d to the output y. Show that as k increases, the disturbances are more and more suppressed.

11.4.2. Operational circuits. Operational circuits are analog electronic networks that closely approximate ideal circuit elements such as gains and integrators. A diagram of an operational circuit is given in Fig. 11.33(a). The circuit contains an active element called *operational amplifier*. It has infinite input impedance and zero output impedance. The input to the operational amplifier is the difference of the two inputs. The amplifier has the transfer matrix H. The magnitude of the frequency response function $\hat{h}(f) = H(j2\pi f)$, $f \in \mathbb{R}$, of the operational amplifier is large (say 10^6) from zero frequency up to a sufficiently high frequency. The other two circuit elements are passive networks consisting of resistors and capacitors, with impedances Z_1 and Z_2 as indicated. The input to the system is the voltage u, while its output is the voltage y. The circuit is often schematically represented as in Fig. 11.33(b).

(a) Show that the closed-loop transfer function H_{cl} of the circuit is the same as that of the feedback configuration of Fig. 11.34.

(b) Show that if the closed-loop system is stable, over those frequencies where the magnitude of the loop frequency response function

$$\hat{l}(f) = \frac{HZ_1}{Z_1 + Z_2}(j2\pi f), \qquad f \in \mathbb{R},$$

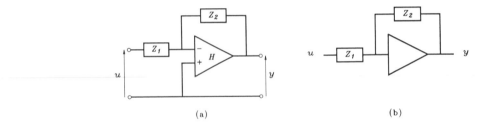

Figure 11.33.   Left: operational circuit. Right: schematic representation.

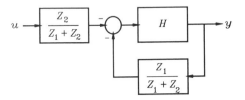

Figure 11.34.   Equivalent feedback configuration.

is much larger than 1 the closed-loop frequency response function $\hat{h}_{cl}(f)$ $= H_{cl}(j2\pi f), f \in \mathbb{R}$, is closely approximated by

$$\hat{h}_{cl}(f) \approx -\frac{Z_2(j2\pi f)}{Z_1(j2\pi f)}, \qquad f \in \mathbb{R}.$$

(c) In particular, show that the circuit of Fig. 11.35(a) approximately represents a *gain* with IO relation

$$y = -\frac{R_2}{R_1}u,$$

while the circuit of Fig. 11.35(b) approximately represents an *integrator* with IO relation

$$\dot{y} = -\frac{1}{RC}u.$$

Note the minus sign in both IO relationships.

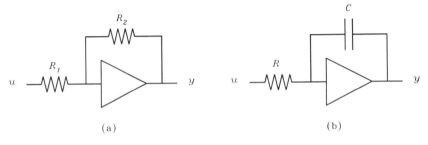

Figure 11.35.   Operational circuits. Left: a pure gain. Right: an integrator.

11.4.3. Integral control. In the linear time-invariant feedback system of Fig. 1..36, the transfer function L contains a factor $1/s$, that is, $\lim_{s\to 0} sL(s)$ is nonzero. Since a factor $1/s$ represents integration, this is referred to as *integral control*. Suppose that the closed-loop system is stable. Prove that if the external input r is a unit step, the error signal $e = r - y$ asymptotically approaches zero as time goes to infinity. Thus, integral control results in a zero steady-state error in the response to step inputs. As a result, also low-frequency inputs are tracked accurately.

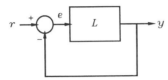

Figure 11.36.   Linear time-invariant feedback system.

11.4.4. Cruise control system. Consider the cruise control system of Example 11.1.1, whose diagram is given in Fig. 11.2. The output of the system is taken as the normalized speed w. In terms of w the motion of the car is described by the differential equation

$$\dot{w} = \alpha(u - w^2), \tag{1}$$

with the constant α as given in 4.8.1.

The cruise controller consists of a pure integrator, described by

$$\dot{u} = ke, \tag{2}$$

with the value of the gain k to be determined, and $e = w_r - w$. The function of the cruise controller is to maintain the speed w of the car at the reference value r.

(a) Show that if the external input w_r has the constant value R_o, the differential equations (1) and (2), supplemented with the equation

$$e = w_r - w,$$

have a constant equilibrium solution $w(t) = W_o = R_o$ and $u(t) = U_o$. What is the equilibrium solution for e? Express U_o in R_o.

(b) Let $w_r = R_o + \tilde{r}$, $u = U_o + \tilde{u}$ and $w = W_o + \tilde{w}$, and show that under the assumption that $\tilde{r}$, $\tilde{u}$ and $\tilde{w}$ are small, the system may be described by the linearized equations

$$\dot{\tilde{u}} = ke,$$

$$e = \tilde{r} - \tilde{w},$$

$$\dot{\tilde{w}} = \alpha(\tilde{u} - 2W_o\tilde{w}).$$

(c) Show that the linearized system may be represented by the block diagram of Fig. 11.37, where H and G are transfer functions given by

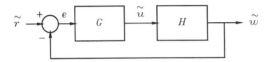

Figure 11.37. Linearized cruise control system.

$$H(s) = \frac{\alpha}{s + a}, \qquad G(s) = \frac{k}{s},$$

with $a = 2\alpha W_o$.

(d) Determine the closed-loop characteristic polynomial Q_{cl} of the system. Plot the locations of the closed-loop characteristic roots as a function of the gain k. Show that the closed-loop system is stable for all positive k.

(e) The closed-loop system has two closed-loop characteristic roots, which are either both real or form a complex-conjugate pair. The characteristic root that is closest to the origin is called the *dominant* characteristic root, because it dominates the response of the closed-loop system. The larger the distance of the dominant characteristic root from the origin is, the larger is the bandwidth of the closed-loop system. If the dominant characteristic root is one of a complex-conjugate pair, then the response of the system becomes overly oscillatory if the imaginary part of the dominant root is much larger than its real part. For the feedback system at hand, show that if the gain k is chosen as $k = a^2/2\alpha$ then the bandwidth of the closed-loop system is maximal within the constraint that the imaginary part of the dominant characteristic root pair does not exceed the real part.

(f) Suppose that the gain is selected according to the rule developed in (e) for the linearized system corresponding to $W_o = \frac{1}{2}$ (i.e., the reference speed is half the top speed). Show that this results in $k = \alpha/2$. Suppose that this setting for the gain is maintained at *all* speeds W_o. Investigate how the closed-loop characteristic roots of the linearized closed-loop system vary with the reference speed W_o. Discuss the corresponding changes in the closed-loop behavior.

11.4.5. Equivalent two-degree-of-freedom feedback configurations. Figure 11.38 shows two "two-degree-of-freedom" feedback configurations that may be used to control a plant. Figure 11.38(a) is the basic configuration of Fig. 11.8(a), with the plant, forward, and return compensator all linear time-invariant systems with transfer functions H, F, and G, respectively. Figure 11.38(b) is a unit feedback configuration that is often thought to be equivalent to that of (a), with a forward compensator with transfer function K, and a "prefilter" with transfer function L.

(a) Derive the closed-loop transfer functions from the external input r and the disturbance d to the output y for both block diagrams.

(b) Show that the closed-loop transfer functions are identical for the two configurations if and only if

$$K = FG, \qquad L = 1/G.$$

(c) Show that even if these conditions are satisfied the closed-loop characteristic roots of the two configurations are *not* identical, and hence the dynamic behaviors of the two configurations differ. In particular, show that the closed-loop characteristic roots of the configuration (b) consist of the closed-loop characteris-

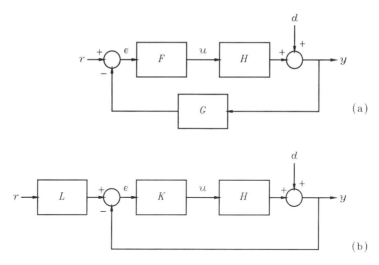

Figure 11.38. Two two-degree-of-freedom feedback configurations.

tic roots of the configuration (a) *together* with the characteristic roots of the prefilter.

11.4.6. Root loci. In the feedback system of Fig. 11.36, let L represent the transfer function of any of the following discrete-time systems. Determine the loci in the complex plane of the closed-loop characteristic roots as a function of the gain constant k for $k \geq 0$. For what values of k is the closed-loop system stable?

(a)

$$L(z) = \frac{kz}{z - a},$$

with a real and $0 \leq a < 1$.

(b)

$$L(z) = \frac{kz^2}{(z - a)(z - b)},$$

with a and b real and $0 \leq a < 1, 0 \leq b < 1$.

(c)

$$L(z) = kz^{-4}$$

11.4.7. Nyquist stability criterion. Apply the Nyquist stability criterion to determine the stability of the continuous-time feedback system of Fig. 11.36, where the transfer function L is given as follows:

(a)

$$L(s) = \frac{k}{s + 1},$$

with k a positive constant.

(b)

$$L(s) = \frac{k}{s},$$

with k a positive constant. *Hint:* Replace $L(s)$ with $k/(s + \epsilon)$, and let the positive constant ϵ approach 0.

(c)

$$L(s) = ke^{-sT},$$

with k and T positive constants. The Nyquist criterion applies to nonrational functions of this type.

11.5 COMPUTER EXERCISES

The exercises in this section deal with various feedback systems.

11.5.1. Cruise controller. We consider the cruise control system as in 11.4.4 and describe the motion of the car by the differential equation

$$\dot{w} = \alpha(u - w^2),$$

with u the throttle position and w the normalized speed of the car.

(a) *Integral control.* First we consider a control scheme with integrating action as in 11.4.4, where the cruise controller generates the throttle position as

$$\dot{u} = ke,$$

with $e = w_r - w$ the speed error signal and k a constant to be determined. In 11.4.4 we found by linearization about a fixed operating point that the gain

$$k = \frac{a^2}{2\alpha}$$

maximizes the bandwidth of the closed-loop system without making its response oscillatory. The constant a is given by $2\alpha W_o$, where W_o is the car speed corresponding to the operating point about which the system is linearized. Use the numerical value $\alpha = 1/10 \ [\text{s}^{-1}]$ in what follows. Choose k such that the response of the linearized cruise control system is optimized at $W_o = \frac{1}{2}$ (i.e., at half the top speed). If the initial conditions are $w(0) = W_o$, $u(0) = U_o = W_o^2$, and the external input is $w_r(t) = W_o$, then the system starts and remains in an equilibrium state. Simulate the response (of the nonlinear system) from these initial conditions to a constant external input $w_r(t) = R_o$, where R_o slightly differs from

W_o. Do this for $W_o = \frac{1}{2}$ and a smaller and a larger value of W_o, e.g. $W_o = 0.2$ and $W_o = 0.9$. Plot the response of the speed w and throttle position u. If the throttle position exceeds the range $[0, 1]$, include a limiter so that values of u outside this range are reduced to 0 or 1, whichever value is closer. Comment on the results. *Hint:* Simulate over a period of 100 [s]. If a Runge-Kutta 2 integration scheme is used, a step size of 1 [s] seems to be a suitable value.

(b) *Proportional and integral control.* As noted in 11.1.1, with pure integral control it takes the cruise control system quite a long time to reach the desired speed. This difficulty may be remedied by introducing *proportional* control in addition to the integral control. This means that the input is taken as

$$u(t) = u(0) + k_1 \int_0^t e(\tau)\, d\tau + k_2 e(t), \qquad t \geq 0.$$

The first two terms on the right-hand side constitute the integral of the error, but the third term, which is new, is proportional to the error. Select experimentally values of k_1 and k_2 that give a good response for $W_o = 0.5$ and $R_o = 0.6$. For the values of k_1 and k_2 thus found repeat the simulations for $W_o = 0.2$ and 0.9 and suitable corresponding values of R_o. Comment on the results. *Hint:* Let $u = k_1 z + k_2 e$, where $\dot{z} = e$.

11.5.2. Nyquist plot and stability test. Make Nyquist plots to test the stability of a feedback system with the configuration of Fig. 11.36 for the following open-loop transfer functions L. If the closed-loop system is stable, compute the closed-loop frequency response function corresponding to the closed-loop transfer function $H_{cl} = L/(1 + L)$, and plot its magnitude. Compute the impulse response of the closed-loop system by inverse Fourier transformation. Compute and plot the step response of the closed-loop system.

(a)

$$L(s) = \frac{8}{(s + 1)(s + 2)}.$$

(b)

$$L(s) = \frac{8e^{-0.1s}}{(s + 1)(s + 2)}.$$

(c)

$$L(s) = \frac{1}{1 - 0.5e^{-s}}.$$

(d)

$$L(s) = \frac{2(s + 4)}{(s - 2)(s + 3)}.$$

Hint: The open-loop system is unstable!

11.5.3. Nonlinear feedback system. In the feedback system of Fig. 11.39, the block marked "H" is a linear time-invariant system with transfer function

$$H(s) = \frac{1}{s(s + 1)},$$

while the block marked "f" is memoryless nonlinear with IO map

$$f(e) = k \, \text{sign} \, (e),$$

with k a positive constant gain.

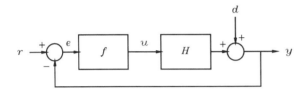

Figure 11.39. A nonlinear feedback system.

(a) Simulate the system with zero external input r and zero disturbance d for various initial conditions and different values of k. Does it look as if the closed-loop system is stable for any positive gain k? *Hints:* Determine a state representation for the linear system. Simulate over the interval $[0, 10]$. With Runge-Kutta 2 integration a step size of 0.1 seems to give adequate results.

(b) Let $k = 1$. Determine the response of the system to a unit step in the external input r and in the disturbance d. Comment on the results. Does it help to increase the gain k to 2?

11.5.4. Control system design. Consider the linear time-invariant feedback system of Fig. 11.40, where the plant has the transfer function

$$H(s) = \frac{1}{(1 + sT_1)(1 + sT_2)},$$

with $T_1 = 1$ [s] and $T_2 = 0.1$ [s]. We consider two possibilities for the transfer function G of the forward compensator: a *pure gain*

$$G(s) = k,$$

and a so-called *lag-lead filter*

$$G(s) = k \frac{1 + sT_3}{1 + sT_4}.$$

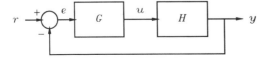

Figure 11.40. A linear time-invariant feedback system.

(a) First consider the pure gain compensator. Show by inspection of the Nyquist plot of the feedback system for $k = 1$ that the closed-loop system is stable for all positive k. Also show that as k increases the gain margin remains constant but the phase margin becomes smaller and smaller. The closed-loop transfer function of the system is given by $H_{cl} = L/(1 + L)$, with $L = HG$. Compute and plot the closed-loop frequency response $\hat{h}_{cl}(f) = H_{cl}(j2\pi f), f \in \mathbb{R}$, for several values of the gain k. Select a value of k for which the bandwidth of the closed-loop system is as large as possible without undesirable peaking of the closed-loop frequency response function. Compute the corresponding impulse response of the closed-loop system by inverse Fourier transformation. From this, compute the step response of the closed-loop system, and plot it.

(b) Repeat all this for the lag-lead compensator, with $T_3 = 1$ [s] and $T_4 = 0.1$ [s]. Show that by selecting a suitable gain k the lag-lead compensator results in a faster closed-loop response.

A Review of Complex Numbers, Sets, and Maps

Complex numbers play an important role in the theory of signals and systems. We therefore summarize in the first part of this supplement the most important properties of complex numbers and their operations. It also turns out to be very useful to employ some simple notions and notations connected with sets and maps. These are reviewed in the remainder of the supplement.

Complex Numbers

A *complex number* z is a pair of real numbers x and y written as

$$z = x + jy.$$

The real number x is called the *real part,* and the real number y the *imaginary part* of z. This is expressed by the notations

$$x = \text{Re}\ (z), \qquad y = \text{Im}\ (z).$$

If $y = 0$ we usually write $z = x$; if $x = 0$ we write $z = jy$.
    Complex numbers have special rules for addition and multiplication. If

$$z_1 = x_1 + jy_1 \qquad \text{and} \qquad z_2 = x_2 + jy_2$$

are two complex numbers, their sum and product are defined as

$$z_1 + z_2 = (x_1 + x_2) + j(y_1 + y_2),$$
$$z_1 \cdot z_2 = (x_1 x_2 - y_1 y_2) + j(x_1 y_2 + x_2 y_1).$$

It follows from the multiplication rule that

$$j^2 = -1.$$

The complex numbers, like the real numbers, constitute what is called a *field*. A field is a set with two operations, in this case addition and multiplication, satisfying a number of hypotheses. These include several commutativity, associativity, and distributivity properties, the existence of a *zero element* and a *unit element*, and that of the *negative* and the *reciprocal* of nonzero elements. The zero element of the complex field is $0 + j0 = 0$, while the unit is $1 + j0 = 1$. The negative of $z = x + jy$ is

$$-z = -x + j(-y),$$

while if $z \neq 0$ its reciprocal is

$$z^{-1} = \frac{x - jy}{x^2 + y^2}.$$

Two complex numbers $z_1 = x_1 + jy_1$ and $z_2 = x_2 + jy_2$ may be subtracted as

$$z_1 - z_2 = z_1 + (-z_2)$$
$$= (x_1 - x_2) + j(y_1 - y_2),$$

and if $z_2 \neq 0$ they may be divided as

$$\frac{z_1}{z_2} = z_1 \cdot z_2^{-1} = \frac{x_1 x_2 + y_1 y_2}{x_2^2 + y_2^2} + j\frac{-x_1 y_2 + x_2 y_1}{x_2^2 + y_2^2},$$

although division is more easily done using the polar representation that is discussed next.

A complex number $z = x + jy$ may be represented by a point in a plane with Cartesian coordinates (x, y) as in Fig. A.1. The plane is called the *complex plane, and*

$$z = x + jy$$

the *Cartesian* representation of the complex number. An alternative way of representing the complex number is in terms of its *polar* coordinates, formed by its *mag-*

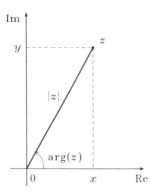

Figure A.1. Complex number as a point in a plane.

nitude $|z|$, and its *argument* arg (z), as indicated in Fig. A.1. The magnitude is also referred to as the *modulus* or *absolute value* of the complex number. The argument is sometimes called the *angle* or *phase* of the complex number. The *polar representation of z* is

$$z = |z|e^{j\arg(z)}.$$

The relations between the real and imaginary parts x and y and the magnitude and argument of $z = x + jy$ are

$$|z| = \sqrt{x^2 + y^2}, \qquad \arg(z) = \begin{cases} \operatorname{atan}\left(\dfrac{y}{x}\right) & \text{if } x > 0, \\[2mm] \operatorname{atan}\left(\dfrac{y}{x}\right) + \pi & \text{if } x \le 0 \text{ and } y > 0, \\[2mm] \operatorname{atan}\left(\dfrac{y}{x}\right) - \pi & \text{if } x \le 0 \text{ and } y \le 0, \\[2mm] \text{undetermined} & \text{if } x = 0 \text{ and } y = 0, \end{cases}$$

$$x = |z| \cos(\arg(z)), \qquad y = |z| \sin(\arg(z)),$$

with atan denoting the arctangent.

Addition and subtraction of complex numbers are most easily performed using the Cartesian representation, while multiplication and division are simplest when the polar forms are available, because

$$z_1 \cdot z_2 = |z_1| \cdot |z_2| e^{j(\arg(z_1) + \arg(z_2))},$$

$$\frac{z_1}{z_2} = \frac{|z_1|}{|z_2|} e^{j(\arg(z_1) - \arg(z_2))}.$$

> The adjective Cartesian derives from the name of the French philosopher, mathematician, and natural scientist René Descartes (1596–1650).

The *complex conjugate* $\overline{z}$ of a complex number $z = x + jy$ is the complex number

$$\overline{z} = x - jy.$$

It is easily verified that

$$z \cdot \overline{z} = |z|^2.$$

Finally, any two complex numbers z_1 and z_2 satisfy the *triangle inequality*

$$|z_1 + z_2| \le |z_1| + |z_2|.$$

A.1.   Example: Operations with complex numbers.   Let z_1 and z_2 be the complex numbers

$$z_1 = 1 + j, \qquad z_2 = 1 + j\sqrt{3},$$

so that

$$z_1 + z_2 = 2 + j(1 + \sqrt{3}), \qquad z_1 - z_2 = j(1 - \sqrt{3}).$$

We have

$$|z_1| = \sqrt{1 + 1} = \sqrt{2}, \qquad |z_2| = \sqrt{1 + 3} = 2,$$
$$\arg(z_1) = \operatorname{atan}(1) = \pi/4, \qquad \arg(z_2) = \operatorname{atan}(\sqrt{3}) = \pi/3,$$

so that the polar representations of z_1 and z_2 are

$$z_1 = \sqrt{2}e^{j\pi/4}, \qquad z_2 = 2e^{j\pi/3}.$$

From the polar representation we find easily

$$z_1 \cdot z_2 = 2\sqrt{2}e^{j7\pi/12}, \qquad \frac{z_1}{z_2} = \frac{1}{2}\sqrt{2}e^{-j\pi/12}. \qquad\blacksquare$$

A.2.   Exercise. Complex exponential and logarithm.
    (a) If $z = x + jy$ is a complex number, the complex exponential is given by

$$e^z = e^x(\cos(y) + j\sin(y)).$$

Prove

$$|e^z| = e^x, \qquad \arg(e^z) = y,$$

$$e^{z_1+z_2} = e^{z_1}e^{z_2}.$$

(b) If $z \neq 0$ is a complex number, the complex (natural) logarithm of z is

$$\log(z) = \log(|z|) + j\arg(z).$$

Prove that if $z_1 z_2 \neq 0$

$$\log(z_1 z_2) = \log(z_1) + \log(z_2),$$

and

$$\log\left(\frac{z_1}{z_2}\right) = \log(z_1) - \log(z_2). \qquad \blacksquare$$

Sets

We denote by

$$X = \{a, b, c, \cdots\}$$

the set X whose elements are $a, b, c, \cdots$. The familiar notations $x \in X$ and $x \notin X$ are assumed to be known, as well as the inclusion symbols $\subset$ and $\supset$ and the elementary set operations union $\cup$ and intersection $\cap$. We use the following notations for some well-known sets:

$\mathbb{N}$ the set of all natural numbers
$\mathbb{Z}$ the set of all integers,
$\mathbb{R}$ the set of all real numbers,
$\mathbb{C}$ the set of all complex numbers.

By $\{x \in X \mid P(x)\}$ we denote the subset of X that consists of elements $x \in X$ for which the proposition $P(x)$ holds.

A.3.   Example: The set of nonnegative integers.   The set

$$\mathbb{Z}_+ := \{x \in \mathbb{Z} \mid x \geq 0\}, \tag{1}$$

consists of all nonnegative integers. $\qquad \blacksquare$

The notation $a := b$ used in (1) indicates "assign the meaning of b to a", or "a equals b by definition."

The set $X_1 \times X_2 \times \cdots \times X_N$, called the *product set* of the sets $X_1, X_2, \cdots$, X_N, is the set of N tuples $(x_1, x_2, \cdots, x_N)$ with $x_1 \in X_1, x_2 \in X_2, \cdots, x_N \in X_N$. The product set $X \times X \times \cdots \times X$, with X repeated N times, is written as X^N.

A.4.   Example: The product sets $\mathbb{R}^N$ and $\mathbb{C}^N$. The sets $\mathbb{R}^N = \mathbb{R} \times \mathbb{R} \times \cdots \times \mathbb{R}$, with $\mathbb{R}$ repeated N times, and $\mathbb{C}^N = \mathbb{C} \times \mathbb{C} \times \cdots \times \mathbb{C}$, again with $\mathbb{C}$ repeated N times, consist of all N tuples $(x_1, x_2, \cdots, x_N)$ with $x_1, x_2, \cdots$, x_N real or complex numbers, respectively. Sometimes it is useful to arrange the Ntuple $(x_1, x_2, \cdots, x_N)$ as a column vector

$$\begin{bmatrix} x_1 \\ x_2 \\ \cdots \\ x_N \end{bmatrix}.$$

An equivalent but space saving notation for this column vector is $\text{col}(x_1, x_2, \cdots, x_N)$. The set consisting of all such column vectors is also denoted $\mathbb{R}^N$ or $\mathbb{C}^N$, depending on whether the elements are real or complex.   ■

If a set has finitely many elements, then it is called a *finite* set. If it has countably infinitely many elements, then the set is *countably infinite*. In all other cases it is *uncountable*.

A.5.   Example: Bits.   *Bits* are the elements of the finite set $\mathbb{B} := \{0, 1\}$.   ■

A.6.   Example: Bytes.   Four-bit *bytes* are the elements of the finite product set $\mathbb{B}^4 = \{0000, 0001, 0010, \cdots, 1111\}$.   ■

Typical examples of countably infinite sets are the set $\mathbb{N}$ of all natural numbers and the set $\mathbb{Z}$ of all integers. The best known examples of uncountable sets are the set of real numbers $\mathbb{R}$ and the set of complex numbers $\mathbb{C}$.

A.7.   Example: Bit streams.   An infinite sequence of zeros and ones, such as $01001100 \cdots$ is called a *bit stream*. The set of *all* bit streams is uncountable. To see this, note that any bit stream can be taken as the mantissa, in binary form, of a number between 0 and 1. As the numbers between 0 and 1 are uncountable, so are the bit streams.   ■

Maps

A *map* ϕ from the set X to the set Y assigns to every element $x \in X$ a unique element $\phi(x) \in Y$. We write

$$\phi: X \to Y,$$

and call $\phi(x)$ the *image* of $x \in X$ under ϕ or the *value* of ϕ at x. A map is some-times also referred to as a *function*, an *operator*, or a *transformation*, depending on the context. The set X is called the *domain* of the map ϕ. The set of all images of elements in X under ϕ is the *range* of the map.

A.8.   Example: Real function of a real variable.   Let $\phi: \mathbb{R} \to \mathbb{R}$ be the func-tion defined by $\phi(x) = x^2$. The function has domain $\mathbb{R}$ and range $\mathbb{R}_+$ $:= \{x \in \mathbb{R} \mid x \geq 0\}$.   ∎

If the range of the map ϕ is the whole set Y, then we say that ϕ is an *onto* or *surjec-tive* map. If no two different elements of X have the same image, that is, if for $x_1 \in X$ and $x_2 \in X$

$$\phi(x_1) = \phi(x_2) \quad \text{implies} \quad x_1 = x_2,$$

then ϕ is said to be a *one-to-one* or *injective* map. If ϕ is both surjective and injec-tive it is called *bijective*.

A.9.   Examples: Injective, surjective, and bijective maps.
(a) The map $\phi: \mathbb{R} \to \mathbb{R}$ defined by $\phi(x) = x^2$ is neither surjective (because its range is not the entire set $\mathbb{R}$) nor injective (because $-x$ and x map to the same num-ber).

(b) As in 2.3.4, define the *entier* function int: $\mathbb{R} \to \mathbb{Z}$ such that int (x) is the largest integer N with the property that $N \leq x$. The map int: $\mathbb{R} \to \mathbb{Z}$ is surjective (because its range is the whole set $\mathbb{Z}$) but not injective (because every $x \in \mathbb{R}$ such that $N \leq x < N + 1$ has the same image N.)

(c) Finally, let $\rho: \mathbb{R} \to \mathbb{R}$ be defined by $\rho(x) = x^3$. As ρ is both injective and surjective, it is bijective.   ∎

If ϕ and ψ are maps

$$\phi: X \to Y, \qquad \psi: U \to V,$$

with $V \subset X$, then the *composition* of ϕ and ψ is the map $\phi \circ \psi: U \to Y$ given by

$$(\phi \circ \psi)(u) = \phi(\psi(u)) \qquad \text{for all} \quad u \in U.$$

The operation $\circ$ is called *map composition*.

A.10.   Example: Composition of maps.   Let $\phi: \mathbb{R} \to \mathbb{R}$ be the map defined by $\phi(x) = x^2$, and $\psi: \mathbb{R} \to \mathbb{Z}$ the map given by $\psi(x) = $ int (x). Then $\phi \circ \psi: \mathbb{R} \to \mathbb{R}$ is the composite map defined by $(\phi \circ \psi)(x) = [$int $(x)]^2$, while $\psi \circ \phi: \mathbb{R} \to \mathbb{Z}$ is the composite map given by $(\psi \circ \phi)(x) = \text{int}\,(x^2)$. Note that the maps $\phi \circ \psi$ and $\psi \circ \phi$ are *not* the same: the composition operation in general is not commutative.   ∎

A bijective map $\phi: X \to Y$ has an *inverse* map

$$\phi^{-1}: Y \to X,$$

that satisfies

$$(\phi^{-1} \circ \phi)(x) = x \qquad \text{for all} \quad x \in X,$$
$$(\phi \circ \phi^{-1})(y) = y \qquad \text{for all} \quad y \in Y.$$

A.11.   Example: A map and its inverse.   The maps defined in Examples A.9(a) and (b) have no inverses because they are not bijective. The map $\rho: \mathbb{R} \to \mathbb{R}$ given by $\rho(x) = x^3$ of A.9(c) is bijective. Its inverse $\rho^{-1}: \mathbb{R} \to \mathbb{R}$ is given by $\rho^{-1}(y) = y^{1/3}$.   ∎

Power Sets

The set of all maps from the set X into the set Y is denoted by Y^X and is called a *power set*. This notation is effective in denoting sets of signals.

A.12.   Examples: Power sets.
  (a) The power set

$$\mathbb{C}^{\{0,\,1,\,\cdots,\,N-1\}}$$

is the set of all maps

$$\{0,\,1,\,\cdots,\,N-1\} \to \mathbb{C}.$$

Thus, an element x of this power set is a map that assigns a complex number $x(i)$ to each $i \in \{0,\,1,\,\cdots,\,N-1\}$. Hence, specifying the map x is the same as specifying the N tuple

$$(x(0),\,x(1),\,\cdots,\,x(N-1)).$$

It follows that $\mathbb{C}^{\{0,\,1,\,\cdots,\,N-1\}}$ is identical to the product set $\mathbb{C}^N$, which also consists of all N tuples of complex numbers.

  (b) Similarly, $\mathbb{C}^{\mathbb{Z}}$, the set of all maps $x: \mathbb{Z} \to \mathbb{C}$, may be identified with the set of all infinite sequences of complex numbers of the form $x = (\cdots,\,x(-1),\,x(0),$

$x(1), \cdots)$, with $x(i) \in \mathbb{C}$ for each $i \in \mathbb{Z}$. As the set $\mathbb{C}^{\mathbb{Z}}$ appears frequently, we refer to it by the special notation

$$\ell = \mathbb{C}^{\mathbb{Z}}.$$

(c) Finally, the power set $\mathbb{C}^{\mathbb{R}}$ is the set of all complex-valued functions of a real variable. A typical element x of $\mathbb{C}^{\mathbb{R}}$ is a function $x(t)$, $t \in \mathbb{R}$, where $x(t) \in \mathbb{C}$ for each $t \in \mathbb{R}$. As also the set $\mathbb{C}^{\mathbb{R}}$ frequently makes its appearance, we denote it as

$$\mathscr{L} = \mathbb{C}^{\mathbb{R}}. \qquad\qquad\qquad \blacksquare$$

Supplement B

A Review of Linear Spaces, Norms and Inner Products

In this supplement we review the basic definitions of linear spaces, norms, and inner products.

Linear Spaces

Before giving the definition of a linear space we recall that the set of reals $\mathbb{R}$ and the set of complex numbers $\mathbb{C}$, with the usual operations of addition and multiplication, both form a *field*. A field, roughly, is a set with two binary operations, called addition and multiplication, having all the usual properties such as commutativity, associativity and distributivity. Besides $\mathbb{R}$ and $\mathbb{C}$ also the set $\underline{N} = \{0, 1, \cdots, N - 1\}$, with N a prime and addition and multiplication defined modulo N, is a field. Elements of a field are called *scalars*. The sum of two scalars α and β is denoted as $\alpha + \beta$, their product as $\alpha \cdot \beta$ or $\alpha\beta$. The unit of the field is written as 1 and its zero as 0.

Linear spaces are sets where the operations of *multiplication* of any element *by a scalar* and *addition* of any two elements are defined, satisfying a number of properties.

B.1. **Definition: Linear space.** Consider a field $\mathcal{F}$ and the triple $(X, +, \cdot)$, where X is a set, $+$ a binary operation

$$+ : X \times X \to X,$$

called *addition,* and $\cdot$ a binary operation

$$\cdot : \mathscr{F} \times X \to X,$$

called *multiplication by a scalar*. The image of $(x, y) \in X \times X$ under $+$ is denoted $x + y$, and the image of $(\alpha, x) \in \mathscr{F} \times X$ under $\cdot$ is denoted $\alpha \cdot x$ or αx. Then, $(X, +, \cdot)$ is a linear space over the field $\mathscr{F}$ if the following conditions are satisfied:

(a) Addition of two elements of X has the following properties:

 (i) $(x + y) + z = x + (y + z)$ for all $x, y, z \in X$ (associativity),
 (ii) $x + y = y + x$ for all $x, y \in X$ (commutativity),
 (iii) there exists a zero element $\theta \in X$ so that $x + \theta = x$ for all $x \in X$,
 (iv) for every $x \in X$ there exists an element $-x$ so that $x + (-x) = \theta$.

(b) Multiplication of an element of X by a scalar has the following properties:

 (i) $\alpha(\beta x) = (\alpha\beta)x$ for all $\alpha, \beta \in \mathscr{F}$ and all $x \in X$ (associativity),
 (ii) $1 \cdot x = x$ and $0 \cdot x = \theta$ for all $x \in X$,
 (iii) $\alpha(x + y) = \alpha x + \alpha y$ for all $\alpha \in \mathscr{F}$ and all $x, y \in X$ (distributivity),
 (iv) $(\alpha + \beta)x = \alpha x + \beta x$ for all $\alpha, \beta \in \mathscr{F}$ and all $x \in X$ (distributivity). ∎

Customarily the zero element θ is written as 0, not to be confused with the zero of the field $\mathscr{F}$. When it is clear what $+$, $\cdot$, and $\mathscr{F}$ are, usually the set X itself is referred to as a linear space. The elements of linear spaces are often called *vectors,* and linear spaces are sometimes referred to as *vector spaces*.

The sets $\mathbb{R}^N$ and $\mathbb{C}^N$ are the best known examples of linear spaces.

B.2. Example: The linear spaces $\mathbb{R}^N$ and $\mathbb{C}^N$. Given two elements $x = (x_1, x_2, \cdots, x_N)$ and $y = (y_1, y_2, \cdots, y_N)$ of $\mathbb{R}^N$ or $\mathbb{C}^N$, define addition in the usual way as $x + y = (x_1 + y_1, x_2 + y_2, \cdots, x_N + y_N)$, and multiplication by a scalar as $\alpha x = (\alpha x_1, \alpha x_2, \cdots, \alpha x_N)$. Then $\mathbb{R}^N$ is a linear space over $\mathbb{R}$, and $\mathbb{C}^N$ a linear space over $\mathbb{C}$. ∎

A *subspace* U of a linear space X is a subset of X that is itself a linear space over the same field as X. To check whether a subset U is a subspace it is necessary to verify *closure* (i.e., whether $x \in U$ and $y \in U$ implies $x + y \in U$ and $x \in U$, $\alpha \in \mathscr{F}$ implies $\alpha x \in U$).

B.3. Example: Span. Let $x_1, x_2, \cdots, x_K$ be elements of $\mathbb{C}^N$. The set $\{x_1, x_2, \cdots, x_K\}$ is not a subspace of $\mathbb{C}^N$, but it may be made into one by taking all linear combinations of the vectors $x_1, x_2, \cdots, x_K$. We thus obtain the set

$$\text{span} (x_1, x_2, \cdots, x_K) := \left\{ x \in \mathbb{C}^N \mid (\exists \alpha_1, \alpha_2, \cdots, \alpha_K \in \mathbb{C}) \quad x = \sum_{i=1}^{K} \alpha_i x_i \right\},$$

which is a subspace. It is called the subspace that is *spanned* by the vectors x_1, $x_2, \cdots, x_K$. This technique of generating subspaces is quite general. ∎

The reason why we are interested in linear spaces is that all the sets of complex-valued time signals we deal with constitute linear spaces.

B.4.   Example: The linear signal space $\mathbb{C}^{\mathbb{T}}$.   The triple $(\mathbb{C}^{\mathbb{T}}, +, \cdot)$, with $\mathbb{T}$ a time axis, addition defined pointwise as in 2.3.14 and multiplication by a scalar likewise defined pointwise, is a linear space over the complex numbers. In particular, the signal sets $\ell_{\underline{N}}$, ℓ_+, ℓ, $\ell_{\underline{N}}(T)$, $\ell_+(T)$, $\ell(T)$, $\mathscr{L}[a, b]$, $\mathscr{L}_+$, and $\mathscr{L}$ defined in Section 2.2 all are linear spaces.

Norms

The *norm* is a measure for the "size" of a signal.

B.5.   Definition: Norm and normed space.   Let X be a linear space over the field $\mathscr{F}$ of real or complex scalars. A function

$$\| \cdot \| \colon X \to \mathbb{R}$$

that maps X into $\mathbb{R}$ is called a *norm* on X if it satisfies the following conditions:

(a) $\| x \| > 0$ for all $x \in X$ (nonnegativity),
(b) $\| x \| = 0$ if and only if $x = 0$ (positive-definiteness),
(c) $\| \lambda x \| = | \lambda | \cdot \| x \|$ for all $\lambda \in \mathscr{F}$ and all $x \in X$ (homogeneity with respect to scaling),
(d) $\| x + y \| \le \| x \| + \| y \|$ for all $x \in X$ and $y \in X$ (triangle inequality).

In (c), $| \lambda |$ denotes the absolute value of the scalar λ. The pair $(X, \| \cdot \|)$ is called a *normed* linear space. ∎

Fig. B.1 illustrates the norm and the triangle inequality geometrically.
        The *p*-norm on the signal spaces ℓ and $\mathscr{L}$ is defined in 2.4.3.

B.6.   Summary: Normed signal spaces.   The signal spaces ℓ_p and $\mathscr{L}_p$ are normed linear subspaces of ℓ and $\mathscr{L}$. ∎

B.7.   Exercise: Proof of B.6.   Prove B.6. *Hint*. To prove that ℓ_∞ is a subspace of ℓ and $\mathscr{L}_\infty$ a subspace of $\mathscr{L}$ is straightforward, by using the triangle equality for

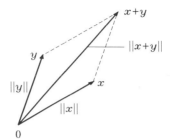

Figure B.1. Norm and the triangle inequality.

complex numbers. To prove that ℓ_2 and ℓ_1 are subspaces of ℓ, use *Minkowski's inequality for sums*. This inequality states that for any $x = (\ \cdots\ ,\ x_{-1},\ x_0,\ x_1,\ x_2,\ \cdots\ ) \in \ell$ and $y = (\ \cdots\ ,\ y_{-1},\ y_0,\ y_1,\ y_2,\ \cdots\ ) \in \ell$ and for $1 \leq p < \infty$,

$$\left(\sum_{i=-\infty}^{\infty} |x_i \pm y_i|^p \right)^{1/p} \leq \left(\sum_{i=-\infty}^{\infty} |x_i|^p \right)^{1/p} + \left(\sum_{i=-\infty}^{\infty} |y_i|^p \right)^{1/p}.$$

To prove that $\mathcal{L}_1$ and $\mathcal{L}_2$ are subspaces of $\mathcal{L}$, use Minkowski's inequality for *integrals*, which takes the form

$$\left(\int_{-\infty}^{\infty} |x(t) \pm y(t)|^p\, dt \right)^{1/p} \leq \left(\int_{-\infty}^{\infty} |x(t)|^p\, dt \right)^{1/p} + \left(\int_{-\infty}^{\infty} |y(t)|^p\, dt \right)^{1/p}$$

for any $x, y \in \mathcal{L}$, and any $1 \leq p < \infty$.   ∎

Inner Product

The inner product of two elements of a linear space is axiomatically defined as follows.

B.8.   Definition: Inner product.   Let X be a linear space over the field $\mathcal{F}$ of real or complex numbers. Then a map

$$\langle\, \cdot\, ,\, \cdot\, \rangle\colon X \times X \to \mathcal{F}$$

is called an *inner product* on X if it has the following properties.

(a) $\langle x, y \rangle = \overline{\langle y, x \rangle}$ for all $x, y \in X$. The overbar denotes the complex conjugate (conjugate symmetry),

(b) $\langle \alpha x + \beta y, z \rangle = \alpha \langle x, z \rangle + \beta \langle y, z \rangle$ for all $x, y, z \in X$ and all $\alpha, \beta \in \mathcal{F}$ (bilinearity),

(c) $\langle x, x \rangle$ is real and nonnegative for all $x \in X$,

(d) $\langle x, x \rangle = 0$ if and only if $x = 0$.

A linear space X that has an inner product is called an *inner product space*.   ∎

The Cauchy-Schwarz inequality plays an important role.

B.9.   Summary: The Cauchy-Schwarz inequality.   Let X be an inner product space over the real or complex field $\mathcal{F}$. Then, for any $x, y \in X$

$$|\langle x, y \rangle|^2 \leq \langle x, x \rangle \langle y, y \rangle.$$

Equality holds if and only if there exists an $\alpha \in \mathcal{F}$ such that $x = \alpha y$ or $y = \alpha x$.

B.10.   Proof.   If y is the zero element of X then the theorem holds trivially. If y is not the zero element then for any scalar $\alpha \in \mathcal{F}$ it follows by B.8(c) and repeated application of B.8(a) and (b) that

$$0 \leq \langle x - \alpha y, x - \alpha y \rangle$$
$$= \langle x, x \rangle - \alpha \overline{\langle x, y \rangle} - \overline{\alpha}\langle x, y \rangle + |\alpha|^2 \langle y, y \rangle.$$

If in particular $\alpha = \langle x, y \rangle / \langle y, y \rangle$, the inequality to be proven follows immediately. The remainder of the proof is left to the reader.   ∎

Inner product spaces automatically have a norm.

B.11.   Summary: Natural norm.   Let $(X, \langle \cdot, \cdot \rangle)$ be an inner product space. Then the map $\|\cdot\|: X \to \mathbb{R}$ defined by

$$\|x\| = \langle x, x \rangle^{1/2}$$

is a norm on X, called the *natural norm*.   ∎

B.12.   Exercise: Proof of B.11.   Prove B.11.   ∎

An Introduction to the Theory of Generalized Signals

In this supplement we present a brief outline of *distribution theory*, which is the mathematical foundation of the theory of generalized functions. *Regular* functions f with domain $\mathbb{R}$ and range $\mathbb{C}$ are specified *pointwise* (i.e., given $t \in \mathbb{R}$ the value $f(t)$ of f at t is well-defined). It turns out that regular functions cannot describe physical phenomena that occur "instantaneously," such as charging a capacitor with wires without resistance or notions such as point masses and charges. This encourages enlarging the set of functions so that they include *singular* functions. These are "functions" that cannot be defined pointwise but only *indirectly,* by specifying their "effect" on a set of *test functions*. Together the regular and singular functions form the *generalized* functions.

Distributions

Distribution theory deals with regular and singular functions in a unified way. The starting point is the observation that a regular function alternatively may be characterized as a *linear functional,* which specifies the effect of the function on a well-defined set of test functions. The set of all linear functionals also includes functionals that do *not* correspond to regular functions. By definition, such functionals correspond to *singular* functions.

The functionals that are considered are called *distributions,* and this is why the investigation of singular functions amounts to distribution theory. Operations on distributions, such as addition, multiplication by a scalar, and differentiation, are defined in such a way that for *regular* distributions, that is, distributions corresponding to regular functions, the operations are equivalent to the conventional operations on the functions.

We start our brief review of distribution theory by explaining what are linear functionals.

C.1.   Definition: Linear functionals.   Let X be a linear space over the field $\mathbb{C}$. Then a *linear functional* $\mathbf{f}$ on X is a map $\mathbf{f}: X \rightarrow \mathbb{C}$ such that

$$\mathbf{f}(\alpha_1 x_1 + \alpha_2 x_2) = \alpha_1 \mathbf{f}(x_1) + \alpha_2 \mathbf{f}(x_2)$$

for all $\alpha_1, \alpha_2 \in \mathbb{C}$ and all $x_1, x_2 \in X$.   ∎

We illustrate the definition with an example.

C.2.   Example: Linear functional on $\mathbb{C}^N$.   Consider the functional $\mathbf{f}$ defined on $\mathbb{C}^N$ by

$$\mathbf{f}(x) = \sum_{i=1}^{N} \alpha_i x_i \qquad \text{for any} \quad x = (x_1, x_2, \cdots, x_N) \in \mathbb{C}^N,$$

with $\alpha_1, \alpha_2, \cdots, \alpha_N$ fixed complex numbers. It is easy to verify that $\mathbf{f}$ is a linear functional over the field $\mathbb{C}$.   ∎

Generalized functions are described by means of linear functionals on the *set of test functions* $\mathscr{D}$, which is defined as follows.

C.3.   Definition: The set of test functions $\mathscr{D}$.   The *set of test functions* $\mathscr{D}$ consists of all functions in $\mathscr{L}$ that are zero outside some finite interval and may be differentiated arbitrarily often.   ∎

A function ϕ whose derivative $D^k\phi$ exists for every $k \in \mathbb{Z}_+$ is called *smooth.*

C.4.   Example: Test function.   An example of a test function is the function $\phi \in \mathscr{D}$ given by

$$\phi(t) = \begin{cases} e^{-\frac{1}{1-t^2}} & \text{for } |t| < 1, \\ 0 & \text{for } |t| \geq 1, \end{cases}$$

as shown in Fig. C.1.   ∎

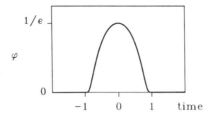

Figure C.1.   Example of a test function $\phi \in \mathcal{D}$.

Distributions are now defined as follows.

C.5.   **Definition: Distribution.**   A *distribution* is a linear functional on the set of test functions $\mathcal{D}$.   ∎

More precisely, a distribution is a *continuous* linear functional on $\mathcal{D}$, but we do not dwell on this point.

In the sequel we denote distributions by bold English or Greek characters, usually lower case. In the following examples two typical instances of distributions are considered.

C.6.   **Example: A distribution.**   The functional $\mathbf{f}$ on $\mathcal{D}$, defined by

$$\mathbf{f}(\phi) = \int_a^b \phi(t)\, dt, \qquad \phi \in \mathcal{D},$$

with a and b fixed real numbers, is linear. Hence $\mathbf{f}$ is a distribution.   ∎

C.7.   **Definition: Delta distribution.**   The linear functional $\boldsymbol{\delta}$ on $\mathcal{D}$, defined by

$$\boldsymbol{\delta}(\phi) = \phi(0), \qquad \phi \in \mathcal{D},$$

is a distribution, called the *delta distribution*.   ∎

By way of example, the value of $\boldsymbol{\delta}$ at the function ϕ given in C.4 is

$$\boldsymbol{\delta}(\phi) = e^{-\frac{1}{1-t^2}}\bigg|_{t=0} = e^{-1}.$$

The distributions of C.6 and C.7 differ in one essential respect. The distribution of C.6 is of the form

$$\mathbf{f}(\phi) = \int_{-\infty}^{\infty} f(t)\phi(t)\, dt, \qquad \phi \in \mathcal{D}, \tag{1}$$

where the function f is given by

$$f(t) = \begin{cases} 1 & \text{for } a \leq t < b, \\ 0 & \text{otherwise.} \end{cases}$$

The delta distribution of C.7, however, *cannot* be represented in the form (1). Apparently some but not all distributions may be represented in the integral form (1). This introduces the following dichotomy among distributions.

C.8.   Definition: Regular and singular distributions.   A distribution **f** is called *regular* if there exists a function $f \in \mathscr{L}$ such that

$$\mathbf{f}(\phi) = \int_{-\infty}^{\infty} f(t)\phi(t) \, dt, \qquad \phi \in \mathscr{D}.$$

Otherwise, **f** is called *singular*. If **f** is regular, then we say that the regular function $f \in \mathscr{L}$ *represents* the distribution.   ∎

As was pointed out before, distributions are denoted by bold characters. The function that *represents* a distribution is usually denoted by the same character but in Italic typeface. Although if a distribution **f** is singular, there exists *no* function $f \in \mathscr{L}$ such that

$$\mathbf{f}(\phi) = \int_{-\infty}^{\infty} f(t)\phi(t) \, dt, \qquad \phi \in \mathscr{D},$$

we nevertheless often write **f** *symbolically* as an integral and call the "function" f that enters into it a *singular* function. The singular functions together with the regular functions are called *generalized* functions.

We thus write the delta distribution **δ** of C.7 symbolically as

$$\boldsymbol{\delta}(\phi) = \int_{-\infty}^{\infty} \delta(t)\phi(t) \, dt = \phi(0), \qquad \phi \in \mathscr{D}.$$

The singular function δ is called the *delta function* or *δ-function*.

C.9.   Remark: Operating with the δ-distribution on a function not in $\mathscr{D}$ **.**   Because the outcome of $\boldsymbol{\delta}(\phi)$ only depends on the value of ϕ at 0, we may also operate with the δ-function on functions ϕ that are not in $\mathscr{D}$ (i.e., are not zero outside some finite interval and are not smooth). When applying the δ-function, the only requirement is that ϕ be continuous at 0. We thus have

$$\int_{-\infty}^{\infty} \delta(t) f(t) \, dt = f(0)$$

for any $f \in \mathscr{L}$ that is continuous at 0.   ∎

Basic Operations

In what follows we extend the various operations on and among regular time signals, such as addition, scaling, time translation, and differentiation to distributions and generalized signals, taking care, of course, that these extensions are *consistent* with the definitions for regular signals. For each new definition we first verify consistency. Next we show the effect of the operation on the δ-distribution and -function, and thus generate all the distributions and singular signals needed in this text.

Before proceeding, we define *equality* of two distributions.

C.10.   **Definition: Equality of distributions.**   Two distributions $\mathbf{f}$ and $\mathbf{g}$ are *equal* if

$$\mathbf{f}(\phi) = \mathbf{g}(\phi) \qquad \text{for all} \quad \phi \in \mathscr{D}. \qquad\qquad\blacksquare$$

Addition and Multiplication by a Scalar

The definitions of addition of distributions and multiplication of a distribution by a scalar are straightforward.

C.11.   **Definition. Addition of distributions and multiplication by a scalar.**   If $\mathbf{f}$ and $\mathbf{g}$ are distributions and α and β scalars in $\mathbb{C}$, then the distribution $\alpha\mathbf{f} + \beta\mathbf{g}$ is defined by

$$(\alpha\mathbf{f} + \beta\mathbf{g})(\phi) = \alpha\mathbf{f}(\phi) + \beta\mathbf{g}(\phi), \qquad \phi \in \mathscr{D}.$$

If $\mathbf{f}$ and $\mathbf{g}$ are represented by the regular *or* singular functions f and g, respectively, then we write $\alpha f + \beta g$ for the generalized function that represents $\alpha\mathbf{f} + \beta\mathbf{g}$. $\blacksquare$

Note that defining a distribution means specifying its value for each test function ϕ in the set of test functions $\mathscr{D}$.

C.12.   **Consistency proof: Addition and multiplication by a scalar.**   If $\mathbf{f}$ and $\mathbf{g}$ are *regular* distributions represented by the regular functions $f \in \mathscr{L}$ and $g \in \mathscr{L}$, respectively, then the distribution $\alpha\mathbf{f} + \beta\mathbf{g}$ as defined in C.11 is a regular distribution represented by $\alpha f + \beta g \in \mathscr{L}$. This is easy to see because

$$\begin{aligned}
(\alpha\mathbf{f} + \beta\mathbf{g})(\phi) &= \alpha\mathbf{f}(\phi) + \beta\mathbf{g}(\phi) \\
&= \alpha \int_{-\infty}^{\infty} f(t)\phi(t)\,dt + \beta \int_{-\infty}^{\infty} g(t)\phi(t)\,dt \\
&= \int_{-\infty}^{\infty} (\alpha f + \beta g)(t)\phi(t)\,dt, \qquad \phi \in \mathscr{D}.
\end{aligned}$$

$\blacksquare$

Now that addition and multiplication have been defined, it is easy to see that the set of all distributions forms a *linear space*. Correspondingly, also the set of generalized functions, consisting of the set of regular functions $\mathscr{L}$ together with the singular functions, is a linear space. Abusing notation, we sometimes denote this space as the continuous-time signal space $\mathscr{L}$. Signals in this space that are equal in the sense of distributions are considered to be one and the same signal.

Time Scaling and Time Translation

Time scaling of continuous-time signals may be defined by the *time scaling operator* μ^α, with α a nonzero real number, that transforms any signal x in $\mathscr{L}$ to the signal $\mu^\alpha x$ in $\mathscr{L}$, given by

$$(\mu^\alpha x)(t) = x(\alpha t), \qquad t \in \mathbb{R}.$$

This leads to the following definition of time scaling for distributions.

C.13.   Definition: Time scaling of distributions.   *Time scaling* the distribution $\mathbf{f}$ by the nonzero real constant α results in the distribution $\mu^\alpha \mathbf{f}$, defined by

$$(\mu^\alpha \mathbf{f})(\phi) = \frac{1}{|\alpha|} \mathbf{f}(\mu^{1/\alpha}\phi) \qquad \text{for all} \quad \phi \in \mathscr{D}.$$

If $\mathbf{f}$ is represented by the regular *or* singular function f, then we write $\mu^\alpha f$, or $f(\alpha t)$, $t \in \mathbb{R}$, for the generalized function that represents $\mu^\alpha \mathbf{f}$.   ∎

C.14.   Consistency proof: Time scaling.   If $\mathbf{f}$ is a regular distribution represented by the function f, time scaling $\mathbf{f}$ by α according to C.13 results in the distribution $\mu^\alpha \mathbf{f}$ given by

$$(\mu^\alpha \mathbf{f})(\phi) = \frac{1}{|\alpha|} \int_{-\infty}^{\infty} f(t)(\mu^{1/\alpha}\phi)(t) \, dt = \frac{1}{|\alpha|} \int_{-\infty}^{\infty} f(t)\phi(t/\alpha) \, dt$$

for all $\phi \in \mathscr{D}$. The change of variable $t/\alpha = \tau$ results in

$$(\mu^\alpha \mathbf{f})(\phi) = \int_{-\infty}^{\infty} f(\alpha\tau)\phi(\tau) \, d\tau = \int_{-\infty}^{\infty} (\mu^\alpha f)(\tau)\phi(\tau) \, d\tau$$

for all $\phi \in \mathscr{D}$. This proves that $\mu^\alpha \mathbf{f}$ is represented by $\mu^\alpha f$, as required for consistency.   ∎

C.15.   Example: Time scaling of the δ-function.   Let us see what time scaling by α does to the delta distribution and the δ-function. Application of the definition of a time scaled distribution yields

$$(\mu^\alpha \boldsymbol{\delta})(\phi) = \frac{1}{|\alpha|}\boldsymbol{\delta}(\mu^{1/\alpha}\phi) = \frac{1}{|\alpha|}(\mu^{1/\alpha}\phi)(0) = \frac{1}{|\alpha|}\phi(0/\alpha)$$

$$= \frac{1}{|\alpha|}\phi(0)$$

for all $\phi \in \mathcal{D}$. It follows that

$$\mu^\alpha \boldsymbol{\delta} = \frac{1}{|\alpha|}\boldsymbol{\delta}.$$

In terms of generalized functions,

$$\delta(\alpha t) = \frac{1}{|\alpha|}\delta(t), \qquad t \in \mathbb{R}. \hspace{3em} \blacksquare$$

C.16.   Exercise: Time reversing a $\boldsymbol{\delta}$-function. Show that $\delta(-t) = \delta(t)$, $t \in \mathbb{R}$. $\blacksquare$

Time translation of continuous-time signals may be defined by the *back shift operator* σ^θ, with θ a real number, that transforms any time signal x in $\mathcal{L}$ to the back shifted signal $\sigma^\theta x$ in $\mathcal{L}$, given by

$$(\sigma^\theta x)(t) = x(t + \theta), \qquad t \in \mathbb{R}.$$

Time translation of a distribution is defined as follows.

C.17.   Definition: Time translation of a distribution. *Time translating* the distribution $\mathbf{f}$ by the real number θ results in the *back shifted* distribution $\sigma^\theta \mathbf{f}$, defined by

$$(\sigma^\theta \mathbf{f})(\phi) = \mathbf{f}(\sigma^{-\theta}\phi) \qquad \text{for all} \quad \phi \in \mathcal{D}.$$

If $\mathbf{f}$ is represented by the regular or singular function f, then we write $\sigma^\theta f$, or $f(t + \theta)$, $t \in \mathbb{R}$, for the generalized function that represents $\sigma^\theta \mathbf{f}$. $\blacksquare$

C.18.   Consistency proof: Time translation of a distribution. We verify that back shifting a regular distribution $\mathbf{f}$ represented by the regular function f by θ results in a regular distribution $\sigma^\theta \mathbf{f}$ represented by the back shifted function $\sigma^\theta f$. Indeed, if $\mathbf{f}$ is regular it follows from the definition of time translation that

$$(\sigma^\theta \mathbf{f})(\phi) = \mathbf{f}(\sigma^{-\theta}\phi) = \int_{-\infty}^{\infty} f(t)(\sigma^{-\theta}\phi)(t)\, dt = \int_{-\infty}^{\infty} f(t)\phi(t - \theta)\, dt$$

for all $\phi \in \mathcal{D}$. By the change of variable $t - \theta = \tau$ it follows that

$$(\sigma^\theta \mathbf{f})(\phi) = \int_{-\infty}^{\infty} f(\tau + \theta)\phi(\tau)\, d\tau = \int_{-\infty}^{\infty} (\sigma^\theta f)(\tau)\phi(\tau)\, d\tau$$

for all $\phi \in \mathcal{D}$, which proves that $\sigma^\theta \mathbf{f}$ is represented by $\sigma^\theta f$.   ∎

C.19.   Example: Time translation of a δ-function.   Let us see what happens when the delta distribution and the δ-function are translated. By the definition of time translation we have

$$(\sigma^\theta \boldsymbol{\delta})(\phi) = \boldsymbol{\delta}(\sigma^{-\theta}\phi) = (\sigma^{-\theta}\phi)(0) = \phi(-\theta)$$

for all $\phi \in \mathcal{D}$. In terms of generalized functions,

$$\int_{-\infty}^{\infty} \delta(t + \theta)\phi(t)\, dt = \phi(-\theta).$$

The formula shows that the δ-function that is back shifted by θ singles out the value of the test function ϕ at time $-\theta$ rather than at time 0.   ∎

Multiplication by a Function

The *product* of two distributions, in general, is not defined. An exception occurs when one of the distributions is regular and represented by a *smooth* function (i.e., a function that may be differentiated as often as desired).

C.20.   Definition: Multiplication of a distribution by a smooth function.   If $\mathbf{f}$ is a distribution, and $\mathbf{g}$ is a regular distribution represented by a smooth regular function g, then the product $\mathbf{fg}$ is the distribution defined by

$$(\mathbf{fg})(\phi) = \mathbf{f}(g\phi) \qquad \text{for all} \quad \phi \in \mathcal{D}.$$

If $\mathbf{f}$ is represented by the regular or singular function f, then we write fg for the generalized function that represents $\mathbf{fg}$.   ∎

C.21.   Consistency proof: Product of a distribution and a smooth function.   We verify the consistency of the definition by letting $\mathbf{f}$ be a regular distribution represented by the regular function f. Then if the distribution $\mathbf{g}$ is regular and represented by the smooth function g, by C.20 the product $\mathbf{fg}$ is defined by

$$(\mathbf{fg})(\phi) = \mathbf{f}(g\phi) = \int_{-\infty}^{\infty} f(t)[g(t)\phi(t)]\, dt = \int_{-\infty}^{\infty} [f(t)g(t)]\phi(t)\, dt$$

for all $\phi \in \mathcal{D}$, which confirms that $\mathbf{fg}$ is represented by fg.   ∎

C.22.   **Example: Product of the δ-function and a smooth function.**   Suppose that **f** is the delta distribution. Then, if **g** is regular and represented by the smooth function g,

$$(\delta \mathbf{g})(\phi) = \delta(g\phi) = g(0)\phi(0) = g(0)\delta(\phi)$$

for all $\phi \in \mathcal{D}$. It follows that $\delta \mathbf{g} = g(0)\delta$. In terms of generalized functions,

$$g(t)\delta(t) = g(0)\delta(t), \qquad t \in \mathbb{R}.$$

Note that the product $g\delta$ only depends on the value $g(0)$ of g at 0.   ∎

Differentiation

We next discuss *differentiation* of distributions. A striking property of distributions and generalized functions is that they *always* have derivatives of *any* order.

C.23.   **Definition: Differentiation of distributions.**   The *derivative* of a distribution **f** is denoted $D\mathbf{f}$, $\mathbf{f}'$ or $\mathbf{f}^{(1)}$, and defined by

$$(D\mathbf{f})(\phi) = -\mathbf{f}(D\phi) \qquad \text{for all} \quad \phi \in \mathcal{D}.$$

If **f** is represented by the regular or singular function f, then we write Df, f' or $f^{(1)}$ for the generalized function that represents $D\mathbf{f}$.   ∎

It follows from Definition C.23 that the derivative $D\mathbf{f}$ of a distribution always exists. Repeated application of the definition leads to the *nth derivative* $D^n\mathbf{f}$ of **f**, with n any nonnegative integer, given by

$$(D^n\mathbf{f})(\phi) = (-1)^n\mathbf{f}(D^n\phi) \qquad \text{for all} \quad \phi \in \mathcal{D}.$$

We sometimes denote $D^n\mathbf{f}$ by $\mathbf{f}^{(n)}$.

C.24.   **Consistency proof: Differentiation of a distribution.**   As usual, we verify that the definition reduces to the conventional notion in case **f** is a regular distribution, represented by a differentiable regular function f. It follows that

$$(D\mathbf{f})(\phi) = -\mathbf{f}(D\phi) = -\int_{-\infty}^{\infty} f(t)(D\phi)(t)\, dt = -\int_{-\infty}^{\infty} f(t)\frac{d\phi(t)}{dt}\, dt$$

for all $\phi \in \mathcal{D}$. Integration by parts results in

$$(D\mathbf{f})(\phi) = -f(t)\phi(t)\Big|_{t=-\infty}^{\infty} + \int_{-\infty}^{\infty} \frac{df(t)}{dt}\phi(t)\,dt$$

$$= \int_{-\infty}^{\infty} \frac{df(t)}{dt}\phi(t)\,dt$$

for all $\phi \in \mathcal{D}$. Here we used the fact that ϕ is zero outside some finite interval so that $\phi(-\infty) = \phi(\infty) = 0$. It follows that $D\mathbf{f}$ is represented by Df, as required.   ∎

An interesting application of the differentiation formula is that the unit step has a derivative, in the distribution sense, which is the delta function.

C.25.   Example: The δ-function is the derivative of the unit step.   The unit step is the distribution **1** defined by

$$\mathbf{1}(\phi) = \int_{-\infty}^{\infty} \mathbb{1}(t)\phi(t)\,dt = \int_{0}^{\infty} \phi(t)\,dt \qquad \text{for all} \quad \phi \in \mathcal{D}.$$

Following C.23, the derivative of this distribution is the distribution defined by

$$(D\mathbf{1})(\phi) = -\mathbf{1}(D\phi) = -\int_{0}^{\infty} (D\phi)(t)\,dt = -\int_{0}^{\infty} \frac{d\phi(t)}{dt}\,dt = \phi(0)$$

$$= \boldsymbol{\delta}(\phi)$$

for all $\phi \in \mathcal{D}$. As a result, $D\mathbf{1} = \boldsymbol{\delta}.$ In terms of generalized functions we have

$$\frac{d}{dt}\mathbb{1}(t) = \delta(t), \qquad t \in \mathbb{R}. \qquad\qquad ∎$$

Definition C.24 allows differentiating δ-functions as often as desired.

C.26.   Example: The nth derivative of the delta distribution and the δ-function.   The nth derivative of the delta distribution is defined by

$$(D^{n}\boldsymbol{\delta})(\phi) = (-1)^{n}\boldsymbol{\delta}(D^{n}\phi) = (-1)^{n}\frac{d^{n}\phi(t)}{dt^{n}}\bigg|_{t=0}$$

for all $\phi \in \mathcal{D}$. We usually write the nth derivative of the δ-distribution as

$$D^{n}\boldsymbol{\delta} = \boldsymbol{\delta}^{(n)},$$

and represent it by the singular function $\delta^{(n)}$. In terms of generalized functions we have

$$\int_{-\infty}^{\infty} \phi(t)\delta^{(n)}(t) \, dt = (-1)^n \frac{d^n \phi(t)}{dt^n}\bigg|_{t=0}.$$ ∎

C.27.   Exercise. Product of $\delta^{(n)}$ and a smooth function.

(a) Show that if **g** is a regular distribution, represented by a smooth regular function g,

$$\mathbf{g}\boldsymbol{\delta}' = g(0)\boldsymbol{\delta}' - g'(0)\boldsymbol{\delta}.$$

In terms of generalized functions,

$$g(t)\delta'(t) = g(0)\delta'(t) - g'(0)\delta(t), \qquad t \in \mathbb{R}.$$

(b) Show by induction that if **g** is a regular distribution represented by the smooth function g,

$$\mathbf{g}\boldsymbol{\delta}^{(n)} = \sum_{k=0}^{n} (-1)^k \binom{n}{k} g^{(k)}(0)\boldsymbol{\delta}^{(n-k)},$$

or, in terms of generalized functions,

$$g(t)\delta^{(n)}(t) = \sum_{k=0}^{n} (-1)^k \binom{n}{k} g^{(k)}(0)\delta^{(n-k)}(t), \qquad t \in \mathbb{R}.$$ ∎

The converse of differentiation is of course integration.

C.28.   Definition: Indefinite integral of distributions.   If **f** is a distribution, then any distribution **F** such that $D\mathbf{F} = \mathbf{f}$ is called an *indefinite integral* of **f**.   ∎

C.29.   Exercise: Indefinite integral.

(a) Suppose that **f** is a regular distribution represented by the regular function f, and let F be an indefinite integral (in the regular sense) of f. Show that the distribution **F** represented by F is an indefinite integral of the distribution **f**.

(b) Prove that if **F** is an indefinite integral of **f**, also $\mathbf{c} + \mathbf{F}$, with **c** the regular distribution represented by the constant $c \in \mathbb{C}$, is an indefinite integral of **f**.

(c) Show that the unit step $\mathbb{1} \in \mathcal{L}$ is an indefinite integral of the δ-function. ∎

Infinite Sums

By scaling, adding, shifting, and differentiating δ-functions, a wide variety of generalized functions may be obtained, which suit most of our purposes in this text, provided also *infinite* sums are allowed. In particular we want to consider signals such as the "infinite comb," given by

$$w_P(t) = \sum_{n=-\infty}^{\infty} \delta(t + nP), \qquad t \in \mathbb{R},$$

and depicted in Fig. C.2. To handle this we need define the *convergence* of a sequence of distributions.

w_P

$-P \quad 0 \quad P \quad 2P$ time   **Figure C.2.** The infinite comb.

C.30.   Definition: Convergence of a sequence of distributions and of an infinite sum of distributions.

(a) *Convergence*. A sequence of distributions $\mathbf{f}_n$, $n = 1, 2, 3, \cdots$, *converges* to a distribution $\mathbf{f}$ if $\mathbf{f}_n(\phi) \to \mathbf{f}(\phi)$ as $n \to \infty$ for any $\phi \in \mathcal{D}$. If a sequence $\mathbf{f}_n$, $n = 1, 2, 3, \cdots$, converges to $\mathbf{f}$, then we write

$$\lim_{n \to \infty} \mathbf{f}_n = \mathbf{f}.$$

(b) *Infinite sums*. The infinite sum

$$\sum_{n=1}^{\infty} \mathbf{f}_n$$

is said to exist if the sequence of partial sums

$$\mathbf{s}_N = \sum_{n=1}^{N} \mathbf{f}_n, \qquad N = 1, 2, \cdots,$$

converges as $N \to \infty$. In this case

$$\sum_{n=1}^{\infty} \mathbf{f}_n = \mathbf{s},$$

where $\mathbf{s} = \lim_{N \to \infty} \mathbf{s}_N$.   ∎

C.31.   Example: Infinite comb.   The infinite comb

$$\mathbf{w}_P = \sum_{n=-\infty}^{\infty} \sigma^{nP} \boldsymbol{\delta},$$

with P a real number, is a well-defined distribution, because the sequence of partial sums

$$\sum_{n=-N}^{N} (\sigma^{nP}\boldsymbol{\delta})(\phi) = \sum_{n=-N}^{N} \boldsymbol{\delta}(\sigma^{-nP}\phi) = \sum_{n=-N}^{N} \phi(-nP)$$

converges as $N \to \infty$ for any $\phi \in \mathcal{D}$. In fact, because ϕ is zero outside some finite interval, the sum only has a *finite* number of nonzero terms. In terms of generalized functions we may represent the infinite comb as

$$w_P(t) = \sum_{n=-\infty}^{\infty} \delta(t + nP), \qquad t \in \mathbb{R}.$$ ∎

What follows has important applications.

C.32.   Summary: Convergence of a sequence of derivatives and term-by-term differentiation and integration.

(a) *Convergence of a sequence of derivatives.* Suppose that the sequence of distributions $\mathbf{f}_n$, $n = 1, 2, 3, \cdots$, converges to a distribution $\mathbf{f}$. Then the sequence $D\mathbf{f}_n$, $n = 1, 2, 3, \cdots$, converges to $D\mathbf{f}$.

(b) *Term-by-term differentiation and integration.* Suppose that

$$\mathbf{f} = \sum_{n=1}^{\infty} \mathbf{f}_n$$

exists. Then,

$$D\mathbf{f} = \sum_{n=1}^{\infty} D\mathbf{f}_n.$$

Likewise, if $\mathbf{F}_n$ is an indefinite integral of $\mathbf{f}_n$ for $n = 1, 2, 3, \cdots$, and

$$\mathbf{F} = \sum_{n=1}^{\infty} \mathbf{F}_n$$

exists, then $\mathbf{F}$ is an indefinite integral of $\mathbf{f}$.   ∎

C.33.   Example: Comb of derivatives.
The comb $\mathbf{w}_P^{(k)}$ of kth order derivatives of delta distributions may be obtained by differentiating the infinite comb $\mathbf{w}_P$ term by term, that is,

$$\mathbf{w}_P^{(k)} = D^k \left(\sum_{n=-\infty}^{\infty} \sigma^{nP}\boldsymbol{\delta} \right) = \sum_{n=-\infty}^{\infty} D^k \sigma^{nP}\boldsymbol{\delta} = \sum_{n=-\infty}^{\infty} \sigma^{nP} D^k \boldsymbol{\delta}$$

$$= \sum_{n=-\infty}^{\infty} \sigma^{nP}\boldsymbol{\delta}^{(k)}.$$

In generalized function form,

$$w_P^{(k)}(t) = \sum_{n=-\infty}^{\infty} \delta^{(k)}(t - nP), \qquad t \in \mathbb{R}.$$ ∎

C.34.   Exercise: Indefinite integral of the infinite comb.   Let int be the entier function defined in 2.3.4. Show that the function W_P defined by

$$W_P(t) = \text{int}\,(t/P), \qquad t \in \mathbb{R},$$

is an indefinite integral of the infinite comb w_P.   ∎

Periodic Distributions

The graph of the infinite comb of Fig. C.2 has a periodic appearance. Periodicity of distributions is defined as follows.

C.35.   Definition: Periodic distribution.
(a) *Periodic distribution*. The distribution **f** is *periodic* if $\sigma^P \mathbf{f} = \mathbf{f}$ for some $P \in \mathbb{R}$ with $P \neq 0$.

(b) *Period of a periodic distribution*. The periodic distribution **f** has *period P* if P is the smallest positive number such that $\sigma^P \mathbf{f} = \mathbf{f}$.   ∎

C.36.   Exercise: Consistency proof of periodicity of distributions.   Prove that the regular distribution **f** represented by the regular function f is periodic if and only if f is periodic. Show also that the period of **f** equals the period of f.   ∎

C.37.   Example: The infinite comb.   The infinite comb $\mathbf{w}_P$ of Example C.31 is periodic since

$$\sigma^P \mathbf{w}_P = \sigma^P \left(\sum_{n=-\infty}^{\infty} \sigma^{nP} \boldsymbol{\delta} \right) = \sum_{n=-\infty}^{\infty} \sigma^{(n+1)P} \boldsymbol{\delta} = \sum_{k=-\infty}^{\infty} \sigma^{kP} \boldsymbol{\delta} = \mathbf{w}_P.$$

Here we use the fact that time translation of infinite sums may be done term-by-term, and substituted $k = n + 1$. The period of $\mathbf{w}_P$ is P.   ∎

Convolution

We complement Section 3.5 with a discussion of the *convolution* of distributions and generalized signals. Like the other operations on distributions, convolution of distributions need be defined indirectly.

C.38.   Definition: Convolution of distributions.   The *convolution* of two distributions **f** and **g**, if it exists, is the distribution $\mathbf{f} * \mathbf{g}$ defined by

$$(\mathbf{f} * \mathbf{g})(\phi) = \mathbf{f}(\psi),$$

for all $\phi \in \mathcal{D}$, where ψ is given by

$$\psi(t) = \mathbf{g}(\sigma^t \phi), \qquad t \in \mathbb{R}.$$

As before, σ is the back shift operator. If $\mathbf{f}$ and $\mathbf{g}$ are represented by the regular or singular functions f and g, respectively, we write $f * g$ for the generalized function that represents $\mathbf{f} * \mathbf{g}$. ∎

The definition shows that to determine the effect of the convolution $\mathbf{f} * \mathbf{g}$ on a test function ϕ, first the function ψ is obtained pointwise for each $t \in \mathbb{R}$ by applying $\mathbf{g}$ to the back shifted function $\sigma^t \phi$, and then $\mathbf{f}$ is applied to ψ.

C.39.   **Consistency proof: Convolution.**   We show that if $\mathbf{f}$ and $\mathbf{g}$ are regular distributions represented by regular functions f and g such that $f * g$ exists, $\mathbf{f} * \mathbf{g}$ is regular and represented by $f * g$. If $\mathbf{g}$ is regular, then the function ψ in C.38 is given by

$$\psi(t) = \mathbf{g}(\sigma^t \phi) = \int_{-\infty}^{\infty} g(\tau)(\sigma^t \phi)(\tau) \, d\tau = \int_{-\infty}^{\infty} g(\tau)\phi(\tau + t) \, d\tau$$

$$= \int_{-\infty}^{\infty} g(\theta - t)\phi(\theta) \, d\theta, \qquad t \in \mathbb{R},$$

where we substituted $\tau + t = \theta$. It may be proved that $\mathbf{f}(\psi)$ is well-defined for any test function ϕ and as a result,

$$(\mathbf{f} * \mathbf{g})(\phi) = \mathbf{f}(\psi) = \int_{-\infty}^{\infty} f(t)\psi(t) \, dt$$

$$= \int_{-\infty}^{\infty} f(t)\left(\int_{-\infty}^{\infty} g(\theta - t)\phi(\theta) \, d\theta \right) dt$$

$$= \int_{-\infty}^{\infty} \left(\int_{-\infty}^{\infty} f(t)g(\theta - t) \, dt \right) \phi(\theta) \, d\theta$$

$$= \int_{-\infty}^{\infty} (f * g)(\theta)\phi(\theta) \, d\theta.$$

This shows that $\mathbf{f} * \mathbf{g}$ is represented by $f * g$. ∎

C.40.   **Example: The delta function is the unit of the convolution.**   Suppose that $\mathbf{f}$ is any distribution, and let $\mathbf{g}$ be the delta distribution. Then, for any $\phi \in \mathcal{D}$ the function ψ is given by

$$\psi(t) = \delta(\sigma^t \phi) = (\sigma^t \phi)(0) = \phi(t), \qquad t \in \mathbb{R},$$

so that $\psi = \phi$ and, hence, $\psi \in \mathcal{D}$. It follows that $\mathbf{f} * \boldsymbol{\delta}$ is defined by

$$(\mathbf{f} * \boldsymbol{\delta})(\phi) = \mathbf{f}(\psi) = \mathbf{f}(\phi) \qquad \text{for all} \quad \phi \in \mathcal{D}.$$

Consequently,

$$\mathbf{f} * \boldsymbol{\delta} = \mathbf{f}$$

for any distribution $\mathbf{f}$. In terms of generalized functions,

$$f * \delta = f$$

for any regular or singular function f. Written in full generalized function notation we have

$$\int_{-\infty}^{\infty} f(t - \tau)\delta(\tau)\, d\tau = \int_{-\infty}^{\infty} \delta(\tau)f(t - \tau)\, d\tau = f(t), \qquad t \in \mathbb{R}.$$

The delta function is what is called the *unit* of the convolution.   ∎

On the basis of Definition C.38 of the convolution of distributions and generalized functions, it may be proved that all the properties 3.5.2 of the continuous-time convolution also apply to the generalized convolution.

C.41.   Example: Convolution with derivatives of delta functions.   We use several of the properties of the convolution to obtain some formulas involving convolution with derivatives of delta functions. Let n and m be nonnegative integers. Then, if f and g are generalized signals, by the differentiation property and the commutativity of the convolution we have

$$D^{n+m}(f * g) = D^n(D^m(f * g)) = D^n(f * (D^m g)) = (D^n f) * (D^m g).$$

In particular, since $f * \delta = f$, it follows that

$$f * \delta^{(n)} = D^n f,$$

for all n in $\mathbb{Z}_+$ and any generalized signal f. Also, since $\delta * \delta = \delta$,

$$\delta^{(n)} * \delta^{(m)} = \delta^{(n+m)}$$

for all n and m in $\mathbb{Z}_+$. In full generalized signal notation we have

$$\int_{-\infty}^{\infty} f(t - \tau)\delta^{(n)}(\tau)\, d\tau = \frac{d^n}{dt^n}f(t), \qquad t \in \mathbb{R},$$

for every generalized signal f, and

$$\int_{-\infty}^{\infty} \delta^{(n)}(t - \tau)\delta^{(m)}(\tau)\, d\tau = \delta^{(n+m)}(t), \qquad t \in \mathbb{R},$$

for all nonnegative n and m.   ∎

C.42.   Exercise: Integration via convolution. The convolution provides a way to obtain the integral of a distribution.

(a) Let **1** be the distribution represented by the unit step $\mathbb{1}$ and **c** the regular distribution represented by the constant $c \in \mathbb{C}$. Prove that if

F = 1 ∗ f + c

exists, it is an indefinite integral of **f** for any **c**.

(b) Suppose that f is a regular function such that the definite integral

$$F(t) = \int_{-\infty}^{t} f(\tau)\, d\tau, \qquad t \in \mathbb{R}, \tag{2}$$

exists for all $t \in \mathbb{R}$. Prove that F represents the distribution **1 ∗ f**.

(c) By analogy, we write the generalized function F that represents the definite integral **F = 1 ∗ f**, if it exists, in function form as (2). Prove that

$$\int_{-\infty}^{t} \delta(\tau)\, d\tau = \mathbb{1}(t), \qquad t \in \mathbb{R}.$$ ∎

C.43.   Remark: Existence of the convolution. The convolution of two distributions may or may not exist, just as the convolution of two regular functions does not always exist. Like for regular convolutions (see 3.4.5) sufficient conditions are available for the existence of the convolution of distributions depending on their supports. A distribution **f** has *bounded support* if there exists a bounded interval $[a, b]$ such that $\mathbf{f}(\phi) = 0$ for every test function $\phi \in \mathcal{D}$ with support *outside* this interval. Similarly, the distribution **f** is *right one-sided* if its support is right-one sided (i.e., there exists a semi-infinite interval $[a, \infty)$ such that $\mathbf{f}(\phi) = 0$ for every test function ϕ with support outside this interval). Left one-sided distributions are defined analogously.

If **f** and **g** are two distributions, then we have:

(a) If **f** or **g** has bounded support, then **f** * **g** exists. If **f** and **g** both have bounded support, then also **f** * **g** has bounded support.

(b) If **f** and **g** are both one-sided (both left or both right), then **f** * **g** exists and is also one-sided (in the same direction as **f** and **g**.)

(c) If **f** or **g** is regular and represented by a smooth function with bounded support, then **f** * **g** exists, is regular, and may be represented by a smooth function.   ∎

Tempered Signals

In Section 7.4 the CCFT of finite-energy continuous-time signals is studied. There are many useful signals, however, that do not belong to the space $\mathcal{L}_2$, such as the constant signal, the unit step, and the δ-function, and may easily figure as inputs to linear systems. We therefore extend the CCFT to a much wider class of signals than $\mathcal{L}_2$, namely, the signals *of polynomial growth*, also called *tempered* signals. This class includes both singular and regular signals. The constant signal, the unit step, and δ-function belong to it, but signals that increase faster than polynomials, such as exponentially increasing signals, do not.

Regular tempered signals include all continuous-time signals of polynomial growth, that is, signals x for which there exist a nonnegative integer N and real constants β and γ such that

$$|x(t)| \leq \beta |t|^N + \gamma, \qquad t \in \mathbb{R}.$$

Singular tempered signals need be introduced by using the notion of *distribution*. A tempered signal is the generalized signal defining a *tempered distribution*. A tempered distribution is a linear functional on the set of test functions $\mathcal{S}$ of *rapid decay*.

C.44.   Definition: Test functions of rapid decay.   The set $\mathcal{S}$ of test functions of rapid decay consists of all smooth functions on $\mathbb{R}$ that decay faster than any polynomial, that is, ϕ belongs to $\mathcal{S}$ if it is smooth and there exist a nonnegative number N and a real constant α such that

$$|t^N \phi(t)| \leq \alpha, \qquad t \in \mathbb{R}. \qquad\qquad\qquad \blacksquare$$

A smooth function of bounded support obviously is of rapid decay and, hence, belongs to $\mathcal{S}$. It follows that the set of test functions $\mathcal{D}$ of bounded support of C.3 is contained in $\mathcal{S}$.

C.45.   Example: Test function of rapid decay.   The "Gaussian bell" is the smooth function ϕ defined by

$$\phi(t) = \frac{1}{\sigma \sqrt{2\pi}} e^{-t^2/2\sigma^2}, \qquad t \in \mathbb{R},$$

with σ a positive real number. Because ϕ has unbounded support, it does not belong to $\mathcal{D}$, but it clearly is an element of $\mathcal{S}$.   ∎

Tempered distributions are defined as linear functionals on the set of test functions $\mathcal{S}$ of rapid decay. *Tempered* generalized signals are the generalized functions that represent tempered distributions.

C.46.   Definition: Tempered distributions and tempered generalized signals.   A *tempered* distribution is a linear functional on the set of test functions $\mathcal{S}$. The generalized signal representing a tempered distribution is called a *tempered* signal.   ∎

Because the set of test functions $\mathcal{D}$ of bounded support is contained in the set of test functions $\mathcal{S}$ of rapid decay, every linear functional on $\mathcal{S}$ is also a linear functional on $\mathcal{D}$. As a result, every tempered distribution is a distribution as defined in C.5, and every tempered generalized signal is a generalized signal. The converse is not true: not every generalized signal is tempered.

C.47.   Examples: Tempered distributions and signals.
      (a) *Regular signal of polynomial growth.* Every regular signal x of polynomial growth defines a tempered distribution **x** given by

$$\mathbf{x}(\phi) = \int_{-\infty}^{\infty} x(t)\phi(t)\, dt, \qquad \phi \in \mathcal{S}.$$

      (b) *The δ-function and its derivatives.* The delta distribution, defined by

$$\boldsymbol{\delta}(\phi) = \int_{-\infty}^{\infty} \delta(t)\phi(t)\, dt = \phi(0), \qquad \phi \in \mathcal{S},$$

is a tempered distribution. Hence, the δ-function is a tempered singular signal. Similarly, all derivatives of delta functions $\delta^{(k)}$ are tempered singular signals.

      (c) *Periodic generalized signals.* Every *periodic* generalized signal is tempered.

      (d) *Exponentially increasing signals are not tempered.* Exponentially increasing signals, such as the signal x defined by $x(t) = e^{\alpha t}$, $t \in \mathbb{R}$, with Re $(\alpha) \neq 0$, are generalized signals, but not tempered.   ∎

The CCFT of Tempered Signals

Preparatory to the definition of the CCFT of signals of polynomial growth we state the following important fact.

C.48.   Summary: The CCFT of test functions of rapid decay.   The CCFT is a bijection from the set of test functions $\mathscr{S}$ of rapid decay to itself.   ∎

The statement implies that if ϕ belongs to $\mathscr{S}$, then its CCFT $\hat{\phi}$ also belongs to $\mathscr{S}$ and vice-versa. It allows us to define the CCFT of generalized signals of polynomial growth as follows.

C.49.   Definition: The generalized CCFT.
   (a) *CCFT of a tempered distribution*. The generalized CCFT of the tempered distribution $\mathbf{x}$ is the tempered distribution $\hat{\mathbf{x}}$ defined by

$$\hat{\mathbf{x}}(\phi) = \mathbf{x}(\hat{\phi}) \qquad \text{for all} \quad \phi \in \mathscr{S},$$

with $\hat{\phi} \in \mathscr{S}$ the CCFT of ϕ.

   (b) *CCFT of a tempered generalized signal*. Suppose that the tempered distribution $\mathbf{x}$ is represented by the generalized signal x. Then the generalized CCFT $\hat{x}$ of x is the tempered generalized signal that represents the CCFT $\hat{\mathbf{x}}$ of $\mathbf{x}$.   ∎

As usual with operations on generalized signals, the CCFT of a generalized signal is defined by its "effect" on suitable test functions. For regular signals the definition reduces to the usual one.

C.50.   Consistency proof.   We prove that if $x \in \mathscr{L}_2$ is a regular signal, then the definition C.49 of the CCFT of x results in the usual CCFT. The regular signal x defines the tempered distribution $\mathbf{x}$ given by

$$\mathbf{x}(\phi) = \int_{-\infty}^{\infty} x(t)\phi(t)\, dt, \qquad \phi \in \mathscr{S}.$$

According to C.49, the CCFT $\hat{\mathbf{x}}$ of $\mathbf{x}$ is the distribution given by

$$\hat{\mathbf{x}}(\phi) = \mathbf{x}(\hat{\phi}) = \int_{-\infty}^{\infty} x(t)\hat{\phi}(t)\, dt = \int_{-\infty}^{\infty} x(t)\left(\int_{-\infty}^{\infty} \phi(f)e^{-j2\pi ft}\, df \right) dt, \qquad \phi \in \mathscr{S}.$$

Note that when writing $\hat{\phi}$ as the CCFT of ϕ for convenience we interchanged the usual roles of f and t. We now apply *Fubini's theorem*, which states, roughly, that the order of a repeated integral may be reversed provided the result is a convergent integral. This yields

$$\hat{\mathbf{x}}(\phi) = \int_{-\infty}^{\infty} \left(\int_{-\infty}^{\infty} x(t) \phi(f) e^{-j2\pi ft} \, df \right) dt$$

$$= \int_{-\infty}^{\infty} \left(\int_{-\infty}^{\infty} x(t) \phi(f) e^{-j2\pi ft} \, dt \right) df$$

$$= \int_{-\infty}^{\infty} \left(\int_{-\infty}^{\infty} x(t) e^{-j2\pi ft} \, dt \right) \phi(f) \, df$$

$$= \int_{-\infty}^{\infty} \hat{x}(f) \phi(f) \, df, \qquad \phi \in \mathcal{S}.$$

This confirms that the generalized CCFT $\hat{\mathbf{x}}$ of $\mathbf{x}$ is a regular tempered distribution represented by the regular CCFT $\hat{x}$.   ∎

Guido Fubini (1897–1943) was an Italian mathematician who taught in Turin until he emigrated to the United States in 1938.

By way of example, we consider the generalized CCFT of some signals that have no CCFT in the ordinary sense.

C.51.   Example: Generalized CCFTs.

(a) *CCFT of the δ-function*. Suppose that

$$x(t) = \delta(t), \qquad t \in \mathbb{R}.$$

Then the generalized CCFT $\hat{\mathbf{x}}$ of the tempered distribution $\mathbf{x} = \boldsymbol{\delta}$ is defined by

$$\hat{\mathbf{x}}(\phi) = \mathbf{x}(\hat{\phi}) = \int_{-\infty}^{\infty} \delta(t) \hat{\phi}(t) \, dt = \hat{\phi}(0) = \int_{-\infty}^{\infty} \phi(f) \, df$$

$$= \int_{-\infty}^{\infty} 1 \cdot \phi(f) \, df, \qquad \phi \in \mathcal{S}.$$

This shows that $\hat{\mathbf{x}}$ is a regular tempered distribution represented by

$$\hat{x}(f) = 1, \qquad f \in \mathbb{R}.$$

Thus, the CCFT of the δ-function is the constant 1.

(b) *CCFT of a constant*. Next, suppose that x is the constant signal

$$x(t) = 1, \qquad t \in \mathbb{R},$$

which has no CCFT in the ordinary sense. The generalized CCFT $\hat{\mathbf{x}}$ of the tempered distribution $\mathbf{x}$ represented by x is given by

$$\hat{\mathbf{x}}(\phi) = \mathbf{x}(\hat{\phi}) = \int_{-\infty}^{\infty} 1 \cdot \hat{\phi}(f) \, df = \int_{-\infty}^{\infty} \hat{\phi}(f) \, df, \qquad \phi \in \mathscr{S}.$$

The latter expression simply is the inverse CCFT of $\hat{\phi}$ evaluated at 0, so that

$$\hat{\mathbf{x}}(\phi) = \phi(0) = \int_{-\infty}^{\infty} \delta(f)\phi(f) \, df, \qquad \phi \in \mathscr{S}.$$

This shows that the CCFT of the constant signal x is the δ-function

$$\hat{x}(f) = \delta(f), \qquad f \in \mathbb{R}. \qquad\qquad\qquad\qquad\qquad \blacksquare$$

C.52.   Exercise: CCFT of derivatives of δ-functions.   Prove that the CCFT of the kth derivative delta function $\delta^{(k)}$ is the regular tempered signal $(j2\pi f)^k$, $f \in \mathbb{R}$.
$\qquad\qquad\qquad\qquad\qquad\qquad\qquad\qquad\qquad\qquad\qquad\qquad\qquad\qquad\qquad\qquad\qquad\qquad\blacksquare$

The *inverse* CCFT of a tempered signal is easily established, namely by reversing Definition C.49.

C.53.   Summary: Inverse generalized CCFT.
    (a) *Inverse CCFT of a tempered distribution.* The inverse generalized CCFT of the tempered distribution $\hat{\mathbf{x}}$ is the tempered distribution $\mathbf{x}$ defined by

$$\mathbf{x}(\phi) = \hat{\mathbf{x}}(\check{\phi}) \qquad \text{for all} \quad \phi \in \mathscr{S},$$

with $\check{\phi}$ the inverse CCFT of ϕ.

    (b) *Inverse CCFT of a tempered generalized signal.* Suppose that the tempered distribution $\hat{\mathbf{x}}$ is represented by the tempered generalized signal $\hat{x}$. Then the generalized inverse CCFT x of $\hat{x}$ is the tempered generalized signal that represents the inverse CCFT $\mathbf{x}$ of $\hat{\mathbf{x}}$.
$\qquad\blacksquare$

By way of example, we establish that application of the inverse CCFT to the CCFTs found in Example C.51 restores the original time signals.

C.54.   Examples: Inverse generalized CCFT.
    (a) *Inverse CCFT of a constant.* Suppose that $\hat{x}$ is the constant signal

$$\hat{x}(f) = 1, \qquad f \in \mathbb{R}.$$

By C.53, the inverse CCFT **x** of the tempered distribution $\hat{\mathbf{x}}$ defined by $\hat{x}$ is given by

$$\mathbf{x}(\phi) = \hat{\mathbf{x}}(\check{\phi}) = \int_{-\infty}^{\infty} \hat{x}(f)\check{\phi}(f)\, df = \int_{-\infty}^{\infty} \check{\phi}(f)\, df, \qquad \phi \in \mathscr{S}.$$

The latter expression is the CCFT ϕ of $\check{\phi}$ evaluated at 0, so that

$$\mathbf{x}(\phi) = \phi(0) = \int_{-\infty}^{\infty} \delta(t)\phi(t)\, dt, \qquad \phi \in \mathscr{S}.$$

This proves that the inverse CCFT of the constant 1 is the δ-function, in agreement with C.51(a), where we found that the CCFT of the δ-function is the constant 1.

(b) *Inverse CCFT of the δ-function.* By C.53, the inverse CCFT **x** of the tempered distribution $\hat{\mathbf{x}} = \boldsymbol{\delta}$ is given by

$$\mathbf{x}(\phi) = \int_{-\infty}^{\infty} \delta(f)\check{\phi}(f)\, df = \check{\phi}(0) = \int_{-\infty}^{\infty} \phi(t)\, dt, \qquad \phi \in \mathscr{S}.$$

This confirms that the inverse CCFT of the δ-function is the constant 1, as expected from C.51(b), where we found that the CCFT of the constant 1 is the δ-function. ∎

C.55.   Exercise: Inverse CCFT of derivatives of δ-functions.   Show by application of the inverse CCFT that for every nonnegative integer k the generalized CCFT of the tempered signal $(-j2\pi t)^k$, $t \in \mathbb{R}$, is $\delta^{(k)}(f), f \in \mathbb{R}$.   ∎

The generalized CCFT possesses all the important properties the ordinary CCFT has, to wit, the *linearity, convolution, shift,* and *differentiation* properties and their converses. Also the symmetry properties carry over.

The identities of *Plancherel* and *Parseval* hold of course for regular signals that belong to $\mathscr{L}_2$, but do not make sense for regular signals not in $\mathscr{L}_2$ or for singular signals, because singular signals cannot always be multiplied and never can be squared.

The convolution property, which implies that convolution transforms into a product, and vice-versa, only applies if the convolutions and products actually exist.

C.56.   Example: CCFT of a harmonic signal.   We apply the shift property to determine the CCFT of the harmonic signal

$$z(t) = e^{j2\pi f_o t}, \qquad t \in \mathbb{R},$$

with f_o a real number. The shift property implies that if the CCFT of x is $\hat{x}$, the CCFT of $x(t)e^{j2\pi\phi t}, f \in \mathbb{R}$, is $\hat{x}(f - \phi), f \in \mathbb{R}$. Writing

$$z(t) = e^{j2\pi f_o t} \cdot 1, \qquad t \in \mathbb{R},$$

and recalling from C.51(b) that the CCFT of the constant 1 is $\delta(f), f \in \mathbb{R}$, it follows that the CCFT of the harmonic z is

$$\hat{z}(f) = \delta(f - f_o), \qquad f \in \mathbb{R}.$$

The interpretation of this result is that a complex harmonic with frequency f_o has all its frequency content concentrated at f_o.   ∎

Supplement D

Jordan Normal Form

The discussion of the modal transformation of linear time-invariant state difference or differential systems of Section 5.6 is based on the assumption that the $N \times N$ matrix A has N linearly independent eigenvectors. If A does not have N linearly independent eigenvalues (i.e., A is defective), then it cannot be diagonalized, but it can be transformed into a near-diagonal form, called the *Jordan normal form*. We summarize the main results.

D.1. Summary: Jordan normal form. Suppose that the $N \times N$ matrix A has L mutually different eigenvalues $\lambda_1, \lambda_2, \cdots, \lambda_L$, such that the eigenvalue λ_i has multiplicity N_i in the characteristic polynomial $\det(\lambda I - A)$. Then there exists a nonsingular $N \times N$ matrix V, which may be partitioned into L blocks of columns as

$$V = [V_1, V_2, \cdots, V_L],$$

such that

$$V^{-1}AV = J.$$

J consists of L diagonal blocks

$J = \text{diag}(J_1, J_2, \cdots, J_L)$.

For each $i \in \{1, 2, \cdots, L\}$ the block V_i has N_i columns and the block J_i has dimensions $N_i \times N_i$. The block J_i may be subpartitioned into L_i diagonal sub-blocks, with $L_i \leq N_i$, as

$$J_i = \text{diag}(J_{i1}, J_{i2}, \cdots, J_{iL_i}), \tag{1}$$

where each sub-block J_{ij} is of the form

$$J_{ij} = \begin{bmatrix} \lambda_i & 1 & 0 & \cdots & 0 & 0 \\ 0 & \lambda_i & 1 & \cdots & 0 & 0 \\ \cdots & \cdots & \cdots & \cdots & \cdots & \cdots \\ 0 & 0 & 0 & \cdots & \lambda_i & 1 \\ 0 & 0 & 0 & \cdots & 0 & \lambda_i \end{bmatrix}.$$

J is called the *Jordan normal form* of the matrix A. ■

Camille Jordan (1838–1922) was a French mathematician who taught at the Ecole Polytechnique in Paris.

In the case of a defective matrix, the Jordan normal form J takes the place of the diagonal matrix Λ with the eigenvalues of A as diagonal elements. The Jordan normal form also has the eigenvalues of A on the main diagonal, but on the first diagonal above the main diagonal it has sequences of ones interrupted by zeros. An example of a matrix that is in Jordan form, with two sub-blocks, is

$$J = \left[\begin{array}{ccc|cc} 2 & 1 & 0 & 0 & 0 \\ 0 & 2 & 1 & 0 & 0 \\ 0 & 0 & 2 & 0 & 0 \\ \hline 0 & 0 & 0 & 3 & 1 \\ 0 & 0 & 0 & 0 & 3 \end{array} \right].$$

From $J = V^{-1}AV$ it follows that

$$AV = VJ.$$

This suggests the following way of determining the transformation matrix V and the Jordan normal form of A. Denote the columns of V as $v_1, v_2, \cdots, v_N$. Then, from the form of J it follows with $AV = VJ$ that

$$Av_j = \lambda v_j + \gamma_j v_{j-1}, \qquad j = 2, 3, \cdots, N, \tag{2}$$

where λ is an eigenvalue of A, and γ_j is either 0 or 1, depending on whether or not the ith column of J has a 1 above the main diagonal.

Let us subpartition the block V_i of V corresponding to the subpartitioning (1) of J_i as

$$V_i = [V_{i1}, V_{i2}, \cdots, V_{iL_i}].$$

Then γ_i is zero whenever the corresponding column v_i is the *first* column of a sub-block.

If $\gamma_i = 0$ the vector v_i satisfies $Av_i = \lambda v_i$ and, hence, is an eigenvector of A. Thus, the *first* columns of the sub-blocks of V_i may be found by determining the eigenvectors of A corresponding to the eigenvalue λ. The remaining columns of V successively follow by (2) with $\gamma_j = 1$. These remaining columns are known as the *generalized eigenvectors* of the matrix A. An eigenvector v_i followed by a number of generalized eigenvectors satisfying (2) with $\gamma_j = 1$ is called a *chain*. We illustrate the procedure with an example.

D.2. **Example: Jordan normal form.** Consider the matrix A as obtained in Example 5.6.12(b) but suppose that $R_1 = R_2 = 1$, $C_1 = C_2 = 2$, and $L_1 = L_2 = \frac{1}{2}$. The matrix then takes the form

$$A = \begin{bmatrix} -2 & \frac{1}{2} & 0 & 0 \\ -2 & 0 & 0 & 0 \\ 0 & 0 & -2 & \frac{1}{2} \\ 0 & 0 & -2 & 0 \end{bmatrix},$$

and its characteristic polynomial is

$$\det(\lambda I - A) = (\lambda^2 + 2\lambda + 1)(\lambda^2 + 2\lambda + 1) = (\lambda + 1)^4.$$

Hence, the matrix A has the single eigenvalue -1 with multiplicity 4. It is easily verified that corresponding to this eigenvalue A has two linearly independent eigenvectors, which may be chosen as $\mathrm{col}(1, 2, 0, 0)$ and $\mathrm{col}(0, 0, 1, 2)$. Thus, choose the first column of V as

$$v_1 = \begin{bmatrix} 1 \\ 2 \\ 0 \\ 0 \end{bmatrix}.$$

Then the second column v_2 of V by (2) must satisfy

$$Av_2 = -v_2 + v_1.$$

It may be found that this equation has the solution

$$v_2 = \begin{bmatrix} 0 \\ 2 \\ 0 \\ 0 \end{bmatrix}.$$

To this solution one may add any multiple of v_1, but we leave it as it is. It may also be verified that

$$Av_3 = -v_3 + v_2$$

has no solution so that the first chain consisting of an eigenvector followed by generalized eigenvectors consists of v_1 and v_2.

We now start a new chain by setting v_3 equal to the second eigenvector we found, namely,

$$v_3 = \begin{bmatrix} 0 \\ 0 \\ 1 \\ 2 \end{bmatrix}.$$

Solution of

$$Av_4 = -v_4 + v_3$$

results in

$$v_4 = \begin{bmatrix} 0 \\ 0 \\ 0 \\ 2 \end{bmatrix}.$$

This completes the computation of V, which together with its inverse takes the form

$$V = \begin{bmatrix} 1 & 0 & 0 & 0 \\ 2 & 2 & 0 & 0 \\ 0 & 0 & 1 & 0 \\ 0 & 0 & 2 & 2 \end{bmatrix}, \qquad V^{-1} = \begin{bmatrix} 1 & 0 & 0 & 0 \\ -1 & \frac{1}{2} & 0 & 0 \\ 0 & 0 & 1 & 0 \\ 0 & 0 & -1 & \frac{1}{2} \end{bmatrix}.$$

The Jordan normal form of A is

$$
J = \begin{bmatrix} -1 & 1 & 0 & 0 \\ 0 & -1 & 0 & 0 \\ 0 & 0 & -1 & 1 \\ 0 & 0 & 1 & -1 \end{bmatrix}.
$$
                                                                            ∎

Given the Jordan normal form of a matrix A it is comparatively simple to determine integral powers of A and its exponential, needed to compute transition matrices of linear time-invariant discrete- or continuous-time systems.

D.3.   Summary: Powers and exponentials of matrices.   Consider the matrix A as in D.1. Then,

(a) $A^n = VJ^nV^{-1}$ for all $n \in \mathbb{Z}_+$.

(b) $e^{At} = Ve^{Jt}V^{-1}$ for all $t \in \mathbb{R}$.

(c) $J^n = \mathrm{diag}(J_1^n, J_2^n, \cdots, J_L^n)$ and
    $J_i^n = \mathrm{diag}(J_{i1}^n, J_{i2}^n, \cdots, J_{iL_i}^n)$ for all $n \in \mathbb{Z}_+$.

(d) $e^{Jt} = \mathrm{diag}(e^{J_1 t}, e^{J_2 t}, \cdots, e^{J_L t})$ and
    $e^{J_i t}, = \mathrm{diag}(e^{J_{i1} t}, e^{J_{i2} t}, \cdots, e^{J_{iL_i} t})$ for all $t \in \mathbb{R}$.

(e) For all $n \in \mathbb{Z}_+$,

$$
J_{ij}^n = \begin{bmatrix} \lambda_i^n & \binom{n}{1}\lambda_i^{n-1} & \binom{n}{2}\lambda_i^{n-2} & \cdots & \binom{n}{N_{ij}-1}\lambda_i^{N_{ij}-1} \\ 0 & \lambda_i^n & \binom{n}{1}\lambda_i^{n-1} & \cdots & \binom{n}{N_{ij}-2}\lambda_i^{N_{ij}-2} \\ \cdots & \cdots & \cdots & \cdots & \cdots \\ 0 & 0 & 0 & \cdots & \lambda_i^n \end{bmatrix},
$$

where N_{ij} is the dimension of J_{ij}.

(f) For all $t \in \mathbb{R}$,

$$
e^{J_{ij}t} = e^{\lambda_i t} \begin{bmatrix} 1 & t & \dfrac{t^2}{2!} & \cdots & \dfrac{t^{N_{ij}-1}}{(N_{ij}-1)!} \\ 0 & 1 & t & \cdots & \dfrac{t^{N_{ij}-2}}{(N_{ij}-2)!} \\ \cdots & \cdots & \cdots & \cdots & \cdots \\ 0 & 0 & 0 & \cdots & 1 \end{bmatrix}.
$$
                                                                            ∎

The computation of powers of A and its exponential reduces to the computation of powers and the exponential of its Jordan sub-blocks J_{ij}, for which simple formulas exist. In D.3, $\binom{n}{j}$ denotes the binomial coefficient

$$\binom{n}{j} = \begin{cases} \dfrac{n!}{j!(n-j)!} & \text{for } j = 0, 1, \cdots, n, \\ 0 & \text{otherwise.} \end{cases}$$

D.4. **Example: Powers and exponential of a matrix.** In Example D.2 we found the Jordan normal form J of a 4×4 matrix A as

$$J = \begin{bmatrix} -1 & 1 & 0 & 0 \\ 0 & 1 & 0 & 0 \\ 0 & 0 & -1 & 0 \\ 0 & 0 & 0 & -1 \end{bmatrix}.$$

By D.3 and by using the fact that $\binom{n}{n-1} = n$ we obtain the nth power of J as

$$J^n = \begin{bmatrix} (-1)^n & n(-1)^n & 0 & 0 \\ 0 & (-1)^n & 0 & 0 \\ 0 & 0 & (-1)^n & n(-1)^n \\ 0 & 0 & 0 & (-1)^n \end{bmatrix}, \qquad n \in \mathbb{Z}_+.$$

The exponential of Jt is

$$e^{Jt} = e^{-t} \begin{bmatrix} 1 & t & 0 & 0 \\ 0 & 1 & 0 & 0 \\ 0 & 0 & 1 & t \\ 0 & 0 & 0 & 1 \end{bmatrix}, \qquad t \in \mathbb{R}.$$

We use the latter result to compute the exponential of the matrix At as

$$e^{At} = V e^{Jt} V^{-1} = \begin{bmatrix} 1 & 0 & 0 & 0 \\ 2 & 2 & 0 & 0 \\ 0 & 0 & 1 & 0 \\ 0 & 0 & 2 & 2 \end{bmatrix} e^{-t} \begin{bmatrix} 1 & t & 0 & 0 \\ 0 & 1 & 0 & 0 \\ 0 & 0 & 1 & t \\ 0 & 0 & 0 & 1 \end{bmatrix} \begin{bmatrix} 1 & 0 & 0 & 0 \\ -1 & \frac{1}{2} & 0 & 0 \\ 0 & 0 & 1 & 0 \\ 0 & 0 & -1 & \frac{1}{2} \end{bmatrix}$$

$$= e^{-t} \begin{bmatrix} 1-t & t/2 & 0 & 0 \\ -2t & 1+t & 0 & 0 \\ 0 & 0 & 1-t & t/2 \\ 0 & 0 & -2t & 1+t \end{bmatrix}, \qquad t \in \mathbb{R}. \qquad \blacksquare$$

Supplement E

Proofs

This supplement contains proofs of various results in Chapters 4, 7, and 8.

Proofs for Chapter 4

The proof for Chapter 4 relate to BIBO and CICO stability.

E.1.   Proof of 4.6.3: BIBO stability of initially-at-rest difference and differential systems.

Discrete-time case.   By 4.5.1, the initially-at-rest difference system described by the constant coefficient differential equation $Q(\sigma)y = P(\sigma)u$ is a convolution system with impulse response h given by

$$ h(n) = \sum_{i=1}^{N} \alpha_i y_i(n-1)\mathbb{1}(n-1) + \sum_{i=0}^{M-N} \beta_i \Delta(n+i), \qquad n \in \mathbb{Z}. $$

By 3.6.2, a convolution system is BIBO stable if and only if its impulse response has finite action $\|h\|_1$. Thus, for the BIBO stability of the initially-at-rest system we need to check when h has finite action. By 4.4.1, each basis solution y_i may be chosen to be of the form $n^i \lambda^n$, $n \in \mathbb{Z}_+$, where λ is a characteristic root of the system. If $|\lambda| < 1$, then the basis solution converges exponentially to zero. If *all* characteristic roots have magnitude less than 1, then the impulse response h consists of exponentially converging terms only and, hence, has finite action, so that the initially-at-rest system is BIBO stable.

Suppose that λ^* is a characteristic root that *cancels* against a root of P (with full multiplicity m^*). This means that λ^* is a characteristic root but *not* a pole of the system. We show that the basis solutions corresponding to this root do not appear in the impulse response h and, hence, do not affect the BIBO stability of the initially-at-rest system. Define the polynomials Q^* and P^* as the polynomials Q and P, respectively, with all common factors corresponding to the root λ^* *canceled*. Then the difference system $Q^*(\sigma)y = P^*(\sigma)u$ has a uniquely defined impulse response h^*, which does not contain any terms corresponding to the root λ^*. Now

$$Q^*(\sigma)h^* = P^*(\sigma)\Delta$$

implies

$$(\sigma - \lambda^*)^{m^*}Q^*(\sigma)h^* = (\sigma - \lambda^*)^{m^*}P^*(\sigma)\Delta,$$

which in turns implies $Q(\sigma)h^* = P^*(\sigma)\Delta$. Since h^* satisfies the initial conditions for the initially-at-rest system, it follows that h^* is the impulse response of the initially-at-rest system, so that $h^* = h$. This proves that h does not contain any terms corresponding to the root λ^*.

Thus, a sufficient condition for the impulse response h to have finite action is that all the poles of the system have magnitude strictly less than 1. It may be shown that *all* poles have a nonzero corresponding term in the impulse response, so that the sufficient condition is also necessary.

Continuous-time case. By 4.5.1, the impulse response of the initially-at-rest differential system described by the constant coefficient differential equation $Q(D)y = P(D)u$ is of the form

$$h(t) = \sum_{i=1}^{N} \alpha_i y_i(t)\mathbb{1}(t) + \sum_{i=1}^{M-N} \beta_i \delta^{(i)}(t), \qquad t \in \mathbb{R}.$$

If the degree of P is greater than that of Q (i.e., $M - N > 0$), then the impulse response contains derivatives of delta functions and, hence, has infinite action. This is why for BIBO stability the degree of P should be less than or equal to that of Q. The proof why the poles of the system need have strictly negative real part follows the same line as that for the discrete-time case. Because the basis solutions are of the form $t^i e^{\lambda t}$, $t \in \mathbb{R}_+$, convergence is determined by the sign of the real part $\text{Re}(\lambda)$ of the poles.   ∎

E.2.   Proof of 4.6.13: CICO stability of convolution systems.   The proof of 4.6.13 is given for the discrete-time case. That for continuous-time systems is similar, with integrals replacing the sums.

Denoting the impulse response of the convolution system as h, the IO relationship of the system takes the form $y = h * u$. If (u_1, y_1) and (u_2, y_2) are any two IO pairs it follows by linearity that $y_1 - y_2 = h * (u_1 - u_2)$. Thus, writing $u_1 - u_2 =: \bar{u}$ and $y_1 - y_2 =: \bar{y}$ we have $\bar{y} = h * \bar{u}$. Hence, from the definition, the system is CICO stable if and only if (i) it is BIBO stable and (ii) $\|\bar{u}\|_\infty < \infty$ and $\bar{u}(t) \to 0$ imply that $\bar{y}(t) \to 0$.

Necessity. BIBO stability obviously is a necessary condition for CICO stability.

Sufficiency. Suppose that the system is BIBO stable. Then we need prove that $\|\bar{u}\|_\infty < \infty$ and $\bar{u}(n) \to 0$ as $n \to \infty$ imply that $\bar{y}(n) \to 0$. If $\bar{u}(n) \to 0$, for every $\epsilon_1 > 0$ there exists an n_1 such that $|\bar{u}(n)| \le \epsilon_1$ for all $n \ge n_1$. It follows that

$$
\begin{aligned}
|\bar{y}(n)| &= \left| \sum_{k \in \mathbb{Z}} h(n-k)\bar{u}(k) \right| \\
&= \left| \sum_{k < n_1} h(n-k)\bar{u}(k) + \sum_{k \ge n_1} h(n-k)\bar{u}(k) \right| \\
&\le \sum_{k < n_1} |h(n-k)| \cdot |\bar{u}(k)| + \sum_{k \ge n_1} |h(n-k)| \cdot |\bar{u}(k)| \\
&\le \sum_{k < n_1} |h(n-k)| \cdot \|\bar{u}\|_\infty + \sum_{k \ge n_1} |h(n-k)| \cdot \epsilon_1, \qquad n \in \mathbb{Z}.
\end{aligned}
$$

Substituting $n - k = m$ it follows that

$$
\begin{aligned}
|\bar{y}(n)| &\le \sum_{m > n - n_1} |h(m)| \cdot \|\bar{u}\|_\infty + \sum_{m \le n - n_1} |h(m)| \cdot \epsilon_1 \\
&\le \sum_{m > n - n_1} |h(m)| \cdot \|\bar{u}\|_\infty + \|h\|_1 \cdot \epsilon_1, \qquad n \in \mathbb{Z}. \tag{1}
\end{aligned}
$$

Because by assumption the system is BIBO stable, $\|h\|_1$ is finite by 3.6.2(a). It follows that for every $\epsilon_2 > 0$ there exists an n_2 such that

$$
\sum_{m > n_2} |h(m)| \le \epsilon_2 \qquad \text{for} \quad n \ge n_2.
$$

Thus, from (1) we obtain

$$
|\bar{y}(n)| \le \epsilon_2 \cdot \|\bar{u}\|_\infty + \|h\|_1 \cdot \epsilon_1 \qquad \text{for} \quad n \ge n_1 + n_2, n \in \mathbb{Z}.
$$

As a result, by choosing $n_1 + n_2$ large enough, $|\bar{y}(n)|$ may be made as small as desired for $n \ge n_1 + n_2$, and hence $\bar{y}(n)$ approaches 0 as n goes to infinity. This proves that if the convolution system is BIBO stable, then it is also CICO stable. ∎

Proofs for Chapter 7

We continue with the proofs of some of the properties of the various Fourier transforms of Chapter 7.

E.3.   Proof of 7.3.6: Convolution and converse convolution properties of the DDFT and CDFT.

(a′) *Convolution property of the CDFT.* We prove the convolution property for the CDFT; that for the DDFT is similar. We may express the CDFT of the cyclical convolution $z = x \odot y$ of two continuous-time signals x and y defined on $[0, P)$ as

$$\hat{z}(f) = \int_0^P \left(\int_0^P x((t - \tau) \bmod P) y(\tau) \, d\tau \right) e^{-j2\pi ft} \, dt, \qquad f \in \mathbb{Z}(F),$$

with $F = 1/P$. Interchanging the order of integration we obtain

$$\hat{z}(f) = \int_0^P \left(\int_0^P x((t - \tau) \bmod P) e^{-j2\pi ft} \, dt \right) y(\tau) \, d\tau, \qquad f \in \mathbb{Z}(F). \tag{2}$$

The integral inside the large parentheses may be written as

$$\int_0^P x((t - \tau) \bmod P) e^{-j2\pi ft} \, dt$$

$$= \int_0^\tau x(t - \tau + P) e^{-j2\pi ft} \, dt + \int_\tau^P x(t - \tau) e^{-j2\pi ft} \, dt, \qquad 0 \le \tau < P.$$

Substitution of $t - \tau + P = \theta$ into the first integral on the right-hand side and $t - \tau = \theta$ into the second together with the fact that fP is an integer results in

$$\int_0^P x((t - \tau) \bmod P) e^{-j2\pi ft} \, dt$$

$$= \int_{-\tau+P}^P x(\theta) e^{-j2\pi f(\theta + \tau - P)} \, d\theta + \int_0^{P-\tau} x(\theta) e^{-j2\pi f(\theta + \tau)} \, d\theta$$

$$= \left(\int_0^P x(\theta) e^{-j2\pi f\theta} \, d\theta \right) e^{-j2\pi f\tau} = \hat{x}(f) e^{-j2\pi f\tau}, \qquad 0 \le \tau < P.$$

With this it follows from (2) that

$$\hat{z}(f) = \int_0^P \hat{x}(f) e^{-j2\pi f\tau} y(\tau) \, d\tau = \hat{x}(f) \hat{y}(f), \qquad f \in \mathbb{Z}(F),$$

which proves that circular convolution transforms into multiplication.

(b′) *Converse convolution property of the CDFT*. To show that multiplication transforms into convolution we work in reverse direction. Consider

$$(\hat{x} * \hat{y})(f) = F \sum_{\phi \in \mathbb{Z}(F)} \hat{x}(f - \phi)\hat{y}(\phi)$$

$$= F \sum_{\phi \in \mathbb{Z}(F)} \left(\int_0^P x(t)e^{-j2\pi(f-\phi)t}\, dt \right) \hat{y}(\phi), \qquad f \in \mathbb{Z}(F).$$

Interchanging the order of summation and integration we obtain

$$(\hat{x} * \hat{y})(f) = \int_0^P x(t)\left(F \sum_{\phi \in \mathbb{Z}(F)} \hat{y}(\phi)e^{j2\pi\phi t} \right) e^{-j2\pi ft}\, dt$$

$$= \int_0^P x(t)y(t)e^{-j2\pi ft}\, dt, \qquad f \in \mathbb{Z}(F),$$

which shows that $\hat{x} * \hat{y}$ is the CDFT of $x \cdot y$.   ∎

E.4.   Proof of 7.3.7: Shift and converse shift properties of the DDFT and CDFT.

(a′) *Shift property of the CDFT*. The shift property is only proved for the CDFT. The proof for the DDFT is analogous. The shift property may easily be proved by simple substitution. Consider the CDFT of z given by

$$z(t) = x((t + \theta) \bmod P), \qquad t \in [0, P).$$

We have

$$\hat{z}(f) = \int_0^P x((t + \theta) \bmod P)e^{-j2\pi ft}\, dt, \qquad f \in \mathbb{Z}(F),$$

with $F = 1/P$. Suppose for the time being that $0 \le \theta < P$. Then substitution of $t + \theta = \tau$ yields

$$\hat{z}(f) = \int_\theta^{\theta+P} x(\tau \bmod P)e^{-j2\pi f(\tau-\theta)}\, d\tau,$$

$$= e^{j2\pi f\theta}\left(\int_\theta^P x(\tau)e^{-j2\pi f\tau}\, d\tau + \int_P^{\theta+P} x(\tau - P)e^{-j2\pi f\tau}\, d\tau \right), \qquad f \in \mathbb{Z}(F).$$

Substitution of $\tau = t$ in the first term on the right-hand side and $\tau - P = t$ in the second yields, together with the fact that fP is an integer and, hence, $e^{j2\pi fP} = 1$,

$$\hat{z}(f) = e^{j2\pi f\theta}\left(\int_{\theta}^{P} x(t)e^{-j2\pi ft}\, dt + \int_{0}^{\theta} x(t)e^{-j2\pi ft}\, dt\right)$$

$$= e^{j2\pi f\theta}\hat{x}(f), \qquad f \in \mathbb{Z}(F).$$

This proves the shift property. If θ does not lie in the interval $[0, P)$ one may write $\theta = kP + \theta'$, with k an integer and $\theta' \in [0, P)$, and repeat the proof with minor modification.

(b') *Converse shift property of the CDFT.* The CDFT of the signal z given by

$$z(t) = e^{j2\pi\phi t}x(t), \qquad t \in [0, P),$$

with $\phi \in \mathbb{Z}(F)$, is

$$\hat{z}(f) = \int_{0}^{P} e^{-j2\pi ft}e^{j2\pi\phi t}x(t)\, dt = \int_{0}^{P} e^{-j\pi(f-\phi)t}x(t)\, dt$$

$$= \hat{x}(f - \phi), \qquad f \in \mathbb{Z}(F).$$

This proves the converse shift property.   ■

E.5.   Proof of 7.3.9:   Differentiation property of the CDFT. Suppose that the time signal x defined on the finite time axis $[0, P)$ is cyclically continuous and has derivative $z = Dx$. It follows by partial integration that the CDFT $\hat{z}$ of the derivative is given by

$$\hat{z}(f) = \int_{0}^{P} Dx(t)e^{-j2\pi ft}\, dt = x(t)e^{-j2\pi ft}\Big|_{0}^{P} + j2\pi f \int_{0}^{P} x(t)e^{-j2\pi ft}\, dt$$

$$= j2\pi f \cdot \hat{x}(f), \qquad f \in \mathbb{Z}(F).$$

This proves the differentiation property.   ■

E.6.   Derivation of the DCFT.   The derivation of the DCFT follows by reversing the roles of time and frequency in the CDFT. Let $z \in \mathcal{L}_2[0, P)$ be a finite-energy signal defined on the time axis $[0, P)$. Then by the CDFT we have

$$z(t) = F \sum_{f \in \mathbb{Z}(F)} \hat{z}(f)e^{j2\pi ft}, \qquad t \in [0, P),$$

where $F = 1/P$ and $\hat{z} \in \ell_2(F)$ is given by

$$\hat{z}(f) = \int_{0}^{P} z(t)e^{-j2\pi ft}\, dt, \qquad f \in \mathbb{Z}(F).$$

Interchanging f and t on the one hand and z and $\hat{z}$ on the other, we obtain

$$\hat{z}(f) = F \sum_{t \in \mathbb{Z}(F)} z(t)e^{j2\pi ft}, \qquad f \in [0, P),$$

$$z(t) = \int_0^P \hat{z}(f)e^{-j2\pi ft}\, df, \qquad t \in \mathbb{Z}(F).$$

The latter equality shows that the signal z, which belongs to $\ell_2(F)$ and, hence, has finite energy, may be expanded in a continuum of harmonics with frequencies ranging over a finite interval. It remains to rearrange so that we obtain the result in the form of 7.4.1. Setting $P = 1$ and hence $F = 1$, substituting $z(t) = x(-t)$, $t \in \mathbb{Z}$, and replacing $\hat{z}$ with $\hat{x}$ results in

$$\hat{x}(f) = \sum_{t \in \mathbb{Z}} x(-t)e^{j2\pi ft}, \qquad f \in [0, 1),$$

$$x(-t) = \int_0^1 \hat{x}(f)e^{-j2\pi ft}\, df, \qquad t \in \mathbb{Z}.$$

The DCFT and its inverse now follow by replacing t with $-n$ in both expressions, so that

$$\hat{x}(f) = \sum_{n \in \mathbb{Z}} x(n)e^{-j2\pi fn}, \qquad f \in [0, 1),$$

$$x(n) = \int_0^1 \hat{x}(f)e^{j2\pi fn}\, df, \qquad n \in \mathbb{Z}. \qquad \blacksquare$$

Proofs for Chapter 8

In the remainder of this supplement we present the proofs of several of the properties of the z- and Laplace transforms as listed in Section 8.4. Most of the proofs are given either for the discrete- or for the continuous-time case. The proof for the alternate case in each instance is closely parallel, with integrals replacing sums or vice-versa.

E.7.   **Proof of 8.4.2: Convolution property of the z- and Laplace transforms.**   We prove the convolution properties of the two- and one-sided Laplace transforms.

(a') *Convolution property of the two-sided Laplace transform.* Let x and y be continuous-time signals whose convolution $w = x * y$ exists. The two-sided Laplace transform W of z is given by

$$W(s) = \int_{-\infty}^{\infty} \left(\int_{-\infty}^{\infty} x(t-\tau)y(\tau)\, d\tau \right) e^{-st}\, dt.$$

We invoke *Fubini's theorem,* which states that if a repeated integral exists, the order of integration may be reversed. It follows that

$$W(s) = \int_{-\infty}^{\infty} \left(\int_{-\infty}^{\infty} x(t-\tau)y(\tau)e^{-st}\, dt \right) d\tau$$

$$= \int_{-\infty}^{\infty} \left(\int_{-\infty}^{\infty} x(t-\tau)e^{-st}\, dt \right) y(\tau)\, d\tau.$$

Substitution of $t - \tau = \theta$ in the inside integral results in

$$W(s) = \int_{-\infty}^{\infty} \left(\int_{-\infty}^{\infty} x(\theta)e^{-s\theta}\, d\theta \right) y(\tau)e^{-s\tau}\, d\tau = X(s) \int_{-\infty}^{\infty} y(\tau)e^{-s\tau}\, d\tau$$

$$= X(s) \cdot Y(s),$$

which proves the convolution property for the two-sided Laplace transform. For W to exist we need both X and Y to exist, so that the existence region of W is the intersection of the existence regions of X and Y. If the intersection is empty, the convolution does not exist.

(b′) *Convolution property of the one-sided Laplace transform.* Suppose that x and y are continuous-time signals, and let $w = x\mathbb{1} * y\mathbb{1}$. Since the supports of $x\mathbb{1}$ and $y\mathbb{1}$ are both contained in $[0, \infty)$, by 3.5.5 the support of w is also contained in $[0, \infty)$, so that w is zero for negative times. By the convolution property of the two-sided Laplace transform,

$$W = \mathcal{L}(x\mathbb{1}) \cdot \mathcal{L}(y\mathbb{1}) = X_+ \cdot Y_+,$$

where W is the two-sided Laplace transform of w. Because w is zero for negative times, its one-sided Laplace transforms W_+ equals its two-sided transform W. It follows that

$$W_+ = X_+ \cdot Y_+,$$

which proves the convolution property for the one-sided Laplace transform.   ■

E.8.   Proof of 8.4.3: Shift properties of the z- and Laplace transforms.   We prove the shift properties for the discrete-time case.

(a) *Shift property of the two-sided z-transform.* The two-sided z-transform of the shifted discrete-time signal

$$w(n) = x(n + k), \qquad n \in \mathbb{Z},$$

is

$$W(z) = \sum_{n=-\infty}^{\infty} x(n + k)z^{-n}.$$

Substitution of $n + k = m$ results in

$$W(z) = \sum_{m=-\infty}^{\infty} x(m)z^{-(m-k)} = \left(\sum_{m=-\infty}^{\infty} x(m)z^{-m} \right) z^k$$

$$= z^k X(z),$$

where X is the two-sided z-transform of x. W exists if X exists, so that the existence region of W coincides with that of X.

   (b) *Shift property of the one-sided z-transform.* The one-sided z transform of the backward shifted signal

$$w(n) = x(n + 1), \qquad n \in \mathbb{Z},$$

is

$$W_+(z) = \sum_{n=0}^{\infty} x(n + 1)z^{-n}.$$

By substitution of $n + 1 = m$ it follows that

$$W_+(z) = \sum_{m=1}^{\infty} x(m)z^{-(m-1)} = \left(\sum_{m=0}^{\infty} x(m)z^{-m} - x(0) \right) z$$

$$= zX_+(z) - zx(0),$$

where X_+ is the one-sided z-transform of x.
   Similarly, the one-sided z-transform of the forward shifted signal

$$v(n) = x(n - 1), \qquad n \in \mathbb{Z},$$

is

$$V_+(z) = \sum_{n=0}^{\infty} x(n - 1)z^{-n}.$$

By the substitution $n - 1 = m$ we obtain

$$V_+(z) = \sum_{m=-1}^{\infty} x(m)z^{-(m+1)} = x(-1) + \left(\sum_{m=0}^{\infty} x(m)z^{-m} \right)z^{-1}$$

$$= z^{-1}X_+(z) + x(-1).$$

The existence regions of W_+ and V_+ are the same as that of X_+. This completes the proof of the shift properties of the one-sided z transform.

(c) *Converse shift property of the z-transform.* The two-sided z transform of the signal

$$w(n) = a^n x(n), \qquad n \in \mathbb{Z},$$

is

$$W(z) = \sum_{n=-\infty}^{\infty} a_n x(n)z^{-n} = \sum_{n=-\infty}^{\infty} x(n)\left(\frac{z}{a}\right)^{-n}$$

$$= X\left(\frac{z}{a}\right).$$

$W(z)$ exists for those z such that z/a is in the existence region of X.

The proof of the converse shift property of the one-sided z-transform is essentially the same.   ■

E.9.   Proof of 8.4.5: Differentiation properties of the Laplace transform.

(a) *Differentiation property of the two-sided Laplace transform.* The two-sided Laplace transform of the derivative $w = Dx$ of the continuous-time signal x is

$$W(s) = \int_{-\infty}^{\infty} \frac{dx(t)}{dt} e^{-st} \, dt.$$

It follows by partial integration that

$$W(s) = x(t)e^{-st}\Big|_{-\infty}^{\infty} + s \int_{-\infty}^{\infty} x(t)e^{-st} \, dt$$

$$= sX(s),$$

where X is the two-sided Laplace transform of X. The product $x(t)e^{-st}$ vanishes at $t = \pm\infty$ because of the assumed existence of $X(s)$. The existence region of W coincides with that of X.

(b) *Differentiation property of the one-sided Laplace transform.* By partial integration it follows that the one-sided Laplace transform of the derivative $w = Dx$ of x is

$$W_+(s) = \int_{0^-}^{\infty} \frac{dx(t)}{dt} e^{-st}\, dt = x(t)e^{-st}\Big|_{0^-}^{\infty} + s\int_{0^-}^{\infty} x(t)e^{-st}\, dt$$

$$= sX_+(s) - x(0^-), \qquad s \in \mathscr{E}_+,$$

where X_+ is the one-sided Laplace transform of x. The existence region of W_+ is the same as that of X_+.   ∎

E.10. Proof of 8.4.7: Converse differentiation property of the z- and Laplace transforms.

We prove the converse differentiation property of the two-sided z-transform. Suppose that X is the two-sided z-transform of x with existence region $\mathscr{E}$. Term-by-term differentiation of

$$X(z) = \sum_{n=-\infty}^{\infty} x(n)z^{-n}, \qquad z \in \mathscr{E},$$

results in

$$\frac{dX(z)}{dz} = \sum_{n=-\infty}^{\infty} x(n)(-n)z^{-n-1}, \qquad z \in \mathscr{E},$$

so that

$$z\frac{dX(z)}{dz} = \sum_{n=-\infty}^{\infty} [-nx(n)]z^{-n}, \qquad z \in \mathscr{E}.$$

The proof of the converse differentiation property of the one-sided z-transform is similar.   ∎

E.11. Proof of 8.4.8: Integration properties of the Laplace transform.

(a) *Integration property of the two-sided Laplace transform.* Because

$$(\mathbb{1} * x)(t) = \int_{-\infty}^{\infty} \mathbb{1}(t-\tau)x(\tau)\, d\tau = \int_{-\infty}^{t} x(\tau)\, d\tau, \qquad t \in \mathbb{R},$$

it follows that the integrated signal y defined by

$$y(t) = \int_{-\infty}^{t} x(\tau)\, d\tau, \qquad t \in \mathbb{R},$$

may be expressed as

$$y = \mathbb{1} * x.$$

Application of the convolution property of the two-sided Laplace transform results in

$$Y(s) = \frac{1}{s} \cdot X(s) = \frac{X(s)}{s}.$$

Because the existence region of the Laplace transform $1/s$ of the unit step is Re $(s) > 0$, the existence region of the Laplace transform Y of y is $\{s \in \mathscr{C} \mid$ Re $(s) > 0\}$, with $\mathscr{C}$ the existence region of x.

(b) *Integration property of the one-sided Laplace transform.* Consider the signal y defined by

$$y(t) = \int_{-\infty}^{t} (x \cdot \mathbb{1})(\tau) \, d\tau = \begin{cases} 0 & \text{for } t < 0, \\ \int_{0}^{t} x(\tau) \, d\tau & \text{for } t \geq 0, \end{cases} \quad t \in \mathbb{R}.$$

Note that because y is zero for negative times its one-sided Laplace transform Y_+ equals its two-sided Laplace transform Y. Also, the two-sided Laplace transform of $x \cdot \mathbb{1}$ is the one-sided transform X_+ of x. By application of the integration property of the two-sided Laplace transform it thus follows that

$$Y_+(s) = \frac{\mathscr{L}(x\mathbb{1})(s)}{s} = \frac{X_+(s)}{s}, \qquad s \in \{s \in \mathscr{C}_+ \mid \text{Re } (s) > 0\},$$

where $\mathscr{C}_+$ is the existence region of X_+.   ∎

E.12.   Proof of 8.4.9: Initial and final value properties of the one-sided z- and Laplace transforms.

(a) *Initial value property of the one-sided z-transform.* To prove the initial value property, we write the one-sided z-transform X_+ of x as

$$X_+(z) = x(0) + x(1)z^{-1} + x(2)z^{-2} + \cdots, \qquad |z| > \rho,$$

with ρ the positive number that determines the existence region of X_+. Letting $|z|$ approach ∞ yields the desired result.

(a′) *Initial value property of the one-sided Laplace transform.* In the continuous-time case we make the initial value property plausible as follows. We have

$$sX_+(s) = s \int_{0}^{\infty} x(t)e^{-st} \, dt, \qquad \text{Re } (s) > \sigma, \tag{3}$$

with σ the real number that determines the existence region of the one-sided Laplace transform X_+ of x. If s is very large and real, the function $se^{-st}\mathbb{1}(t)$, $t \in \mathbb{R}$, may be considered an approximation to the δ-function, so that in the limit $s \to \infty$ the right-hand side of (3) approaches $x(0^+)$.

(b) *Final value property of the one-sided z-transform.* The complete proof of the final value property is beyond the scope of this text. Consider the discrete-time case. Given the signal x, the idea of the proof is to define another signal $\tilde{x}$ as

$$\tilde{x}(n) = x(n) - x(\infty), \qquad n \in \mathbb{Z},$$

that is, $\tilde{x}$ is the signal x with the limit $x(\infty)$ subtracted. The result is that $\tilde{x}$ approaches zero as time goes to infinity. The one-sided z-transform $\tilde{X}$ of x is given by

$$\tilde{X}_+(z) = X_+(z) - \frac{z}{z-1} x(\infty), \qquad |z| > 1,$$

where we use the assumption that X_+ exists for $|z| > 1$. It follows that

$$zx(\infty) = (z-1)X(z) - (z-1)\tilde{X}(z).$$

The difficult part of the proof is to show that because $\tilde{x}(n) \to 0$ as $n \to \infty$, $\tilde{X}(z)$ is finite at $z = 1$. Once this has been established, the final value property follows easily by letting z approach 1.   ∎

Bibliography

The following books roughly cover the same material as the present text. The list is by no means complete.

R. A. GABEL and R. A. ROBERTS, *Signals and Linear Systems,* 3rd. ed., John Wiley, New York, 1987.

T. H. GLISSON, *Introduction to System Analysis*. McGraw-Hill, New York, 1985.

C. L. LIU and J. W. S. LIU, *Linear Systems Analysis*. McGraw-Hill, New York, 1975.

H. P. NEFF, JR., *Continuous and Discrete Linear Systems*. Harper and Row, New York, 1984.

A. V. OPPENHEIM and A. S. WILLSKY, with I. T. YOUNG, *Signals and Systems*. Prentice Hall, Englewood Cliffs, N.J., 1983.

H. W. SCHÜSSLER, *Netzwerke, Signale und Systeme,* vols. 1 and 2. Springer-Verlag, Berlin, 1981, 1984.

W. M. SIEBERT, *Circuits, Signals, and Systems*. MIT Press, Cambridge, Mass., 1986.

More advanced texts on signals and systems follow:

T. KAILATH, *Linear Systems*. Prentice Hall, Englewood Cliffs, N.J., 1980.

A. PAPOULIS, *Signal Analysis*. McGraw-Hill, New York, 1977.

For further reading on digital signal processing we refer to

C. S. BURRUS and T. W. PARKS, *DFT/FFT and Convolution Algorithms.* John Wiley, New York, 1985.

M. T. JONG, *Methods of Discrete Signal and System Analysis.* McGraw-Hill, New York, 1982.

S. L. MARPLE, JR., *Digital Spectral Analysis With Applications.* Prentice Hall, Englewood Cliffs, N.J., 1987.

A. V. OPPENHEIM and R. W. SHAFER, *Digital Signal Processing.* Prentice Hall, Englewood Cliffs, N.J., 1975.

For further reading on communication systems, the following books may be consulted:

S. BENEDETTO, E. BIGLIERI, and V. CASTELLANI, *Digital Transmission Theory.* Prentice Hall, Englewood Cliffs, N.J., 1987.

A. B. CARLSON, *Communication Systems,* 3rd ed., McGraw-Hill, New York, 1986.

Some well-known texts in the area of automatic control follow:

G. F. FRANKLIN, J. D. POWELL, and A. EMAMI-NAEINI, *Feedback Control of Dynamic Systems.* Addison-Wesley, Reading, Mass., 1986.

B. C. KUO, *Automatic Control Systems,* 5th ed., Prentice Hall, Englewood Cliffs, N.J., 1987.

SIGSYS
A Signals and Systems Interpreter
TUTORIAL

Description. SIGSYS is an interpreter that has been designed to generate, manipulate and display time and frequency signals. A wide and flexible range of operations is supported.

Contents

---

0. Using SIGSYS

For instructions how to install and start
SIGSYS on a personal computer under PC-
DOS or MSDOS, consult the READ.ME
file on the disk that contains SIGSYS. The
READ.ME file also includes recent information, such as changes in and additions to
this tutorial.

After the program has been started a
prompt appears, and SIGSYS is ready to
accept commands.

SIGSYS is an interpreter, and command lines are executed as soon as they
have been concluded by pressing the
⟨ENTER⟩ key.

1. Data Types

SIGSYS supports the following data types:

> real or complex scalar,
>
> real or complex signal,
>
> real or complex polynomial.

Data may be identified by any name consisting of at most 8 lower or upper case letters or the figures 0 to 9. The first character
must be a letter. Thus, x, x1, X and
period are all legal names.

2. Entering Scalars

Real scalars may be entered by a simple assignment of the form

    a=1.234

The special font used here indicates text
typed on the keyboard. The number on the
right-hand side may be in fixed-point format or in scientific format such as
.1234e01. Complex scalars may be entered using the permanent variable j,
which equals the square root of -1. For instance,

    b=1+2*j

defines b as a complex scalar. Another useful permanent variable is pi, whose value
is obvious.

The permanent scalar variable rand
returns a random number. By default the
random variable is normally distributed
with mean zero and standard deviation 1.
The command

    rand N(a,b)

changes the mean to a and the standard deviation to b. On the other hand, the command

    rand U(a,b)

makes the distribution uniform on the interval (a, b). Finally, the command

    rand seed a

sets the seed of the random number generator equal to a.

One command or several commands
separated by a semicolon may be entered
on a single line, such as

    a=1.234;b=1+2*j

Spaces and tabs may be inserted to improve
the layout, such as in

    a=1.234; b=1+2*j

The value of any variable is displayed by
simply typing its name. Thus,

    a

shows the value of a on the screen.

2.1. Exercise. Enter the scalars a and b
as above. Display their values.   ∎

2.2. Exercise. Generate a few random
numbers. See what happens if you change
the distribution.   ∎

By default, SIGSYS displays numerical data
in a floating point or scientific notation (depending on which is more compact) with

seven digits precision. The precision of the display may be changed to 15 digits with the command

```
format long
```

and changed back to seven digits with

```
format short
```

Also, by default SIGSYS displays complex numbers in Cartesian form. The display may be changed to polar form by the command

```
format polar
```

In this format, the complex number $1 + j$ is shown as

```
1.414214*exp(j*0.7853982)
```

The display may be changed back to Cartesian form by the command

```
format cartesian
```

The single command

```
format
```

reverts the display to default.

2.3. Exercise. Define the variables a and b as in 2.1, and display them in polar and Cartesian format and with long and short precision. ∎

3. Elementary Data Management and Command Line Editing

At any time during a session, the command

```
who
```

shows a list of the variables that have been created so far, together with their types. The command

```
clear
```

clears all variables, while

```
clear a,x,A
```

clears a, x and A only. It is useful to know that a session may be peacefully ended at any time by typing

```
exit
```

and that SIGSYS accepts the keyboard interrupts

⟨CTRL⟩S   halt execution,
⟨CTRL⟩Q   resume execution,
⟨CTRL⟩C   cancel the current operation.

⟨CTRL⟩S means that the CTRL key is pressed together with the (lower case) s key.

At any point during a session all variables and their values may be saved on disk by typing for instance

```
save data
```

This creates a file on disk called DATA.DAS. The extension .DAS is added by SIGSYS, and no extension should be specified. If the file name "data" is omitted, the default file name SIGSYS is used. In file names no distinction is made between upper and lower case characters. Saved data may be retrieved by typing

```
load data
```

where any existing SIGSYS data file name (with extension .DAS) may be used instead of "data". The variables in the data file that is loaded are merged with the existing variables. Variables with duplicate names are overwritten by the file that is loaded.

3.1. Exercise. Display the variables you have created so far and clear some of them. Save your variables, exit SIGSYS and enter it again, and load the saved data. ∎

The rest of this section may be skipped on first reading.

If no more memory is available for storing variables, SIGSYS displays an error message. Free space may be created by clearing variables. The stored variables may be reorganized to make more efficient use of the available space by the command

```
pack
```

This command first saves all variables on disk in a temporary file, then clears all variables, next reloads the variables from file and finally erases the temporary file.

The effectiveness of the ⟨CTRL⟩C command is increased by issuing the DOS command BREAK ON before starting SIGSYS (or by issuing the shell command !BREAK ON from SIGSYS—see 18.) This facility slows down execution; it may be switched off by the DOS command BREAK OFF.

By default, .DAS files are written to and read from the current disk and sub-directory. The command chdir ("change directory") may be used to change this. Thus,

```
chdir c:\sigsys\data
```

changes the default directory to C:\SIGSYS\DATA. The same may be done *before* starting SIGSYS by the command (from DOS)

```
SET SIGSYS=C:\SIGSYS\DATA
```

where of course any other valid path name may be used.

SIGSYS has a simple but effective facility for command line editing. While typing a command line, the cursor may be moved back and forth by the ← and → keys, and the ⟨INS⟩, ⟨DEL⟩ and ⟨BACK-SPACE⟩ keys have their usual functions. By means of the ↑ and ↓ keys commands that recently have been issued may be recalled and displayed. They may be re-issued, possibly after editing, by pressing the ⟨ENTER⟩ key.

3.2. Exercise. Experiment with the command line editing facility. ■

4. Operations Among Scalars

The following operations among scalars are defined

a^b	exponentiation,
-a	unary minus,
a*b	multiplication,
a/b	division,
a+b	addition,
a-b	subtraction,
a mod b	reduction of a modulo b.

The operations apply both to real and complex numbers. If a and b are complex, the modulo operation is applied to the real and imaginary parts separately and returns a complex number, that is, a mod b is the complex number with real part Re (a) mod Re (b) and imaginary part Im (a) mod Im (b)

The scalars may be entered by name or by value. Operations may be combined to algebraic expressions. The usual precedence rules apply. Parentheses may be used to change the natural precedences.

4.1. Exercise. To compute the root to the base 10 of *j* type

```
j^(1/10)
```
■

4.2. Exercise. Compute

```
11+12j mod 3+4j
```

See what the outcome of 5 mod 0 is. ■

Typing an expression without assigning it to a variable produces the result directly on the screen. The result is stored in a variable named ans. The previous value of ans is overwritten.

4.3. Example. Typing

```
2+5
ans/7
```

results in successive displays of 7 and 1. After completion, ans has the value 1. ■

5. Functions of Scalars

A number of standard functions have been implemented. They are the following

real(a)	real part,
imag(a)	imaginary part,
abs(a)	absolute value,
arg(a)	argument,
conj(a)	complex conjugate,
exp(a)	exponential,
log(a)	natural logarithm,
log10(a)	logarithm to the base 10,
sqrt(a)	square root,
sin(a)	sine,
cos(a)	cosine,
tan(a)	tangent,
asin(a)	arcsine,
acos(a)	arccosine,
atan(a)	arctangent,
sinc(a)	sin(a)/a,
sign(a)	sign,
floor(a)	largest integer less than or equal to a,
ceil(a)	smallest integer greater than or equal to a,
round(a)	nearest integer to a,
step(a)	1 for $a \geq 0$, 0 otherwise,
ramp(a)	a for $a \geq 0$, 0 otherwise,
rect(a)	1 for $-1/2 \leq a < 1/2$, 0 otherwise,
trian(a)	$1 - |a| \leq 1$, for $|a| \leq 1$, 0 otherwise,
sat(a)	a for $|a| \leq 1$, sign(a) otherwise.

The operations are all defined for real and complex argument a. If the argument a is complex, the functions sign, floor, ceil, round, step, ramp, rect, trian and sat operate on the real and imaginary parts separately, and return a complex number. Thus, sign(0.1-5*j), for instance, returns the complex number 1-j.

5.1. Exercise. Take a=pi/2 and b= exp(j*a). Check that the real part, the imaginary part, the absolute value and the argument of b are as expected. Check some of the other functions as well. ■

Note. Complex scalars with imaginary part equal to zero are automatically converted to real scalars.

6. Creating Time Signals

A time signal is handled by the program as a finite-time discrete-time signal. A signal is represented by the following attributes:

> The *increment,* which is the distance between the sampling instants.
>
> The *domain,* characterized by two real numbers, denoting the lower and the upper limit of the time axis on which the signal is defined. The difference between the upper and lower limit is an integral multiple of the increment.
>
> A *list* of real or complex numbers, representing the signal values at the sampling instants.

A direct way of creating a time signal is by using the cat function. Cat is a contraction of *concatenation.* The command

    x=cat(2,3,1)

creates a real signal x, whose increment is 1, whose domain is [0, 2] and that takes the successive values 2, 3, and 1. The cat function uses the increment inc, which is a permanent variable with default value 1. The list of arguments of the cat function (which may have arbitrary length) specifies the successive signal values. The first signal value is placed at the time origin. An exclamation mark may be used to change the location of the time origin. Thus,

    y=cat(5,4!3,2,1)

creates a signal with increment 1, domain $[-1, 3]$ and successive values $y(-1) = 5$, $y(0) = 4$, $y(1) = 3$, $y(2) = 2$ and $y(3) = 1$.

6.1. Exercise.   Create the signals x and y. Display them.   ∎

The cat function may not only be used to define signals from their sampled values, but also to concatenate signals that already exist. The general format of the command is

    u=cat(a,b,...,m!n,...,z)

Here a, b, ..., z are either scalars or signals. The signals must all have the same increment. The new signal is formed by concatenation of the signal and scalar values. The exclamation mark indicates the location of the origin. If m is a scalar, this signal value is placed at the origin. If m is a signal, the origin of m becomes the origin for u. If no exclamation mark is present, the signal starts at time 0.

6.2. Exercise.   Define x=cat(2,3,1) and y=cat(5,4!3,2,1) as before. Check that cat(x,y!) is the same as cat(2,3,1,5,4!3,2,1) and that cat(x,y) is cat(2,3,1,5,4,3,2,1). What is cat(x!y)?   ∎

Once a signal has been defined, four functions may be used to extract some of its attributes:

inc(x)	returns the increment of x,
num(x)	returns the number of samples of x,
low(x)	returns the lower limit of the domain of x,
up(x)	returns the upper limit of the domain of x.

6.3. Exercise.   Verify that inc(y), num(y), low(y) and up(y), with y as in

Exercise 6.2, result in the expected answers.   ∎

There are two important permanently defined time signals, named dot and dotplus. In Section 9 it is explained how they may be used to generate a wide class of signals. The signal dot has increment inc (with default value 1), two-sided domain

$$\left[-\frac{num}{2}inc, \left(\frac{num}{2} - 1\right)inc \right]$$

(provided num is even), and signal values specified by $dot(t) = t$. Here num is another permanent variable with default value 10. Displaying dot results in something like

```
signal specifiers
num=10
inc=1
domain=[-5,4]
signal data
signal axis          signal value
```

signal axis	signal value
−5	−5
−4	−4
−3	−3
−2	−2
−1	−1
0	0
1	1
2	2
3	3
4	4

If num is odd, the signal dot has domain

$$\left[-\frac{num-1}{2}inc, \frac{num-1}{2}inc \right].$$

The signal dotplus has increment inc, right one-sided domain

[0, (num-1)inc],

and signal values specified by dot-plus(*t*) = *t*.

It will be seen in Section 9 that the main function of the signals dot and dot-plus is to define a time axis. A more flexible way of doing this is with the command a:b:c, with a, b and c three real numbers with b positive and c ≥ a. The command

    x=a:b:c

creates a signal with increment b and domain [a, c] such that x(*t*) = *t*. If necessary, a is rounded up and c is rounded down so that a and c are integral multiples of b. The real scalars a, b and c may be called by name or by value. Thus, the command

    x=2:0.5:4

creates a signal x with increment 0.5, domain [2, 4] and successive values y(2) = 2, y(2.5) = 2.5, y(3) = 3, y(3.5) = 3.5 and y(4) = 4.

6.4. Exercise. Display dot, dotplus, and the signal 2:0.5:4. ∎

A further permanent time signal is delta. It has the same increment and domain as dot (and hence is two-sided), while its signal values are given by delta(0) = 1/inc and delta(*t*) = 0 for *t* ≠ 0.

Besides dot, dotplus, and delta another permanent time signal is noise. The command

    z=noise

generates a signal z with increment inc and two-sided time axis, whose successive values equal random numbers. The successive random signal values are generated as rand/√inc (see the description of the permanent scalar rand). The command

    z=noiseplus

generates a similar random signal but on a right one-sided time axis with increment inc and one-sided domain.

6.5. Exercise. Generate the noise signal z. See what happens if you change the distribution. ∎

Note. Complex signals that have their imaginary part equal to zero are automatically converted to real signals.

7. Changing the Default Values

The scalars j and pi are permanent and cannot be changed. In addition to the permanent variables inc and num that we already met there is a third permanent variable, called INC. INC is the increment for frequency signals, which are discussed in Section 12. The default value of inc is 1, that of num is 10 and that of INC is 0.1. The default values of inc and num can be changed by simple assignment statements such as

    inc=.01

or

    num=100

Whenever a new value is assigned to inc, num or INC, the new values for inc, num and INC are all displayed.

The meaning of the permanent variables is the following. That of inc, the increment for time signals, is clear. The variable num defines the number of sampling instants of the standard time signals dot and dotplus. Together, inc and num define the time axes of the standard time signals dot and dotplus and via these of all signals derived from dot and dotplus. How this is done is discussed in Section 9. Given inc and num, the domain of dot is taken (more or less) symmetrically with respect to the origin (imitating a two-sided time axis) and that of dotplus at the right-hand side of the origin (imitating a right one-sided time axis).

7.1. Exercise. Change `inc` to .01 and `num` to 100. Verify that the lower limit `low(dot)` and the upper limit `up(dot)` and those of `dotplus` are as expected. Try also `num = 101`.   ∎

There are three `set` commands that may be used to set some of the defaults automatically. They are mainly useful when simultaneously working with time signals and their Fourier transforms (see Section 17). The commands are the following

`set INC`	make `INC` equal to $1/(\text{num}\cdot\text{inc})$,
`set inc`	make `inc` equal to $1/(\text{num}\cdot\text{INC})$,
`set num`	round up `num` to the next integral power of 2.

8. Operations Among and on Signals

The following pointwise operations among signals are defined:

`x^y`	pointwise exponentiation,
`-x`	pointwise unary minus,
`x*y`	pointwise multiplication,
`x/y`	pointwise division,
`x+y`	pointwise addition,
`x-y`	pointwise subtraction.
`x mod y`	pointwise reduction of x modulo y.

In each of these cases x and y must have the same increments. The domain of the signal that results from any of the binary operations is the smallest closed interval that includes the domains of the operands. Signal values of the operands outside their domains are set equal to zero. If x or y is a scalar (called by name or by value), it is taken as a constant signal. If x and y are complex-valued, then the modulo reduction operates on the real and imaginary parts separately, as in the case of scalars.

8.1. Exercise. Define `x=cat(2,3,1)` and `y=cat(5,4!3,2,1)` as before. Compute x + y and verify the result. Also check some of the other operations from the list. Is `dot mod 4` what you expect it to be? How about `(-1)^dot`?   ∎

Two important binary operations between signals are not defined pointwise. They are

`x**y`	fast convolution,
`x oo y`	fast cross-correlation.

The spaces before and after `oo` are important, like those before and after `mod`. The signals x and y need have the same increments. Both x and y are taken to be zero outside their domains. The domains of the convolution and the cross-correlation of x and y are made equal to their supports.

The fast convolution `x**y` and cross-correlation `x oo y` use the fast Fourier transform as an intermediate step. When the command `x**y` is invoked, first the discrete Fourier transforms of the signals x and y are computed, next the transforms are multiplied, and finally the inverse discrete Fourier transform of the product is calculated. Before the transforms are determined the signals x and y are padded with sufficiently many zeros. At the end of the computation the domain of the result `x**y` is trimmed to the correct interval. For the correlation `x oo y` a similar algorithm is employed.

The commands

`x*.*y`	direct convolution,
`x o.o y`	direct cross-correlation,

also perform convolution and cross-correlation, respectively, but do it directly by first computing the convolution or cross-correlation sum applied to the lists of signal values, and multiplying the result by the (common) increment of x and y. For long signals (over 100 samples or more) fast convolution and correlation are much quicker than the direct algorithms.

8.2. Exercise. Define `x=cat(2,3,1)` and `y=cat(5,4!3,2,1)` as before. Compute their convolution z using the fast and direct algorithms. Verify that `x**y = y**x`. ∎

8.3. Exercise. Set num equal to a fairly large number (e.g. num = 256), and create some time signal x (e.g. `x=dot`). Compute the convolution of x with itself by the fast and the direct algorithms, and compare the computation times as well as the results. ∎

Two further commands are available to compute the cyclical convolution of two time signals. They are

    x cc y   fast cyclical convolution,
    x c.c y   direct cyclical convolution,

Before computing the cyclical convolution of two signals, any parts of the signals defined for negative times are truncated. Next, for positive times the shortest signal is padded with zeros so that the signals have the same length. The two resulting signals are the starting point for the direct cyclical convolution. For fast cyclical convolution the signals are further padded with zeros to a length consisting of a number of samples that is an integral power of 2, and the fast Fourier transform is used. As a result, fast and direct cyclical convolution may not produce the same outcomes.

9. Functions and Functionals of Signals

All the functions that are introduced in Section 5 also work when the argument a is a signal. The evaluation is pointwise. Thus, assuming that inc = 1 and num = 10,

```
x=sin(dot)
```

creates a signal x with increment 1, domain $[-5, 4]$ and signal values $\sin(-5)$, $\sin(-4)$, $\cdots$, $\sin(0)$, $\cdots$, $\sin(4)$. In this way, using the standard functions of Section 5 and the permanent time signal dot, a variety of well-known elementary time signals may be created, such as real and complex harmonic signals, steps, ramps, and pulses.

9.1. Exercise. Verify that

```
num=100; inc=1/num
period=up(dot)-low(dot)+inc
x=sin(2*pi*dot/period)
```

creates a single period of the sinusoid. Plot the signal using the command

```
plot x
```

Plotting signals is discussed in Section 13. ∎

9.2. Exercise. Check that

```
num=20; inc=1
z=rect((dot-2)/4)
```

generates a rectangular pulse starting at time 0 and ending at time 3. ∎

9.3. Exercise. Generate a step of size 10 starting at time 2 on the same time axis as in 9.2. ∎

The following unary operations on signals are not pointwise:

int(z)	integral,
der(z)	derivative,
sum(x)	sum (synonymous with int),
dif(z)	difference (synonymous with der),
del(z)	delay by one increment,
inv(z)	inverse (z real only),
sort(z)	sort,
hist(z)	histogram (z real only).

These operations are explained in what follows.

If z is a signal, then x=int(z) is another signal representing the running integral of z, while y=der(z) is its derivative. The signal int(z) is obtained by simple Euler integration and der(z) by taking differences and dividing by the increment. The signals sum(z) and dif(z) represent the corresponding operations for discrete-time signals but actually are synonyms for int and der. The signal x=del(z) is defined on the same axis as z and given by $x(t) = z(t - iz)$ with t ranging over the domain of z and iz the increment of z.

9.4. Exercise. Let z be the rectangular pulse of Exercise 9.2 and look at the signals int(z), der(z), and del(z). ∎

Suppose that z is a real, monotonically increasing or decreasing signal. Then the command x=inv(z) creates a signal x that is the inverse of z taken as a map from its domain to its range. The domain of x is [min(z), max(z)] (see below for the definition of max and min), while the number of increments of x equals that of z. If z is not monotonically increasing or decreasing, then it is truncated to the initial monotonically increasing or decreasing part. The inverse x of z is computed on a uniformly sampled signal axis, and its values are obtained by linear interpolation.

9.5. Exercise. Compute x=inv(dot) and verify the result. Also, compute z=exp(0:0.1:1) and x=inv(z). Define e=exp(1) and compare x with the signal y=log(1:(e−1)/10:e). Explain. ∎

The command sort(z) returns a signal on the same axis as z but whose signal values are rearranged in order of increasing magnitude. Duplicate signal values are retained. If z is complex, then the real and imaginary parts are sorted separately.

If z is a real signal, then the command h=hist(z) generates a histogram of z. The number of sample points of the signal h according to a well-known rule of thumb is chosen as $10 \log_{10}(\text{num})$, rounded to the nearest integer. Its domain is [min(z), max(z)] (see below), and h(t) *is* the number of signal values of z in the interval $[t - ih/2, t + ih/2)$, with ih the increment of h.

9.6. Exercise. Generate the signal x of Exercise 9.1. Compute and plot sort(x) and hist(x). ∎

The following functionals may be used to determine certain signal characteristics:

mean(z)	airthmetic mean,
rms(z)	root mean square value,
ampl(z)	maximum of the magnitude of z,
max(z)	maxima of real and imaginary parts of z,
argmax(z)	values at which real and imaginary parts of z take their first maximum,
min(z)	minima of real and imaginary parts of z,
argmin(z)	values at which real and imaginary parts of z

take their first mini-
mum,

zero(z)   values at which real and
imaginary parts of z first
cross zero,

<x, y>   inner product of the sig-
nals x and y.

If the signal z is complex, then `mean(z)`, `max(z)`, and `min(z)` are computed as complex numbers whose real parts are the mean, maximum, and minimum of the real part of z and whose imaginary parts are the mean, maximum, or minimum of the imaginary part of z, respectively.

   Similarly, the numbers `argmax(z)`, `argmin(z)`, and `zero(z)` are computed as complex numbers indicating the times or frequencies at which the real and imaginary part of z take their maximum or minimum or cross zero, respectively. If several maxima, minima, or zero crossings exist the *smallest* time or frequency is taken. The values of `argmin(z)` and `argmax(z)` are taken at sampling instants but those of `zero(z)` are obtained by linear interpolation of adjacent sampling instants.

9.7. Exercise. Generate x as in 9.1 and check whether `mean(x)`, `rms(x)`, `ampl(x)`, `max(x)`, `argmax(x)`, `min(x)`, `argmin(x)`, and `zero(x)` are as expected. Define `period` as in 9.1, and let

   `y= exp(j*2*pi*dot/period)`

Check if `mean(y)`, `rms(y)`, `ampl(y)`, `max(y)`, `argmax(y)`, `min(y)`, `argmin(y)`, and `zero(y)` are as expected. How about ⟨x,y⟩?   ∎

10. Pointwise Evaluation and Composition of Signals

Let x be an existing signal and t a real scalar. Then the command

   `x(t)`

results in the evaluation and display of the signal x at time t. If the time t does not coincide with a sampling instant, then `x(t)` is evaluated by linear interpolation between adjacent sampling instants. Thus,

   `dotplus(1.6)`

displays the value 1.6.

   The same construction may be used in an assignment. If x is an existing signal, t a real scalar and c a scalar, then the command

   `x(t)= c`

assigns the value c to x at time t. If t is not a sampling instant, then it is rounded to the nearest sampling instant. If x does not exist, then the command is rejected. If t is outside the domain of x then the domain is automatically expanded. Missing signal values are set equal to zero.

   If x is again an existing signal and a another existing real signal, then the command

   `x(a)`

results in the composition of the two signals as follows. The increment and the domain of the signal x(a) that is created are the increment and the domain of a. The successive signal values of x(a) are evaluated as $x(a)(t) = x(a(t))$ with t ranging over the time axis of a. If for some value of t the value of a(t) does not coincide with a sampling instant of x, then the value of x(a(t)) is obtained by linear interpolation between adjacent sampling instants. If for some value of t the value of a(t) is outside the domain of x, then x(a(t)) is set equal to zero.

   If x is a scalar (called either by name or by value), then x(a(t)) is set equal to x for each t in the domain of a. Thus, y=1(dot)

results in a signal y that has the constant
value 1 on the axis of dot.

    The composition operation is mainly
intended to change time scales on existing
signals and to shift them. Suppose that x is
the rectangular pulse generated in Exercise
9.2. Then,

    x(dot/2); x(2*dot)

generates two rectangular pulses that have
double the length respectively half the
length of the original pulse. The command

    x(dot-8)

shifts the pulse so that it starts at time 8
rather than at time 0.

    Another use of composition is to
generate periodic signals. If x is an existing
signal and P some positive real scalar, then

    x(dotplus mod P)

generates a periodic repetition of the part
of x defined on [0, P).

10.1. Exercise.   Define x as in Exercise
9.2. Look at the signals

    x(dot)
    x(dotplus)
    x(dot/2)
    x(2*dot)
    x(dot-8)
    x(dot mod 6)
    (1+j)(dot)

and explain what you find.   ■

11. Change of Domain of a Signal

Suppose that x is an existing signal. It is
possible to change the domain of x by one
of the following four commands:

    y=x[a:b]
    y=x(a:b]
    y=x[a:b)
    y=x(a:b)

Here a and b are real scalars. The commands generate a signal y that equals x but
whose domain is extended or restricted to
the interval delimited by a and b, which is
taken (half) open or (half) closed as indicated. Thus

(a	sets the lower limit of the domain equal to the smallest sampling in- stant greater than a,
[a	sets the lower limit of the domain equal to the smallest sampling in- stant greater than or equal to a,
b)	sets the upper limit of the domain equal to the largest sampling instant less than b,
b]	sets the upper limit of the domain equal to the largest sampling instant less than or equal to b.

If the domain of y is extended beyond that
of x, then the missing signal values are set
equal to zero.

    If in any of the four commands a or
b is replaced with ., then the *current* lower
or upper limit is taken. By way of example,
let x=cat(1,2!1). Then

    x=x(-5:5)

makes x equal to cat(0,0,0,1,2!1,0,0,
0), but

    x=x[-5:5]

results in x being equal to cat(0,0,0,0,
1,2!1,0,0,0,0). The command

    x=x(.:.)

reduces x to cat(2!), while

    x=x[.:.]

leaves x unchanged.

11.1. Exercise. Letting `x=cat(1,2!1)` as above, verify that

`z=x[low(dot):up(dot)]`

leads to the same result as

`z=x(dot)`   ■

11.2. Exercise. Generate and plot the signal x that is given by

$$x(t) = \begin{cases} 1 & \text{for } t < 0, \\ \cos(2\pi t/P) & \text{for } t \geq 0, \end{cases}$$

on the default time axis dot with inc = 1, num = 20 and $P = 10$.   ■

12. Creating and Manipulating Frequency Signals

Frequency signals are only distinguished from time signals by their default increment, which equals INC. Frequency signals may be created directly using the CAT function. This function works precisely the same way as the cat function except that it assigns INC to the increment of the signals it creates.

Frequency signals may be generated indirectly by using the standard frequency signals DOT and DOTPLUS in the same way as dot and dotplus are used for time signals. DOT has increment INC (default value 1/10), domain

$$\left[-\frac{\text{num}}{2} \text{INC}, \left(\frac{\text{num}}{2} - 1 \right) \text{INC} \right]$$

if num is even and

$$\left[-\frac{\text{num} - 1}{2} \text{INC}, \frac{\text{num} - 1}{2} \text{INC} \right]$$

if num is odd, and signal values specified by DOT(f) = f with f ranging over all sam-

pling frequencies in the domain. DOTPLUS has increment INC, domain

$$[0, (\text{num} - 1)\text{INC}]$$

and signal values similarly defined by DOT-PLUS(f) = f.

The permanent frequency signal DELTA is defined on the time axis of DOT and given by DELTA(0) = 1/INC, DELTA(f) = 0 for $f \neq 0$.

All functions and operations defined for time signals also work for frequency signals.

12.1. Exercise. Create the frequency signal z defined by

$$Z(f) = \frac{1}{1 + j2\pi f}, \quad 0 \leq f < 10,$$

with increment INC = .05. Compute the magnitude and argument of z as a function of frequency.   ■

13. Graphic Display of Signals

The command

`plot x`

with x the name of a signal or a signal expression, provides a plot of the signal. The real and imaginary parts are plotted in different frames. If the signal is real, then the frame for the imaginary part is not displayed and the whole screen is used for the real part.

When the command plot is invoked, the screen is changed to graphic mode and the normal text display disappears. The text display may be recalled by pressing the ⟨ESC⟩ key. Typing

`plot`

without argument restores the last plot, if available.

A hardcopy dump of the graphics

screen on a printer connected to the computer may be made by holding down the shift key and pressing the ⟨PRTSCR⟩ key.

Scaling of the plots is automatic, the axes are labeled by default, and the sample points of the plot are connected by straight lines. Several signals may be displayed in the same frame by a command such as

```
plot x,y,z
```

If the format status is polar, the plot command provides plots of the magnitude and argument of x rather than of the real and imaginary parts.

13.1. Exercise.   Generate a single period of a sine and two periods of a cosine as

```
inc=.01; num=101
x=sin(2*pi*dotplus)
y=sin(4*pi*dotplus)
z=x+j*y
```

Display the signals by

```
plot x
⟨ESC⟩ plot y
⟨ESC⟩ plot z
⟨ESC⟩ plot x,y
```

Look at some of the plots in polar format (see 2).   ∎

The command

```
polarplot x
```

provides a plot (in a single frame) of the imaginary part of x versus its real part. Several plots can be made in the same frame as for the plot command.

13.2. Exercise.   Compute z as in Exercise 13.1 and give the command

```
polarplot z
```

Note that actually the signal x is plotted against the signal y.   ∎

Curves in a plot may be marked with a number or letter. For instance,

```
plot x"x"
```

marks the plot of x at regular distances with an "x," while

```
plot x"1",y"2"
```

marks the plot of x with "1" and that of y with "2." The command

```
plot x"."
```

plots x point by point, without connecting lines. The command format

```
plot x"|"
```

finally, is meant for displaying discrete-time signals, and plots each point as a vertical bar located at the correct horizontal coordinate whose height equals the vertical coordinate.

13.3. Exercise.   Display the signals of Exercise 13.1 with different markings and as discrete-time signals.   ∎

If a color monitor is available, curves may be plotted in color. For instance, the command

```
plot x"/r"
```

plots the signal x in red. The available colors are

/b	blue
/c	cyan
/g	green
/r	red
/y	yellow

Default is white.

Once a plot is on the screen, it may be relabeled, titled, or a grid may be drawn in. After reverting to text mode the command

```
title "Legend"
```

writes "Legend" at the top of the current plot, which may be recalled by the command plot. The commands

```
hlabel •"legend"
vlabel "legend"
vlabelt "legend"
vlabelb "legend"
```

overwrite the default labels beneath the horizontal axis, next to the vertical axes, next to the vertical axis of the top plot and that of the bottom plot, respectively. The commands

```
text x,y,"legend"
textt x,y,"legend"
textb x,y,"legend"
```

write "legend" in both frames, the top frame and the bottom frame, respectively, at the location (x, y), where (0, 0) is the lower left corner of the frame and (1, 1) the upper right.
The command

```
grid
```

draws grid lines on the current plot. The grid is extinguished by repeating the command.
The command

```
haxis xmin,xmax
```

freezes the left-hand and right-hand side ends of the horizontal scale at xmin and xmax, respectively. This command has to be given *before* issuing a plot command, and cannot be used to change the scale of the current plot. Similarly, the commands

```
vaxis ymin,ymax
vaxist ymin,ymax
vaxisb ymin,ymax
```

freeze the scales for both plots, the top plot or the bottom plot, respectively. The commands

```
axis
haxis
vaxis
vaxist
vaxisb
```

display *all* scale settings, respectively those

for the horizontal and vertical axes separately. The commands

```
set axis
set haxis
set vaxis
set vaxist
set vaxisb
```

revert all scalings to automatic, respectively those for the horizontal and vertical axes separately.

13.4. Exercise.   Compute z as in Exercise 13.1 and see what the effect is of typing

```
plot z
⟨ESC⟩ haxis 0,.5
plot z
⟨ESC⟩ hlabel "time"
vlabelt "x"
vlabelb "y"
haxis −2,2
vaxis −2,2
polarplot z
```

Put the legend "Lissajoux figure" both at the top of the plot and somewhere inside the frame by using the commands title and text. Add a grid as well.   ∎

The command

```
plotfile ⟨file name⟩
```

finally, creates a plot file of the current plot (i.e., the last that appeared on the screen). The file name should not have an extension; SIGSYS gives it the extension .PLT. If no file name is given, the default name SIGSYS.PLT is assigned. The file may be printed from DOS by the command copy SIGSYS.PLT prn: assuming that the file name is SIGSYS.PLT.

14. Entering Polynomials

Polynomials may be entered by using the indeterminate variable sigma. Typing

```
p=1.5+(2+3*j)*sigma+sigma^2
```

defines the polynomial

$$p = 1.5 + (2 + 3j)\sigma + \sigma^2.$$

Displaying p results in something like

```
(1.5+j*0)*σ^0+
(2+j*3)*σ^1+
(1+j*0)*σ^2
```

Note that the indeterminate variable is displayed as σ, which does not exist on the keyboard. The indeterminate variable sigma may be redefined, for instance to the handier s, by typing

```
indeter=s
```

The display remains in terms of σ, however. The current name of the indeterminate variable may be displayed by simply typing

```
indeter
```

If p has been defined as a polynomial, the command

```
p[n]=c
```

with n an integer and c a real or complex scalar, sets the coefficient of the term with the nth power equal to c. If n is greater than the degree of the polynomial, the degree is automatically increased.

If the leading coefficient of a polynomial equals zero, then its degree is automatically decreased by 1. This is repeated until the leading coefficient differs from zero. Similarly, if all the coefficients of a complex polynomial have zero imaginary part, then the polynomial is converted to a real polynomial.

14.1. Exercise. Enter the polynomial $p = 1 + 2\sigma + 3\sigma^2 + 4\sigma^4 + \sigma^5$. Change the leading coefficient to 0 and see how the degree of the polynomial decreases to 4. ∎

15. Operations Among Polynomials

The following operations among polynomials are defined:

p^n	(with n an integer) exponentiation,
−p	unary minus,
p*q	multiplication of polynomials,
p/a	(with a scalar) division by a scalar,
p+q	addition of polynomials,
p−q	subtraction of polynomials.

15.1. Exercise. Compute the polynomials $p_1 = (1 + \sigma)^{10}$, $p_2 = (1 - \sigma)^{10}$ and determine $p = p_1 + p_2$. ∎

16. Functions and Evaluation of Polynomials

For polynomials there are three attribute functions, namely,

degree(p)	the degree of p,
roots(p)	the roots of p,
p[n]	(with n an integer) the coefficient of the term with power n.

The roots roots(p) of p are returned as a *signal* with increment 1 and domain [0, degree(p) − 1]. The roots are given in order of increasing magnitude.

Two further functions convert polynomials into other polynomials:

p'	replaces the indeterminate σ with $-\sigma$,
conj(p)	replaces the coefficients with their complex conjugates.

Finally, the command

```
p(a)
```

with a a scalar, returns the value of the polynomials at $\sigma = a$. When a is a signal `p(a)` is a signal that is obtained by pointwise substitution of the signal a.

16.1. Exercise. Enter the polynomials $p = 1 + \sigma + \sigma^2$ and $q = 1 + \sigma^4$. Compute the roots of p and q. Determine the leading coefficient of the product polynomial pq. Compute and plot the frequency signal Z defined by

$$Z(f) = \frac{1 + \sigma + \sigma^2}{1 + \sigma^4}$$

with $\sigma = j2\pi f$, where f ranges over [0, 2) with increment INC = .05. *Hint:* Set INC = .05, num = 2/INC,

```
s=j*2*pi*DOTPLUS
```

and simply compute `Z=p(s)/q(s)`. ∎

17. Fourier Transformation

SIGSYS provides the following Fourier transforms:

fft	Fast Fourier Transform,
ifft	inverse Fast Fourier Transform,
fc	Fourier coefficients,
fs	Fourier series,
ddft	discrete-to-discrete Fourier transform,
iddft	inverse discrete-to-discrete Fourier transform,
cdft	continuous-to-discrete Fourier transform,
icdft	inverse continuous-to-discrete Fourier transform,
dcft	discrete-to-continuous Fourier transform,
idcft	inverse discrete-to-continuous Fourier transform
ccft	continuous-to-continuous Fourier transform,
iccft	inverse continuous-to-continuous Fourier transform.

The various transforms differ only with respect to

The *domain* of the signals they operate on, and

a *scaling factor*.

All transforms are based on the Fast Fourier Transform. One consequence of this is that all signals that are transformed need be defined on a number of sample points that is an integral power of 2. The domain of signals that are defined on a different number of sample points is automatically extended; missing signal values are set equal to zero.

Two types of domain are used:

"One-sided." The domain of the signal is

$$[0, (N - 1)T],$$

where N is the number of sample points of the signal for nonnegative times, and T the increment of the signal. Signal values for negative times are discarded. If N is not an integral power of 2 it is increased to the next integral power of 2 and the signal is supplemented with zeros.

"Two-sided." The domain of the signal is

$$\left[-\frac{N}{2}T, \left(\frac{N}{2} - 1\right)T \right],$$

where N is the number of sample points of the signal and T its increment. If N is not an integral power of 2 it is increased to one and the signal is supplemented with zeros.

The various transforms are based on the formula

$$\hat{x}(f) = c \sum_{t \in D} \hat{x}(t)e^{-j2\pi ft}, \qquad f \in \hat{D},$$

while the inverse transforms are given by

$$x(t) = \hat{c} \sum_{f \in \hat{D}} \hat{x}(f)e^{j2\pi ft}, \qquad t \in D.$$

D is the domain of the time-signal and $\hat{D}$ that of the transform, while c and $\hat{c}$ are constants. Given a time signal with N sample points and increment T, the increment of its transform is $F = 1/NT$. Given a transform with N sample points and increment F, the increment of the inversely transformed time signal is $T = 1/NF$. D and $\hat{D}$ both contain N points, and are one- or two-sided depending on the transformation.

For the transform pair fft/ifft (Fast Fourier Transform and its inverse) the constants are $c = 1$ and $\hat{c} = 1/N$, according to the usual convention. Both D and $\hat{D}$ are one-sided.

For the transform pair fc/fs (Fourier coefficients/Fourier series) the constants are $c = 1/N$ and $\hat{c} = 1$, so that the transform fc generates the Fourier coefficients (in the alternative form of 6.4.5) and the inverse transform fs reconstructs the time signal from the Fourier coefficients. D is one-sided and $\hat{D}$ two-sided.

For all remaining Fourier transform pairs the constants are $c = T$ and $\hat{c} = F$. The domains and coefficients of all transform pairs are indicated in the table.

We look at an example. Suppose that a time signal z is generated by the sequence of commands

```
num=200;  inc=.01
z=rect(dot)
```

The time signal is a rectangular pulse with duration from $-.5$ to $.5$, defined on the time axis $[-1, 1)$, with increment 0.01 and 200 sample points. The command

```
Z=ccft(z)
```

first expands the number of sample points from 200 to 256, then extends the domain two-sidedly from $[-1, 1)$ to $[-1.28, 1.28)$ while padding with zeros, and finally computes the continuous-to-continuous Fourier transform (approximating the integral by a sum) on 256 points, with increment $F = 1/(256 \cdot 0.01) = 0.390625$, on the two-sided frequency domain $[-128F, 128F) = [-50, 50)$. The subsequent command

```
zz=iccft(Z)
```

ATTRIBUTES FOR THE VARIOUS FOURIER TRANSFORM PAIRS

Transform	Inverse	Name	c	$\hat{c}$	D	$\hat{D}$
fft	ifft	Fast Fourier Transform	1	$1/N$	one-sided	one-sided
fc	fs	Fourier coefficients	$1/N$	1	one-sided	two-sided
ddft	iddft	DDFT	T	F	one-sided	one-sided
dcft	idcft	DCFT	T	F	two-sided	one-sided
cdft	icdft	CDFT	T	F	one-sided	two-sided
ccft	iccft	CCFT	T	F	two-sided	two-sided

leaves the number of sample points and the domain of z unaffected, and returns zz as the inverse CCFT (approximating again the integral by a sum) on 256 points, with increment $T = 1/NF = 0.01$, on the two-sided domain $[-1.28, 1.28]$. Because the original signal z has its support entirely inside its domain, the signal zz is identical to the original signal z.

We continue the example. Given the time signal z as originally defined, the command

```
Z=fc(z)
```

discards the signal values at negative times, changes the number of sample points from 100 to 128 (the next integral power of 2 up from 100), redefines the domain of the signal to [0, 1.28) and supplements the signal with zeros. The command considers the resulting signal as a single period of a periodic signal and computes its Fourier coefficients. The coefficients are stored as a frequency signal with 128 sample points, frequency increment $F = 1/(128 \cdot 0.01) = 0.78125$ and domain $[-64F, 64F) = [-50, 50]$. The command

```
zz=fs(Z)
```

sums the Fourier series whose coefficients are stored in z on 128 sample points, with time increment $T = 1/(128 \cdot 0.78125) = 0.01$ on the one-sided domain [0, 1.28). The signal zz equals the original signal z only for nonnegative times.

17.1. Exercise. Go through the steps indicated above and verify the results. ∎

18. Macros and Shell Escape

SIGSYS may handle macros. Macros should be prepared as ASCII files, and given the extension .MAS. Suppose that MACRO1.MAS is a macro file name. When the command

```
macro1
```

is given (without the extension, and no distinction is made between upper and lower case characters), SIGSYS searches the disk for a file of that name with the extension .MAS, reads it line by line, and presents each line to the interpreter as a command line. A macro can call other macros.

Macros are searched for in the default directory, or the directory set externally or by the `chdir` command as explained in Section 3.

Macros may be internally documented using the % character. Any text on a command line following the character % up to the end of the line or the first separator ; is ignored, including the % character itself. The % may be preceded by any number of spaces.

By way of example, a macro that shows a Lissajoux figure follows.

```
%Macro DEMO
%Plot of a Lissajoux figure
num=200; inc=1/num
%Add 1 point
%to obtain a complete figure:
num=num+1
x=sin(2*pi*dotplus*4)
y=sin(2*pi*dotplus*5)
z=x+j*y
polarplot z
```

The command

```
pause
```

may be inserted in a macro file. When this command is encountered, execution stops, and resumes when the ⟨ENTER⟩ key is pressed.

A useful facility in creating and modifying macros is the *shell escape*

```
!
```

A single ! on a command line shifts control to the "shell," in this case DOS, while keeping SIGSYS resident. In this mode, an editor may be invoked to create or modify a macro. After finishing with editing, typing

```
exit
```

(from DOS) returns control to SIGSYS. The escape ! may also be followed directly on the command line by a DOS command. Thus,

```
!edit macro1.mas
```

invokes an editor called "edit" to work on the file MACRO1.MAS (note that DOS automatically converts lower case letters to capitals.) After exiting from the editor control is automatically returned to SIGSYS. As another example, the command

```
!dir
```

displays the contents of the current directory.

Note that because SIGSYS stays resident, the memory available for any program invoked via the shell escape is much less than it would normally be.

18.1. Exercise. Create the macro file DEMO1.MAS that will display a Lissajoux figure, and run it. ∎

19. Integration of Differential Equations

SIGSYS offers a unique facility for the integration of a set of differential equations. Consider for instance solving the Volterra equation

$$\dot{x} = x(y - 1),$$

$$\dot{y} = y(1 - x),$$

on the interval [0, 10] with an integration step size of .04 and initial conditions $x(0) = 2$, $y(0) = 2$. This is accomplished by the following sequence of commands, which are best collected and executed as a macro.

```
x=2; y=2
integrate from 0 to 10 stepsize .04
@x=x*(y−1)
@y=y*(1−x)
write x,y
end
```

Before entering the integration loop the initial conditions need be set. This is done in the first line. Inside the integration loop the notation @ may be read as the differential operator. For the interpreter the symbol @ marks the variables that need be integrated.

The `write` command indicates the signals whose values have to be retained during integration and stored upon completion. After completion of the command, x and y are signals with domain [0, 10] and increment .04 containing the solution of the differential equation.

By default, integration is performed with the standard second-order Runge-Kutta method. Modification of the `integrate` command to

```
integrate RKx · · ·
```

with $x = 1$, $x = 2$ or $x = 4$ results in the use of the standard first-, second-, or fourth-order Runge-Kutta method, while

```
integrate AB · · ·
```

uses the second-order Adams-Bashforth method, initialized with a single second-order Runge-Kutta step.

If `stepsize` is not specified, the current value of `inc` is used as a default for the step size. If neither the integration range nor the step size are specified, integration is performed on the default time axis `dot-plus`. Thus, the second command line may for instance be modified to

```
integrate from 0 to 10
```

or to

```
integrate AB
```

In the example, upon exit of the integration loop x and y, which are initially defined as scalars, are overwritten by the signals included in the `write` command. If this command is omitted, then x and y on exit have scalar values equal to their values at the final time.

The integration loop may also involve auxiliary variables and externally defined signals. For instance, the differential equation

$$\dot{x} = -x + u,$$

with u an existing time signal, may be integrated on the time axis `dotplus` as

```
x=0; t=0
integrate
@t=1
@x=-x+u(t)
write x
end
```

By integrating t along with x, each time the right-hand side of the differential equation for x is called the external input u is evaluated at the correct time instant.

Finally, consider the integration of the differential equation

$$\frac{dw}{dt} = \alpha (u - w^2),$$

where u is given by

$$u = k(w_o - w).$$

Here w might represent the normalized speed of a car, and u the throttle position, which follows by proportional feedback of the difference $w_o - w$ between a constant reference speed w_o and the actual speed w. We integrate this as follows:

```
alpha=.1; k=1; wo=0.9
w=0.2
integrate from 0 to 100 stepsize 1
u=k*(wo-w)
@w=alpha*(u-w*w)
write u,w
end
```

Note that the command u=k*(wo-w) must precede the integration command to ensure that @w is evaluated for the correct value of u. Upon completion of the command, the solutions for both u and w are stored over the entire time interval.

19.1. Exercise. Try the three examples presented in this section. ∎

The integration loop may prematurely be terminated with the `break` command. If we modify the last example to

```
alpha=.1; k=.1; wo=.9; w=.2
integrate from 0 to 100 stepsize 1
u=k*(wo-w)
@w=alpha*(u-w*w)
if w>.55 break end
write u,w
end
```

integration is stopped as soon as the speed w exceeds the value 0.55.

20. Iteration

Sets of iterative equations may conveniently be solved by using the `iterate` command. Suppose by way of example that we wish to solve the simultaneous difference equations

$$x_1(n + 1) = x_2(n),$$
$$x_2(n + 1) = x_1(n) + x_2(n),$$

for $n = 0, 1, 2, \cdots, 20$, with the initial conditions $x_1(0) = 0$, $x_2(0) = 1$. This may be done as follows in a way that is similar to integration:

```
x1=0; x2=1
iterate from 0 to 20
@x1=x2
@x2=x1+x2
write x1,x2
end
```

The iteration loop is repeated as many times as indicated in the from ⋯ to statement. Each time the loop is executed the variables marked with @ are temporarily stored and given their definitive values *after* the loop has been completed once. The values of the variables specified in the write statement are retained during the iteration process and stored as signals on the

axis defined by the from · · · to statement
with increment one. If the from · · · to
part is omitted, iteration is performed on
the axis 0, 1, · · · , num.
    Like integration, iteration may be in-
terrupted by using the break command.
Modification of the example to

```
num=20
x1=0;  x2=1
iterate
@x1=x2
@x2=x1+x2
if x1>1000 break end
write x1,x2
end
```

causes the iteration process to stop as soon
as x1 exceeds the value 1000.

20.1. Exercise.   See how this iteration
works.   ■

21. Command Flow Control

SIGSYS has several facilities for command
flow control. They are mainly useful within
macros. The first is the if command,
which takes the form

```
if ⟨condition⟩
⟨compound statement⟩
end
```

The four basic elements of this statement
may be separated by spaces, commas, or
newlines. An example is

```
if real(x)>0 x=0 end
```

In ⟨condition⟩ the following comparators
may be used:

==	equal to,
>	greater than,
>=	greater than or equal to,
<	less than,
<=	less than or equal to,
~=	not equal to

Furthermore, the logical operators

&	"and,"
|	"or," and
~	"not"

are permitted.
    The ⟨compound statement⟩ may en-
compass several statements. An example is

```
if x>0 & y¯=0
x=1;  y=0
end
```

The if statement may include an else op-
tion, in the form

```
if ⟨condition⟩
⟨compound statement⟩
else
⟨compound statement⟩
end
```

By way of example consider

```
if x>=0 x=1
else x=-1
end
```

More flow control is possible with a while
loop. It has the general form

```
while ⟨condition⟩
⟨compound statement⟩
end
```

The ⟨compound statement⟩ is repeatedly executed until ⟨condition⟩ becomes false.
Thus, in

```
t=0
while t<10
t=t+1
end
```

the value of t is increased from 0 to 10.
While loops may be interrupted with the
break command. In

```
t=0;  x=1
while t<10
t=t+1;  x=2*x
if x>1000 break end
end
```

the loop is broken when x becomes larger than 1000.

Finally, a `for` loop may be executed. It has the general form

```
for ⟨scalar variable⟩
in ⟨signal variable⟩
⟨compound statement⟩
end
```

In this command, the ⟨compound statement⟩ is repeated as many times as ⟨signal variable⟩ has sample points, with ⟨scalar variable⟩ successively taking the values assumed by ⟨signal variable⟩. For instance,

```
for t in dot
z(t)=x(t)+x(t)
end
```

is a way of adding the time signals x and y on the time axis dot. A more useful example is

```
x=0(0:1:100)
for t in 0:10:100
x(t)=1
end
```

which creates x as a comb on the signal axis 0:1:100 with period 10. As a final example consider

```
x=0
for v in x
X=X+abs(v)
end
```

The final value of x is the sum of the absolute values taken by the signal x.

Also a `for` loop may be interrupted with the `break` command.

21.1 Exercise. Test the various examples in this section. ∎

22. User-Defined Functions and Procedures

SIGSYS allows the user to define functions. The general form is

```
function ⟨name⟩ ((a. list))
⟨compound statement⟩
return ⟨expression⟩
end
```

where "a. list" stands for "argument list." We look at an example. The series of commands

```
function crt(x)
return x^(1/3)
end
```

defines a function similar to `sqrt` that may be used to compute cube roots of scalars and signals. Calling `crt(8)` returns the value 2. Once defined, a function may be called as often as desired. Another example is

```
function Max(x,y)
if real(x)>=real(y)
  r=real(x)
  else r=real(y) end
if imag(x)>=imag(y)
  i=imag(x)
  else i=imag(y) end
return r+j*i
end
```

The function `Max` computes the maximum of the real and imaginary parts of two scalars or signals.

All variables within a function definition are local. This means that externally defined variables are not available.

Besides functions, also *procedures* may be defined. The general form is

```
procedure ⟨name⟩ ((a. list))
⟨compound statement⟩
end
```

where again "a. list" stands for "argument list." The arguments in the list may be input or output parameters. For instance, after defining

```
procedure swap(x,y)
u=x; x=y; y=u
end
```

the command

swap(a,b)

causes the values of a and b to be interchanged.

22.1 Exercise. Define and test the functions crt and Max and the procedure swap. ∎

23. Export and Import of Signals

Signals may be exported as ASCII files using the commands

export ⟨s. name⟩ as ⟨f. name⟩

where "s. name" stands for "signal name" and "f. name" for "file name." The file name should have no extension: SIGSYS gives it the extension .DAT. If

as ⟨f. name⟩

is omitted, the file name is the same as the signal name. The file is written to the current directory (see 3.)

SIGSYS writes the signal data in the file in three column format, with the first column the signal axis, the second column the real part, and the third column the imaginary part of the signal values. Thus, part of a .DAT file could look like

0	1	1
0.1	1.5	0.5
0.2	3	1
0.3	4.5	1
0.4	6	1.5

If the signal is real, the third column is omitted.

Exported .DAT files may be used as input to other programs, such as text processing and graphical programs. The converse of the export command is the command

import ⟨f. name⟩ as ⟨s. name⟩

with "f. name" standing for "file name", and "s. name" for "signal name." No ex-

tension should be specified; SIGSYS looks for a file with the extension .DAT. If

as ⟨s. name⟩

is omitted, the signal name is the file name. The import command converts a two or three column ASCII file as created by the export command or otherwise to a real or complex signal.

The export and import commands may be modified to

lexport   export in list format

limport   import from list format

In files in list format the first column (i.e., the signal axis), is omitted. Imported files in list format are defined as signals on a one-sided time axis with the default increment inc.

24. Help and Diary Facilities and Diagnostics

SIGSYS offers an on-line help facility. Typing

help

displays a list of all available commands, one of which may be selected by using the cursor keys. By pressing the ⟨RETURN⟩ key, information concerning the command is shown on the left-half part of the screen. Additional related items may be selected with the cursor and displayed on the right-half part of the screen by again using the ⟨RETURN⟩ key. The help command may be terminated by using the ⟨ESC⟩ key. Help may alternatively be invoked, even in the middle of typing a command line, by using the

⟨ALT⟩H

combination (i.e., by using the H key while pressing the ⟨ALT⟩ key).

The diary facility makes it possible to keep a record of all text that appears on the screen. Typing diary ⟨file name⟩

opens a file with the specified name, and all text that subsequently appears on the screen is written to the file. Plots are not recorded. The file name should be given without extension; SIGSYS gives the file the extension .LOG. If ⟨file name⟩ is omitted, then the default name SIGSYS.LOG is used. The command

diary off

temporarily suspends recording the screen text. Recording may be resumed by typing

diary on

The command

diary close

closes the file.

If during the compilation of a SIGSYS command line an error is detected, an arrow appears below the command line, approximately indicating the location of the error in the command line. Moreover, a concise error message is shown.

25. Synopsis

Syntax element	Meaning	Section
→, ←, ↑, ↓	command line editing	3
!	origin mark for concatenation	6
!	shell escape	18
!BREAK OFF	turn off DOS abort feature	3
!BREAK ON	turn on DOS abort feature	3
&	logical "and"	21
%	comment follows	18
'	mirror polynomial	16
( )	precedence	4
(:)	change domain	11
(:]	change domain	11
*	multiply	4, 8, 17
**	fast convolution	8
.	direct convolution	8
+	add	4, 8, 15
−	unary minus, subtract	4, 8, 15
/	divide	4, 8, 15
::	define signal axis	6
;	command separator	2
<	less than	21
< , >	inner product	9
<ALT> H	invoke the help utility	24
<BACKSPACE>	delete preceding character	3
<CTRL> C	cancel current operation	3
<CTRL> Q	resume execution	3
<CTRL> S	halt execution	3
<DEL>	delete current character	3
<INS>	toggle insert mode	3
<PRTSCR>	print screen	13
<=	less than or equal to	21
=	assigment operator	2
==	equal to	21

Syntax element	Meaning	Section
>	greater than	21
>=	greater than or equal to	21
@	differential operator, store temporarily	19, 20
abs	absolute value	5
acos	arccosine	5
ampl	amplitude	9
ans	value of last unassigned variable	4
arg	argument	5
argmax	argument of maximum	9
argmin	argument of minimum	9
asin	arcsine	5
atan	arctangent	5
axis	return settings for plot scales	13
BREAK OFF	turn off DOS abort feature (external command)	3
BREAK ON	turn on DOS abort feature (external command)	3
break	break a loop	19, 20, 21
CAT	concatenate frequency signals	12
cat	concatenate time signals	6
cc	fast cyclical convolution	8
c.c	direct cyclical convolution	8
ccft	CCFT	17
cdft	CDFT	17
ceil	round up	5
chdir	change directory	3, 18
clear	clear all or some variables	3
conj	complex conjugate	5, 16
cos	cosine	5
dcft	DCFT	17
ddft	DDFT	17
degree	degree of polynomial	16
del	delay by one increment	9
DELTA	unit frequency impulse	12
delta	unit time impulse	6
der	derivative	9
diary	start a diary	24
diary close	close the diary	24
diary off	suspend writing to the diary	24
diary on	resume writing to the diary	24
dif	difference	9
DOT	permanent frequency signal	12
dot	permanent time signal	6
DOTPLUS	permanent frequency signal	12
dotplus	permanent time signal	6
else	else	21
end	end of statement	19, 20, 21, 22
exit	terminate SIGSYS, exit from DOS shell	3, 18
exp	exponential	5
export	export signal	23
fft	Fast Fourier Transform	17
floor	round down	5
for	for loop	21
format	revert to default formats	2

Syntax element	Meaning	Section
format cartesian	Cartesian format	2
format long	long precision	2
format polar	polar format	2
format short	short precision	2
function	define function	22
grid	insert grid	13
haxis	freeze or show horizontal scale	13
help	help utility	24
hist	histogram	9
hlabel	label horizontal axis	13
iccft	inverse CCFT	17
icdft	inverse CDFT	17
idcft	inverse DCFT	17
iddft	inverse DDFT	17
if	if	21
ifft	inverse FFT	17
imag	imaginary part	5
import	import signal file	23
INC	default increment frequency signal	7, 12
inc	default increment time signal	6
indeter	redefine or display indeterminate variable	14
integrate	integrate differential equation	19
int	integrated signal	9
inv	inverse	9
iterate	iterate	20
j	$\sqrt{-1}$	2
lexport	export in list format	23
limport	import in list format	23
load	load a data file	3
log	natural logarithm	5
log10	logarithm to the base 10	5
low	lower limit of domain	5
max	maximum value	9
mean	mean value	9
min	minimum value	9
mod	reduction modulo	4, 8
noise	white noise	6
noiseplus	white noise	6
num	number of sample points	6
o.o	direct correlation	8
oo	fast correlation	8
pack	reorganize memory	3
pause	pause	18
pi	π	2
plot	plot a signal	13
plotfile	create a plot file	13
polarplot	polar plot of a signal	13
procedure	define procedure	22
ramp	ramp function	5
rand N	change distribution to normal	2
rand seed	set seed random number generator	2
rand U	change distribution to uniform	2

Syntax element	Meaning	Section
rand	random number	2
real	real part	5
rect	rectangular pulse	5
return	return function value	22
rms	root mean square value	9
roots	roots of polynomial	16
round	round to nearest integer	5
sat	saturation function	5
save	save all data	3
SET SIGSYS=	set default directory (external command)	3, 18, 23
set axis	unfreeze all scalings	13
set haxis	unfreeze horizontal scaling	13
set INC	set increment frequency signal	7
set inc	set increment time signal	7
set num	set number of sample points	7
set vaxisb	unfreeze vertical scaling of the bottom plot	13
set vaxist	unfreeze vertical scaling of the top plot	13
set vaxis	unfreeze vertical scaling	13
sigma	default indeterminate variable	14
sign	signum function	5
sin	sine	5
sinc	sinc function	5
sort	sort function	9
sqrt	square root	5
step	step function	5
stepsize	integration step size	20
sum	summed equal	9
tan	tangent	5
text	insert text in plot	13
textb	insert text in bottom plot	13
textt	insert text in top plot	13
title	display title	13
trian	triangular function	5
up	upper limit of domain	6
vaxisb	freeze or show vertical axis of the bottom plot	13
vaxist	freeze or show vertical axis of the top plot	13
vlabelb	label vertical axis of the bottom plot	13
vlabelt	label vertical axis of the top plot	13
vaxis	freeze or show vertical scales	13
vlabel	label vertical axes	13
while	while	21
who	display variables and free memory space	3
write	store signal in integration or iteration loop	19, 20
zero	zero crossing	9
[:)	change domain	11
[:]	change domain	11
[ ]	extract a coefficient of a polynomial	16
^	exponentiation	4, 8, 15
|	logical "or"	21
~=	not equal to	21

Index

Note: The SIGSYS Tutorial has its own index, named *Synopsis* (p. 777).